Nucleic Acid and Monoclonal Antibody Probes

Infectious Disease and Therapy

Series Editors

Brian E. Scully, M.B., B.Ch.

College of Physicians
and Surgeons
Columbia University
New York, New York

Harold C. Neu, M.D.

College of Physicians
and Surgeons
Columbia University
New York, New York

Volume 1 Parasitic Infections in the Compromised Host, *edited by Peter D. Walzer and Robert M. Genta*

Volume 2 Nucleic Acid and Monoclonal Antibody Probes: Applications in Diagnostic Microbiology, *edited by Bala Swaminathan and Gyan Prakash*

Additional Volumes in Preparation

Nucleic Acid and Monoclonal Antibody Probes

Applications in Diagnostic Microbiology

edited by

BALA SWAMINATHAN
Purdue University
West Lafayette, Indiana

GYAN PRAKASH
Mesa Diagnostics, Inc.
Albuquerque, New Mexico

MARCEL DEKKER, INC. **New York and Basel**

ISBN 0-8247-8023-X

MARCEL DEKKER, INC.
270 Madison Avenue, New York, New York 10016

Current printing (last digit):
10 9 8 7 6 5 4 3 2 1

PRINTED IN THE UNITED STATES OF AMERICA

To
Mangal, Savita, Shivaani, and Nikhil

Series Introduction

Marcel Dekker, Inc., has for many years specialized in the publication of high-quality monographs in tightly focused areas in a variety of medical disciplines. These have been of great value to both the practicing physician and the research scientist as sources of detailed and up-to-date information presented in an attractive format. During the last decade, there has been a veritable explosion in knowledge in the various fields related to infectious diseases and clinical microbiology. Antimicrobial resistance, antibacterial and antiviral agents, AIDS, Lyme disease, infections in immunocompromised patients, and parasitic diseases are but a few of the areas in which an enormous amount of significant work has been published. This new Infectious Disease and Therapy series will cover carefully chosen topics which should be of interest and value to the practicing physician, the clinical microbiologist, and the research scientist.

Brian E. Scully, M.B., B.Ch.
Harold C. Neu, M.D.

Foreword

The infectious diseases include more acute and life-threatening illnesses than any of the other classic disciplines of medicine such as cardiology and gastroenterology. Intoxications such as tetanus and botulism are very serious illnesses, as are such infections as meningitis, brain abscess, epiglottitis, anaerobic myonecrosis (gas gangrene), and necrotizing fasciitis. Commonly encountered infections, however—pneumonia, bacteremia (particularly when complicated by septic shock or endocarditis), and peritonitis—may also carry a significant mortality, especially in the very young or old and in the individual who is immunocompromised or who has other significant medical illness. A number of infections may be caused by a large variety of organisms, the therapy for which may differ widely. For example, pneumonia may involve a large number of different bacteria, including *Mycoplasma*, chlamydiae, rickettsiae, and mycobacteria, viruses, fungi, and parasitic agents. Often the clinical characteristics, epidemiology, general laboratory studies, and so forth, cannot clearly differentiate between a number of various etiological agents that might be involved. Conventional diagnostic tests may take days to weeks for definitive answers; however, preliminary results are very helpful in guiding therapy.

It is not always feasible or safe to use multiple antimicrobial agents in order to cover most reasonable etiological possibilities, even for the period of time until the specific etiology can be determined. It has been shown conclusively, in many serious infections, that the sooner one treats the patient with effective antimicrobials (and other appropriate therapy), the better the prognosis for survival without sequelae. Early diagnosis permits early use of the best drugs, minimizes the need for surgical therapy (or abortion), prevents or minimizes tissue destruction and abscess formation, and may prevent the spread of infection to others. The more rational use of drugs facilitated by early diagnosis minimizes use of unnecessary toxic agents or combinations, minimizes development of microbial resistance to drugs, and results in cost savings. Even early negative results redirect the diagnostic appraisal toward fastidious or unusual organisms or noninfectious processes and may prevent unnecessary use of antimicrobial drugs.

Aside from the utility of nucleic acid and monoclonal antibody probes in rapid specific diagnosis of infectious diseases, these reagents may be used to study microbial pathogenesis and to develop therapeutic or preventive interventions. Immunologically active, functionally defective mutants may serve effectively as toxoids or vaccines. Recombinant techniques may be used to identify protective peptides which might then be produced synthetically.

The editors are to be complimented for arranging this exciting reference and utilitarian text. They have assembled a first-class group of contributors and have organized the book beautifully to provide excellent coverage of the field with minimal overlap between chapters. The final chapter on recent developments provides a late update, something of particular value in a fast-moving field. Nevertheless, the rest of the book has such good, basic material in it that it will be useful for years to come.

Sydney M. Finegold, M.D.
Wadsworth VA Medical Center and
UCLA School of Medicine
Los Angeles, California

Preface

The impact of biotechnology on diagnostic microbiology is just beginning to be felt at the user level, in hospital laboratories and clinics, industrial quality control laboratories, and other settings where microbiological analyses are routinely performed. For years, the microbiologist has been envious of his counterparts working in chemical laboratories because the chemists have had modern analytical tools to perform their analyses with increased sensitivity, objectivity, and ease. The microbiologist has had to be content with conventional culture of microorganisms in liquid media and on agar plates. Their identification by conventional methods is by means of a battery of biochemical tests performed in liquid media, followed by serotyping by the use of polyclonal antisera derived from rabbits and carefully and laboriously absorbed with appropriate antigens to remove cross-reactive immunoglobulins. Rapid detection methods are desperately needed in diagnostic microbiology to improve patient care in clinical microbiology and to allow the timely institution of appropriate remedial measures in microbiological quality control in industrial processes. Rapid noncultural diagnostic methods are of particular significance for the rapid detection of slow-growing (e.g., *Mycobacterium*) or difficult-to-culture (e.g., viruses) microorganisms. Rapid noncultural techniques will be of great value in identifying microorganisms that have very slow metabolic processes. Further, the detection of specific toxins produced by microorganisms (e.g., heat-labile and heat-stable toxins of enterotoxigenic *Escherichia coli*) has conventionally been performed by costly, laborious animal response tests, such as the rabbit ileal loop assay and the suckling mouse assay. New methods are needed that are specific, sensitive, and easy to perform.

Diagnostic microbiology is presently undergoing profound changes as a result of application of biotechnology to detection and identification of microorganisms that cause human or animal disease or cause product spoilage. The two major technologies that have fostered the revolution in diagnostic microbiology are (1) nucleic acid hybridization and (2) monoclonal antibody technology. Nucleic acid probe-based detection systems are just being marketed. Currently available commercial nucleic probe systems include those for ***Salmonella*** and ***Listeria*** (Gene-Trak), ***Legionella*** (Gen-Probe), ***Campylobacter*** (New England Nuclear), enterotoxigenic (heat-labile and heat-stable toxins) ***Escherichia coli*** (New England Nuclear), and ***Mycoplasma*** (Gen-Probe). Monoclonal antibody-based in vitro diagnostic kits have been commercially available since the early 1980s and are now available for a wide spectrum of bacterial and viral pathogens and microbial toxins. The market for monoclonal antibody-based in vitro diagnostic kits is expected to grow over $1 billion by 1990.

There is a voluminous body of literature detailing the basic developments in molecular biology and research applications. However, there are relatively few books that deal exclusively with the application of hybridoma technology or DNA hybridization to the detection of pathogenic microorganisms, and no one book deals with both techniques. This book is aimed at filling this void and is primarily directed at the end users of the products of biotechnology for the rapid identification of pathogenic microorganisms in the real world: clinical laboratory and public health laboratory microbiologists, clinicians, supervisors, and directors, and personnel and managers of industrial microbiological quality control laboratories. Also, upper-level undergraduate students, graduate students, and researchers entering this field should find this book to be a useful reference. We hope that this book will put the new technology in perspective for our target audience and make them comfortable with and receptive to the evolving technology, acquaint them with the shortcomings of the new technologies, and allow them to make rational decisions about the choice of the appropriate technology for their situation. We expect that this book will stimulate those involved in the development of assays for diagnostic microbiology to make new and improved diagnostic kits.

One of the most frequently debated topics among diagnostic microbiologists is: Which technique is better, monoclonal antibody-based immunoassay or nucleic acid hybridization? This is a complex issue, and there is no single, unequivocal answer available at this point. Nucleic acid probe-based assays have been shown to be highly specific and sensitive in limited laboratory trials and in evaluations by reference laboratories; however, new and unforeseen problems may arise when such assays are used in the field on "real" specimens. For an assay technique to be widely accepted, it should have comparable sensitivity and specificity to the currently used standard technique, it should be easy to perform and automate, and it should be free from safety concerns. Monoclonal antibody-based immunoassays presently hold an edge over nucleic acid hybridization assays with respect to these requirements. Further, monoclonal antibody-based immunoassays provide for the direct detection of a toxin in an assay, whereas the nucleic acid probe-based test can provide only an indirect answer in such a situation. A major research effort is underway in commercial and public research laboratories to improve nucleic hybridization technology through such means as nonradioactive probe development, increased sensitivity through gene amplification, and rapid reaction times by optimizing the kinetics of hybridization reactions. This massive effort is expected to bear results in the near future. Also, new formats of hybridization tests are being proposed where the identification of a pathogenic bacterium and its antibiotic sensitivity profile can be ascertained in one step.

Ultimately, it must be realized that both techniques can complement one another. For example, amino acid sequences of immunogens can be deduced from the nucleic acid sequence of the appropriate cloned gene and hydropathy plots derived from such amino acid sequences can be used to identify the most hydrophobic region of the protein. Such regions are known to be highly antigenic, and monoclonal antibodies can be prepared against synthetic peptides containing the antigenic regions for use in diagnostics. Similarly, monoclonal antibodies have been prepared against DNA:RNA and DNA:DNA hybrids. Such monoclonal antibodies are used to detect nucleic acid hybridization reactions. Thus, we can see the two technologies merging into each other in the future so that the strengths of each are augmented in the achievement of the final objective: the development of highly specific, highly sensitive, easy-to-perform assays.

The book is organized in two parts. Part I deals with nucleic acid-based typing and detection of microorganisms. The first chapter, by Dr. John Johnson, provides an overview of the basic principles of nucleic acid hybridization. A clear understanding of the basic principles is crucial to the development and evaluation of nucleic acid probe-based test systems. The chapter by Dr. Jeffry Leary and Dr. Jerry Ruth details the state of the art in nonisotopic detection systems for nucleic acid hybridization. DNA probes to ribosomal RNA are particularly attractive because of the greater sensitivity that can be achieved by this technique. Dr. James Hogan's chapter deals with DNA probes to ribosomal RNA of some clinically significant microorganisms.

Detection is only one facet of nucleic acid technology. Bacterial taxonomy has been revolutionized by nucleic acid hybridization, as can be seen from the current edition of Bergey's *Manual of Systematic Bacteriology*. Dr. Don Brenner describes this important application in the next chapter. The following chapter, by Dr. Scott Holmberg and Dr. Kaye Wachsmuth, fascinatingly details the application of plasmid profiling in molecular epidemiology of bacterial diseases. Specific applications of nucleic acid technology are dealt with in the ensuing eight chapters. The final chapter in Part I, by Dr. John Washington and

Dr. Gail Woods, summarizes the application of nucleic acid probes in diagnostic microbiology and provides specific examples of the application of this technology in the detection of viruses. Because of space limitations, it was deemed infeasible to include individual chapters on the detection of medically important viruses in this book.

Part II of the book deals with monoclonal antibody-based immunoassays, and its organization is similar to that of Part I. Methodology for generating hybridomas and characterization of monoclonal antibodies are covered in the first chapter, by Dr. Richard Goldsby. The next chapter, by Dr. David Nau, covers the application of high-performance liquid chromatography to the purification of monoclonal antibodies. Specific applications of monoclonal antibody-based immunoassays are covered in the next 10 chapters. The concluding chapter, by Dr. John Herrmann, summarizes the present status of monoclonal antibody technology in infectious disease diagnosis.

New developments in nucleic acid and monoclonal antibody probe technologies, which were reported after work on the book began and which are expected to significantly impact diagnostic microbiology, are discussed in Part III.

We were privileged to work with a group of authors who are recognized leaders in their respective areas and who willingly gave their time to contribute to this volume despite their busy schedules. We are forever in their debt.

This book would not have become a reality without the support, encouragement, and assistance of several of our peers and colleagues. Graham Garratt of Marcel Dekker, Inc., provided the impetus for the initiation of this project. Dr. Don Brenner and Dr. Sydney Finegold encouraged us to take on this project and helped us in recruiting prospective contributors. Ms. Marian Davis provided valuable assistance with literature search. Ms. Pat Aveline helped with correspondence and word processing. Kaberi Das and Sandhya Sathe assisted with the editing of manuscripts. Elaine Grohman of Marcel Dekker, Inc., was ready to help whenever we needed her. We are indebted to all those mentioned above and several others who willingly helped us in our endeavors to put this manuscript together.

Bala Swaminathan
Gyan Prakash

Contributors

Don J. Brenner Molecular Biology Laboratory, Meningitis and Special Pathogens Branch, Division of Bacterial Diseases, Center for Infectious Diseases, Centers for Disease Control, Atlanta, Georgia

Mary A. Buesing Department of Preventive Medicine, Uniformed Services University of the Health Sciences, Bethesda, Maryland

David R. Bundle Immunochemistry Section, Division of Biological Science, National Research Council of Canada, Ottowa, Ontario, Canada

Michael C. Chen[a] Eukaryotic DNA Cloning, BioTechnica Diagnostics, Inc., Cambridge, Massachusetts

John W. Cherwonogrodzky Agriculture Canada, Animal Diseases Research Institute (Nepean), Nepean, Ontario, Canada

Marie B. Coyle Departments of Laboratory Medicine and Microbiology, University of Washington, Seattle, Washington

Michael S. Curiale Integrated Genetics, Inc., Framingham, Massachusetts

Present affiliation:
[a]DNA Cloning and Expression, American BioTechnologies, Inc., Cambridge, Massachusetts.

Deborah E. Dixon-Holland[a] Department of Food Science and Human Nutrition, Michigan State University, East Lansing, Michigan

Suzanne M. Eklund Operations, BioTechnica Diagnostics, Inc., Cambridge, Massachusetts

Peter Feng[b] Department of Microbiology, IGEN, Inc., Rockville, Maryland

Cynthia K. French Research and Development, BioTechnica Diagnostics, Inc., Cambridge, Massachusetts

Michael G. Gabridge Bionique Laboratories, Inc., Saranac Lake, New York

Steven J. Geary Research and Development, Bionique Laboratories, Inc., Saranac Lake, New York

Richard A. Goldsby Department of Biology, Amherst College, and Department of Veterinary and Animal Science, University of Massachusetts at Amherst, Amherst, Massachusetts

Patricia Guerry Department of Infectious Diseases, Naval Medical Research Institute, Bethesda, Maryland

Mary Catherine Harris Department of Pediatrics, University of Pennsylvania School of Medicine, and The Children's Hospital of Philadelphia, Philadelphia, Pennsylvania

John E. Herrmann Division of Infectious Diseases, University of Massachusetts Medical School, Worcester, Massachusetts

James J. Hogan Gen-Probe, Inc., San Diego, California

Scott D. Holmberg[c] Enteric Diseases Branch, Division of Bacterial Diseases, Center for Infectious Diseases, Centers for Disease Control, Atlanta, Georgia

Present affiliations:
[a]Research and Development, Neogen Corporation, Lansing, Michigan.
[b]Department of Microbiology, U.S. Food and Drug Administration, Washington, District of Columbia.
[c]AIDS Program, Epidemiology Branch, Center for Infectious Diseases, Centers for Disease Control, Atlanta, Georgia.

John L. Johnson Department of Anaerobic Microbiology, Virginia Polytechnic Institute and State University, Blacksburg, Virginia

Charles Kaspar[a] Department of Microbiology, Iowa State University, Ames, Iowa

Changmin Kim Department of Foods and Nutrition, Purdue University, West Lafayette, Indiana

Lynn C. Klotz[b] BioTechnica Diagnostics, Inc., Cambridge, Massachusetts

Joan S. Knapp Sexually Transmitted Diseases Laboratory Program, Center for Infectious Diseases, Centers for Disease Control, U.S. Public Health Service, Department of Health and Human Services, Atlanta, Georgia

Arend H. J. Kolk N. H. Swellengrebel Laboratory of Tropical Hygiene, Royal Tropical Institute, Amsterdam, The Netherlands

Dennis J. Kopecko Department of Bacterial Immunology, Walter Reed Army Institute of Research, Washington, District of Columbia

Jeffry J. Leary Anti-infectives, Smith, Kline & French Laboratories, King of Prussia, Pennsylvania

Rance B. Le Febvre[c] National Animal Disease Center, Agricultural Research Service, U.S. Department of Agriculture, Ames, Iowa

Sheila A. Lukehart Division of Infectious Diseases, Department of Medicine, University of Washington School of Medicine, Seattle, Washington

David M. Lyerly Department of Anaerobic Microbiology, Virginia Polytechnic Institute and State University, Blacksburg, Virginia

David R. Nau Research Specialty Products Division, J. T. Baker, Inc., Phillipsburg, New Jersey

---

Present affiliations:
[a]Fisheries Research Branch, U.S. Food and Drug Administration, Dauphin Island, Alabama.
[b]Biotechnology consultant, Cambridge, Massachusetts.
[c]Department of Veterinary Microbiology, University of California at Davis, Davis, California.

Malcolm B. Perry Division of Biological Sciences, National Research Council of Canada, Ottawa, Ontario, Canada

James J. Pestka Department of Food Science and Human Nutrition, Michigan State University, East Lansing, Michigan

Carol J. Phelps Department of Anaerobic Microbiology, Virginia Polytechnic Institute and State University, Blacksburg, Virginia

Gyan Prakash Mesa Diagnostics, Inc., Albuquerque, New Mexico

Bhanu P. Ram Research and Development, Idetek, Inc., San Bruno, California

Ayoub Rashtchian[a] Integrated Genetics, Inc., Framingham, Massachusetts

Nalin Rastogi Unité de la Tuberculose et des Mycobactéries, Institut Pasteur, Paris, France

W. Stuart Riggsby Department of Microbiology and Graduate Program in Cellular, Molecular, and Developmental Biology, University of Tennessee, Knoxville, Tennessee

Marilyn C. Roberts Department of Pathobiology, University of Washington, Seattle, Washington

Fran A. Rubin Department of Bacterial Immunology, Walter Reed Army Institute of Research, Washington, District of Columbia

Jerry L. Ruth Molecular Biosystems, Inc., San Diego, California

Eugene D. Savitt[b] Department of Periodontology, Forsyth Dental Center, Boston, Massachusetts

Susanne L. Simon BioTechnica Diagnostics, Inc., Cambridge, Massachusetts

Richard S. Stephens[c] Department of Biomedical and Environmental Health Science, University of California–Berkeley, Berkeley, California

Present affiliations:
[a]Life Technologies, Inc., Gaithersburg, Maryland.
[b]Department of Dental Research, BioTechnica Diagnostics, Inc., Cambridge, Massachusetts.
[c]Departments of Laboratory Medicine and Pharmaceutical Chemistry and the Francis I. Proctor Foundation, University of California–San Francisco, San Francisco, California.

Bala Swaminathan[a] The Food Sciences Institute, Purdue University, West Lafayette, Indiana

Alex B. Thiermann[b] National Animal Disease Center, Agricultural Research Service, U.S. Department of Agriculture, Ames, Iowa

Karen K. Vaccaro BioTechnica Diagnostics, Inc., Cambridge, Massachusetts

Kaye Wachsmuth Enteric Bacteriology Section, Center for Infectious Diseases, Centers for Disease Control, Atlanta, Georgia

John A. Washington Department of Microbiology, Cleveland Clinic Foundation, Cleveland, Ohio

Tracy Dale Wilkins Department of Anaerobic Microbiology, Virginia Polytechnic Institute and State University, Blacksburg, Virginia

Gail L. Woods[c] Department of Microbiology, Cleveland Clinic Foundation, Cleveland, Ohio

E. Pamela Wright[d] Department of Medical Microbiology, University of Amsterdam, Amsterdam, The Netherlands

Present affiliations:

[a]Epidemic Investigations Laboratory, Meningitis and Special Pathogens Branch, Center for Infectious Diseases, Centers for Disease Control, Atlanta, Georgia.

[b]Agricultural Research Service, National Program Staff, U.S. Department of Agriculture, Beltsville, Maryland.

[c]Department of Pathology and Microbiology, University of Nebraska Medical Center, Omaha, Nebraska.

[d]Department of International Education and Training, Royal Tropical Institute, Amsterdam, The Netherlands.

Contents

II. MONOCLONAL ANTIBODY-BASED IMMUNOASSAYS

Nucleic Acid and Monoclonal Antibody Probes

I

NUCLEIC ACID DETECTION AND CHARACTERIZATION

1
Nucleic Acid Hybridization
Principles and Techniques

JOHN L. JOHNSON
Virginia Polytechnic Institute and State University, Blacksburg, Virginia

I. INTRODUCTION

One of the oldest disciplines in science is that of classification and the corresponding use of classifications for identification. Historically, the classification of organisms has been based upon phenotypic properties that range from size, shape, color, and other physical traits to the transformation of specific compounds, the presence of specific enzymic activities, and the presence of specific antigens. More recently, the molecules that contain or transmit the genetic information in a cell, deoxyribonucleic acid (DNA) and ribonucleic acid (RNA), have been subjected to detailed comparisons, and the results have been used in classification. Nowhere have these nucleic acid studies had more impact than in microbiology.

The unique physical property of DNA and RNA—the complementary strand separation and reassociation of DNA and the hybridization between RNA and its template strand of DNA—has enabled investigators to compare the nucleotide sequences of one organism with those of another. During the past 25 years, microbiologists have directed most of their studies toward elucidating the taxonomic and phylogenetic relationships among the various groups of microorganisms, viruses, and plasmids. With this accumulated data base as background, investigators are in a position to use DNA, from selected organisms, viruses, or plasmids, as labeled probes to identify complementary DNA or RNA isolated from unidentified organisms. These methods can also be used to detect the presence of specific organisms in mixed populations. The purpose of this chapter is to describe the nature of the hybridization reactions and to discuss the important variables that must be considered in designing experiments and in interpreting the results.

II. HISTORICAL BACKGROUND

History is a somewhat relative term, in that the more rapidly events or advances occur, what is considered the historical past becomes less distant. This is truly so with nucleic acid research. What is discovered one year often is in general use by other investigators by the following year. Without trying to make anybody feel old, I wish to highlight some of the classic experiments and studies that have contributed to the present understanding of nucleic acid structure and the technology for its use in classification and identification.

A. Early Experiments

Experiments carried out in Paul Doty's laboratory at Harvard in the late 1950s and early 1960s ushered in the era of DNA reassociation and RNA hybridization (1,2). While using antibiotic resistance to investigate bacterial transformation, Doty and his associates observed that as DNA samples were subjected to increasingly high temperatures before the transformation experiment, the number of transformants decreased because of the "denaturation" of the DNA. However, if the DNA samples were allowed to cool slowly before the transformation step, the transforming capability of the DNA returned to a level comparable with unheated "native" DNA and, thus, was considered to have been "renatured." These studies were extended to the use of a UV spectrophotometer with a heated cuvette chamber, which enabled the investigators to correlate denaturation with the hyperchromic shift (melting curve) of the DNA as it went from a double-stranded to a single-stranded form. Additional observations from the spectrophotometric experiments included (a) The midpoint of the DNA melting curve (T_m) correlated with the mole percentage guanine plus cytosine (mol% G+C) of the DNA (3); (b) DNA renaturation could also be measured spectrophotometrically; (c) The rate of renaturation was a function of the DNA concentration, salt concentration of the buffer, and the temperature; and (d) DNA from higher organisms tended to renature more slowly than DNA from lower organisms.

B. Use of Heavy Isotopes of Nitrogen

The first successful experiments detecting the reassociation between two preparations of DNA involved the use of heavy isotopes of nitrogen (4). Two DNA preparations were isolated from *Bacillus subtilis*. Cells for the first preparation were grown on ^{14}N (normal or light isotope) and for the second preparation on a ^{15}N (heavy isotope)-enriched nitrogen source. The difference in nitrogen density cause the DNA preparations to band at different positions when centrifuged in CsCl. This property is illustrated in panels A and B in Figure 1. Thus, mixtures of heavy and light isotope-labeled DNA can be separated by this

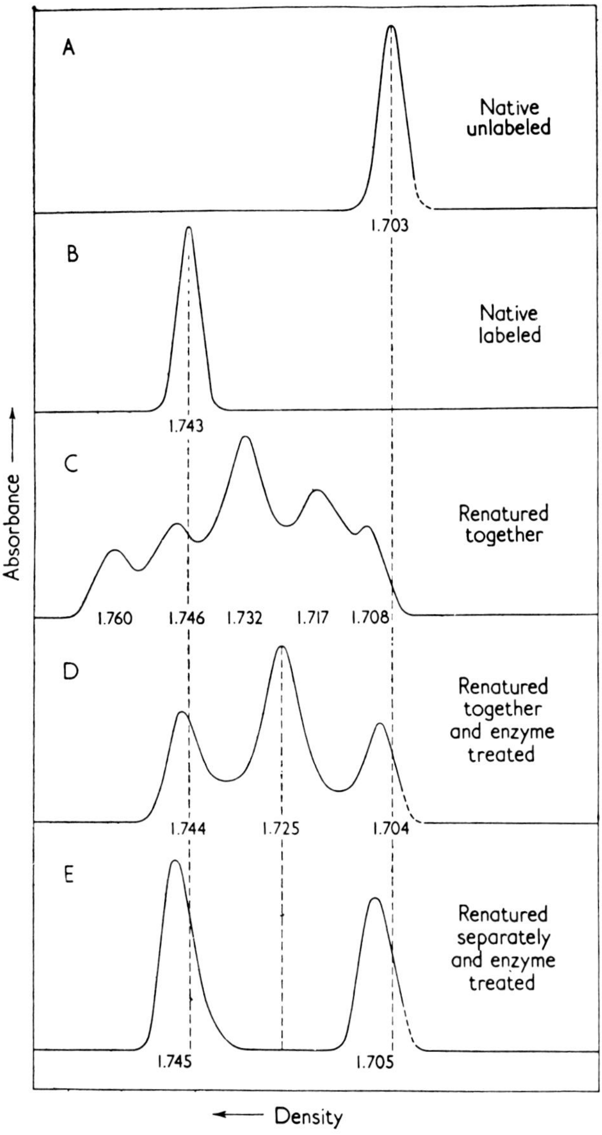

Figure 1 The second-order reassociation kinetics of DNA, as demonstrated using mixtures of ^{14}N and ^{15}N density-labeled DNA and an analytical ultracentrifuge. See text for details (*Source*: Ref. 4).

technique. When such a DNA mixture is heated and fast cooled, upon centrifugation both bands shift to higher densities because of single-strandedness (this is not illustrated in Fig. 1). If the two DNA preparations are mixed together, denatured, and then allowed to reassociate before CsCl centrifugation, several bands will form (Figure 1, panel C). The multiple bands are due to partial duplexes, in which part of a fragment is double-stranded and part is still single-stranded, resulting in intermediate densities. When the reassociated DNA mixture is digested with *Escherichia coli* phosphodiesterase before centrifugation, single-stranded regions are removed, and the banding pattern shown in Figure 1, panel D, is obtained. The major duplex peak is midway between the densities of the individually reassociated peaks. Moreover, the sizes of the peaks are in a 1:2:1 ratio which demonstrates that the complementary strands completely separated from each other and that there was a random reassociation of complementary DNA fragments. When each denatured DNA preparation was allowed to reassociate and then treated with phosphodiesterase, each shifted back to the respective double-stranded density (Figure 1, panel E). This experiment demonstrated a method for comparing the DNA from one organism with that of another; however, the experiments were difficult and slow, and the results were only qualitative. In addition, the method worked best for only small genomes such as those of viruses (5).

C. Immobilized Nucleic Acids

A method for immobilizing denatured DNA was developed at the Carnegie Institute in Washington (6,7). High-molecular-weight (HMW) DNA was denatured, mixed with molten agar, and the mixture rapidly cooled. The agar was then cut into small pieces by passage through a screen placed at the bottom of a syringe. The single strands of denatured (HMW) DNA remained trapped in the agar matrix and were unable to reassociate with one another. Radioactively labeled low-molecular-weight DNA fragments were then mixed with the agar particles. These freely diffused in and out of the agar particles and readily formed duplexes with the immobilized DNA. After the reassociation step, the agar and DNA complexes were washed free of the unreassociated labeled DNA fragments. The labeled DNA fragments that had formed duplexes with the immobilized DNA were then eluted from the agar by using a combination of low salt and high temperature. The extent of reassociation was determined by measuring the radioactivity. This was the first "easy" method for comparing DNAs from different organisms. A number of organisms could be readily compared, and the results were reasonably quantitative. The optimization of many of the reassociation characteristics was done with this procedure.

Nitrocellulose membranes were first used to bind DNA and RNA-DNA hybrids by Nygaard and Hall (8). Gillespie and Spieglman (9) took advantage of this property and first immobilized DNA on nitrocellulose membranes and demonstrated quantitative hybridization of ribosomal RNA (rRNA) to the immobilized DNA. Although extremely useful in RNA hybridization studies, the nonspecific binding of DNA to the membranes precluded the use of labeled DNA probes. Denhardt (10) discovered that after the unlabeled DNA had been immobilized on the nitrocellulose membrane, the remaining binding sites on the membrane could be covered by incubating the membrane in a solution containing a mixture of bovine serum albumin, ficoll, and polyvinylpyrrolidone. This mixture has become universally known as the "Denhardt prehybridization mixture." In 1975, Southern (11) took the membrane immobilization of nucleic acids a step further by first separating DNA fragments by electrophoresis in agarose gels and then, after denaturing the fragments, transferring them to nitrocellulose membranes. This was rapidly followed by a variation of the transfer procedure, introduced by Thomas (12), by which messenger RNAs (mRNA) could be transferred from gels onto nitrocellulose membranes. The nucleic acid immobilization technology introduced by these individuals represents one of the major milestones in molecular biology.

The kinetics of the reassociation and hybridization reactions were important considerations in all of the aforementioned studies, but because of its importance in designing experiments and tests, reaction kinetics will be discussed in detail in the next section. However, from a historical perspective, the studies of Britten and Kohne (13), Wetmur and Davidson (14), Britten and Davidson (15), and Galau and associates (16,17) have provided much of the basic information about the kinetics of these reactions.

III. REASSOCIATION AND HYBRIDIZATION KINETICS

One of the more important considerations in using DNA reassociation and RNA hybridization techniques for organism identification or for the detection of unique genes, is the reaction kinetics. The rate-limiting step in DNA reassociation or RNA hybridization is *nucleation*, which is the alignment of the required number of bases to be the focal point from which the rest of the bases along the two complementary fragments can form hydrogen bonds. The rate of the reaction is dependent upon a complex interaction of several components that include the type of DNA or RNA, the concentrations of the nucleic acids, the type and concentration of salt in the reaction buffer, the concentrations and kinds of certain organic molecules that may be included in the reaction mixture, and the temperature at which the reaction occurs. With an understanding of these variables, one can predict how long a reaction must be incubated to approach completion, how much reassociation will occur among the probe DNA fragments as opposed to the reaction between the probe fragments and the target fragments, the minimal amount of DNA or RNA (i.e., the number of cells) that is required for detection when using a specific experimental design and, in general, aid in designing optimal detection systems. These experimental variables will be discussed in detail.

A. Solution Kinetics

Reassociation and hybridization reactions occurring in solution are the easiest to describe in mathematical terms, thus these will be discussed in some detail. The reassociation of DNA (from double-stranded genomes) follows second-order reaction kinetics, as initially demonstrated in the heavy isotope experiments illustrated in Figure 1. Experiments designed to detect the presence of a particular nucleic acid usually involve a mixture of at least two nucleic acid preparations. The labeled nucleic acid, often referred to as the *probe* or *tracer*, is usually at a low concentration. The unlabeled nucleic acid, referred to as the *driver* or *target* nucleic acid, is usually at a high concentration. In designing experiments, the kinetics of both the probe and the driver nucleic acid must be considered.

1. *DNA Reassociation*

The reassociation of DNA (as described for an organism that has double-stranded DNA) follows second-order reaction kinetics. Here, there is an equal mixture of complementary DNA fragments that are removed from the reacting population of fragments by forming duplexes. Figure 2 illustrates the kinetics for a reassociation reaction containing one kind of DNA. When more than one kind of DNA are included (for example, probe and driver) the kinetics of both must be considered individually.

The traditional form of the second-order rate equation [Eq. (1)] has been rearranged by Britten and Kohne (13) [Eq. (2)] and is used for generating what has come to be known as "C_0t curves."

$$1/C - 1/C_0 = k \tag{1}$$

$$C/C_0 = 1/(1 + kC_0t) \tag{2}$$

The concentration of single-stranded DNA (in moles nucleotides per liter) at zero time is represented by C_0, C is the concentration of single-stranded DNA at time t, (t = seconds) and k represents the reassociation rate constant. The rate constant units are $L \cdot mol^{-1} \cdot t^{-1}$. When the DNA is half reassociated, C/C_0 in Eq. (2) equals 1/2, and $kC_0t = 1$. Therefore, at one-half reassociation, $k = 1/C_0t$ and $C_0t = 1/k$. A general log plot of C_0t values is shown in Figure 2. There is a near linear region on the curve that extends over 2 logs. This is because any significant change in C/C_0 occurs between $1/(1 + 0.1)$, i.e., 90% still denatured, and $1/(1 + 10)$, i.e., only 9% still denatured.

Although one usually works with DNA in terms of milligrams per milliliter (mg/ml), these values are readily converted to moles nucleotides per liter by di-

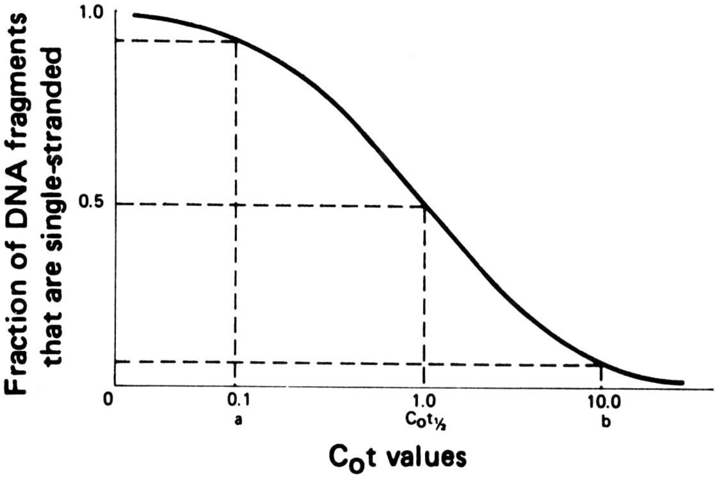

Figure 2 Generalized C_0t plot. At C_0t values less than, or equal, to (*a*), $1/(1 + C_0t)$ values are 0.9 or greater. At C_0t values equal or greater than (*b*), $1/(1 + C_0t)$ values are less than 0.1 (*Source*: Ref. 18).

viding them by 331 (the average molecular weight of the deoxyribonucleotide sodium salts). Approximate C_0t values can also be estimated by dividing the optical density at 260 nm of denatured DNA (at room temperature) by 2 and multiplying by reassociation time in hours (13).

2. RNA Hybridization

When the driver nucleic acid is RNA or single-stranded DNA (for example, viruses such as ϕX174 or M13), the reaction between it and the probe nucleic acid follows pseudo-first-order reaction kinetics. In this reaction, there is essentially no change in the concentration of the driver nucleic acid throughout the reaction period. This is because there are no complementary fragments among the driver fragments and the number of hybrids between the probe and the driver fragments will not measurably reduce the concentration of the driver. Equation (3) is a pseudo-first-order equation used by Galau and co-workers (16) in characterizing the reassociation kinetics of ϕX174 DNA.

$$\frac{C_{RF}}{C_{RF_0}} = e^{-k_{pFO}C_{(+)}t} \tag{3}$$

Here, the double-stranded replicating form (RF) DNA was used as the probe. The concentration of RF DNA at any time t equals C_{RF}, C_{RF_0} is the concentration the RF probe DNA at zero time, $C_{(+)}$ is the concentration of the single-strand (plus strand) driver DNA, and k_{PF0} is the pseudo-first-order rate constant in $L \cdot mol^{-1} \cdot t^{-1}$. Graphic representation of data using this equation is shown in Figure 3 (16).

The plot is the same as that for the C_0t curve seen in Figure 2, in that the fraction of unreassociated probe (C_{RF}/C_{RF_0}) is plotted against the log of the driver concentration times time ($C_{(+)}t$). Again the rate constant can be readily calculated at one-half reassociation or hybridization, where $-k_{PF0}$ will equal $0.693/C_{(+)}t$. The dotted line in panel b of Figure 3 is the second-order reassociation curve that results from using RF DNA as the driver (i.e., from panel a), which helps to illustrate the differences between the two reactions. The pseudo-first-order curve is sharper, extending over only about 1.5 logs and the rate is about twice that of the second-order reaction. It should also be noted that nearly identical results were obtained when (+)strand RNA was used in place of the (+)strand DNA for the driver nucleic acid.

When labeled RNA is used as the probe for hybridization with DNA, the rate constant for RNA-DNA hybrid formation is 3- to 4.5-fold lower than the rate constant for DNA reassociation (17,19). Although clearly documented, the reasons for these differences are not understood.

B. Kinetics Involving Immobilized Nucleic Acids

Although the kinetic principles are the same for reassociation and hybridization reactions when the driver nucleic acid is immobilized, the concentration values for the immobilized nucleic acids are impossible to determine. For example, where a given amount of DNA is immobilized on a nitrocellulose membrane, the measured reassociation reactions are occurring only at the surface of the membrane. Therefore, the effective concentration will be a function of the size of the membrane and the volume of the reassociation mixture. As a result, optimization of these systems has been more empirical. It is important to remember that the probe will be following solution kinetics, i.e., a double-stranded DNA probe will be reassociating with its own complementary fragments in solution as a function of its concentration and genomic complexity. Thus, the concentration of probe DNA fragments having the potential of forming duplexes with the im-

---

Figure 3 Comparison of second-order and pseudo-first-order reassociation kinetics. Hybridization of labeled replicating-form (RF) DNA of ϕX174 with (*a*) an excess of unlabeled RF DNA (second-order), (*b*) an excess of unlabeled (+)strand DNA (pseudo-first-order). The dashed line in *b* is the second-order curve from panel *a*. (*c*) An excess of (+)strand RNA (pseudo-first-order). The dashed line in *c* is the pseudo-first-order curve from panel *b* (*Source*: Ref. 16).

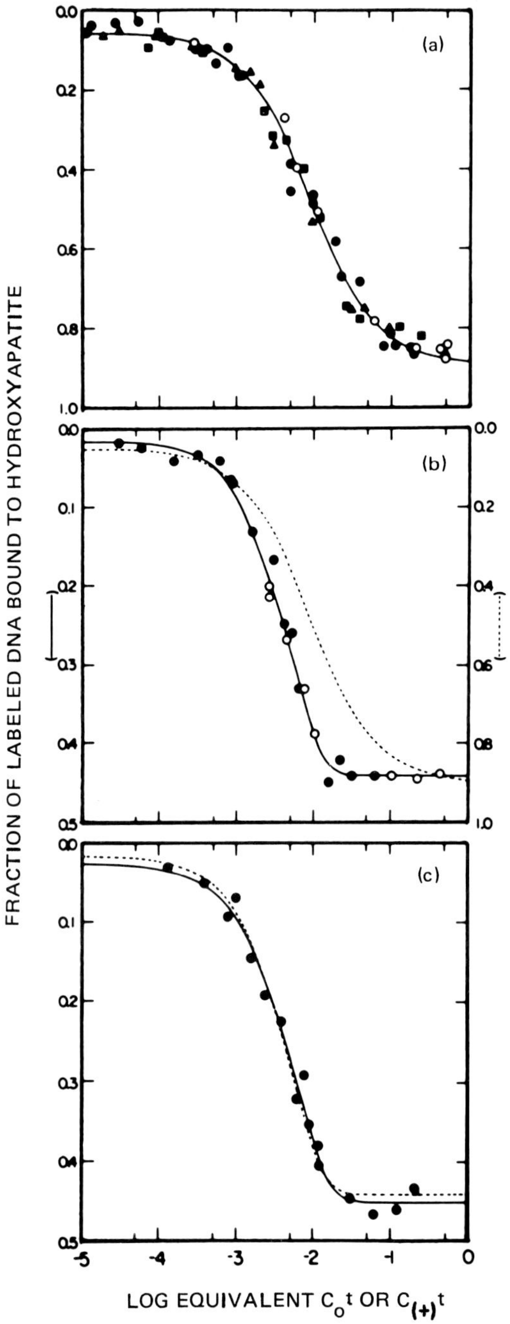
FRACTION OF LABELED DNA BOUND TO HYDROXYAPATITE
LOG EQUIVALENT C_0t OR $C_{(+)}t$
(a)
(b)
(c)
0.0
0.2
0.4
0.6
0.8
1.0
0.1
0.3
0.5
-5
-4
-3
-2
-1
0

mobilized DNA is steadily decreasing during the reassociation period. With a single-stranded probe, on the other hand, the concentration of probe will be decreased only by those fragments that have duplexed or hybridized with the immobilized nucleic acid. McCarthy (20), demonstrated the effect of variations in both the concentrations of labeled DNA in solution and the amounts of unlabeled DNA immobilized on membranes. It should be pointed out that these early experiments were conducted with rather low specific-activity DNA, for which the concentrations of labeled DNA were substantial (i.e., from 0.65–14 μg/ml). It was also demonstrated with the early experiments that it is very difficult to reassociate immobilized DNA to completion with labeled DNA. This is because as the concentration of the labeled DNA in solution is increased, the solution reassociation of the probe DNA competes with the rate of reassociation between the labeled and immobilized DNA. As a result, the probe DNA only asymptotically approaches saturation of the immobilized DNA. On the other hand, when labeled rRNA is used as the probe, the rRNA cistrons are readily saturated with this probe (9). This is due to both the high copy number of the rRNA molecules per unit weight of rRNA and to the fact that it is single-stranded. With probes that use the high-specific-activity DNA and rRNA preparations that are readily available today (i.e., 10^6–10^8 cpm/μg), one can use very low concentrations of labeled probe (i.e., 0.01-0.5 μg/ml), so that within a reasonable time frame, free solution hybridization does not influence the rate of probe hybridization with the immobilized DNA. These probes are very sensitive for detecting complementary sequences, and the amount of signal will be directly proportional to both the amount of probe and the amount of DNA immobilized on the membrane.

Nitrocellulose membranes were the first to be extensively used for immobilizing nucleic acids. In recent years, diazotized paper has been introduced for the covalent attachment of nucleic acids, and membranes prepared from nylon have also become popular.

IV. EFFECTS OF EXPERIMENTAL VARIABLES ON REASSOCIATION AND HYBRIDIZATION RATES

The aforegiven reaction kinetics are dependent upon a number of experimental variables; if any are altered, the rates will also change. Some of the major factors will be considered in the following discussion.

A. Genome Complexity, Fragment Size, and Nucleic Acid Concentrations

The effect of genome size or complexity on the rate of DNA reassociation was clearly demonstrated by Britten and Kohne (13) in their classic paper intro-

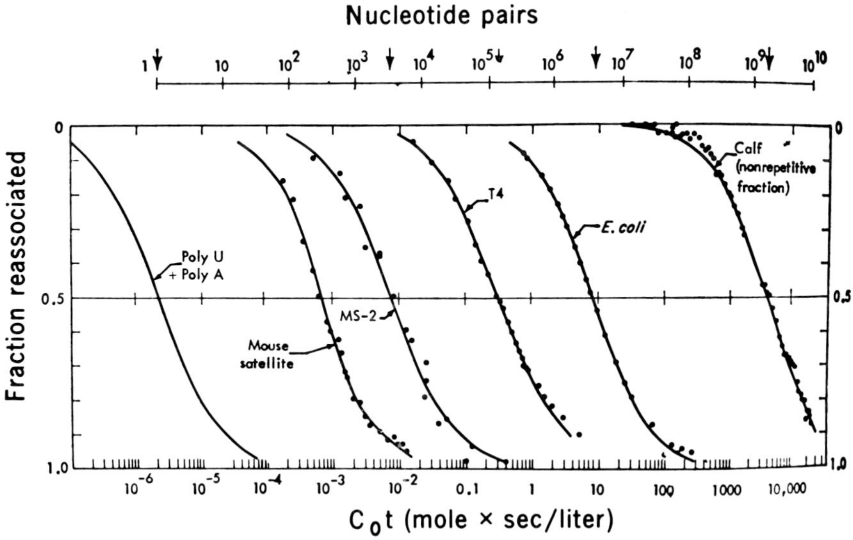

Figure 4 Reassociation of double-stranded nucleic acids from various sources. All of the rates are normalized to those that would be observed at 0.18 M sodium-ion concentration (*Source*: Ref. 13).

ducing the use of C_0t curves. Some of their C_0t curves are reproduced in Figure 4. The rate of reassociation is inversely proportional to the size of the genome, i.e., there will be fewer gene copies per weight of DNA (e.g., per μg) for a large genome than for a small one. The genome sizes are given in nucleotide base pairs (bp) per genome, and are on the scale over the graph. Two genomes that differ in size by a factor of 10 will have reassociation rates that differ by the same factor. Therefore, there are some general guides that one can follow in designing detection or identification systems. In this respect I wish to consider genomes that range in size from bacteria to small viruses and plasmids. The most obvious fact is that for obtaining comparable results, one needs much less DNA for a virus or a plasmid than for a bacterium. However, with small genomes, one must be more concerned about self-reassociation of the probe DNA. The extreme condition would be using a cloned fragment (1-2 kb in length) for detecting that sequence in bacterial DNA. These relationships are illustrated in Figure 4; most bacterial DNA preparations will have a reassociation rate comparable with that of *E. coli*, but the self-reassociation rate for the probe DNA would lie somewhere between the rate for mouse satellite DNA and MS-2 phage, i.e., about 4000 times faster. In this instance, a total DNA excess of 2000-4000 would result only in a 1:1 ratio of probe fragments to target fragments in the unlabeled DNA preparation.

The self-reassociation rate of the probe DNA would be about half that of the probe and target fragment mixture. In this type of an experiment it would be easy to have a situation in which the probe may, in fact, function as a driver. Therefore, one must compromise between using a highly labeled probe, so that low concentrations can be used, and a reduced reaction rate between the probe and the target DNA, which also results from the lower probe concentration.

The fragment size of DNA determines the amount of reassociation that occurs with each effective nucleation; therefore, the longer the fragments, the faster the reassociation (14). In practice, DNA preparations from the organisms being compared are fragmented (usually by sonication or shearing in a French pressure cell) to a similar extent, so that fragment size is no longer a variable. One must be aware of potential differences when comparing results from one laboratory with those from another. When DNA preparations are being immobilized on membranes, fragment sizes are not as often considered because, usually, HMW DNA is denatured and immobilized. This is, in part, a carryover from the days of agar gel immobilization, when the DNA fragments had to be large or they would leach out of the agar but, also, because they cannot reassociate with other HMW fragments, one obtains maximum reassociation with each effective nucleation. If solution reassociation of probe DNA is a factor for a particular application, larger fragment sizes will, of course, enhance it. The length of the probe DNA also needs to be considered if only a part of the probe sequence has partial sequence similarity with the sequences of interest in the target DNA immobilized on the membrane. If there is too much unrelated flanking DNA, it may destabilize the duplexes of interest.

Another aspect of fragment size that is rather interesting, is the specificity of reassociation for very short fragments. When they used oligonucleotides prepared by the hydrolysis of bacterial, phage, and mouse DNA, McConaughy and McCarthy (21) found that species-specific reassociation occurred with fragments that were longer than 14 bases. Similar results were obtained by Niyogi (22) who used oligoribonucleotides with T-phage DNAs. The specificity of these short fragments are clearly shown by the 15-base oligonucleotide primers that are used for dideoxy-DNA sequencing reactions.

B. Temperature of the Reaction Mixtures

The temperature at which reassociation or hybridization reactions are carried out is extremely important, because it will affect both the rate and the specificity of the reactions. The specific temperature requirements are interrelated with the mole percentage guanine plus cytosine (mol% G+C) of the DNA or RNA, whether the reaction is between DNA or between DNA and RNA, and other components of the reaction mixture, including salts and organic molecules. In this section, only the nucleic acid effects will be considered.

Marmur and Doty (23) observed that the optimal reassociation rate for DNA in 2X SSC (SSC = standard saline citrate; 0.15 M NaCl, 0.015 M sodium citrate, pH 7.0) was about 25°C below the midpoint of the optical melting profile (T_m) of the DNA. This has been a very useful guideline for a wide range of studies because the reaction is reasonably specific, yet can detect similarities between reasonably divergent sequences. Raising the temperature to, say, 15°C below the melting temperature will increase the specificity to the point that only highly conserved sequences will be able to form duplexes. More diverged sequences can be detected by lowering the reassociation temperature to 30 to 40°C below the T_m for the test organism with the highest mol% G+C. However, the background will be higher because of an increase of nonspecific associations between the higher mol% G+C components.

A cell contains several classes of RNA, which include messenger RNA (mRNA), ribosomal RNA (rRNA) and transfer RNA (tRNA). An important property of RNA is that during hybridization at high temperatures (65-70°C), it will hydrolyze quite rapidly (24), with a corresponding decrease in hybridization efficiency. Formamide is commonly added to the hybridization reaction mixture (see under Sect. IV.C.2) to decrease the amount of hydrolysis, because it has both a stabilizing effect on the RNA and also enables one to use lower hybridization temperatures. An interesting property of rRNA (and probably of other types as well) is that in the presence of high concentrations of formamide the optimal hybridization conditions and the thermal stability of DNA-RNA hybrids differ from that of the corresponding DNA duplexes. At high formamide concentrations, RNA hybrids are progressively more stable than comparable DNA duplexes, and the optimal hybridization rates occur at higher temperatures relative to DNA duplex formation (25,26). The optimal hybridization temperature for rRNA in the presence of 40% or more formamide is only about 5°C below the midpoint of the melting profile of the hybrids (27,28).

C. Reassociation and Hybridization Reaction Components

In addition to the nucleic acids, other organic molecules and salts are important components of the reassociation or hybridization reaction mixtures.

1. Type and Concentration of Salt

Denatured DNA fragments do not reassociate (and RNA will not hybridize) in low ionic strength buffers. This is because of the repelling effect of the negatively charged phosphate groups. Increasing the ionic strength tends to repress this repelling effect by shielding the phosphate groups. Sodium has been, by far, the most used cation, with others, such as potassium, cesium, rubidium, and lithium, used only occasionally. Although chloride is the most commonly used anion, many, including both inorganic and organic anions, have been employed in these experiments.

Initially, Marmur and Lane (2) observed that maximum reassociation occurred at a KCl concentration of about 0.4 M and that similar results were obtained using 2X SSC, which is about 0.4 M Na^+, although for sodium, the maximum rate of hybridization is approached at about 6X SSC (1.2 M Na^+). The C_0t curve studies of Britten and Kohne (13) were carried out in 0.12 M sodium phosphate buffer (pH 6.8) and, as a result, the C_0t values are defined for this ionic strength (i.e., 0.18 M Na^+). When other salts or different ionic strengths are used, rate corrections are applied to obtain equivalent C_0t (EC_0t) values. For an example of these correction factors, see Table 1 in Britten et al. (29), in which are listed EC_0t values for various concentrations of phosphate buffer. The EC_0t value for 6X SSC is about 10, i.e., the C_0t values at one-half reassociation, is about 10 times less than for 0.12 M phosphate buffer.

The anionic component of a salt exerts the greatest effect on the properties of the salt. Although most salts tend to shield the phosphate groups in the DNA fragments, resulting in greater thermal stability of native DNA as the ionic strength is increased (for example, the T_m in 1X SSC is about 15.3°C higher than in 0.1X SSC), the more chaotropic salts will tend to lower the thermal stability at high concentrations. The relative chaotropic properties of several anions are listed as follows (30):

$$CCl_3COO^- \gg CNS^- > CF_3COO^- > ClO_4{}^- > I^- > CH_3COO^- > Br^-, Cl^-, CHOO^-$$

Some of these salts have been used for the development of low-temperature, or high-reaction rate, reassociation, and hybridization reaction mixtures (31,32).

2. *Organic Molecules and Solvents*

Organic compounds are added to reaction mixtures to either lower the thermal stability of the duplexes or hybrids, or to increase the rate of reassociation or hybridization.

Several of the organic compounds that depress the thermal stability of DNA duplexes and RNA hybrids include formamide (26,33-35), urea (35,36), and dimethyl sulfoxide (37). These allow one to do the hybridization experiments at lower temperatures so that there is less thermal scission of the nucleic acid strands. This is important for eukaryotic DNAs, for which the reassociation reactions must proceed for a long time because of the large genome size, and for RNAs, which are more thermolabile.

A second effect that these compounds have on the reaction mixtures is that they increase the viscosity, which decreases the reassociation or hybridization rates.

Formamide. Formamide has been the most widely used compound for lowering the thermal stabilities of duplexes and hybrids. It lowers the thermal stability of DNA (and, thus, the required temperature for reassociation) by about 0.6°C for each percentage of formamide in the mixture. Another interesting property of formamide is that its effect on the thermal stability of RNA hybrids is not as pronounced as that on DNA duplexes. This effect is shown in Figure 5 for *E. coli* DNA and rRNA, where both nucleic acids are in the 52-54 mol% G+C range (27). At formamide concentrations in the 50-80% range, conditions can be obtained in which rRNA hybridization proceeds in the absence of DNA reassociation (27,28). Conditions for this type of experiment can be met for organisms with DNA having a G+C content of 50 mol% or less, but may not be possible for organisms having DNA with a significantly greater mol% G+C content. This is because the G+C content of rRNA cistrons is similar for most bacteria and, as a result, the differences in thermal stability will be less pronounced for organisms having DNA with a high G+C content.

Urea. High concentrations of urea, up to 8 M, in high salt buffers have also been used for nucleic acid homology experiments done at low temperatures (35,36). Urea lowers the melting temperature of DNA at a rate of 2.25°C per molar concentration (35).

Dimethyl Sulfoxide. The use of dimethyl sulfoxide (DMSO) for lowering the thermal stability of nucleic acid hybrids was introduced by Legault-Démare et al. (37). The thermal stability of DNA is lowered by about 0.46°C for each percentage of DMSO.

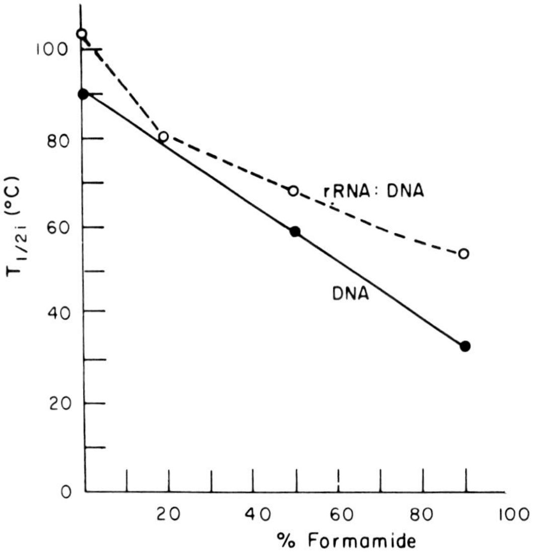

Figure 5 The effect of formamide concentration on the thermal stabilities of rRNA-DNA hybrids and DNA-DNA duplexes of *E. coli*. The $T_{1/2i}$ values are the temperatures at which one-half of the labeled DNA or rRNA had irreversibly disassociated from the immobilized DNA (*Source*: Ref. 27).

Phenol. Mixtures of phenol and chaotropic salts have been incorporated into reassociation mixtures by Kohne et al. (38). These mixtures, maintained as emulsions by rapid shaking during the reassociation step not only lower the melting temperatures but also dramatically increase reassociation rates. The interactions of these components are rather complex. For example, with a mixture consisting of 0.9 ml of 2 M NaSCN, plus 0.1 ml 90% phenol, reassociation of bacterial DNAs that ranged from 32 to 67 mol% G+C, all occurred at room temperature. In addition, the rate constant values were dependent upon DNA concentration and also upon the intensity of the shaking. The effect of DNA concentration on the rate constant was unexpected, with the faster rates occurring at the lower concentrations. The rate constants were directly proportional to the shaking intensity. The most probable mechanism involved in this system is that the single-stranded DNA is somehow concentrated by the interaction with the emulsion. The DNA may be concentrated in the phenol-aqueous interface. If this is the mechanism, the faster shaking rates would result in finer emulsions and, thus, more surface areas. Perhaps, with the higher DNA concentrations, the effective concentrations at this interface area become so great that aggregates or semi-precipitates form that depress the apparent reassociation rates. This is a procedure that should be considered if one is working with small amounts of DNA from large genomes. The major problem that I see with the procedure is the difficulty of standardization; for example, standardized shaking intensities would seem to be very hard to achieve, as would even, perhaps, the DNA concentrations and the 90% phenol preparations.

Dextran Sulfate and Polyethylene Glycol. The addition of 10% dextran sulfate to reassociation solutions can increase the reassociation rate 10-fold (32). Similar, or even greater rate increases, have been observed for hybridization or reassociation with immobilized DNA (39) and with cytological preparations (40). Amasino (41) has observed that the addition of 10% polyethylene glycol enhances the reassociation or hybridization rates with immobilized DNA even more than dextran sulfate. This enhancement is more pronounced with single-stranded than with double-stranded probes. The effect of these compounds appear to be that of tying up water, such that the effective concentrations of the reactants become greater, increasing the reaction rates.

V. DUPLEX AND HYBRID SPECIFICITIES

Through much of the preceding discussion, the specificity of the reassociation or hybridization reaction products have been only implied or alluded to indirectly. In this section, the effect of base-pair mismatches on the formation of DNA duplexes and RNA hybrids and on their reaction rates and thermal stabilities will be covered.

When a preparation of fragmented native DNA is denatured, allowed to reassociate, and then denatured a second time, the T_m of the second melting profile is nearly the same as that of the first. This indicates that the base pairings formed during the reassociation were as precise and complete as in the native DNA. The same type of experiment can be carried out using radioactive probe DNA or RNA, hybridized with an excess of unlabeled DNA. The thermal stability of these duplexes or hybrids can be determined by placing a tube containing them into a water bath, then increasing the temperature of the bath stepwise, and measuring the dissociation of the labeled fragments. If the unlabeled DNA has been immobilized on a nitrocellulose membrane, the membrane (after being washed free of unhybridized probe) can be transferred from one temperature to the next. If the duplexes have been formed in solution, they can be adsorbed to hydroxyapatite at a buffer concentration that will retain only double-stranded DNA. If the hydroxyapatite is then subjected to increasing temperatures, the fragments will release when the duplexes dissociate. By using these procedures, the midpoint of the melting profiles can be reasily determined, and they are often given the symbol $T_{m(e)}$, indicating the elution aspect of the assay. Although the curves from which one obtains the $T_{m(e)}$ values are very similar to the curves for the T_m values (i.e., midpoint of the optical melting curve), the $T_{m(e)}$ values will be 2° or 3°C higher.

Initial thermal stability experiments revealed that the maximum temperature difference between homologous and heterologous $T_{m(e)}$ values [referred to as $\Delta T_{m(e)}$ values] was about 20°C, when the reassociation step was carried out at about 25°C below the T_m of the homologous reactions. By increasing the reassociation temperature, the more stable heterologous duplexes or hybrids will be selected for, and the maximum $\Delta T_{m(e)}$ values will be less. When the reassociation temperature is lowered, duplexes of lower stability are retained, and the $\Delta T_{m(e)}$ values can be larger. However, at the lower reassociation temperatures, the fraction of nonspecific complexes will increase for both the homologous and heterologous reactions.

It is easiest to consider the role of mismatched base pairs on duplex thermal stabilities at a salt concentration of 1X SSC. Extrapolating with the Marmur and Doty T_m equation (3) to 100% A+T, the T_m is 69.3°C, and to 100% G+C, the T_m is 110°C, or about 0.69 and 0.11°C for each percentage mismatch of each of the respective base pairs. Taking an average of these values, the value would be 0.9°C per base-pair mismatch for organisms in the 50 mol% G+C range. If one assumes that a base-pair mismatch does not contribute to (but does not inhibit) the thermal stability of the duplex (i.e., that it does not have a negative effect, such as repulsion), then the base-pair mismatch can be estimated by dividing the $\Delta T_{m(e)}$ by 0.9, i.e., a maximum base-pair mismatch of about 22%.

The effect of mispaired bases was first investigated by Bautz and Bautz (42) and Kotaka and Baldwin (43) by using synthetic polymers. The results of these studies indicated a decrease in thermal stability of about 1°C for each percentage of base-pair mismatch. Ullman and McCarthy (44,45), who used alkaline deamination of cytosine in *E. coli* DNA to disrupt guanine-cytosine base pairing, obtained estimates of 2.2°C decrease in thermal stability for each percentage loss in base pairing. Therefore, under optimal reassociation conditions (i.e., 25°C below the T_m of the native DNA), the maximum extent of mispairing allowed in heterologous duplexes is somewhere from 10 to 20%.

The thermal stabilities of rRNA hybrids have usually been measured with labeled rRNA and membrane-immobilized DNA. Here, because one is measuring thermal stability differences within a single molecule in which some sequences may be highly conserved, it is best to digest the hybrids with RNase before doing the thermal elution steps. This will digest out the nonpairing loops so that the strand separation of the less-stable regions is not dependent upon the more-stable regions (46,47).

VI. TECHNIQUES

The reassociation and hybridization techniques discussed in this section will be those that are currently most amenable to the rapid detection of total genomes, specific DNA fragments, or RNAs.

A. Labeled Probes

Nucleic acids can be labeled in vivo by growing the organism in the presence of ^{3}H-, ^{14}C-, ^{32}P-, or ^{33}P-labeled precursor; however, this can be a rather expensive route that results in a lot of radioactive waste, and the specific activities of the nucleic acids often are not high enough for most of the detection methods described in this book. It is more common to label in vitro by introducing labeled precursors either enzymatically or catalytically.

1. Enzymatic Labeling

Nick translation is the most widely used procedure for introducing radioactive- (usually ^{32}P or ^{35}S) or biotin-labeled deoxyribonucleotides into DNA. This procedure involves the generation of single-strand breaks (nicks) in the DNA with DNase I, followed by nucleotide removal and replacement by *E. coli* DNA polymerase I (48-51). Kits patterned after the procedure of Rigby et al. (50) are available from several companies. These come with complete instructions for labeling 1 or 2 μg amounts of DNA and are easy to use.

The nick translation procedure is designed primarily for HMW DNA, for which the DNase I/polymerase I ratio is such that the resynthesized, labeled fragments are several hundred bases long. The labeling of a short fragment of cloned DNA, which has been digested out of a vector and purified, would result in much shorter fragments because of the polymerase reaching the ends of the fragments. If a substantial proportion of these labeled fragments are very short, the reaction rate would be slow (because of the relatively short length that would become double-stranded after each nucleation event), and the resulting duplexes or hybrids may be less stable than probe DNA prepared with HMW DNA. To avoid the generation of short, labeled fragments from small-insert DNA, Feinberg and Vogelstein (52,53) have used short oligonucleotide primers and the large fragment of *E. coli* DNA polymerase I for synthesizing nearly full-length complementary strands of denatured, cloned fragments. Another approach for labeling cloned fragments has been to use coliphage T4 DNA polymerase to label by replacement synthesis (54).

End-labeling has been used to label nucleic acids at either the 3′- or 5′-ends. The 3′-ends can be labeled with either cordycepin triphosphate or dideoxyadenosine triphosphate using polynucleotide transferase (55,56). Labeled ATP and polynucleotide kinase are used for labeling the 5′-ends of DNA fragments (57,58). Again, kits are available for carrying out these reactions.

Highly labeled RNA probes can be prepared by transcription of DNA cloned into specially constructed vectors (59-61). These vectors have either a SP6, T3, or T7 phage RNA polymerase promoter region located adjacent to the cloning region, and the specific phage RNA polymerase is then used to transcribe the cloned insert by using labeled precursors. Occasionally, one of the promoter regions is located at one end of the cloning region and another at the other end, allowing one to transcribe from either strand. Kits are also available from a number of companies for generating these probes.

2. *Catalytic Labeling*

Iodine can be catalytically bonded to the cytosine residue in DNA or RNA in the presence of thallium chloride ($TlCl_3$) at pH 4.8 (62-66).

Biotin can also be attached to nucleic acids by a catalytic process. Forster et al. (67) have described a photocatalytic procedure for the covalent attachment of biotin to nucleic acids. A kit for this has also been developed and is commercially available.

B. Reassociation and Hybridization in Solution

The reassociation reactions may vary greatly in terms of reaction components, as discussed previously. Keeping the reaction volumes small will result in the highest effective concentrations of the nucleic acids.

1. *Detection*

There are two approaches available for detecting probe nucleic acids that have formed double-stranded complexes in solution. The first is by selective adsorption to hydroxyapatite, and the second is by selective degradation of noncomplexed probe DNA with single-strand specific nucleases.

Hydroxyapatite. Hydroxyapatite (HA) has been used most extensively to physically separate single-stranded probe DNA from probe DNA that has formed complexes with unlabeled target DNA or RNA. The double-stranded nucleic acids will adsorb to HA at a phosphate buffer concentration of 0.12-0.14 M (the complete buffer will usually consist of 0.14 M sodium phosphate, pH 6.8, plus 0.2-0.5% sodium dodecyl sulfate; and all carried out at a temperature of about 60°C), whereas the unreacted single-stranded probe DNA will not. The double-stranded DNA is desorbed from the HA at a buffer concentration of about 0.2-0.25 M phosphate. The HA may be placed in columns (68-70) or used in a batch procedure by centrifugation (71).

The interaction of a number of factors must be taken into consideration when performing these experiments. Many have been discussed individually earlier. One needs to know if the hybridization reaction has gone to completion. This is particularly important if one is distinguishing between organisms that have a significant amount of nucleotide sequence similarity. For example, if the reassociation is not complete, very similar organisms would appear to be more distantly related then they really are. On the other hand, if one is trying to detect the presence of a unique organism (i.e., with no closely related organisms in the samples), then partial reactions would represent a titration of the number of the organisms present in the sample. The relative sizes of the probe and target DNA fragments can have a substantial effect on adsorption to HA. If the probe fragments are small relative to the target fragments, most of the probe fragment would be in a double-stranded form, whereas, if the reverse were true, any duplexed region along the length of the probe fragment would cause the whole fragment to be adsorbed to the HA. The reassociation temperature is a particularly important factor in these experiments; at low temperatures (i.e., 30-40°C below the T_m of the reference nucleic acid), a greater fraction of the probe DNA will form complexes with the heterologous target nucleic acids, than at higher temperatures (15-25°C below the T_m). Another important consideration is the loading capacity of the HA; it is best to have a reasonable excess of HA-binding capacity (3- to 10-fold). Sometimes, the secondary structure of the nucleic acids is important. The best example of this is the use of rRNA. There are two problems with rRNA. The first is that there are many double-stranded regions within a single molecule, such that it will bind to HA almost as efficiently as double-stranded DNA. Second, highly conserved regions in the rRNA genes occur in most groups of organisms, and hybridization in these regions would cause the entire molecule to be adsorbed to the HA. Here, digesting the reaction mixture with ribonuclease or S_1 nuclease before adsorbing it to the HA has reduced these problems.

S_1 Nuclease. The single-strand-specific nucleases represent a very simple method for differentiating between double- and single-stranded DNA. The single-stranded DNA is simply degraded, whereas double-stranded DNA is not attacked.

Although S_1 nuclease from *Aspergillus oryzae* was the first used (72,73) and has been used almost universally, nucleases from other organisms, including *Staphylococcus aureus* (74) and from the mung bean (75) also exhibit preferences for single-strand substrates.

The S_1 nuclease has two requirements for specificity: the presence of zinc cations and a pH of about 4.6. The enzyme is extremely stable and has been used at temperatures up to 75°C (76). Variations of the protocol described by Crosa and coauthors (77) are probably the most commonly used.

The reassociation or hybridization procedures can be the same as those used for the HA analysis just discussed, except that polyvalent anions, such as phosphate, should be avoided because they will tie up the zinc and probably alter the pH. In general, the assay involves diluting the reassociation mixture in S_1 nuclease buffer, adding a set amount of single-stranded DNA to provide the enzyme with ample substrate, and then incubating at a set temperature for a specific period (18). The S_1-resistant fragments are usually precipitated and collected on membrane or glass-fiber filters (18); they have also been selectively bonded to DEAE ion exchange paper (78,79). There are other potential procedures for separating the S_1-resistant fragments from the degraded material, such as gel filtration, but they have not been exploited to any extent.

The effects of most of the reassociation variables are the same for both HA and S_1 assay systems. However, there is one major difference. Because only double-stranded DNA is S_1-resistant, localized mismatched regions and overhanging fragment ends are degraded. As a result, the initial reassociation rates are not precisely second-order (80), and the reassociation temperature has much less influence on the experimental results. For example, at low hybridization temperatures, weakly associated fragments will adsorb to HA and result in a high apparent similarity value, but these weakly associated complexes are still susceptible to S_1 muclease. As a result, at very low reassociation temperatures, a significant fraction of even homologous complexes may be S_1-sensitive. Figure 6 shows the effect of reassociation temperature on the DNA homology values, the same results were obtained whether the reassociation was at 60°C or 75°C. The figure also illustrates that similar results are obtained when the S_1 nuclease-resistant fragments are precipitated with trichloroacetic acid (TCA) or bound to ion-exchange paper.

The HA and membrane competition results reflect similar reassociation temperature effects and homology values. The relationships between the S_1 nuclease procedure and the membrane competition and HA procedures have also been compared in detail by Grimont and his associates (82,81). These are illustrated in Figure 7 and 8.

Ribonucleases. When the probe nucleic acid is RNA, ribonucleases will degrade the unhybridized or unreassociated probe RNA. Although most of these assays have been done in conjunction with target DNA immobilized on membrane

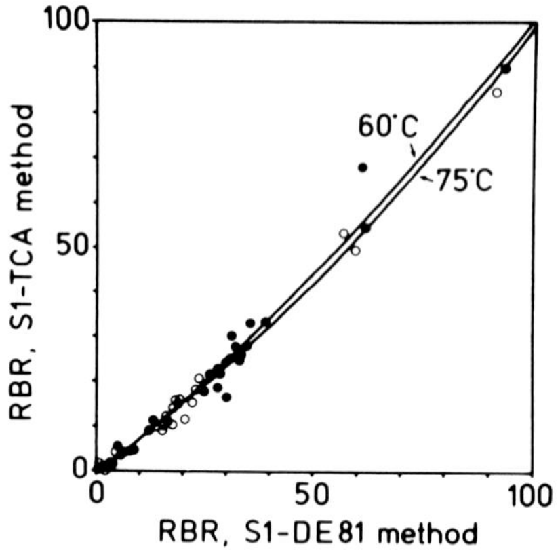

Figure 6 Comparison of trichloroacetic acid precipitation and binding to ion-exchange paper for the detection of S_1 nuclease-resistant labeled DNA fragments in DNA homology experiments. Also illustrated in this figure, is the insensitivity of the S_1 procedure to reassociation temperature. The same results were obtained for both the 60°C and 75°C reassociation temperatures (*Source*: Ref. 81).

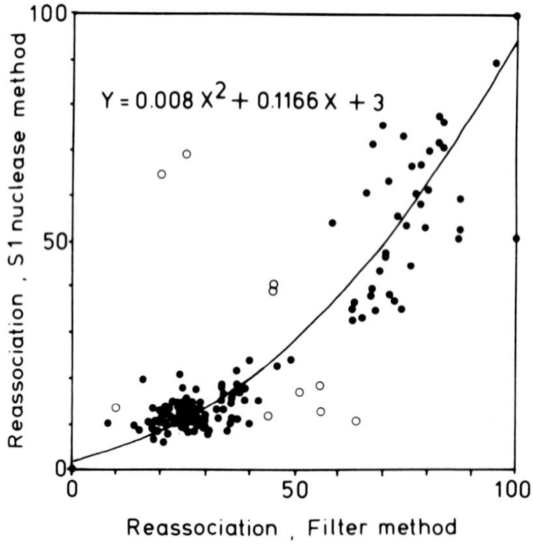

Figure 7 Comparison of the S_1 nuclease and membrane competition DNA homology methods (*Source*: Ref. 82).

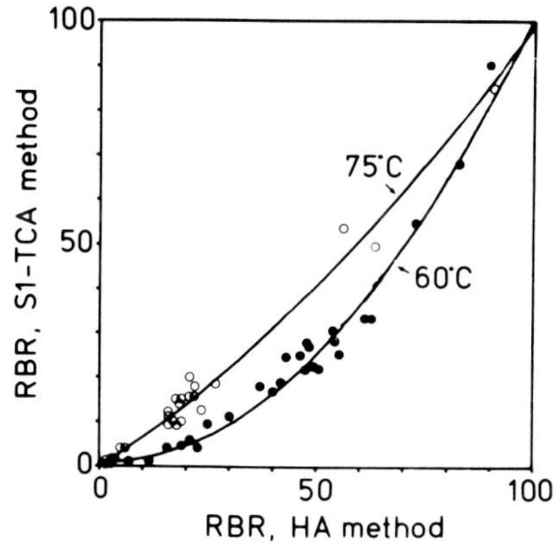

Figure 8 Comparison of the S_1 nuclease and hydroxyapatite (HA) DNA homology methods (*Source*: Ref. 81).

filters, for which an RNasing step may not be required, the potential remains for hybridization or reassociation with complementary DNA in solution.

C. Reassociation and Hybridization by Use of Immobilized Nucleic Acids

The immobilization of nucleic acids to a membrane is the first, and a very important, step for these experiments. The original Gillespie and Spiegelman protocol (9) for immobilizing DNA on nitrocellulose membranes incorporated high salt concentrations (0.9 M NaCl-0.09 M sodium citrate) for the initial trapping of the DNA on the membrane, followed by baking in an oven to make the bonding irreversible. Thomas (12) was able to get RNA to bond to nitrocellulose membranes by increasing the salt concentrations to 3.0 M NaCl-0.03 M sodium citrate. Recent modifications include the use of the chaotropic salt NaI (83-86). When used at room temperature, bonding is specific for mRNA, and when used at 50-70°C, both mRNA and DNA are bonded with a high degree of efficiency. The covalent bonding of DNA or RNA to diazobenzyloxymethyl paper (DBM paper) has enabled investigators to denature hybridized probe nucleic acid and then use the membrane-bound nucleic acid in another hybridization reaction (38,86,87). The most recently introduced type of membrane has been positvely charge-modified nylon 66. The nylon membrane has a greater capacity than either nitrocellulose or DBM paper and the nucleic acids bond with low-ionic-strength buffer. All of these products are available from several companies, which also supply detailed protocols.

1. Detection

Membrane-bound DNA duplexes and RNA hybrids are the easiest to detect because all that is required is that the unbound probe be washed away. Although one will find many variations in procedures in the literature, membrane washing will usually be carried out in a salt solution of moderate concentration, such as 2X SSC, containing 0.1-0.5% sodium dodecyl sulfate and perhaps some of the Dendhardt preincubation mixture or casein. The background will usually be lower if the washing is done (at least the initial buffer changes) at the same temperature as the reassociation. Probes that are radioactively labeled can be detected and measured by exposing them to x-ray film and the differences in density patterns estimated by eye, or they can be quantitated by densitometry measurements or, with dot-blots, they can also be cut out and measured in a scintillation counter. For biotinated probes, colorimetric assays involving avidin-linked enzyme reactions are used.

VII. CONCLUSIONS

The physical properties of nucleic acids, together with differences in genomic sizes, have enabled investigators to design DNA reassociation and RNA hybridization experiments with which they can detect specific nucleic acid sequences at the levels of gene product, gene, species, and species cluster identification. The following chapters in this section will provide details for employing these properties for specific identification problems.

REFERENCES

1. Doty, P., J. Marmur, J. Eigner, and C. Schildkraut, Strand separation and specific recombination in deoxyribonucleic acids: Physical chemical studies. *Proc. Natl. Acad. Sci. USA 46*:461–476 (1960).
2. Marmur, J., and D. Lane, Strand separation and specific recombination in deoxyribonucleic acids: Biological studies. *Proc. Natl. Acad. Sci. USA 46*: 453–461 (1960).
3. Marmur, J., and P. Doty, Determination of the base composition of deoxyribonucleic acid from its thermal denaturation temperature. *J. Mol. Biol. 5*:109–118 (1962).
4. Schildkraut, C. L., J. Marmur, and P. Doty, The formation of hybrid DNA molecules and their use in studies of DNA homologies. *J. Mol. Biol. 3*: 595–617 (1961).
5. Schildkraut, C. L., L. L. Wierzchowski, J. Marmur, D. M. Green, and P. Doty, A study of the base sequence homology among the T series of bacteriophages. *Virology 18*:43–55 (1962).
6. Hoyer, B. H., B. J. McCarthy, and E. T. Bolton, A molecular approach in the systematics of higher organisms. *Science 144*:959–967 (1964).

7. McCarthy, B. J., and E. T. Bolton, An approach to the measurement of genetic relatedness among organisms. *Proc. Natl. Acad. Sci. USA 50*:156–164 (1963).
8. Nygaard, A. P., and B. D. Hall, A method for the detection of RNA–DNA complexes. *Biochem. Biophys. Res. Commun. 12*:98–104 (1963).
9. Gillespie, D., and S. Spiegelman, A quantitative assay for DNA–RNA hybrids with DNA immobilized on a membrane. *J. Mol. Biol. 12*:829–842 (1966).
10. Denhardt, D. T., A membrane-filter technique for the detection of complementary DNA. *Biochem. Biophys. Res. Commun. 23*:641–646 (1966).
11. Southern, E. M., Detection of specific sequences among DNA fragments separated by gel electrophoresis. *J. Mol. Biol. 98*:503–517 (1975).
12. Thomas, P. S., Hybridization of denatured RNA and small DNA fragments transferred to nitrocellulose. *Proc. Natl. Acad. Sci. USA 77*:5201–5205 (1980).
13. Britten, R. J., and D. E. Kohne, Repeated sequences in DNA. *Science 161*: 529–540 (1968).
14. Wetmur, J. G., and N. Davidson, Kinetics of renaturation of DNA. *J. Mol. Biol. 31*:349–370 (1968).
15. Britten, R. J., and E. H. Davidson, Studies on nucleic acid reassociation kinetics: Empirical equations describing DNA reassociation. *Proc. Natl. Acad. Sci. USA 73*:415–419 (1976).
16. Galau, G. A., R. J. Britten, and E. H. Davidson, Studies on nucleic acid reassociation kinetics: Rate of hybridization of excess RNA with DNA, compared to the rate of DNA renaturation. *Proc. Natl. Acad. Sci. USA 74*:1020–1023 (1977).
17. Galau, G. A., M. J. Smith, R. J. Britten, and E. H. Davidson, Studies on nucleic acid reassociation: Retarded rate of hybridization of RNA with excess DNA. *Proc. Natl. Acad. Sci. USA 74*:2306–2310 (1977).
18. Johnson, J. L., DNA homology procedures. In *Manual of Methods for General Microbiology*, P. Gerhardt (ed.), American Society for Microbiology, pp. 450–472 (1981).
19. Melli, M., C. Whitfield, K. V. Rao, M. Richardson, and J. O. Bishop, DNA–RNA hybridization in vast DNA excess. *Nature 231*:8–12 (1971).
20. McCarthy, B. J., Arrangement of base sequences in deoxyribonucleic acid. *Bacteriol. Rev. 31*:215–229 (1967).
21. McConaughy, B. L., and B. J. McCarthy, The interaction of oligodeoxyribonucleotides with denatured DNA. *Biochim. Biophy. Acta 149*:180–189 (1967).
22. Niyogi, S. K., The influence of chain length and base composition on the specific association of oligoribonucleotides with denatured deoxyribonucleic acid. *J. Biol. Chem. 244*:1576–1581 (1969).
23. Marmur, J., and P. Doty, Thermal renaturation of deoxyribonucleic acids. *J. Mol. Biol. 3*:585–594 (1961).
24. Gillespie, S., and D. Gillespie, Ribonucleic acid–deoxyribonucleic acid hybridization in aqueous solutions and in solutions containing formamide. *Biochem. J. 125*:481–487 (1971).

25. Birnstiel, M. L., B. H. Sells, and I. F. Purdom, Kinetic complexity of RNA molecules. *J. Mol. Biol. 63*:21–39 (1972).
26. Schmeckpeper, B. J., and K. D. Smith, Use of formamide in nucleic acid reassociation. *Biochemistry 11*:1319–1326 (1972).
27. Casey, J., and N. Davidson, Rates of formation and thermal stabilities of RNA:DNA and DNA:DNA duplexes at high concentrations of formamide. *Nucleic Acid Res. 4*:1539–1552 (1977).
28. Vogelstein, B., and D. Gillespie, RNA–DNA hybridization in solution without DNA reannealing. *Biochem. Biophys. Res. Commun. 75*:1127–1132 (1977).
29. Britten, R. J., D. E. Graham, and B. R. Neufeld, Analysis of repeating sequences by reassociation. In *Methods in Enzymology*, L. Grossman and Moldave, K. (eds.). Academic Press, New York, vol. XXIX, part E, pp. 363–406 (1974).
30. Hamaguchi, K., and E. P. Geiduschek, The effect of electrolytes on the stability of the deoxyribonucleate helix. *J. Am. Chem. Soc. 84*:1329–1338 (1962).
31. Chien, Y.-H., and N. Davidson, RNA:DNA hybrids are more stable than DNA:DNA duplexes in concentrated perchlorate and trichloroacetate solutions. *Nucleic Acids Res. 5*:1627–1637 (1978).
32. Wetmur, J. G., Acceleration of DNA renaturation rates. *Biopolymers 14*: 2517–2524 (1975).
33. Bonner, J., G. Kung, and I. Bekhor, A method for the hybridization of nucleic acid molecules at low temperature. *Biochemistry 6*:3650–3653 (1967).
34. McConaughy, B. L., C. D. Laird, and B. J. McCarthy, Nucleic acid reassociation in formamide. *Biochemistry 5*:3289–3294 (1969).
35. Hutton, J. R., Renaturation kinetics and thermal stability of DNA in aqueous solutions of formamide and urea. *Nucleic Acids Res. 4*:3537–3555 (1977).
36. Kourilsky, P., S. Manteuil, M. H. Zamansky, and F. Gros, DNA–RNA hybridization at low temperature in the presence of urea. *Biochem. Biophys. Res. Commun. 41*:1080–1087 (1971).
37. Legault-Démare, B. Desseaux, T. Heyman, S. Séror, and G. P. Ress, Studies on hybrid molecules of nucleic acids. I. DNA–DNA hybrids on nitrocellulose filters. *Biochem. Biophys. Res. commun. 28*:550–557 (1967).
38. Kohne, D. E., S. A. Levison, and M. J. Byers, Room temperature method for increasing the rate of DNA reassociation by many thousandfold: The phenol emulsion reassociation technique. *Biochemistry 16*:5329–5341 (1977).
39. Wahl, G. M., M. Stern, and G. R. Stark, Efficient transfer of large DNA fragments from agarose gels to diazobenzyloxymethyl-paper and rapid hybridization by using dextran sulfate. *Proc. Natl. Acad. Sci. USA 76*: 3683–3687 (1979).
40. Lederman, L., E. S. Kawasaki, and P. Szabo, The rate of nucleic acid annealing to cytological preparations is increased in the presence of dextran sulfate. *Anal. Biochem. 117*:158–163 (1981).

41. Amasino, R. M., Acceleration of nucleic acid hybridization rate by polyethylene glycol. *Anal. Biochem. 152*:304–307 (1986).
42. Bautz, E. K. V., and F. A. Bautz, The influence of noncomplementary bases on the stability of ordered polynucleotides. *Proc. Natl. Acad. Sci. USA 52*:1476–1481 (1964).
43. Kotaka, T., and R. L. Baldwin, Effects of nitrous acid on the dAT copolymer as a template for DNA polymerase. *J. Mol. Biol. 9*:323–339 (1964).
44. Ullman, J. S., and B. J. McCarthy, Alkali deamination of cytosine residues in DNA. *Biochim. Biophys. Acta 294*:396–404 (1973).
45. Ullman, J. S., and B. J. McCarthy, The relationship between mismatched base pairs and the thermal stability of DNA duplexes. *Biochim. Biophys. Acta 294*:416–424 (1973).
46. De Ley, J., and J. De Smedt, Improvements of the membrane filter method for DNA:rRNA hybridization. *Antonie van Leeuwenhoek J. Microbiol. 41*: 154–168 (1975).
47. Johnson, J. L., and B. Harich, Comparisons of procedures for determining ribosomal ribonucleic acid similarities. *Curr. Microbiol. 9*:111–120 (1983).
48. Chelm, B. K., and R. B. Hallick, Changes in the expression of the chloroplast genome of *Euglena gracilis* during chloroplast development. *Biochemistry 15*:593–599 (1976).
49. Kelly, R. B., N. R. Cozzarelli, M. P. Deutscher, I. R. Lehman, and A. Kornberg, Enzymatic synthesis of deoxyribonucleic acid. XXXII. Replication of duplex deoxyribonucleic acid by polymerase at a single strand break. *J. Biol. Chem. 245*:39–45 (1970).
50. Rigby, P. W. J., M. Dieckmann, C. Rhodes, and P. Berg, Labeling deoxyribonucleic acid to high specific activity in vitro by nick translation with DNA polymerase I. *J. Mol. Biol. 113*:237–251 (1977).
51. Maniatis, T., E. F. Fritsch, and J. Sambrook, *A Laboratory Manual.* Cold Spring Harbor Laboratory, Cold Spring Harbor, N.Y., pp. 114–121 (1982).
52. Feinberg, A. P., and B. Vogelstein, A technique for radiolabeling DNA restriction endonuclease fragments to high specific activity. *Anal. Biochem. 132*:6–13 (1983).
53. Feinberg, A. P., and B. Vogelstein, Addendum—A technique for radiolabeling DNA restriction endonuclease fragments to high specific activity. *Anal. Biochem. 137*:266–267 (1984).
54. Deen, K. C., T. A. Landers, and M. Berninger, Use of T4 polymerase replacement synthesis for specific labeling of plasmid-cloned inserts. *Anal. Biochem. 135*:456–465 (1983).
55. Tu, C.-P. D., and S. N. Cohen, 3′-end labelling of DNA with [α-^{32}P] cordycepin-5′-triphosphate. *Gene 10*:177–183 (1980).
56. Yousaf, S. I., A. R. Carroll, and B. E. Clarke, A new improved method for 3′-end labelling DNA using [α-^{32}P] ddATP. *Gene 27*:309–313 (1984).
57. Lillehaug, J. R., R. K. Kleppe, and K. Kleppe, Phosphorylation of double-stranded DNAs by T4 polynucleotide kinase. *Biochemistry 15*:1858–1865 (1976).
58. Richardson, C. C., Phosphorylation of nucleic acid by an enzyme from T4

bacteriophage-infected *Escherichia coli. Proc. Natl. Acad. Sci. USA 54*: 158-165 (1965).
59. Green, M. R., T. Maniatis, and D. A. Melton, Human β-globin pre-mRNA synthesized in vitro is accurately spliced in *Xenopus* oocyte nuclei. *Cell 32*: 681-694 (1985).
60. Lynn, D. A., L. M. Angerer, A. M. Bruskin, W. H. Klein, and R. C. Angerer, Localization of a family of mRNAs in a single cell type and its precursors in sea urchin embryos. *Proc. Natl. Acad. Sci. USA 80*:2656-2660 (1983).
61. Melton, D. A., P. A. Krieg, M. R. Rebagliati, T. Maniatis, K. Zimm, and M. R. Green, Efficient in vitro synthesis of biologically active RNA and RNA hybridization probes from plasmids containing a bacteriophage SP6 promoter. *Nucleic Acids Res. 12*:7035-7056 (1984).
62. Chan, H. C., W. T. Ruyechan, and J. G. Wetmur, In vitro iodination of low complexity nucleic acids without chain scission. *Biochemistry 15*:5487-5490 (1976).
63. Commerford, S. L., Iodination of nucleic acids in vitro. *Biochemistry 11*: 1993-1999 (1971).
64. Orosz, J. M., and J. G. Wetmur, In vitro iodination of DNA. Maximizing iodination while minimizing degradation; use of buoyant density shifts for DNA-DNA hybrid isolation. *Biochemistry 13*:5467-5473 (1974).
65. Selin, Y. M., B. Harich, and J. L. Johnson, Preparation of labeled nucleic acids (nick translation and iodination) for DNA homology and rRNA hybridization experiments. *Curr. Microbiol. 8*:127-132 (1983).
66. Tereba, A., and B. J. McCarthy, Hybridization of ^{125}I-labeled ribonucleic acid. *Biochemistry 12*:4675-4679 (1973).
67. Forster, A. C., J. L. McInnes, D. C. Skingle, and R. H. Symons, Nonradioactive hybridization probes prepared by the chemical labeling of DNA and RNA with a novel reagent, photobiotin. *Nucleic Acids Res. 13*:745-761 (1985).
68. Lachance, M.-A., Simple method for determination of deoxyribonucleic acid relatedness by thermal elution in hydroxyapatite microcolumns. *Int. J. Syst. Bacteriol. 30*:433-436 (1980).
69. Miyazawa, Y., and C. A. Thomas, Nucleotide composition of short segments of DNA molecules. *J. Mol. Biol. 11*:223-237 (1965).
70. Sibley, C. G., and J. E. Ahlquist, The phylogeny and relationships of the ratite birds as indicated by DNA-DNA hybridization. In *Evolution Today*, Proceedings of the Second International Congress of Systematic and Evolutionary Biology, G. G. E. Scudder and Reveal, J. L. (eds.). Hunt Institute for Botanical Documentation, Carnegie-Mellon University, Pittsburg, pp. 301-335 (1981).
71. Brenner, D. J., G. R. Fanning, A. V. Rake, and K. E. Johnson, Batch procedure for thermal elution of DNA from hydroxyapatite. *Anal. Biochem. 28*:447-459 (1969).
72. Ando, T., A nuclease specific for heat-denatured DNA isolated from a product of *Aspergillus oryzae*. *Biochim. Biophys. Acta 114*:158-168 (1966).

73. Sutton, W. D., A crude nuclease preparation suitable for use in DNA reassociation experiments. *Biochim. Biophys. Acta 240*:522–531 (1971).
74. Kacian, D. L., and S. Spiegleman, Use of micrococcal nuclease to monitor hybridization reactions with DNA. *Anal. Biochem. 58*:534–540 (1974).
75. Johnson, P. H., and M. Laskowski Sr., Mung bean nuclease. I. Resistance of double stranded deoxyribonucleic acid and susceptibility of regions rich in adenosine and thymidine to enzymatic hydrolysis. *J. Biol. Chem. 245*: 891–898 (1970).
76. Barth, P. T., and N. J. Grinter, Assay of deoxyribonucleic acid homology using a single-strand-specific nuclease at 75°C. *J. Bacteriol. 121*:434–441 (1975).
77. Crosa, J. H., D. J. Brenner, and S. Falkow, Use of single-strand-specific nuclease for analysis of bacterial and plasmid deoxyribonucleic acid homo- and heteroduplexes. *J. Bacteriol. 115*:904–911 (1973).
78. Maxwell, I. H., J. Van Ness, and W. E. Hahn, Assay of DNA–RNA hybrids by S_1 nuclease digestion and adsorption to DEAE cellulose filters. *Nucleic Acids Res. 5*:2033–2038 (1978).
79. Popoff, M., and C. Coynault, Use of DEAE-cellulose filters in the S_1 nuclease method for bacterial deoxyribonucleic acid hybridization. *Ann. Microbiol. (Paris) 131A*:151–155 (1980).
80. Smith, M. J., R. J. Britten, and E. H. Davidson, Studies on nucleic acid reassociation kinetics: Reactivity of single-stranded tails in DNA–DNA renaturation. *Proc. Natl. Acad. Sci. USA 72*:4805–4809 (1975).
81. Grimont, P. A. D., M. Y. Popoff, F. Grimont, C. Coynault, and M. Lemelin, Reproducibility and correlation study of three deoxyribonucleic acid hybridization procedures. *Curr. Microbiol. 4*:325–330 (1980).
82. Bouvet, P. J. M., and P. A. D. Grimont, Taxonomy of the genus *Acinetobacter* with the recognition of *Acinetobacter baumannii* sp. nov., *Acinetobacter haemolyticus* sp. nov., *Acinetobacter johnsonii* sp. nov., and *Acinetobacter junii* sp. nov. and emended descriptions of *Acinetobacter calcoaceticus* and *Acinetobacter lwoffi. Int. J. Syst. Bacteriol. 36*:228–240 (1986).
83. Bresser, J., and D. Gillespie, Quantitative binding of covalently closed circular DNA to nitrocellulose in NaI. *Anal. Biochem. 129*:357–364 (1983).
84. Bresser, J., H. R. Hubbell, and D. Gillespie, Biological activity of mRNA immobilized on nitrocellulose in NaI. *Proc. Natl. Acad. Sci. USA 80*:6523–6527 (1983).
85. Bresser, J., J. Doering, and D. Gillespie, Quick-blot: Selective mRNA or DNA immobilization from whole cells. *DNA 2*:243–253 (1983).
86. Seed, B., Attachment of nucleic acids to nitrocellulose and diazonium-substituted supports. *Genet. Eng. 4*:91–102 (1982).
87. Alwine, J. C., D. J. Kemp, and G. R. Stark, Method for detection of specific RNAs in agarose gels by transfer to diazobenzyloxymethyl-paper and hybridization with DNA probes. *Proc. Natl. Sci. USA 74*:5350–5354 (1977).

2
Nonradioactive Labeling of Nucleic Acid Probes

JEFFRY J. LEARY
Smith, Kline & French Laboratories, King of Prussia, Pennsylvania

JERRY L. RUTH
Molecular Biosystems, Inc., San Diego, California

I. INTRODUCTION

Nucleic acid hybridization probes have historically been labeled with high-energy radioisotopes. Although generally short-lived and somewhat hazardous, the best radioisotopic labels can be detected in extremely small quantities, have no measurable effect on hybridization quality or kinetics, and can be quantitated in relatively short times by scintillation counting or autoradiographic techniques. It is these latter qualities that are the major challenges for nonradioactive labels: to be sensitive, ideally quantitative, and have no detrimental effect on hybridization.

Although considered a relatively new tool in molecular biology, the use of nonisotopic labels for the detection of hybridization events dates back to the early 1970s. Davidson and colleagues (1,2) used RNA probes, chemically labeled with biotin, to visualize reiterated ribosomal and transfer RNA genes in polytene chromosomes with election microscopy. By 1980, methods using other labels and RNA as probes began to appear. These nonisotopic probes were limited in use because of poor sensitivity compared with isotopic probes and because of the relatively small amounts of purified RNA available.

In subsequent work, one of the primary forces driving the research on nonisotopic labels was the need for hybridization assays that can be used routinely outside of the specialized molecular biology laboratory. Of particular interest is the diagnosis of infectious and genetic diseases. The clinical use of hybridization probes requires additional considerations of stability, cost, ease of use, and assay time. All of these directly impact upon the choice and use of the nonisotopic label.

The quest for nonradioactive labels that fulfill all these criteria has largely paralleled the path forged by protein-based assays a decade earlier, and can be generally broken down into two categories:

1. *Indirect primary labels* that require secondary recognition after hybridization, such as biotin and antigens
2. *Direct primary labels* attached covalently to the nucleic acid probe, such as enzymes, fluorophores, and lumiphors

The choice of label can often be constrained by the method of probe production. The backbone and sequence of the nucleic acid probe are derived by one of two methods: (a) cloning and enzymatic replication of naturally occurring nucleic acids (*cloned probes*); and (b) chemical synthesis of oligodeoxynucleotide fragments (*synthetic probes*). Each method of production requires quite different considerations in the choice and use of the nonisotopic labels. Additionally, the introduction of the label into the nucleic acid backbone can also be accomplished both enzymatically and chemically.

The nature of the signal, the hybridization format, and the sophistication of any instrumentation desired also, to some degree, dictate the choice of nonisotopic label. Although insoluble dye products of enzymatic reactions may be detected visually, soluble dyes may require the use of a spectrophotometer. Adequate fluorescent detection requires devices ranging from simple fluorometers to sophisticated time-resolution, photon-counting instrumentation. Luminescence detection requires luminometers, or at least photosensitive film equipment. To quantify any hybridization event, some level of instrumentation is required, and this, in turn, will affect the choice of label.

The following discussions attempt to review the major nonisotopic labels for nucleic acid probes. This includes methods devised for incorporation of label into nucleic acid probes and some discussion on how the labels fulfill the criteria of stability, sensitivity, and ease of use. The labeling of the nucleic acid probe is categorized primarily by type of probe (cloned or synthetic), then by method of probe labeling (enzymatic or chemical). Examples, advantages, and disadvantages of each labeling scheme will be briefly presented, with a summary of detection methods and sensitivities. Topics peripheral to probe labels, such as strategies of probe design; new hybridization formats, such as solution or sandwich assays; or specific protocols for use of nucleic acid probes, are outside the scope of this chapter.

II. LABELING OF ISOLATED OR CLONED NUCLEIC ACID PROBES

A. By Enzymatic Methods

1. *Indirect Primary Labels*

After 3′-biotin labeling of RNA was reported (1,2), the first broadly applicable method for labeling DNA with biotin was described in 1981 by Langer et al. (3), who used biotinylated analogues of thymidine triphosphate. Biotinylated C-5 alkyl-2′-deoxyuridine triphosphate (bio-4-dUTP) was synthesized (3) using the C-5 mercuration and palladium alkylation reactions of Bergstrom and Ruth (4). This process produced an allylamino-dUTP that was subsequently labeled with a biotin-*N*-hydroxysuccinimide ester, a reagent developed earlier for biotinylation of proteins (5). Although many bulky nucleotide analogues are not good substrates for nucleic acid polymerases, bio-4-dUTP was recognized as a thymidine analogue during nick translation by many DNA polymerases, including polymerases from *Escherichia coli*. Although the incorporation rate was significantly lower than that of thymidine 5′-triphosphate (TTP), nick translation with bio-4-dUTP in place of TTP still proceeded effectively. Product DNA that was highly biotinylated retained its hybridization characteristics, with only moderate decreases in melting temperature (3). Analogues with a longer spacer-arm between biotin and the base (bio-11-dUTP and bio-16-dUTP) were constructed and were used in a similar manner to overcome the inability of avidin to effectively bind immobilized *bio-4* biotinylated DNA (6–8). Antibodies to biotin, however, bound immobilized DNA made from bio-4-dUTP or the longer-spacer analogues equally well. The longer 11- and 16-atom spacer analogues are incorporated efficiently into DNA (6) but at a reduced rate relative to bio-4-dUTP, with a final incorporation generally in the range of 50 biotinylated bases per kilobase (Kb) of DNA. Refinement (9) of the nick translation has produced more homogenous and fully biotinylated DNA, although the avidin-binding capacity of biotinylated DNA does not appear to increase after more than 10–15% of the thymidine residues (or 25–40 bases per kilobase) are replaced. Similar analogues of dCTP (*bio-11-dCTP*) have also been synthesized and incorporated by *E. coli* DNA polymerases to obtain DNA biotinylated at cytosine bases (10).

Enzymatic methods that incorporate biotin into DNA tails have been described, such as using terminal transferase to incorporate bio-dUTP onto the 3′-end of the DNA fragment (11). In general, sensitivity and reproducibility of these methods are less than those of nick translation. Cleavable analogues of bio-dUTP, such as *bio-12-SS-dUTP*, have been incorporated into DNA to allow affinity separations (12), and they have some advantages over the use of iminobiotin, a biotin analogue whose binding to avidin can be reversed by acidic pHs.

Biotin has been incorporated into RNA by similar methods. The biotinylated riboside monomers (bio-UTPs) are very poor substrates for most RNA polymerases (3); this is not surprising because thymine-base analogues are generally not naturally occurring in RNA. Although bio-UTP can be incorporated into RNA by T7 (3) or SP6 (13,14) RNA polymerases, a more effective approach appears to be the incorporation of a ribonucleoside triphosphate with an amine spacer attached (C-5 aminoalkyl-UTP), followed by subsequent biotinylation (13,14).

Other biotinylated deoxynucleoside triphosphate analogues designed to be enzymatically incorporated into nucleic acids also have been described. Analogues of dCTP and dATP which are biotinylated at the N-4 and N-6 exocyclic amines, respectively, have been synthesized by Gebeyhu and co-workers (15) and incorporated into DNA by using *E. coli* DNA polymerases. In this procedure, the N-4-substituted dCTP analogues were synthesized by N-4 transamination of the cytosine nucleosides or nucleotides by use of bisulfite in the presence of diaminoalkanes. This resulted in cytidine or deoxycytidine adducts with aminoalkyl groups attached to N-4. This process was originally described (16,17) for cytidine bases in mRNA and tRNA and later, in more detail, at the monomer level (18) by Draper and others.

One of the original uses (3) envisioned for biotin-labeled hybridization probes was purification or collection of hybrids by affinity chromatography on avidin resins, rather than using the biotin label, as such, for detection. This application of the biotin label has recently been exploited for the purification of proteins involved in nucleic acid metabolism, such as transcription factors and splicing complex proteins (19,20). Biotinylated nucleic acids have also been used for selective enrichment of specific DNA or RNA sequences (21) and to determine orientation of DNA sequences (22).

Indirect primary labels, other than biotin, have also been incorporated into nucleic acids by enzymatic methods. An analogue of ATP, 8-[*N*(2,4-dinitrophenyl)-6-aminohexyl] amino-ATP (DNP-rATP), is a substrate for deoxynucleotidyl terminal transferase and, as such, can be incorporated into DNA by 3′-end labeling or nick translation (23). Alternatively, the DNP group can be chemically attached to nucleic acids that have been enzymatically modified with 8-aminohexylamino-ATP to produce similar DNP-labeled probes (23). The dinitrophenyl (DNP) group can then serve as an antigenic recognition site for nonradioactive detection using anti-DNP antibodies. Using in vivo enzymatic labeling, another antigenic tag, 5-bromo-2′-deoxyuridine (BrUdR), can be incorporated into M13 DNA by growing thymine-requiring strains of *E. coli* in the presence of BrUdR; the subsequently isolated brominated DNA is used as a probe and is detected by using monoclonal antibodies (24).

2. *Direct Primary Labels*

Direct primary labels have also been incorporated into probes enzymatically. Fluorescent nucleoside triphosphates have been synthesized and incorporated into DNA by polymerases (25) for use in fluorescent DNA sequencing by dideoxy termination. In this procedure, only one labeled analogue is incorporated at the 3′-end; general application to labeling DNA probes has not yet been demonstrated, and it may be difficult because most polymerizing enzymes recognize fluorescently labeled analogues poorly, if at all. A more proven enzymatic method for labeling nucleic acids with fluorescent groups is to incorporate primary amine groups into the probe by nick translation with use of aminoalkyl-dNTP or -NTP analogues (3,14,15,23); fluorescent groups can then be efficiently attached by reaction of reactive fluorophores, such as fluorescein isothiocyanate (FITC), with the amino-derivatized RNA or DNA. Fluorescent groups such as fluorescein (26), rhodamine (26), and bimanes (27) have also been enzymatically introduced onto the 3′-terminal ribose of nucleic acids using RNA ligases and modified nucleoside diphosphates.

B. Labeling of Cloned Nucleic Acids by Chemical Methods

Enzymatic reactions that incorporate labeled nucleotides into probes are much more suited to the molecular biology laboratory than to the manufacture of diagnostic products for general use. Chemical approaches offer alternate methods for the introduction of a broader variety of labels into nucleic acid probes.

1. *Indirect Primary Labels*

Methods that take advantage of unique hydroxyl and phosphate reactive groups at the 3′- or 5′-end of nucleic acid molecules have long been known (1,2). Biotin has been attached to 3′-ribose moieties in nucleic acids (RNA or RNA-tailed DNA) by oxidation of the 2′,3′-*cis* diol to a dialdehyde followed by condensation with amines (2). Fluorophores have been added to the 3′-terminus of RNA by use of similar approaches (28). Because these methods add a single label, sensitivity is generally much lower than with enzymatically labeled probes.

Hapten labels have been introduced chemically throughout nucleic acids by direct modification with activated haptens. When DNA is reacted with *N*-acetoxy-*N*-2-acetylaminofluorene (AAAF) or its 7-iodo derivative (AAIF), the major site of reaction is known to be, almost exclusively, at C-8 of the guanine residues thereby giving about 5% modified bases (29). The resulting AAAF- or AAIF-labeled nucleic acids can be effective hybridization probes with use of antibodies to detect the hybrid (29), although AAAF modification of guanine may have a drastic effect on melting characteristics (30). Haptens, such as sulfonyl groups, can also be chemically introduced into DNA (31) at the C-6 of cytosines by treatment of the nucleic acid with a bisulfite/methylhydroxylamine mixture, and the resulting sulfonylated probes can be detected with enzyme-labeled antibodies.

Mercury has also been used to chemically incorporate labels into nucleic acids. Hopman and colleagues (32) mercurated DNA and RNA at the C-5 of cytosine or uracil bases by established methods (33a,b) and used the resulting products as probes. Sulfhydryl ligands were labeled with biotin, fluorescers, or haptens, and then bound to the mercurated nucleic acids after hybridization and washes (32). Detection of the labeled ligand was accomplished by standard methods. Although mercurated nucleotides and nucleic acids can be unstable in the presence of sulfhydryl groups (34), the mercurated probes were used successfully in dot-blot and in situ hybridizations (32).

Manning et al. (1) employed cytochrome C as a DNA-binding component that could be labeled with protein-modifying agents such as biotin esters, and they successfully biotinylated probes with this method. The cytochrome C was covalently bonded to the probe by formaldehyde fixation, although this linkage later proved unstable in hybridization. Renz (35) reported variations of this approach, that used biotinylated histone as the DNA-binding component, cross-linking the histone to DNA with glutaraldehyde. Syvanen and co-workers (36) bound the single-stranded binding protein (SSB) of *E. coli* to M13 single-stranded DNA, also cross-linked the complex with glutaraldehyde, then biotinylated the protein complex. Al-Hakim and Hull (37) later published an analysis of several variations of the polyanion/biotinylation technique. In this study, both aldehyde fixation and bifunctional cross-linkers were used to conjugate biotinylated histone, cytochrome C, polyethyleneimine, and polymin G35 to single- and double-stranded DNA. Of these approaches, biotinylated polymin G35 cross-linked to DNA with a diepoxyoctane or bis(sulfosuccinimidyl)-suberate gave the best hybridization results (37). Because *N*-hydroxysuccinimidyl (NHS) esters are generally not reactive with unmodified DNA, it is unclear how the biotin-polyamines are attached by such esters of suberate; Al-Hakim and Hull suggest the N-7 of guanine, without explanation (37).

A reagent for direct biotin labeling of hybridization probes was developed by Forster and colleagues (38). This photoactive reagent (photobiotin) offers one of the simplest methods of chemical labeling for cloned nucleic acids and shows a sensitivity (38) comparable with enzymatic biotinylation. Sheldon et al. (39) also used a photoreactive agent, biotinylated psoralen, to chemically label double-stranded sections of M13 DNA with biotin. Photolabeling with these agents is accomplished by mixing photobiotin (38) or biotinylated psoralen (39) with the nucleic acid to be labeled, then irradiating the mixture with a sun lamp for a few minutes. Stability of the reagents may be a minor problem, and the site(s) of modification on DNA is not always clear. Photolabeling with biotin, as with most of the other chemical methods, must be controlled such that only a small degree of modification is achieved to ensure that hybridization or hybrid stability is not substantially affected (38).

Viscidi et al. (40) modified cytosine bases in single-stranded nucleic acids by a transamination reaction used earlier by Draper (16,18) to attach fluorescent molecules to RNA. The transamination reaction was catalyzed by bisulfite, and produces N-4 aminoethylcytosine residues in the nucleic acid. The aminoethylcytosines were subsequently labeled with biotin by reaction with biotinylaminocaproyl-*N*-hydroxysuccinimide. Probes labeled in this fashion were similar to enzymatically biotinylated probes in terms of sensitivity, but the modification appears to inhibit hybridization because modified bacteriophage DNA could not completely reanneal after the modification. This may be due to the substantial N-4 deamination of cystosine to uracil in the presence of aqueous bisulfite (18). Because these analogues have been shown to be mutagenic (41) and the N-4 position is involved in the G:C base pair, it is apparent that the modification perturbs base pairing. This is also consistent with the observation (18) that nucleic acids must be single-stranded before the N-4 aminoalkylation can occur.

2. Direct Primary Labels

Cytidine residues in RNA can also be aminoalkylated at N-4 by use of aqueous bisulfite and diaminoalkanes (16-18). The improved method of Draper (18) allows specific modification at unpaired cytosine bases. RNA labeled with fluorescent groups has been constructed by such techniques.

Renz and Kurz (42) devised a method to overcome the normal electrostatic repulsion between nucleic acids and some enzymes. In this approach, alkaline phosphatase or horseradish peroxidase was covalently linked to organic polyamines (polyethyleneimine, polymin G35) in a 1:1 ratio with use of benzoquinone. These positively charged polyamine tails were then allowed to bond electrostatically to DNA, and were then nonspecifically cross-linked using glutaraldehyde. The isolated enzyme-DNA conjugates had a protein/DNA mass ratio of about 30, but they were still able to function as useful colorimetric probes (42).

III. LABELING OF SYNTHETIC OLIGODEOXYNUCLEOTIDE PROBES

A. By Chemical Methods

In 1981, Letsinger and Schott (43) reported the chemical synthesis of a thymidine dinucleotide that had a phenanthracene interculator dye linked to the internucleotide phosphate. The dimer was used to study physical binding to polyA as a possible delivery vehicle for interculating molecules. The synthetic route had little application to longer oligomers and resulted in somewhat unstable phosphoramidate triester products.

1. Indirect Primary Labels for Synthetic Probes

The chemical synthesis of nonisotopically labeled oligodeoxynucleotides for use as diagnostic probes was first reported by Ruth and colleagues (44a,b). In this approach, a C-5 aminoalkyl 2′-deoxyuridine analogue was synthesized, the amine group was blocked, and the nucleoside was chemically converted to its 3′-phosphite form (45). By use of standard phosphoramidite methods (46), this "linker-arm" thymidine analogue was then incorporated synthetically into oligonucleotides in place of one or more thymidines in the chosen sequence. Normal work-up provided defined sequence oligomers with one or more 12-atom amine linker arms attached at specific sites in the probe. Synthetic probes of up to 27 bases in length were reported with up to four internal linker-arm bases (44a,b). Multiple nonisotopic labels such as biotin, FITC, and Texas Red were then chemically attached to the linker-arm oligonucleotide. The same approach was later used to attach alkaline phosphatase or other enzymes directly to synthetic probes (47).

Almost all other chemical approaches to labeling oligonucleotides with biotin have focused on end labeling through a 5′- or 3′-terminus, rather than internally. Kempe and colleagues (48) synthesized oligomers on a solid support by standard chemistries, chemically phosphorylated the 5′-hydroxy group of blocked oligomers, and then condensed biotinylamidoethanol to the support-bound oligomer. Conventional workup provided DNA of up to five bases in length, biotinylated at the 5′-phosphate with a three-atom spacer (48). A similar approach was chosen by Chollet and Kawashima (49), who phosphorylated 15- to 40-base unblocked oligodeoxynucleotides at the 5′-end and condensed the oligomer with diamine to form a 5′-aminoalkyl phosphoramidate. The amino-oligomer was then biotinylated chemically to give a single biotin attached to 5′-phosphoramidate through a four- to eight-atom spacer.

Chu and Orgel (50) also developed chemical methods for 5′-end labeling of nucleic acids with biotin. DNA containing a 5′-phosphate is activated by reaction with carbodiimide and imidazole, then reacted with diaminoalkanes to give 5′-amino-terminated phosphoramidates (51). The amino-oligomers are then biotinylated to give 5′-biotin-phosphoramidate oligonucleotides of a structure identical with that reported by Chollet and Kawashima (49). Wachter et al. (52) used similar condensing agents to attach aminoalkyl groups to the 5′-hydroxy of unprotected oligonucleotides through a carbamate, rather than phosphate, linkage; the 5′-aminoalkyl groups were then biotinylated by conventional means. Connolly (53) and Agrawal et al. (54) condensed protected aminoalkylphosphoramidites to 5′-hydroxy of blocked oligonucleotides during oligomer synthesis; deprotection gave 5′-aminoalkylphosphate diester analogues that were then biotinylated to give labeled oligomers similar in structure to those of Kempe et al. (48). Fluorescent probes were similarly derived (53,54). With the use of phosphotriester chemistry, Tanaka and colleagues (55) have also reported 5′-aminoalkylphosphates. Kremsky et al. (56) also chose to link functional groups, such as aldehydes and carboxylic acids, to the 5′-phosphate, then reacted the carbonyl groups with biotin hydrazide to obtain biotinylated oligomers.

Chemical modification of bases in unprotected oligodeoxynucleotides has also been reported. Sanford and Krugh (30) studied the reaction of *N*-acetoxy-2-acetylaminofluorene (AAAF) on a synthetic nine-base DNA oligomer. The product was an oligomer with AAAF at C-8 of a centrally located guanine. Hybridization to complementary oligonucleotides was studied, and a drastic effect on melting behavior was noted as the melting temperature was decreased from 53° ± 3°C to 35° ± 3°C. Psoralen adducts also have been directly cross-linked to specific thymine bases in oligodeoxynucleotides by a photofixation/reversal method reported by Gamper et al. (57). The psoralen-modified oligomers are then hybridized and specifically photocross-linked to complementary DNA to provide covalent hybrids. The use of biotinylated psoralens (39) or photobiotin (38) may provide biotin-labeled oligomers by a similar mechanism.

Rashtchian and co-workers (58) incorporated N-4 aminoalkylcytosine bases into synthetic oligomers, choosing to put the modified base at the 5′-terminal unit. The amino-oligomer was then biotinylated to allow detection by avidin complexes. No synthetic details have been published.

2. *Direct Primary Labels for Synthetic Probes*

In addition to linker-arm oligomers labeled internally (44a,b,45), fluorescent labels have also been attached to oligonucleotide probes through 5′- or 3′-end labeling. Smith et al. (59) chose to fluorescently label at the 5′-terminus of oligodeoxynucleotides using a 5′-aminothymidine incorporated during the oligomer synthesis. The purified 5′-amino-oligomer was then labeled with a single fluorophore such as FITC or Texas Red. Modifications of this method are the basis for successful fluorescent sequencing of DNA by ddNTP techniques (60).

Sproat and co-workers (61) also reported the synthesis of 5′-amino-oligodeoxynucleotides by using all four 5′-amino-deoxynucleoside bases. Similar to Smith et al. (59) the 5′-amino-oligomers could then be labeled with fluorophores or biotin. The synthesis of 5′-mercapto-oligonucleotides and fluorescently labeled products were also reported by Sproat et al. (62).

Connolly (63) chemically attached active phosphites containing blocked thioalkyl chains to the 5′-hydroxy-terminus of oligomers during chemical synthesis. The deblocked 5′-thio-oligomer was then labeled with a thio-reactive fluorophore, such as naphthylmaleiimide. Linkers of two carbon atoms in length between the 5′-phosphate and thio group were reported unstable, but three- and six-carbon linkers appeared to be acceptable (63). Oligonucleotides with 3′-thioalkyl groups also have been made and labeled with fluorophores such as fluorescein (64).

Laboratories other than Ruth et al. (44,45) have used C-5 aminoalkyl derivatives of thymidine to functionalize DNA oligomers. Haralambidis and colleagues (65) also synthesized blocked C-5 aminoalkyl-2′-deoxyuridine phosphoramidite analogues, and incorporated these into oligodeoxynucleotides during chemical synthesis, resulting in primary amine linker arms attached to thyminelike bases. These amine linker arms were subsequently FITC-labeled. In a similar manner, Gibson and Benkovic (66) synthesized aminoalkyl thymidine analogues and incorporated these into synthetic oligomers; the oligomers were then labeled with azidonitrobenzoate for use in photocross-linking studies.

Fluorescent base analogues have also been incorporated into oligodeoxynucleotides during chemical synthesis of the oligomer. Pyridopyrimidine analogues of cytosine nucleoside have been synthesized (67), blocked, and incorporated at chosen sites into oligonucleotides by using phosphotriester monomers and dimers (68). The resulting pyridopyrimidine-containing oligomers are mildly fluorescent, but quenching occurs when hybridized. Interestingly, during hybridization, the pyridopyrimidine cytosine analogues base pair with cytosine rather than guanine bases (68).

The most practical synthetic probes appear to be oligonucleotides covalently attached to alkaline phosphatase. By use of amine linker-arm oligomers (44a,b, 45) Jablonski et al. (47) conjugated amino-oligonucleotides to alkaline phosphatase with di-*N*-hydroxysuccidimidyl suberate. This yielded a 1:1 enzyme/probe conjugate linked through a 19-atom spacer to an internal thymidine analogue. No loss in enzyme activity was observed, and good sensitivities were reported. Li and colleagues (69) later described a different approach for attaching alkaline phosphatase to oligonucleotide probes. Oligomers with 3′ "tails" of four extraneous bases were constructed which contained an N-4 amino-alkylamino-derivatized deoxycytidine base at the 3′-terminus. The aminoalkyl-cytosine (69) was then thio functionalized and cross-linked to bromoacetyl-modified alkaline phosphatase. This method resulted in a 40% loss of enzyme activity and subsequent poorer hybridization sensitivities, perhaps as a result of multiple sites of bromoacetylation.

B. Labeling of Oligodeoxynucleotides by Enzymatic Methods

Synthetic probes have also been nonisotopically labeled using enzymatic methods. To date, the labels have been limited to biotin (11,70) or dinitrophenyl (23), both of which are indirect labels.

Murasugi and Wallace (70) used *E. coli* DNA polymerase I and bio-11-dUTP to add a single biotin label at the 3′-end of a 23mer hybridized to an overhanging 16mer template. The primer extension reaction was limited to the addition of a single base, by lack of specific dNTP. A single-stranded biotinylated probe was isolated from the unlabeled template by denaturing electrophoresis. With a similar approach, Riley et al. (11) used terminyl deoxynucleotidyl transferase (TdT) and bio-11-dUTP to incorporate multiple biotinylated bases into the 3′-end of a synthetic 25mer.

Vincent and colleagues (23) also used terminal transferase to incorporate a dinitrophenyl label into a 35-base pair DNA fragment using 8-(2,4-dinitrophenylaminohexyl)amino-ATP (DNP-rATP) as a substrate. The labeling appears to proceed poorly, and significant degradation from contaminating exonuclease was noted.

To date, enzymatic labeling of oligonucleotides has been limited to 3′-end labels. It would appear possible, however, to incorporate internal labels at specific sites by extension of known primer template procedures (70). Also, incorporation of biotin- or hapten-labeled ribonucleoside triphosphates into specific sequence RNA oligomers appears feasible because techniques for making unlabeled RNA transcripts from complementary oligodeoxynucleotides without using primers have been reported (71).

IV. GENERAL METHODS OF DETECTION

All of the labels described have been used in nucleic acid hybridizations. However, relatively few have been routinely used to detect hybridization events by nonisotopic means. Biotin, haptens, and direct enzyme labels are the most frequently reported labels, and all have been used extensively in dot-blot hybridizations. Table 1 summarizes selected labels, the extent of label incorporation, how the probes were detected, and the reported sensitivities. The relative signal contribution of each individual label is also noted.

A. Detection of Indirect Primary Labels

Indirect primary labels can be described as reporter groups that are covalently attached to the nucleic acid probe and require posthybridization recognition by proteins. Indirect primary labels that have been used for nucleic acids are biotin and haptens, such as dinitrophenyl, acetylaminofluorene analogues (AAF, AAAF, AAIF), and brominated or sulfonylated pyrimidines. The unusual forms of nucleic acid hybrids that can be recognized by antibodies, such as Z-DNA and RNA:DNA hybrids, are also included in this group. The method of detection is essentially independent of how the probe is labeled, but sensitivity of detection may depend very strongly upon the labeling method.

1. Biotin Labels

Biotinylated hybridization probes were originally detected in situ by electron-microscopy (1,2) after complexation with an avidin-ferritin complex, but the

Table 1 Detection Limits of Nonradioactive Probes

Probe type	Labeling agent[a]	Ref.	Label/100 bases[b]	Probe size (in bases)
A. Cloned	^{32}P	–	10–20	8,000
	Bio-11-dUTP	88	5–15	4,360
		74	5–15	23,600
		89	12	48,500
		7a,b	5–15	9,460
		7a,b	5–15	9,460
	N^4-bio-dCTP	15	3–10	4,360
	Biotin–SSB	36	28–42	8,920
	Biotin–polyamine	37	35	7,250ss
	Photobiotin	38	2–7	7,250ss
	Bisulfite/biotin	40	3–10	48,500
	Biotin–psoralen	39	10	8,300
	Biotin–histone	35	2	5,000
	AAIF	29	2–5	7,700
	AAIF	92	5–7	8,300
	AP–polyamine	42	24	2,300
	HRP–polyamine	42	24	2,300
	DNP–rATP	23	2	48,500
	Br–dU	24	5–25	7,560
	Mercury	32	3	–
B. Synthetic oligomer	^{32}P	44a,b		25
	AP (internal)	47		20–25
	Biotin (internal)	44a,b		22–27
	AP (3′-end)	69		22–24
	3′Bio-dUTP	70		23
		11		24
	5′-Biotin	54		17
		50		16
		49		20
		48		18

[a] AP = alkaline phosphatase; HRP = horseradish peroxidase; β-gal = β-galactosidase, Ab = antibody to label; TNP-SH = trinitrophenylthiol ligand; Eu = europium; TRF = time-resolved fluorometry; lumin = luminescence; SSB = single strand binding protein; DNP = dinitrophenyl; AAIF = *N*-acetoxy-*N*-2-acetyl-7-iodofluorene.

[b] Reported values; number from which others are calculated.

[c] Based on reported sizes of targets.

[d] Approximate total number of target molecules detected by hybridization.

[e] Relative signal per label is inversely proportional to (molecules of target detected) times (labels per molecule probe).

Average labels/ molecule	Detection method[a]	Reported sensitivities			Relative[e] signal/ label
		pg	Detected (amol)[c]	Molecules[d] ($\times 10^{-6}$)	
1,200	Autorad (18-hr)	0.5	0.1	0.06	420
440	HRP; lumin	0.25[b]	0.1	0.06	1,700
2,360	AP; color	3[b]	0.2	0.1	130
5,820	β-gal; lumin	10	0.32	0.2	30
950	Poly AP; color	4[b]	0.65	0.4	80
950	AP; color	32[b]	5.2	3	10
300	AP; color	0.25[b]	0.1	0.06	1,700
3,120	AP; color	–	2[b]	1	10
4,900	AP; color	0.05[b]	0.04	0.02	306
510	AP; color	0.5	6[b]	4	20
3,400	AP; color	2[b]	0.06	0.04	210
830	HRP; color	–	0.2	0.1[b]	360
100	AP; color	100[b]	30	19	16
310	Ab-AP; color	1–5[b]	0.6	0.4	250
500	Eu; TRF	20	0.74	0.5	230
550	Color	1–5[b]	1.8	1	50
550	Color	20[b]	13	8	7
970	Ab-AP; color	500[b]	16	10	3
1,130	Ab-AP; color	1[b]	0.2	0.12	210
–	TNP-SH; color	4[b]	–	–	–
1	Autorad (18-hr)	20[b]	8	4	7,500
1	Color	8	1.7	1[b]	30,000
1	AP; color	50[b]	10	6	5,000
1	Color	80	24	14	2,100
1	AP; color	–	500[b]	300	100
4–6	AP; color	1000[b]	400	200	300
1	AP; color	1000	1,000	600	50
1	AP; color	–	400[b]	200	30
1	AP; color	5000[b]	2,000	1,200	25
1	AP; color	10^5	50,000	30,000	1

useful scope was limited. Langer et al. (3) initially suggested the use of antibiotin antibodies. Although biotin antibodies were used, with some success (8), in both filter-based hybridizations and in situ, avidin or streptavidin, used in conjunction with biotinylated horseradish peroxidase (8) or alkaline phosphatase (7a,b) to precipitate insoluble dyes, generally have proved superior. Nonradioactive detection of nucleic acid probes labeled enzymatically or chemically with biotin have been reported in Southern blot (40,72,73), dot blot (7a,b,15,37–40,49,50,70,74-76) and in situ (6,8,76-87) formats.

As summarized in Table 1, most of these systems use combinations of streptavidin and alkaline phosphatase to achieve reported sensitivities of down to about 0.1 attomoles (amol; 10^{-19} moles, or about 60,000 molecules) of target nucleic acid. This is usually accomplished in a dot-blot format using dye deposition. Biotin-based systems have also been detected with horseradish peroxidase (HRP) (88) or β-galactosidase (89) to generate luminescent signals, with reported sensitivities similar to color detection. Biotin can also be detected spectrophotometrically by avidin-peroxidase systems (58,90), or by using europium bound to streptavidin to provide time-resolved fluorometric detection (91).

Longer cloned probes labeled with biotin can also be fluorescently detected with FITC-labeled avidin, but applications have largely been limited to in situ formats (78).

The reported sensitivities of biotinylated synthetic probes have varied widely and appear to depend highly on the site of biotin attachment to the oligomer. When the biotin is attached to the 5′- or 3′-termini of synthetic probes, color detection limits are generally about 400 amol (2.4×10^8 molecules) of target, regardless of labeling method (11,48-50,54,70). Rashtchian et al. (58) spectrophotometrically detected biotinylated 50-base oligonucleotide probes by using streptavidin-peroxidase conjugates after RNA–DNA hybrid capture by antibodies; this system was able to detect 70,000 bacteria, or an estimated 1×10^8 molecules (170 amol) of ribosomal RNA. However, when biotin is attached to oligodeoxynucleotides through a 19-atom internal linker-arm base Ruth et al. (44a,b,45) reported that down to 10 amol of target can be detected. The precise reason for the sensitivity of the latter method is unknown, but it may reflect the length and mixed hydrophilic character of the biotin attachment, as well as internal position in the oligomer.

2. *Hapten Labels*

Probes that have been chemically labeled with AAIF have been detected by AAIF antibodies (29,92,93) coupled to alkaline phosphatase and peroxidase. Reported dot-blot sensitivities with dye deposition are 0.5–1 amol, or generally 2–10 times less sensitive than the best reported biotin systems, although recent side-by-side comparisons indicate that AAIF-modified probes can be more sensitive than biotinylated probes (93). Time-resolved fluorescence detection of AAIF-probes with use of europium has been reported (92) and is also capable of detecting 0.5–1 amol of target. Brominated nucleic acids (24) have also been detected down to similar levels with antibody-based alkaline phosphatase systems.

Other methods of detecting nucleic acids with indirect primary labels have been reported to be less sensitive. Antibodies to dinitrophenyl (DNP) modification have been used (23) to detect down to about 10 amol (1×10^{-17} mol) of target nucleic acid.

3. *Nontraditional Indirect Labels*

Nucleic acid hybrids have also been detected when no traditional nonradioactive labels are present. The following three methods have been grouped with indirect labels because posthybridization recognition by proteins is required for detection.

Vary and co-workers (94) have used long (8000 bases) polyadenosine tails on the 3′-terminus of a synthetic 50-base oligonucleotide as a probe. After the polyA-tailed probe was hybridized and displaced (95) into solution, the polyA was degraded to ATP and measured by bioluminescence. Although theoretical sensitivity is subattomole, only femtomole (10^{-15} mol) sensitivities have been demonstrated, to date, with this system, when either ^{32}P or bioluminescence is used.

Yehle and colleagues (96) used RNA–DNA hybrids themselves as labels. Biotinylated DNA probes hybridized to ribosomal RNA targets in solution were captured by immobilized streptavidin, then detected spectrophotometrically with β-galactosidase conjugated to antibody against RNA–DNA hybrids. This system was capable of detecting about 1000 bacteria, or an estimated 10^6 (2×10^{-18} mol) ribosomal RNA targets.

Chu et al. (97) have suggested that midivariant RNA can be an effective reporter because it can be amplified exponentially by Qβ RNA polymerase. The initial report demonstrated that 5′-biotinylated RNA was still a substrate for the replicase (97), but no nonradioactive detection was described.

B. Detection of Direct Primary Labels

Direct primary labels can be described as reporter groups that are covalently attached to the probe and can be measured directly. Direct labels used, to date, include alkaline phosphatase, horseradish peroxidase, fluorescent groups, and chemiluminescent groups. With the possible exception of fluorescent groups, each of the direct labels does require posthybridization chemical treatment (addition of substrates, peroxides, other) to produce a signal; fluorescence detection may or may not require chemical treatment. Reports of nucleic acid probes detected by direct labels have been far fewer.

1. Enzyme Labels

Alkaline phosphatase and horseradish peroxidase conjugated directly to DNA probes can be detected colorimetrically by adding the appropriate substrates after the hybridization is complete. Renz and Kurz (42) were able to colorimetrically detect 1-5 pg (2×10^{-18} mol) of target in a Southern blot format using cloned probes nonspecifically cross-linked to alkaline phosphatase through polyamines. Surprisingly, a 22-base oligonucleotide labeled with a single alkaline phosphatase demonstrated equal sensitivity (47). This undoubtedly is due to a combination of factors, including the ability of the latter method (47) to maintain essentially full enzyme activity throughout conjugation and hybridization; the former method (42) requires more vigorous chemical cross-linking and harsher hybridization conditions, resulting in substantial loss of phosphatase activity.

2. Fluorescent Labels

Many laboratories have described the synthesis of fluorescently labeled nucleic acids. Practical use, however, has not been demonstrated on a routine basis because of the rigors of fluorescence detection integrated into hybridization assays. In dot-blot or Southern blot formats, fluorescent detection of probes suffers from the need for small volumes and sophisticated instrumentation. The practical detection limit of fluorescein is 10^{-12} M in solution. Theoretically, a 5-kb cloned probe containing 500-1000 fluoresceins should be capable of easily detecting 10^{-19} mol (0.1 amol) of complementary sequences in a 100 μl solution or on filters, a sensitivity that would rival colorimetric methods. In practice, fluorescent quenching, low hybridization efficiencies of longer probes, inherent high fluorescent backgrounds, and the current inability to produce high-quality probes with well-defined fluorescent groups have, thus far, limited fluorescent detection to the research laboratory.

Fluorescent systems that use nonradioactive energy-transfer approaches have successfully combined appropriate energy-transfer donors and acceptors, either on separate oligonucleotide probes (98) or on the same oligomer (99). Although energy transfer techniques have basic advantages of lower backgrounds and also offer the possibility of true solution hybridization and detection, sensitivities are poorer than for direct fluorescence because some energy loss (and, therefore, signal loss) must occur during transfer.

V. ADDITIONAL CONSIDERATIONS

A. Stability of Labeled Probes

1. Label Stability During Storage

Almost all the labels described are relatively stable when properly stored in aqueous solution and have shelf lives of 1-2 years or more. As with all nucleic acids, care must be taken to avoid nuclease activity, and proteins used as direct or indirect labels must be protected from proteinase activities or bacterial contamination. Labels that may require additional considerations are: thiol-sensitive labels such as bio-SS-dUTP (12) or mercaptoethylphosphates (63); photosensitive labels (38,39,57) such as psoralens; mercurated nucleic acids (32) which may slowly decompose (34); and any label cross-linked with aldehyde linkers (1, 36,37,42,56) because the resulting imine linkages slowly hydrolyze in aqueous solutions unless chemically reduced. Labels linked to nucleic acids using phosphoramidate diesters (49-51) will also hydrolyze slowly in aqueous solutions, particularly in mild acid.

2. Label Stability During Hybridization

Current hybridization techniques invariably require elevated temperatures (42-80°C), time (minutes to days), usually detergents and, for many, 20-50% organic solvents such as formamide. Indirect labels have not been reported to be as sensitive to these treatments. However, many enzymes that could be very useful as direct labels, such as β-galactosidase or luciferases, cannot withstand the temperatures, detergents, or organic solvents. Alkaline phosphatase and horseradish peroxidase are somewhat hardier (42,47). Enzymes are particularly useful when coupled to synthetic probes (47) because time, temperature, and organic solvents can be drastically reduced or eliminated to preserve enzyme activity.

B. Effect of Label On Hybridization

The effect of most labels on hybridization is difficult to determine, especially with cloned probes, and has not been thoroughly investigated. When using cloned probes, biotin enzymatically incorporated using bio-dUTP has been shown (3) to decrease melting temperatures only 3-5°C, a moderate effect considering the extensive biotinylation. Biotinylated N-4 cytosine bases (15,40,58) appear to perturb hybridization because sites of hydrogen bonding are modified, and DNA-containing N-4 alkyl cytosines cannot reanneal completely (15); the direct effects on selectivity have not been studied. Photobiotin (38) and biotin-psoralen (39) nonspecifically cross-link to nucleic acids and, as such, may have an effect on hybridization; any effect has been minimized by either restricting the amount of label (38) or by labeling only nonhybridizing portions of M13 DNA (39). Nonspecific cross-linking of biotin-conjugated (37) or enzyme-conjugated (42) polyamines to nucleic acids has been suggested (37) to occur at N-7 of guanines; if true, this modification would not be expected to significantly affect hybridization.

Internal linker arms conjugated to biotin or even to enzymes do not appear to alter hybridization characteristics (44a,b,47). Although initial reports (47) suggested that the internal attachment of alkaline phosphatase to oligonucleotide probes lowered the probe melting temperature by up to 10°C, more detailed studies have since shown the effect to be less than 2°C (J. Ruth and E. Jablonski, unpublished results). Modification of oligonucleotides by a single AAAF, however, drastically decreases melting behavior (30). Fluorescent pyridopyrimidine analogues of cytosine appear to base pair better with cytosine than with guanosine (68).

C. Detectability of The Label

Ultimately, any effect of label, or the method of attachment, on hybridization should be reflected in both the sensitivity and selectivity of the probe. As one measure of this effect, the detectability, or relative signal contribution, of each individual label in a given probe can be derived. As indicated in Table 1, relative contribution of each label can be defined as inversely proportional to the number of labels per probe molecule times the minimum number of target molecules needed for a positive signal. A 9.46-kb probe, labeled with 10 biotins per 100 bases, has about 950 biotins per probe molecule; if 3×10^6 molecules of target can be detected, each biotin contributes a signal equivalent to (950) $(3 \times 10^6) = 2.85 \times 10^9$ targets, on an average. Relative to an arbitrary standard (3×10^{10} targets = relative signal of 1), a label signal equivalent to 2.85×10^9 targets has a relative signal, per label, of 10. The relative signal undoubtedly reflects a variety of factors, including hybridization efficiency and any effect of the label on hybridization. It is interesting to note certain comparisons, however. For example, biotin-psoralen-labeled cloned probes (39) (relative signal = 360) result in about the same signal per biotin as biotin-polyamine-labeled cloned probes (37) (relative signal = 306) or 3′-biotin oligonucleotides (11,70) (relative signal = 100–300). Three groups (42,47,69) have reported probes with alkaline phosphatase attached as a direct label; interestingly, the relative signal generated by each alkaline phosphatase varied widely between the methods, ranging from 50 for cloned probes (42), to 2100 for 3′-labeled oligomers (69), to 30,000 for internally labeled oligomers (47). This represents as much as 600-fold difference in relative signal from the same label.

VI. SUMMARY

Nonisotopic labels for nucleic acid probes have generally paralleled labeling schemes described earlier for immunoassays, with the most successful systems continuing to use direct or indirect enzyme reporters, such as alkaline phosphatase, horseradish peroxidase, or β-galactosidase. This has allowed nonisotopic detection of hybridizations to develop rapidly from nonexistence less than 10 years ago, to current techniques capable of detecting as few as 20,000 target molecules. Defined in the context of traditional analytical chemistry, this represents a sensitivity equal to detecting 0.1 part per trillion in a 1-ml aqueous solution, a remarkable feat by almost any standard.

Commercial application to clinical diagnosis of microbial agents is still in its infancy, however. Potential future developments including microtiter-based formats, sandwich assays, complete hybridization and detection in solution, and overall assay simplification will continue to overcome the remaining barriers to widespread clinical use of nonradioactive probes. (For reviews of nonradioactive DNA probes, hybridization strategies, and detection methods see Refs. 100a,b).

REFERENCES

1. Manning, J. E., N. D. Hershey, T. R. Broker, M. Pelligrini, H. K. Mitchell, and N. Davidson, A new method of in situ hybridization. *Chromosoma (Berl.) 53*:107–117 (1975).
2. Broker, T. R., L. M. Angerer, P. H. Yen, N. D. Hershey, and N. Davidson, Electron microscopic visualization of tRNA genes with ferritin-avidin:biotin labels. *Nucleic Acids Res. 5*:363–384 (1978).
3. Langer, P. R., A. A. Waldrop, and D. C. Ward, Enzymatic synthesis of biotin-labeled polynucleotides: Novel nucleic acid affinity probes. *Proc. Natl. Acad. Sci. USA 78*:6633–6637 (1981).
4. Bergstrom, D. E., and J. L. Ruth, Synthesis of C-5 substituted pyrimidine nucleosides via organopalladium intermediates. *J. Am. Chem. Soc. 98*: 1587–1589 (1976).
5. Heitzman, H., and F. M. Richards, Use of the avidin–biotin complex for specific staining of biological membranes in electron microscopy. *Proc. Natl. Acad. Sci. USA 71*:3537–3541 (1974).
6. Brigati, D. J., D. Myerson, J. J. Leary, B. Spalholz, S. Z. Travis, C. K. Y. Fong, G. H. Hsiung, and D. C. Ward, Detection of viral genomes in cultured cells and paraffin-embedded tissue sections using biotin-labeled hybridization probes. *Virology 126*:32–50 (1983).

7a. Leary, J. J., D. J. Brigati, and D. C. Ward, Rapid and sensitive colorimetric method for visualizing biotin-labeled DNA probes hybridized to DNA or RNA immobilized on nitrocellulose: Bio-blots. *Proc. Natl. Acad. Sci. USA 80*:4045–4049 (1983).

7b. Leary, J. J., D. J. Brigati, and D. C. Ward, A sensitive colorimetric method for visualizing biotin-labeled probes hybridized to DNA or RNA on nitrocellulose. In Chromosomes and Cancer: From Molecules of Man, J. Rowley and Ultmann, J. (eds.). Academic Press, New York, pp. 274–290 (1983).

8. Hsu, S. M., L. Raine, and H. Fanger, Use of avidin–biotin–peroxidase complex (ABC) in immunoperoxidase techniques. A comparison between ABC and unlabeled antibody (PAP) procedures. *J. Histochem. Cytochem. 29*:577–580 (1981).
9. Koch, J., S. Kolvraa, and L. Bolund, An improved method for labelling of DNA probes by nick-translation. *Nucleic Acids Res. 14*:7132 (1986).
10. Kline, S. A., and D. Ward, Synthesis and evaluation of a biotin-substituted dCTP analogue. *Fed. Proc. 43*:2050 (1984).

11. Riley, L. K., M. E. Marshall, and M. S. Coleman, A method for biotinylating oligonucleotide probes for use in molecular hybridizations. *DNA 5*:333–337 (1986).
12. Shimkus, M. L., J. Levy, and T. A. Herman, A chemically cleavable biotinylated nucleotide: Usefulness in the recovery of protein DNA complexes from avidin affinity columns. *Proc. Natl. Acad. Sci. USA 82*:2593–2597 (1985).
13. McCracken, S., Preparation of RNA transcripts using SP6 RNA polymerase. *BRL Focus 7*(2):5–8, 11–12 (1985).
14. Luehrsen, K. R., and M. P. Baum, In vitro synthesis of biotinylated RNA probes from A-T rich templates: Problems and solutions. *Biotechniques 5*:660–670 (1987).
15. Gebeyhu, G., P. Y. Rao, P. SooChan, D. A. Simms, and L. Klevan. Novel biotinylated nucleotide analogs for labeling and colorimetric detection of DNA. *Nucleic Acids Res. 15*:4513–4534 (1987).
16. Draper, D. E., and L. M. Gold, *Biochemistry 19*:1774–1781 (1980).
17. Schulman, L. H., H. Pelka, and S. A. Reines, *Nucleic Acids Res. 9*:1203–1217 (1981).
18. Draper, D. E., Attachment of reporter groups to specific, selected cytidine residues in RNA using a bisulfite-catalyzed transamination reaction. *Nucleic Acids Res. 12*:989–1002 (1984).
19. Kasher, M. S., D. Pintel, and D. C. Ward, Rapid enrichment of HeLa transcription factors IIIB and IIIC by using affinity chromatography based on avidin-biotin interaction. *Mol. Cell. Biol. 6*:3117–3127 (1986).
20. Grabowski, P. J., and P. Sharp, Affinity chromatography of splicing complexes: U2, U5, and U4 + U6 small nuclear ribonuclearprotein particles in the spliceosome. *Science 233*:1294–1299 (1986).
21. Welcher, A. A., A. R. Torres, and D. C. Ward, Selective enrichment of specific DNA, cDNA, and RNA sequences using biotinylated probes, avidin, and copper-chelate agarose. *Nucleic Acids Res. 14*:10027–10044 (1986).
22. Theveny, B., and B. Revet, DNA orientation using specific avidin-ferritin biotin end labeling. *Nucleic Acids Res. 15*:947–958 (1987).
23. Vincent, C., P. Tchen, M. Cohen-Solal, and P. Kourilsky, Synthesis of 8-(2,4-dinitrophenyl-2,6-aminohexyl) amino-adenosine 5′-triphosphate: Biological properties and potential uses. *Nucleic Acids Res. 10*:6787–6796 (1982).
24. Sakamoto, H., F. Traincard, T. Vo-Quang, T. Ternynck, J-L. Guesdon, and S. Avrameas, 5-Bromodeoxyuridine in vivo labeling of M13 DNA, and its use as a nonradioactive probe for hybridization experiments. *Mol. Cell. Probes 1*:109–120 (1987).
25. Prober, J. M., G. L. Trainor, R. J. Dam, F. W. Hobbs, C. W. Robertson, R. J. Zagursky, A. J. Cocuzza, M. A. Jensen, and K. Baumeister, A system for rapid DNA sequencing with fluorescent chain-terminating dideoxynucleotides. *Science 238*:336–341 (1987).
26. Richardson, R. W., and R. I. Gumport, Biotin and fluorescent labeling of RNA using T4 RNA ligase. *Nucleic Acids Res. 11*:6167–6184 (1983).

27. Cosstick, R., L. W. McLaughlin, and F. Eckstein, Fluorescent labeling of tRNA and oligodeoxynucleotides using T4 RNA ligase. *Nucleic Acids Res. 12*:1791–1810 (1984).
28. Bauman, J. G. J., J. Wiegant, and P. Van Duijn, Cytochemical hybridization with fluorochrome-labeled RNA. *J. Histochem. Cytochem. 29*:227–237 (1981).
29. Tchen, P., R. P. P. Fuchs, E. Sage, and M. Leng, Chemically modified nucleic acids as immunodetectable probes in hybridization experiments. *Proc. Natl. Acad. Sci. USA 81*:3466–3470 (1984).
30. Sanford, D. G., and T. R. Krugh, *N*-Acetoxy-2-acetylaminofluorene modification of a deoxyoligoribonuceotide duplex. *Nucleic Acids Res. 13*: 5907–5917 (1985).
31. Lebacq, P., D. Squalli, M. Duchenne, P. Pouletty, and M. Joannes, A new sensitive non-isotopic method using sulfonated probes to detect picogram quantities of specific DNA sequences on blot hybridization. *J. Biochem. Biophy. Meth. 15*:255–266 (1988).
32. Hopman, A. H. N., J. Wiegant, G. I. Tesser, and P. van Duijn, A non-radioactive in situ hybridization method based on mercurated nucleic acid probes and sulfhydryl–hapten ligands. *Nucleic Acids Res. 14*:6471–6488 (1986).
33a. Dale, R. M. K., E. Martin, D. C. Livingston, and D. C. Ward, Direct covalent mercuration of nucleotides and polynucleotides. *Biochemistry 14*: 2447–2457 (1975).
33b. Dale, R. M. K., and D. C. Ward, Mercurated polynucleotides: New probes for hybridization and selective polymer fractionation. *Biochemistry 14*: 2458–2469 (1975).
34. Van Broeckhoven, C., and R. De Wachter, The reactions of mercurated pyrimidine nucleotides with thiols and with hydrogen sulfide. *Nucleic Acids Res. 5*:2133–2151 (1978).
35. Renz, M., Polynucleotide-histone H-1 complexes as probes for blot hybridizations. *EMBO J. 2*:817–822 (1983).
36. Syvanen, A.-C., M. Alanen, and H. A. Soderlund, Complex of single-strand binding protein and M13 DNA as hybridization probe. *Nucleic Acids Res. 13*:2789–2802 (1985).
37. Al-Hakim, A. H., and R. Hull, Studies towards the development of chemically synthesized non-radioactive biotinylated nucleic acid hybridization probes. *Nucleic Acids Res. 14*:9965–9976 (1986).
38. Forster, A. C., J. L. McInnes, D. C. Skingle, and R. H. Symons, Non-radioactive hybridization probes prepared by the chemical labelling of DNA and RNA with a novel reagent, photobiotin. *Nucleic Acids Res. 13*: 745–762 (1985).
39. Sheldon, E. L., D. E. Kellog, R. Watson, C. H. Levenson, and H. A. Erlich, Use of nonisotopic M13 probes for genetic analysis: Application to HLA class II loci. *Proc. Natl. Acad. Sci. USA 83*:9085–9089 (1986).
40. Viscidi, R. P., C. J. Connelly, and R. H. Yolken, Novel chemical method for the preparation of nucleic acids for non-radioactive hybridization. *J. Clin. Microbiol. 23*:311–317 (1986).

41. Nomura, A., K. Negishi, and H. Hayatsu, Direct-acting mutagenicity of N^4-aminocytidine derivatives bearing alkyl groups at the hydrazino nitrogens. *Nucleic Acids Res. 13*:8893–8899 (1985).
42. Renz, M., and C. Kurz, A colorimetric method for DNA hybridization. *Nucleic Acids Res. 12*:3435–3444 (1984).
43. Letsinger, R. L., and M. E. Schott, Selectivity in binding a phenanthridium-dinucleotide derivative to homopolynucleotides. *J. Am. Chem. Soc. 103*:7394–7396 (1981).
44a. Ruth, J. L., Chemical synthesis of nonradioactively labeled DNA hybridization probes. *DNA 3*:123 (1984).
44b. Ruth, J. L., and R. N. Bryan, Chemical synthesis of modified oligonucleotides and their utility as nonradioactive hybridization probes. *Fed. Proc. 43*:2048 (1984).
45. Ruth, J. L., C. A. Morgan, and A. Pasko, Linker arm nucleotide analogs useful in oligonucleotide synthesis. *DNA 4*:93 (1985).
46. Matteucci, M. D., and M. H. Caruthers, Synthesis of deoxyoligonucleotides on a polymer support. *J. Am. Chem. Soc. 103*:3185–3191 (1981).
47. Jablonski, E., E. W. Moomaw, R. H. Tullis, and J. L. Ruth, Preparation of oligodeoxynucleotide-alkaline phosphatase conjugates and their use as hybridization probes. *Nucleic Acids Res. 14*:6115–6128 (1986).
48. Kempe, T., W. I. Sundquist, F. Chow, and S. Hu, Chemical and enzymatic biotin-labeling of oligodeoxyribonucleotides. *Nucleic Acids Res. 13*:45–57 (1985).
49. Chollet, A., and E. H. Kawashima, Biotin-labeled synthetic oligodeoxyribonucleotides: Chemical synthesis and uses as hybridization probes. *Nucleic Acids Res. 11*:659–669 (1985).
50. Chu, B. C. F., and L. E. Orgel, Detection of specific DNA sequences with short biotin-labeled probes. *DNA 4*:327–331 (1985).
51. Chu, B. C. F., G. M. Wahl, and L. E. Orgel, Derivatization of unprotected oligonucleotides. *Nucleic Acids Res. 11*:6513–6528 (1983).
52. Watcher, L., J. Jablonski, and K. L. Ramachandran, A simple and efficient procedure for the synthesis of 5′-aminoalkyl oligodeoxynucleotide. *Nucleic Acids Res. 14*:7985–7994 (1986).
53. Connolly, B. A., The synthesis of oligonucleotides containing a primary amino group at the 5′-terminus. *Nucleic Acids Res. 15*:3131–3139 (1987).
54. Agrawal, S., C. Cristodoulous, and M. J. Gait, Efficient methods for attaching nonradioactive labels to the 5′-ends of synthetic oligodeoxyribonucleotides. *Nucleic Acids Res. 14*:6227–6245 (1986).
55. Tanaka, T., T. Sakata, K. Fujimoto, and M. Ikehara, Synthesis of oligodeoxyribonucleotide with aliphatic amino or phosphate group at the 5′-end by the phosphoptriester method on a polystyrene support. *Nucleic Acids Res. 15*:6209–6224 (1987).
56. Kremsky, J. N., J. L. Wooters, J. P. Dougherty, R. E. Meyers, M. Collins, and E. L. Brown, Immobilization of DNA via oligonucleotides containing aldehyde or carboxylic acid group at the 5′-terminus. *Nucleic Acids Res. 15*:2891–2909 (1987).

57. Gamper, H. B., G. D. Cimino, S. T. Issacs, M. Ferguson, and J. C. Hearst, Reverse Southern hybridization. *Nucleic Acids Res. 14*:9943–9954 (1986).
58. Rashtchian, A., J. Eldridge, M. Ottaviani, M. Abbot, G. Mock, D. Lovern, J. Klinger, and G. Parsons, Immunological capture of nucleic acid hybrids and application to nonradioactive DNA probe assays. *Clin. Chem. 33*: 1526–1530 (1987).
59. Smith, L. M., S. Fung, M. W. Hunkapiller, T. J. Hunkapiller, and L. E. Hood, The synthesis of oligonucleotides containing an alphatic amino group at the 5′-terminus: Synthesis of fluorescent DNA primers for use in DNA sequence analysis. *Nucleic Acids Res. 13*:2399–2412 (1985).
60. Connell, C., S. Fung, C. Heiner, J. Bridgham, V. Chakerian, E. Heron, B. Jones, C. Menchen, W. Mordan, M. Raff, M. Recknor, L. Smith, J. Springer, S. Woo, and M. Hunkapiller, Automated DNA sequence analysis. *Biotechniques 5*:342–348 (1987).
61. Sproat, B. A., B. Beijer, and P. Rider, The synthesis of protected 5′-amino-2′, 5′-dideoxyribonucleoside 3′-*O*-phosphoramidites; applications of 5′-amino-oligodeoxyribonucleotides. *Nucleic Acids Res. 15*:6181–6196 (1987).
62. Sproat, B., B. Beijer, P. Rider, and P. Neuner, The synthesis of protected 5′-mercapto-2′,5′-dideoxyribonucleoside-3′-*O*-phosphoramidites; uses of 5′-mercapto-oligodeoxyribonucleotides. *Nucleic Acids Res. 15*:4837–4848 (1987).
63. Connolly, B. A., Chemical synthesis of oligonucleotides containing a free sulphydryl group and subsequent attachment of thiol-specific probes. *Nucleic Acids Res. 13*:4485–4502 (1985).
64. Zuckermann, R., D. Corey, and P. Schultz, Efficient methods for attachment of thiol specific probes to the 3′-ends of synthetic oligodeoxyribonucleotides. *Nucleic Acids Res. 15*:5305–5321 (1987).
65. Haralambidis, J., M. Chai, and G. Tregear, Preparation of base-modified nucleosides suitable for non-radioactive label attachment and their incorporation into synthetic oligodeoxyribonucleotides. *Nucleic Acids Res. 15*:4857–4876 (1987).
66. Gibson, K., and S. J. Benkovic, Synthesis and application of derivatizable oligonucleotides. *Nucleic Acids Res. 15*:6455–6467 (1987).
67. Bergstrom, D. E., H. Inoue, and P. A. Reddy, *J. Org. Chem. 47*:2174–2178 (1982).
68. Inoue, H., A. Imura, and E. Ohtsuka, Synthesis and hybridization of dodecadeoxyribonucleotides containing a fluorescent pyridopyrimidine deoxynucleoside. *Nucleic Acids Res. 13*:7119–7128 (1985).
69. Li, P., P. P. Medon, D. C. Skingle, J. A. Lanser, and R. H. Symons, Enzyme-linked synthetic oligonucleotide probes: Non-radioactive detection of enterotoxigenic *E. coli* in faecal specimens. *Nucleic Acids Res. 15*:5275–5287 (1987).
70. Murasugi, A., and R. B. Wallace, Biotin-labeled oligonucleotides: Enzymatic synthesis and use as hybridization probes. *DNA 3*:269–277 (1984).
71. Sharmeen, L., and J. Taylor, Enzymatic synthesis of RNA oligonucleotides. *Nucleic Acids Res. 15*:6705–6711 (1987).

72. Garbutt, G. J., J. T. Wilson, G. S. Schuster, J. J. Leary, and D. C. Ward, Use of biotinylated probes for detecting sickle cell anemia. *Clin. Chem. 31*: 1203–1206 (1985).
73. Koch, J., N. Gregerson, S. Kolvraa, and L. Bolund, The use of biotinylated hybridization probes for routine analysis of unique sequences in DNA. *Nucleic Acids Res. 14*:7133 (1986).
74. Chan, V. T.-W., K. A. Fleming, and J. O'D. McGee, Detection of subpicogram quantities of specific DNA sequences on blot hybridization with biotinylated probes. *Nucleic Acids Res. 13*:8083–8091 (1985).
75. Dorman, M. A., C. D. Blair, J. K. Collins, and B. J. Beaty, Detection of bovine herpesvirus 1 DNA immobilized on nitrocellulose by hybridization with biotinylated DNA probes. *J. Clin. Microbiol. 22*:990–995 (1985).
76. Gardner, L., Nonradioactive DNA labeling: Detection of specific DNA and RNA sequences on nitrocellulose and in situ hybridization. *Biotechniques 1*:38–43 (1983).
77. Unger, E. R., L. R. Budgeon, D. Meyerson, and D. J. Brigati, Viral diagnosis by in situ hybridization. Description of a rapid simplified colorimetric method. *Am. J. Surg. Pathol. 10*:1–8 (1986).
78. Pinkel, D., T. Straume, and J. W. Gray, Cytogenetic analysis using quantitative, high sensitivity, fluorescence hybridization. *Proc. Natl. Acad. Sci. USA 83*:2934–2938 (1986).
79. Manuelidis, L., In situ detection of DNA sequences using biotinylated probes. *BRL Focus 7*:408 (1985).
80. Burns, J., V. T. W. Chan, J. A. Jonasson, K. A. Fleming, S. Taylor, and J. O'D McGee, Sensitive system for visualizing biotinylated DNA probes hybridised in situ: Rapid sex determination of intact cells. *J. Clin. Pathol. 38*: 1085–1092 (1985).
81. Singer, R. H., J. G. Lawrence, and C. Villnave, Optimization of in situ hybridization using isotopic and non-isotopic detection methods. *Biotechniques 4*:230–250 (1986); errata *4*:314 (1986).
82. Landegent, J. E., N. J. in de Wal, G.-J. B. van Ommen, F. Baas, J. J. M. de Vijlder, P. van Duijn, and M. van der Ploeg, Chromosomal localization of a unique gene by non-autoradiographic in situ hybridization. *Nature 317*:175–177 (1985).
83. Trask, B., G. van den Engh, J. Landegent, N. J. in de Wal, and M. van der Ploeg, Detection of DNA sequences in nuclei in suspension by in situ hybridization and dual beam flow cytometry. *Science 230*:1401–1403 (1985).
84. Meyerson, D., R. C. Hackman, and J. D. Meyers, Diagnosis of cytomegaloviral pneumonia by in situ hybridization. *J. Infect. Dis. 150*:272–277 (1984).
85. Smith, G. H., D. J. Doherty, R. B. Stead, C. M. Gorman, D. E. Graham, and B. H. Howard, Detection of transcription and translation in situ with biotinylated molecular probes in cells transfected with recombinant DNA plasmids. *Anal. Biochem. 156*:17–24 (1986).
86. Unger, E. R., J. J. Leary, D. C. Ward, and D. J. Brigati, Application of

nucleic acid hybridization in clinical virology. In *Microbial Antigen Diagnosis*, K. Wicher (ed.). CRC Press, Boca Raton (1987).
87. Garson, J. A., J. A. van den Berghe, and J. T. Kemshead, Novel non-isotopic in situ hybridization technique detects small (1 kb) unique sequences in routinely G-banded human chromosomes: Fine mapping of *N-myc* and *B-NGF* genes. *Nucleic Acids Res. 15*:4761–4770 (1987).
88. Matthews, J. A., A. Batki, C. Hynds, and L. J. Kricka, Enhanced chemiluminescent method for the detection of DNA dot-hybridization assays. *Anal. Biochem. 151*:205–209 (1985).
89. Nagata, Y., H. Yokota, O. Kosuda, K. Yokoo, K. Takemura, and T. Kikuchi, Quantification of picogram levels of specific DNA immobilized in microtiter wells. *FEBS Lett. 183*:379–382 (1985).
90. Syvanen, A.-C., M. Laaksonen, and H. Soderlund, Fast quantification of nucleic acid hybrids by affinity-based hybrid collection. *Nucleic Acids Res. 14*:5037–5048 (1986).
91. Dahlen, P., A.-C. Syvanen, P. Hurskainen, M. Kwiatkowski, C. Sund, J. Ylikoski, H. Soderlund, and T. Lovgren, Sensitive detection of genes by sandwich hybridization and time-resolved fluorometry. *Mol. Cell. Probes 1*:159–168 (1987).
92. Syvanen, A.-C., P. Tchen, M. Ranki, and H. Soderlund, Time-resolved fluorometry; a sensitive method to quantify DNA-hybrids. *Nucleic Acids Res. 14*:1017–1028 (1986).
93. Donovan, R. M., C. E. Bush, W. R. Peterson, L. H. Parker, S. H. Cohen, G. W. Jordan, K. M. W. Brink, and E. Goldstein, Comparison of non-radioactive DNA hybridization probes to detect human immunodeficiency virus nucleic acid. *Mol. Cell. Probes 1*:359–366 (1988).
94. Vary, C. P. H., F. J. McMahon, F. P. Barbone, and S. E. Diamond, Non-isotopic detection methods for strand displacement assays of nucleic acids. *Clin. Chem. 32*:1696–1701 (1986).
95. Ellwood, M. S., M. Collins, E. F. Fritsch, J. I. Williams, S. E. Diamond, and J. G. Brewen, Strand displacement applied to assays with nucleic acid probes. *Clin. Chem. 32*:1631–1636 (1986).
96. Yehle, C. O., W. L. Patterson, S. J. Boguslawski, J. P. Albarella, K. F. Kip, and R. J. Carrico, A solution hybridization assay for ribosomal RNA from bacteria using biotinylated DNA probes and enzyme-labeled antibody to DNA:RNA. *Mol. Cell. Probes 1*:177–193 (1987).
97. Chu, B. C. F., F. R. Kramer, and L. E. Orgel, Synthesis of an amplifiable reporter RNA for bioassays. *Nucleic Acids Res. 14*:5591–5603 (1986).
98. Heller, M. J., and L. E. Morrison, In *Rapid Detection and Identification of Infectious Agents*, D. Kingsbury and Falkow, S. (eds.). Academic Press, New York, pp. 245–256 (1985).
99. Heller, M. J., E. Hennesy, J. L. Ruth, and E. Jablonski, Fluorescent energy transfer oligonucleotide probes. *Fed. Proc. 46*:1968 (1987).
100a. Matthews, J. A., and L. J. Kricka, Analytical strategies for the use of DNA probes. *Anal. Biochem. 169*:1–25 (1988).
100b. Jablonski, E., In *DNA Probes for Infectious Diseases*, F. Tenover (ed.). CRC Press, Boca Raton, Chapter 2 (1988).

3
DNA Probes to Ribosomal RNA

JAMES J. HOGAN
Gen-Probe, Inc., San Diego, California

I. INTRODUCTION

The discussion in Dr. Brenner's chapter (Chapt. 4) on DNA hybridization and the importance of rapid identification of disease-causing organisms is excellent. This chapter on DNA probes to ribosomal RNA will first concentrate on why ribosomal RNA is an excellent taxonomic tool and, second, on the importance of designing testing protocols that can be easily adopted by clinical laboratories with a minimum of effort or equipment. Development of new assay protocols that are more desirable in the eyes of the clinical laboratory are being implemented as rapidly as possible. New generations of tests are striving for easier sample handling, coupled with features such as nonisotopic labels and heightened sensitivity. Gen-Probe's first generation of tests were designed to be used with virtually no new training or replacement of existing equipment. An example of this is the tissue culture bacterial contamination assay that was developed primarily to detect mycoplasmas. These are cell-wall-less bacteria and are extremely difficult to detect with standard techniques. The second generation of tests were designed to be performed with a reduction of hands-on time for the technician and to be used directly on the patient specimen without the need of laborious sample preparation procedures. The initial second-generation tests were directed toward respiratory pathogens, with the samples being either throat swabs or sputum samples. The bacteria that cause legionellal, mycobacterial, and mycoplasmal pneumonias were the three target organisms. A final description of the actual assays and performance characteristics for both genus- and species-specific probes will then be presented.

II. RIBOSOMAL RNA, A TAXONOMIC RULER

Why is the ribosomal RNA (rRNA) such a good target nucleic acid for bacterial identification purposes with DNA probes? Before discussing this question, a brief background on ribosomes would be appropriate.

Prokaryotic ribosomes and eukaryotic ribosomes are referred to as *70S* particles and *80S* particles, respectively. These broad classifications, which are based upon sedimentation characteristics, are just handy labels because the actual sizes of ribosomes can be much smaller than 70S, whereas there are eukaryotic examples that fall between the 70S and 80S values, and even examples that are larger than 80S.

The ribosome is made up of two subunits, the *30S* and *50S*, from the prokaryotic 70S, and the *40S* and *60S* subunits from the eukaryotic 80S. These subunits are multicomponent and have both RNA and protein fractions that also span a large range of sizes. For convenience, we refer to the prokaryotic 30S subunit rRNA as 16S and to the 50S subunit RNAs as 23S and 5S. The eukaryotic rRNA counterpart to the 16S is the 18S rRNA in the 40S subunit. In the eukaryotic 60S subunit there are three rRNAs with the 28S and 5.8S molecules corresponding to the 23S from prokaryotic 50S subunits (1,2) and the 5S rRNAs from both prokaryotic and eukaryotic subunits being classed together (3). The ribosomal proteins, in general, show little homology between organisms, with only a few displaying antigenic cross-reactions. We will be concerned with only the larger rRNAs in this discussion. These have the conservation and divergence of nucleotide sequence necessary for DNA probe design that can be useful in a clinical setting, these are the 16S and 23S rRNAs.

The rRNA was, for many years, considered just a structural framework on which the ribosomal proteins were supported. The basic thrust of ribosome research was concerned with the assembly of the *Escherichia coli* ribosomal subunits and the identification of which of the 55 ribosomal proteins were responsible for the different functions inherent in protein synthesis. The primary functional questions centered around transfer RNA (tRNA) binding; messenger RNA (mRNA) binding, initiation, elongation, and termination factors; subunit association; and the actual formation of peptide bonds between the amino acids that is referred to as *peptidyltransferase* activity.

The attention of many ribosomologists was centered around reconstitution experiments between the 21 ribosomal proteins and the 16S rRNA found in the *E. coli* 30S subunit. It was possible to do single-protein omission experiments and chemical modification of these proteins to look for functional results in an vitro assay systems. In the course of this protein modification work, it was found that chemical modification of the rRNA also had functional consequences. It was found that tRNA binding and subunit association were both affected by only a few specific base modifications out of the 1542 bases in 16S RNA. The sequences that were modified were identified, but little sense could be made of the modification patterns because the complete rRNA sequence was unknown. This led to the sequencing of the *E. coli* 16S rRNA, which was possible with the new DNA-sequencing technology (4-5). The *E. coli* 23S rRNA was then also sequenced and the results of functional perturbations by chemical modification of this rRNA were also mapped onto the primary structure (6).

At this time, independent work based upon T_1 maps (which entails digestion of the rRNAs with the guanine-specific nuclease T_1, then two-dimensional separation of the fragments, and subsequent sequencing of the T_1 fragments) led to crucial taxonomic correlations between related organisms. As more and more 16S rRNAs were mapped with this technique, evolutionary relationships were found to closely follow the conservation patterns of certain rRNA sequences (7). In fact, by analysis of these T_1 maps, Woese and co-workers (8) determined that there was a third kingdom, the archaebacterial, which was composed of many bacteria that, in retrospect, did not fit into the prokaryotic kingdom, even by classic criteria. This work by Woese, when combined with the hybridization techniques of Kohne (9,10), set the foundations for the ability to make DNA probes to identify organisms down to the species level by analysis of rRNA sequences.

These and other workers realized that an important aspect of the ribosome function lay in the rRNA, not just in the proteins.

The next stage of endeavor was to examine how the ribosomal 16S RNA was folded and what relationship this secondary structure had to the observation of taxonomic relatedness. By using a comparative approach, coupled with the chemical modification results, it was determined that many functionally important sites were located in unpaired regions and were phylogentically highly conserved sequences. Fairly complete secondary structures based upon comparative analysis have been determined, and the most variable regions seem to be in helical regions.

The realization that rRNA was a phylogenetically useful tool had been determined by an extension of DNA relatedness techniques involving hybridization criteria (10). With the understanding that the structure of the rRNA was a primary component necessary for ribosomal function, it became clear that both conservation of structure as well as certain conserved sequences were equally important. The basic secondary structure of the 16S *E. coli* rRNA is shown in Figure 1. For a more complete treatment of rRNA conservation patterns there are many detailed reviews (11-15). The conclusion from the primary structure and from these secondary structures is that some sequences are relatively invarient throughout all life, and these are in single-stranded regions. But the characteristic feature most conserved in all ribosomes is the secondary structure, rather than the primary sequences. If the helical regions must be conserved in structure, but not necessarily in sequence, then compensatory mutations must

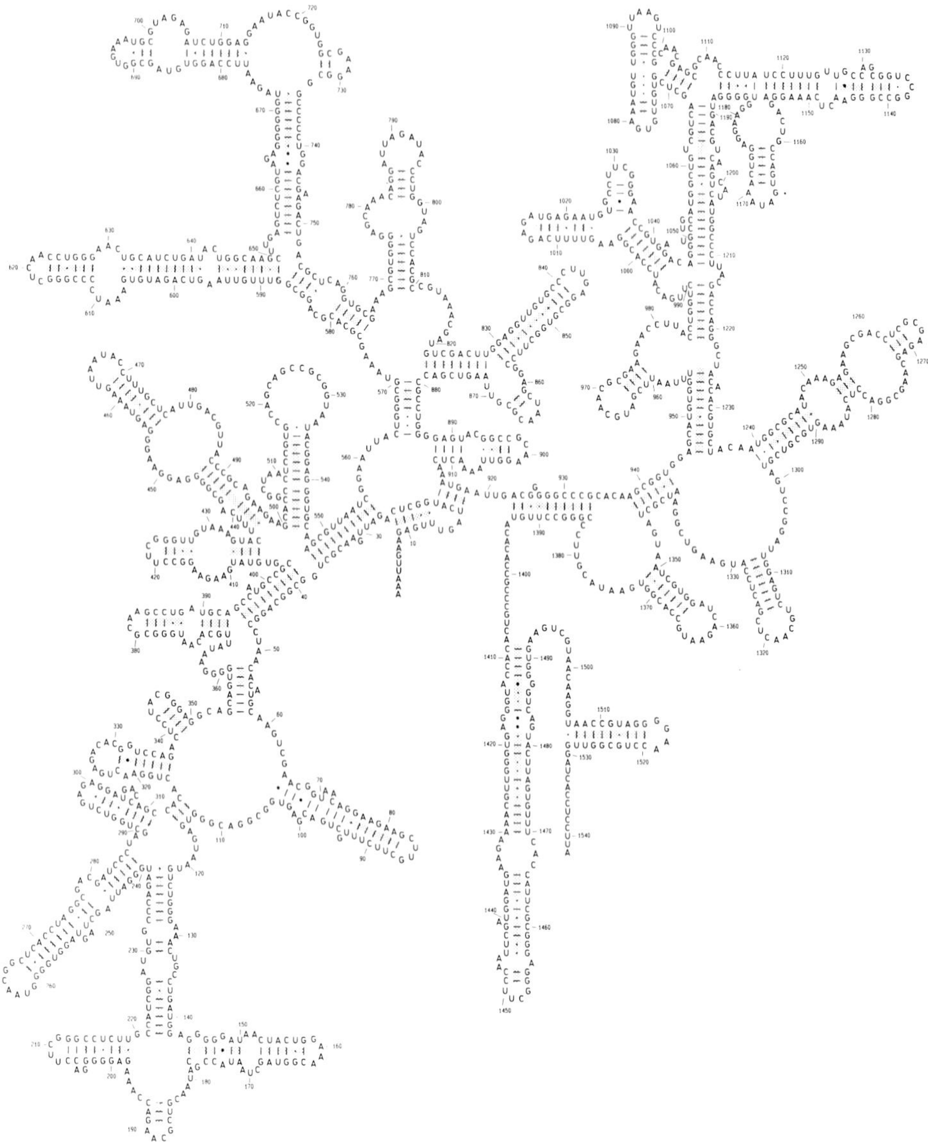

Figure 1 Secondary structure of *E. coli* rRNA (Source: Ref. 12).

occur on complementary strands to maintain these structures. As can be seen in Figure 1, the distance between opposing bases in the helical regions can be hundreds of nucleotides in the primary sequence, which implies that even if one region, in the DNA, that codes for the rRNA is a mutational hot spot, a compensatory mutation necessary for maintenance of structure and, thereby, function might be hundreds of nucleotides away and, therefore, highly unlikely. These two-site mutational restraints obviously place limits on the amount of randomly accumulated mutations to be found in helical regions of the RNA. This means that even though sequences do vary from one another in different organisms, the closer the organisms are to one another, the better chance of locating similar differences in relation to distant organism. Random changes in the rRNA are, therefore, selected against because the rRNA secondary structure is an integral part of the protein synthesis machinery. Any changes in the fidelity of this protein synthesis machinery would have far-reaching consequences in the cell's function.

III. ASSAY FORMAT

The following factors are considered optimal when designing DNA probes for clinical assays:

1. Variable specificity: species, genus, all bacterial, all life
2. Sensitivity: 10^3-10^6 organisms
3. Time: 2-4 hr
4. Simplicity: no more than four steps

All of these factors must be balanced against one another to make a useful diagnostic test for the laboratory. The first factor is determined by the desired organisms to be detected. The second is determined primarily by the patient specimen with a trade-off with 3. The third is determined by both 2 and the need for a same-day result being given to the doctor. The fourth factor is determined by required efficiency and expertise in the clinical laboratory. An example of balancing the different variables would be the length of time of the hybridization versus sensitivity of the assay. By using longer hybridization times, sensitivity could be increased. Alternatively, sensitivity could also be increased by using more of the patient sample and concentrating the organisms by either filtration or centrifugation. Other experimental variables involve the size of patient specimen, which may be small. If sensitivity is then to be increased, the hybridization time must be increased, or faster rate-acceleration systems must be used if the assay results are to be available in a few hours. Ultimately, the specific activity of the reporter group on one's probe, coupled with the number of organisms normally found in the patient sample, will determine the sensitivity assuming factors 3 and 4 are fixed. Now let us switch to actual probe assays to see how the formats of our kits reflect the goals of factors 1-4.

IV. MYCOPLASMA DETECTION

Gen-Probe's first kit was designed for use by laboratories that use tissue culture systems and need a rapid and inexpensive test for the detection of mycoplasma-type contaminations. These contaminations are refractory to normal methods used to detect bacterial organisms. The mycoplasmas are cell-wall-less organisms that are very difficult and time-consuming to culture. Most tissue culture users are research laboratories and have equipment slightly different from standard clinical laboratories, i.e., scintillation counters versus gamma counters. This was why this kit uses a tritium-labeled probe that is detected by a scintillation counter.

Our mycoplasma probes are primarily directed toward mycoplasmas and acheleoplasmas, with a capability of detecting other bacterial contaminants with a two- to fourfold reduction of sensitivity, depending on the evolutionary divergence from these primary target organisms. Going down our list from 1 through 4 for this application will give insight into designing a DNA probe assay.

The broad specificity of this probe for detection of bacterial contaminants in tissue culture is at the kingdom level, with an increased sensitivity toward hard-to-detect organisms. Our probe could be chosen to react with only organisms from the *Mycoplasma* group but tissue culture users are desirous of knowing of any bacterial contamination, so we chose a more broadly reacting probe for this application. The primary criteria for this probe's specificity is that it will not react with any eukaroytic samples, which it does not.

The titer of a sample determines what sensitivity is needed for a probe. With mycoplasmal infections of tissue culture, the titer is usually 10^7-10^8 organisms per milliliter of medium in an established infection, which is well within the sensitivity range of our probes. Our new mycoplasmal test kit, soon to be available, has been reformulated to give two orders of magnitude greater sensitivity. This will enable early detection of developing contaminations in one's tissue culture system. The earlier detection and subsequent removal of the contaminated lines will help reduce spread of the contamination. The primary source of tissue culture contamination is from other contaminated lines, so earlier detection and elimination of the contaminated lines is desirable.

One could save time and money in the laboratory if one could have a rapid feedback on tissue culture contaminants. With this in mind, we set the time for our assay at 2 hr. This is easily done with our four-step protocol, sample preparation, hybridization, hybrid isolation, and detection. The supernatant from the tissue culture line is spun down to pellet the mycoplasmas; depending on the available equipment, this supernatant can be from 250 ml down to 1 ml. In this hybridization system, we can detect 10^6 organisms in 1 hr, which is our desired sensitivity and time. The assay is incubated in a standard water bath at 72°C in a 7-ml scintillation vial. By performing the entire assay in the detection vessel, no transfers of the sample are necessary for processing. A solution containing the lysis solution and the DNA probe directed against the rRNA is added, and the temperature is raised to accelerate hybridization. Once hybridization is complete, a hydroxyapatite suspension is added that selectively binds only hybridized probe (16). Washes are performed with a clinical centrifuge and the scintillation cocktail is added directly to the pelleted hydroxyapatite. This is a test-tube version of the column method normally used in DNA hybridization studies described by Dr. Brenner in Chapter 4. Counting is performed in a standard scintillation counter in a few minutes. Recent publications evaluating this test illustrate the specificity and sensitivity of this assay on a broad range of tissue culture contaminants (17).

V. CLINICAL APPLICATIONS

A test for the detection of *Legionella* spp. had been approved by the FDA for both direct assays and colony confirmation. A recent review on this assay (18) also show the desired characteristics of a good clinical diagnostic procedure, i.e., good sensitivity and specificity, with a 2-hr completion time, with a minimum of steps. This probe has been labeled with radioactive iodine, for most clinical laboratories have access to a gamma counter because radioactive iodine assay systems are routinely used. The two newest probes with FDA approval are a genus-specific probe for *Mycobacterium* to be used for colony confirmation, and a species-specific probe for *Mycoplasma pneumoniae*. These respiratory pathogens historically have been extremely difficult to detect in a rapid manner, which is why we decided to develop these tests first. In the following sections, I will present some assay formats and a brief review of some characteristics of the kits. I will first present the data on the *M. pneumoniae* probe, which is particularly useful for detection directly from patient specimens in under 2 hr. Next I will present performance data on a genus-specific probe for *Mycobacterium* species, which has been altered to narrow the specificity, deemed desirable by the medical professional.

A. *Mycoplasma pneumoniae*

Until now, DNA probes have been limited in clinical assays by two factors: sensitivity and ease of use. Many research techniques would give excellent results, but the equipment, time, expense, required expertise, and amount of hands-on time make the technique impractical for general clinical use. This is not true for our assay system. As can be seen in Figure 2, the assay for *Mycoplasma*

MYCOPLASMA ASSAY

STEP	PURPOSE
• Throat swab speciman in transport medium	
• Centrifuge 10 min. in microcentrifuge. Decant supernatant.	Concentrate mycoplasma
• Add lysis solution. Suspend pellet. Transfer to Gamma counter tube.	Lyse organisms. Release ribosomal RNA. Destroy inhibitors.
• Add accelerator and probe. Heat 60 min at 72°C.	Hybridize probe to rRNA.
• Add separation solution. Heat 5 min at 72°C.	Absorb hybrid to hydroxyapatite.
• Centrifuge 2 min. Decant.	Separate bound from unbound probe.
• Add wash solution. Heat 5 min. at 72°C. Centrifuge 2 min. Decant.	Wash hydroxyapatite pellet.
• Count.	Determine amount of hybrid formed.

Figure 2 Assay protocol for identification of *M. pneumoniae*.

pneumoniae begins with the throat swab being transported to the laboratory in a standard transport medium. Next the transport medium is quickly spun down, the cells lysed, and nuclease inhibitors are added to preserve the target. It is transferred to the final detection vessel, here, a gamma-counter tube, and the probe and rate accelerators are added. The additions can be made with automatic pipettes for speed and accuracy. The hybridization occurs in 1 hr in a standard water bath at 72°C. The separation solution is then added and the procedure is finished with a quick-wash step. The time spent after hybridization is approximately 10 min, with the beginning steps before hybridization being only 15 min. This means the results are available within 2 hr, with less than 30-min hands-on time.

The results are displayed in Figures 3-5. Figures 3 and 4 show two separate studies, one conducted with students and the other with pediatric patients. The mean, the standard deviation, and the range are displayed. As can be seen, the signal/noise ratio leaves nothing to interpretation in determining the presence

		Gen-Probe Assay +	Gen-Probe Assay −
Culture*	+	100	0
	−	0	100

	Gen-Probe Assay Ratio		
	Mean	S.D.	Range
Culture Positives	19.6	3.1	11.8-27.7
Culture Negatives	1.2	0.2	0.9-1.6

Figure 3 Results from comparative study 1. Throat swab samples, collected from university students, were cultured on modified PPLO broth agar plates. Identification of *M. pneumoniae* was confirmed by guinea pig hemadsorption and antibody inhibition of growth.

		Gen-Probe Assay +	Gen-Probe Assay −
Culture*	+	94	0
	−	0	94

	Gen-Probe Assay Ratio		
	Mean	S.D.	Range
Culture Positives	22.8	3.8	17.5-29.7
Culture Negatives	1.1	0.1	0.8-1.5

Figure 4 Results from comparative study 2. Throat swab samples, collected from pediatric patients, were cultured on modified PPLO broth agar plates after overnight incubation in modified PPLO broth. Identification of typical colonies as *M. pneumoniae* was confirmed by the guinea pig hemadsorption method.

Gen-Probe Assay

		+	−
M. pneumoniae	2 × 10⁶	51	0
	1 × 10⁵	49	0
	0	0	100

Commercial Culture Method**

		+	P	I	−
M. pneumoniae	2 × 10⁶	48	0	3	0
	1 × 10⁵	28	12	9	0
	0	0	0	13	87

Figure 5 Results from comparative study 3. Four throat swabs were collected, in 4 ml of transport medium, from patients with pharyingitis or with respiratory infection, or from control volunteers. Two 0.9-ml aliquots of each sample were inoculated with either 2×10^6 or 1×10^5 *M. pneumoniae* cells. One 0.9-ml aliquot and one uninoculated 0.9-ml aliquot were assayed with the Gen-Probe *Mycoplasma pneumoniae* Rapid Detection System. The remaining matched inoculated and uninoculated 0.9-ml aliquots were cultured with Mycotrim-RS, a commercial culture method available from Hana Media, Inc. ** The results obtained with the commercial culture method are reported as follows: +, positive (typical colonies observed with appropriate medium color change); P, presumptive positive (only color change observed); I, inconclusive (overgrown by contaminating oral flora); −, negative (no colonies or color change observed).

of *M. pneumoniae*. Figure 5 shows a comparative evaluation of our test at 1×10^5 organisms, which is easily obtained with our method. The standard method that involves culture takes days to perform and has interpretation problems at the 10^5-organism level, which is common in patient samples. The actual sensitivity of this DNA probe assay is below 1×10^5 organisms. Table 1 shows some of the organisms tested against this probe with the signal/noise expressed as a ratio. From the earlier figures it is clear that a positive specimen usually has a signal/noise ratio greater than 10-fold. The cross-reactions with all other known mycoplasmas are less than the 3.0 ratio for the positive criteria, except for one recently discovered organism, *M. genitalium*. This organism now has been isolated only from the human urogenital tract in rare cases and has not been isolated from the respiratory tract (19–20). For a more in-depth study of the performance parameters of this probe, a recent publication is available (21).

Table 1 Reactivity of the *M. pneumoniae* Probe With Other Bacteria

ATCC Identification	Gen-Probe assay ratio	Gen-Probe assay +/-[a]
Actinomyces israelii	1.0	-
Bacteroides fragilis	1.3	-
Bifidobacterium breve	1.3	-
Bordetella bronchiseptica	1.4	-
Clostridium innocuum	1.0	-
C. pasteurianum	2.7	-
C. perfringens	1.1	-
C. ramosum	1.5	-
Corynebacterium xerosis	1.1	-
Erysipelothrix rhusiopathiae	1.1	-
Escherichia coli	1.0	-
Flavobacterium meningosepticum	1.1	-
Haemophilus influenzae	1.2	-
Klebsiella pneumoniae	1.0	-
Lactobacillus acidophilus	1.2	-
Legionella pneumophila	1.0	-
Listeria monocytogenes	1.0	-
Moraxella osloensis	1.1	-
Mycobacterium tuberculosis	1.9	-
Neisseria meningitidis	1.0	-
Pasteurella multocida	1.4	-
Peptococcus magnus	1.0	-
Propionibacterium acnes	1.0	-
Pseudomonas aeruginosa	1.1	-
Staphylococcus aureus	1.0	-
Streptococcus faecalis	1.2	-
S. mitis	1.3	-
S. pneumoniae	1.4	-
S. pyogenes	1.0	-

[a]A positive sample produces a ratio ≥ 3.0; a negative sample produces a ratio <3.

Step 1 Sample Preparation

In a Biological Safety Cabinet:
Add 100 μl of mycobacterial cell suspension, turbidity to match a #3 McFarland nephalometer standard,

or

Add 100 μl of a concentration-decontamination sediment to Reagent I tube (containing glass beads and buffer).
Disrupt cells in a prepared ultrasonic cleaner for 10 min.

Step 2 Hybridization

Add 1 ml of Reagent II, Probe Solution (containing ^{125}I labeled probe and buffers) to Reagent I tube.
Incubate at 72°C for 2 hr.

Step 3 Separation

Add 5 ml of Reagent III, Separation Solution.
Incubate at 72°C for 5 min.
Centrifuge at 2000 rpm for 2 min. Decant supernate.
Add 5 ml of Reagent IV, Wash Solution.
Centrifuge 2000 rpm for 2 min. Decant supernate.

Step 4 Detection

Place tube in gamma counter. Count for 5 min.

$$\text{Calculate RATIO} = \frac{\text{sample cpm} - \text{background cpm}}{\text{negative control cpm} - \text{background cpm}}$$

Figure 6 Sample preparation protocol for detection of the genus *Mycobacterium*.

B. *Mycobacterium* Genus Probe

A DNA probe that has genus-specific properties is directed toward mycobacterial rRNA. As can be seen in Figure 6, the sample preparation is slightly different from the preceding assay in the initial patient sample. The standard concentrated sediment specimen is added directly to the gamma tube, which eliminates any further transfers. The hybridization runs for 2 hr, which is currently necessary for the desired sensitivity. The processing after hybridization is also rapid and simple and follows the proceeding format. The probe detects members of the *M. tuberculosis* complex, as can be seen in Table 2. Species-specific probes in this genus have been designed; they have excellent discrimination characteristics and are available. An abbreviated presentation of the specificity and sensitivity of our genus-level probe is shown in Figure 7.

Table 2 *Mycobacterium* spp. Relatedness

Organism	% Hyb normalized to Mtb at 100%
M. tuberculosis (avir)	100.0
M. tuberculosis (vir)	100.3
M. bovis	100.2
M. bovis (BCG)	94.3
M. africanum	93.4
M. gastri	83.2
M. ulcerans	79.9
M. haemophilum	79.2
M. intracellulare	78.9
M. kansasii	78.7
M. marinum	78.0
M. avium	76.5
M. scrofulaceum	76.1
M. malmoense	72.3
M. gordonae	63.8
M. asiaticum	62.5
M. simiae	53.5
M. thermoresistible	53.3
M. nonchromogenicum	43.8
M. triviale	43.6
M. smegmatis	42.2
M. terrae	40.1
M. shimoidei	38.2
M. fortuitum	35.3
M. flavescens	34.4
M. xenopi	34.1
M. vaccae	33.9
M. szulgai	33.4
M. chelonae	33.4
M. phlei	28.9

		ATCC +	ATCC −
Gen-Probe	+	90	1
	−	0	74

Agreement = 99.4%
Specificity = 98.7%
Sensitivity = 100%

		Centers for Disease Control +	Centers for Disease Control −
Gen-Probe	+	184	0
	−	1	15

Agreement = 99.5%
Specificity = 100%
Sensitivity = 99.5%

		Combined +	Combined −
Gen-Probe	+	274	1
	−	1	89

Agreement = 99.5%
Specificity = 98.9%
Sensitivity = 99.6%

Figure 7 Results of genus-specific identification of *Mycobacterium* with Gen-Probes for this organism.

VI. CONCLUSIONS

In summary, the rRNA has proved to be an excellent taxonomic target. It also is a good target because of its high copy number of 10^3–10^4 copies per cell, which greatly improves sensitivity over that of a gene in the DNA that is present only in a few copies. As can be seen in the actual probe formats, no extensive processing of the nucleic acid is necessary to make it available for probe binding, as is found for duplex DNA. Also, because there is not a complementary strand competing for the probes binding site (the rRNA is single-stranded), the hybridization is easy to perform. With the new improvements in labeling of probes and the new accelerated rate systems, the clinical applicability of DNA probes to rRNA have been greatly increased, and they will be even more sensitive in the future. Already other areas besides the clinical laboratory, such as food testing and water quality, are being developed with very promising results. Sensitivities in the 10^3 organism range are possible, and our nonisotopic probes have been formated to be used in clinical assays. The rapid identification of pathogenic bacteria will dramatically reduce lag times in treatment and will enable the doctors to use more specific treatment regimens early in the treatment of patients.

REFERENCES

1. Noller, H. F., J. Kop, V. Wheaton, J. Brosius, R. R. Gutell, A. M. Kopylov, F. Dohme, W. Herr, D. A. Stahl, R. Gupta, and C. R. Woese, *Nucleic Acids Res. 9*:6167-6189 (1981).
2. Nazar, R. N., *FEBS Lett. 119*:212-214 (1980).
3. Garret, R. A., S. Douthwaite, and H. F. Noller, *Biochem. Sci. 6*:137-138 (1981).
4. Sanger, F., S. Nicklen, and A. R. Coulsen, *Proc. Natl. Acad. Sci. USA 74*: 5463-5467 (1977).
5. Maxam, A. M., and W. Gilbert, *Proc. Natl. Acad. Sci. USA 74*:560-564 (1977).
6. Noller, H. F., *Annu. Rev. Biochem. 53*:119-162 (1984).
7. Fox, G. E., E. Stakebrandt, R. B. Hespell, J. Gibson, J. Maniloff, T. A. Dyer, R. S. Wolfe, W. E. Balch, R. S. Tanner, L. A. Magrum, L. B. Zablem, R. Blakemore, R. Gupta, L. Bonen, B. J. Lewis, D. A. Stahl, K. R. Luehrsen, K. N. Chen, and C. R. Woese, *Science 209*:457-463 (1980).
8. Woese, C. R., *Sci. Am. 244*:98-122 (19xx).
9. Britten, R. J., and D. E. Kohne, *Carnegie Inst. Wash. Year Book 65*:78-106 (1966).
10. Kohne, D. E., A. G. Steigerwalt, and D. J. Brenner, In *Legionella*, C. Thornsberry (ed.). American Society for Microbiology, Washington, pp. 107-108 (1984).
11. Noller, H. F., and C. R. Woese, *Science 212*:403-411 (1981).
12. Woese, C. R., R. R. Gutell, R. Gupta, and H. F. Noller, *Microbiol. Rev. 47*:621-669 (1983).
13. Noller, H. F., *Annu. Rev. Biochem. 53*:119-162 (19xx).
14. Gutell, R. R., B. Weiser, C. R. Woese, and H. F. Noller, *Prog. Nucleic Acids Res. Mol. Biol. 32*:156-216 (1985).
15. Brimacombe, R., P. Maly, and C. Swieb, *Prog. Nucleic Acids Res. Mol. Biol. 28*:1-48 (1983).
16. Bernardi, G., *Nature 206*:779 (1965).

17. McGarrity, G. J., and H. Kotani, *Exp. Cell Res. 163*:273–278 (1986).
18. Wilkinson, H. W., J. S. Sampson, and B. B. Plikaytis, *J. Clin. Microbiol. 23*: 217–220 (1986).
19. Tully, J. G., D. Taylor-Robinson, D. L. Rose, R. M. Cole, and J. M. Bore, *Int. J. Syst. Bacteriol. 33*:387–396 (1983).
20. Lind, K., B. O. Lindhardt, H. J. Schutten, J. Blom, and C. Christiansen, *J. Clin. Microbiol. 20*:1036–1043 (1984).
21. Shaw, S. B., L. B. Weiner, T. J. Fultz, T. G. Lawrence, L. A. Boas, J. G. Putnam, D. K. Cabanas, J. A. McMillan, C. L. Page, S. H. McDonough, M. S. Hoppe, L. Poe, D. E. Kohne, and D. L. Kacian, *N. Engl. J. Med.* (submitted).

RECENT REFERENCES ADDED IN PROOF

Hanna, B. A., and R. Gonzalez, Evaluation of Gen-Probe DNA hybridization system for the identification of *M. tuberculosis* and *M. avium/intracellulare. Diagnostic Microbiol. Infect. Dis.* (1987).

Jonas, V., J. J. Hogan, K. M. Young, and R. N. Bryan. Evaluation of an isotopic DNA probe assay for direct detection. Gen-Probe, Inc., San Diego, CA, ASM (1988).

Kiehn, T. E., and F. F. Edwards. Rapid identification using a specific DNA probe of *Mycobacterium avium* complex from patients with acquired immunodeficiency syndrome. *J. Clin. Microbiol. 25*:1551–1552 (1987).

Putbrese, S. C., F. A. Meier, B. A. Johnson, R. R. Brookman, and H. P. Dalton. Comparison of a non-isotopic DNA probe and ELISA for detecting *Chlamydia trachomatis* in clinical samples. Medical College of Virginia, Richmond, ASM (1988).

Stolzenbach, F., L. A. P. Phillips, Y. Y. Yang, R. K. Enns, and M. S. You. Highly specific DNA probes for the detection and differentiation of *Campylobacter* species. Gen-Probe, Inc., San Diego, CA, ASM (1988).

4

DNA Hybridization for Characterization, Classification, and Identification of Bacteria

DON J. BRENNER
Centers for Disease Control, Atlanta, Georgia

I. INTRODUCTION

It seems appropriate that this volume should appear almost exactly on the Silver Anniversary of the experiments that showed the specificity of deoxyribonucleic acid (DNA) and enabled its use for the detection, characterization, classification, taxonomy, and evolution of bacteria. We have become used to hearing about first-, second-, third-, and fourth-generations of antibiotics and computers. Applications of DNA for the detection and classification of bacteria can also be divided into generations. The first generation is the use of total DNA relatedness to identify bacteria, detect new species, define the biochemical limits of new and existing species, and to properly classify organisms at the species level. A second generation includes the use of specific gene probes to rapidly detect the presence of bacteria or bacterial virulence factors in cultures or directly in foods, tissues, or body fluids, and the use of plasmid profiles and restriction endonuclease analyses to detect and determine the transmission of pathogenic strains of bacteria in outbreaks of disease, in the contamination of foods, or the environment. The third generation consists of comparative sequencing of, or relatedness between, genes that are preferentially conserved [the genes that specify ribosomal ribonucleic acid (rRNA) or 5S RNA or the RNAs themselves] to determine evolutionary divergence in bacteria at the levels of kingdom, superfamily, family, and perhaps genus.

The thrust of this chapter will be the first-generation applications of DNA hybridization. Second- and third-generation applications will be mentioned only briefly because detailed treatments are given in other chapters. Methodology will also be introduced, but it is covered in detail elsewhere. To understand and use applications of DNA hybridization in all three generations, it is essential to have some appreciation for the principles, practices, and rules of bacterial classification, taxonomy, and nomenclature. These subjects will, therefore, be approached in some detail.

Bacterial DNAs are double-stranded molecules with relative molecular masses in the range of 1×10^9 and 8×10^9 D. The two strands are held together specifically by hydrogen bonds formed between only adenine (A) and thymine (T) and between guanine (G) and cytosine (C). The hydrogen bonds can be broken (denatured) by heating or by treatment with base. Single DNA strands will reform the double-stranded molecule when incubated with complementary single-stranded sequences. This process is called reassociation, reannealing, double-strand formation, binding, or hybridization. Single-stranded DNA (or RNA) will reassociate specifically with, and only with, completely or partially complementary DNA (or RNA) strands. The reversible denaturation of DNA and its ability to reassociate with complementary sequences from the same (*homologous*) strain or from different (*heterologous*), but still complementary (*related*) strains or species, are the bases for all nucleic acid relatedness studies and for all nucleic acid probe studies. For example, a sequence composed of GACTAGACTG would reassociate with its complementary sequence CTGATCTGAC and with the predominantly complenetary sequence CTGA*A*CTGAC, but not with the noncomplementary sequence AGTTGTGCTG.

II. HYBRIDIZATION METHODS

Marmur and Doty first showed that native DNA could be reversibly heat denatured (1), that the temperature at which denaturation occurred was dependent upon the guanine plus cytosine (G+C) content of the DNA (and therefore could be used to determine the G+C content (2), and that single-stranded DNA could hybridize with complementary DNA from another organism (3). Shortly thereafter, Goodman and Rich hybridized transfer RNA (tRNA) to DNAs from a variety of bacteria in the first attempt to determine evolutionary divergence of a single gene (4). Both of these investigations used CsCl gradients to detect hybridized DNA, a tedious and not generally applicable method.

The DNA agar method was the first generally applicable approach for detecting hybridized nucleic acids (5,6). It was replaced by a variety of methods that are still widely used. These are DNA filter methods done by direct (7-9) or by competition hybridization (10), hydroxyapatite method (11-13), S_1 nuclease procedure (14-16), and optical reassociation (17,18). Each of these methods has certain advantages and disadvantages. When properly used, results from each of the methods are highly comparable (14,16-18). Thermal elution profiles to determine divergence in related hybrid nucleotide sequences can be performed with the filter, S_1, and hydroxyapatite methods.

The relative molecular mass of bacterial DNA (genome size) varies not only between different species, but also between strains of a single species. The range is from about 1.0×10^9 D to 8×10^9 D, but the most commonly encountered bacteria have genome sizes of 2.0×10^9 D to 4×10^9 D (19). Differences in genome size of up to 30% were observed among *Escherichia coli* strains (20), and it is likely that comparable differences exist among strains of at least some other species. In certain situations, genome size determinations can be used to great advantage in determining the identity of unknown or previously undescribed strains. It is often useful to do reciprocal DNA relatedness reactions (labeled DNA from strain *A* plus unlabeled DNA from strain *B*, and vice versa). In such reactions, similar relatedness values are obtained when, and only when, the two strains have identical or nearly identical genome sizes (for example, if strain *A* has a genome size 25% smaller than that of strain *B*, and labeled strain *A* DNA is 80% related to unlabeled strain *B* DNA, then in the reciprocal reaction, relatedness will be 60%). Relative genome sizes can be determined in reciprocal hybridization reactions, and estimates of genome size are possible if strains of known molecular mass are included. Alternatively, the genome size of an organism can be determined directly by determining its reassociation kinetics—the time required for reassociation is directly proportional to the genome size (13,18,19).

Determination of the G+C content of DNA is a technique often done before DNA hybridization. The DNA of each species has a characteristic, but not unique G+C content. Therefore, the G+C content cannot be used to identify an unknown isolate, but it is useful for ruling out a wide range of possibilities. Reasonably accurate G+C values are obtainable by thermally denaturing double-stranded DNA in a spectrophotometer (2,18,19). This is possible because single-stranded DNA absorbs more ultraviolet light than does double-stranded DNA and because thermal stability is directly proportional to the G+C content of the DNA. A newer, perhaps more accurate, means of determining G+C content is to use high-performance liquid chromatography (HPLC) to chromatograph 5′-deoxyribonucleotide-monophosphates obtained by nuclease digestion of DNA (21).

III. TAXONOMY, CLASSIFICATION, AND NOMENCLATURE

Before discussing the use of DNA hybridization in the characterization, classification, taxonomy, and identification of bacteria, it is necessary to define these terms and place them in their proper context. The *characteristics* of a strain, or of any group of strains, usually include biochemical, morphological, and cultural properties, as well as DNA relatedness data. These may be supplemented by any other traits of interest, such as serology, antimicrobial susceptibility, plasmid content, toxins, and the like.

Nomenclature is the means through which we define, summarize, and communicate the characteristics of a species. The name *E. coli* conveys a very different set of characteristics than the name *Mycobacterium tuberculosis*. If we do not know the characteristics of an organism, its name provides a means of obtaining them. The naming of bacteria is governed by a set of rules found in the *International Code of Nomenclature for Bacteria* (22). A name proposed for a new species becomes valid only after the organism has been described and named, according to the rules of nomenclature, in the *International Journal of Systematic Bacteriology* (*IJSB*) or, if the organism is described in another journal, the name has been published in the list of validly published new names that appears in each issue of the *IJSB*. All names with standing in the literature as of January 1980 are found in the *Approved List of Bacterial Names* (23). Names published between 1980 and January 1985 are listed in the *IJSB* by Moore et al. (24).

Classification is an orderly arrangement of bacteria into groups. These groups may be formal (governed by the rules of nomenclature): superfamily, family, genus, species, subspecies; or informal: serotypes, biogroups, phage types, lactose fermenters, enteric bacteria, anaerobes, fermenters, chemolithotrophs, halophiles, thermophiles, and such. Aesthetically, if for no other reason, one would like a classification to have a sound scientific basis, but this is frequently not the case. Classification is purpose oriented; thus, a good classification may not be successful, and a successful classification is not necessarily good (25).

Taxonomy is the science of classification. It should, therefore, be based upon the most complete and newest data. Taxonomy should be dynamic. New approaches and data can result in changes in existing classification, in nomenclature, in criteria for identification, and in the recognition of new species. To keep abreast of these changes one should read the *IJSB*. The most comprehensive treatment of bacterial taxonomy is found in *Bergey's Manual of Systematic Bacteriology* (26). Further information on taxonomy, classification, and nomenclature is found in References 22-28.

Identification is the practical use of a classification to isolate and identify bacteria to verify the authenticity or utility of a culture, to isolate and identify the causative agent of a disease, and so forth. The ability to identify an organism depends on how well it is classified, on the use of appropriate and sufficient tests, and on the proper use of these tests.

As a final introductory note, it is useful to dispel some erroneous beliefs about classification and taxonomy.

1. There is no "official" classification.
2. Names do not become official after being accepted by a committee of taxonomists. Names must be validly published, as noted earlier (22,23).
3. A proposed new name does not automatically replace an existing name. If

a name violates the rules of nomenclature (22), a replacement name can be proposed in the *IJSB* and, if accepted by the Judicial Commission of the International Union of Microbiological Societies in a written opinion, the new name replaces the old name. If the change is made based on interpretation of scientific data, its acceptance is not voted on or ruled on, but it is either accepted or rejected by its use in the scientific community. For example, the proposed transfer of *Proteus morganii* to the genus *Morganella* appears to have been generally accepted, whereas proposed transfers of certain *Legionella* species to the genera *Tatlockia* and *Fluorobacter* have not been widely accepted.

IV. SPECIES DEFINITION BASED ON DNA RELATEDNESS

Before the concept of DNA relatedness was established, species were defined subjectively. Single traits, such as host range, pathogenicity, and toxin production, were often used to define species. There was no single definition that could be applied to all groups and, therefore, creation of species was heavily slanted toward the interests of the investigators who described them and whether these workers preferred lumping or splitting organisms. After some 20 years of taxonomic DNA-relatedness work, most investigators agree that a species definition can be made from DNA relatedness data that are essentially equally applicable to all bacteria. In general, the definition of a species is a group of strains in which DNA relatedness is 70% or greater at conditions optimal for reassociation, 60% or more relatedness at stringent reassociation criteria, and 5% or less divergence in related sequences (29,30). The DNA relatedness data, obtained from total DNA or from conserved DNA genes (5S and ribosomal RNA genes), have been used to justify the creation of genera, families, and superfamilies; however, there are no accepted relatedness definitions for taxa at, or above, the level of genus.

Although arbitrary, the definition of species based on DNA relatedness is supported by voluminous data on at least 700 species that cover most major bacterial families. Its use has resulted in solutions to a large number of taxonomic problems and enigmas and in the designation of more than 250 new species.

V. CHARACTERIZATION OF SPECIES

Many DNA relatedness studies start as efforts to characterize the biochemical limits of a given species. The best way to accomplish this is by a three-step process. The first step is to characterize strains biochemically. The second step is to choose appropriate strains on the basis of biochemical, source, geographic origin, virulence factors, and the like, for DNA hybridization studies. The third step is to reassess the biochemical reactions useful in the identification of the species under study and of any new species or biogroups uncovered during the course of the study. The type strain should be included in all studies of named species.

With some species, this approach gives a simple, straightforward result. *Edwardsiella tarda* strains exhibited almost no biochemical variability, but isolates were available from many parts of the world (31). Strains from 11 states and four foreign countries, with diverse serotypes and isolated from human stools, wounds, and meningitis, as well as from reptile, turtle, and rodent sources, were all 82% or more related in both optimal and stringent DNA hybridization studies, with divergence in related sequences averaging 1.5%. These strains were less than 30% related to other species of *Enterobacteriaceae.*

Vibrio cholerae and *E. coli* are species in which the DNA relatedness results obtained were more complex. Historically, only toxigenic, 0 antigen 1 strains were considered to be *V. cholerae* because these were thought to be the only strains capable of causing cholera. It was subsequently shown that certain non-01 strains produced choleratoxin, and that both toxigenic and nontoxigenic strains could cause cholera. Several groups showed that both 01 and non-01 strains were the same species (32-34). Studies on biochemically atypical *V. cholerae* strains indicated that mannose-, mannitol-, and lysine-negative strains; as well as salicin-positive, cellobiose-positive; and bioluminescent strains were *V. cholerae* but, surprisingly, that sucrose-negative strains (Heiberg biogroup 5) belonged to a separate species, *V. mimicus* (34,35).

A more complex situation was encountered in *E. coli* and the four named *Shigella* species (20,36-38). *Escherichia coli* is a normal inhabitant of the human intestine and is widely found in animals and in the environment. *Escherichia coli* strains are also found in disease, causing a large number of urinary tract and septicemic infections, and at least five types of diarrheal disease. The *E. coli* strains also exhibit a great deal of biochemical variability. The four species of *Shigella* cause invasive diarrheal disease. They are much less biochemically active than typical *E. coli* strains, and only *S. sonnei* can be identified solely on the basis of its biochemical reactions (serotyping serves to distinguish between shigellae).

All shigellae, other than *S. boydii* serogroup 13, are a single genetic species. *Shigella boydii* 13 represents a second, closely related species (36). The *E. coli* strains are also in the same genetic species as the shigellae other than *S. boydii* 13 (20,36). This is not terribly surprising to those who have tried to separate nonmotile, non-gas-forming, lactose-negative *E. coli* strains from shigellae. Strains of *E. coli* isolated from a variety of human and nonhuman sources, including strains isolated from four distinct types of gastrointestinal disease were all 80% or more related. Further studies indicated that biotypically atypical *E. coli*-like strains, with one exception, were in fact *E. coli*. These included strains with the atypical reactions shown in Table 1. The exception was a group of

Table 1 Relatedness of Biochemically Atypical *E. coli* Strains and *E. hermannii*[a]

Atypical reactions	% Relatedness to *E. coli* K-12		
	60°C	%D	75°C
E. coli K-12	100	0.0	100
E. coli, adonitol-positive (2)[b]	100	0.0	100
E. coli, citrate-positive (4)	99	N.D.	97
E. coli, citrate-positive, H_2S-positive (1)	100	0.5	100
E. coli, triple decarboxylase-negative (30)	98	1.0	96
E. coli, H_2S-positive (5)	100	0.5	99
E. coli, H_2S-positive, indole-negative (1)	100	N.D.	100
E. coli, H_2S-positive, lactose-negative (1)	96	N.D.	95
E. coli, lactose-negative (1)	92	N.D.	89
E. coli, inositol-positive (2)	92	N.D.	94
E. coli, inositol-positive, H_2S-positive (1)	94	N.D.	100
E. coli, inositol-positive, indole-negative (1)	100	N.D.	100
E. coli, mannitol-negative (5)	94	N.D.	89
E. coli, mannitol-negative, H_2S-positive (1)	99	N.D.	100
E. coli, mannitol-negative, methyl red-negative (3)	96	0.0	96
E. coli, phenylalanine deaminase-positive (3)	100	0.0	96
E. coli, urea-positive (8)	97	3.5	94
E. coli, KCN-positive (2)	100	0.0	100
E. hermannii, KCN-positive, cellobiose-positive, yellow-pigmented (5)	43	13.5	14

[a]Procedures done by the hydroxyapatite method (104). %D, % divergence as determined by thermal elution profiles (104).
[b]Numbers in parentheses indicate the number of strains tested.
Source: Data taken from Ref. 38.

KCN-positive, cellobiose-positive, yellow-pigmented strains that represented a new species, *Escherichia hermannii* (Table 1; 38). This study demonstrated the importance of identifying an organism on the basis of its complete biochemical profile.

The results also illustrate that a classification must be accepted by the scientific community. Here, the name *E. coli* has priority over the four *Shigella* species. Because it seemed that clinical microbiologists would not accept such a taxonomic recommendation because of the confusion it would cause, the recommendation was never made. There are many other examples in which species created for medical or phytopathological reasons were later shown to be a single genetic species. These include *Klebsiella pneumoniae, K. ozaenae*, and *K. rhinoscleromatis* (39); all *Brucella* species (40); *Bordetella pertussis, B. parapertussis*, and *B. bronchiseptica* (41); and a large number of *Erwinia* species (42,43). Most of the confusion in *Erwinia* has been resolved, but the clinically relevant species remain unchanged, except in *Klebsiella* for which three subspecies of *K. pneumoniae* were proposed, but are not used (44), and in *Brucella*, for which a single species was recommended with, as yet, no acceptance (45).

Two additional examples focus on the magnitude of the problem of changing nomenclature in medically important species. In 1967 it was shown that *Neisseria meningitidis* and *N. gonorrhoeae* were the same genetic species (46)–a formal recommendation to consider them both as *N. gonorrhoeae* (which has priority) has not been made. Such a recommendation was made for *Yersinia pestis* and *Y. pseudotuberculosis* which belong to the same genetic species (47). The authors stated that the two names should continue to be used as previously, yet such concern arose among clinical microbiologists that *Y. pestis* might be misidentified that the Judicial Commission rejected this recommendation on the basis of "practical concerns for human welfare" (48).

VI. NEW SPECIES "MASQUERADING" IN ESTABLISHED SPECIES

The use of DNA hybridization to describe new species that were hidden among strains of established species was introduced with the examples of *V. mimicus* (34) and *E. hermannii* (38). Usually, these masqueraders are not suspected, and they are discovered during the course of studies to characterize what are thought to be well-defined species, such as *E. coli* and *V. cholerae*, as in the preceding examples. Other examples are the discovery of *Enterobacter sakazakii*, which is frequently associated with highly fatal cases of neonatal meningitis, during a study of *Enterobacter cloacae* (49), and of at least seven species masquerading under *Yersinia enterocolitica* (50,51). It was unexpectedly found that strains considered to be yellow-pigmented *E. cloacae* were a separate species. Similarly, an in-depth study to define the biochemical limits of *Y. enterocolitica* revealed the additional species *Y. intermedia, Y. frederiksenii, Y. kristensenii, Y. aldovae, Y. rohdei, Y. bercovieri*, and *Y. mollaretii*.

Sometimes the masquerading species are not quite so obvious. A preliminary study of *Serratia liquefaciens* showed borderline relatedness of several strains to the reference strain (52). Further studies revealed three additional, closely related species, *S. plymuthica, S. grimesii*, and *S. proteomaculans* (53). This study points out the necessity for doing thermal stability tests or relatedness determinations at stringent incubation criteria before designating new species, because four strains showing 68% to 75% optimal relatedness (60°C reactions) to *S. liquefaciens* showed 8.8-9.8% divergence in related sequences and 44-51% relatedness at stringent conditions (75°C reactions). Similar results were obtained with some *Yersinia* species (50). *Citrobacter* is another species in which heterogeneity probably exists and which needs to be reexamined (54).

In all of the preceding studies the new species were defined biochemically. Occasionally, new species cannot, at least initially, be separated biochemically. Workers studying *Enterobacteriaceae* have chosen not to name such species. Two current examples are *Hafnia alvei*, in which two DNA hybridization groups are 50% related (52), and the *Enterobacter agglomerans* complex, in which most of the 13 or more DNA hybridization groups are not phenotypically distinguishable (43).

VII. NEW SPECIES IN ESTABLISHED GENERA OR FAMILIES

An excellent way to "discover" new species in established genera or families is that practiced for *Enterobacteriaceae* and *Vibrionaceae* in the Enteric Identification Laboratory and the Molecular Biology Laboratory at the Centers for Disease Control. The method consists of biochemical characterization, computer-assisted identification, comparison of each unidentified strain with biochemical profiles of every strain in the data base by a "strain match" computer program, DNA relatedness tests on phenotypically similar strains, and finally, biochemical differentiation of new species from all previously defined species (55). Strains that cannot be identified are grouped with their closest phenotypic relatives. These groups are given vernacular names as Enteric groups, further characterized by DNA hybridization (55) and, when warranted, described as new species.

Occasionally, single new species were identified within existing or new genera. These include *Edwardsiella ictaluri, Enterobacter taylorae, Escherichia fergusonii, Ewingella americana, Koserella trabulsii, Moellerella wisconsensis, Rahnella aquatilis, Tatemella ptyseos, Vibrio hollisae*, and *V. damsela*. In other cases, Enteric groups contained more than one species: *Cedecea davisae, C. lapagei*, and *C. neteri*; *Kluyvera ascorbata* and *K. cryocrescens*; and *Leminorella grimontii* and *L. richardii*. Except for the *Vibrio* species (56,57), *Koserella* (58), and *Leminorella* (59), these and other new species are reviewed in Reference 55. Data used to propose the new species typically consist of at least three categories: overall biochemical reactions, DNA relatedness between strains of the new species and between the new species and other species, and biochemical reactions that distinguish the new species from its closest phenotypic relatives. An example of such data for *Y. aldovae* is shown in Table 2. It is obvious that *Y. enterocolitica* is the species most closely related to *Y. aldovae* (one strain was

Table 2 Characterization of *Y. aldovae* Biochemically and by DNA Relatedness[a]

A. Relatedness to the type strain of *Y. aldovae*.

Source of unlabeled DNA	% Relatedness, 60°C	%D	% Relatedness, 75°C
Y. aldovae (14)[b]	88	0.5	85
Y. enterocolitica (4)	69	12.5	38
Y. frederiksenii (3)	61	14.0	25
Y. kristensenii (1)	59	11.5	30
Y. intermedia (1)	59	11.5	27
Y. ruckeri (1)	44	15.5	c
Y. pseudotuberculosis (1)	42	14.0	
Other *Enterobacteriaceae* (65)	11–32[d]		

B. Biochemical differentiation of *Y. aldovae* from other yersiniae.

Test	*Y. aldovae*	*Y. enterocolitica*	*Y. frederiksenii*	*Y. kristensenii*	*Y. intermedia*	*Y. ruckeri*	*Y. pseudotuberculosis*	*Y. pestis*
Voges-Proskauer	+	+	+	–	+	–	–	–
Sucrose	–	+	+	–	+	–	–	–
Sorbose	–	+	+	+	+		–	–
L-Rhamose	+	–	+	–	+	–	+	–
Melibiose	–	–	–	–	+	–	+	V
D-Sorbitol	+	+	+	+	+	–	–	–
Ornithine decarboxylase	+	+	+	+	+	+	–	–
Urease	+	+	+	+	+	–	+	–

[a]See Table 1 for DNA methods.
[b]Numbers in parentheses are the number of strains, except for other *Enterobacteriaceae* where it indicates the number of species.
[c]Blank space indicates that test was not done.
[d]Range of relatedness values obtained.
+, 90% or more strains positive within 48 hr; –, 10% or less strains positive; V, between 11% and 89% positive.
Source: Data from Ref. 51.

73% related) and that divergence within related sequences and the relatedness at 75°C were necessary to unequivocally determine that *Y. aldovae* was a separate species. It should also be noted that *Y. pestis* was not tested. Because strains of *Y. pestis* and *Y. pseudotuberculosis* are a single genetic species (90% or more relatedness) (51), only one of them had to be included.

So-called polyphasic (phenotypic and genotypic) characterization is now the method by which most culturable new species are discovered and described. That these new species are shown to be genetically unique is the powerful justification for their description. They do exist and, therefore, must be described and acknowledged by microbiologists. It is not only unscientific, but somewhat unbelievable, that a portion of the medical community asks us to ignore their presence to "keep things simple" and "avoid confusion." To ignore the existence of well-characterized species is untenable; however, whether to consider them in the clinical laboratory remains the choice of each laboratory. The well-established pathogens remain responsible for the overwhelming number of human infections. No doubt, many of the new species are of limited, if any, clinical importance. Yet, if we do not consider them, how can we be sure that we are not misidentifying them as pathogens, and how can we accurately determine their present and future prevalence? The fallacy of this argument (forgetting its scientific absurdity) lies in the etiological agents discovered during the past decade. Should we have ignored the causal agents of legionellosis, Lyme disease, and the subset of strains in existing species that cause toxic shock syndrome and hemolytic uremic syndrome? Should we ignore the new species of *Enterobacteriaceae* involved in neonatal meningitis and in nosocomial infections of compromised patients?

VIII. TOTALLY NEW SPECIES

A surprising number of new bacterial diseases and new pathogenic agents have been detected during the past decade. Many of these are caused by a subset of strains in well-defined species such as enterotoxigenic and enterohemorrhagic *E. coli* strains and toxic shock toxin-producing strains of *Staphylococcus aureus*. Others involve the recognition of known species as pathogens, exemplified by campylobacteria. Still others are recognized as new species in well-defined genera such as *Vibrio* and *Enterobacter*. In others, the family or even the bacterial nature of the causative agent is not certain. The best examples of this last category are *Borrelia burgdorferi*, the causative agent of Lyme disease, and legionellae, the causative agents of legionellosis.

Lyme disease, caused by ticks, is a disease in which a skin lesion, erythema chronicum migrans, develops, often followed by neurological, arthritic, and cardiac complications (60,61). The disease was first reported in 1909, but not until 1982 was an infectious agent isolated. The organism was morphologically a spirochete, but it could not be assigned to a genus on the basis of phenotype. The Lyme spirochete was shown to resemble the genus *Borrelia* in G+C composition. DNA relatedness studies showed that all tested Lyme spirochete strains were the same species, that they were 30-59% related to *Borrelia* species, and that they were negligibly related to species of *Leptospira* and *Treponema*. The organism was subsequently named *B. burgdorferi* (61).

The investigations that led to the discovery and description of *Legionella pneumophila*, and subsequently to other *Legionella* species, illustrate the use of many nucleic acid methods for the characterization of a totally new group of bacteria (29,62). Legionellosis comprises two distinct diseases; a fulminating pneumonia, often referred to as Legionnaires' disease, and a mild, febrile, flulike illness often called Pontiac fever. A large outbreak of pneumonia (221 cases and 34 deaths) occurred at an American Legion convention in 1976. A year of intensive study resulted in the isolation of an infectious agent and also revealed other outbreaks, some preceding the Legionnaires' convention, as well as many sporadic cases of the same disease.

The disease agent was initially referred to as the Legionnaires' disease organism (LDO) because it was isolated by methods used for rickettsiae and appeared similar to *Rochalimaea quintana*, a rickettsial species that can be cultivated on culture media. The first task in classifying the organism was to determine whether it was a bacterium or a rickettsia. *Rochalimaea quintana* was reported to have a genome size (total amount of DNA) of some 1×10^9 D, at least 50% smaller than genome sizes of almost all bacteria. Because the time course of DNA reassociation is directly proportional to its molecular mass, the genome size of the LDO was determined and compared with those of *E. coli* DNA and *R. quintana* DNA (Fig. 1). The LDO DNA reassociated at a rate essentially identical with that of the *E. coli* strain which had a genome size of 2.5×10^9 D and, therefore, appeared to be a bacterium. This observation was shortly confirmed by cellular structure and cell wall composition studies. The organism was then referred to as the Legionnaires' disease bacterium (LDB).

The next question was whether the LDB was related to other bacteria at the family, genus, or species level. The LDB could not be identified at any level on the basis of its biochemical, morphological, and cultural characteristics. It was a fastidious, metabolically inactive, gram-negative, oxidase-variable, aerobic, nonspore-forming rod. The G+C content of LDB DNA was determined to be 39 mol%. With this knowledge, organisms with substantially higher or lower G+C contents could be eliminated from consideration.

The DNA from the LDB was then shown to have little, if any, relatedness to DNAs from a wide variety of pathogenic bacteria, from nonpathogens that bore some resemblance to the LDB, and from *R. quintana*. Although it is impossible to unequivocally state that an organism is unrelated to all other species, without having tested everything, it seemed highly likely that the LDB represented a

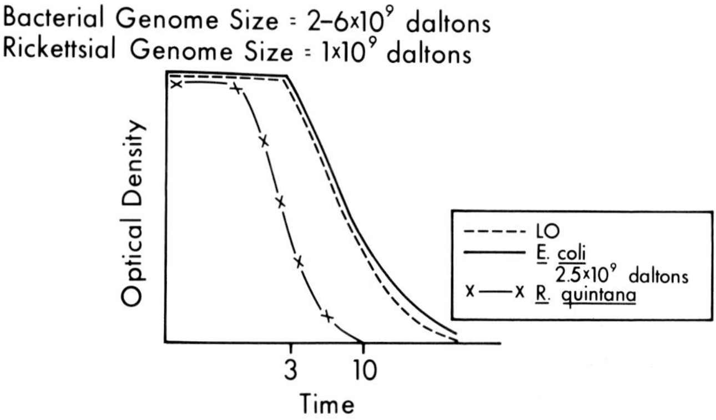

Figure 1 Diagrammatic representation of genome size determination of DNA from the Legionnaires' disease organism. Approximately 50 μg of DNA from the Legionnaires' disease organism strain Philadelphia 1 and from *E. coli* K-12 were denatured and were allowed to reassociate in a spectrophotometer in which the cuvettes were heated to 60°C. As reassociation occurred the optical density (at 260 μM) decreased because double-stranded DNA absorbs less ultraviolet light than single-stranded DNA. The time course of reassociation was compared for these two DNAs and with a literature value for *R. quintana* DNA. Results indicated that the Legionnaires' disease organism had a genome size equal to that of the *E. coli* strain (2.5×10^9 D) and substantially larger than the literature value for *R. quintana*. A typical bacterial genome size is $2–6 \times 10^9$ D, and that of a rickettsia is 1×10^9 D.

previously undescribed species that could not be linked phenotypically or genetically to an existing genus or family. This conclusion was subsequently confirmed on the basis of 16S ribosomal RNA oliogonucleotide cataloging (63) and a fractionated ribosomal RNA gene probe (64).

It remained to determine whether all strains from the Legionnaires' convention outbreak, other outbreaks, sporadic cases, from cases of Pontiac fever, and strains serologically unreactive with the original outbreak strain, represented one or multiple species. Results of these studies (Table 3) showed that all strains were the same species. Therefore, Pontiac fever and Legionnaires' disease were caused by strains of the same species, and multiple serogroups were present in that species. It was then that the new species was formally named and described in the new family *Legionellaceae* and the new genus *Legionella*, as *L. pneumophila* (62). Continuing studies have led to the description of 26 species containing 42 serogroups in the genus *Legionella* (29,65), and there is good evidence that at least 12 additional species exist (D.J. Brenner, unpublished data).

Table 3 DNA Relatedness Between Strains of *L. pneumophila*

Strain	Characteristic	% Relatedness to Philadelphia 1 at 60°C
Philadelphia 1	Legionnaires' Convention strain	100
Philadelphia 2	Legionnaires' Convention strain	96
Philadelphia 3	Legionnaires' Convention strain	94
Philadelphia 4	Legionnaires' Convention strain	93
Pontiac 1	Pontiac fever, Michigan	82
Flint 1	Sporadic case, Michigan	89
Flint 2	Sporadic case, Michigan	98
Burlington 1	Outbreak, Vermont	86
Bellingham 1	Sporadic case, Washington	81
Albuquerque 1	Sporadic case, New Mexico	97
Knoxville 1	Sporadic case, Tennessee	87
Berkeley 1	Sporadic case, California	97
OLDA	Isolated in 1947	90
Togus 1	Serogroup 2, Maine	85
Bloomington 2	Serogroup 3, Indiana	84
Los Angeles 1	Serogroup 4, California	71
Dallas 10	Serogroup 5, Texas	73
Chicago 2	Serogroup 6, Illinois	79
Chicago 8	Serogroup 7, Illinois	74
Concord 3	Serogroup 8, New Hampshire	93

Source: Data from a variety of publications cited in Ref. 29.

IX. NAME CHANGES AND RECLASSIFICATION

It is not uncommon for the same organism to be given two names by investigators working independently of one another. When it is subsequently shown that both names represent the same organism, the name that was first published in the *International Journal of Systematic Bacteriology* has priority, and the other name loses standing in the literature. One recent example of duplicate names is in the genus *Providencia* for which the names *P. rustigianii* (66) and *P. friedericiana* (67) were proposed for what was subsequently shown to be the same organism by DNA hybridization. Here, *P. rustigianii* has priority (55,55a).

Changes in classification usually result from two types of studies. In one, a well-established group is reinvestigated and the data obtained are interpreted to indicate errors in the original classification. In the second, two laboratories interpret data on the same groups of organisms differently, leading to different classifications. In either event the changes in classification are either accepted or rejected by use within the scientific community—they are not "officially" ruled on by the Judicial Commission or any other body.

The first type is exemplified by the genera *Proteus, Providencia*, and *Morganella*. Until 1978, the *Proteeae* contained the two genera *Proteus*, including *P. morganii* and *P. vulgaris*, and *Providencia*. The DNA relatedness studies showed that *Proteus rettgeri* was closer to *Providencia* than to *Proteus*, and that *P. morganii* was less related to *Proteus* and *Providencia* than to most other genera in *Enterobacteriaceae* (68). It was, therefore, proposed that *P. morganii* be transferred to a separate genus, *Morganella*, and that *P. rettgeri* be transferred to *Providencia*. It seems that these changes have been widely accepted by the scientific community.

The second type is exemplified by the family *Legionellaceae*. The first five described species were placed in the genus *Legionella* by one group of investigators (29). A second group interpreted similar DNA relatedness data to justify three genera, *Legionella, Fluoribacter*, and *Tatlockia*, for these five species (69, 70). There are now 26 species, all of which were placed in the genus *Legionella* (29) when described. The second group of investigators has not made recommendations on the other species, but given their interpretation that a genus contains species with 25% or more relatedness, nine genera would be required (29). If 40% relatedness was used to define genera, 14 genera would be required (29). There is no definition for a genus either for total DNA relatedness or for relatedness of conserved genes such as ribosomal RNA (rRNA). Oligonucleotide analysis of rRNA in *Legionella* species appears to support the one-genus interpretation (63).

It has been argued that because no definition exists, if genetic and phenotypic data are seemingly not compatible, a phenotypic definition should take precedence (29,55). Simply stated, a genus consists of phenotypically similar species that are dealt with as a single group at the laboratory bench. This is the case with legionellae where all species are extremely similar biochemically; in fact, it is not now possible to phenotypically identify most species. Further, legionellae have similar cell wall fatty acids, cause the same disease, and are susceptible to the same antibiotics. Thus far, the scientific community has accepted the single genus *Legionella*.

Although no precise definitions exist, the relatedness of ribosomal and 5S RNA sequences (either by hybridization or by comparing nucleotide sequences) has been used to determine the relatedness of organisms and to make taxonomic recommendations at the levels of kingdom and superfamily and, occasionally, at the levels of family and genus. It is important to remember several points relative to this exciting and revolutionary new work that enables us, for the first time, to glimpse at the evolutionary process in bacteria by providing nucleotide sequence data that are equivalent to a fossil record. Designations of various taxonomic levels are made on the basis of sequence similarities, but no definitions exist for the degree of similarity required for any level from genus to kingdom. These designations, therefore, are made arbitrarily on the basis of data interpretation in a single laboratory, and the criteria used will likely vary considerably between laboratories. Owing to the newness of this approach and the large number of comparisons to be made, new and changed designations were often made on the basis of incomplete data (appropriate type strains, species, genera, and sometimes families have not been included in the study). The proposals generated from 5S and 16S rRNA studies are based on science and, therefore, are not legislated but must gain acceptance. Finally, it should be noted that most of the proposed changes, certainly those above the level of family, will have little or no effect on the bench microbiologist.

It should be emphasized that these caveats are not meant to denigrate the tremendous accomplishments made in this field and the ever greater future impact. Total DNA relatedness studies suffered from these and other problems just 10 years ago. We should look forward to future work with rRNA and the exciting findings it will undoubtedly generate.

Because this subject is covered in detail elsewhere, I will attempt only to cite a few examples of the kinds of studies that have been done (71-73). Three separate studies showed that the genus *Pseudomonas*, as presently constituted, contains five very different groups (74-77). These certainly represent different genera and, in some cases, different families and superfamilies; formal taxonomic proposals have not yet been made. A proposal, based on rRNA hybridization, 5S and 16S ribosomal sequences, was made to transfer the genus *Aeromonas* from the family *Vibrionaceae* to the new family *Aeromonadaceae* (78). Studies of sequence and structure of ribosomal RNA have led to descriptions of two new kingdoms, the archaebacteria or urkaryotes (79,80) and the eocytes or "*Eocyta*" (81), neither of which has formally been proposed.

With this greatly enhanced armamentarium, it is not surprising that there has been a substantial proliferation in both the number of medically important species and in the number of total species of bacteria. In 1974, *Enterobacteriaceae* contained 13 genera with 40 species, 13 of them in the genus *Erwinia*. Today, there are 26 published genera containing some 115 species. Similar increases were seen in other groups. *Legionella* did not exist in 1974 and now has 26 named and 12 as yet unnamed species; *Vibrio* had 5 species and now has 33; *Campylobacter* had 3 and now has 14. Virtually all of these species are unique genetically. There are genera for which the number of species has been substantially reduced because many of the named species were found to be scientifically invalid. The genus *Streptomyces*, which contained 463 species and subspecies in 1974 (most named on the basis of producing a given antibiotic; 82), now contains 383 species and subspecies, and continuing work is expected to decrease this number by at least one-third. DNA relatedness has been an effective means of showing that many names, including some on the Approved Lists (23) were synonyms of well-described species. "*Escherichia aurescens*" was shown to be a pigmented *E. coli, Vibrio albensis* was shown to be a luminescent *V. cholerae*, and many of the names given to erwiniae on the basis of source of isolation were shown to be synonymous with previously named species. Thus, the purpose of DNA relatedness studies is to remove unwarranted species as well as to describe valid new species.

X. DNA RELATEDNESS FOR IDENTIFICATION OF CLINICAL ISOLATES

In our CDC laboratory we frequently use DNA hybridization to confirm the identity of unusual or rarely seen bacteria for other reference laboratories. In these instances a specific species is suspected. The DNA from the isolate to be tested is radiolabeled and reacted against unlabeled DNA from type or reference strains of the species suspected to be closely related to it. This has been done to identify atypical strains of *Yersinia* and many other *Enterobacteriaceae*, the first human isolate of *Y. ruckeri*, and of several *Legionella* species, and the first disease outbreak associated with *Campylobacter laridis*. This approach requires the availability of DNA from all species in a given group and is, therefore, not applicable to a clinical laboratory. There is no reason, however, that such a test cannot be initiated in any laboratory that has a specific, limited need to distinguish between a few species. One possibility would be to rule out *Y. pestis*. A modification of the DNA hybridization procedure (taxonomic spot blots) has been used to rapidly identify strains of *Campylobacter cinaedi* and *C. fennelliae*, two species associated with enteric disease in homosexuals (83). Other such specific applications will undoubtedly be forthcoming.

XI. PROBES, PLASMIDS, AND RESTRICTION ENDONUCLEASES

Each of these subjects is covered in detail elsewhere in this volume. Each is briefly mentioned here because they are all actually methods for the characterization and detection of bacteria at the species or infrasubspecific (below species) level.

The concept and application of gene probes for the detection of bacteria and virulence genes is perhaps the most exciting advance in the detection and enumeration of microbial pathogens. In practice, a *gene probe* is any specific

Table 4 Representative Microbial Gene Probes[a]

Probe directed to	Type of probe[b]	Commercially available
Adenovirus 2	Gene	Yes
Cytomegalovirus	Gene	Yes
Epstein-Barr virus	Gene	Yes
Feline leukemia virus	Gene	Yes
Hepatitis A virus	Gene	Yes
Hepatitis B virus	Gene	Yes
Herpes simplex I virus	Gene	Yes
Herpes simplex II virus	Gene	Yes
Human papilloma virus	Gene	Yes
SV-40	Gene	Yes
Bacteria (all species)	Unfractionated ribosomal RNA	No
Bordetella pertussis	Total DNA and gene	No
Bordetella sp.	Total DNA	No
Campylobacter cinaidi	Total DNA	No
C. fennelliae	Total DNA	No
C. jejuni	Gene	Yes
Chlamydia	?	Yes
Enterobacteriaceae (all species)	Fractionated ribosomal RNA	No
Escherichia coli, adhesion factor	Gene	No
E. coli, heat-labile enterotoxin	Gene	Yes
E. coli, heat-stable enterotoxin	Gene	Yes
E. coli + *Shigella*, invasiveness factor	Gene	No
Legionella pneumophila	Total DNA	No
Legionellaceae (all species)	Fractionated ribosomal RNA	Yes
Mycobacterium tuberculosis	Fractionated ribosomal RNA	Yes
Mycoplasma pneumoniae	?	Yes
Mycoplasma (all species)	Unfractionated ribosomal RNA	Yes
Neisseria gonorrhoeae	Gene	No
Salmonella typhi	Gene	No

Table 4 (Continued)

Probe directed to	Type of probe[b]	Commercially available
Salmonella (all species)	Gene	Yes
Vibrio cholerae, choleratoxin	Gene	No
V. parahaemolyticus	Gene	No
Yersinia enterocolitica	Gene	No

[a]No attempt was made to exhaustively survey the literature and there are undoubtedly probes that are not listed.
[b]Gene, isolated gene, gene portion, or genetically defined segment obtained from DNA of the organism or constructed synthetically; Total DNA, all or a portion of total DNA that has not been genetically characterized.

gene, gene region, portion of or total DNA (or RNA) that is used to detect itself (gene, region, or organism). The probe can be used to detect purified DNA (or RNA), DNA from colonies of strains lysed directly on nitrocellulose filter (84), or to detect DNA directly with unprocessed stool material smeared on a nitrocellulose filter (85). Probes have been used to detect a variety of viruses directly in tissue, and one will shortly be able to "probe" for bacteria directly from tissue, blood, urine, sputum, other body fluids and secretions, as well as from water, foods, and other environmental sources. This means that the diagnosis of many pathogenic bacteria will soon be obtainable without the need to wait for biochemical identification and, often, without isolating the organism. This capability already exists for several viruses, as well as for the campylobacteria cited earlier; salmonellae; mycoplasmas; legionellae; *N. gonorrhoeae*; and toxigenic strains of *V. cholerae, E. coli*, and *Y. enterocolitica* (Table 4). This will result in bacterial identification several days to more than a week earlier than at present. A probe is even available that recognizes the presence of all bacteria (at least all eubacteria; D.E. Kohne, personal communication). This probe allows one to detect bacteria in any normally sterile material, including blood and culture media. Probes are also available for an increasing number of antimicrobial resistance genes (see Table 4). We can anticipate the day when these can be used to rapidly initiate specific antimicrobial therapy.

An ideal gene probe is highly specific (detection of only the desired organism or gene with no false-positive reactions) and highly sensitive (detection of all positives, with no false-negative, reactions). The specificity of probes is generally acceptable to excellent, at least with pure cultures. Improvements are necessary in rapid lysis systems for gram-positive bacteria and in the detection of bacteria directly in tissue and body fluids. Three types of probe assay systems are commercially available. There are the "traditional" radioactive assay systems; systems in which DNA is biotinylated and then visualized cytochemically by means of avidin and biotinylated polymers of alkaline phosphatase or other enzymes; and systems in which DNA is immunologically altered and detected by antibodies to the altered region. Sensitivity is directly dependent on the specific activity of the assay system. The sensitivity of probe assays is usually between 10^5 and 10^8 organisms, although sensitivities of as low as 10^3 have been obtained experimentally (D.E. Kohne, personal communication). Fluorescently labeled DNA assays have been reported, and these may ultimately become the methods of choice, yielding sensitivities in the range of 10^2 organisms or less. Probes may be grouped into three classes: (a) those composed of specific genes or portions of genes (either isolated from the organism or made synthetically), or defined genetic regions; (b) those made from conserved genes or portions thereof (5S, 16S, or 23S rRNA); and (c) those consisting of portions or all of total DNA. The choice of the proper class is quite important in designing a probe for a given purpose. Ribosomal RNA sequences, because of their central role in protein production, have been extremely conserved throughout the evolutionary process. Portions of the genes that code for rRNA have remained essentially identical in most bacteria, whereas other portions have diverged to varying degrees. These genes, therefore, are analogous to a panspecific antibody, portions of which react with essentially all organisms and other portions of which are specific for various groups. Unfractionated 16S rRNA gene probes from *E. coli* or *L. pneumophila* have detected all bacteria against which they were tested (D.E. Kohne, D.J. Brenner, unpublished observations). A similar probe made from a *Mycoplasma* rRNA gene is commercially available for the detection of mycoplasmal contamination in tissue culture. The ribosomal RNA gene can be fractionated (analogous to removing cross-reactions in a panspecific antigen by appropriate absorptions) to remove the most conserved sequences and obtain a family or group-specific probe. This has been done successfully for *Legionellaceae* (64) and for *Enterobacteriaceae* (D.E. Kohne, D.J. Brenner, unpublished observations). It seems reasonable to predict that this method can be used to obtain probes to many bacterial families and to some genera.

Probes made from specific genes (including portions of genes and gene regions) are used to detect viruses, some bacteria, and bacterial virulence genes (see Table 4). Among these are many of the virulence genes in *E. coli*. One must be aware of which, if any, organisms contain the same or a very similar gene sequence to the one being probed. For example, some strains of *V. mimicus* contain the choleratoxin gene (34), the heat-labile enterotoxin gene in *E. coli* is similar to the choleratoxin gene in *V. cholerae* (86), and the Vi antigen gene used as a probe for *S. typhi* is also found in *Citrobacter diversus* (87).

All or portions of total DNA may be used to generate probes to detect most bacterial species and some genera. The advantage of this type of probe is its nonreliance on isolating or characterizing genes. Because every species that is properly classified genetically contains at least 30% of DNA not shared by other bacteria, one should be able to produce species-specific probes without much difficulty. It is necessary to know the taxonomy of the group of interest and of related groups that must be tested to assure the specificity of the probe. The genera *Bordetella, Brucella*, and *Shigella* (plus *E. coli*) can be detected because each of these genera contains only one genetic species. Alternatively, these genera and *Y. pestis* and *Y. pseudotuberculosis* cannot be separated by portions of total DNA because their genetic classification is incorrect.

Enteroinvasive *E. coli* strains and *Shigella* strains contain a plasmid that is associated with invasiveness (88,89), and *N. gonorrhoeae* strains contain a plasmid not found in other *Neisseria* species (90). These and similar plasmids found in pathogenic strains of other species can be used as probes.

Strains of many, but certainly not all, bacteria harbor *plasmids*; extrachromosomal DNA elements that replicate autonomously from chromosomal DNA are nonessential for growth. In these strains the number and size of the plasmids can be determined by the distance that they migrate on an agarose gel during electrophoresis, which is directly proportional to their molecular mass. This method, often called *plasmid profile analysis*, can be used to identify and trace epidemic strains (91), and to monitor the existence spread, and change of antibiotic-resistance plasmids in a given species (92) or their dissemination to other species (93). In effect, plasmid profile analysis is a molecular subtyping method, not unlike serotyping or phage typing. It has the advantage of not needing reagents or quality control, and it can be used for many species in which other subtyping systems are either unavailable, too expensive, or too rarely needed to warrant their routine use.

Plasmid profile analysis has been used to characterize a large number of disease outbreaks involving many species of *Enterobacteriaceae* (94), *Pseudomonas aeroginosa* (94), and *L. pneumophila*. Among its successful applications are nosocomial outbreaks caused by *Serratia marcescens, Klebsiella pneumoniae,* and other *Enterobacteriaceae* (91), sequential outbreaks of *K. pneumoniae* in a neonatal care unit (94), *Enterobacter cloacae* outbreaks in burn units (94), salmonellosis in commercially precooked roast beef (94), neonatal sepsis and meningitis caused by *Enterobacter sakazakii* (95), a multistate outbreak of salmonellosis caused by contaminated marijuana (96), hemorrhagic colitis outbreaks caused by a rare serotype of *E. coli* (97) and, more recently, the largest salmonella outbreak in the United States (L.W. Mayer, personal communication).

It is possible that plasmids of similar molecular mass are structurally different. To rule out this possibility, plasmids are treated with one or more of a large number of *restriction endonucleases* that cleave DNA only at specific sites, generating a specific set of fragments (98). This is called *plasmid fingerprinting*.

Restriction endonuclease digestion of chromosomal DNA has been used with apparent success to subtype strains of some organisms that rarely contain plasmids. Restriction endonuclease mapping was found more sensitive than serotyping in characterizing an outbreak of enteritis caused by *Campylobacter jejuni* and linked to cattle (99). In a preliminary survey, not involving a disease outbreak, restriction endonuclease analysis of pig, fowl, bird, and human *Campylobacter coli* strains yielded 41 different patterns with little overlap in patterns obtained from different species (100). In studies on several serotypes of *Leptospira interrogans* (101-103) restriction endonuclease analysis is as sensitive or more sensitive than serotyping in identifying serotypes. Whether or not these observations hold for all serotypes and whether or not the method will be useful in tracing disease source remains to be seen. Phage typing is the method of choice, although not without problems, for subtyping the limited number of serotypes present in strains of *L. monocytogenes*. Because *Listeria* phage typing is not done in the United States, plasmid profile analysis and restriction endonuclease analysis were tried on strains from two outbreaks of listeriosis. Neither method was successful in subtyping the *Listeria* strains. Thus, as powerful as molecular methods can be for epidemiologic surveillance, they are not universally effective.

XII. CONCLUDING REMARKS

This chapter is limited to nucleic acid methods. It must be mentioned that, especially with subtyping and rapid detection, immunological, protein, and enzymatic methods are alternatives to, and sometimes preferable to, nucleic acid methods. Monoclonal antibodies are commercially available for rapid detection of a large number of pathogenic bacteria. Electrophoretic separation and analysis of outer-membrane or whole-cell proteins may result in improved subtyping. Finally, isoenzyme analysis has been successfully used to subtype strains of legionellae, enteric bacteria, and *L. monocytogenes*.

As I hope is clear from the preceding commentary, significant advances have been made in the taxonomy, characterization, classification, and rapid identification of bacteria. We are on the threshold of even greater advances, many of which are already being integrated into clinical laboratories. The clinical laboratory must remain aware of advances in these areas and should be ready to implement the newer techniques as they become available and as they are appropriate to increase the capabilities of the laboratory.

The author requests forgiveness for relying heavily on his own work to exemplify various points. There are other examples that are just as good, and often better, but one obviously feels most comfortable with one's own results. The research for this chapter was completed on December 20, 1985. The reader is encouraged to keep current by reading the literature and by soliciting research update materials from commercial sources.

REFERENCES

1. Marmur, J., and P. Doty, Thermal renaturation of DNA. *J. Mol. Biol. 3*: 585–594 (1961).
2. Marmur, J., and P. Doty, Determination of the base composition of deoxyribonucleic acid from its thermal denaturation point. *J. Mol. Biol. 5*:109–118 (1962).
3. Schildkraut, C. L., J. Marmur, and P. Doty, The formation of hybrid DNA molecules and their use in studies of DNA homologies. *J. Mol. Biol. 3*:595–617 (1961).
4. Goodman, H. M., and A. Rich, Formation of a DNA-soluble RNA hybrid and its relation to the origin, evolution, and degeneracy of soluble RNA. *Proc. Natl. Acad. Sci. USA 48*:2101–2109 (1962).
5. Bolton, E. T., and B. J. McCarthy, A general method for the isolation of RNA complementary to DNA. *Proc. Natl. Acad. Sci. USA 48*:1390–1397 (1962).
6. McCarthy, B. J., and E. T. Bolton, An approach to the measurement of genetic relatedness among organisms. *Proc. Natl. Acad. Sci. USA 50*:156–164 (1963).
7. Nygaard, A. P., and B. D. Hall, A method for the detection of RNA-DNA complexes. *Biochem. Biophys. Res. Commun. 12*:98–104 (1963).
8. Gillespie, D., and S. Spiegelman, A quantitative assay for DNA/RNA hybrids with DNA immobilized on a membrane. *J. Mol. Biol. 12*:829–842 (1965).
9. Denhardt, D. T., A membrane filter technique for the detection of complementary DNA. *Biochem. Biophys. Res. Commun. 23*:641–646 (1966).
10. Johnson, J. L., Genetic characterization. In *Manual of Methods for General Bacteriology*, P. Gerhardt, Murray, R. G. E., Costilow, R. N., Nester, E. W., Wood, W. A., Krieg, N. R., and Phillips, G. B. (eds.). American Society for Microbiology, Washington, pp. 450–472 (1981).
11. Bernardi, G., Chromatography of nucleic acids on hydroxyapatite. *Nature 206*:779–783 (1965).
12. Brenner, D. J., G. R. Fanning, A. V. Rake, and K. E. Johnson, A batch procedure for thermal elution of DNA from hydroxyapatite. *Anal. Biochem.* 447–459 (1969).
13. Britten, R. J., and D. E. Kohne, Nucleotide sequence repetition in DNA. *Carnegie Inst. Wash. Year Book 65*:78–106 (1966).
14. Crosa, J. G., D. J. Brenner, and S. Falkow, The use of a single-strand specific nuclease for the analysis of bacterial and plasmid DNA homo- and heteroduplexes. *J. Bacteriol. 115*:904–911 (1973).
15. Popoff, M., and C. Coynault, Use of DEAE-cellulose filters in the S_1 nuclease method for bacterial deoxyribonucleic acid hybridization. *Ann. Microbiol. (Paris) 131A*:151–155 (1980).
16. Grimont, P. A. D., M. Y. Popoff, F. Grimont, C. Coynault, and M. Lemelin, Reproducibility and correlation study of three deoxyribonucleic acid hybridization procedures. *Curr. Microbiol. 4*:325–330 (1980).

17. De Ley, J., H. Cattoir, and A. Reynaerts, The quantitative measurement of DNA hybridization from renaturation rates. *Eur. J. Biochem. 12*:133–142 (1970).
18. Seidler, R. J., and M. Mandel, Quantitative aspects of deoxyribonucleic acid renaturation: Base composition, state of chromosome replication, and polynucleotide homologies. *J. Bacteriol. 106*:608–614 (1971).
19. Gillis, M., J. De Ley, and M. De Cleene, The determination of molecular weight of bacterial genome DNA from renaturation rates. *Eur. J. Biochem. 12*:143–153 (1970).
20. Brenner, D. J., G. R. Fanning, F. J. Skerman, and S. Falkow, Polynucleotide sequence divergence among strains of *Escherichia coli* and closely related organisms. *J. Bacteriol. 109*:953–965 (1972).
21. Katayama-Fujimura, Y., Y. Komatsu, H. Kuraishi, and T. Kaneko, Estimation of DNA base composition by high performance liquid chromatography of its nuclease P1 hydrolysate. *Agric. Biol. Chem. 48*:3169–3172 (1984).
22. Lapage, S. P., P. H. A. Sneath, E. F. Lessel, V. B. D. Skerman, H. P. R. Seeliger, and W. A. Clark (eds.), *International Code of Nomenclature for Bacteria, 1975 Revision*. American Society for Microbiology, Washington (1976).
23. Skerman, V. B. D., V. McGowan, and P. H. A. Sneath (eds.), Approved lists of bacterial names. *Int. J. Syst. Bacteriol. 30*:225–240 (1980).
24. Moore, W. E. C., E. P. Cato, and L. V. H. Moore, Index of the bacterial and yeast nomenclatural changes published in the *International Journal of Systematic Bacteriology* since the 1980 approved lists of bacterial names (1 January 1980 to 1 January 1985). *Int. J. Syst. Bacteriol. 35*:382–407 (1985).
25. Cowan, S. T., Sense and nonsense in bacterial taxonomy. *J. Gen. Microbiol. 67*:1–8 (1971).
26. Krieg, N. R., and J. G. Holt (eds.), *Bergey's Manual of Systematic Bacteriology*, Vol. 1. Williams & Wilkins Co., Baltimore (1984).
27. Farmer, J. J., III, Should clinical microbiology keep up with taxonomy—yes. *Clin. Microbiol. Newsl. 1*:5–6 (1981).
28. Brenner, D. J. Impact of modern taxonomy on clinical microbiology. *ASM News 49*:58–63 (1983).
29. Brenner, D. J., A. G. Steigerwalt, G. W. Gorman, H. W. Wilkinson, W. F. Bibb, M. Hackel, R. L. Tyndall, J. Campbell, J. C. Feeley, W. L. Thacker, P. Skaliy, W. T. Martin, B. J. Brake, B. S. Fields, H. V. McEachern, and L. K. Corcoran, Ten new species of *Legionella*. *Int. J. Syst. Bacteriol. 35*: 50–59 (1985).
30. Grimont, P. A. D., and M. Y. Popoff, Use of principal component analysis in interpretation of deoxyribonucleic acid relatedness. *Curr. Microbiol. 4*: 337–342 (1980).
31. Brenner, D. J., A. G. Steigerwalt, and G. R. Fanning, Polynucleotide sequence relatedness in *Edwardsiella tarda*. *Int. J. Syst. Bacteriol. 24*:186–190 (1974).
32. Citarella, R. V., and R. R. Colwell, Polyphasic taxonomy of the genus

Vibrio: Polynucleotide sequence relationships among selected *Vibrio* species. *J. Bacteriol. 104*:434–442 (1970).
33. Baumann, P., L. Baumann, M. J. Woolkalis, and S. S. Bang, Evolutionary relationships in *Vibrio* and *Photobacterium*: A basis for a natural classification. *Annu. Rev. Microbiol. 37*:369–398 (1983).
34. Davis, B. R., G. R. Fanning, J. M. Madden, A. G. Steigerwalt, H. B. Bradford, Jr., H. L. Smith, Jr., and D. J. Brenner, Characterization of biochemically atypical *Vibrio cholerae* strains and designation of a new pathogenic species, *Vibrio mimicus. J. Clin. Microbiol. 14*:631–639 (1981).
35. Fanning, G. R., B. R. Davis, A. G. Steigerwalt, I. K. Wachsmuth, F. W. Hickman, J. J. Farmer III, and D. J. Brenner, Biochemical and genetic parameters of *Vibrio cholerae. Abstr. Ann. Meet. Am. Soc. Microbiol. 80*:47 (1980).
36. Brenner, D. J., G. R. Fanning, C. V. Miklos, and A. G. Steigerwalt, Polynucleotide sequence relatedness among *Shigella* species. *Int. J. Syst. Bacteriol. 23*:1–7 (1973).
37. Brenner, D. J., G. R. Fanning, A. G. Steigerwalt, I. Ørskov, and F. Ørskov, Polynucleotide sequence relatedness among three groups of pathogenic *Escherichia coli* strains. *Infect. Immun. 6*:308–315 (1972).
38. Brenner, D. J., B. R. Davis, A. G. Steigerwalt, C. F. Riddle, A. C. McWhorter, S. D. Allen, Y. Saitoh, and G. R. Fanning, Atypical biogroups of *Escherichia coli* found in clinical specimens and the description of *Escherichia hermannii* sp. nov. *J. Clin. Bacteriol. 15*:700–713 (1982).
39. Brenner, D. J., A. G. Steigerwalt, and G. R. Fanning, Differentiation of *Enterobacter aerogenes* from klebsiellae by deoxyribonucleic acid reassociation. *Int. J. Syst. Bacteriol. 22*:193–200 (1972).
40. Hoyer, B. H., and N. B. McCullough, Homologies of deoxyribonucleic acids from *Brucella ovis*, canine abortion organisms, and other *Brucella* species. *J. Bacteriol. 96*:1783–1790 (1968).
41. Kloos, W. E., N. Mohapatra, W. J. Dobrogosz, J. W. Ezzell, and C. R. Manclark, Deoxyribonucleic sequence relationships among *Bordetella* species. *Int. J. Syst. Bacteriol. 31*:173–176 (1981).
42. Brenner, D. J., G. R. Fanning, and A. G. Steigerwalt, Deoxyribonucleic acid relatedness among erwiniae and other enterobacteria. II. Corn stalk rot bacterium and *Pectobacterium chrysanthemi. Int. J. Syst. Bacteriol. 27*: 211–221 (1977).
43. Brenner, D. J., G. R. Fanning, J. K. Leete Knutson, A. G. Steigerwalt, and M. I. Krichevsky, Attempts to classify Herbicola group-*Enterobacter agglomerans* strains by deoxyribonucleic acid hybridization and phenotypic tests. *Int. J. Syst. Bacteriol. 34*:45–55 (1984).
44. Ørskov, I. Genus V. *Klebsiella* Trevisan. In *Bergey's Manual of Systematic Bacteriology*, Vol. 1, N. R. Krieg and Holt, J. G. (eds.). Williams & Wilkins Co., Baltimore, pp. 461–465 (1984).
45. Verger, J.-M., F. Grimont, P. A. D. Grimont, and M. Grayon, *Brucella*, a monospecific genus as shown by deoxyribonucleic acid hybridization. *Int. J. Syst. Bacteriol. 35*:292–295 (1985).

46. Kingsbury, D. T., Deoxyribonucleic acid homologies among species of the genus *Neisseria*. *J. Bacteriol.* *94*:870–874 (1967).
47. Bercovier, H., H. H. Mollaret, J. M. Alonso, J. Brault, G. R. Fanning, A. G. Steigerwalt, and D. J. Brenner, Intra- and interspecies relatedness of *Yersinia pestis* by DNA hybridization and its relatedness to *Yersinia pseudotuberculosis*. *Curr. Microbiol.* *4*:225–229 (1980).
48. Judicial Commission of the International Committee on Systematic Bacteriology. Opinion 60. Rejection of the name *Yersinia pseudotuberculosis* subsp. *pestis* (van Loghem) Bercovier et al. 1981 and conservation of the name *Yersinia pestis* (Lehmann and Neumann) van Loghem 1944 for the plague bacillus. *Int. J. Syst. Bacteriol.* *35*:540 (1985).
49. Farmer, J. J., III, M. A. Asbury, F. W. Hickman, D. J. Brenner, and The *Enterobacteriaceae* Study Group, *Enterobacter sakazakii*: A new species of "*Enterobacteriaceae*" isolated from clinical specimens. *Int. J. Syst. Bacteriol.* *30*:569–584 (1980).
50. Brenner, D. J., J. Ursing, H. Bercovier, A. G. Steigerwalt, G. R. Fanning, J. M. Alonso, and H. H. Mollaret, Deoxyribonucleic acid relatedness in *Yersinia enterocolitica* and *Yersinia enterocolitica*-like organisms. *Curr. Microbiol.* *4*:195–200 (1980).
51. Bercovier, H., A. G. Steigerwalt, A. Guiyoule, G. Huntley-Carter, and D. J. Brenner, *Yersinia aldovae* (formerly *Yersinia enterocolitica*-like group X2): A new species of *Enterobacteriaceae* isolated from aquatic ecosystems. *Int. J. Syst. Bacteriol.* *34*:166–172 (1984).
52. Steigerwalt, A. G., G. R. Fanning, M. A. Fife-Asbury, and D. J. Brenner, DNA relatedness among species of *Enterobacter* and *Serratia*. *Can. J. Microbiol.* *22*:121–137 (1976).
53. Grimont, P. A. D., K. Irino, and F. Grimont, The *Serratia liquefaciens*–*S. proteomaculans*–*S. grimesii* complex: DNA relatedness. *Curr. Microbiol.* *7*:63–68 (1982).
54. Crosa, J. H., A. G. Steigerwalt, G. R. Fanning, and D. J. Brenner, Polynucleotide sequence divergence in the genus *Citrobacter*. *J. Gen. Microbiol.* *83*:271–282 (1974).
55. Farmer, J. J., III, B. R. Davis, F. W. Hickman-Brenner, A. McWhorter, G. P. Huntley-Carter, M. A. Asbury, C. Riddle, H. G. Wathen-Grady, C. Elias, G. R. Fanning, A. G. Steigerwalt, C. M. O'Hara, G. K. Morris, P. B. Smith, and D. J. Brenner, Biochemical identification of new species and biogroups of *Enterobacteriaceae* isolated from clinical specimens. *J. Clin. Microbiol.* *21*:46–76 (1985).
55a. Hickman-Brenner, F. W., G. R. Fanning, H. E. Müller, and D. J. Brenner, Priority of *Providencia rustigianii* Hickman-Brenner, Farmer, Steigerwalt, and Brenner 1983 over *Providencia friedericiana* Müller 1983. *Int. J. Syst. Bacteriol.* *36*:565 (1986).
56. Hickman, F. W., J. J. Farmer III, D. G. Hollis, G. R. Fanning, A. G. Steigerwalt, R. E. Weaver, and D. J. Brenner, Identification of *Vibrio hollisae* sp. nov. from patients with diarrhea. *J. Clin. Microbiol.* *15*:395–401 (1982).
57. Love, M., D. Teebken-Fisher, J. E. Hose, J. J. Farmer III, F. W. Hickman,

and G. R. Fanning, *Vibrio damsela*, a marine bacterium, causes skin ulcers on the damselfish *Chromis punctipinnus*. *Science 214*:1139–1140 (1981).

58. Hickman-Brenner, F. W., G. P. Huntley-Carter, G. R. Fanning, D. J. Brenner, and J. J. Farmer III, *Koserella trabulsii*, a new genus and species of *Enterobacteriaceae* formerly known as Enteric group 45. *J. Clin. Microbiol. 21*:39–42 (1985).
59. Hickman-Brenner, F. W., M. P. Vohra, G. P. Huntley-Carter, G. R. Fanning, V. A. Lowery III, D. J. Brenner, and J. J. Farmer III, *Leminorella*, a new genus of *Enterobacteriaceae*: Identification of *Leminorella grimontii* sp. nov. and *Leminorella richardii* sp. nov. found in clinical specimens. *J. Clin. Microbiol. 21*:234–239 (1985).
60. Schmid, G. P., A. G. Steigerwalt, S. E. Johnson, A. G. Barbour, A. C. Steere, I. M. Robinson, and D. J. Brenner, DNA characterization of the spirochete that causes Lyme disease. *J. Clin. Microbiol. 20*:155–158 (1984).
61. Johnson, R. C., G. P. Schmid, F. W. Hyde, A. G. Steigerwalt, and D. J. Brenner, *Borrelia burgdorferi* sp. nov.: Etiologic agent of Lyme disease. *Int. J. Syst. Bacteriol. 34*:496–497 (1984).
62. Brenner, D. J., A. G. Steigerwalt, and J. E. McDade, Classification of the Legionnaires' disease bacterium: *Legionella pneumophila* genus novum, species nova of the family *Legionellaceae* familia nova. *Ann. Intern. Med. 90*:656–658 (1979).
63. Ludwig, W., and E. Stackebrandt, A phylogenetic analysis of *Legionella*. *Arch. Microbiol. 135*:45–50 (1983).
64. Kohne, D. E., A. G. Steigerwalt, and D. J. Brenner, Nucleic acid probe specific for members of the genus *Legionella*. In *Legionella*, C. Thornsberry, Balows, A., Feeley, J. C., and Jakubowski, W. (eds.). Proc. Second Intl. Symp., American Society for Microbiology, Washington, pp. 107–108 (1984).
65. Thacker, W. L., B. B. Plikaytis, and H. W. Wilkinson, Identification of 22 *Legionella* species and 33 serogroups with the slide agglutination test. *J. Clin. Microbiol. 21*:779–782 (1985).
66. Hickman-Brenner, F. W., J. J. Farmer III, A. G. Steigerwalt, and D. J. Brenner, *Providencia rustigianii*: A new species in the family *Enterobacteriaceae* formerly known as *Providencia alcalifaciens* biogroup 3. *J. Clin. Microbiol. 17*:1057–1060 (1983).
67. Müller, H. E., *Providencia friedericiana*, a new species isolated from penguins. *Int. J. Syst. Bacteriol. 33*:709–715 (1983).
68. Brenner, D. J., J. J. Farmer III, G. R. Fanning, A. G. Steigerwalt, P. Klykken, H. G. Wathen, F. W. Hickman, and W. H. Ewing, Deoxyribonucleic acid relatedness of *Proteus* and *Providencia* species. *Int. J. Syst. Bacteriol. 28*:269–282 (1974).
69. Brown, A., G. M. Garrity, and R. M. Vickers, *Fluoribacter dumoffii* (Brenner et al.) comb. nov. and *Fluoribacter gormanii* (Morris et al.) comb. nov. *Int. J. Syst. Bacteriol. 31*:111–115 (1981).
70. Garrity, G. M., A. Brown, and R. M. Vickers, *Tatlockia* and *Fluoribacter*: Two new genera of organisms resembling *Legionella pneumophila*. *Int. J. Syst. Bacteriol. 30*:609–614 (1980).

71. Fox, G. E., E. Stakebrandt, R. B. Hespell, J. Gibson, J. Maniloff, T. A. Dyer, R. S. Wolfe, W. E. Balch, R. S. Tanner, L. A. Magrum, L. B. Zablem, R. Blakemore, R. Gupta, L. Bonen, B. J. Lewis, D. A. Stahl, K. R. Luehrsen, K. N. Chen, and C. R. Woese, The phylogeny of prokaryotes. *Science 209*: 457–463 (1980).
72. Stackebrandt, E., and C. R. Woese, The phylogeny of prokaryotes. *Microbiol. Sci. 1*:117–122 (1984).
73. Lake, J. A., Evolving ribosome structure: Domains in archaebacteria, eubacteria, eocytes, and eukaryotes. *Ann. Rev. Biochem. 54*:507–530 (1985).
74. Palleroni, N. J., R. Kunisawa, R. Contopoulou, and M. Doudoroff, Nucleic acid homologies in the genus *Pseudomonas. Int. J. Syst. Bacteriol. 23*: 333–339 (1973).
75. De Ley, J., Phylogenetic relationships between gram-negative bacteria. *INSERM 114*:205–210 (1983).
76. De Vos, P., M. Goor, M. Gillis, and J. De Ley, Ribosomal ribonucleic acid cistron similarities of phytopathogenic *Pseudomonas* species. *Int. J. Syst. Bacteriol. 35*:169–184 (1985).
77. Woese, C. R., P. Blanz, and C. M. Hahn, What isn't a pseudomonad: The importance of nomenclature in bacterial classification. *Syst. Appl. Microbiol. 5*:179–195 (1984).
78. Colwell, R. R., M. T. MacDonell, and J. De Ley, Proposal to recognize the family *Aeromonadaceae* fam. nov. *Int. J. Syst. Bacteriol. 36*:473–477 (1986).
79. Woese, C. R., and G. E. Fox, Phylogenetic structure of the prokaryotic domain: The primary kingdoms. *Proc. Natl. Acad. Sci. USA 74*:5088–5090 (1977).
80. Woese, C. R., Archaebacteria. *Sci. Am. 244*:98–122 (1981).
81. Lake, J. A., E. Henderson, M. Oakes, and M. W. Clark, Eocytes: A new ribosome structure indicates a kingdom with a close relationship to eukaryotes. *Proc. Natl. Acad. Sci. USA 81*:3786–3790 (1984).
82. Buchanan, R. E., and N. E. Gibbons (eds.), *Bergey's Manual of Determinative Bacteriology*, 8th ed. Williams & Wilkins Co., Baltimore (1974).
83. Totten, P. A., C. L. Fennell, F. C. Tenover, J. M. Wezenberg, P. L. Perine, W. E. Stamm, and K. K. Holmes, *Campylobacter cinaedi* (sp. nov.) and *Campylobacter fennelliae* (sp. nov): Two new *Campylobacter* species associated with enteric disease in homosexual men. *J. Infect. Dis. 151*: 131–139 (1985).
84. Grunstein, M., and D. S. Hogness, Colony hybridization: A method for the isolation of cloned DNAs that contain a specific gene. *Proc. Natl. Acad. Sci. USA 72*:3961–3965 (1975).
85. Moseley, S. L., I. Huq, A. R. M. A. Alim, M. So, M. Samadpour-Motalebi, and S. Falkow, Detection of enterotoxigenic *Escherichia coli* by DNA colony hybridization. *J. Infect. Dis. 142*:892–898 (1980).
86. Moseley, S. L., and S. Falkow, Nucleotide sequence homology between the heat-labile enterotoxin gene of *Escherichia coli* and *Vibrio cholerae* deoxyribonucleic acid. *J. Bacteriol. 144*:444–446 (1980).
87. Rubin, F. A., D. J. Kopecko, K. F. Noon, and L. S. Baron, Development

of a DNA probe to detect *Salmonella typhi. J. Clin. Microbiol. 22*:600–605 (1985).
88. Harris, J. R., I. K. Wachsmuth, B. R. Davis, and M. L. Cohen, High-molecular-weight plasmid correlates with *Escherichia coli* invasiveness. *Infect. Immun. 37*:1295–1298 (1982).
89. Sansonetti, P. J., H. d'Hauteville, C. Écobichon, and C. Pourcel, Molecular comparison of virulence plasmids in *Shigella* and enteroinvasive *Escherichia coli. Ann. Microbiol. (Paris) 134A*:295–318 (1983).
90. Totten, P. A., K. K. Holmes, H. H. Handsfield, J. S. Knapp, P. L. Perine, and S. Falkow, DNA hybridization technique for the detection of *Neisseria gonorrhoeae* in men with urethritis. *J. Infect. Dis. 148*:462–471 (1983).
91. Schaberg, D. R., L. S. Tompkins, and S. Falkow, Use of agarose gel electrophoresis of plasmid deoxyribonucleic acid to fingerprint gram-negative bacilli. *J. Clin. Microbiol. 13*:1105–1108 (1981).
92. Brunner, F., A. Margadant, R. Peduzzi, and J.-C. Piffaretti, The plasmid pattern as an epidemiological tool for *Salmonella typhimurium* epidemics: Comparison with the lysotype. *J. Infect. Dis. 148*:7–11 (1983).
93. Rubens, C. E., W. E. Farrar Jr., Z. A. McGee, and W. Schaffner, Evolution of a plasmid mediating resistance to multiple antimicrobial agents during a prolonged epidemic of nosocomial infections. *J. Infect. Dis. 143*:170–181 (1981).
94. Farrar, W. E. Jr., Molecular analysis of plasmids in epidemiologic investigation. *J. Infect. Dis. 148*:1–6 (1983).
95. Muytjens, H. L., H. C. Zanen, H. J. Sondercamp, L. A. Kallee, J. J. Farmer III, and I. K. Wachsmuth, Neonatal meningitis and sepsis due to *Enterobacter sakazakii. J. Clin. Microbiol. 18*:115–120 (1983).
96. Taylor, D. N., I. K. Wachsmuth, Y. Shangkuan, E. V. Schmidt, T. J. Barrett, J. S. Schrader, C. S. Scherach, H. B. McGee, R. A. Feldman, and D. J. Brenner, Salmonellosis associated with marijuana: A multistate outbreak traced by plasmid fingerprinting. *N. Engl. J. Med. 306*:1249–1253 (1982).
97. Wells, J. G., B. R. Davis, I. K. Wachsmuth, L. W. Riley, R. S. Remis, R. Sokolow, and G. K. Morris, Laboratory investigation of outbreaks of hemorrhagic colitis associated with a rare *Escherichia coli* serotype. *J. Clin. Microbiol. 18*:512–520 (1983).
98. Thompson, R., S. G. Hughes, and P. Broda, Plasmid identification using specific endonucleases. *Mol. Gen. Genet. 133*:141–149 (1974).
99. Bradbury, W. C., A. D. Pearson, M. A. Marko, R. V. Congi, and J. L. Penner, Investigation of a *Campylobacter jejuni* outbreak by serotyping and chromosomal restriction endonuclease analysis. *J. Clin. Microbiol. 19*:342–346 (1984).
100. Kakoyiannis, C. K., P. J. Winter, and R. B. Marshall, Identification of *Campylobacter coli* isolates from animals and humans by bacterial restriction endonuclease DNA analysis. *Appl. Environ. Microbiol. 48*:545–549 (1984).
101. Robinson, A. J., P. Ramadass, A. Lee, and R. B. Marshall, Differentiation of subtypes within *Leptospira interrogans* serovars *hardjo, balcanica* and

tarossovi, by bacterial restriction-endonuclease DNA analysis (BRENDA). *J. Med. Microbiol. 15*:331–338 (1982).

102. Marshall, R. B., P. J. Winter, and R. Yanagawa, Restriction endonuclease DNA analysis of *Leptospira interrogans* serovars *icterohaemorrhagiae* and *hebdomadis. J. Clin. Microbiol. 20*:808–810 (1984).
103. Thiermann, A. B., A. L. Handsaker, S. L. Moseley, and B. Kingscote, New method for classification of leptospiral isolates belonging to serogroup Pomona by restriction endonuclease analysis: Serovar *kennewicki. J. Clin. Microbiol. 21*:585–587 (1985).
104. Brenner, D. J., A. C. McWhorter, J. K. Leete Knutson, and A. G. Steigerwalt, *Escherichia vulneris*: A new species of *Enterobacteriaceae* associated with human wounds. *J. Clin. Microbiol. 15*:1133–1140 (1982).

RECENT REFERENCES ADDED IN PROOF

Grimont, F., and P. A. D. Grimont, Ribosomal ribonucleic acid gene restriction patterns as potential taxonomic tools. *Ann. Inst. Pasteur/Microbiol. 137B*: 165–175 (1986).

Marx, J. L., Multiplying genes by leaps and bounds. *Science 140*:1408–1410 (1988).

Wayne, L. G., D. J. Brenner, R. R. Colwell, P. A. D. Grimont, O. Kandler, M. I. Krichevsky, L. H. Moore, W. E. C. Moore, R. G. E. Murray, E. Stackebrandt, M. P. Starr, and H. G. Trüper, Report of the Ad Hoc Committee on reconciliation of approaches to bacterial systematics. *Int. J. Syst. Bacteriol. 37*:463–464 (1987).

5
Plasmid and Chromosomal DNA Analyses in the Epidemiology of Bacterial Diseases

SCOTT D. HOLMBERG and KAYE WACHSMUTH
Center for Infectious Diseases, Centers for Disease Control, Atlanta, Georgia

I. INTRODUCTION

New and simplified techniques of molecular biology have had a major impact on the epidemiology of bacterial pathogens. Before 1980, virtually no investigation of pathogenic bacteria in the hospital or the community relied on plasmid or chromosomal DNA analysis to "fingerprint" organisms. Today, investigations of epidemics caused by *Staphylococcus, Legionella, Enterobacter, Klebsiella, Salmonella, Campylobacter, Vibrio* species, and many other bacterial pathogens are considered incomplete if such DNA analysis, or a more sophisticated technique, is not used to typify organisms.

We think that there are three general reasons why this sudden revolution occurred. First, modifications in plasmid DNA isolation by Meyers et al. (1) and by Birnboim and Doly (2) made plasmid DNA analysis and restriction endonuclease digestion of plasmid DNA easy, simple techniques that are within the capabilities of most research and many nonresearch laboratories. Second, the technique of DNA analysis was reproducible and applicable to a wide range of organisms (3). Finally, as this chapter points out, the value of genetically typing the organisms being investigated, rather than relying on analysis of their phenotypic characteristics, became immediately apparent to epidemiologists (4,5). Our knowledge of bacterial pathogens and their modes of transmission has expanded considerably in the last several years.

Plasmid analysis is much more commonly used than chromosomal DNA analysis, but both rely on the same basic technique. Simply described, bacterial broth culture is grown overnight (more than 4 hr; 2). Outer membranes of the harvested cells are disrupted by EDTA, the cell walls are removed by treatment with lysozyme, and internal membranes are lysed with detergent. The chromosome is denatured at alkaline pH and cleared along with cellular debris by centrifugation for several minutes. Plasmid DNA is further purified by precipitation in cold ethanol.

Next, plasmid DNAs are electrophoresed through a 0.7-0.8% agarose gel, and DNAs are separated by their molecular mass during migration toward the anode. Larger, heavier plasmid DNAs migrate less rapidly than smaller plasmids: we have generally found an inverse log-log relationship between the size of plasmids in megadaltons (Md) and the number of millimeters they have progressed down the gel. The gel is stained with ethidium bromide which binds to the DNA and fluoresces as distinct "bands" that constitute the plasmid profile or fingerprint when illuminated under ultraviolet light (Fig. 1). Plasmid DNAs of known molecular mass are included as standards on each gel, and their relative gel mobility is used to calculate the mass of the unknown plasmids (1).

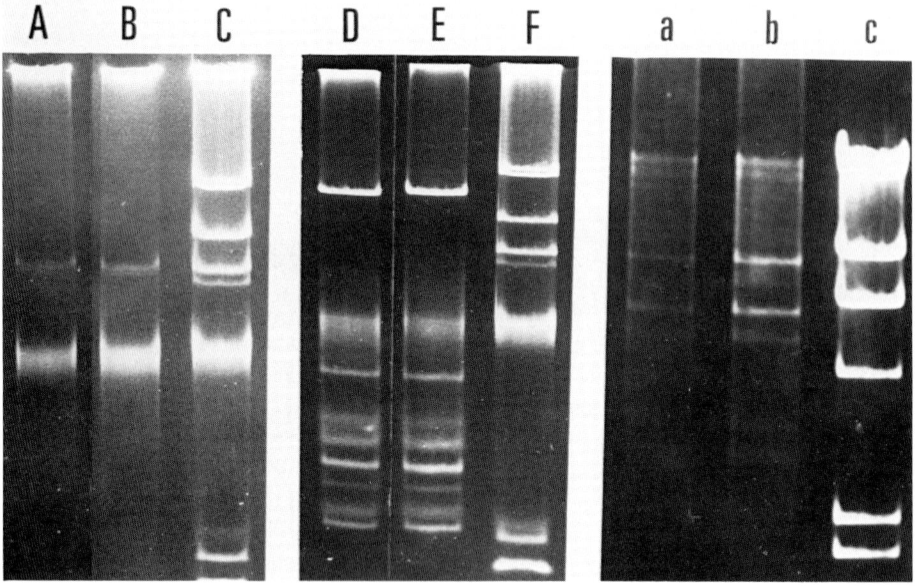

Figure 1 Agarose gel electrophoresis of plasmid DNA isolated from *S. typhimurium* strains (A–F) and *Hin*dIII digest fragments of plasmid DNAs (*a–c*). A and a, human isolate from farm 1; B and b, bovine isolate from farm 1; C, molecular weight control; c, λ-phage DNA (control); D, human isolate from farm 2; E, bovine isolate from farm 2; F, molecular mass control (*Source*: from Refs. 40 and 41).

Plasmid DNA can be further characterized by digestion with restriction endonucleases. These enzymes usually have unique DNA target sequences of four to six bases and cleave the double-stranded DNA at specific sites within the target. Identical plasmids have identical targets and are cleaved into identical DNA fragments (see Fig. 1). Like the parent plasmids, fragments are separated on agarose gel by electrophoresis and produce profiles when, after staining, they are visualized under ultraviolet light.

Chromosomal DNAs can also be digested by restriction endonucleases to give reproducible electrophoretic gel patterns. Generally, chromosomal DNA analysis is somewhat more difficult than plasmid profile analysis because the many bands produced by digestion of chromosomal DNA are often difficult to compare; thus, chromosomal DNA analysis is usually reserved for those cases in which the investigated organism does not have plasmids or in which another molecular technique, such as hybridization, is used.

II. NOSOCOMIAL OUTBREAKS

The first uses of plasmid analysis by epidemiologists were not, strictly speaking, for epidemic investigations; rather, characteristics of resistance plasmids (R-plasmids) were examined. In perhaps the first such use of plasmid analysis, McGowan and co-workers investigated nosocomial infections with gentamicin-resistant *Staphylococcus aureus* during the period October 1976 through August 1977 (6). Gentamicin-resistant *S. aureus*, that had not been recovered in the hospital before that period, were recovered from 22 high-risk infants, 20 adults on general medical and surgical services, and 12 patients in the burn unit. The isolates were of two different phage types but had the same antimicrobial resistance; two of five representative isolates (both phage groups) had apparently identical plasmids as visualized on the agarose electrophoresis gel. In addition, *Eco*RI restriction endonuclease fragments appeared to be common between plasmids of isolates from patients on different services at different times. These investigators concluded that spread of gentamicin resistance probably involved the dissemination of both a single R-plasmid and multiple R-plasmids mediating resistance to gentamicin.

Jaffe and coinvestigators explored the possibility that a single plasmid that mediated gentamicin resistance might have transferred in vivo and interspecies, between *S. aureus* and *S. epidermidis* in a neonatal special care nursery (7). In addition to being the same size by electron microscopy, R-plasmids isolated from the *S. aureus* and the *S. epidermidis* and their fragments by endonuclease digestion with enzymes *Hae*III, *Eco*RI, *Xba*I, and *Hin*dIII appeared to be identical on agarose gel. Although this study could not prove the interspecies transfer of R-plasmids nor the direction in which such transfer might have taken place, it did lend credence to the hypothesis that R-plasmid transfer between

S. aureus and *S. epidermidis* occurs in nature. This is epidemiologically and clinically important in that antimicrobial-resistance engendered in one organism by use of an antimicrobial(s) can be transferred to other organisms not exposed to the antimicrobial(s).

Cohen et al. (8) examined a 32-Md plasmid mediating resistance to several antimicrobials and recovered from *S. aureus* infecting 31 patients and personnel in a Kentucky hospital in a 7-month period in 1978 and 1979. As determined by plasmid analysis and colony hybridization, this same R-plasmid was recovered from *S. epidermidis*. These multiresistant *S. epidermidis* had been first detected a year before the *S. aureus* outbreak. Thus, the emergence of antimicrobial resistance in *S. aureus* appeared to result from genetic transfer from *S. epidermidis* and from the interhospital spread of resistant staphylococci.

Another study of R-plasmids in *Pseudomonas aeruginosa* and *Serratia marcescens* over a 5-year period in a Tennessee hospital (9) found transfer of a gentamicin-resistance-coding transposable DNA sequence (transposon) between 100- and 9.8 Md R-plasmids. This transposition resulted in a 105-Md R-plasmid that was recovered from strains of *Serratia* and *Klebsiella* isolated in 1976-1977. The investigators concluded that nosocomial outbreaks caused by different bacterial genera were, in fact, related by the presence of plasmids with a common transposable DNA sequence.

The use of plasmid or chromosomal DNA analysis as a epidemiological tool to identify persons infected by an epidemic organism or to identify the source of an epidemic pathogen (10) is an issue separate from the microbiological factors responsible for the emergence and spread of antimicrobial resistance. Schaberg and co-workers found that the total and types of plasmids recovered from nosocomial gram-negative rods seemed to be excellent markers for particular strains (11). To see if plasmid content was an effective epidemiological tool, they collected strains from seven large outbreaks of nosocomial infections in several cities (12). Results of plasmid analysis of the epidemic strains were then compared with phenotypic characteristics of the organisms—biotype, serotype, bacteriocin type, and antibiograms. Invariably, plasmid DNA of isolates from the same outbreak were identical with one another and different from coisolate control strains. Thus, plasmid DNA analysis appeared to be a highly specific way to fingerprint strains of the same genus. The implications of this finding were apparent to these and other nosocomial outbreak investigators. Typing techniques for *S. aureus* and *S. epidermidis*, such as phage typing and biotyping, did not appear to be as specific in distinguishing staphylococcal strains of the same clone.

An investigation of a multiple-antimicrobial-resistant strain of *S. aureus* in a Seattle hospital in 1980 and 1981 demonstrated some of the complexities of how a particular strain evolved and is transmitted (13). By plasmid analysis of *S. aureus* recovered from patients in a burn unit, it became apparent that 34 patients—half of whom later died—had acquired their infections from a patient transferred from another hospital. In addition to demonstrating the identity of isolates from index and subsequent cases, the investigators analyzed *S. aureus* from the burn unit over 15 months. Concurrent with a trial of chloramphenicol in cases of bowel sepsis in the burn unit, the epidemic *S. aureus* acquired a 2.9-Md plasmid that mediated resistance to chloramphenicol and cephalothin. Plasmids were spontaneously lost and added to the original epidemic strain and, in all, the investigators identified four epidemic strains. Plasmid profile analysis was useful in identifying patient and personnel reservoirs. (The outbreak was finally controlled after rifampin was combined with vancomycin in treating those infected with the epidemic strains.) Thus, in this investigation, plasmid analysis was effectively used to identify persons infected with epidemic *S. aureus*, to follow the natural evolution of *S. aureus* in a hospital, and to control the outbreak.

An investigation of methicillin-resistant *S. aureus*, transmitted from an index case to 17 hospital patients and three hospital personnel, compared the utility of four epidemiological markers—antimicrobial resistance pattern, phage type, production of aminoglycoside-inactivating enzymes, and plasmid profile (14). Phage typing was poorly reproducible, and identification of the enzyme was difficult and not useful during the outbreak. Rifampin-resistance was useful in screening *S. aureus* for the epidemic strain, but plasmid DNA analysis was generally easy to perform and more useful than antimicrobial-resistance as an epidemiological marker.

Subsequent investigations of nosocomial staphylococcal epidemics have generally confirmed the epidemiological utility of plasmid profile analysis. For example, a recent report of *S. epidermidis*, isolated from a valve prosthesis (15), showed that, although epidemiological evidence implicated a common source for the 20 isolates obtained, 14 of these exhibited variations in their antimicrobial susceptibility pattern (*antibiograms*). When these same strains were tested by phage typing, they were all of the same type. However, plasmid profile analysis showed a prototypic pattern from which various subtypes varied by addition or loss of plasmids mediating antimicrobial resistance. Thus, although phage typing alone would have been adequate to specify epidemiologically related organisms, plasmid analysis provided the most specific information about the epidemic strain. In any event, the authors concluded that antimicrobial susceptibility patterns were not a reliable way of tracing this epidemic.

Antimicrobial resistance of recovered organisms is now frequently used to screen quickly large numbers of organisms before plasmid or chromosomal DNA analysis is used. When it is suspected that the plasmid(s) of an epidemic organism is an R-plasmid(s), it makes sense to first screen suspect organisms to determine if they have the same antimicrobial resistance(s). For example, such an approach was recently applied to 109 *S. aureus* obtained from geographically dispersed hospitals in Australia (16). The authors noted that these staphylococci had similar patterns of antimicrobial resistance—specifically, to gentamicin, kanamycin, tobramycin, and chloramphenicol—and found that these *S. aureus* commonly had three plasmids—a 15- to 22-Md—R-plasmid, a 3-Md R-plasmid, and a 1-Md cryptic plasmid. Restriction endonuclease digestion showed similar DNA fragments in plasmids from strains of *S. aureus* recovered from all the hospitals. Although this study was not truly an "epidemiological" investigation, the results do suggest the wide dissemination of a precursor *S. aureus*, its derivatives now recoverable from geographically separate Australian hospitals.

Nosocomial outbreaks of Legionnaires' disease have frequently been investigated using plasmid and chromosomal DNA analysis techniques. In one study (17), chromosomal DNA was extracted from *Legionella pneumophila* serogroup 1 from three patients and their environment. Endonuclease restriction showed DNA digests in *Legionella* from the patients to be identical with one another and with *Legionella* isolated from the hot-water supply of the hospital. However, although the hot-water plumbing would account for all of the cases, it is not clear whether or not other environmental isolates were tested; thus, there is a possibility that hot water was not the exclusive source of the patients' infections. Although a substantial number of reports indicate that inhalation of aerosols in hot showers may infect susceptible patients with *L. pneumophila*, some of the reports that use plasmid or chromosomal analysis lack analysis of comparison (control) environmental isolates. For example, analysis of plasmids from *L. pneumophila* serogroups 1 and 10 recovered from hot water and from 19 immunocompromised patients showed these plasmids to be identical (18). However, because no other environmental sources were cultured and because the source of the infections was not determined by histories of exposures obtained from patients (or their records) and from control (comparison) persons, this should be viewed as good circumstantial, but poor epidemiological, evidence of hot water as the source of the legionella infections in the patients studied.

Studies of *Legionella* species have been reported in which plasmid analysis was useful in distinguishing the source of the organism. An investigation of *L. pneumophila* serogroup 1 from two patients, the air-conditioning cooling tower, and the two hot-water tanks of the hospital where the study was conducted showed that all the serogroup 1 strains contained the same two 21- and 48-Md plasmids (19). However, restriction-endonuclease analysis of the smaller plasmid revealed identical *Eco*R1 and *Hae*III fragments only in isolates from the patients and the hot-water towers but not in isolates from the cooling tower. These findings implicate hot, rather than cold, water as the source of the patients' infections. Plasmid analysis of random clinical and environmental isolates of *L. pneumophila* serogroup 1 from a water source in an American city showed the multiplicity of different plasmid profiles found in *Legionella* from even a single source (20).

Plasmid DNA analysis also has been applied to studies of nosocomial enterobacter infections. Markowitz and associates (21) retrospectively analyzed silver sulfadiazine- and mafenide acetate-resistant *Enterobacter cloacae* isolates from a 1976 outbreak in a burn unit. Epidemic strains of the *E. cloacae* had identical plasmid profiles that showed four plasmid bands ranging in size from 2 to 66 Md. Comparison (control) strains of susceptible *E. cloacae* from 1982 had distinctly different plasmid profiles. Another study of eight cases of neonatal meningitis caused by *E. sakazaki* (formerly yellow-pigmented *E. cloacae*) by Muytjens et al. in the Netherlands (22) also relied on plasmid profile analysis to distinguish epidemic strains. Retrospective analysis of plasmids recovered from these organisms, which had been collected over a 6-year period, showed that a cluster of four patient strains in one hospital had almost identical plasmid profiles; however, two strains of *E. sakazakii* isolated from formula after the cluster of four cases had a different plasmid profiles. It appeared that one of the plasmids had evolved by losing a transposable DNA element (transposon) of about 2 Md.

John and coinvestigators (23) studied two sequential epidemics of multiply resistant *Klebsiella pneumoniae* in a pediatric intensive care unit in 1979. Both epidemics apparently resulted from strains with similar antimicrobial resistances but with different plasmid profiles. Plasmids of approximately 105 Md appeared to be the same in both outbreak strains, but they were subsequently shown to yield different fragments after endonuclease digestion with *Bam*HI and *Ava*I. In addition to the early use of more sophisticated methods of "molecular epidemiology" to understand nosocomial infections, the results of this study suggest that plasmid profile analysis alone might be misleading if identical plasmids are not further studied by endonuclease restriction or other methods discussed elsewhere (e.g., gene probes, blot hybridization). In this study, multiple plasmids were recovered from the two outbreak strains, making their disparity clear. However, had only the single 105-Md R-plasmid been isolated and if the authors had not done further research, they might have erroneously concluded that the two outbreak strains were identical.

Usually, plasmids isolated from strains of *Enterobacteriaceae* causing infection are sufficiently numerous and diverse to allow easy distinction between strains. Such was the case in a study by Williams and colleagues (24) of a *Citrobacter diversus* outbreak in a hospital nursery. Two cases of neonatal *C. diversus* meningitis and sepsis were followed by the recovery of the epidemic organism from the rectums and umbilicuses of 11 neonates. *Citrobacter* of the same plasmid profile was recovered from hand and rectal cultures of two epidemiologically suspect nurses who attended the infants shortly after birth. A second cluster of four infants colonized with the epidemic strain was associated with a mother who was a carrier of the same *C. diversus*, and the outbreak was stopped only after strict control measures were instituted.

In addition to the investigation of epidemic organisms introduced into hospitals, some molecular epidemiological investigations have implicated hospitals as the source of bacterial pathogens in the community. Riley and co-workers (25) in Brazil studied 470 children with multiply resistant *Salmonella typhimurium*; the children had visited eight outpatient clinics from March 1981 to May 1982. Plasmid DNA analysis of strains recovered from 28 children revealed four distinct plasmid profiles, each of which was temporally clustered (Fig. 2). In addition to identifying the source of community *S. typhimurium* as pediatric clinics, results of this study suggested that so-called endemic salmonellosis may actually represent the integration of large numbers of small epidemics of microbiologically different strains of the same *Salmonella* species.

Nosocomial infections are often spread from person to person, as seen in an investigation of antibiotic-associated (pseudomembranous) colitis involving a distinctive strain of *Clostridium difficile* (26). Wüst and colleagues recovered 16 strains of *C. difficile* from 15 patients during a nosocomial outbreak and were able to show that at least 12 were phenotypically identical by several techniques—crossed immunoelectrophoresis for detecting extracellular antigens and toxins, polyacrylamide gel electrophoresis for analyzing soluble proteins, assays for cytotoxicity, and antimicrobial susceptibility. All *C. difficile* isolates cultured from June 1979 to January 1980 had the same large plasmid, and plasmid profile analysis alone reliably indicated epidemiologically associated and phenotypically identical strains.

The opposite conclusion resulted from a study of cryptic plasmids among 123 isolates of *Providencia stuarti* from a hospital ward (27). Hawkey and co-workers (27) found that only 40 strains contained cryptic plasmids and these varied among *P. stuarti* isolated from closely related sources; some epidemiologically related *P. stuarti* were all of serotype 063. When considering cryptic plasmids in *P. stuarti*, there appeared to be no particular benefit from plasmid profile analysis. These investigators concluded that plasmid analysis was probably most useful for typifying bacteria that are apparently similar by phenotypic markers such as antimicrobial resistance.

In another study in which *Providencia* were resistant to antimicrobials (i.e., bore R-plasmids) plasmid analysis was epidemiologically useful. Shlaes and Curie (28) found an endemic gentamicin-resistance-mediating R-plasmid, apparently spread among different species of the family *Enterobacteriaceae*, recovered from patients in a spinal cord injury unit. The investigators studied the R-plasmids from 22 plasmid-bearing strains of *Enterobacteriaceae* (including *Providencia* and *Acinetobacter*) and found a 36-Md gentamicin-resistant R-plasmid in them. *Pseudomonas aeruginosa* strains recovered in the prospective study had different plasmids. Gentamicin resistance was apparently spread among the *Enterobacteriaceae* by a conjugative R-plasmid.

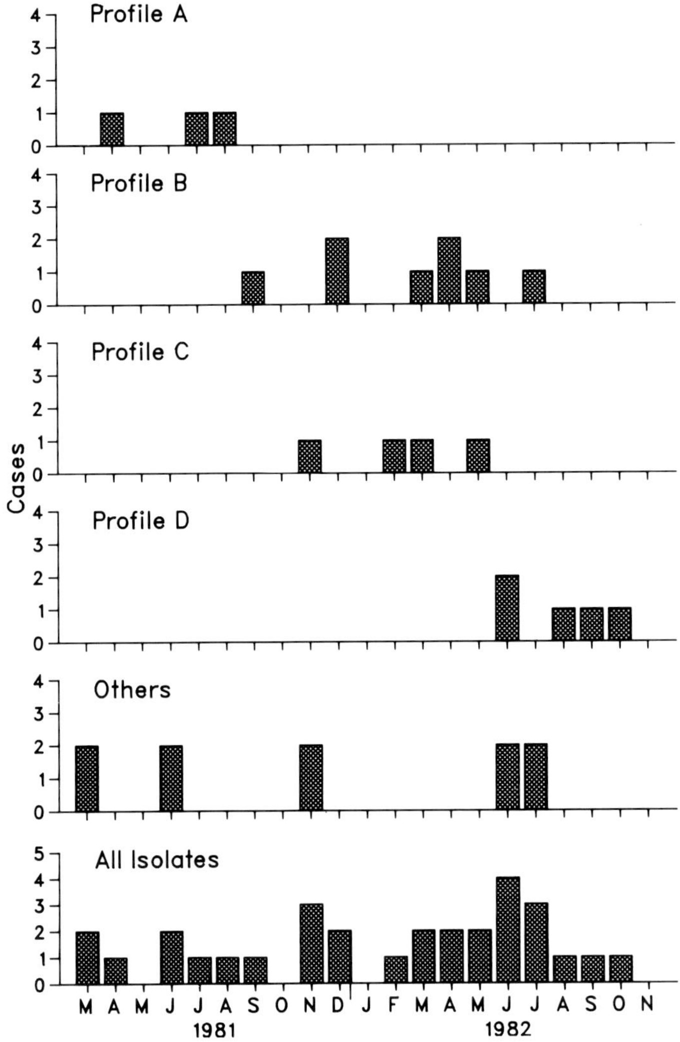

Figure 2 Plasmid DNA profile patterns and months of isolation of *S. typhimurium* strains from São Paulo children with a history of a previous hospital admission (*Source*: from Ref. 25).

Nosocomial *P. aeruginosa* outbreaks can be investigated by plasmid analysis, as performed by Parrott et al. (29). They were able to demonstrate that *P. aeruginosa* peritonitis or wound infection developed in five persons who had been swabbed with poloxamer-iodine solution. The *P. aeruginosa* recovered from unopened poloxamer-iodine bottles was the same by plasmid profile analysis (and by serotype and antimicrobial susceptibility) as organisms recovered from the infected patients. As a result of the investigation, the manufacturer voluntarily withdrew the iodophor disinfectant, and more stringent manufacture of such iodine-containing compounds was urged.

III. COMMUNITY OUTBREAKS

Molecular epidemiology has caused major changes in investigations of community outbreaks and the results of those investigations in the past several years. These studies have mainly been of enteric pathogens, i.e., *Salmonella*, enterotoxigenic *E. coli, Campylobacter*, and *Vibrio cholerae*. Plasmid and chromosomal DNA analysis has yielded some remarkable insights into the epidemiology of these and other bacteria.

Perhaps the first successful epidemiological study to rely heavily on molecular techniques was an investigation of a multistate outbreak of *Salmonella muenchen* in January and February of 1981 (30). Taylor and colleagues studied an apparent increase in *S. muenchen* by examining 85 isolates from Ohio, Michigan, Georgia, and Alabama. Initial epidemiological investigation failed to find a common food source, and almost all patients were young adults. Further scrutiny revealed that patients had been smoking marijuana, and marijuana samples from some households yielded as many as 10^7 *S. muenchen* per gram. Plasmid fingerprinting showed that marijuana-associated *S. muenchen* contained two small plasmids (3.1- and 7.4-Md), and such isolates were recovered from areas as separate as California and Massachusetts. The investigators speculated that the marijuana may have originated in Colombia or Jamaica, and that high counts of fecal coliforms in cannabis samples indicated contamination from animal feces. Aside from discerning a nationwide epidemic by the use of plasmid analysis, the investigators noted that the plasmid fingerprint remained essentially unchanged over a broad geographic area and after passage through several human hosts.

In Britain, Rowe and Threlfall (31) found that an increase in multiple-drug-resistant *S. typhimurium* in cattle and humans since 1977 had followed the sequential acquisition of resistance plasmids in the bovine host by one strain of *S. typhimurium*, phage type 204. Type 204 acquired additional antimicrobial resistances (to ampicillin, kanamycin, and streptomycin) with the acquisition of an R-plasmid. This plasmid also coded for restriction of *S. typhimurium* typing bacteriophages and, thereby, converted *S. typhimurium* phage type 204 to phage type 193 (32). During 1978 and 1979 these multiply resistant strains of types 204 and 193 spread to Europe, apparently by the export of infected calves from Britain. In this series of investigations, as in the one previously described, national and international patterns of spread of *Salmonella* were discerned with the aid of plasmid DNA analysis.

The utility of plasmid profile analysis in salmonella epidemiology was noted in an *S. drypool* outbreak in Texas traced to commercially prepared Mexican food (33) and in differentiating *S. muenster* isolates in Canada (34). However, the reliability of plasmid analysis compared with other systems for typing *Salmonella*—specifically, bacteriophage lysis pattern (phage type) or antimicrobial susceptibility pattern (sometimes called antibiogram)—was unproved. In the United States, phage typing is limited to a few common serotypes such as *S. typhimurium* and must be performed at the Centers for Disease Control (CDC). The specificity and sensitivity of antimicrobial susceptibility patterns in differentiating *Salmonella* of the same serotype was unknown as well.

We (35) compared the phage types, antimicrobial resistance patterns, and plasmid profiles of 20 groups of isolates received at CDC from *S. typhimurium* outbreaks between 1975 and 1982 to determine the most useful laboratory method for identifying epidemiologically related isolates of this organism. In 18 (90%) of the 20 outbreaks, epidemiologically related isolates were identified as being the same by each of the three methods. However, in a subgroup of nine outbreaks in which *S. typhimurium* isolates unrelated to the outbreak were submitted for comparison, outbreak isolates were differentiated from such control isolates six times (67%) by phage typing alone, four times (44%) by antimicrobial susceptibility testing alone, and eight times (89%) by plasmid profile analysis alone. Epidemic isolates were susceptible to all antimicrobials tested, non-phage-typable, or without plasmids in 14 (70%), 1 (5%), and 3 (15%), respectively, of the 20 outbreaks. Thus, plasmid analysis appeared to be at least as specific as phage typing in identifying epidemiologically related isolates of *S. typhimurium* and in differentiating them from non-epidemic-related (control) specimens; both techniques appeared to be superior to antimicrobial susceptibility testing alone (Table 1).

Similar results were obtained in a smaller study by Brunner et al. (36). These investigators compared lysotyping (phage typing), antimicrobial resistance patterns (14 antibiotics), and biotyping with plasmid profile analysis during a limited epidemic of *S. typhimurium*. Again, plasmid profile determination was as useful and accurate as phage typing, and antimicrobial susceptibility patterns were far less so; biotyping was quite unreliable.

A demonstration of the ability of plasmid analysis to identify geographically disparate but related strains of *Salmonella* is illustrated in an investigation by Riley and co-workers (37). They studied an outbreak of *S. newport* in New Jersey and Pennsylvania, in July and August of 1981, and found that infections

Table 1 Characteristics of *S. typhimurium* Strains from Three Outbreaks of Disease in the United States

State/date of outbreak	Source of isolates	No. of specimens	Antimicrobial resistances[a]	Plasmid profile (MD)	Phage group[b]
Ohio/1976	Hospital patients	9	Su, Sm, Km, Te, Am, Cb, Cf	68, 15, 6, 3.8	Res 135
	Dietary worker	1	Su, Sm, Km, Te, Am, Cb, Cf	68, 15, 6, 3.8	NST
Virginia/1978	Family	5	None	71	NST
	Ice cream	1	None	71	NST
Pennsylvania/1980	Children	7	None	4.7, 2.9, 2.4, 2.2	NST
	Meat	1	None	4.8, 3.0, 2.6, 2.4	NST
	Control	1	None	None	Res 10

[a]Su, sulfadiazine; Sm, streptomycin; Km, kanamycin; Te, tetracycline; Am, ampicillin; Cb, carbenicillin; Cf, cephalothin.
[b]Res, resembles; NST, no specific type.
Source: from Ref. 35.

with the epidemic *S. newport* were epidemiologically related to the consumption of a brand of precooked roast beef. Plasmid profile analysis showed that supposedly "sporadic" *S. newport* strains in the two states were the same as those isolated during the outbreak, and that patients with sporadic strains had also eaten the same brand of precooked roast beef. It would have been impossible to trace the introduction and transmission of this bacterial clone by antimicrobial susceptibility testing because the *S. newport* was not antimicrobial-resistant. In addition, this pattern of spread would not have been easily identified by phage lysis of the *S. newport* strains because a phage-lysis typing system for *S. newport* was (and is) not readily available in the United States.

Salmonella is a pathogen transmitted almost exclusively by undercooked or mishandled foods of animal origin (e.g., beef, pork, chicken, eggs, and raw milk). Antimicrobial-resistant strains now account for approximately 25% of *Salmonella* isolated from humans, and the proportion of resistant strains has been increasing in the last 30 years. This increase has been roughly concomitant with the use of tetracycline, penicillin, and other antimicrobials at low doses in the feed of animals to promote their growth and to prevent disease. Antibiotics used in human medicine are frequently added to animal feed, and the use of antimicrobials for veterinary purposes accounts for almost half of all antimicrobials produced in the United States yearly. Some experts have been concerned that the use of antibiotics in animal feeds selects for resistant bacteria, which make their way through the food chain to cause infections and serious disease in humans. The putative pathway from farm to dinner table has been exceedingly difficult to discern in the individual case because of the complexities of modern marketing and distribution of animal foods. Two recent studies that relied on plasmid DNA analysis have elucidated this suspected relationship between antimicrobials in animal feeds and antimicrobial-resistant salmonellal infections in humans.

In 1982, O'Brien and others reported the results of a complex study of *Salmonella* from animals and humans in the United States (38). They obtained from animal and human reference laboratories *S. typhimurium, S. newport*, and *S. dublin*, tested these serotypes for antimicrobial resistance, and extracted plasmids from them. Within each serotype, computer analysis showed that restriction endonuclease digestion of plasmids yielded fragments from animal and human isolates that were often identical or nearly so. Human salmonella cases were not apparently related except for their common plasmids and appeared to be clustered in time, but geographically dispersed, as is seen in spread of organisms by animal food products. This study demonstrated that R-plasmids were apparently extensively shared by animal and human salmonellae and suggested that the source of antimicrobial-resistance in *Salmonella* species and other enteric pathogens (*Campylobacter*, pathogenic *E. coli*) was the farm.

In a more specific study, Holmberg and co-workers (39), in early 1983, identified 18 persons in four Midwestern states who were infected with *S. newport* in which ampicillin, carbenicillin, and tetracycline resistances were mediated by a 38-kb (24-Md) R-plasmid. Twelve of the patients had been taking penicillin derivatives for medical problems other than diarrhea in the 24–48 hr before the onset of salmonellosis. These illnesses were severe in that 11 patients were hospitalized for an average of 8 days, and one had a fatal nosocomial infection. The investigators compared plasmid profiles of 162 *S. newport* isolates from humans (six-state area) and animals (continental United States) and examined computer records of meat distribution. The *S. newport* had a distinct plasmid fingerprint from all but one of the 162 animal and human *S. newport* during the 18 months before the cluster of human cases. The animal isolate that was identical with the strain in the human outbreak was recovered from a dairy herd on a farm neighboring a beef herd in South Dakota; the beef herd had received subtherapeutic chlortetracycline added to its feed. The timing and geographic distribution of the cases coincided with the slaughter and distribution of the beef herd. Patients with *S. newport* had either received meat directly from and eaten at the beef farm or, outside of South Dakota, had purchased it from supermarkets apparently supplied with meat from the herd. This sequence of events would not likely occur by chance. The investigators concluded that antimicrobial-resistant organisms of animal origin cause serious human illness and that there was a need for more prudent use of antimicrobials in both humans and animals.

Plasmid analysis has remained an important component of all recent investigations of this subject. For example, Ølsvik et al. (40,41; see Fig. 1) used plasmid profile analysis to demonstrate the transmission of *S. typhimurium* from cattle on a Norwegian farm to humans directly in contact with the ill animals. (These investigators found a 24-Md plasmid that appeared to be similar to that found in the *S. newport* in the investigation described earlier. However, endonuclease restriction by *Hin*dIII showed the plasmids from the two investigations to be dissimilar.) On a second farm, these investigators isolated tetracycline-resistant *S. typhimurium* strains possessing eight plasmids from both human and animals; this multiple-plasmid-bearing, tetracycline-resistant *Salmonella* species was also recovered from animals at a third farm that had no known contact between this farm and the other two. Results of this study indicate the extent to which salmonellae strains are shared between animals and the people tending them.

Plasmid profiles have also been found to be identical in *E. coli* strains from animals and humans, suggesting that, as for *Salmonella*, substantial overlap exists between animal and human reservoirs of *E. coli*. Specific plasmids have been recovered from nonpathogenic *E. coli* transferred from animal to human reservoirs

(42). In another study, chloramphenicol-resistance-mediating R-plasmids recovered from enteropathogenic *E. coli* in piglets and a human infant were almost identical by restriction endonuclease digestion (43).

Endonuclease restriction of chromosomal DNA from nine enterotoxigenic *E. coli* (ETEC) from an outbreak of human diarrhea showed that these were the same, but they were different from seven comparison (control) specimens of nontoxigenic strains from animals and humans (44). Interestingly, the nine outbreak strains, although epidemiologically associated, had different H serotypes. (Variability of the H antigen has not occurred in *E. coli* serotypes from any outbreak analyzed at CDC.) This finding tends to confirm the usefulness of DNA analysis to typify *E. coli* outbreak strains.

Even when pathogenic *E. coli* have unusual serotypes, plasmid analysis is frequently used and valuable. Two recent investigations illustrate this point. Two outbreaks of hemorrhagic colitis, a newly recognized syndrome characterized by bloody diarrhea, severe abdominal pain, and little or no fever, occurred in 1982 (45). Nine of 20 cases, but no controls, excreted a rare *E. coli* serotype, 0157:H7. Wells and others examined the plasmid content of outbreak strains and found these to be closely related (46). *Escherichia coli* 0157:H7 recovered from patients in Oregon and Michigan had the same 72-Md with a very small, less than 1-Md, plasmid as determined by *Hin*dIII endonuclease digestion.

In another study, enterotoxigenic *E. coli* 027:H20 were recovered from persons in five states who had eaten imported semisoft French brie cheese and, subsequently, developed watery diarrhea and abdominal cramps. Initially, MacDonald and co-workers identified three clusters of intestinal illness in 45 persons who had attended office parties in Washington, D.C. (47). Plasmid profile analysis showed the ETEC to contain a large 100-Md plasmid that bound radiolabeled DNA probe for heat-stable *E. coli* enterotoxin. This same plasmid was recovered from *E. coli* 027:H20 in Illinois, Wisconsin, Georgia, and Colorado in symptomatic persons who had eaten the same brand of brie cheese.

As with outbreaks caused by *Salmonella* and *E. coli*, plasmid analysis has been useful in investigating those caused by *Campylobacter jejuni*. Although some serotyping systems of *Campylobacter* have been described (48,49), no single method to characterize these recently recognized, very common agents of gastrointestinal illness has been acceptable. To address the issue of whether or not plasmid analysis–combined with other techniques such as serotyping and antimicrobial susceptibility patterns–might adequately typify outbreak-related isolates of *C. jejuni*, Bopp et al. examined 31 *Campylobacter* strains from 11 outbreaks (50). Of the 31 strains, 19 (61%) possessed plasmid DNA, and those that had a common 38-Md tetracycline-resistance plasmid shared common nucleic acid sequences. In the 11 outbreaks, three (27%) were caused by *C. jejuni* of multiple serotypes; in two of these three outbreaks, plasmid profile analysis was also indicative of multiple strains. These investigators concluded that plasmid profile analysis and antimicrobial susceptibility patterns might be potentially useful epidemiologically as supplements to serotyping.

The limited availability of serotype reagents has prompted epidemiologists to analyze DNA from plasmids and chromosomes to identify outbreak-related strains of *Campylobacter*. Tenover and co-workers (51) analyzed plasmids from three outbreaks of campylobacteriosis. In the first outbreak, which occurred in 11 members of a college sorority, three of four analyzed strains had a 41-kb plasmid; in the second, occurring in a mother and her three young children, all four patients had *C. jejuni* that contained 28- and 38-kb plasmids; and, in an outbreak involving a mother, her two sons, and a newly acquired puppy, all isolates contained 31-, 35-, and 38-kb plasmids. Plasmid fingerprinting was apparently useful in all three outbreaks.

Even though Penner and coworkers can serotype *C. jejuni* on the basis of an extractable thermostable antigen (48), these investigators have also used chromosomal restriction endonuclease digestion to aid in the analysis of a laboratory-acquired case of *C. jejuni* enteritis (52). The frequently passaged laboratory strain was identical, by both serotyping and chromosomal analysis, with *C. jejuni* isolated from the laboratory worker. These workers then compared serotyping and restriction endonuclease digestion of chromosomal DNA in investigating an outbreak of *C. jejuni* enteritis that involved 29 persons (53). Milk from cattle was suspected as the source of the outbreak. Chromosomal restriction patterns of all the human isolates and some of the isolates from 20 cattle were identical, corroborating the suspected link between human and animal. However, the single milk isolate had a different serotype and chromosomal digestion pattern and, thus, this milk could not be implicated by laboratory analysis.

Because of the suspected and now well-recognized association between the consumption of other undercooked or raw animal foods (meat, milk, cheese) and campylobacter infection (54), Bradbury and others have reported their results of two studies that analyzed plasmid DNA in *C. jejuni* and *C. coli* from humans and animals. In the first study (55), serologically defined strains of *C. jejuni* (40 strains) and *C. coli* (17 strains) were examined for their plasmid DNA content. Because 11 (36%) *C. jejuni* isolates had stable plasmids, the investigators concluded that plasmid extraction and analysis might be useful in studying the epidemiology of *Campylobacter* strains. However, some might interpret this low rate of recovery of plasmids from *C. jejuni* as an argument that a specific and rapid typing test other than plasmid profile analysis is necessary. The same workers (56) found plasmid carriage in 116 (53%) of 200 *C. jejuni* and *C. coli* in healthy and diarrheic animals. However, this study was not able to establish a clear correlation between plasmid carriage by recovered campylobacters and the health status of animals. Among the conflicting results, these workers found a higher plasmid carriage rate in *C. coli* from healthy (45/61, 74%) than from diarrheic (3/10, 30%) pigs, and a higher plasmid carriage rate in *C. jejuni* from diarrheic (10/32, 31%) than from healthy (3/22, 14%) cattle.

In sum, the usefulness of DNA analysis in campylobacter outbreaks is limited by the generally low carriage rate of plasmids and the complexities of analyzing the many bands created by restriction endonuclease digestion of chromosomal DNA. These techniques are apparently useful, but less so than for studying the epidemiology of *Salmonella* and *E. coli*.

Molecular techniques have been applied successfully to the epidemiology of *Vibrio cholerae* 01, the causative agent of cholera, and other non-01 *V. cholerae* in the Gulf Coast. Kaper et al. (57) studied enterotoxigenic strains of *V. cholerae* 01, biotype eltor, isolated from a case of cholera in Texas in 1973, an outbreak of cholera in Louisiana in 1978 (58), and Louisiana sewage samples from 1980 and 1981. These investigators digested chromosomal DNA from these strains and showed that, in addition to the identity of resultant digest patterns, a radioactive probe for *E. coli* heat-labile enterotoxin (LT) hybridized with DNA sequences both from *V. cholerae* 01 and *V. cholerae* non-01 isolated from Gulf coastal waters. (Probes for LT will hybridize with gene sequences coding for cholera toxin production.) In addition to demonstrating the identity of *V. cholerae* 01 in the US Gulf Coast from 1973 to 1980, they concluded that a second reservoir of cholera toxin genes existed in *V. cholerae* non-01 in Louisiana.

Subsequently, Blake and others recovered *V. cholerae* 01, biotype eltor, serotype inaba, from an American tourist who had visited Cancun, Mexico, in June 1983 (59). In addition to being of the same biotype and serotype and having other phenotypic features (hemolytic activity, bacteriophage sensitivity pattern) in common with the earlier isolates from the US Gulf Coast, extracted chromosomal DNA from the *V. cholerae* 01 from this tourist had gene sequences for cholera toxin production apparently identical with those found in earlier *V. cholerae* 01 in the United States. These findings lent credence to the contention that the causative agent of cholera has probably persisted for decades in the Gulf of Mexico, causing uncommonly detected disease in Mexico and the United States.

Rarely seen since the beginning of the seventh cholera pandemic in the 1960s, recent isolates of classical *V. cholerae* 01 might originate from precursor strains of *V. cholerae* 01 eltor or from nontoxigenic strains of classical *V. cholerae* 01. To address this issue, Cook and others (60) examined the plasmid profiles, the location of cholera toxin subunit A genes, and the presence of defective VcAl prophage genome in *V. cholerae* 01, biotype classical, in Bangladesh in 1982. They found that the newer classical strains typically had two plasmids (21- and 3-Md), whereas strains of *V. cholerae* 01, biotype eltor, recovered in Bangladesh before 1982 and strains of nontoxigenic *V. cholerae* 01, principally isolated in the United States, had no and zero to three plasmids, respectively. The results of plasmid profile analysis and other laboratory investigations led these workers to conclude that neither the eltor nor the classical nontoxigenic 01 strains were precursors to the classical toxigenic strains recovered in Bangladesh since 1982. Also, classical biotypes from Japan in 1916 and from Asia until the 1960s were all identical by these genetic derminants. This suggests that a persistent single clone of classical *V. cholerae* 01 is responsible for the rare classical toxigenic strains isolated in recent years.

In Bangladesh, plasmid profile analysis has also been applied to 136 shigella isolates (61). Many different plasmid profiles can be found within each species, indicating that many different strains of *Shigella* are responsible for illness (shigellosis, bacillary dysentery) in that country.

The use of plasmid and chromosomal DNA analyses has been largely, but not exclusively, devoted to the epidemiology of bacterial enteric pathogens in the community. Pappenheimer and Murphy in a recent study (62) demonstrated that chromosomal DNA from a toxigenic *Corynebacterium diphtheriae mitis* in a baby with membranous tonsillitis yielded different endonuclease digestion fragments of DNA from digests of *C. diphtheriae* in diphtheria outbreaks elsewhere. (These investigators then used other molecular techniques, namely, hybridization with labeled nick-translated corynephage probe, to demonstrate that spread of the *tox* gene via prophage might result in the conversion of resident *tox*-negative strains in the nasopharynx to diphtheria-causing organisms.)

Restriction endonuclease digestion of chromosomal DNA of *N. gonorrhoeae* has been advocated as an epidemiological tool because auxotyping and serological classification are tedious and complicated. Falk et al. (63) showed that seven samples from the same gonococcal clone had identical restriction patterns and that these remained stable through 41 passes in vitro. Also, *N. meningitidis* and *N. gonorrhoeae* may frequently carry a 2.6-Md and other plasmids with different molecular masses (64,65). However, to our knowledge, plasmid DNA analysis has not been widely applied to the epidemiological investigation of cases or outbreaks of these *Neisseria* species. Chromosomal DNA analysis has been performed recently on meningococci from three infected schoolchildren with symptoms and 36 carriers (66) in northern Norway; chromosomal DNA fingerprinting showed that epidemiologically related isolates from schoolchildren in 2 of 15 small villages were identical or nearly so, and distinct from epidemiologically unrelated isolates.

Meissner and Falkinham have examined plasmid DNA from clinical and environmental isolates of *Mycobacterium avium, M. intracellulare*, and *M. scrofulaceum* (67). Because plasmids were frequently recovered from mycobacteria from clinical (56%) and aerosol (75%) samples, but infrequently from mycobacteria in soil (5%), dust (7%), sediment (6% or less), and water (25%), these investigators suggested that mycobacteria-laden aerosols were a likely source of human infection. However, because no specific epidemiological associations were looked for, this conclusion must be viewed cautiously.

Daly has advocated the enumeration of plasmids and restriction endonuclease digestion analysis in studying the epidemiology of *Haemophilus ducreyi* (68). A number of studies have looked at the molecular characteristics of β-lactamase-producing plasmids in *H. ducreyi* (69-71) but not at the molecular epidemiology of the organism.

IV. SUMMARY

This review of recent (i.e., to July 1986) advances in molecular epidemiology based on plasmid and chromosomal DNA analyses covers a wide range of investigations of a large number of different bacteria. However, from all of these investigations, certain conclusions can be made.

1. Plasmid and chromosomal DNA analysis is rapid, practical, simple, and applicable to the epidemiology of many different organisms (72).
2. Because plasmid and chromosomal DNA analyses are based on genotype, these analyses seem to be more reliable than those based on phenotypic characteristics, such as antimicrobial resistance, serotype, bacteriophage lysis, biochemical characteristics, and so forth. Also, because plasmid and chromosomal DNA code for phenotype, analysis of DNA provides insight into the biology and evolution of the organism(s) being studied.
3. The usual limitation of plasmid analysis occurs when organisms, such as *Campylobacter*, do not commonly carry plasmids. Chromosomal digestion patterns are always somewhat limited by the large number of digest fragments (bands) shown on agarose electrophoresis, making comparison of the many fragments difficult.
4. Molecular epidemiological techniques and results, as described here, have been useful in better defining the spread of bacterial "clones" through hospital and community (73).
5. Gene probes are now also useful for epidemic investigations (74,75), but not yet widely available. However, the "recombinant DNA industry" is working on time- and money-saving innovations to make genetic probes practical for the modern nonresearch laboratory.
6. Finally, the use of plasmid or chromosomal DNA analysis is not, in itself, molecular epidemiology. Without proper control specimens or data, analysis of randomly collected isolates produces little useful information. Plasmid and chromosomal DNA analyses remain most useful epidemiologically as typing methods to identify associated bacterial isolates in well-designed and conducted investigations.

REFERENCES

1. Meyers, J. A., D. Sanchez, L. P. Elwell, and S. Falkow, Simple agarose gel electrophoretic method for the identification and characterization of plasmid deoxyribonucleic acid. *J. Bacteriol. 127*:1529–1537 (1976).
2. Birnboim, H. C., and J. Doly, A rapid alkaline extraction procedure for screening recombinant plasmid DNA. *Nucleic Acids Res. 7*:1513–1523 (1979).
3. Aber, R. C., and D. C. Mackel, Epidemiologic typing of nosocomial microorganisms. *Am. J. Med. 70*:899–905 (1981).
4. Farrar, W. E., Jr., Molecular analysis of plasmids in epidemiologic investigations. *J. Infect. Dis. 148*:1–6 (1983).
5. Wachsmuth, K., Genotypic approaches to the diagnosis of bacterial infections: Plasmid analyses and gene probes. *Infect. Control 6*:100–109 (1985).
6. McGowan, J. E., Jr., P. M. Terry, T.-S. R. Huang, C. L. Houk, and J. Davies, Nosocomial infections with gentamicin-resistant *Staphylococcus aureus*: Plasmid profile analysis as an epidemiologic tool. *J. Infect. Dis. 140*:864–872 (1979).
7. Jaffe, H. W., H. M. Sweeney, C. Nathan, R. A. Weinstein, S. A. Kabins, and S. Cohen, Identity and interspecific transfer of gentamicin-resistance plasmids in *Staphylococcus aureus* and *Staphylococcus epidermidis. J. Infect. Dis. 141*:738–747 (1980).
8. Cohen, M. L., E. S. Wong, and S. Falkow, Common R-plasmids in *Staphylococcus aureus* and *Staphylococcus epidermidis* during a nosocomial *Staphylococcus aureus* outbreak. *Antimicrob. Agents Chemother. 21*: 210–215 (1982).
9. Rubens, C. E., W. E. Farrar Jr., Z. A. McGee, and W. Schaffner, Evolution of a plasmid-mediating resistance to multiple antimicrobial agents during a prolonged epidemic of nosocomial infections. *J. Infect. Dis. 143*:170–181 (1981).
10. Parisi, J. T., and D. W. Hecht, Plasmid profiles in epidemiologic studies of infections by *Staphylococcus epidermidis. J. Infect. Dis. 141*:637–643 (1980).
11. Tompkins, L. S., D. R. Scaberg, and S. Falkow, Use of agarose gel electrophoresis of plasmid DNA as an epidemiologic marker of bacterial strains. In *Current Chemotherapy and Infectious Diseases*, J. D. Nelson and Grassi, C. (eds.). American Society for Microbiology, Washington, pp. 726–728 (1980).
12. Schaberg, D. R., L. S. Tompkins, and S. Falkow, Use of agarose gel electrophoresis of plasmid deoxyribonucleic acid to fingerprint gram-negative bacilli. *J. Clin. Microbiol. 13*:1105–1108 (1981).
13. Locksley, R. M., M. L. Cohen, T. C. Quinn, L. S. Tompkins, M. B. Coyle, J. M. Kirihara, and G. W. Counts, Multiply antibiotic-resistant *Staphylococcus aureus*: Introduction, transmission, and evolution of nosocomial infection. *Ann. Intern. Med. 97*:317–324 (1982).
14. Archer, G. L., and C. G. Mayhall, Comparison of epidemiologic markers

used in the investigation of an outbreak of methicillin-resistant *Staphylococcus aureus* infections. *J. Clin. Microbiol. 18*:395-399 (1983).
15. Mickelson, P. A., J. J. Plorde, K. P. Gordon et al., Instability of antibiotic resistance in a strain of *Staphylococcus epidermidis* isolated from an outbreak of prosthetic valve endocarditis. *J. Infect. Dis. 152*:50-58 (1985).
16. Lyon, B. R., J. L. Iuorio, J. W. May, and P. A. Skurray, Molecular epidemiology of multiresistant *Staphylococcus aureus* in Australian hospitals. *J. Med. Microbiol. 17*:79-89 (1984).
17. van Ketel, R. J., J. ter Schegget, and H. C. Zanen, Molecular epidemiology of *Legionella pneumophila* serogroup 1. *J. Clin. Microbiol. 20*:362-364 (1984).
18. Meenhorst, P. L., A. L. Reingold, D. G. Groothuis et al., Water-related nosocomial pneumonia caused by *Legionella pneumophila* serogroups 1 and 10. *J. Infect. Dis. 152*:356-364 (1985).
19. Nolte, F. S., C. A. Conlin, A. J. Roisin, and S. R. Redmond, Plasmids as epidemiological markers in nosocomial Legionnaires' disease. *J. Infect. Dis. 149*:251-256 (1984).
20. Maher, W. E., J. F. Plouffe, and M. F. Para, Plasmid profiles of clinical and environmental isolates of *Legionella pneumophila* serogroup 1. *J. Clin. Microbiol. 18*:1422-1423 (1983).
21. Markowitz, S. M., S. M. Smith, and D. S. Williams, Retrospective analysis of plasmid patterns in a study of burn unit outbreaks of infection due to *Enterobacter cloacae. J. Infect. Dis. 148*:18-23 (1983).
22. Muytjens, H. L., H. C. Zanen, H. J. Sonderkamp, L. A. Kollée, I. K. Wachsmuth, and J. J. Farmer III, Analysis of eight cases of neonatal meningitis and sepsis due to *Enterobacter sakazakii. J. Clin. Microbiol. 18*:115-20 (1983).
23. John, J. F., Jr., K. T. McKee, Jr., J. A. Twitty, and W. Schaffner, Molecular epidemiology of sequential nursery epidemics caused by multiresistant *Klebsiella pneumoniae. J. Pediatr. 102*:825-830 (1983).
24. Williams, W. W., J. Mariano, M. Spurrier et al., Nosocomial meningitis due to *Citrobacter diversus* in neonates: New aspects of the epidemiology. *J. Infect. Dis. 150*:229-235 (1984).
25. Riley, L. W., B. S. O. Ceballos, L. R. Trabulsi, M. R. Fernandes de Toledo, and P. A. Blake, The significance of hospitals as reservoirs for endemic multiresistant *Salmonella typhimurium* causing infection in urban Brazilian children. *J. Infect. Dis. 150*:236-241 (1984).
26. Wüst, J., N. M. Sullivan, U. Hardegger, and T. D. Wilkins, Investigation of an outbreak of antibiotic-associated colitis by various typing methods. *J. Clin. Microbiol. 16*:1096-1101 (1982).
27. Hawkey, P. M., P. M. Bennett, and C. A. Hawkey, Cryptic plasmids in hospital isolates of *Providencia stuarti. J. Med. Microbiol. 18*:277-284 (1984).
28. Shlaes, D. M., and C. A. Currie, Endemic gentamicin resistance R factors on a spinal cord injury unit. *J. Clin. Microbiol. 18*:236-241 (1983).
29. Parrott, P. L., P. M. Terry, E. N. Whitworth, L. W. Frawley, R. S. Coble,

I. K. Wachsmuth, and J. E. McGowan Jr, *Pseudomas aeruginosa* peritonitis associated with contaminated poloxamer-iodine solution. *Lancet 2*:683-685 (1982).
30. Taylor, D. N., I. K. Wachsmuth, Y.-H. Shangkuan et al., Salmonellosis associated with marijuana: A multistate outbreak traced by plasmid fingerprinting. *N. Engl. J. Med. 306*:1249-1253 (1982).
31. Rowe, B., and E. J. Threlfall, Multiply-resistant clones of *Salmonella typhimurium* in Britain: Epidemiologic and laboratory aspects. In *Molecular Biology, Pathogenicity, and Ecology of Bacterial Plasmids*, S. B. Levy, Cloves, R. C., and Koenig, E. L. (eds.). Plenum Press, New York, pp. 567-573 (1981).
32. Willshaw, G. A., E. J. Threlfall, L. R. Ward, A. S. Ashley, and B. Rowe, Plasmid studies of drug-resistant epidemic strains of *Salmonella typhimurium* belonging to phage types 204 and 193. *J. Antimicrob. Chemother. 6*:763-773 (1980).
33. Riley, L. W., and M. L. Cohen, Plasmid profiles and salmonella epidemiology (letter). *Lancet 1*:573 (1982).
34. Bezanson, G. S., R. Khakhria, and D. Pagnutti, Plasmid profiles of value in differentiating *Salmonella muenster* isolates. *J. Clin. Microbiol. 17*:1159-1160 (1983).
35. Holmberg, S. D., I. K. Wachsmuth, F. W. Hickman-Brenner, and M. L. Cohen, Comparison of plasmid profile analysis, phage typing, and antimicrobial susceptibility testing in characterizing *Salmonella typhimurium* isolates from outbreaks. *J. Clin. Microbiol. 19*:100-104 (1984).
36. Brunner, F., A. Margadant, R. Peduzzi, and J.-C. Piffaretti, The plasmid profile as an epidemiologic tool for *Salmonella typhimurium* epidemics: Comparison with the lysotype. *J. Infect. Dis. 148*:7-11 (1983).
37. Riley, L. W., G. T. DiFerndinando, Jr., T. M. DeMelfi, and M. L. Cohen, Evaluation of isolated cases of salmonellosis by plasmid profile analysis: Introduction and transmission of a bacterial clone by precooked roast beef. *J. Infect. Dis. 148*:12-17 (1983).
38. O'Brien, T. F., J. D. Hopkins, E. S. Gilleece et al., Molecular epidemiology of antibiotic resistance in *Salmonella* from animals and human beings in the United States. *N. Engl. J. Med. 307*:1-6 (1982).
39. Holmberg, S. D., M. T. Osterholm, K. A. Senger, and M. L. Cohen, Drug-resistant *Salmonella* from animals fed antimicrobials. *N. Engl. J. Med. 311*: 617-622 (1984).
40. Olsvik, Ø., H., Sørum, K. Birkness et al., Plasmid characterization of *Salmonella typhimurium* transmitted from animals to humans. *J. Clin. Microbiol. 22*:336-338 (1985).
41. Olsvik, Ø., H. Sørum, K. Wachsmuth, K. Fossum, and J. C. Feeley, Animal-to-human transmission of both sensitive and resistant *Salmonella typhimurium* demonstrated by plasmid profiling (letter). *Lancet 1*:172-173 (1985).
42. Levy, S. B., G. B. Fitzgerald, and A. B. Macone, Spread of antibiotic-

resistant plasmids from chicken to chicken and from chicken to man. *Nature 260*:40–42 (1976).
43. Jorgensen, S. T., Relatedness of chloramphenicol resistance plasmids in epidemiologically unrelated strains of pathogenic *Escherichia coli* from man and animals. *J. Med. Microbiol. 16*:165–173 (1983).
44. Marshall, R. B., P. J. Winter, A. J. Robinson, and K. A. Bettelheim, A study of enterotoxigenic *Escherichia coli*, serogroup 0126, by bacterial restriction endonuclease DNA analysis (BRENDA). *J. Hyg. (Camb.) 94*:263–268 (1985).
45. Riley, L. W., R. S. Remis, S. D. Helgerson et al., Hemorrhagic colitis associated with a rare *E. coli* serotype. *N. Engl. J. Med. 308*:681–685 (1983).
46. Wells, J. G., B. R. Davis, I. K. Wachsmuth, L. W. Riley, R. S. Remis, R. Sokolow, and G. K. Morris, Lboratory investigation of hemorrhagic colitis outbreaks associated with a rare *Escherichia coli* serotype. *J. Clin. Microbiol. 18*:512–520 (1983).
47. MacDonald, K. L., M. Eidson, C. Strohmeyer et al., A multistate outbreak of gastrointestinal illness caused by enterotoxigenic *Escherichia coli* in imported semisoft cheese. *J. Infect. Dis. 151*:716–720 (1985).
48. Penner, J. L., and J. N. Hennessy, Passive hemagglutination technique for serotyping *Campylobacter fetus* subsp. *jejuni*. *J. Clin. Microbiol. 12*:732–737 (1980).
49. Lior, H., D. L. Woodward, J. A. Edgar, L. J. Laroche, and P. Gill, Serotyping of *Campylobacter jejuni* by slide agglutination based on heat-labile antigenic factors. *J. Clin. Microbiol. 15*:761–768 (1982).
50. Bopp, C. A., K. A. Birkness, I. K. Wachsmuth, and T. J. Barrett, In vitro antimicrobial susceptibility, plasmid analysis, and serotyping of epidemic-associated *Campylobacter jejuni*. *J. Clin. Microbiol. 21*:4–7 (1985).
51. Tenover, F. C., S. Williams, K. P. Gordon, N. Harris, C. Nolan, and J. J. Florde, Utility of plasmid fingerprinting for epidemiological studies of *Campylobacter jejuni* infections (letter). *J. Infect. Dis. 149*:279 (1984).
52. Penner, J. L., J. N. Hennessy, S. D. Mills, and W. C. Bradbury, Application of serotyping and chromosomal restriction endonuclease digest analysis in investigating a laboratory-acquired case of *Campylobacter jejuni* enteritis. *J. Clin. Microbiol. 18*:1427–1428 (1983).
53. Bradbury, W. C., A. D. Pearson, M. A. Marko, R. V. Congi, and J. L. Penner, Investigation of a *Campylobacter jejuni* outbreak by serotyping and chromosomal restriction endonuclease analysis. *J. Clin. Microbiol. 19*:342–346 (1984).
54. Hopkins, R. S., R. Olmstead, and G. R. Istre, Endemic *Campylobacter jejuni* in Colorado: Identified risk factors. *Am. J. Public Health 74*:249–250 (1984).
55. Bradbury, W. C., and D. L. G. Munroe, Occurrence of plasmids and antibiotic resistance among *Campylobacter jejuni* and *Campylobacter coli* isolated from healthy and diarrheic animals. *J. Clin. Microbiol. 22*:339–346 (1985).

56. Bradbury, W. C., A. D. Pearson, M. A. Marko, R. V. Congi, and J. P. Penner, Investigation of a *Campylobacter jejuni* outbreak by serotyping and chromosomal restriction endonuclease analysis. *J. Clin. Microbiol. 19*: 342–346 (1984).
57. Kaper, J. B., H. B. Bradford, N. C. Roberts, and S. Falkow, Molecular epidemiology of *Vibrio cholerae* in the U.S. Gulf Coast. *J. Clin. Microbiol. 16*:129–134 (1982).
58. Blake, P. A., D. T. Allegra, J. D. Snyder et al., Cholera—a possible endemic focus in the United States. *N. Engl. J. Med. 302*:305–309 (1980).
59. Blake, P. A., K. Wachsmuth, B. R. Davis, C. A. Bopp, B. P. Chaiken, and J. V. Lee, Toxigenic *Vibrio cholerae* 01 strain from Mexico identical to United States isolate (letter). *Lancet 2*:912 (1983).
60. Cook, W. L., K. Wachsmuth, S. R. Johnson, K. A. Birkness, and A. R. Samadi, Persistence of plasmids, cholera toxin genes, and prophage DNA in classical *Vibrio cholerae* 01. *Infect. Immun. 45*:222–226 (1984).
61. Tacket, C. O., N. Shahid, M. I. Huq, A. R. M. A. Alim, and M. L. Cohen, Usefulness of plasmid profiles for differentiation of *Shigella* isolates in Bangladesh. *J. Clin. Microbiol. 20*:300–301 (1984).
62. Pappenheimer, A. M., Jr., and J. R. Murphy, Studies on the molecular epidemiology of diphtheria. *Lancet 2*:923–926 (1983).
63. Falk, E. S., B. Bjorvatn, D. Danielsson, B.-E. Kristiansen, K. Melby, and B. Sørensen, Restriction endonuclease fingerprinting of chromosomal DNA of *Neisseria gonorrhoeae. Acta Pathol. Microbiol. Immunol. Scand. (B) 92*: 271–278 (1984).
64. Ison, C. A., C. M. Bellinger, and A. A. Glynn, Plasmids in throat and genital isolates of meningococci. *J. Clin. Pathol. 37*:1123–1128 (1984).
65. Kalaimathee, K. K., C.-L., Koh, and Y.-F. Ngeow, Plasmid profiles of *Neisseria gonorrhoeae* in peninsular Malaysia. *Microbiol. Immunol. 29*: 921–926 (1985).
66. Kristiansen, B.-E., B. Sørensen, B. Bjorvatn et al., An outbreak of group B meningococcal disease: Tracing the causative strain of *Neisseria meningitidis* by DNA fingerprinting. *J. Clin. Microbiol. 23*:764–767 (1986).
67. Meissner, P. S., and J. O. Falkinham III, Plasmid DNA profiles as epidemiologic markers for clinical and environmental isolates of *Mycobacterium avium, Mycobacterium intracellulare*, and *Mycobacterium scrofulaceum. J. Infect. Dis. 153*:325–331 (1986).
68. Daly, J. A., *Haemophilus ducreyi. Infect. Control 6*:203–205 (1985).
69. Handsfield, H. H., P. A. Totten, C. L. Fennel, S. Falkow, and K. K. Holmes, Molecular epidemiology of *Haemophilus ducreyi* infections. *Ann. Intern. Med. 95*:315–318 (1981).
70. Brunton, J., M. Meier, N. Ehrman, I. MacLean, L. Slaney, and W. L. Albritton, Molecular epidemiology of beta-lactamase-specifying plasmids of *Haemophilus ducreyi. Antimicrob. Agents Chemother. 21*:857–863 (1982).
71. Anderson, B., W. L. Albritton, J. Biddle, and S. R. Johnson, Common β-lactamase-specifying plasmid in *Haemophilus ducreyi* and *Neisseria gonorrhoeae. J. Clin. Microbiol. 25*:296–297 (1984).

72. Wachsmuth, K., Ø. Olsvik, and W. Cook. Genetic approaches to the diagnosis of enteropathogenic bacterial infections. In *Infectious Diarrhoea in the Young. Strategies for Control in Humans and Animals*. S. Tzipori, Barnes, G., Bishop, R., Holmes, I., and Robins-Browne, R. (eds.). Elsevier Science Publishers, Amsterdam, pp. 337–348 (1985).
73. Ørskov, F., and I. Ørskov, Summary of a workshop on the clone except in the epidemiology, taxonomy, and evolution of the *Enterobacteriaceae* and other bacteria. *J. Infect. Dis. 148*:346–357 (1983).
74. Echeverria, P., J. Seriwatana, D. N. Taylor, C. Tirapat, W. Chaicumpa, and B. Rowe, Identification by DNA hybridization of enterotoxigenic *Escherichia coli* in a longitudinal study of villages in Thailand. *J. Infect. Dis. 151*:124–130 (1985).
75. Echeverria, P., J. Seriwatana, O. Sethabutr, and D. N. Taylor, DNA hybridization in the diagnosis of bacterial diarrhea. *Clin. Lab. Med. 5*:447–462 (1985).

6

Whole-Chromosomal DNA Probes for Rapid Identification of *Mycobacterium tuberculosis* and *Mycobacterium avium* Complex

MARILYN C. ROBERTS and MARIE B. COYLE
University of Washington, Seattle, Washington

I. INTRODUCTION

During the 35 years since isoniazid therapy became available, the United States has experienced a steady decline in the incidence of tuberculosis. However the number of reported cases has yet to fall below 21,000/year (1). Developing countries have had a much slower decline in the incidence of tuberculosis. These countries have an estimated risk of tuberculosis 20-50 times higher than that in developed countries. Although estimates of the worldwide incidence of tuberculosis varies considerably, it appears that approximately 15 million individuals have active disease and over 2 million die of this disease each year. The influx of Southeast Asian refugees between 1978 and 1981 accounted for a large number of the new cases of tuberculosis in the United States and slowed the downward trend of tuberculosis statistics during that period. The Centers for Disease Control recently reported that the expected decline in cases was not realized in 1985. In Florida, where a large proportion of the patients with acquired immunodeficiency syndrome (AIDS) are Haitian, 10% of these patients had tuberculosis at the time AIDS was diagnosed or had had it within 18 months preceding the AIDS diagnosis. There was also a correlation in New York City between the presence of AIDS and a history of tuberculosis (1). As a result of these statistics, the current slowing of the decline in tuberculosis cases is thought to be due to the rise in AIDS cases during this period (1).

II. MYCOBACTERIAL DISEASES

There are 54 recognized species in the genus *Mycobacterium* (2-4). Only about 10 are pathogenic or opportunistic for man or animals, the rest are nonpathogenic saprophytic organisms that can be found as contaminants in patient specimens but are thought not to cause disease (4,5). Most pathogenic species grow very slowly, and isolation and identification from clinical specimens can take over 12 weeks (2). The diagnosis of mycobacterial disease can be confirmed quickly in those specimens that have acid-fast organisms on the initial smear, but the species identification requires additional time to allow for sufficient growth to yield reliable results from conventional biochemical tests (2).

Mycobacterium tuberculosis is the most common cause of mycobacterial disease in the United States. The recommended drug regimens for *M. tuberculosis* infections are not harmful for those patients whose disease is caused by mycobacteria other than tubercle (MOTT) bacilli, therefore, all acid-fast positive smears are presumed to be *M. tuberculosis* until proven otherwise. *Mycobacterium tuberculosis* is highly pathogenic and tuberculosis requires isolation of the patient, whereas diseases due to MOTT bacilli do not require isolation because of the absence of any evidence for transmission from person to person (3,6). The source of infection with MOTT bacilli is thought to be environmental, whereas *M. tuberculosis* and *M. leprae* have not been isolated from the environment (6).

The incidence of clinically significant infections with *M. avium* complex has markedly increased in recent years, particularly with the advent of AIDS patients who often acquire disseminated mycobacterial infections that are caused by *M. avium* complex (7-11). The disease is usually disseminated and there is evidence to suggest that the infection occurs late in the course of the AIDS (12-14). Over 80% of the isolates taken from AIDS patients have been identified in the Centers for Disease Control as *M. avium* complex as compared with 9% identified as *M. tuberculosis* (1). In addition, a recent report by du Moulin et al. (15) showed that a fivefold increase in disease caused by the *M. avium* complex has occurred in the general population of Massachusetts between 1972 through 1983. They suggested that disease due to MOTT bacilli may increase in the general population as the number of cases of *M. tuberculosis* declines. Similar data has been published from British Columbia where they have documented a sixfold increase in infections with MOTT bacilli between 1960 and 1981 and a steady decrease in *M. tuberculosis* infections (16). Published reports of the incidence of mycobacterial pulmonary disease due to MOTT bacilli in the United States ranged from 0.5 to 30% with *M. avium* complex organisms predominating (17-22). Pulmonary disease caused by MOTT bacilli is indistinguishable clinically, radiologically, and histologically from tuberculosis (21-23). Species of MOTT bacilli can infect more than one site, and even

M. gordonae, which is usually classified as a saprophyte, has occasionally been implicated in infections of the respiratory tract, bursa, prosthetic aortic valve, and disseminated disease (4,5).

III. MYCOBACTERIAL IDENTIFICATION

The introduction of the BACTEC radiometric culture system for mycobacteria has reduced the time of detection from patient specimens (24-26). The BACTEC detects growth of mycobacteria by release of radioactive CO_2 during metabolism of [^{14}C] palmitic acid in a broth medium. The Haroborview Mycobacterial Laboratory has found that the BACTEC system has reduced the time of detecting *M. tuberculosis* from smear-positive respiratory specimens and smear-negative nonrespiratory specimens by 50% when compared with the conventional media. For smear-negative respiratory specimens, the BACTEC system provides only a minimal improvement when compared with the conventional media (27). A more striking improvement in time of detection was seen with smear-positive specimens containing the *M. avium* complex. The mean detection time was 12 days by BACTEC and 32-38 days with conventional media (27). The BACTEC system includes a differential growth test that distinguishes *M. tuberculosis* from the MOTT bacilli. A strain is identified as a member of the *M. tuberculosis* complex by their failure to grow in BACTEC medium containing *p*-nitroacetylamino-β-hydroxypropiophenone (NAP). All MOTT bacilli are resistant to this compound (2,24). Species identification of MOTT bacilli requires subculturing onto a solid medium before characterization by conventional biochemical methods.

The *M. avium* complex includes *M. avium* and *M. intracellulare*, which are indistinguishable by biochemical tests. Some authors have suggested that *M. scrofulaceum* should be classified as a member of the *M. avium* complex and have coined the term MAIS complex for strains within the *M. avium-intracellulare-scrofulaceum* group. This proposal was based on the similarity of biochemical reactions, surface antigens, drug-resistance patterns, and variable pigmentation (2,4,5). The species *M. avium* and *M. intracellulare* were initially distinguished on the basis of chick virulence tests and later by serotyping in which the *M. avium* species comprised serotypes 1, 2, and 3, whereas *M. intracellulare* included serotypes 4 through 28. More recently, DNA homology analysis has been used to look at the genetic relationship of the various species within the genus. Baess (28) found that the MAIS group was composed of three DNA homology groups. The first group included the three serotypes of *M. avium* and some of the virulent serotypes of *M. intracellulare*. The second group comprised avirulent and virulent serotypes of *M. intracellulare*, whereas the third group included *M. scrofulaceum* and the virulent serotype 9 of *M. intracellulare*. The DNA homologies of strains within groups ranged between 71 and 100%, which is the normal range for strains within a species. Homology among the three groups ranged from 33 to 56%. These data suggest that more work is needed to sort out the taxonomic status and criteria for the three groups. Members of each DNA homology group can be associated with disease in man. For the remainder of the paper, these three species will be placed together under the collective term of MAIS complex. These three groups share <25% DNA homology with *M. tuberculosis* (29).

The *M. tuberculosis* complex includes *M. tuberculosis, M. bovis, M. africanum*, and *M. microti* (30,31). DNA studies show that these four species share 85–100% of their sequences in common (31). Similarly, serological tests and structural homologies of proteins suggest that all four could be classified as a single species. The *M. avium* complex strains are generally resistant to the drugs normally prescribed for *M. tuberculosis* infections. The treatment for disseminated infections with *M. avium* complex frequently includes experimental drugs that can be toxic and are not provided to the physician unless the mycobacterial isolate has been identified as *M. avium* complex (32). In Seattle, between 1983 and 1984 there were 39 confirmed or presumed cases of AIDS. Specimens from 19 of these patients were submitted for culture and 13 (71%) were positive for mycobacteria. Eleven (58%) were positive for *M. avium* complex and two (13%) for *M. tuberculosis*. Similar experiences have been reported in other parts of the country (15). It has also been noted that Haitians with AIDS are more likely to have *M. tuberculosis* than *M. avium* (33). Consequently, the need for rapid identification of these mycobacteria has been intensified by their role in opportunistic but treatable infections in AIDS patients.

We decided to examine the potential of DNA probes in a dot-blot assay for rapid identification of mycobacteria. We chose to use whole-chromosomal DNA probes instead of cloned sequences because very little is known about the genetic organization of these bacteria. Furthermore, probes containing cloned single-copy genes for detection of bacteria in dot-blot assays are less sensitive (as much as 100-fold) when compared with multicopy cloned gene probes (34, 35). We have used the dot-blot assay and whole-chromosomal DNA probes for identification of strains in the anaerobic genus *Mobiluncus* (36). We found that organisms removed from agar plates and grown in broth, as well as direct patient specimens could each be used as the source of the bacteria in the dot-blot assay (36,37). The dot-blot test employing purified DNA has been used to compare *M. leprae* with other members of the genus (38). As a result of the *M. leprae* report and our experience with other organisms, we started with pure DNA and progressed to whole-cell lysates as the source of the DNA.

A. Methods

In 1985, 172 mycobacterial strains were received and tested in the Mycobacteria Laboratory at Seattle's Harborview Medical Center. Fifty-nine percent were *M.*

tuberculosis, 26% *M. avium* complex, 10% *M. gordonae*, and the rest were other species of *Mycobacterium*. The probe strains listed in Table 1 included an attenuated *M. bovis* GCG strain (ATCC 27291) that has been shown to be 100% related to *M. tuberculosis* (29). The probe for the *M. avium* complex was prepared from three strains representing one strain from each of the three DNA homology groups; *M. avium* serotype 1 (ATCC 15769), *M. intracellulare* serotype 14 (ATCC 25169), and *M. scrofulaceum* serotype 9 (ATCC 19073) (28). The fourth probe was *M. gordonae* (ATCC 14470), the third most common species isolated in 1985 and the most common nonpathogen isolated in the laboratory (see Table 1). All strains were purchased from The American Type Culture Collection and were confirmed biochemically in the laboratory (39).

The probe strains were grown in 1 L of Middlebrook 7H11 broth containing 0.05% Tween 80 and 5% albumin-glucose-catalase enrichment. They were grown at 36°C. When the cultures were in log phase (turbidity equal to a 0.5 McFarland standard) D-cycloserine (500 μg/ml) and ampicillin (1 mg/ml) were added to enhance their susceptibility to lysis (40). Cultures were incubated for an additional 16-24 hr, harvested by centrifugation, and lysed with the method of Spiegel and Roberts (41). The ability of ethanol to inactivate mycobacteria was determined with the five ATCC reference strains and four clinical *M. tuberculosis* isolates. Bacterial pellets from 50 ml of a broth culture grown to a 0.5 McFarland standard were mixed with 1.0 ml of 70% ethanol. At 7, 15, 30, 45, and 60 min, 0.1 ml was plated on Middlebrook 7H11 medium and was added to BACTEC vials without antibiotics. The plates and vials were held for 8 and 6 weeks, respectively, to assure that no growth would occur. All nine strains tested were unable to grow after 15 min of ethanol treatment. The pellets were usually dried to remove the ethanol before the cells were lysed.

Purified DNA was labeled with ^{32}P by the nick translation procedure of Maniatis et al. (42). DNA (0.1 μg) was labeled using [^{32}P] adenosine-5′-triphosphate and [^{32}P] thymidine-5′-triphosphate. The three DNAs for the labeled *M.*

Table 1 DNA Probes Used for Dot Blot Assay

Positive DNA probe hybridization	Identified as
M. bovis	*M. tuberculosis*
M. gordonae	*M. gordonae*
M. avium	
M. intracellulare	MAIS complex
M. scrofulaceum	
No reaction	Other *Mycobacterium* sp.

avium complex probe were prepared, either individually and then combined, or the DNAs were mixed together and then labeled. Both methods gave identical results in dot-blot hybridizations. In general, the latter method was used because of convenience. These organisms have a high G+C content and as a result hybridization of the dot blots was carried out in 50% formamide at 50°C overnight and washed at 60°C for 30 min. The higher temperatures were required to prevent background interference. Under these conditions, the *M. bovis* probe reacted with only *M. bovis* and *M. tuberculosis*, the *M. gordonae* probe with *M. gordonae* and the MAIS probe with all isolates identified as *M. avium, M. intracellulare*, or *M. scrofulaceum*. No cross-hybridization was seen, even when the x-ray films were incubated for 2 weeks before development (normal incubation time for pure DNA was 16 hr). When we were using purified DNA we could routinely detect 100 pg, which corresponds roughly to 10^4 bacteria, although the sensitivity ranged from 100 down to 10 pg depending on the experiment.

Initially, dot blots were prepared with purified DNA samples; however, once we had established that the probes were specific for each group we used cell suspensions of broth cultures grown to log phase and treated with D-cycloserine and ampicillin before harvesting. Because of the infectious nature of dried mycobacteria, the cell pellets were resuspended in 70% ethanol and incubated 15 min, to inactivate the cultures before spotting and lysing. The ethanol was either diluted 1:2 in water or removed by drying before the cells were spotted and lysed. The ethanol did not interfere with cell lysis. Lysing consisted of placing dried nitrocellulose filters into a solution 0.5 M NaOH and 1.5 M NaCl at 60°C for 30 sec followed by incubation at room temperature in the same solution for 10 min. The samples were then treated with two changes of 1.0 M Tris-HCl (pH 7.5) followed by 1.0 M Tris-HCl plus 1.5 M NaCl, extracted twice with chloroform and treated with 1.0 M Tris-HCl plus 1.5 M NaCl and baked in a vacuum oven at 80°C for 2 hr (36,37). Antibiotic-treated cells lysed more completely than untreated cells as judged by the intensity of the hybridization and the extent to which we could dilute the specimen and still detect a reaction. However, the untreated cells provided a strong signal with their homologous probe. Therefore, we examined the sensitivity of the probes by using strains grown on agar plates without the addition of antibiotics.

B. Results

Sixty-one clinical strains were grown on plates, spotted, lysed, and probed with each of the three probes (Table 2). Fifty-seven of the 61 (93%) strains were correctly identified as *M. tuberculosis*, MAIS complex, *M. gordonae*, or other mycobacteria. Three *M. tuberculosis* strains and one *M. avium* strain did not hybridize to any of the three probes. These strains were regrown in Middlebrook

Table 2 Effectiveness of DNA Probes for Identification of *Mycobacterium* spp. Grown in Middlebrook 7H11 Agar and Tested with DNA Probes

Clinical strains	Reaction with DNA probe to *M. bovis*	MAIS complex	*M. gordonae*
M. tuberculosis $n = 33$	30 (3)[a]	0	0
M. avium $n = 15$	0	14 (1)	0
M. scrofulaceum $n = 6$	0	6	0
M. gordonae $n = 6$	0	0	6
M. kansasii $n = 1$	0	0	0

[a]The number of misidentified isolates is shown in parenthesis; the isolates were correctly identified when they were grown in broth with antibiotics.

7H11 broth, treated with antibiotics, and retested. All four isolates were correctly identified (see Table 2). Common respiratory isolates including *Candida albicans, Staphylococcus aureus, S. epidermidis, Streptococcus viridans, S. pneumoniae, Proteus mirabilis, Lactobacillus sp. Enterobacter aerogenes, Escherichia coli, Klebsiella sp.*, and *Pseudomonas aeruginosa*, were tested against the three probes. These common contaminates exhibited no cross-reaction with the probes and none was observed even after extended (2 weeks) incubation of the x-ray film.

We were interested in determining whether or not positive BACTEC cultures could be used as the source of the bacteria because this would greatly reduce the time required for identification. Initially, ATCC strains were inoculated into BACTEC bottles and grown to a growth index of 100. The cultures were pelleted by centrifugation, and a 5-μl sample was spotted and read in a scintillation counter. The ^{14}C level associated with the pellets was very high (1000–20,000 counts/5 μl). This amount of radiolabel was able to expose the x-ray film and, thus, interfere with detection of the ^{32}P-radiolabeled probes used in the dot-blot assay. However, the type of spot made by the ^{14}C was different in appearance from that made by ^{32}P; the ^{14}C spot could be blocked or greatly reduced, depending on the total radioactivity, by placing tape over the area. Washing the cells reduced the radioactivity associated with the pellet at least 10- to 100-fold. Thus, the BACTEC cultures for dot blot tests were routinely washed three times before inactivation in ethanol. Initial experiments with the homologous probe and BACTEC-grown strains were successful. At this point we began screening positive BACTEC cultures as they became available from the clinical laboratory.

We tested 63 BACTEC-positive cultures by use of this approach and were able to correctly identify 70% of the *M. tuberculosis* strains and 56% of the MAIS complex strains with the DNA probes (Table 3). Fourteen of the BACTEC cultures were incubated with antibiotics before processing. When these were examined separately, 13 (93%) were correctly identified. The only strain missed was an *M. gordonae* isolate that did not hybridize with any of the probes and was identified as another species. The DNA probe test from BACTEC cultures takes a maximum of 2 weeks to complete because BACTEC detects relatively low numbers of organisms. The DNA probe provided a substantial improvement over the conventional identification by biochemical analysis, which required an average of 6 weeks for *M. avium* complex isolates.

We found that whole-chromosomal DNA probes could be used to correctly identify pure cultures of mycobacteria. The assay correctly identified 93% of the clinical strains grown on a solid medium. Strains that failed to hybridize under these conditions were correctly identified when they were grown in broth that contained antibiotics to enhance lysis. The DNA probe assay of cells from colonies or 50-ml cultures required a 4- to 17-hr exposure of the film. The entire test required 36–48 hr after the cells were harvested, including the preparation time.

The overall identification of BACTEC specimens with the DNA probes was 59%. We found that by modifying BACTEC growth conditions we could increase the sensitivity of the test to 93%. Further studies have suggested that other

Table 3 Effectiveness of DNA Probes for the Identification of *Mycobacterium* spp. Grown in BACTEC Broth and Tested with DNA Probes

	Probe		
Clinical strains	*M. bovis*	*M. avium*	*M. gordonae*
M. tuberculosis $n = 40$ (6)[a]	28 (6)	0	0
M. avium complex $n = 21$ (6)	0	13 (6)	0
M. gordonae $n = 1$ (1)	0	0	0
Other species $n = 7$ (1)	0	0	0

[a]The number of cultures treated with ampicillin and D-cycloserine before processing is shown in parenthesis.

changes in the growth conditions may improve the sensitivity of the test above the 93% obtained with antibiotic treatment of cells.

C. Discussion

In January 1987, the FDA approved the release of three ^{125}I DNA probes that are complementary to the species-specific regions of the ribosomal RNA (rRNA) of *M. tuberculosis, M. avium*, and *M. intracellulare*. These probes, which were developed by Gen-Probe (San Diego, California), should be very sensitive for detection of low numbers of organisms because there are roughly 10,000 copies of the rRNA per cell.

All three tests are done by preparation of standardized bacterial cell suspension, lysis of the cells by sonication, incubation of the cells with the specific probe for 1 hr in a 72°C water bath, removal of the unhybridized probe, and counting the pellets in a gamma counter. At this time, there has been only a few evaluations of the Gen-Probe system (43-44). Drake et al. (43) tested 134 strains representing the MAIS complex and 66 other species of mycobacteria and found that the Gen-Probe system identified all of the 134 MAIS complex strains and none of the 66 other species. When Kiehn et al. (44) examined 76 positive mycobacterial cultures grown on solid medium, all of the 33 MAIS complex strains reacted with the Gen-Probe DNA probes. Only 19 of 23 strains grown in BACTEC broth yielded positive reactions; however, when these four strains were grown on solid medium they were correctly identified by the probes.

The Centers for Disease Control evaluated the Gen-Probe kit for the identification of 240 stock strains representing the *M. tuberculosis* complex (45). They found one strain, which by conventional biochemical reactions was not believed to be a member of the *M. tuberculosis* complex, that hybridized with the *M. tuberculosis*, whereas all the other strains had a 100% correlation between biochemical tests and DNA probe.

The Mycobacterial Laboratory at Harborview, in Seattle, found that the Gen-Probe probes can be used with BACTEC-grown cultures if the auramine-stained BACTEC-broth contains >200 acid-fast bacilli per low-power field. Unfortunately, this does not correlate with the Growth Index. Cultures that contain fewer organisms have a high probability of yielding false-negative results because of an insufficient biomass. As with the whole-chromosomal DNA probes, considerable time is saved when BACTEC-grown cultures are used for testing with the Gen-Probe kit, as compared with using cultures grown on solid medium or doing conventional biochemical tests. However, we have encountered a number of isolates with an unusual tiny colonial morphology that biochemically resemble the MAIS complex but are not recognized by any of the Gen-Probe DNA probes. It is not yet known if these discrepancies are due to the inadequacies of the biochemical tests or the probes. One disadvantage of the Gen-Probe kit is the short half-life of the ^{125}I-label. We estimate that, in our laboratory, the use of the kit increases the cost of identification of the positive mycobacterial cultures by 50% if the Gen-Probe kit is used on a twice-weekly schedule.

IV. CONCLUSION

DNA probes hold great promise for the future of rapid identification of *Mycobacterium* species in the clinical laboratory. However, improvement in the shelf-life and use of nonradioactive labels would greatly expand the usefulness of this technology.

ACKNOWLEDGMENT

This work was supported by a grant from the University of Washington Graduate School. We gratefully acknowledge the assistance of the technologists from Harborview Medical Center and the University of Washington.

REFERENCES

1. *Morbidity and Mortality Weekly Report*, Tuberculosis–United States, 1985–and the possible impact of human T-lymphotropic virus type III/ lymphadenopathy-associated virus infection. *35*:74–76 (1986).
2. Tsukamura, M., A review of the methods of identification and differentiation of mycobacteria. *Rev. Infect. Dis. 3*:841–861 (1981).
3. Wayne, L. G., Mycobacterial taxonomy: A search for discontinuities. *Ann. Microbiol. 129A*:13–27 (1978).
4. Good, R. C., Opportunistic pathogens in the genus *Mycobacterium*. *Annu. Rev. Microbiol. 39*:347–369 (1985).
5. Wayne, L. G., The "atypical" mycobacteria: Recognition and disease association. *CRC Crit. Rev. Microbiol. 12*:185–222 (1985).
6. Brooks, W. B., K. L. George, B. C. Parker, J. O. Falkinham III, and H. Gruft, Recovery and survival of nontuberculosis mycobacteria under various growth and decontamination conditions. *Can. J. Microbiol. 30*:112–117 (1984).
7. Seshi, B., Two cases of AIDS with florid *Mycobacterium avium-intracellulare* infection in the T-cell areas of the spleen. *Hum. Pathol. 16*:964–965 (1985).
8. Berlin, O. B., P. Zakowski, D. A. Bruckner, M. N. Clancy, and B. L. Johnson, Jr., *Mycobacterium avium*: A pathogen of patients with acquired immunodeficiency syndrome. *Diagn. Microbiol. Infect. Dis. 2*:213–218 (1984).
9. Damsaker, B., and E. J. Bottone, *Mycobacterium avium–Mycobacterium intracellulare* from the intestinal tracts of patients with the acquired immunodeficiency syndrome: Concepts regarding acquisition and pathogenesis. *J. Infect. Dis. 151*:179–181 (1985).

10. Elliot, J. L., W. L. Hoppes, M. S. Platt, J. G. Thomas, I. P. Patel, and A. Gansar, The acquired immunodeficiency syndrome and *Mycobacterium avium-intracellulare* bacterium in a patient with hemophilia. *Ann. Intern. Med. 98*:290–293 (1983).
11. Offenstadt, G., P. Pinta, P. Hericord, M. Jagueux, F. Jean, and P. Amstutz, Multiple opportunistic infection due to AIDS in a previously healthy black woman from Zaire. *N. Engl. J. Med. 308*:775 (1983).
12. Macher, A. M., J. A. Kovacs, V. Gill, G. D. Roberts, J. Ames, C. H. Park, S. Straus, H. C. Lane, J. E. Parrillo, A. S. Fauci, and H. Masur, Bacteremia due to *Mycobacterium avium-intracellulare* in the acquired immunodeficiency syndrome. *Ann. Intern. Med. 99*:782–785 (1983).
13. Zakowski, P., S. Fligiel, O. G. W. Berlin, and L. Johnson Jr., Disseminated *Mycobacterium avium-intracellulare* infections in homosexual men dying of acquired immunodeficiency. *J. Am. Med. Assoc. 248*:2980–2982 (1982).
14. Roth, R. I., R. L. Owen, D. F. Keren, and P. A. Volberding, Intestinal infection with *Mycobacterium avium* in acquired immune deficiency syndrome: Histological and clinical comparison with Whipple's disease. *Dig. Dis. Sci. 30*:497–504 (1985).
15. Du Moulin, G. C., I. H. Sherman, D. C. Hoaglin, and K. D. Stottmeier, *Mycobacterium avium* complex, an emerging pathogen in Massachusetts. *J. Clin. Microbiol. 22*:9–12 (1985).
16. Isaac-Renton, J. L., E. A. Allen, C. W. Chao, S. Grzybowski, E. I. Whittaker, and W. A. Black, Isolation and geographic distribution of *Mycobacterium* other than *M. tuberculosis* in British Columbia 1972–81. *Can. Med. Assoc. J. 133*:573–576 (1985).
17. Wiesenthal, A. M., K. E. Powell, J. Kopp, and J. W. Spindler, Increase in *Mycobacterium avium* complex isolations among patients admitted to a general hospital. *Publ. Health Rep. 97*:61–65 (1984).
18. Wolinsky, E., Nontuberculosis mycobacteria and associated diseases. *Am. Rev. Resp. Dis. 119*:107–159 (1979).
19. Meissner, G., and W. Anz, Sources of *Mycobacterium-avium* complex infection resulting in human diseases. *Am. Rev. Resp. Dis. 116*:1057–1064 (1977).
20. Grange, J. M., *Mycobacterium avium. Eur. J. Respir. Dis. 65*:399–401 (1984).
21. Rosenzweig, D. Y., Pulmonary mycobacterial infections due to *Mycobacterium intracellulare-avium* complex: Clinical features and course in 100 consecutive cases. *Chest. 75*:115–119 (1979).
22. Rosenzweig, D. Y., and D. P. Schlueter, Spectrum of clinical disease in pulmonary infection with *Mycobacterium avium-intracellulare. Rev. Infect. Dis. 3*:1046–1051 (1981).
23. Kiehn, T. E., F. F. Edwards, P. Brannon, A. Y. Tsang, M. Maio, J. W. M. Gold, E. Whimbey, B. Wong, J. K. McClatchy, and D. Armstrong, Infections caused by *Mycobacterium avium* complex in immunocompromised patients: Diagnosis by blood culture and fecal examination, antimicrobial susceptibility tests, and morphological and seroagglutination characteristics. *J. Clin. Microbiol. 21*:1168–173 (1985).

24. Gross, W. M., and J. E. Hawkins, Radiometric selective inhibition tests for differentiation of *Mycobacterium tuberculosis, Mycobacterium bovis*, and other mycobacteria. *J. Clin. Microbiol. 21*:565–568 (1985).
25. Roberts, G. D., N. L. Goodman, L. Heifets, N. W. Larsh, J. K. McClatchy, M. R. McGinnis, S. H. Iddigi, and P. Wright, Evaluation of the BACTEC radiometric method for recovery of mycobacteria and drug susceptibility testing of *Mycobacteria tuberculosis* from acid-fast smear-positive specimens. *J. Clin. Microbiol. 18*:689–696 (1983).
26. Morgan, M. A., C. D. Horstmeier, D. R. DeYoung, and G. D. Roberts, Comparison of a radiometric method (BACTEC) and conventional culture media for recovery of mycobacteria from smear-negative specimens. *J. Clin. Microbiol. 18*:384–388 (1983).
27. Kirihara, J. M., S. Hillier, and M. B. Coyle, Improved detection times of *Mycobacterium avium* complex and *M. tuberculosis* with the BACTEC radiometric system. *J. Clin. Microbiol. 22*:841–845 (1985).
28. Baess, I., Deoxyribonucleic acid relatedness between different serovars of *Mycobacterium avium, Mycobacterium intracellulare* and *Mycobacterium scrofulaceum. Acta Pathol. Microbiol. Immunol. Scand. Sect. B 91*:201–203 (1983).
29. Baess, I., Deoxyribonucleic acid relatedness among species of slowly-growing mycobacteria. *Acta Pathol. Microbiol. Scand. Sect. B 87*:221–226 (1979).
30. Collins, D. M., and G. W. De Lisle, DNA restriction endonuclease analysis of *Mycobacterium tuberculosis* and *Mycobacterium bovis* BCG. *J. Gen. Microbiol. 130*:1019–1021 (1984).
31. Imaeda, T., Deoxyribonucleic acid relatedness among selected strains of *Mycobacterium tuberculosis, Mycobacterium bovis, Mycobacterium bovis BCG, Mycobacterium microti*, and *Mycobacterium africanum. Int. J. Syst. Bacteriol. 35*:147–150 (1985).
32. Davidson, P. T., V. Khanijo, M. Goble, and T. S. Moulding, Treatment of disease due to *Mycobacterium intracellulare. Rev. Infect. Dis. 3*:1052–1059 (1981).
33. Pitchenik, A. E., C. Cole, B. W. Russell, M. A. Fischl, T. J. Spira, and D. E. Snider, Tuberculosis, atypical mycobacteriosis, and the acquired immune deficiency syndrome among Haitian and non-Haitian patients in South Florida. *Ann. Intern. Med. 101*:641–654 (1984).
34. Horn, J. E., T. Quinn, M. Hammer, L. Palmer, and S. Falkow, Use of nucleic acid probes for the detection of sexually transmitted infectious agents. *Diagn. Microbiol. Infect. Dis. 4*:101S–109S (1986).
35. McLafferty, M. A., D. R. Weiss, L. A. Sapiain, and E. L. Hewlett, Development of a DNA probe for identification of *Bordetella pertussis. Abstr. Annu. Meet. Am. Soc. Microbiol. 86*:56 (1986).
36. Roberts, M. C., S. L. Hillier, F. D. Schoenknecht, and K. K. Holmes, Nitrocellulose filter blots for species identification of *Mobiluncus curtisii* and *Mobiluncus mulieris. J. Clin. Microbiol. 20*:826–827 (1984).

37. Roberts, M. C., S. L. Hillier, F. D. Schoenknecht, and K. K. Holmes, Comparison of Gram stain, DNA probe, and culture for the identification of species of *Mobiluncus* in female genital specimens. *J. Infect. Dis. 152*:74-77 (1985).
38. Athwal, R. S., S. S. Deo, and T. Imaeda, Deoxyribonucleic acid relatedness among *Mycobacterium leprae, Mycobacterium lepraemurium* and selected bacteria by dot blot and spectrophotometric deoxyribonucleic acid hybridization assays. *Int. J. Syst. Bacteriol. 34*:371-375 (1984).
39. Sommers, H. M., and R. C. Good, *Mycobacterium*, In *Manual of Clinical Microbiology*, 4th ed., E. H. Lennette, Balows, A., Hausler, W. J., Jr., and Shadomy, H. J. (eds.). American Society for Microbiology, Washington, pp. 216-248 (1985).
40. Crawford, J. T., M. D. Cave, and J. H. Bates, Characterization of plasmids from strains of *Mycobacterium avium-intracellulare. Rev. Infect. Dis. 3*: 949-952 (1981).
41. Spiegel, C. A., and M. Roberts, *Mobiluncus* gen. nov., *Mobiluncus curtisii* subsp. *curtisii* sp. nov., *Mobiluncus curtisii* subsp. *holmesii* subsp. nov., and *Mobiluncus mulieris* sp. nov., curved rods from the human vagina. *Int. J. Syst. Bacteriol. 34*:177–184 (1984).
42. Maniatis, T., A. Jeffery, and D. G. Kleid, Nucleotide sequence of the rightward operator of phage lambda. *Proc. Natl. Acad. Sci. USA 72*:184–188 (1979).
43. Drake, T. A., J. A. Hindler, O. G. W. Berlin, and D. A. Bruckner. Rapid identification of *Mycobacterium avium* complex in culture using DNA probes. *J. Clin. Microbiol. 25*:1442-1445 (1987).
44. Kiehn, T. E., and F. F. Edwards, Rapid identification using a specific DNA probe of *Mycobacterium avium* complex from patients with acquired immunodeficiency syndrome. *J. Clin. Microbiol. 25*:1551-1552 (1987).
45. Silcox, V. A., M. M. Floyd, and D. L. Woodley, *Abstr. Annu. Meet. Am. Soc. Microbiol. 87*:133 (1987).

7

Restriction Endonuclease Analysis and Other Molecular Techniques in Identification and Classification of *Leprospira* and Other Pathogens of Veterinary Importance

ALEX B. THIERMANN* and RANCE B. LE FEBVRE†
National Animal Disease Center, Agricultural Research Service, U.S. Department of Agriculture, Ames, Iowa

I. INTRODUCTION

Leptospirosis is the most widespread contemporary zoonotic disease and one of the most important reproductive diseases in livestock worldwide. The disease is caused by many antigenically distinct, but morphologically similar, spirochetes of the genus *Leptospira*.

Leptospires, like many other bacteria, are classified according to serological methods. Currently, serotyping (1) is based on the comparison of surface antigens of isolates with those of reference strains. Live leptospires are reacted with homologous antisera, and with antisera specific for reference strains, by the microscopic agglutination test (MAT; 2). The reliability of this method is subject to nutritional and environmental influences on the culture being typed. In 1967, Kmety (3) proposed a system of antigenic factor analysis, which used specifically absorbed sera, to determine the taxonomy of leptospires. However, this method is complex, and results are difficult to reproduce in different laboratories. Recently, Terpstra and associates (4) have proposed the classification of leptospires within the Hebdomadis serogroup using a complex set of monoclonal antibodies. Although this method is more accurate than previous serotyping, it also depends on nutritional and environmental factors of the cultures.

Present affiliations:
*Agricultural Research Service, National Program Staff, U.S. Department of Agriculture, Beltsville, Maryland.
†Department of Veterinary Microbiology, University of California at Davis, Davis, California.

The use of biotechnology for a better understanding of leptospires at a genetic level is quite recent. Restriction endonuclease analysis (REA) has been used in classification studies. All field isolates submitted to our laboratory have been classified by REA. The DNA homology has been employed in phylogenetic and epidemiological studies, and specific DNA probes have been developed as sensitive diagnostic and taxonomic tools.

II. RESTRICTION ENDONUCLEASE ANALYSIS

The application of REA to the classification of leptospires was first proposed by Marshall et al. (5). Since then, after improvement of the DNA extraction procedure and the resolution of the restricted fragments, most of the reference strains belonging to 18 serogroups of *Leptospira interrogans* have been examined and classified by REA in our laboratory.

Chromosomal DNA was extracted from leptospiral cultures according to methods previously described (6,7). The REA was conducted by digesting 2 μg of purified DNA with 4-5 units of restriction enzymes, under conditions recommended by the manufacturer. The digested DNA preparations were then electrophoresed at 60 V for 16 hr in gels consisting of 0.7 or 0.5% agarose in tris-borate buffer. The gels were then stained with ethidium bromide and photographed under ultraviolet light (6). Over 20 restriction enzymes were screened for cleavage of leptospiral DNA. Complete digestions and best resolutions were obtained with *Eco*RI, *Hha*I, *Bgl*II, *Hind*III, and *Hpa*II. The REA studies were first concentrated on the taxonomic groups that include serovars important to the livestock industry.

A. Serogroup Pomona

The Pomona serogroup consists of seven serovars: *pomona, monjakov, kennewicki, mozdok, tropica, proechymis*, and *tsaratsova*. Serovars *pomona* and *kennewicki* are common parasites of domestic animals. On the basis of serotyping, these two are indistinguishable from each other and, thus, both are identified as serovar *pomona*. Because of the apparent similarity, serovar *kennewicki* has been recently eliminated from the Pomona serovar list. With REA, all seven serovars show distinct patterns; however, only subtle differences can be observed between serovars *tsaratsova* and *mozdok* (Fig. 1).

When examining North American isolates from swine, cattle, wildlife, and human origin serotyped as *pomona*, all showed an REA pattern that matched that of *kennewicki* and not *pomona* (Fig. 2). However, all *pomona* isolates from Northern Ireland gave a common REA pattern that matched that of the *pomona* reference strain (Fig. 2).

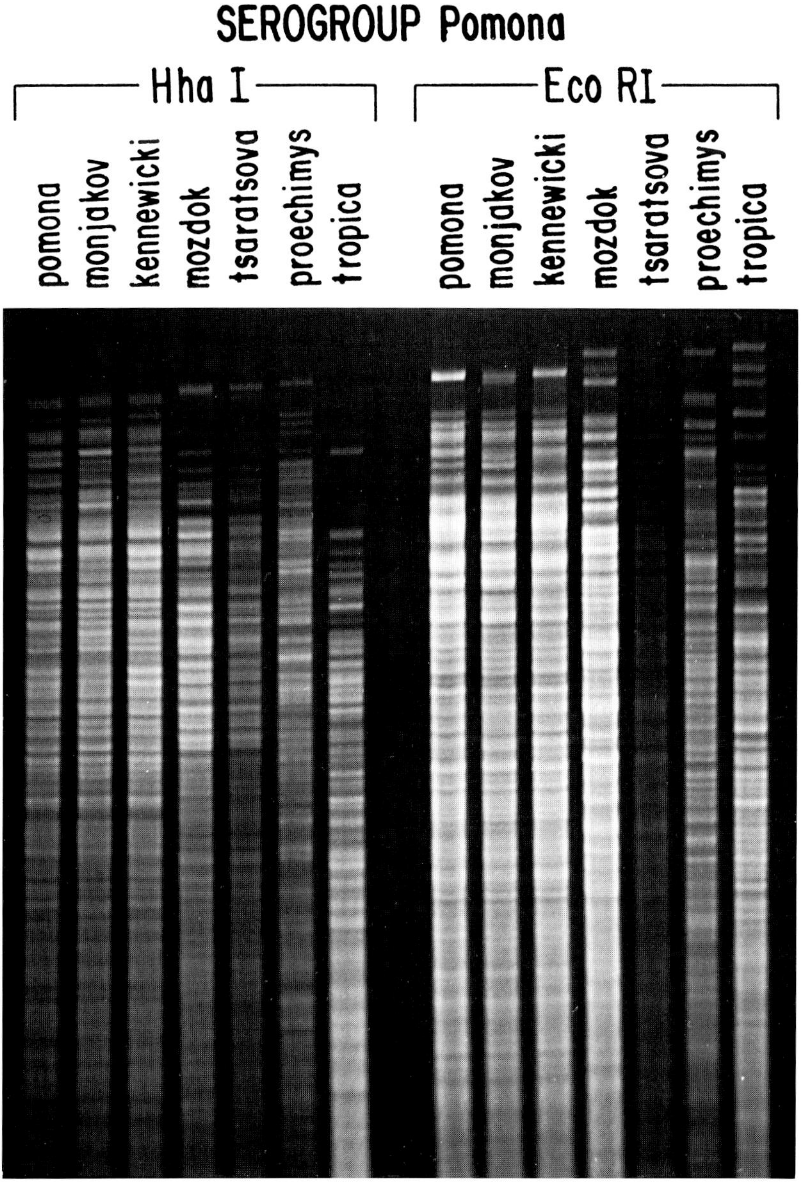

Figure 1 Electrophoretic patterns of chromosomal DNA from serovar reference strains in serogroup Pomona digested with enzymes *Hha*I and *Eco*RI (0.7% agarose gel).

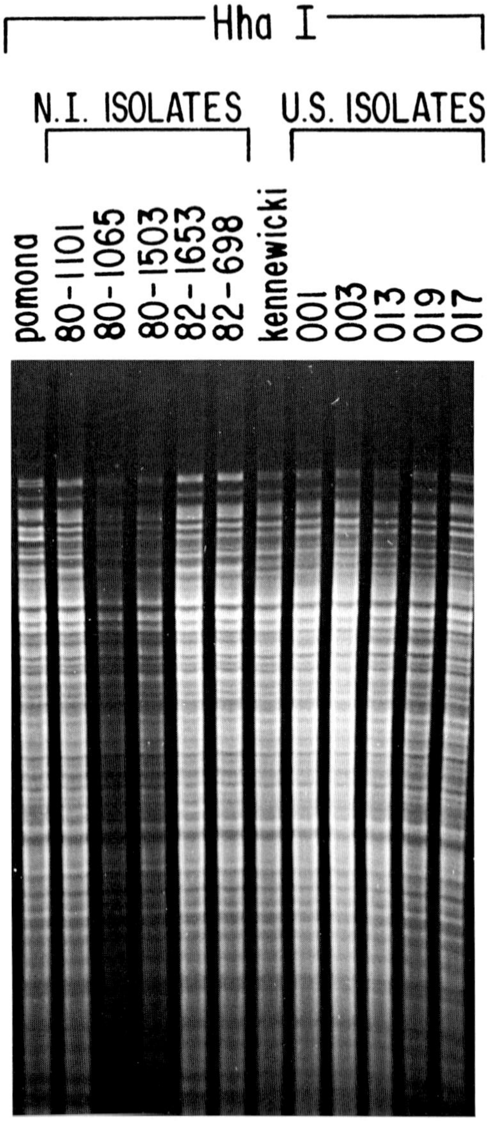

Figure 2 Electrophoretic patterns of chromosomal DNA from serovar reference strains *pomona* and *kennewicki*, and from Northern Ireland and United States isolates in serogroup Pomona digested with enzyme *Hha*I (0.7% agarose gel).

Although consistent, the differences in REA pattern between *pomona* and *kennewicki* when digested with *Eco*RI, *Hha*I, or *Bgl*II are subtle. However, when digested with enzymes *Hha*I and *Bgl*II simultaneously, the differences in pattern become more evident (Fig. 3). These REA pattern differences between North American *kennewicki* and Northern Ireland *pomona* isolates correlate with epizoologic findings. Serovar *pomona*, in Northern Ireland, infects primarily wildlife and causes reproductive disorders in horses (8). Conversely, in North America serovar *kennewicki* although capable of infecting cattle and wildlife, primarily infects swine, causing reproductive disorders and chronic shedding. These findings are quite critical because *kennewicki* has been omitted from the reference list as a result of its apparent serological identity with *pomona*. It would be interesting to examine North American horse isolates by REA to determine if serovar *pomona* is present, but no such isolates are available. Conversely, the absence of serological evidence of *pomona*-like endemic infections in swine in Northern Ireland and the rest of the United Kingdom, indicates that *kennewicki* is not present.

B. Serogroup Sejroe

The Sejroe group consists of 17 serovars: *balcanica, caribe, gorgas, haemolytica, hardjo, istrica, medanensis, nyanza, polonica, recreo, ricardi, roumanica* (iassy), *ruparupae, saxkoebing, sejroe, trinidad*, and *wolffi*. Whereas most reference strains had distinctive REA patterns when cleaved with restriction enzymes *Hha*I, *Bgl*II, or *Eco*RI (Fig. 4), some reference strains could not be differentiated in this manner. When enzymes *Hha*I, *Eco*RI, or *Bgl*II were used, three sets of identical REA patterns were observed among the following reference strains: (a) *wolffi* and *roumanica*, (b) *istrica* and *nyanza*, and (c) *polonica* and *sejroe* (Fig. 5). In light of conflicting serological results involving these organisms, these similarities were not unexpected.

The most crucial findings from REA analysis were observed when examining North American isolates serotyped as *balcanica* and *hardjo*. Serovar *hardjo* is the most important pathogen of cattle with a worldwide distribution, establishing permanent residence in cattle and possibly sheep, but it has no recognized wildlife carrier. It causes reproductive disorders and agalactia and is maintained in the cattle population through chronic carriers. Serovar *balcanica* was recently described in cattle in Florida (9).

When examined by REA, the North American *hardjo* and *balcanica* isolates had a common pattern that differed from that of either reference strain. Because this unique REA pattern could not be matched with that of any known reference strain, it was identified as genotype hardjo-bovis (10). This pattern showed some resemblance to that of *balcanica*, and was very different from that of the *hardjo* reference strain (hardjoprajitno; Fig. 4). Minor differences were

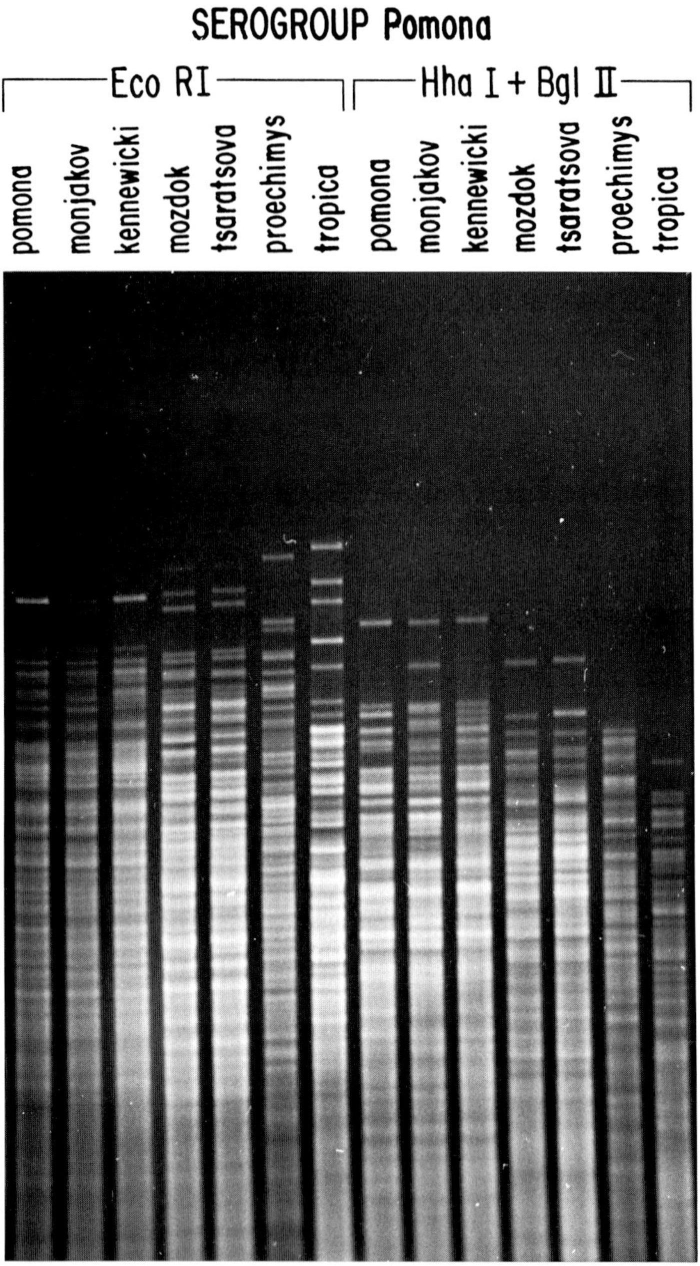

Figure 3 Electrophoretic patterns of chromosomal DNA from serovar reference strains in serogroup Pomona digested with enzymes *Eco*RI and, simultaneously, with *Hha*I and *Bgl*II (0.5% agarose gel).

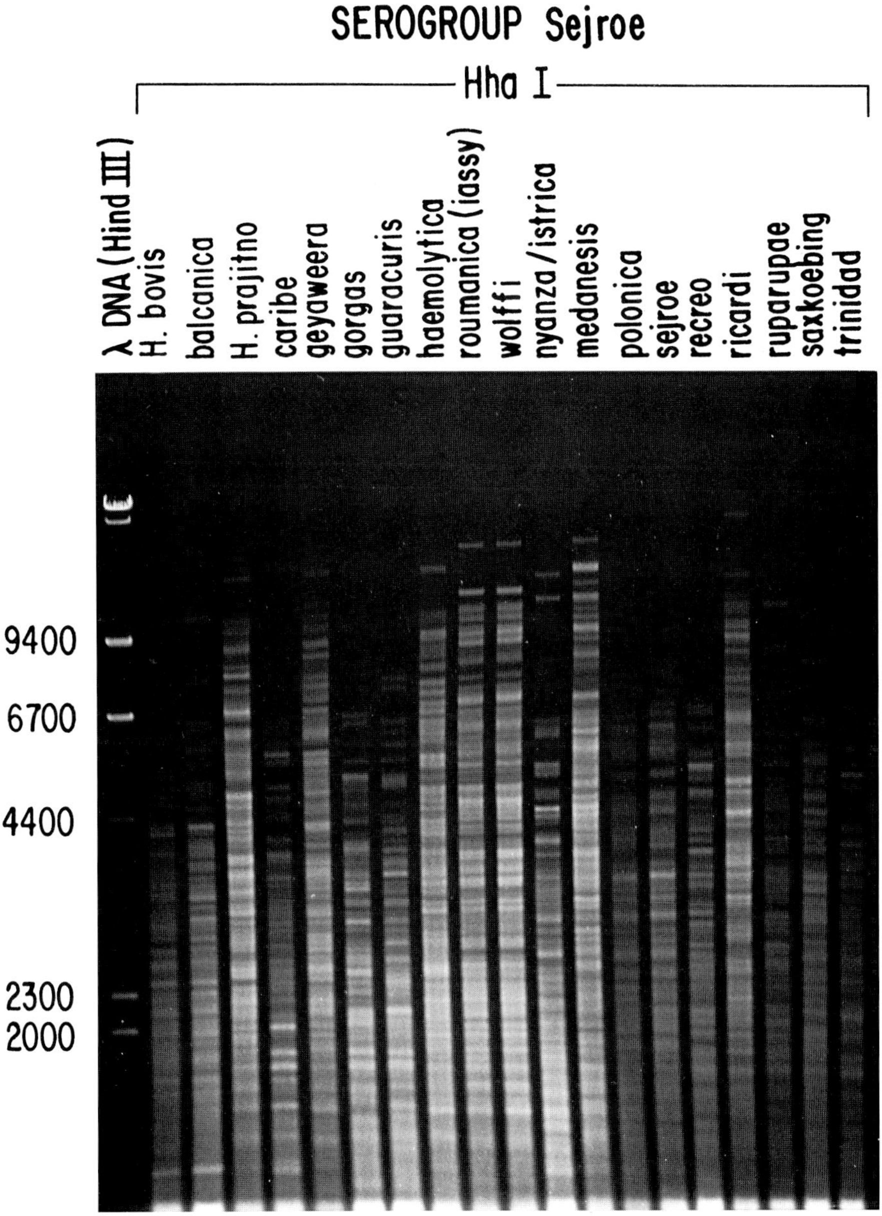

Figure 4 Electrophoretic patterns of chromosomal DNA from hardjo-bovis, and serovar reference strains in serogroup Sejroe digested with enzyme *Hha*I (0.5% agarose gel).

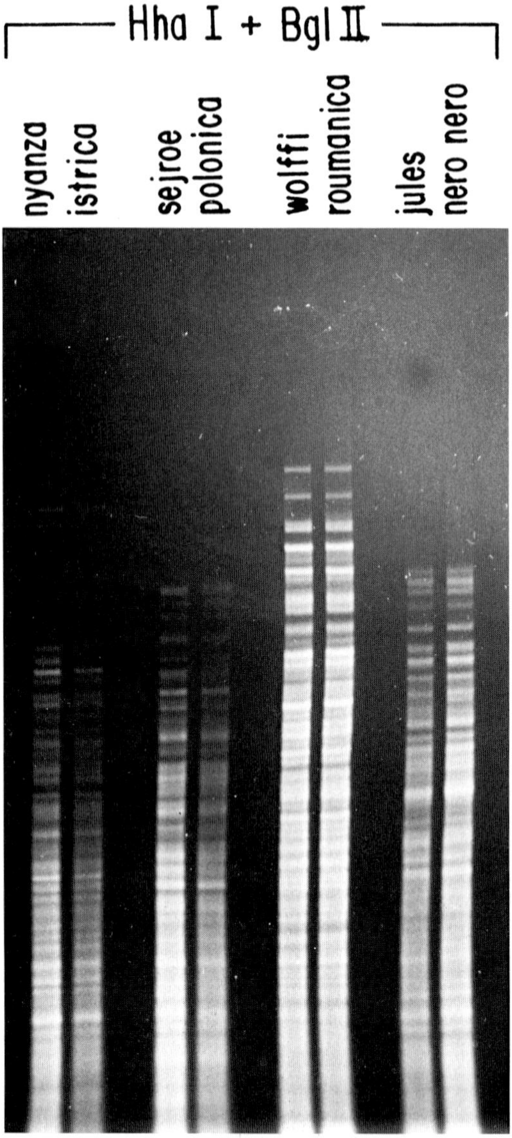

Figure 5 Electrophoretic patterns of chromosomal DNA from serovar reference strains *nyanza, istrica, sejroe, polonica, wolffi*, and *roumanica* in serogroup Sejroe, and serovar reference strains *jules* and *nero nero* in serogroup Hebdomadis, after simultaneous digestion with enzymes *Hha*I and *Bgl*II (0.5% agarose gel).

observed among some hardjo-bovis isolates when cleaved with enzyme *Hha*I. These differences did not correlate with the differences observed with serotyping. These subtypes were identified from the REA study and classified as A, B, and C (Fig. 6). Subtype A is the most common and is found throughout North America, Europe, and South America. Subtype B was found only among Florida *hardjo* and *balcanica* isolates, and subtype C is represented by only one isolate from Nebraska. The differences between these subtypes are limited to migration differences of one high-molecular-weight fragment and, therefore, could be due to single base-pair substitutions or deletions. Thus, although typed serologically as *hardjo* and *balcanica*, none of the North American bovine isolates genetically resembled these reference strains. The REA findings on what were typed as *balcanica* isolates are not unexpected. The presence of *balcanica* in Florida has been questioned on epizootiological grounds. Serovar *balcanica* is a common parasite of opossums in New Zealand and Australia and occasionally affects cattle (11), but it does not establish persistent infections in cattle. However, the Florida isolates serotyped as *balcanica* that have been described only in cattle, despite the presence of an abundant native opposum, in fact, appear to be hardjo-bovis by REA.

When *hardjo* isolates from the United Kingdom were examined by REA, two different patterns were found: one matched that of North American hardjo-bovis type A, and one matched that of *hardjo* reference strain (Fig. 7). There appear to be differences in the type of disease and degree of urinary shedding caused by these two *hardjo* types. More epidemiological studies are needed for a better understanding of these differences.

The recently described serotyping method using a battery of monoclonal antibodies (4) cannot differentiate between these two different types of *hardjo* because monoclonal antibodies specific for *hardjo* appear to be directed against those antigens shared by hardjoprajitno and hardjo-bovis. Thus, REA is the only method now available that is capable of differentiating the two *hardjo* types.

C. Serogroup Mini

The Mini group consists of six serovars: *beye, georgia, mini, perameles, szwajizak*, and *tabaquite*. The members of this group are found mostly in wildlife. This group gained recognition after the isolation of leptospires serotyped as *szwajizak* from cattle in Oregon (12). This resulted in pressure from the vaccine industry to include this new serovar in livestock vaccines.

When this group was examined by REA with enzymes *Eco*RI and *Hha*I, distinct REA patterns were observed for each of the six reference strains. The two Oregon isolates showed a pattern identical with that of reference strain *georgia*, and not *szwajizak*, as suggested by serotyping (Fig. 8).

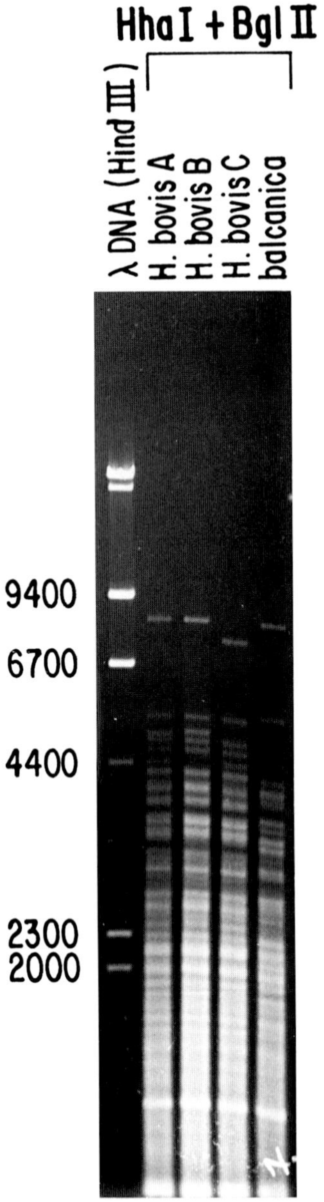

Figure 6 Electrophoretic patterns of chromosomal DNA from genotypes hardjobovis A, B, and C, and serovar *balcanica* in serogroup Sejroe, simultaneously digested with enzymes *Hha*I and *Bgl*II (0.5% agarose gel).

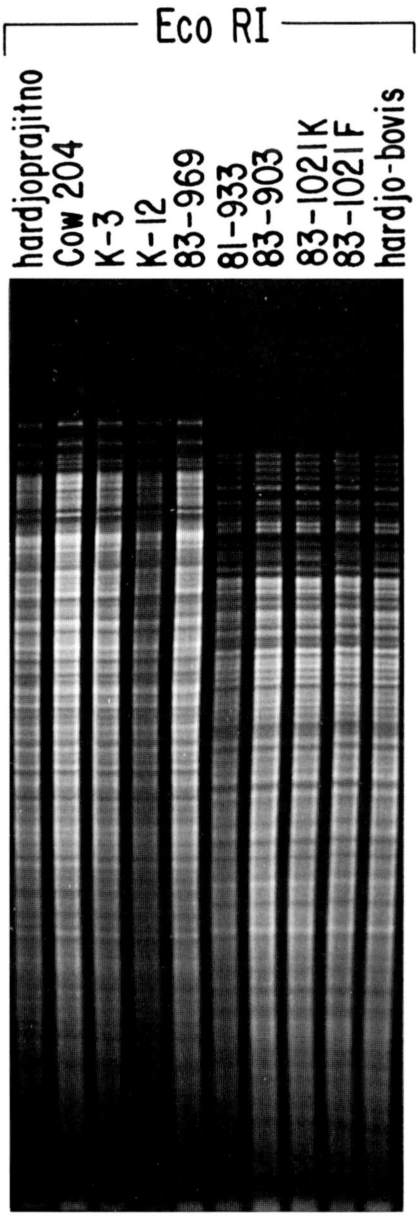

Figure 7 Electrophoretic patterns of chromosomal DNA from serovar reference strains *hardjo* (lane 1), Northern Ireland isolates (lanes 2–9), and hardjo-bovis (lane 10) in serogroup Sejroe digested with enzyme *Eco*RI (0.7% agarose gel).

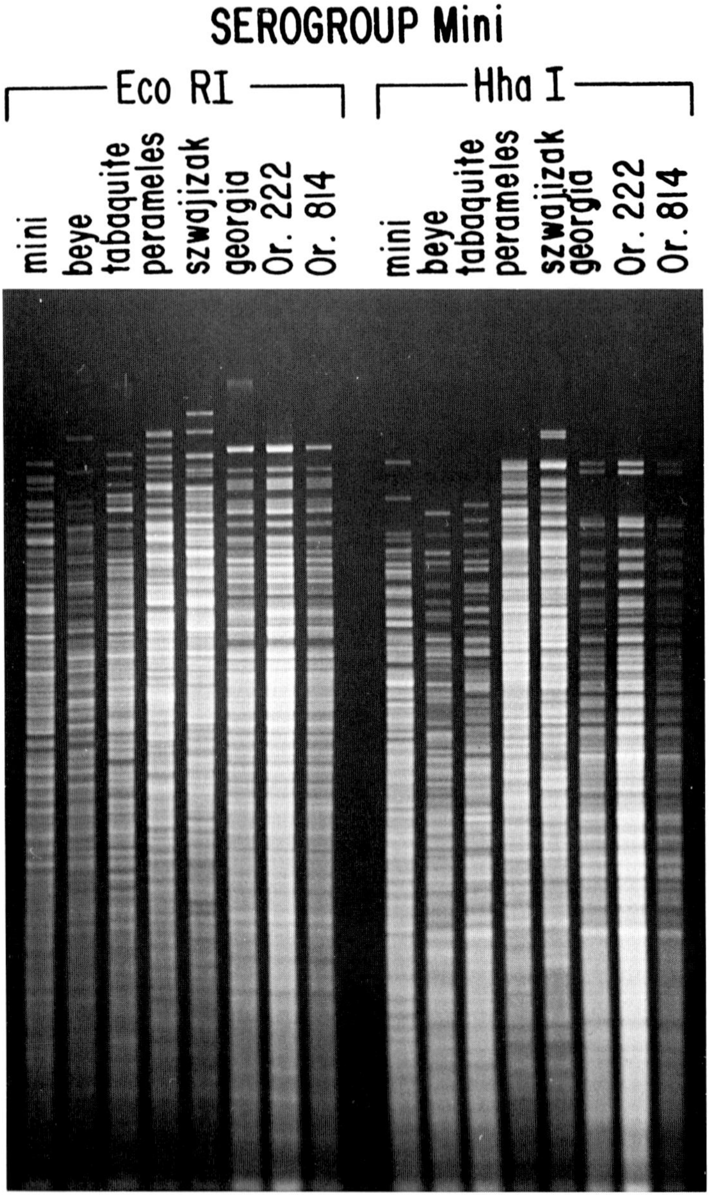

Figure 8 Electrophoretic patterns of chromosomal DNA from serovar reference strains in serogroup Mini, and Oregon isolates 222 and 814, digested with enzymes *Eco*RI and *Hha*I (0.7% agarose gel).

Again, on epizootiological grounds, the findings obtained by REA are no surprise. Of this serogroup, only serovar *georgia* is native to the United States (13). It is carried by raccoons, opossums, and skunks, which are commonly found in Oregon, and it may have been introduced into cattle by contact with infected wildlife, as happened with serovar *grippotyphosa* (14). Serovar *szwajizak* is known to be transmitted in Israel among rodents and, occasionally, cattle and humans.

D. Serogroup Australis

The Australis group consists of 13 serovars: *australis, bangkok, bratislava, fugis, hawaii, jalna, lora, muenchen, nicaragua, peruviana, pina, ramisi*, and *soteropolitana*. Some members of this group (*bratislava, lora*, and *muenchen*) have been implicated in reproductive disorders of swine and other domestic animals in Europe. Recently, two *bratislava* isolates were obtained from swine reproductive tracts in the United States (15). The final identification, by serotyping of isolates belonging to this group, has been difficult and controversial. In this group, the REA has also revealed characteristic patterns for each reference strain when the enzymes *Hha*I, *Eco*RI, and *Bgl*II are used (Fig. 9).

E. Serogroup Canicola

The Canicola group also consists of 13 serovars: *bafani, benjamin, bindjei, broomi, canicola, galtoni, jonsis, kamituga, kuwait, malaya, portlandvere, schueffneri*, and *sumneri*. This group is better known because of serovar *canicola*, which commonly infects dogs and, occasionally, other domestic species. Characteristic patterns were revealed by these serovars with REA. Serovars *canicola* and *portlandvere* exhibited similar patterns when digested with *Eco*RI (Fig. 10). However, a clear difference was observed between these two serovars when digested with *Hha*I (Fig. 11), *Hpa*II, and *Hha*I and *Bgl*II (Fig. 12). So far, all United States' isolates serotyped as *canicola* that have been examined by REA have exhibited an REA pattern that matches *portlandvere* and not *canicola* (Fig. 12).

F. Serogroup Grippotyphosa

In this group we recognize six serovars: *canalzonae, grippotyphosa, ratnapura, rattus, valbuzzi*, and *vanderhoedeni*. Until recently, two different reference strains (Andaman and Moskva V) have been recognized for serovar *grippotyphosa*. Strain Andaman has been eliminated. When examined by REA, all reference strains exhibited characteristic patterns, including strains Andaman and Moskva V of serovar *grippotyphosa* (Fig. 13). Similarities were observed between serovars *grippotyphosa* (Moskva V) and *vanderhoedeni*. The United States'

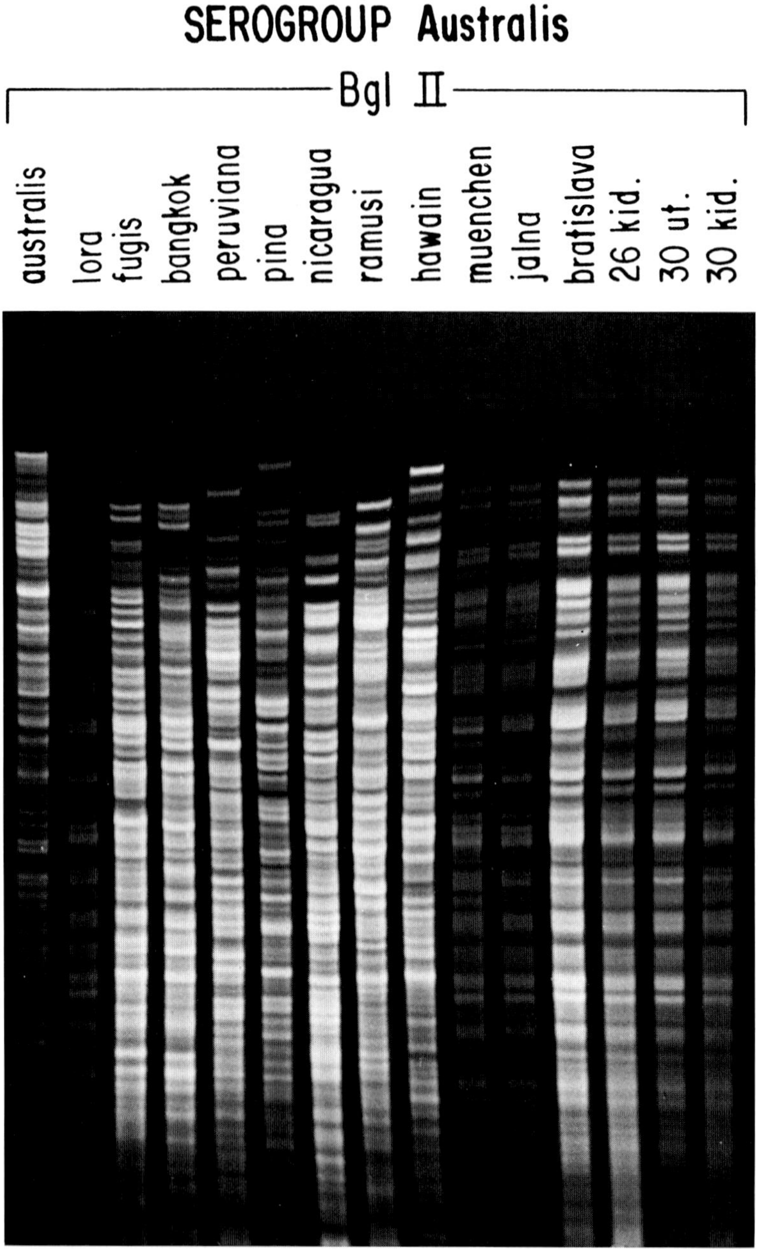

Figure 9 Electrophoretic patterns of chromosomal DNA from serovar reference strains in serogroup Australis, and United States' isolates 26 kid., 30 ut., and 30 kid., digested with enzyme *Bgl*II (0.5% agarose gel).

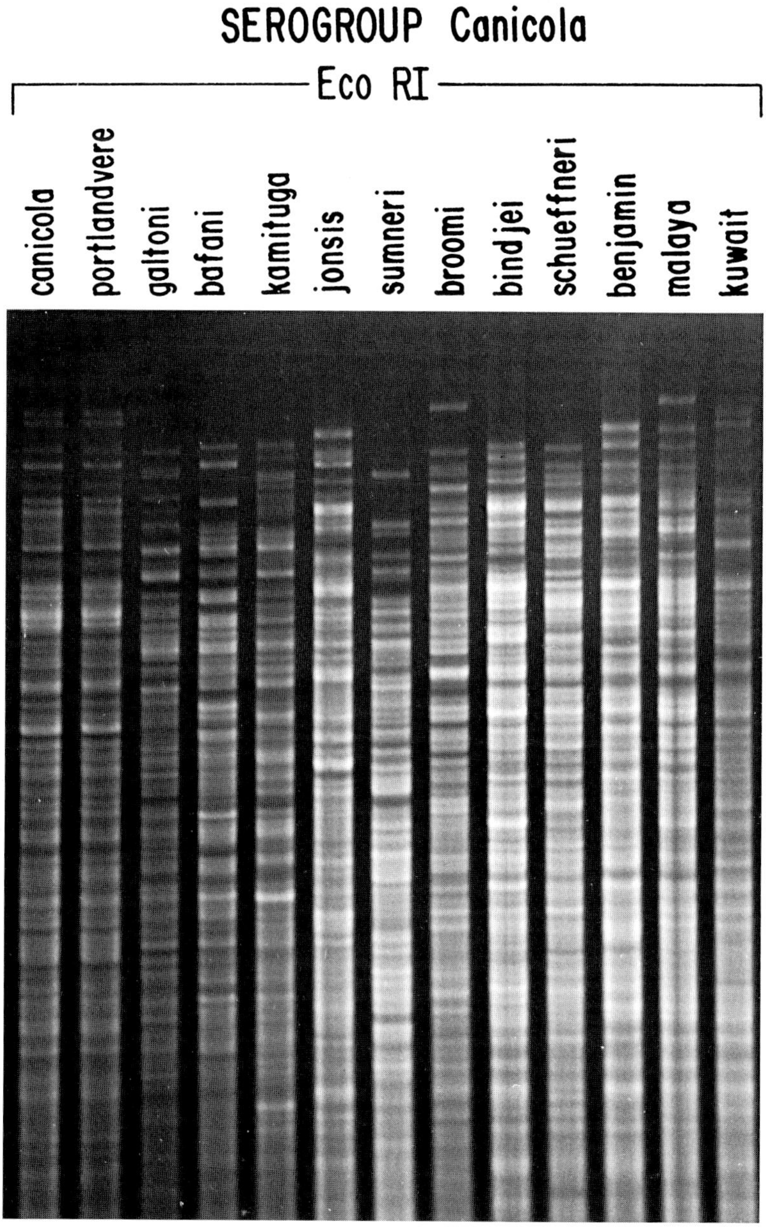

Figure 10 Electrophoretic patterns of chromosomal DNA from serovar reference strains in serogroup Canicola digested with enzyme *Eco*RI (0.7% agarose gel).

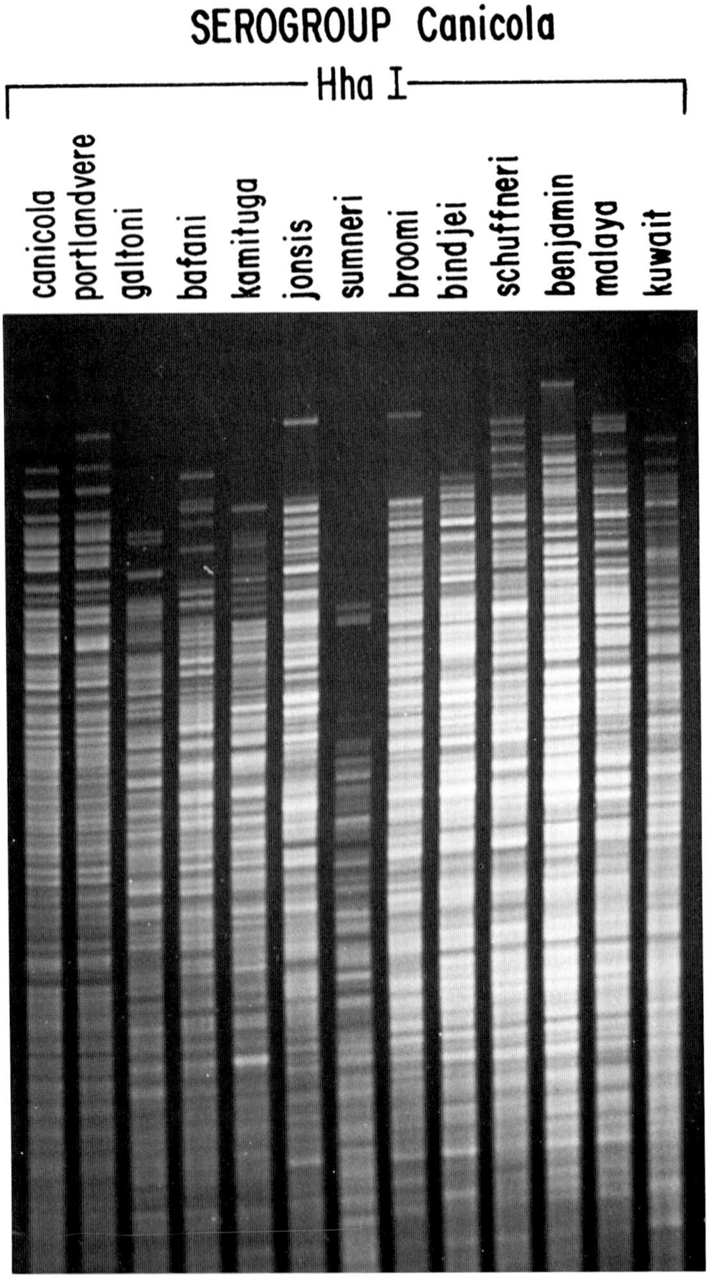

Figure 11 Electrophoretic patterns of chromosomal DNA from serovar reference strains in serogroup Canicola digested with enzyme *Hha*I (0.7% agarose gel).

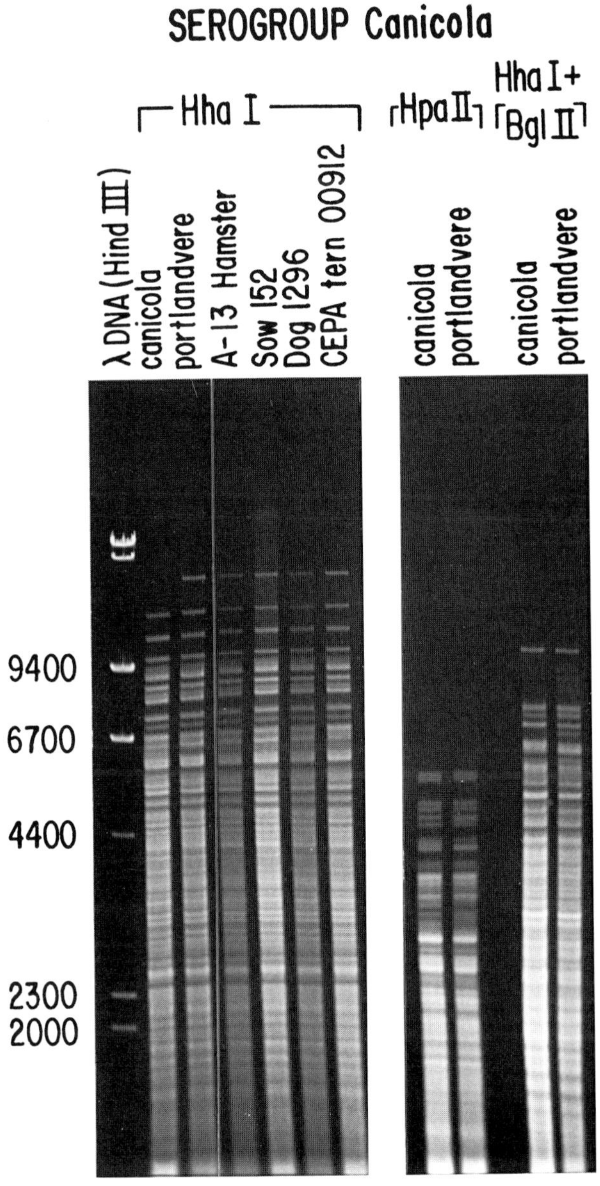

Figure 12 Electrophoretic patterns of chromosomal DNA from serovar reference strains *canicola* and *portlandvere*, and field isolates in serogroup Canicola digested with enzyme *Hha*I, and serovar reference strains *canicola* and *portlandvere* digested with enzymes *Hpa*II and, simultaneously, *Hha*I and *Bgl*II (0.5% agarose gel).

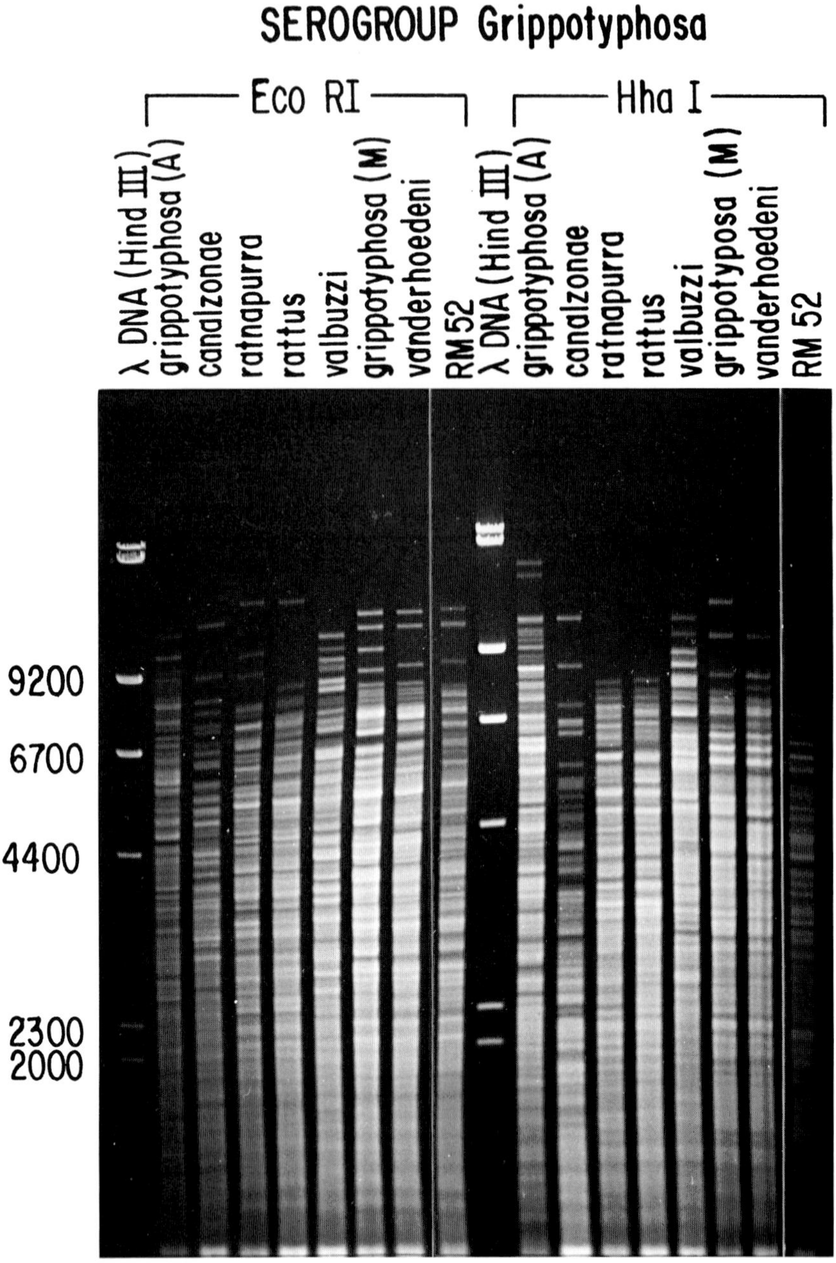

Figure 13 Electrophoretic patterns of chromosomal DNA from serovar reference strains in serogroup Grippotyphosa and isolate RM52 digested with enzymes *Eco*RI and *Hha*I (0.5% agarose gel).

isolates serotyped as *grippotyphosa* exhibit REA patterns that more closely resemble *vanderhoedeni*. These findings were consistent when enzymes *Eco*RI, *Hha*I, and *Bgl*II were used.

G. Serogroup Icterohaemorrhagiae

The Icterohaemorrhagiae group includes 20 different serovars: *birkini, borgvere, budapest, copenhageni, dakota, gem, icterohaemorrhagiae* (RGA), *icterohaemorrhagiae* (Ictero I), *lai, machuguengai, mankarso, monymusk, mwogolo, naam, ndambari, ndahambukuje, sarmin, smithi, tonkini*, and *weaveri*.

Because of unexpected results observed when examining these strains by REA, new reference strains were requested from two different reference centers [Centers for Disease Control (CDC), Atlanta, Georgia and Royal Tropical Institute (RTI), Amsterdam, The Netherlands]. After examining the patterns obtained from digestions with enzymes *Eco*RI and *Hha*I, the following consistent results were obtained: serovars *icterohaemorrhagiae* (RGA), *icterohaemorrhagiae* (Ictero I), *copenhageni, budapest, monymusk*, and *machuguengai* exhibited identical patterns; serovar *ndambari* and *dakota* (CDC origin) also exhibited identical patterns. When comparing reference strains of the same serovar but different origins (CDC and RTI), different patterns were observed for the following serovars: *smithi, dakota, ndahambukuje*, and *mwogolo* (Fig. 14). These differences among reference strains of the same serovars could be a common occurrence, but they could not be demonstrated before the use of REA. So far, this group has been the source of most discrepancies between REA and serotyping.

H. Other *Leptospira* Serogroups

The following additional serogroups were examined by REA, resulting in identifiable patterns for each serovar: Autumnalis (Fig. 15), Bataviae (Fig. 16), Butembo (Fig. 15), Cynopteri (Fig. 17), Djasimin (Fig. 15), Hebdomadis (Fig. 18), Louisiana (Fig. 15), Panama (Fig. 17), Shermani (Fig. 17), and Tarassovi (Fig. 19). The REA has been shown to be a more sensitive and accurate method for the classification of leptospires and possibly other bacterial pathogens of veterinary importance. By and large, it does not contradict previous classifications that are based on serological techniques; however, it identifies differences of epizootiological importance that could not be detected by previous techniques. Therefore, we propose the classification of leptospires by REA once the organisms have been identified within a serogroup by conventional serotyping. The newly proposed genotype will identify an organism or group of organisms whose REA pattern can be differentiated from that of all other members in its serogroup by two or more restriction enzyme digestions. However, differences in REA pattern, even when demonstrable with only one enzyme, must be reported.

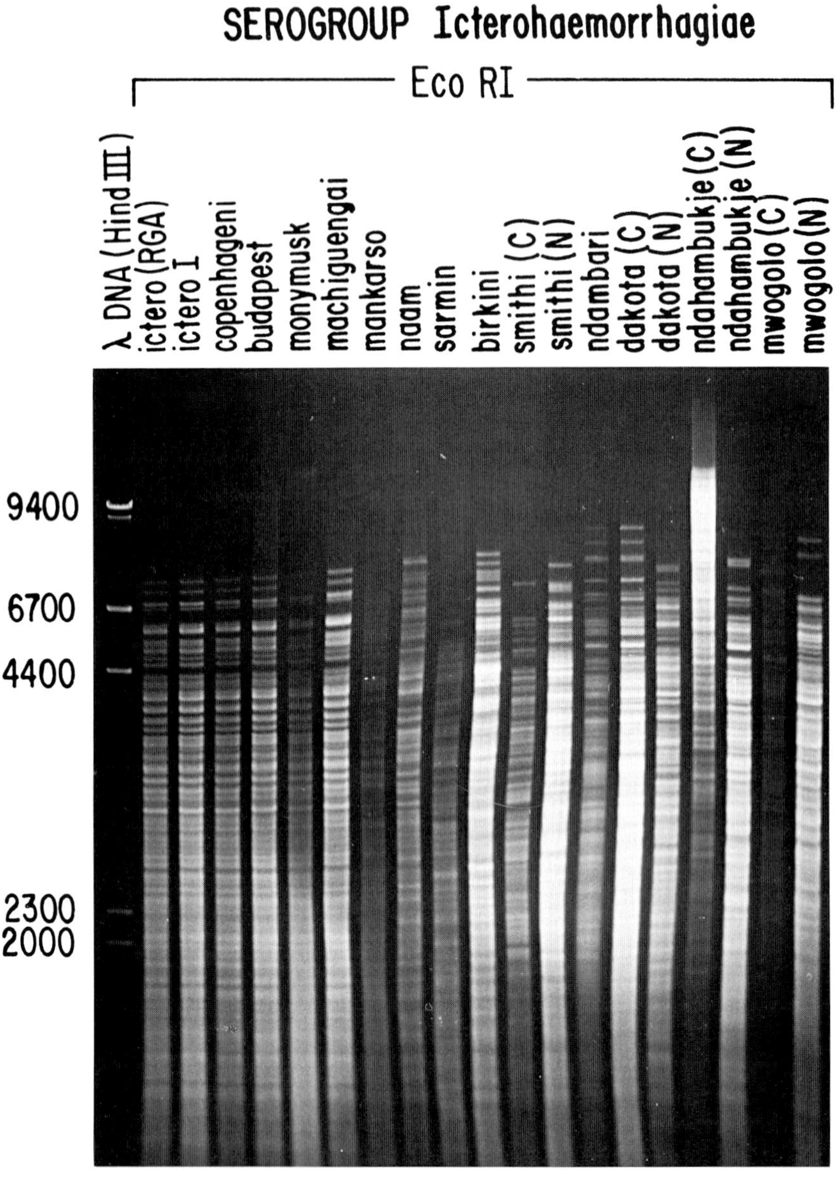

Figure 14 Electrophoretic patterns of chromosomal DNA from seovar reference strains in serogroup Icterohaemorrhagiae digested with enzyme *Eco*RI (0.7% agarose gel). C = strains obtained from the Center for Disease Control, Atlanta, Georgia; N = strains obtained from the Royal Tropical Institute, Amsterdam, The Netherlands.

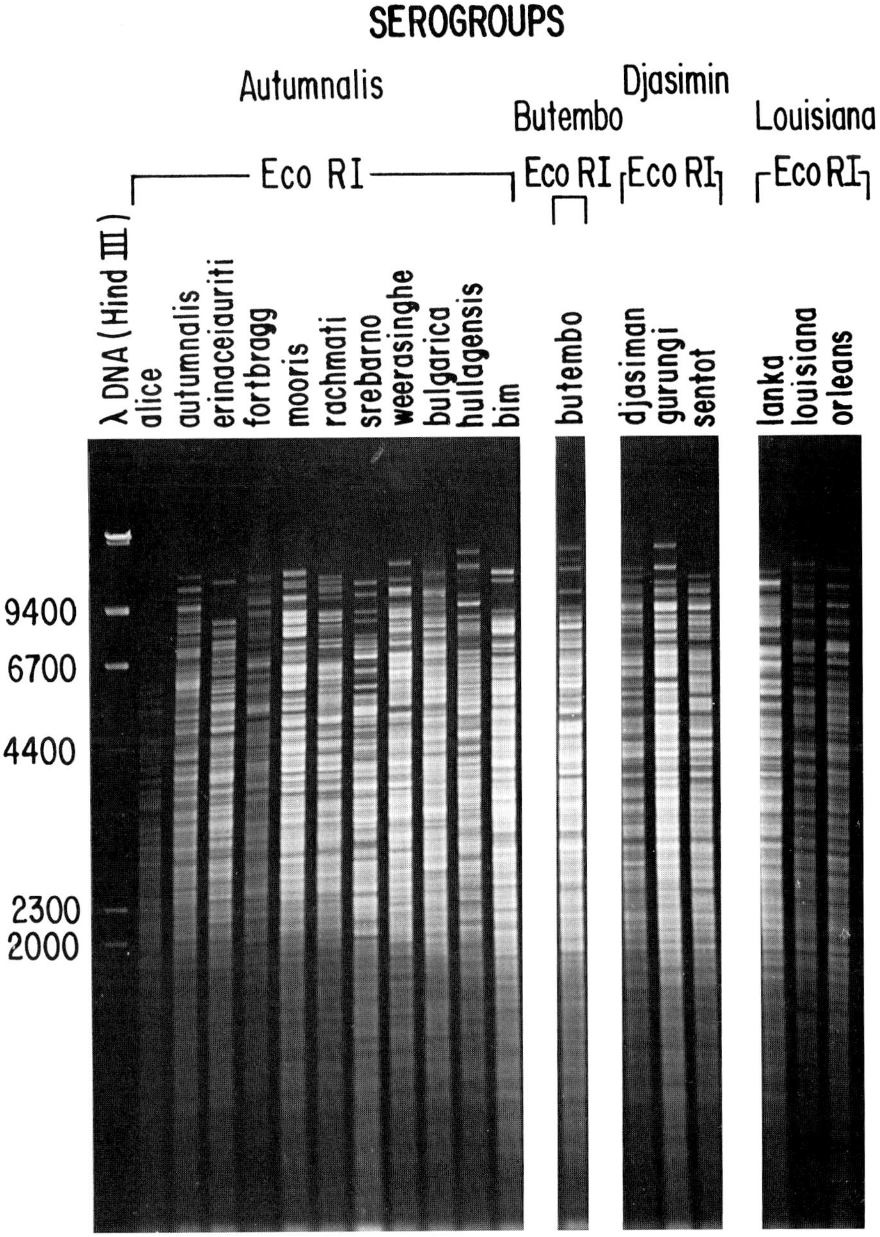

Figure 15 Electrophoretic patterns of chromosomal DNA from serovar reference strains in serogroups Autumnalis, Butembo, Djasimin, and Louisiana digested with enzyme *Eco*RI (0.7% agarose gel).

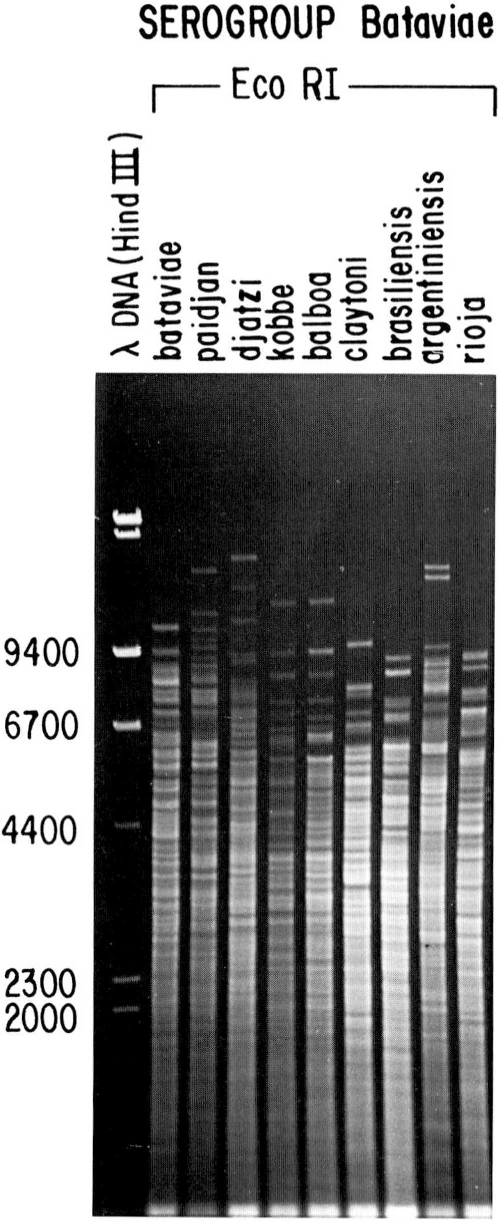

Figure 16 Electrophoretic patterns of chromosomal DNA from serovar reference strains in serogroup Bataviae digested with enzyme *Eco*RI (0.5% agarose gel).

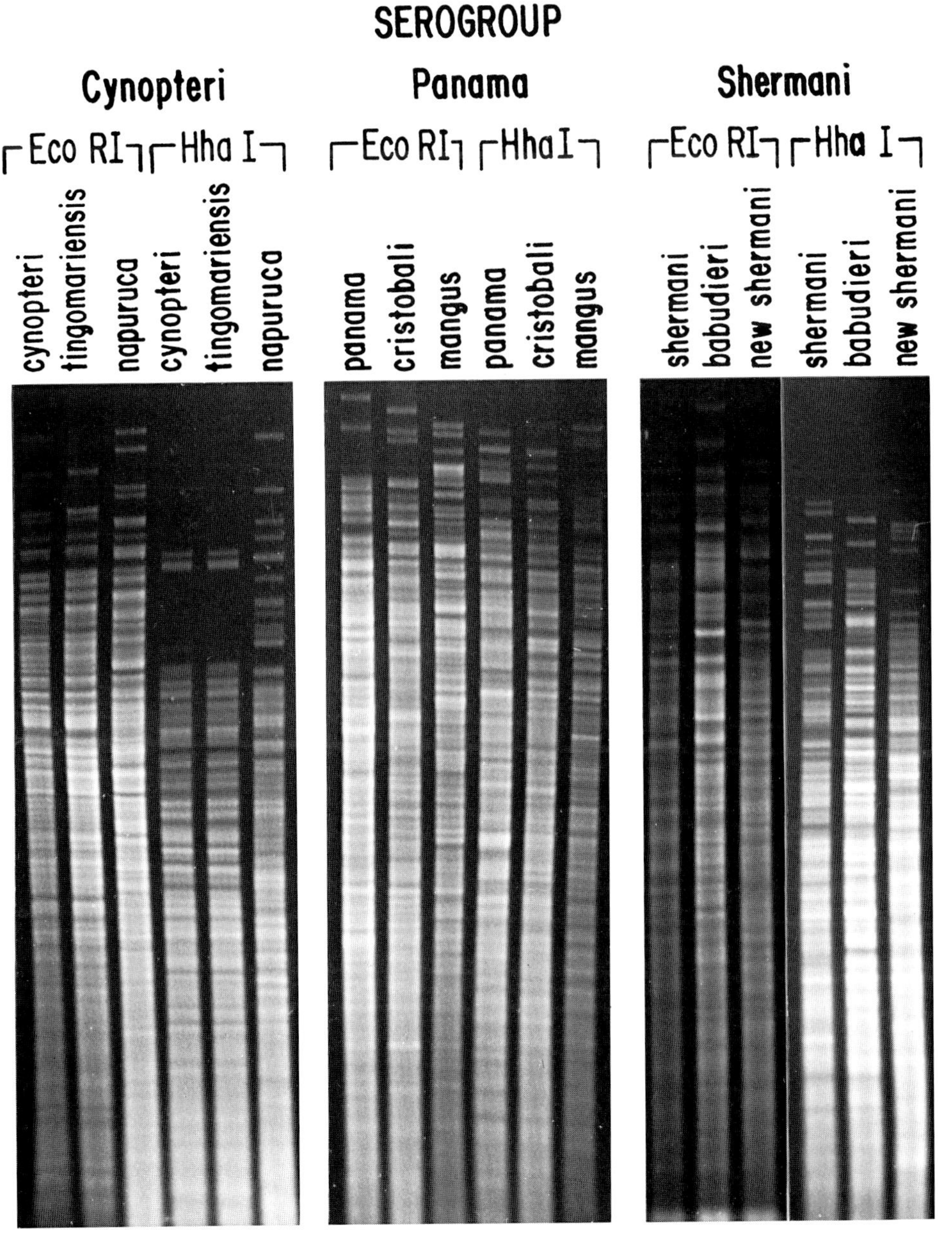

Figure 17 Electrophoretic patterns of chromosomal DNA from serovar reference strains in serogroups Cynopteri, Panama, and Shermani digested with enzymes *Eco*RI and *Hha*I (0.7% agarose gel).

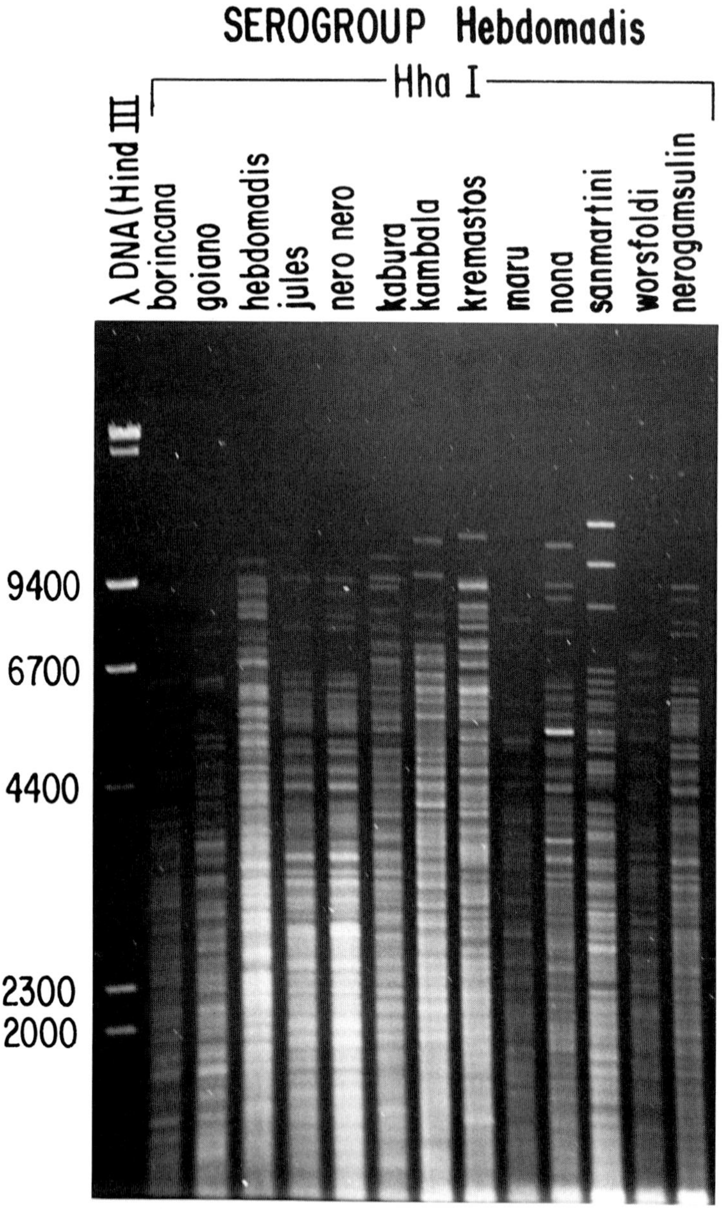

Figure 18 Electrophoretic patterns of chromosomal DNA from serovar reference strains in serogroup Hebdomadis digested with enzyme *Hha*I (0.5% agarose gel).

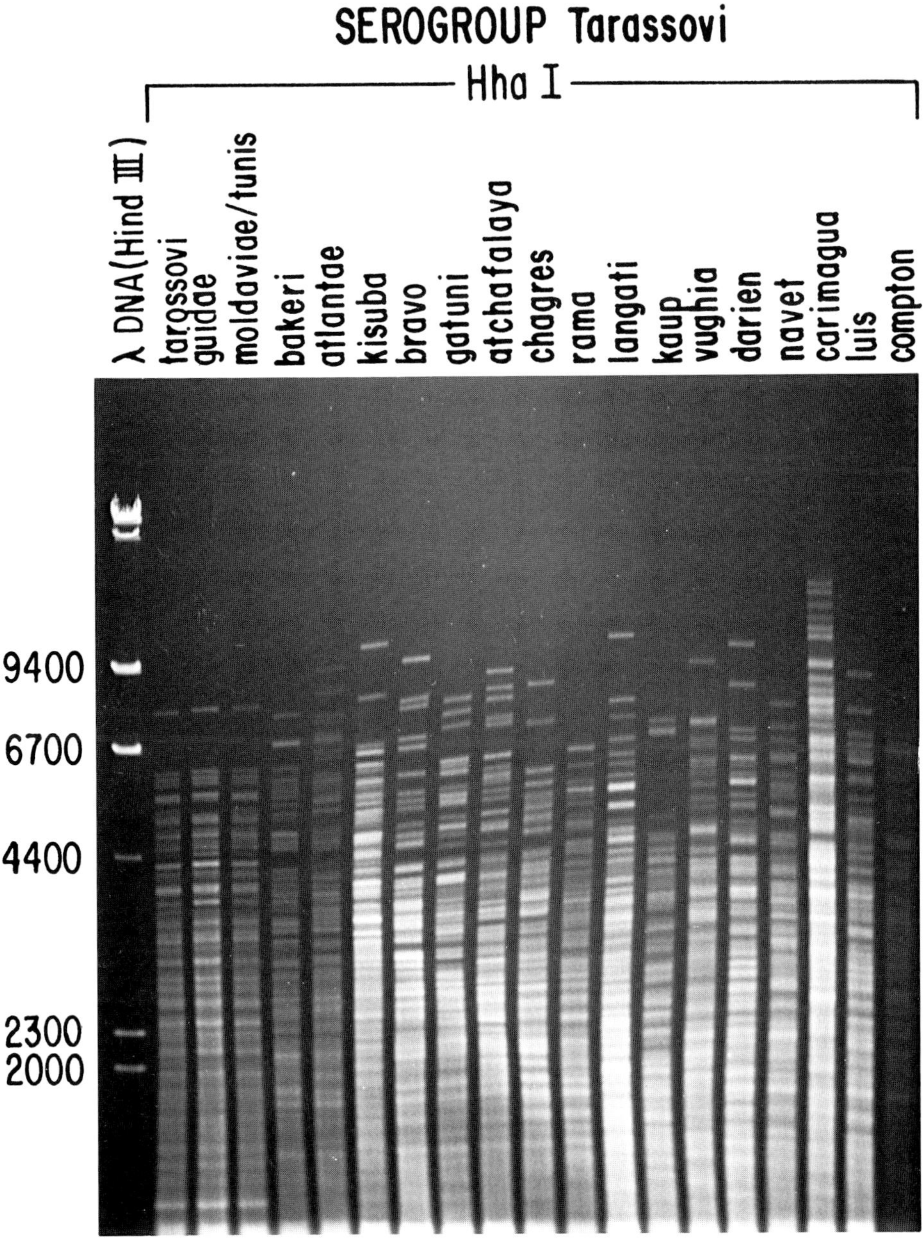

Figure 19 Electrophoretic patterns of chromosomal DNA from serovar reference strains in serogroup Tarassovi digested with enzyme *Hha*I (0.5% agarose gel).

Great stability of the leptospiral genome has been observed, as demonstrated when comparing patterns of fresh field isolates with those of long-term, laboratory-adapted strains. All previous classification methods for leptospires are based on the analysis of surface antigens and are, therefore, subject to the influence of environmental and nutritional factors on the expression of such components by the culture being classified.

I. Other Bacteria of Veterinary Importance

The REA has also been successfully applied to other veterinary pathogens. Work has been conducted by Songer (16) on *Corynebacterium pseudotuberculosis*, the

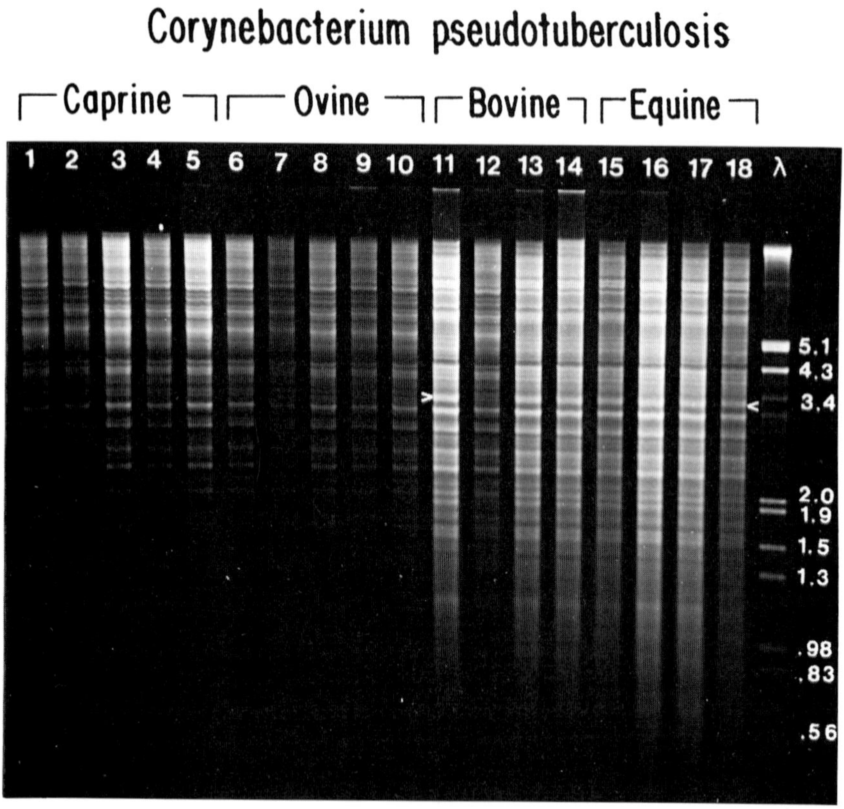

Figure 20 Electrophoretic patterns of chromosomal DNA of *C. bacterium pseudotuberculosis* isolates digested with *Eco*RV. Lanes 1–10 and 12 represent nitrate reductase-negative isolates, and lanes 11 and 13–18 represent reductase-positive isolates. Lane 19 represents λ-phage DNA digested with *Eco*RI and *Hind*III, simultaneously (0.7% agarose gel).

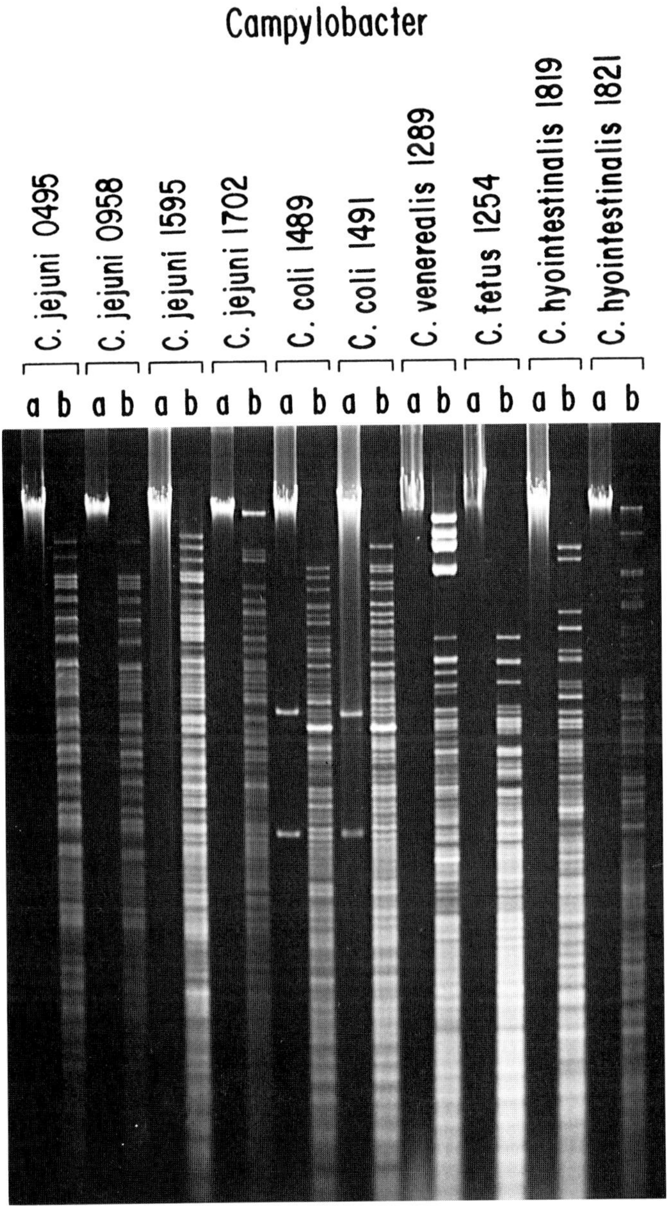

Figure 21 Electrophoretic patterns of chromosomal DNA of *C. jejuni, C. coli, C. venerealis, C. fetus*, and *C. hyointestinalis* isolates. Lanes *a* represent undigested DNA demonstrating plasmids in *C. coli* 1489 and *C. coli* 1491. Lanes *b* represent digestions with enzyme *Hha*I (0.7% agarose gel).

causative agent of caseous lymphadenitis in sheep and goats, and chronic pectoral and inguinal abscesses in horses and cattle. Given the ability to reduce nitrate, these organisms have been divided into two biovars (17): *equi* (usually isolated from horses, producing nitrate reductase), and *ovis* (isolated from sheep and goats, and not producing nitrate reductase). Isolates from cattle have been a mixture of both biovars.

Examination of isolates from sheep, goats, horses, and cattle by REA was carried out to determine if differences between biovars are detectable at a genotypic level. Best resolution of electrophoretic patterns was obtained when using enzymes *Eco*RV and *Pst*I. The differences in patterns correlated perfectly with the phenotypic classification of biovars (Fig. 20; 16).

Similar REA studies have also been reported for *Brucella* (18), *Chlamydia* (19), mycobacteria (20), and *Campylobacter* (21,22). Bryner has examined five *Campylobacter* species by REA. Using enzyme *Hha*I they observed distinctive patterns in *C. jejuni, C. coli, C. hyointestinalis, C. fetus*, and *C. venerealis* (Fig. 21).

Recently, with the use of sensitive laser densitometry and newly developed computer programs, these REA patterns can be recorded, stored, and compared in greater detail. This new technology has allowed a better interpretation and exchange of these genetic fingerprints.

III. DNA HOMOLOGY STUDIES

As noted in the prior discussion, REA has proved to be a very valuable tool in delineating previously unknown but important differences among pathogenical leptospires. To further characterize these leptospires phylogenetically, DNA homology studies were performed. Closely related leptospires should share a high level of similar DNA sequences and, thus, should reflect a high level of hybridization of their respective DNAs. Conversely, leptospires with divergent phylogenetic lines would be expected to share fewer common DNA sequences and, thus, would hybridize to a lesser degree.

Homology studies of the genomes of living organisms have been accepted as important to bacterial classification for more than 20 years. There are several advantages to be gained by basing classification on genetic relatedness.

1. A more unifying concept of a bacterial species is possible.
2. Classifications based on genetic relatedness tend not to be subject to frequent or radical changes.
3. Reliable identification schemes can be prepared after organisms have been classified on the basis of genetic relatedness.
4. Information can be obtained that is useful for understanding how the various bacterial groups have evolved and how they can be arranged according to their ancestral relationships (23).

It is the unique properties of nucleic acids that provide important taxonomic information. One of the first features to be explored was the mole-percentage (mol%) of guanine plus cytosine (G+C). However, this information alone does not lend itself to grouping organisms taxonomically because although closely related organisms will have similar G+C mole-percentages, unrelated organisms may also have similar G+C content. The explanation for this is that DNA base composition does not reflect the actual organization of the nucleotides in the linear DNA molecule. Several methods have been developed to assay actual sequence homologies shared by different organisms (24-33). We chose a slightly different approach than these. By using a modification of the Gillespie and Spiegelman procedure (24), in which restriction enzymes, Southern blotting (34), and autoradiography were used, we were able to visualize differences in degrees of homology as well as the DNA fragments base-paired to the hybrid genome (35). Chromosomal DNA of leptospires of interest were digested with the restriction endonuclease *Eco*RI and fractionated on a 0.7% agarose gel at 60 V for 16 hr. After electrophoresis, the gel was stained with ethidium bromide and photographed under ultraviolet illumination. The DNA in the gel was transferred to nitrocellulose following the procedure of Southern (34). Homology studies were performed by hybridization of transferred DNA with radiolabeled *Eco*RI-digested genomic DNA from a specific leptospire. After hybridization, stringent washes were performed to remove essentially all radiolabeled DNA, except for paired hybrids. Autoradiography revealed the extent to which the genomes were homologous and which fragments were shared in common.

The *hardjo* isolates were analyzed first to resolve the enigma presented earlier. This disparity in DNA restriction sites and subsequent fragment sizes suggested major divergences in the overall sequence organization of the genome of these organisms, but it was not conclusive. Figure 22 illustrates the degree of homology between the genomes of the three North American bovine isolates (hardjo-bovis type A, B, and C), serovar *balcanica* reference strain (of the same Sejroe serogroup), and *hardjo* reference strain hardjoprajitno. The probe used here was ^{32}P-labeled *Eco*RI digest of hardjo-bovis type A genomic DNA. Although it is apparent from the autoradiograph that there is a high degree of DNA homology between the three hardjo-bovis isolates and the serologically unrelated *balcanica* strain, very little homology is seen between the hardjoprajitno reference strain and the hardjo-bovis probe. When the same Southern blot is probed with radio-labeled hardjoprajitno genomic DNA (Fig. 23), little base pairing is observed between the hardjo-bovis isolates and the hardjoprajitno reference strain.

In an attempt to determine the degree of homology of serologically unrelated leptospires, the genomes of *pomona* and *kennewicki*, two serovars in the Pomona group, and a saprophytic *illini* were restricted with *Eco*RI and Southern blotted along with hardjo-bovis isolate type A and hardjoprajitno. When a ^{32}P-labeled *Eco*RI-digested hardjoprajitno DNA was employed, strong homology was

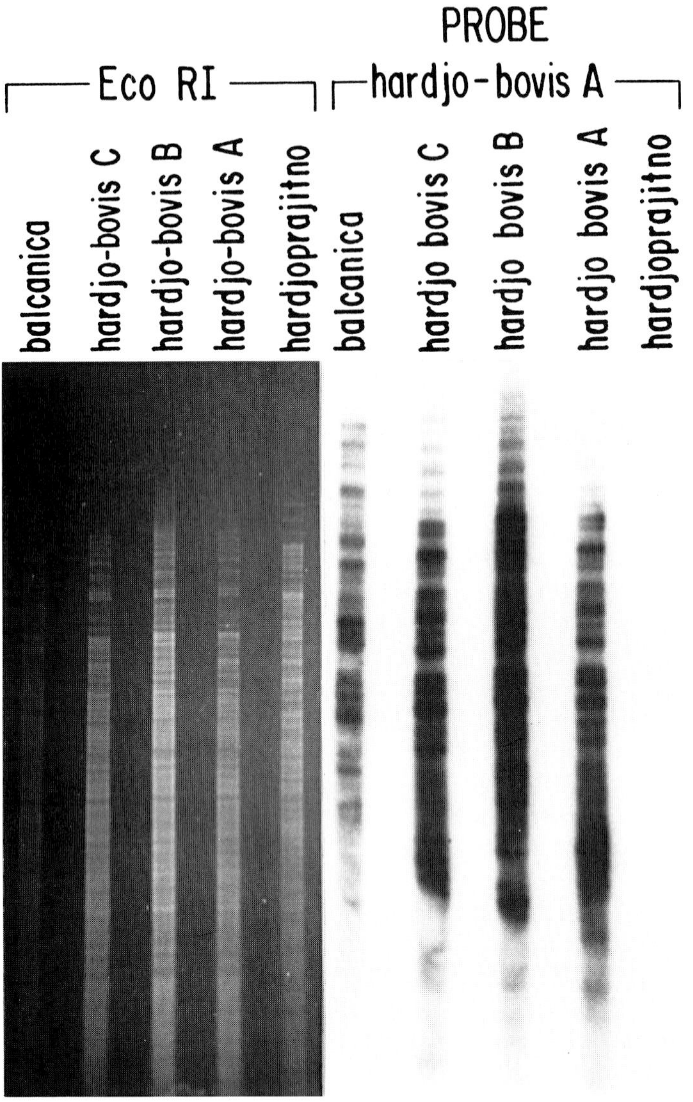

Figure 22 Electrophoretic and hybridization patterns of chromosomal DNA from serogroup Sejroe probed with *Eco*RI-digested and ^{32}P-labeled hardjo-bovis A. Lane 1 represents serovar *balcanica*; lanes 2–4 hardjo-bovis (C, B, and A, respectively); and lane 5 reference strain hardjoprajitno (0.7% agarose gel).

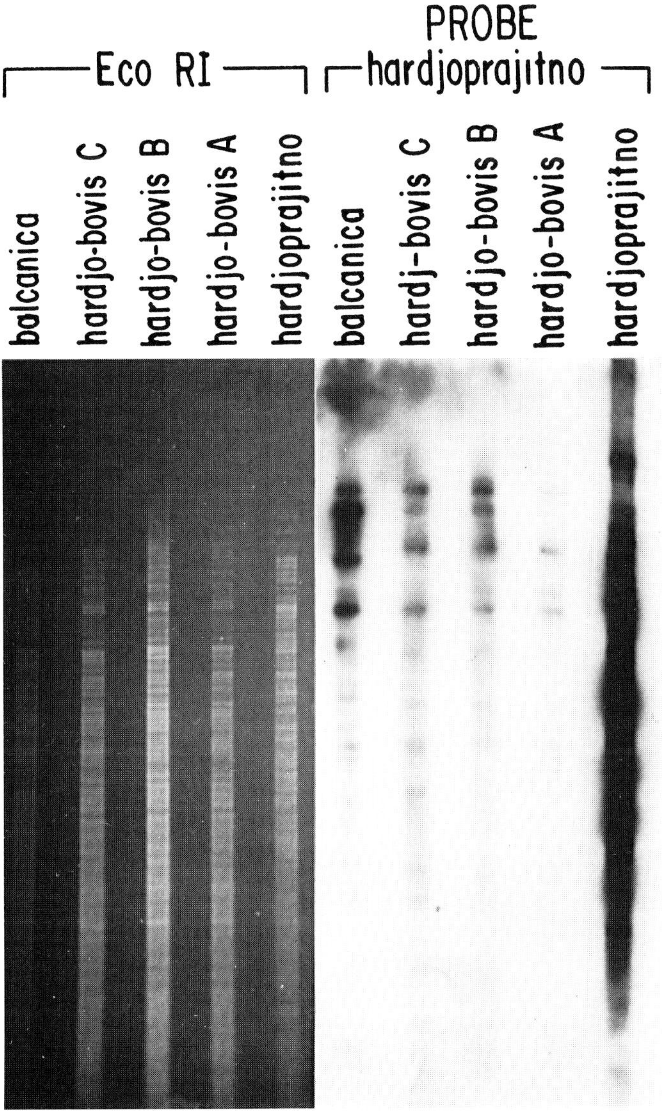

Figure 23 Electrophoretic and hybridization patterns of chromosomal DNA from serogroup Sejroe probed with *Eco*RI-digested and ^{32}P-labeled hardjoprajitno DNA. Lane 1 represents serovar *balcanica*; lanes 2–4 hardjo-bovis (C, B, and A, respectively); lane 5 reference strain hardjoprajitno (0.7% agarose gel).

observed between the DNA of labeled hardjoprajitno and that of serovars *pomona* and *kennewicki* reference strains (Fig. 24). No base pairing was observed with the *illini* strain and, again, minimal homology was detected with hardjo-bovis A.

As stated earlier, DNA-DNA hybridization studies serve as the most accurate assay to determine overall genetic similarity between species. Organisms having a high degree of homology of nucleotide sequences can be classified as being closely related phylogenetically. A difference of 3% in base composition reflected a vast difference in nucleotide sequences of leptospiral DNA (36). Similarities and differences of leptospiral genomes can be correlated to genetic relatedness of serovars (37).

It is not clear how to resolve the problem presented by these results. Hardjoprajitno and the North American *hardjo* isolates are indistinguishable by conventional serological assays now used in leptospiral identification. Restriction endonuclease analysis and DNA homology studies have shown them to be quite disparate genetically. When hardjoprajitno was compared with different Pomona serovars, the converse of the preceding finding was observed; *pomona* and *hardjo* are serologically unrelated, and yet share a high degree of homologous base sequence. Although it is not clear how to organize these data into a workable taxonomy, it is quite clear that reclassification of leptospires is needed to meet the demands of more accurate and sensitive taxonomic assays.

IV. DNA PROBES

The use of genetic probes, either RNA or DNA, as diagnostic tools is rapidly becoming an important field for clinical research. The application has been facilitated by the development of methods that allow nonradioactive labeling of nucleic acids, such as the use of biotin (38-40). As a rule, nucleic acid probes are more sensitive and specific than most immunological tests. Assays with nucleic acid probes are relatively fast and easy, which also contributes to their desirability in diagnostics.

The development of nucleic acid probes for detection of leptospires was approached in two ways. First, it was important to design a probe that was specific for *L. interrogans*, because in early diagnosis of the disease, the final identity of the causative serovar is not as important as the rapid diagnosis of leptospirosis. In previous experiments, it was noted that labeled genomic DNA from either *pomona* or *hardjo* serovars had no homology with nonpathogenic leptospires or with *Escherichia coli*. Radiolabeled genomic DNA from hardjo-bovis type A was used to test the sensitivity of detection of leptospires by dot-blot analysis. As few as 10 organisms have been detected with this method (unpublished data). Work is currently underway to determine if any homology exists between the leptospiral genomic probes and other bacteria. We are also investigating the level of sensitivity of genomic DNA labeled with biotin.

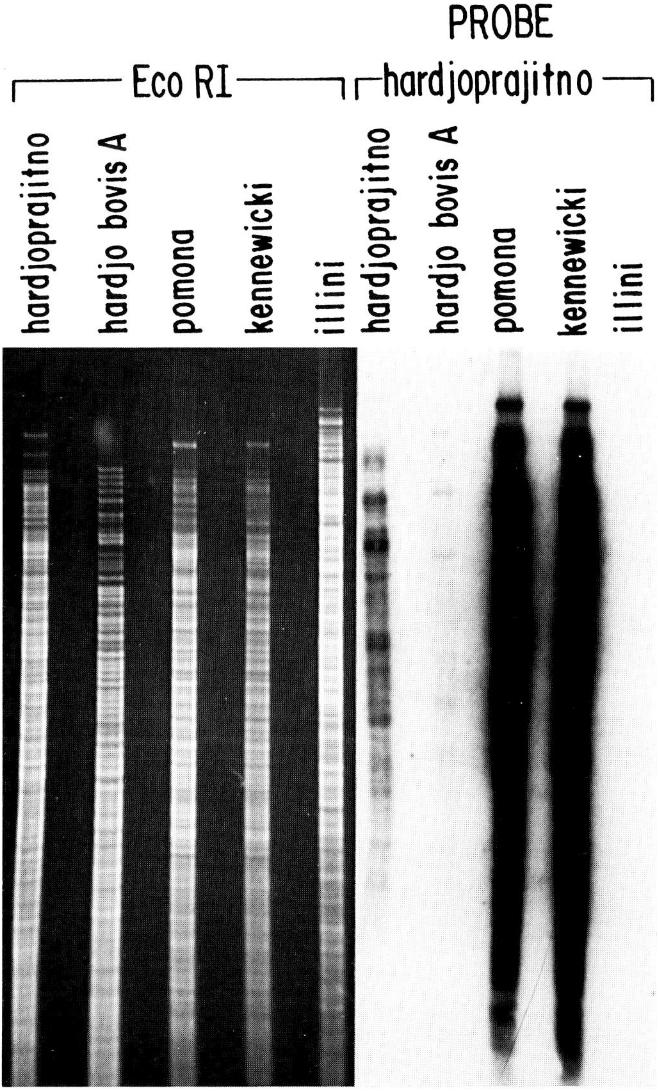

Figure 24 Electrophoretic and hybridization patterns of chromosomal DNA from serogroups Sejroe, Pomona, and Illini probed with *Eco*RI-digested and ^{32}P-labeled hardjoprajitno DNA of serogroup Sejroe. Lane 1 represents reference strain hardjoprajitno; lane 2 hardjo-bovis A of serogroup Sejroe; lanes 3–4 serovars *pomona* and *kennewicki*, respectively, of serogroup Pomona; lane 5 serovar *illini* of serogroup Illini (0.7% agarose gel).

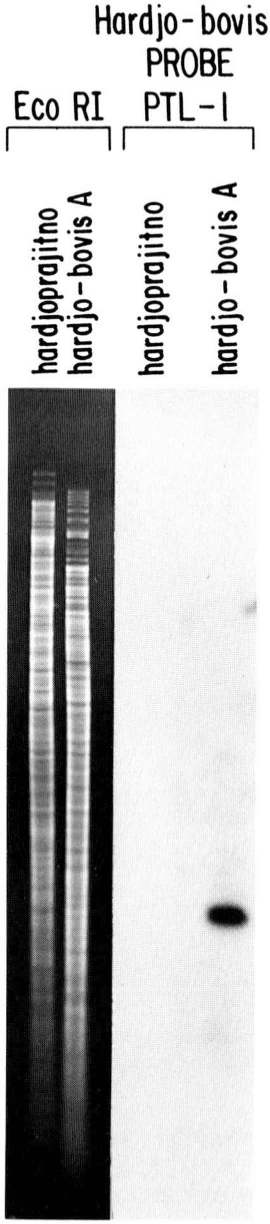

Figure 25 Electrophoretic and hybridization patterns of chromosomal DNA from serogroup Sejroe probed with (pTL1), a DNA probe isolated from hardjo-bovis A and ^{32}P-labeled. Lane 1 represents hardjoprajitno, and lane 2 represents hardjo-bovis A (0.7% agarose gel).

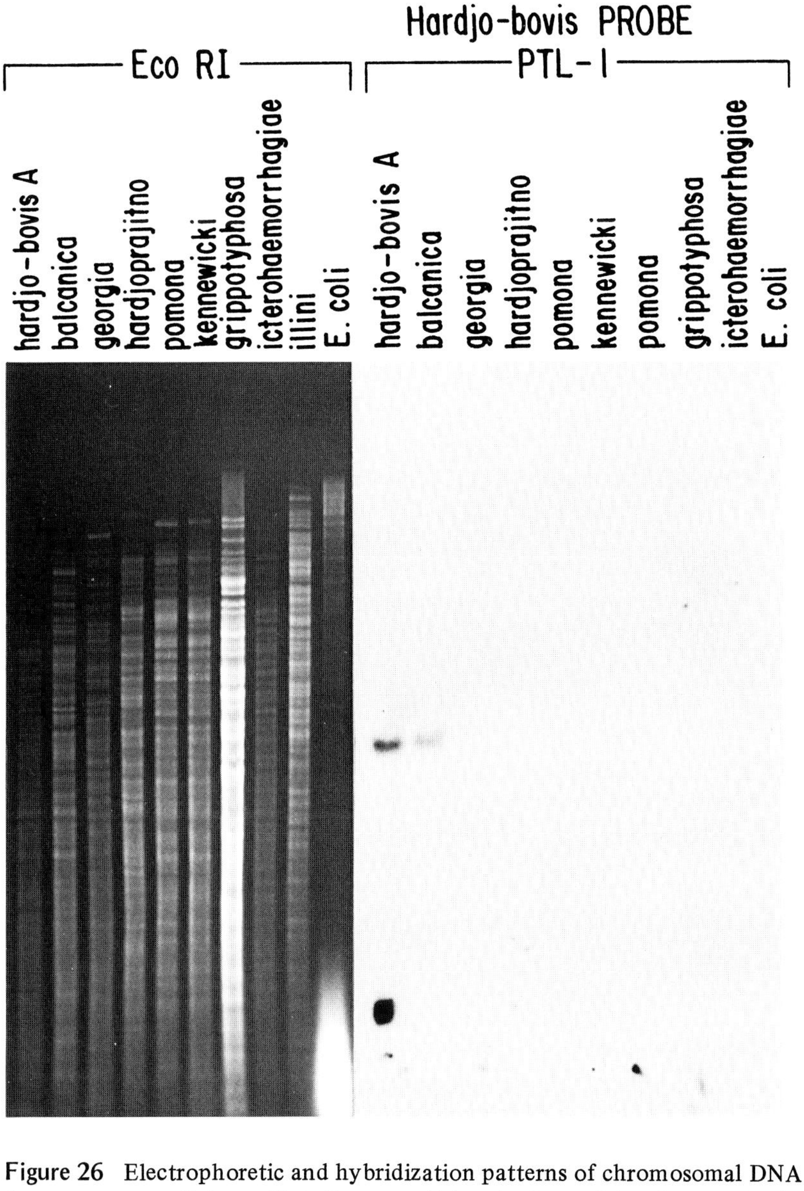

Figure 26 Electrophoretic and hybridization patterns of chromosomal DNA from serogroups Sejroe, Mini, Pomona, Grippotyphosa, Icterohaemorrhagiae, and Illini probed with (pTL1), a DNA probe isolated from hardjo-bovis A and ^{32}P-labeled. Lane 1 represents hardjo-bovis A from serogroup Sejroe; lane 2 serovar *balcanica* from serogroup Sejroe; lane 3 serovar *georgia* from serogroup Mini; lane 4 reference strain hardjoprajitno of serogroup Sejroe; lanes 5 and 6 serovars *pomona* and *kennewicki*, respectively, of serogroup Pomona; lane 7 serovar *grippotyphosa* of serogroup Grippotyphosa; lane 8 serovar *icterohaemorrhagiae* of serogroup Icterohaemorrhagiae; lane 9 serovar *illini* from serogroup Illini; and lane 10 *E. coli* (0.7% agarose gel).

The second approach taken for the development of nucleic acid probes for leptospirosis was to isolate DNA fragments, unique for a given serovar, to be used for more precise diagnosis and for characterization of isolates. With the use of information from the homology studies, we purified DNA fragments from hardjo-bovis A that were not hybridized by hardjoprajitno DNA. These fragments were electroeluted from the agarose gel and cloned into (pUC8) plasmids. Several recombinant plasmids were found to contain fragments unique for hardjo-bovis, and one of these has been further characterized. The recombinant (pTL1) contains a hardjo-bovis *Eco*RI fragment of about 4-kb pairs. When radiolabeled and used to probe Southern blots of hardjoprajitno and hardjo-bovis DNA, only a single band in the hardjo-bovis lane was homologous (Fig. 25). The (pTL1) plasmid was further tested against several other pathogenic leptospiral genomes (Fig. 26), and only a low degree of homology was detected with a *balcanica* strain. For diagnostic use, the hybridization of the probe to *balcanica* does not pose a problem because it is a strain apparently not present in the United States. Probes unique for hardjoprajitno have also been isolated and cloned. Probes unique for *pomona* and *kennewicki* are being developed.

Our understanding of leptospiral taxonomy and diagnosis has increased dramatically over the last few years because of the studies of these organisms at the genomic level. In our laboratory, rapid isolation of leptospiral chromosomal DNA and subsequent restriction enzyme analysis has largely supplanted the serotyping tests previously used.

ACKNOWLEDGMENTS

The authors wish to acknowledge the contributions of Dr. J. W. Bryner Jr. of the National Animal Disease Center in Ames, Iowa and Dr. J. G. Songer of the Department of Veterinary Science, University of Arizona, Tucson, Arizona as well as the technical assistance of John W. Foley and Annette L. Handsaker (from the laboratory of Thiermann and Le Febvre), and Lori A. Pollet (from the laboratory of Bryner).

REFERENCES

1. Dikken, H., and E. Kmety, Serological typing method of leptospires, In *Methods in Microbiology*, Vol. 2, Bergan, and Norris (eds.). Academic Press, London, pp. 259–312 (1978).
2. Cole, J. R., Jr., C. R. Sulzer, and A. R. Pursell, Improved microtechnique for the leptospiral microscopic agglutination test. *Appl. Microbiol. 25*: 976–980 (1973).
3. Kmety, E., Faktoren analyse von Leptospiren de Icterohaemorrhagiae und einiger verwandter Serogruppen. *Biol. Pr. 13*:000–000 (1967).

4. Terpstra, W. J., H. Korver, J. van Leeuwen, P. R. Klatser, and A. H. Kolk, The classification of sejroe group serovars of *Leptospira interrogans* with monoclonal antibodies. *Zentralbl. Bakteriol. Hyg. A 259*:498–506 (1985).
5. Marshall, R. B., B. E. Wilton, and A. J. Robinson, Identification of *Leptospira* serovars by restriction endonuclease analysis. *J. Med. Microbiol. 14*: 163–166 (1981).
6. Thiermann, A. B., A. L. Handsaker, S. L. Moseley, and B. Kingscote, New method for classification of *Leptospiral* isolates belonging to serogroup Pomona by restriction endonuclease analysis: Serovar *kennewicki*. *J. Clin. Microbiol. 21*:585–587 (1985).
7. Le Febvre, R. B., J. W. Foley, and A. B. Thiermann, Rapid and simplified protocol for isolation and characterization of leptospiral chromosomal DNA for taxonomy and diagnosis. *J. Clin. Microbiol. 22*:606–608 (1985).
8. Ellis, W. A., D. G. Bryson, J. J. O'Brien, and S. D. Neil, Leptospiral infection in aborted equine foetuses. *Equine Vet. J. 15*:321–324 (1983).
9. White, F. H., C. R. Sulzer, and R. W. Engel, Isolations of *Leptospira interrogans* serovars *hardjo*, *balcanica*, and *pomona* from cattle at slaughter. *Am. J. Vet. Res. 43*:1172–1173 (1982).
10. Thiermann, A. B., and W. A. Ellis, Identification of leptospires of veterinary importance by restriction endonuclease analysis. In *Proceedings, The Present State of Leptospirosis Diagnosis and Control*, The Commission of European Communities. Martines Nijhoff, The Hague, pp. 91–104 (1986).
11. Mackintosh, C. F., R. B. Marshall, and J. C. Thompson, Experimental infection of sheep and cattle with *Leptospira* interrogans serovar *balcanica*. *N. Z. Vet. J. 29*:15–19 (1981).
12. Glosser, J. W., C. R. Sulzer, and G. C. Reynolds, Isolation of leptospiral serotype *szwajizak* from dairy cattle in Oregon. *Proc., Annu. Meet. US Anim. Health Assoc. 78*:119–125 (1974).
13. Galton, M. M., G. W. Gorman, and E. G. Shotts Jr., A new leptospiral serotype in the Hebdomadis serogroup. *Public Health Rep. 75*:917--921 (1960).
14. Hanson, L. E., H. C. Ellinghausen Jr., and R. Marlowe, Isolation of *Leptospira grippotyphosa* from a cow following an abortion. *Proc., Soc. Exp. Biol. Med. 117*:495–497 (1964).
15. Ellis, W. A., and A. B. Thiermann, Isolation of *Leptospira* interrogans serovar *bratislava* from sows in Iowa. *Am. J. Vet. Res. 47*:1458–1460 (1986).
16. Songer, J. G., G. B. Olson, M. M. Marshall, K. Beckenbach, and L. Kelley, Phenotypic and genotypic variation in *Corynebacterium pseudotuberculosis*. *7th West. Conf. Food Anim. Vet. Med.*, March 17–19 (1986).
17. Biberstein, E. L., H. D. Knight, and S. Jang, Two biotypes of *Corynebacterium pseudotuberculosis*. *Vet. Rec. 80*:691–692 (1971).
18. O'Hara, M. J., D. M. Collins, and G. W. De Lisle, Restriction endonuclease analysis of *Brucella ovis* and other *Brucella* species. *Vet. Microbiol. 10*: 425–429 (1985).

19. Peterson, E. M., and L. M. de la Maza, Characterization of *Chlamydia* DNA by restriction endonuclease cleavage. *Infect. Immun. 41*:604–608 (1983).
20. Collins, D. M., and G. W. De Lisle, DNA restriction endonuclease analysis of *Mycobacterium tuberculosis* and *Mycobacterium bovis* BCG. *J. Gen. Microbiol. 130*:1019–1021 (1984).
21. Kakoyiannis, C. K., P. J. Winter, and R. B. Marshall, Identification of *Campylobacter coli* isolates from animals and humans by bacterial restriction endonuclease DNA analysis. *Appl. Environ. Microbiol. 48*:545–549 (1984).
22. Collins, D. M., and D. E. Ross, Restriction endonuclease analysis of *Campylobacter* strains with particular reference to *Campylobacter fetus* subsp. *fetus. J. Med. Microbiol. 18*:117–124 (1984).
23. Johnson, J. L., Nucleic acids in bacterial classification. In *Bergey's Manual of Systematic Bacteriology*, Vol. 1, Krieg, N. R. and Holt, J. C. (eds.). Williams & Wilkins, Baltimore, pp. 8–11 (1984).
24. Hoyer, B. H., B. J. McCarthy, and E. T. Bolton, A molecular approach in the systematics of higher organisms. *Science 144*:959–967 (1964).
25. Gillespie, D., and S. Spiegelman, A quantitative assay for DNA–RNA hybrids with DNA immobilized on a membrane filter. *J. Mol. Biol. 12*:809–842 (1965).
26. Denhardt, D. T., A membrane-filter technique for the detection of complementary DNA. *Biochem. Biophys. Res. Commun. 83*:641–646 (1966).
27. Johnson, J. L., Genetic characterization. *Manual of Methods for General Bacteriology*, Gerhardt, P. et al. (eds.). American Society for Microbiology, Washington, pp. 450–472 (1981).
28. DeLey, J., H. Caltoir, and A. Reynaerts, The quantitative measurement of DNA hybridization from renaturation rates. *Eur. J. Biochem. 12*:133–142 (1970).
29. Bernardi, G., Chromatography of nucleic acids on hydroxyapatite. I. Chromatography of native DNA. *Biochim. Biophys. Acta 174*:423–434 (1969).
30. Bernardi, G., Chromatography of nucleic acids on hydroxyapatite. II. Chromatography of denatured DNA. *Biochim. Biophys. Acta 174*:435–448 (1969).
31. Miyazawa, Y., and C. A. Thomas, Composition of short segments of DNA molecules. *J. Mol. Biol. 11*:223–237 (1965).
32. Brenner, D. J., G. R. Fanning, A. V. Rake, and K. E. Johnson, Batch procedure for thermal elution of DNA from hydroxyapatite. *Anal. Biochem. 28*:447–459 (1969).
33. Crosa, J. H., D. J. Brenner, and S. Falkow, Use of single-strand specific nuclease for analysis of bacterial and plasmid deoxyribonucleic acid homo- and hetero-duplexes. *J. Bacteriol. 115*:904–911 (1973).
34. Southern, E., Detection of specific sequences among DNA fragments separated by gel electrophoresis. *J. Mol. Biol. 98*:503–518 (1975).

35. LeFebvre, R. B., and A. B. Thiermann, DNA homology studies of leptospires of serogroups Sejroe and Pomona from cattle and swine. *Am. J. Vet. Res. 47*:959–963 (1986).
36. Haaple, D. K., M. Rogul, L. B. Evans, and A. D. Alexander, Deoxyribonucleic acid base composition and homology studies of leptospira. *J. Bacteriol. 98*:421–428 (1969).
37. Bundle, J. J., M. Royal, and A. D. Alexander, Deoxyribonucleic acid hybridization among selected leptospiral serotypes. *Int. J. Syst. Bacteriol. 24*:205–214 (1974).
38. Langer, P. R., A. A. Wardrop, and D. C. Ward, Enzymatic synthesis of biotin-labeled polynucleotides: Novel nucleic acid affinity probes. *Proc. Natl. Acad. Sci. USA 78*:6633–6637 (1981).
39. Landegent, J. E., N. Jansen in de Wal, G.-J. B. Van Ommen, F. Baas, J. J. M. de Vijlder, P. van Duijn, and M. Van der Ploeg, Chromosomal localization of a unique gene by non-autoradiographic in situ hybridization. *Nature 317*:175–177 (1985).
40. Tchen, P., R. P. P. Fuchs, E. Sage, and M. Leng, Chemically modified nucleic acids as immunodetectable probes in hybridization experiments. *Proc. Natl. Acad. Sci. USA 81*:3466–3470 (1984).

8

Nucleic Acid Probes for Detection of Clinically Significant Bacteria

FRAN A. RUBIN and DENNIS J. KOPECKO
Walter Reed Army Institute of Research, Washington, District of Columbia

I. INTRODUCTION

A. Why Nucleic Acid Probes Are Advantageous

Defined nucleic acid segments used to detect complementary sequences within a sample (i.e., genetic probes) have been utilized successfully over the past decade, mainly by molecular biologists and also by taxonomists and epidemiologists. These genetic probes have allowed the cloning, by recombinant DNA technology, of desired bacterial chromosomal gene regions in basic genetic studies and for the detection of certain pathogenic traits (e.g., shigellalike toxin production) or pathogenic organisms in prospective and retrospective epidemiological studies. As a result of the recent advances in DNA hybridization detection technology (see Chap. 2), the development of nucleic acid probes that are specific for a particular bacterial pathogen, and the great practical utility of probe technology, several nucleic acid probes for clinical diagnostic use have already become commercially available and more are expected in the coming several years.

Clinical and quality control microbiologists are tasked with detecting and identifying microorganisms, with the specific objectives of disease prevention, epidemiological surveillance of disease outbreaks, and the diagnostic identification of specific disease agents. Nucleic acid probes offer considerable promise to the quality control microbiologist for detecting food, water, or other samples contaminated with potential disease agents. The special problems and the unique probes constructed for the analysis of food and water are discussed in Chapter 9 of this volume. The present chapter will focus on the epidemiological and diagnostic uses of nucleic acid probes. Detection of a bacterial pathogen typically involves the isolation and identification of viable bacterial cells from a clinical sample. Oftentimes, the disease causative species is mixed with the normal flora in specimens (e.g., in stool); the pathogen(s) must be differentiated from the total bacterial population. Suspect, purified bacterial colonies are then subjected to diagnostic biochemical tests to obtain bacteriological identification. Finally, agglutination reactions in specific antisera or other assays are then employed to define a particular serotype or serogroup (1,2). This process, which generally requires 36-72 hr for the specific identification of a bacterial pathogen, is oftentimes necessary to establish a diagnosis and select the appropriate antibiotic regimen.

Recent studies indicate that nucleic acid probes have the potential to be used for the identification of clinically significant organisms with a high degree of specificity and sensitivity in relatively simple assays that are much more rapid than classic bacteriological methods. In addition, it is probable that nucleic acid probes can be used to detect nonviable organisms or viable bacteria in phagocytic cells, thus enabling the diagnosis of disease when typical bacteriological analysis fails, such as sometimes occurs with blood or stool cultures. Also, genetic probes can be used to differentiate between pathogenic and nonpathogenic strains of a particular bacterial species, which often requires lengthy and costly bioassays (e.g., the guinea pig keratoconjunctivitis assay to identify enteroinvasive *Escherichia coli* can be avoided by using a nucleic acid probe specific for the invasion gene region of these organisms; see later section on probes for *Shigella* species (3a).

Useful genetic probes for viral and protozoan disease agents are discussed elsewhere in this text. It is our intention to review the presently available information on nucleic acid probes for many clinically important pathogenic bacteria. Rather than simply list every potential genetic probe that has been reported in the literature, we will focus on those probes that have been tested for broad application and those probes that appear to have the greatest practical potential and that have not been discussed elsewhere in this book.

B. Nucleic Acid Hybridization and Detection Technology—A Brief Overview

Bacterial genes are encoded in the DNA of the chromosome or of plasmids (i.e., extrachromosomal genetic elements). The salient features of this DNA are two complementary polynucleotide chains, each composed of covalently linked, repeating nucleotide units of four different bases, organized normally into a regular double helix. The two chains are cross-linked by hydrogen bond base pairing; two bonds between the complementary bases adenine (A) and thymine (T) and three bonds between the complementary bases guanine (G) and cytosine (C). Finally, the stereochemical result of this specific base pair cross-linking between DNA strands is that the polynucleotide chains run in opposite directions relative to the sugar-phosphate linkages within each chain (Fig. 1). Double-stranded bacterial DNA within a clinical specimen can be reduced to single-strand chains under appropriate denaturing conditions, and the single strands can be fixed to solid support material (e.g., a nylon membrane). Fixed (or soluble) single strands of sample nucleic acid can be reacted under reannealing (i.e., hybridization) conditions with single-stranded DNA or RNA probes (e.g., ranging in size from about 15 to several thousand nucleotide bases) that are complementary to a segment of the bacterial DNA. Either DNA or the corresponding RNA sequence can effectively serve as a probe. Hybridization conditions can be chosen so that the reannealed DNA has greater than 90% homology (i.e., less than 10% base-pair mismatches), thus, the reaction is highly specific. Hybridization reaction time, dependent upon factors such as concentration of reactant nucleic acids and temperature of reaction, can range from a few minutes to 24 hr or more. Obviously, for greatest usefulness, probes that can be prepared in high concentrations and that could be used in rapid hybridization reactions (e.g., 5-30 min) would be most beneficial. Under ideal laboratory conditions, with a pure bacterial culture of *Salmonella typhi*, as few as 10^2-10^3 cells could be detected with a radiolabeled DNA probe (3b). Thus, it appears possible that future highly sensitive, hybridization detection systems will allow one to detect 100 cells, or fewer, of a pathogenic organism in a clinical specimen within 1-3 hr.

How are clinical samples prepared for a probe hybridization reaction? Generally, small quantities of the clinical specimen or bacterial colonies are spotted onto hybridization support filters (e.g., 25-50 sample spots per 82-mm diameter filter). Bacteria within each spotted sample are often amplified by placing the filter onto a nutrient agar medium and allowing the bacteria to multiply for 2-16 hr. Filter-borne specimens, either directly spotted or growth amplified, are prepared for hybridization by lysing the bacteria to release nucleic acid. The nucleic acid is denatured by treatment at alkaline pH and single-stranded nucleic acid is preferentially bound to the support filter membrane.

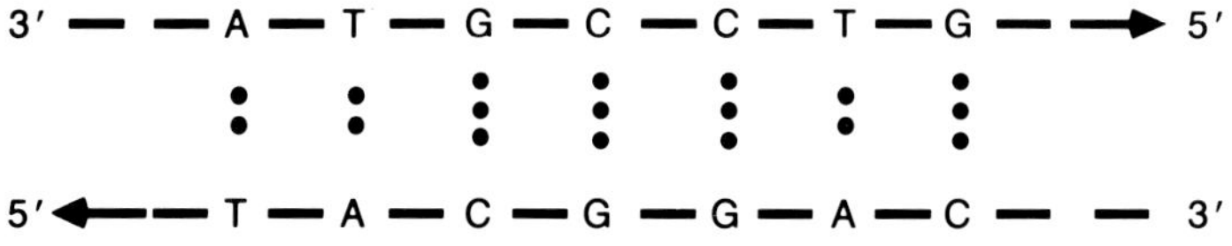

Figure 1 Schematic linear depiction of two nucleic acid strands cross-linked by hydrogen bond base pairing depicted by either two or three dots. Hybridization can occur between DNA and DNA, RNA and RNA, or DNA and RNA single strands, as long as the nucleotide sequences are complementary.

Protein denaturation or protease treatment is used in some procedures to increase the sensitivity of the reaction by removing contaminating material. Finally, the fixed-sample, single-stranded nucleic acid is reacted with soluble, labeled probe, single-stranded nucleic acid to allow reannealing (i.e., hydrogen-bond formation) between complementary sequences. A labeled probe that has reacted with any sample spot can be detected by standard assay for radioactivity or enzymatic activity, depending on the nature of the labeled probe. This procedure has been termed *colony blot hybridization* and is generally applicable to many different types of clinical specimens.

What problems presently exist as barriers to the aforementioned practical application of nucleic acid probes for diagnostic use? First, radiolabeled probes are expensive, have short useful half-lives, require sophisticated, expensive equipment for detection, and are not practical for widespread adaptation. Relatively inexpensive, but highly sensitive, nonradioactive detection systems must be developed to replace radiolabeling of probe DNA. Labeling systems such as biotinylation of probe DNA which, after hybridization, can be detected with avidin conjugates linked to various colorimetrically detectable enzymes, such as alkaline phosphatase, have already been developed and are currently being modified for increased detection sensitivity. Second, optimal conditions need to be established for directly detecting pathogenic organisms within a clinical sample (e.g., stool appears to have some properties that reduce the sensitivity of a probe 10^2 to 10^3-fold) (F. Rubin, unpublished data). Third, for rapid diagnosis, nucleic acid probes must be available in high concentrations and in pure form (i.e., free of contaminating, nonspecific nucleic acid fragments). Small probes, ranging from 15 to about 150 base pairs (bp), can now be synthetically made in large quantity and purified by gel chromatography. Larger probes that exist as part of a recombinant plasmid can present considerable difficulties in obtaining pure, highly concentrated preparations. Although important considerations, these technical problems are surmountable and should be functionally solved in the short term.

C. Nucleic Acid Probe Development—How Are Genetic Probes Made?

The specificity of a nucleic acid probe is based on certain DNA sequences of an organism that are unique to that bacterial pathogen. Recombinant DNA technology has provided methods for gene cloning that have resulted in the construction of recombinant plasmids with inserts of particular DNA regions that can be purified for use as genetic probes. In general, regions of bacterial chromosomal or plasmid DNA that encode virulence-associated traits (e.g., production of toxins or adherence factors, genes controlling tissue invasion) or antibiotic resistance have been chosen for use as genetic probes. Cloned gene regions and restriction endonuclease-fragmented subsets of these DNA regions are then analyzed for specificity and sensitivity in DNA probe hybridization reactions to choose the most suitable genetic probe (e.g., 3b). Finally, the DNA sequence of some genes (e.g., the genes encoding *E. coli* toxins) have been determined and used to construct synthetically small (22-nucleotides long) oligonucleotide probes (4,5). It is, therefore, easy to envision how a toxin gene probe or invasion gene probe can be used to differentiate between a pathogenic and nonpathogenic *E. coli* isolate. However, what constitutes a genus-specific probe? Ribosomal RNA-encoding chromosomal regions are evolutionally distinct in different genera, and portions of these genes have been found to serve as useful genus-specific probes (see Chap. 3). The preceding methods constitute the general approaches taken to construct genetic probes.

Since 1975, genetic probes have been used as experimental tools by molecular biologists and have been developed for use by taxonomists and epidemiologists. This review will focus initially on the earliest developed probes constructed for epidemiological use. However, we will also discuss the recent development of probes made specifically for diagnostic use.

II. NUCLEIC ACID PROBES FOR CLINICALLY SIGNIFICANT BACTERIA

A. Probes for Heat-Labile and Heat-Stable Enterotoxin Genes

1. *Toxin Types and Bioassays*

Since the 1960s, it has been well appreciated that certain *E. coli* serotypes, termed enterotoxigenic *E. coli* (ETEC), are the causative agents of a choleralike diarrhea, often referred to in human disease as traveler's diarrhea. The ETEC strains are a major cause of diarrheal illness in developing countries and among foreign travelers to those areas (6). As the pathogenesis of various intestinal or urinary tract illnesses has been elucidated since 1970, several specific categories of pathogenic *E. coli* serotypes have been defined. In addition to the ETEC strains, there are enteropathogenic *E. coli* (EPEC), enteroinvasive *E. coli* (EIEC), hemorrhagic *E. coli*, and *E. coli* strains that cause urinary tract infections. Identification of many of these pathogenic *E. coli* when mixed with the commensal intestinal *E. coli* presents the clinical microbiologist with a formidable problem. After colony purification and bacteriological identification, serotypic analysis is often used to delineate potential ETEC strains. Proof of enterotoxicity requires subsequent, costly, and time-consuming bioassays that are not easily adaptable to large-scale analysis of numerous clinical bacterial isolates. However, a series of DNA probes for several ETEC-specific enterotoxin genes are now available that simplify tremendously the identification of many ETEC strains. It should be emphasized that these genetic probes distinguish which *E. coli* strains carry the enterotoxin gene(s) but can not demonstrate whether or not the toxin gene is actually expressed.

The hallmark of ETEC strains is their capability to colonize the proximal small bowel and to secrete enterotoxins that stimulate fluid secretion in the animal or human intestine (6). Presently, individual ETEC strains are known to synthesize either a heat-labile enterotoxin, a heat-stable enterotoxin, or both toxin types. Several ETEC-specific enterotoxins have been reported that differ from each other in their mode of action as well as their chemical structure. Some ETEC produce a heat-labile toxin (LT-I), which is a high-molecular-weight antigenic protein that consists of one A subunit and four B subunits held together by noncovalent bonds (7,8). The LT-B subunits bind ganglioside GM1 receptors on the surface of the small intestine epithelial cells (9,10), and appear to be responsible for placement of the toxic (LT-A) part of the molecule. The LT-A peptide crosses the mucosal cell membrane and activates adenylate cyclase in these intestinal epithelial cells, which results in increased intracellular cyclic AMP, followed by net positive secretion of fluid and electrolytes into the bowel lumen and consequent diarrhea (11-14). As one might expect, given the similar pathogenic mechanisms of disease for ETEC and cholera, LT-I is very similar to cholera toxin (CT) in structure and mode of action (9,10,15). Although extensive nucleotide sequence homology has been demonstrated between the plasmid-encoded *E. coli* LT-1 structural genes and the CT structural genes located typically on the *Vibrio cholerae* chromosome (16), these enterotoxins are not completely homologous antigenically (17). In fact, *E. coli* LT-I isolated from humans (LTh-I) and from pigs (LTp-I) have been shown to differ somewhat serologically (18).

Recently, another heat-labile enterotoxin, initially termed LT-like toxin, which was originally described in an *E. coli* isolate from water buffalo, has been renamed type II heat-labile enterotoxin (LT-II). The LT-II toxin induces the rounding of Y1 adrenal cells and the activation of adenylate cyclase, but its activity is not neutralized by antisera prepared against CT or LT-I enterotoxins (19,20). The structural genes for LT-II, which are located on the bacterial chromosome, have recently been cloned and showed no nucleotide sequence homology to LT-I sequences (20).

Some ETEC strains are capable of producing heat-stable enterotoxins (ST) that are poorly immunogenic, low-molecular-weight proteins that cause intestinal fluid secretion by means other than activation of adenylate cyclase. In contrast to LT, the STs from ETEC are structurally very heterogeneous; e.g., several types of ST have been described that differ in mode of action and assay tissue specificity. ST-I is a methanol-soluble enterotoxin that has a biological activity in suckling mice and neonatal piglets (21) that is due to stimulation of intestinal guanylate cyclase, elevation of intracellular levels of cyclic GMP, and consequent fluid and electrolyte efflux (22,23). Two ST-I (also called ST_a)

toxins have been reported. ST-Ia (or STp) was originally isolated from a pig (24), but ETEC that produce STp have been isolated from patients with diarrhea as well as from cows and pigs (25-27). ST-Ib (or STh) was originally isolated from a person with diarrhea in Bangladesh (28), and ETEC that produce STh are almost always isolated from humans rather than animals (25,27). DNA probes for these two ST-I toxins show that they are distinctly different (29).

ST-II (or STb) is a methanol-insoluble enterotoxin that has biological activity in the ligated intestine of older piglets (21). Although this enterotoxin does not induce the formation of cyclic nucleotides in intestinal cells, ST-II appears to cause net fluid efflux by stimulating active secretion of bicarbonate (30). After they cloned the ST-II gene, by recombinant DNA techniques, and determined the nucleotide sequence, Lee et al. (31) reported no homology between the ST-II gene (encodes 71 amino acids) and ST-I gene. A summary of the current LT and ST designations and the bioassays for each toxin type are given in Table 1.

Before nucleic acid probe technology was developed, LT-producing bacterial strains had been identified by the ability of culture filtrates to cause fluid accumulation in rabbit ileal loops after 18 hr (32) or by morphological alteration of Chinese hamster ovary cells (33) or of Y1 adrenal cells (34). ST-I-producing strains, including STp and STh, have been identified by the ability of culture filtrates to cause early fluid accumulation in rabbit ileal loops (32) or by fluid accumulation in the gut of the suckling mouse (35,36), whereas ST-II strains demonstrate fluid accumulation in jejunal loops of pigs only (31). These bioassays are time-consuming, costly, and impractical for the testing of large

Table 1 Current Summary of ETEC Enterotoxins

Enterotoxin designations	Diarrhea stimulated thru	Standard bioassays
LT-I (LTp or LTh)	Adenylate cyclase	CHO cells Y1 adrenal cells 18-hr rabbit ileal loop
LT-II	Adenylate cyclase	CHO cells Y1 adrenal cells
ST-I (STa)		
ST-Ia (STp)	Guanylate cyclase	Suckling mice
ST-Ib (STh)		4-hr rabbit ileal loop
ST-II (STb)	Active bicarbonate secretion	5-hr pig jejunal loop

LT, heat-labile toxin; ST, heat-stable toxin; CHO, Chinese hamster ovary cells. Toxin designations and bioassay responses are described in the text.

numbers of samples. DNA probes, however, have been developed for the improved identification of ETEC. Hybridization reaction with DNA probes makes possible the testing of large numbers of diarrheal samples in a relatively short period, with a relatively simple technique. Hybridization techniques for identifying ETEC have been demonstrated to be at least as sensitive and specific as the traditional method of testing 10 isolated *E. coli* colonies for toxin production in the previously described bioassays (26,27). In addition, DNA probes can be used to detect ETEC directly from clinical specimens or environmental samples without the isolation of purified colonies (discussed later in this section).

2. Probe Development

Because the genetic information for toxin production is generally plasmid encoded in ETEC, DNA probes were developed by recombinant DNA cloning of the toxin genes onto small plasmid vectors; specific fragments of each cloned toxin gene were then purified for use as a specific genetic probe. The following paragraphs summarize hybridization studies that demonstrate the effective use of DNA probes in identifying ETEC. The probes employed in these studies are summarized in Table 2 (20,24,28,31,37).

3. Epidemiological Utility

DNA probes for epidemiological use were first reported in 1980 and used to identify ETEC strains that had been isolated from acutely ill diarrheal patients in Morocco and Bangladesh (39). In this study, one LT-I (~750 bp) or one ST-specific probe was hybridized with the nucleic acid of lysed bacterial colonies fixed on a support filter (the colony blot hybridization method). Probe reactions were compared for sensitivity with standard toxin bioassays, and these two probes were found to be effective in the detection of most, but not all, ETEC strains. Some (~30%) ST only-producing ETEC strains detected by bioassay did not react with this ST-I probe, suggesting the presence of other nonhomologous ST-genes.

Given the success of their initial work, Moseley and co-workers (26), next employed one LT-I probe (approximately 1200-nucleotides long) and two ST probes (STp, 157-nucleotides; STh, 216-nucleotides) separately in a preliminary study of the epidemiology of ETEC in Thailand. Although only 63 ETEC isolates from five population groups were examined, the three probes employed were able to identify all strains. LT^+ strains, analyzed by bioassay, were detected with complete accuracy by the LT-I probe. Most ST^+ isolates reacted with a single ST probe, but five strains reacted with both ST probes, indicating that each carried two ST genes (STp and STh). Strains of ETEC, from rural Thailand, that produce ST only were homologous with only the STh probe, whereas strains isolated in urban Bangkok were homologous with STh, STp, or both probes. Other studies have also demonstrated the specificity of DNA probes for the identification of ETEC in Asia. Isolates of *E. coli* from patients with diarrhea in Bangladesh, Indonesia, the Philippines, and Thailand were tested; all of the

Table 2 Summary of ETEC Probe Studies

Toxin probe designation	Probe from original recombinant plasmid	Dates of studies and refs.
LT-I	(pEWD299) (37)	1980 (39) 1982 (26, 40) 1983 (27, 29, 42, 44) 1984 (25) 1985 (41) 1986 (45a, 46)
LT-II	(pCP3727) (20)	1986 (19)
ST-I		
STp (ST-Ia)	(pRIT10036) (24)	1980 (39) 1982 (26, 40) 1983 (27, 29, 42) 1984 (25) 1985 (41) 1986 (45a)
STh (ST-Ib)	(pSLM004) (28)	1982 (26, 40) 1983 (27, 29, 42) 1984 (25) 1985 (41) 1986 (45a, 46)
ST-II	(pCHL6) (31)	1984 (43) 1985 (41)

The toxins and their designations are described in the text.

984 ETEC, and none of 733 non-ETEC, strains examined hybridized with one or more of the three cloned enterotoxin probes described earlier (29).

In an attempt to evaluate the use of DNA probes to detect ETEC in environmental samples from endemic areas and, thus, to extend their epidemiological studies in Thailand, Echeverria et al. (40) carried out reconstruction experiments in which various amounts of ETEC were added to canal water that contained other bacteria. The ETEC were easily detected by probe hybridization of bacteria filtered from water samples and amplified by overnight growth. In fact, DNA hybridization detection of ETEC was approximately 10^4 times more sensitive than testing 10 individual *E. coli* colonies in standard toxin bioassays (because of the inability to detect *E. coli* at lower concentrations in water that contained other bacteria). In direct analyses of water samples with an LT-I or an STh probe, probe-positive bacteria were found in 9% of 350 samples tested from Thailand. Probe-positive samples occurred about twice as often in water from the homes of patients infected with ETEC than from neighbors' homes. Of 31 probe-positive water specimens, only two specimens were shown by bioassay to contain ETEC, again suggesting that the DNA hybridization assay is considerably more sensitive than standard bioassays for detecting ETEC in water.

To assess further the endemic nature of ETEC in Thailand, bacterial isolates from hand cultures, water, food, and stool specimens were processed and reacted with DNA probes for LT-I, STp, or STh (25). The ETEC strains were found in 30 of 221 (14%) children with diarrhea, in 9% (8/88) of their household contacts, in 8% (8/101) of their neighbors, and only in 2% (32/1379) of inhabitants of 382 homes not associated with ETEC infections. Additionally, ETEC were found significantly more often in water and food and on mothers' hands in homes of children with ETEC-associated diarrhea and in those of their immediate neighbors, than in homes of children without ETEC infections (8/360 versus 3/2290). Drinking water and powdered milk were identified as possible sources of infection in 2 of 18 households with ETEC infections. Similar results were obtained in a more recent longitudinal study of villages in Thailand using DNA probes to detect ETEC (41).

DNA probes have also been used to assess the role of ETEC in childhood diarrhea in Central Africa. Toxin bioassays and probe hybridization methods were equally effective in detecting 24 ETEC strains from 788 *E. coli* isolates obtained from children with diarrhea (42). On the basis of this study, it was concluded that ETEC diarrhea is not a major cause of illness in children under 2 years of age in the Central African Republic.

In 1983, the unusual ST-II gene was cloned by recombinant DNA techniques, and nucleotide sequence analysis revealed no sequence homology with ST-I (31). A 460-bp DNA probe, consisting of an internal segment of the ST-II gene, was found to react with 24 animal isolates and one human isolate (from a patient who had contact with pigs) known to make ST-II toxin, but it did not react with nine ST II-nonproducing ETEC or five non-ETEC strains (31). This ST-II probe was then used to assess the prevalence of ST–II-producing (St-II$^+$) ETEC in pigs, persons, and water at 57 farms in Thailand (43). The ST-II$^+$ ETEC strains were found in 3% (62/2110) of suckling pigs, but 21% of these 62 pigs had diarrhea. Thirty-two percent (181/560) of weaned and 1% (4/457) of adult pigs examined carried ST-II$^+$ ETEC, but none had diarrhea, suggesting that weaned pigs may be a reservoir for these strains. The ST-II$^+$ ETEC were detected in about 5% of clay jars containing water for bathing at these farms and from the rectal swab of only one (at a farm in which 8 of 13 litters of suckling pigs had ST-II$^+$ ETEC-associated diarrhea) of 246 persons living at these pig farms. These data, as well as the results of a more recent study (41), indicate that ST-II$^+$ ETEC may be an enteric pathogen in suckling pigs in Thailand, but it is probably not a significant cause of human disease in that setting. More importantly, weaned pigs might act as a reservoir of infection, and their removal from the suckling pigs' environment may prevent transmission of these bacterial disease agents.

4. *Detection of Enterotoxogenic Escherichia coli in Foods*

The feasibility of detecting ETEC in foods by DNA colony hybridization with a radiolabeled toxin-specific probe was established in 1983 (44). In this study various quantities of ETEC cells were added to scallops that already contained a coliform count of 1.9×10^5 cells/g. Homogenized scallop samples (0.1 ml) were then spread onto nitrocellulose filters, which were incubated for 24 hr on aerobic plate count agar before colony lysis and DNA hybridization procedures were performed. Samples that were contaminated with as few as 15 ETEC per filter before growth amplification were reportedly detected upon audioradiography of the test filters. Although preliminary, these data suggest that DNA probes will be useful in the microbiological testing of foodstuffs. However, the development of nonradioactive hybridization detection systems and the use of synthetic oligonucleotide probes, as well as sample growth amplification techniques, will probably be necessary to make DNA probe hybridization a practical method for testing large numbers of food samples for the presence of pathogenic bacteria.

5. *Oligonucleotide Probes for Enterotoxogenic Escherichia coli*

Recently, two synthetic oligonucleotide probes of 22 bp each were made from the DNA sequences known for the STh and STp genes (4). These radiolabeled probes were tested and were found to be highly specific for ETEC strains encoding either of these ST genes. In other preliminary studies, oligonucleotide probes for LT and ST genes were synthesized, tested, and shown to be as sensitive as bioassays for detection of ETEC (45a,b). The development of oligonucleotide probes for each of the LT or ST genes of ETEC would represent a major step forward in making probe technology widely applicable to epidemiological, diagnostic, and quality control detection of ETEC.

6. *Direct Blotting of Clinical Samples Without Prior Microbiological Identification of Bacteria*

Many of the studies just summarized employed nucleic acid probes to identify ETEC among purified *E. coli* isolates, a process requiring 3 days, or more, from isolation of the initial clinical sample. There is a great need for a rapid, sensitive, specific, and technologically simple method for clinical diagnosis of ETEC infections. Simplification of diagnostic detection procedures can be achieved by inoculating stool samples directly onto a sterile hybridization support filter. Any organisms present in the sample are then allowed to amplify by incubating the inoculated filter on MacConkey agar overnight at 37°C. Growth-amplified samples are then processed for cell lysis, nucleic acid denaturation and fixation to the filter, and hybridization to a tagged (radiolabeled or enzyme-linked) nucleic acid probe. This method of growth amplification of a direct blot of a clinical or environmental sample has been successfully employed for the identification of ETEC in stool (26,28,39,40,46). These studies demonstrate that this direct blot-amplification method is as sensitive as, but much more rapid than, standard microbiological procedures for the identification of ETEC in clinical samples. This procedure offers two other distinct advantages: (a) there is no need to isolate individual *E. coli* from a sample; and (b) a large number of stool samples can be tested easily (e.g., 25-50 samples per 82-mm diameter filter). Although an important step forward, this method will be much more generally applicable with the use of nonradiolabeled probes and reduced growth amplification times.

7. Practical Problems and Lessons

Although DNA probes have proved extremely useful in studies of ETEC-associated diarrhea, some practical problems have arisen in adapting this probe hybridization technology to specific clinical and epidemiological situations. For example, in many of the preceding ETEC studies, approximately 20% of all enterotoxin probe-positive isolates were toxin-negative in all bioassays (25). In one study (40), only 2 of 31 probe-positive water specimens were shown to contain ETEC by bioassay. What are potential reasons for this discrepancy between probe reaction and bioassay results? One explanation is that most ETEC toxin genes are plasmid-borne, and these plasmids may be lost upon subculture, which is necessary to obtain pure cultures for bioassay. Second, there is great difficulty in detecting, by standard bacteriological methods, *E. coli* in water samples that contain much higher numbers of contaminating bacteria. This could explain the discrepancies obtained with water sampling (40). Third, the probe could react with the nucleic acid of dead cells that would not be culturable from the water or stool specimens. A fourth explanation is that the probe is not entirely specific for the enterotoxin gene (i.e., the probe contains contaminating nonspecific DNA sequences). In a recent comparative study aimed at examining the potential for detecting ETEC by DNA probe hybridization technology versus bioassays, only 72% (184/257) of the STh probe-positive colonies produced ST upon bioassay (45). These STh probe-positive, but nontoxigenic isolates, were subsequently shown not to hybridize with shorter versions of the same probe, but they did hybridize with the cloning vector from which the initial larger STh probe was obtained, thereby suggesting that the initial STh probe was probably incompletely purified and contained contaminating plasmid vector sequences. We would like to emphasize here the importance of ensuring that the probe insert fragment is purified away from the cloning vector or any adjacent nonspecific DNA fragments. Single agarose gel electrophoretic separation of a probe fragment from the remainder of the parent recombinant plasmid is usually not sufficient for purification. Appropriate hybridization control experiments should always be conducted to show that the purified toxin gene probe does not contain small amounts of contaminating, nonspecific sequences (i.e., other fragments of the recombinant plasmid carrying the toxin gene). If available, the use of synthetic oligonucleotide probes (from about 15-50 bases long), which do not require purification, is often preferred.

A second problem detected in the preceding epidemiological studies involves the isolation of *E. coli* colonies that are bioassay-positive, but that do not react with the toxin probes tested (26,28,29,39). It seems possible that other undefined toxins present in *E. coli* may have been detected by bioassay but they are not homologous in DNA sequence to the LT-I or ST probes employed in these studies. In fact, Pickett and co-workers (20) recently showed that the LT-II toxin induces rounding of Y1 adrenal cells, but the LT-II genes do not hybridize with LT-I DNA probes. In this study, four ETEC strains obtained from food samples and two strains obtained from the feces of patients with diarrhea were found to produce LT-II toxin and to react with only an LT-II DNA probe.

A third problem surfaced during an epidemiological survey in Thailand (41). Clinical and environmental samples to be tested for ETEC were spotted directly onto nitrocellulose filters that were subjected to overnight incubation on MacConkey agar and subsequently treated for cell lysis, DNA denaturation, and irreversible attachment of the single-stranded nucleic acid to the filter at 65°C at a field laboratory. Additionally, these same samples were characterized bacteriologically, and the resulting *E. coli* were tested for toxigenicity by standard bioassays. However, only 18% of the specimens from which ETEC were identified by bioassay methods were found to hybridize with multiple enterotoxin probes (LT-I, STh, STp, ST-II). Apparently, inconsistent and low oven temperature (insufficient for nucleic acid fixation to the filter), or other sample preparation problems, gave rise to these false-negative results. Alternatively, clinical samples (e.g., stool) may contain substances that interfere with the DNA probe hybridization reaction. The use of hybridization support materials that do not require a high temperature for nucleic acid attachment (e.g., nylon, Whatman 541 paper) and the use of proper probe-positive and probe-negative control samples on each test filter would help alleviate the aforenoted problem(s).

The use of toxin-specific nucleic acid probes to study the epidemiology of ETEC-associated diarrhea has established the tremendous usefulness of hybridization methodology. The results obtained from the preceding studies strongly suggest that nucleic acid probe hybridization technology will be useful in epidemiology, diagnostic microbiology, and food quality control testing programs.

B. Probes for Shiga-Like Toxin (Verotoxin) Genes

O'Brien et al. (47) initially reported that strains of *E. coli* synthesize Shiga-like toxin (SLT; also termed *verotoxin* because of its effect on cultured VERO cells), a cell-associated cytotoxin that is neutralized by antibodies against purified Shiga toxin from *Shigella dysenteriae* serotype 1. In recent more comprehensive studies of many different bacterial strains, O'Brien and co-workers (48,49) showed by bioassay that SLT activity is present in many EPEC, ETEC, enteroinvasive *E. coli* (EIEC), hemorrhagic *E. coli* (EHEC), as well as *E. coli* causing hemolytic uremic syndrome and in *Vibrio cholerae* and *V. parahaemolyticus* strains. The production of high levels of SLT in *E. coli* appears to be associated with increased virulence because strains isolated from healthy humans have been shown to make cytotoxin at lower levels than strains from sources with active disease (50). Furthermore, toxin neutralization by antibodies has defined at least two types of SLT (51). DNA probes have been constructed by cloning SLT genes from two different toxin-converting phages present in the same 0157:H7 *E. coli* strain 933 (52,53a). Hybridization data indicated a distinct heterogeneity between the gene sequences of these biologically similar cytotoxins (51,53a,b). Specific DNA probes for each cytotoxin will be useful epidemiologically in assessing the occurrence of these genes in bacterial isolates.

C. Probes for *Escherichia coli* Colonization

1. Enteropathogenic Escherichia coli

Enteropathogenic *E. coli*, which comprise characteristic O-serogroups, are associated epidemiologically with diarrhea in children but do not elaborate enterotoxins of the heat-labile (LT) or heat-stable (ST) types (54). Electron-microscopic studies of EPEC infection in intestinal biopsies (55) and in animal models (56,57) have revealed a pathognomonic lesion manifested by intestinal brush border alteration and epithelial cell–plasma membrane pedestal formation at the sites of EPEC attachment. The EPEC strains colonize the epithelial mucosa but do not invade epithelial cells. Recent studies have shown that EPEC strains produce a Shiga-like enterotoxin that may be involved in pathogenesis (58). Cravioto et al. (59) observed that most EPEC strains adhere to Hep-2 cells in tissue culture in the presence of mannose (this type of adherence is independent of type 1 fimbriae). The Hep-2 adherence of EPEC strain E2348/69 (0127:H6) was reported by Baldini et al. (60) to be plasmid encoded. This adhesive property (subsequently termed EPEC adherence factor or EAF), which is necessary for virulence in humans (61), appears to be an important virulence factor of EPEC.

A 1-kb pair DNA fragment from within the Hep-2 adherence gene region of plasmid (pMAR2), from strain E2348/69, was recently used as a DNA probe to study the epidemiology of EPEC in Peru, by analysis of the stools of Peruvian infants with or without diarrhea (62). Hybridization of the probe with colony blots of known EPEC and non-EPEC strains demonstrated that the EAF probe was specific and sensitive in detecting strains of *E. coli* that are able to adhere to Hep-2 cells. The frequency of EAF-positive strains, detected equally well either by bioassay or probe hybridization, varied with serotype and resulted in categorizing EPEC strains into two groups. The EPEC I strains are generally adherent to Hep-2 cells (i.e., they are DNA probe-positive) and include the serotypes (055, 0111, 0119, 0127, 0128, and 0142) most commonly associated with this diarrheal syndrome. Another less common set of EPEC strains, termed EPEC II (including, among others, serotypes 044, 086, and 0114) are rarely Hep-2 adherent (i.e., they are DNA probe-negative) and can only be detected by serotype analysis. This EAF probe has proved useful in establishing that there are at least two categories of EPEC strains. However, the development of other DNA probes that can be used to identify all EPEC strains will be necessary to study most effectively the epidemiology of EPEC diarrhea—to assess age-specific and area-specific incidence and to identify modes of transmission.

2. Enterotoxigenic Escherichia coli

In addition to the EPEC Hep-2 adherence factor antigen just described, other colonization factors are present in strains of *E. coli*. Fimbrial or other surface antigens on the bacterial cell of some ETEC and EPEC strains have been reported and are involved in colonization of the intestinal epithelium. Colonization factor antigens CFA/I, CFA/II, CFA/III, and the antigen E8775 have been identified in *E. coli* strains from humans with ETEC diarrhea (63-66). Many of the human colonization factor antigen genes have recently been cloned (67,68), as have animal-derived *E. coli* colonization factor antigen genes (69,70) and, thus, another set of DNA probes are available to detect *E. coli* that encode these specific factors.

D. Probes for the Causative Agents of Dysentery: Enteroinvasive *Escherichia coli* and *Shigella*

All four species of *Shigella*, representing 32 different serotypes, and certain *E. coli* strains (EIEC) falling into at least nine serotypes are capable of causing classic dysentery. This disease is characterized by bacterial invasion of the colonic mucosal epithelium, intracellular multiplication by the bacteria, and focal microulcerations of the colon. The causative bacteria are shed in the stool. Standard microbiological assay for shigellae from stools of infected individuals generally requires 24-48 hr for positive identification, but EIEC isolates are much more difficult to identify. Diagnostic confirmation of a virulent strain requires a subsequent time-consuming and expensive bioassay involving bacterial invasion of the guinea pig corneal epithelium and subsequent provocation of keratoconjunctivitis (the Sereny assay; 71). Thus, the availability of a more rapid and inexpensive screening method for invasive isolates would facilitate both identification and epidemiological studies of these organisms.

1. Nature of the Probes

Invasive isolates of *Shigella* and EIEC have each been shown to carry a large plasmid (120-140 Md) that encodes a functionally and genetically homologous set of genes necessary for epithelial cell penetration (reviewed in Ref. 72).

Several DNA probes have now been constructed by cloning various fragments of these invasion-specific plasmids. As described later, those nucleic acid probes that represent known invasion gene sequences appear to function as highly specific and sensitive DNA probes for the detection of invasive *Shigella* species and EIEC.

In 1984, Boileau and co-workers (73) cloned three *Eco*RI fragments, of uncharacterized function that were shared in cosmid recombinant plasmids that encoded the *S. flexneri* 5 invasion genes, of 7.6, 11.5, and 17 kb in length from a *S. flexneri* serotype 5 invasion plasmid. The 17-kb fragment, but not the other two fragments, hybridized specifically to 40 selected invasive *Shigella* and EIEC strains, but not to 40 negative control strains, including other enteroinvasive bacteria (e.g., *Salmonella, Yersinia, Vibrio*). In a further prospective study, this probe was shown to react positively with all of 452 *Shigella* isolates and with none of 428 gram-negative bacteria isolated from diarrheal stools that were not *Shigella* or EIEC (73). More recently, Small and Falkow (74) cloned a 2.5-kb *Hin*dIII fragment, of unknown function, from the invasion plasmid of an EIEC strain. In very preliminary testing this 2.5-kb probe was found to hybridize with 12 different invasive *Shigella* and EIEC strains, but did not hybridize with eight isolates of avirulent *Shigella* and EIEC or with 14 other negative control bacteria (e.g., *Y. pseudotuberculosis, Y. enterocolitica, Salmonella*).

Both of these probes have been shown to be plasmid-specific, do not hybridize to one another (75), and do not hybridize to the chromosome (73,74). Because of the importance of having an invasion-specific nucleic acid probe, Wood et al. (75) recently conducted a comparison screening assay that used the aforementioned two probes and the Sereny reaction to detect invasive organisms among 42 *Shigella* and 29 *E. coli* clinical isolates. All of 20 Sereny-positive *Shigella* and 19 Sereny-positive EIEC strains reacted with the DNA probes; none of nine nonenteroinvasive *E. coli* isolates reacted with either probe. However, both probes inappropriately reacted with 3 of 23 (13%) Sereny-negative (noninvasive) *Shigella* or EIEC strains. Thus, although these plasmid-specific probes are very useful in identifying *Shigella* and EIEC, it is apparent from these limited studies and other reports (74–76) that these probes are not invasion-specific and can generate a significant number of false-positive results.

Recent studies have revealed that four plasmid-specified outer membrane proteins (*a*, 78 kd; *b*, 57 kd; *c*, 43 kd; *d*, 39 kd) are closely associated with the invasive phenotype (77). Moreover, these four proteins are important immunogens that stimulate significant antibody titers in shigella infected monkeys and humans (78). Three of these four characteristic, invasion-associated, outer membrane proteins (*b, c, d*) have recently been cloned (79a). In an effort to determine if DNA fragments from within these invasion-associated genes can be used as invasion-specific nucleic acid probes, three DNA fragments were tested as probes in hybridization experiments (3a). The three probes are 470 bp (*b* peptide fragment), 1750 bp (*c* peptide fragment), and 870 bp (*d* peptide fragment). In preliminary studies, all three probes behaved identically and reacted with 34 invasive *Shigella* and EIEC strains, but not with eight noninvasive *Shigella* strains nor with more than 300 other nondysenteric, pathogenic enteric bacteria including *Salmonella* spp., EPEC, ETEC, *Citrobacter, Enterobacter*, hemorrhagic *E. coli, Klebsiella, Vibrio cholerae, V. parahaemolyticus, Yersinia* spp., and *Aeromonas*. Theoretically, these invasion-associated gene fragments should act as more suitable invasion-specific probes, and they should not generate the significant number of false-positive results observed with the 17-kb and 2.5-kb probes discussed earlier. In fact, of nine noninvasive EIEC isolates, all of which were observed to hybridize inappropriately with the 17-kb *Eco*RI probe, only one hybridized to the C peptide probe (76; David N. Taylor, personal communication). Thus, although further testing is necessary, these invasion gene probes appear to be very specific for detecting classic invasive dysentery bacilli.

In a recent study of dysentery in Thailand (79b), the dysenteric syndrome (abdominal pain, frequent stools containing blood and leukocytes) was found to be caused by *Shigella* spp., *Campylobacter* spp., *Salmonella* spp., *Pleisiomonas shigelloides, Entamoeba histolytica, V. parahaemolyticus*, or enteroinvasive *E. coli*. Furthermore, *Shigella* and EIEC strains were found to be responsible for approximately 50% of the dysentery studied among 200 Thai children with mucoid or bloody diarrhea. The development and use of probes to detect these other diarrheal disease agents would be extremely beneficial for both diagnostic and epidemiological purposes.

2. *Toward a Practical Field Assay*

The aforenoted dysentery probe development studies were conducted with radiolabeled probe nucleic acid and purified bacterial colonies. A practical assay for detection of dysentery bacilli in the stool demands the use of direct stool blots, without colony purification, and the use of sensitive nonradiolabeled probes that can be detected colorimetrically without the use of costly equipment. Sethabutr et al. (80) biotinylated the 17-kb *Eco*RI probe and found it to react specifically with colony blots of 52 EIEC strains and to be nonreactive with 16 non-EIEC strains examined. Furthermore, when stool samples were growth amplified on MacConkey agar at 37°C overnight and then blotted to filters, 11 of 13 stools from children infected with shigellae or EIEC reacted with the biotinylated probe; stools from 43 children without shigellae or EIEC did not react with the probe. Taylor and co-workers (79b) recently obtained very similar results by reacting a radiolabeled 17-kb *Eco*RI fragment probe with prepared fecal bacterial growth samples. However, in this study only 27 of 88 (31%) stools from children infected with shigellae were found to hybridize with the probe. These authors suggest that shigellae are shed in the stool in much lower numbers than EIEC in infected individuals. These studies demonstrated that EIEC and, sometimes, *Shigella* strains can be identified from stool without colony purification. To be very practical, however, the sensitivity of this type of assay needs considerable improvement.

E. Detection of *Salmonella* species

1. *Probe for Salmonella typhi, the Typhoid Fever Bacillus*

Typhoid fever continues to be a serious health problem in developing countries and is endemic in many areas of the world. A nucleic acid probe for the detection of *S. typhi*, the causative agent of typhoid fever, would be extremely helpful for rapid diagnosis of acutely ill patients, for the detection of typhoid carriers among the population, as well as for epidemiological studies.

All strains of *S. typhi, S. paratyphi* C, and a few atypical *Citrobacter* strains are capable of synthesizing a capsular antigen, termed Vi, that is associated with the virulence of *S. typhi* (81). Two distinct chromosomal genetic loci, *viaA* and *viaB*, have been shown to be involved in the synthesis of this antigen (82). A *C. freundii viaB* locus, contained in a recombinant cosmid, was recently subcloned and various fragments were tested as potential probes in colony blot hybridization assays (3b). An 8.6-kb *Eco*RI fragment was found to be highly specific for strains encoding the Vi capsular antigen, when tested against more than 180 different enteric bacteria including representatives of *Shigella, E. coli*, and *Salmonella* groups A, B, C1, C2, C3, D1, D2, E1, E2, E3, E4, F, G1, G2, H, I, J, K, L, M, N, O, P, Q, R, S, T, U, V, W, X, Y, Z, 51-55, 66, and 67 (3b). Under optimized laboratory conditions, as few as 10^2-10^3 lysed *S. typhi* cells could be detected with this radiolabeled 8.6-kb probe (3b). In more recent studies, this 8.6-kb probe has been employed to examine a variety of microbiologically characterized, gram-negative enteric strains from stool specimens obtained in endemic areas of Lima, Peru and Jakarta, Indonesia. The probe was found to react with fewer than 0.5% of the nontyphoid isolates and reacted positively with more than 99.5% of the *S. typhi* strains analyzed (83a). Thus, this probe is specific for Vi-antigen-encoding bacteria and should be very useful for the diagnostic and epidemiological detection of *S. typhi*. However, the practical utility of this probe will be fully realized when it can be used as a nonradiolabeled probe to analyze stool samples directly, as well as in analyses of blood samples (Rubin, F.A. et al., manuscript in preparation) from acutely ill typhoid patients.

2. *Genus-Specific and Other* Salmonella *Probes*

In contrast to the typhoid-specific probe just described, genus-specific *Salmonella* probes have been prepared from the cloned DNA of *S. typhimurium* strain C23566 and have been used to detect the presence of *Salmonella* spp. in foods (83b). The use of a genus-specific *Salmonella* probe in food analyses is discussed in another chapter of this book.

There have been a number of outbreaks of salmonella-induced gastroenteritis with widespread transmission (84–86) in which the use of a salmonella-specific DNA probe(s) might have been helpful in screening large numbers of patients as well as food and milk samples for the potential sources of infection. For example, *S. dublin* infections are occurring with increased frequency throughout the western United States (87) and the United Kingdom (88), and are significantly correlated with the ingestion of raw milk. Salmonella-specific nucleic acid probes could be used to test large numbers of cows, raw milk samples, and milk-processing equipment to lessen the occurrence of milk-borne salmonellae. Significant control of *Salmonella* spp., the most frequently isolated bacterial pathogen in outbreaks of food-borne disease in the United States, can be achieved by improvements in certain areas of food processing and quality control. However, in addition to the problem of detecting contaminated food products or food-processing machinery, animals themselves are an established reservoir of non-typhoid salmonellae that infect man and other animals (89–92). The use of DNA probes in veterinary medicine and domestic animal breeding could have a substantial impact in identifying etiological agents of disease in animals and in subsequently reducing veterinary-associated human and animal diseases.

There now are no available species-specific nucleic acid probes that can be used directly for the detection of nontyphoid salmonellae in food, water, or clinical samples. However, cloned random chromosomal sequences of *S. dublin* have been used as probes to differentiate between strains of *S. dublin, S. typhimurium*, and *S. enteritidis* in indirect fingerprint analyses of restriction endonuclease-digested, whole-cell DNA (93), as described in Chapter 5 of this book. Although this technique has great epidemiological value, its usefulness is limited by expense and time considerations. However, all three *Salmonella* species mentioned earlier, have been reported to carry a virulence-associated plasmid (94–97), and it seems likely that species-specific probes for direct examination of samples will be developed in the near future from these sequences.

F. Probes for Vibrio Virulence Determinants

1. Probes for Cholera Toxin Genes

Vibrio cholerae is the causative agent of cholera and continues to be a serious public health problem in many areas of the world. An important factor in the control of cholera is rapid detection of *V. cholerae* for diagnosis as well as for the identification of sources of infection. According to current classification systems, 0 group 1 (01) strains of *V. cholerae* are responsible for cholera and all other *V. cholerae* strains are grouped as non-01 (98,99). *Vibrio cholerae* 01 strains adhere to, and multiply on, the human small-intestinal mucosa and elaborate cholera toxin that stimulates a massive fluid and electrolyte efflux. Cholera toxin (CT), which is encoded by the chromosome (100), shares mechanisms of action, subunit structures, and immunological relatedness with ETEC LT-I (9,10,13-17,101). In addition, significant nucleotide sequence homology exists between the genes for CT and LT (102). Not surprisingly, then, an LT-I nucleic acid probe was initially used, under hybridization conditions of reduced stringency, to detect phenotypically toxigenic 01 *V. cholerae* from cholera patients (102). This probe did not hybridize with phenotypically nontoxigenic environmental strains of *V. cholera.* In a separate report, an *E. coli* LT-I probe was also employed in an epidemiological study to analyze the molecular relatedness of the toxin gene regions of *V. cholerae* 01 strains from the United States Gulf Coast (103). When patterns of hybridization of the probe to *V. cholerae* chromosomal restriction endonuclease digests were compared, the toxin gene sequences of *V. cholerae* 01 strains from the United States Gulf Coast area were identical, but were distinctly different from other 01 *V. cholerae* strains isolated throughout the world. More importantly, two strains of enterotoxigenic, non-01 *V. cholerae* isolated from patients with severe diarrhea in Louisiana were probe-positive and had cholera toxin gene regions that differed in the restriction fragment pattern from the United States coastal *V. cholerae* 01 strains.

More recently, a group of *V. cholerae* isolates from Thailand and the Philippines were examined by DNA hybridization by using cloned cholera toxin genes as the DNA probe. All of 339 01 *V. cholerae* isolates from diarrheal patients were homologous with the probe and only 5 of 237 non-01 *V. cholerae* environmental isolates reacted with the probe. None of 44 non-01 *V. cholerae* isolates from persons with diarrhea hybridized with the probe (104). A number of different CT gene probes have now been constructed by recombinant DNA procedures (105-107). Thus, the CT probe offers promise for the rapid identification of the pathogenic, toxigenic 01 *V. cholerae.*

2. *Probes for Hemolysins*

Other species of the family *Vibrionaceae*, in addition to *V. cholerae*, have been associated with human disease. At present, 10 potentially pathogenic species of *Vibrio* are recognized (108). In addition to gastrointestinal diseases, strains of *Vibrio* have been isolated from wound infections and patients with septicemia. The genes encoding the thermostable direct (Kanagawa) hemolysin of *V. parahaemolyticus* have been cloned and recently employed as a probe (109-111). Also, the cytotoxin-hemolysin gene of *V. vulnificus* has been cloned (112), as well as hemolysin genes from *V. cholerae* (113). These cloned gene probes should be helpful in elucidating the mechanisms by which these pathogens cause disease and may have important epidemiological uses (e.g., detecting vibrios in contaminated food or water).

G. Probes for *Yersinia* species

The incidence of human infection caused by *Y. enterocolitica* has increased worldwide over the past 15 years (114-116). In the United States, *Y. enteroco-*

litica is recognized as an important food-borne pathogen (114). The development of genus-specific or species-specific probes for *Yersinia* spp. would undoubtedly help in the diagnosis of and epidemiological control of diseases caused by yersiniae.

Virulence-associated (Vwa) plasmids have been detected in pathogenic strains of *Y. enterocolitica* (117,118), *Y. pestis* (117), and *Y. pseudotuberculosis* (119). These plasmids encode necessary virulence traits (e.g., Ca^{2+} dependency; 120) and share DNA sequence homology. In fact, Hill and co-workers (122,123) have reported the use of plasmid fragments in DNA probe hybridization reactions for the detection of virulent *Y. enterocolitica* in food. Gemski et al. (*Contr. Microbiol. Immunol.*, in press) have cloned from the 42-Md plasmid of a *Y. enterocolitica* 0:8 strain two different DNA segments that act as specific and sensitive DNA probes. Preliminary studies indicate that one probe (4.5 kb) is specific for all yersiniae carrying the (Vwa) plasmid, whereas a second 2.5-kb probe is specific for only *Y. enterocolitica* strains carrying the (Vwa) plasmid. Recently, Isberg and Falkow (121) have cloned a gene from *Y. pseudotuberculosis* that is responsible for epithelial cell invasion. However, the usefulness of this DNA region as a nucleic acid probe for *Yersinia* pathogens has not been reported.

H. Detection of *Neisseria gonorrhoeae*

A rapid and inexpensive assay to detect *N. gonorrhoeae* in clinical samples would go a long way in helping to control this disease of worldwide prevalence. Despite the lack of information on *N. gonorrhoeae* virulence mechanisms, Totten and co-workers (124) initially attempted to see if the 2.6-Md cryptic plasmid isolated from *N. gonorrhoeae* strain NRL8038 could serve as a reasonably specific DNA probe. These studies showed that DNA sequences homologous with this gonococcal plasmid were present in 93% (124 of 134) of the *N. gonorrhoeae* strains tested but were absent in *N. meningitidis, Branhamella catarrhalis*, strains of *Haemophilus*, and 95% (42/44) of commensal neisserial isolates. Furthermore, as few as 100 colony-forming units (cfu) of *N. gonorrhoeae* could be detected by filter hybridization procedures (124).

DNA hybridization has also been employed to detect *N. gonorrhoeae* directly in urethral exudates (124,125). In a study of men with urethritis in Seattle, 89% (63 of 71) of the exudates that were culture-positive for *N. gonorrhoeae* were probe-positive, and all 42 culture-negative exudates were nonreactive with the plasmid probe (124). In a separate study of urethral specimens from men with urethritis from several African and Asian countries, 180 of 216 specimens were found to be probe-positive and culture-positive for *N. gonorrhoeae*, 27 specimens were negative in both tests, and 9 specimens gave discordant results (125). The latter study demonstrated that samples can be spotted onto nitrocellulose filters and transported without loss of sensitivity from the field to a central, well-equipped laboratory for processing. Although this plasmid probe generates 5–10% false-negative reactions, it should serve as an important tool in epidemiological studies of gonorrhea.

I. Probes for Other Clinically Important Bacteria

DNA probes for the detection of *Campylobacter, Legionella*, and various anaerobic bacteria have been developed and are discussed elsewhere in this book.

J. Probes for Antibiotic-Resistance Genes

A large group of DNA probes have been developed that will detect specific antibiotic-resistance genes within a variety of clinically important bacteria. Because of the large amount of antibiotic resistance that exists in bacterial pathogens and the need to assess by rapid techniques the types of resistance genes carried by a particular organism, it was deemed appropriate to mention this series of probes. The present state of antibiotic-resistance gene probes has recently been reviewed by Tenover (126). It is clear that these probes have been, and will continue to be, extremely useful in epidemiological studies aimed at understanding the dissemination and evolution of antibiotic-resistance genes. Although not yet practical, rapid probe analyses of the antimicrobial susceptibility of organisms in clinical specimens represents a future possibility.

III. CONCLUDING REMARKS

A. What is the Present Applicability of Probes in Diagnosis and Epidemiology?

The tremendous utility of nucleic acid probes in epidemiological studies has now been firmly established as a result of many of the studies we have discussed (e.g., 25,26,28,39–43,46,79,80). Epidemiological analyses of the occurrence of certain pathogens (e.g., EIEC), on a statistically meaningful scale, in many areas of the world has not been practical by conventional methods which ultimately depend on costly and time-consuming bioassays. It now appears, however, that nucleic acid probe use will make possible epidemiological surveys that have not heretofore been practical. Also, probe use will greatly facilitate other epidemiological studies that, to date, have been limited by cumbersome bacteriological, serological, or bioassay procedures.

The employment of probes for diagnostic identification of causative disease agents is now possible, but a few practical problems need to be overcome before widespread application of this technology occurs. As opposed to epidemiological applications for which speed is not essential, diagnostic uses of probes will require relatively rapid and direct assays of clinical specimens. The advantages of this technology are that one can detect a pathogen in a mixed bacterial population and, in some instances, even if the organism is nonviable. In addition to being employed in the diagnosis of acutely ill patients, the typhoid fever bacillus probe may prove useful in detecting typhoid carriers. It has not previously been possible, because of cost and other factors, to test entire populations ($> 10^6$ individuals) to detect the carriers that apparently are responsible for maintaining the endemicity of the disease in a particular region. In addition to applications for human disease diagnosis, it seems likely that nucleic acid probes will benefit both disease diagnosis and detection of pathogen carriage in veterinary medical applications. We anticipate that, in the near future, nucleic acid probes for diagnostic use will not supplant current microbiological identification methods but, rather, will serve as an adjunctive, rapid technique for identification of a pathogen in highly industrialized countries. However, in special niches within those countries and in developing nations, probe technology will be used as a primary diagnostic technique and will make a dramatic impact on disease diagnosis.

What are some current limitations to probe use in disease diagnosis and in epidemiological studies? There are still only a limited number of pathogen-specific or virulence trait-specific probes available; more probes need to be developed and tested for specificity and sensitivity in practical hybridization assays. Although not absolutely necessary, the use of sensitive, nonradioactively labeled probes (e.g., biotinylated or enzymatically tagged probes) that have a long shelf-life and that can be detected without the use of expensive equipment would definitely facilitate the broader application of this technology. Third, refinement of clinical sample preparation and hybridization procedures should allow sufficient levels of probe sensitivity to generate a highly reproducible assay that can rival standard microbiological procedures. For example, a recent report indicated that only about 30% of stools, that were confirmed microbiologically as being *Shigella*-positive, could be detected by DNA probe hybridization after fecal growth amplification (79b). These authors suggested that shigellae are shed in low numbers. Alternatively, however, stools may contain some factor that interferes with the hybridization assay. Thus, each probe hybridization assay of a clinical specimen for a particular organism will have to be optimized for a reproducibly sensitive response. Many of the DNA probes reported herein have been developed by using purified DNA fragments obtained from recombinant plasmids. Although this approach has been very useful in the development of probe technology, it is difficult and time-consuming to prepare high concentrations of well-purified probes. As pointed out previously, contaminating adjacent DNA regions can cause spurious hybridization results (45a). The recent availability of automated oligonucleotide synthesizers has made the preparation and use of oligonucleotide probes (e.g., 15-50 bp in length) inexpensive and simple. By sequencing the ends of a large nucleic acid probe, one can obtain short nucleotide sequence information from which an oligonucleotide probe can be synthesized. There are several other advantages to synthesized probes. Short nucleic acid probes hybridize much faster than long probes so that the hybridization and overall assay time can be reduced significantly (e.g., a 25-bp probe can be hybridized completely in 30 min under conditions that would require 6 hr for a 3400-bp probe, at equal concentration, to hybridize; 5). Second, oligonucleotides can be synthesized overnight at very high concentrations. Third, oligonucleotide probes can be labeled during synthesis (5). Also, probes prepared from double-stranded DNA fragments hybridize less efficiently than single-stranded probes, because probe self-annealing decreases the effective concentration available for reaction with the sample. Additionally, oligonucleotide probes make possible the examination of small DNA regions, and under conditions of high hybridization stringency, can be used to detect single base-pair mismatches within a sequence (128). As a practical example, a 415-bp probe for the Kanagawa hemolysin gene of *V. parahaemolyticus* was found to be unable to distinguish all Kanagawa hemolysin-positive (Kp^+) from hemolysin-negative (Kp^-) strains (110). However, small 19- to 21-bp oligonucleotide probes were able to separate Kp^+ from Kp^- strains satisfactorily (111). Thus, future attention to the considerations summarized in the foregoing will undoubtedly lead to increased use of nucleic acid probes in epidemiological and diagnostic applications.

B. Potential for Probe Use in Quality Control

Although not the major focus of this review, the results of several of the studies presented earlier (40,44,83b,122,123) suggest that nucleic acid probes will be useful in specific areas of quality control of foodstuffs, food-handling equipment, products for human use, and environmental resources. Some of the same limitations to probe use mentioned in the previous section also apply here. In addition, there is a great potential for the use of this technology to determine the food source of a disease outbreak as well as reservoirs (e.g., water supply) of infection. The capability of probe technology to test, inexpensively, large numbers of samples with specificity and sensitivity ensures that nucleic acid probes will be an enormous benefit to quality control testing and in the public health arena.

ADDENDUM

Since the completion of this chapter in mid-1987, probes have been reported for the detection of *Listeria monocytogenes* (128), *Pseudomonas aeruginosa* (129), non-01 *Vibrio cholerae* (130) and *E. coli* serotypes causing hemorrhagic colitis and hemolytic uremic syndrome (131). Studies with biotinylated probes (132, 133) and alkaline phosphatase-conjugated oligonucleotide DNA probes (134, 135) demonstrate advances in the development of nonradioactive detection systems. In addition, an RNA transcript probe was recently reported for detection of heat-stable enterotoxigenic *E. coli* (136), as an approach to eliminate preparing probe DNA from recombinant plasmids.

ACKNOWLEDGMENTS

We thank M. Venkatesan, David Taylor, Barbara Murray, and Walter Hill for kindly supplying their unpublished results and J. Newland for supplying unpublished data and for his helpful comments on the manuscript. We are also grateful to Earnestine Durham for her expert help in typing this chapter.

REFERENCES

1. Edwards, P. R., and W. H. Ewing, *Identification of* Enterobacteriaceae, 3rd ed. Burgess Publishing Co., Minneapolis, pp. 146–207 (1972).
2. Lenette, E. H., A. Balows, W. J. Hausler, and J. P. Truant (eds.), *Manual of Clinical Microbiology*, 3rd ed., American Society for Microbiology, Washington (1980).
3a. Venkatesan, M., J. M. Buysse, E. Vandendries, and D. J. Kopecko, Development and testing of invasion-associated DNA probes for detection of *Shigella* spp. and enteroinvasive *Escherichia coli. J. Clin. Microbiol. 26*: 261–266 (1988).
3b. Rubin, F. A., D. J. Kopecko, K. F. Noon, and L. S. Baron, Development of a DNA probe to detect *Salmonella typhi. J. Clin. Microbiol. 22*:600–605 (1985).
4. Hill, W. E., B. A. Wentz, W. L. Payne, J. A. Jagow, and G. Zon, DNA colony hybridization method using synthetic oligonucleotides to detect enterotoxigenic *Escherichia coli*: Collaborative study. *J. Assoc. Off. Anal. Chem. 69*:531–536 (1986).
5. Bryan, R. N., J. L. Ruth, R. D. Smith, and J. M. LeBon, Diagnosis of clinical samples with synthetic oligonucleotide hybridization probes. In *Microbiology–1986*, L. Leive (ed.). American Society for Microbiology, Washington, pp. 113–116 (1986).
6. Sack, R. B., Human diarrheal disease caused by enterotoxigenic *Escherichia coli. Annu. Rev. Microbiol. 29*:333–353 (1975).
7. Clements, J. D., and R. A. Finkelstein, Isolation and characterization of homogeneous heat-labile enterotoxins with high specific activity from *Escherichia coli* cultures. *Infect. Immun. 24*:760–769 (1979).
8. Kunkel, S. L., and D. C. Robertson, Purification and chemical characterization of heat-labile enterotoxin produced by enterotoxigenic *Escherichia coli. Infect. Immun. 25*:586–596 (1979).
9. Donta, S. T., and J. P. Viner, Inhibition of steroidogenic effects of cholera and heat-labile *Escherichia coli* enterotoxins by GM1 ganglioside: Evidence for a similar receptor site for the two toxins. *Infect. Immun. 11*:982–985 (1975).

10. Holmgren, J., Comparison of the tissue receptors for *Vibrio cholerae* and *Escherichia coli* enterotoxins by means of gangliosides and natural cholera toxoid. *Infect. Immun. 8*:851–859 (1973).
11. Evans, D. J., Jr., L. C. Chen, G. T. Curlin, and D. G. Evans, Stimulation of adenyl-cyclase by *Escherichia coli* enterotoxin. *Nature 236*:137–138 (1972).
12. Kantor, H. S., P. Tao, and S. L. Gorbach, Stimulation of intestinal adenyl cyclase by *Escherichia coli* enterotoxin: Comparison of strains from an infant and an adult with diarrhea. *J. Infect. Dis. 129*:1–9 (1974).
13. Moss, J., and S. H. Richardson, Activation of adenylate cyclase by heat-labile *Escherichia coli* enterotoxin: Evidence for ADP-ribosyl transferase activity similar to that of choleragen. *J. Clin. Invest. 62*:281–285 (1978).
14. Zenser, T. V., and J. F. Metzger, Comparison of the action of *Escherichia coli* enterotoxin on the thymocyte adenylate cyclase-cyclic adenosine monophosphate system to that of cholera toxin and prostaglandin E_1. *Infect. Immun. 10*:503–509 (1974).
15. Gill, D. M., and S. H. Richardson, Adenosine diphosphate-ribosylation of adenylate cyclase catalyzed by heat-labile enterotoxin of *Escherichia coli*: A comparison with cholera toxin. *J. Infect. Dis. 141*:64–70 (1980).
16. Yamamoto, T., T. Nakazawa, T. Miyata, A. Kaji, and T. Yokota, Evolution and structure of two ADP-ribosylation enterotoxins, *Escherichia coli* heat-labile toxin and cholera toxin. *FEBS Lett. 169*:241–246 (1984).
17. Lindholm, L., J. Holmgren, M. Wikstrom, U. Karlsson, K. Andersson, and N. Lycke, Monoclonal antibodies to cholera toxin with special reference to cross-reactions with *Escherichia coli* heat-labile enterotoxin. *Infect. Immun. 40*:570–576 (1983).
18. Tsuji, T., S. Taga, T. Honda, Y. Takeda, and T. Miwatani, Molecular heterogeneity of heat-labile enterotoxins from human and porcine enterotoxigenic *Escherichia coli. Infect. Immun. 38*:444–448 (1982).
19. Guth, B. E. C., C. L. Pickett, E. M. Twiddy, R. K. Holmes, T. A. T. Gomes, A. A. M. Lima, R. L. Guerrant, B. D. G. M. Franco, and L. R. Trabulsi, Production of type-II heat-labile enterotoxin by *Escherichia coli* isolated from food and human feces. *Infect. Immun. 59*:587–589 (1986).
20. Pickett, C. L., E. M. Twiddy, B. N. Belisle, and R. K. Holmes, Cloning of genes that encode a new heat-labile enterotoxin of *Escherichia coli. J. Bacteriol. 165*:348–352 (1986).
21. Burgess, N. M., R. J. Bywater, C. M. Cowley, N. A. Mullan, and P. M. Newsome, Biological evaluation of a methanol-soluble, heat-stable *Escherichia coli* enterotoxin in infant mice, pigs, rabbits and calves. *Infect. Immun. 21*:526–531 (1978).
22. Field, M., L. H. Graf, W. J. Laird, and P. L. Smith, Heat-stable enterotoxin of *Escherichia coli*: In vitro effects on guanylate cyclase activity, cyclic AMP concentration and ion transport in small intestine. *Proc. Natl. Acad. Sci. USA 75*:2800–2804 (1978).
23. Hughes, J. M., F. Murad, B. Chang, and R. L. Guerrant, Role of cyclic

AMP in the action of heat-stable enterotoxin of *Escherichia coli. Nature 271*:755–756 (1978).
24. Lathe, R., P. Hirth, M. DeWilde, N. Harford, and J. P. Lecocq, Cell-free synthesis of enterotoxin of *E. coli* from a cloned gene. *Nature 284*:473–474 (1980).
25. Echeverria, P., J. Seriwatana, U. Leksomboon, C. Tirapat, W. Chaicumpa, and B. Rowe, Identification by DNA hybridization of enterotoxigenic *Escherichia coli* in homes of children with diarrhoea. *Lancet 1*:63–66 (1984).
26. Moseley, S. L., P. Echeverria, J. Seriwatana, C. Tirapat, W. Chaicumpa, T. Sakulaipeara, and S. Falkow, Identification of enterotoxigenic *Escherichia coli* by colony hybridization using three enterotoxin gene probes. *J. Infect. Dis. 145*:863–869 (1982).
27. Utomporn, P., J. Seriwatana, and P. Echeverria, Identification of enterotoxigenic *Escherichia coli* isolated from swine with diarrhea in Thailand by colony hybridization using three enterotoxin gene probes. *J. Clin. Microbiol. 18*:1242–1231 (1983).
28. Moseley, S. L., J. N. Hardy, M. I. Huq, P. Echeverria, and S. Falkow, Isolation and nucleotide sequence determination of a gene encoding a heat-stable enterotoxin of *Escherichia coli. Infect. Immun. 39*:1167–1174 (1983).
29. Seriwatana, J., P. Echeverria, J. Escamilla, R. Glass, I. Huq, R. Rockhill, and B. J. Stoll, Identification of enterotoxigenic *Escherichia coli* in patients with diarrhea in Asia with three enterotoxin gene probes. *Infect. Immun. 42*:152–155 (1983).
30. Weikel, C. S., H. N. Nellans, and R. L. Guerrant, In vivo and in vitro effects of a novel enterotoxin, STb, produced by *Escherichia coli. J. Infect. Dis. 153*:893–901 (1986).
31. Lee, C. H., S. L. Moseley, H. W. Moon, S. C. Whipp, C. L. Gyles, and M. So, Characterization of the gene encoding heat-stable toxin II and preliminary molecular epidemiology studies of enterotoxigenic *Escherichia coli* heat-stable toxin II producers. *Infect. Immun. 42*:264–268 (1983).
32. Evans, D. G., D. J. Evans, Jr., and N. F. Pierce, Differences in the response of rabbit small intestine to heat-labile and heat-stable enterotoxins of *Escherichia coli. Infect. Immun. 7*:873–880 (1973).
33. Guerrant, R. L., L. L. Brunton, T. C. Schnaitman, I. Rebhun, and A. G. Gilman, Cyclic adenosine monophosphate and alteration of Chinese hamster ovary cell morphology: A rapid, sensitive in vitro assay for the enterotoxins of *Vibrio cholerae* and *Escherichia coli. Infect. Immun. 10*:320–327 (1974).
34. Sack, D. A., and R. B. Sack, Test for enterotoxigenic *Escherichia coli* using Y1 adrenal cells in miniculture. *Infect. Immun. 11*:334–336 (1975).
35. Dean, A. G., Y. C. Ching, R. G. Williams, and J. B. Harden, Test for *Escherichia coli* enterotoxin using infant mice: Application in a study of diarrhea in children in Honolulu. *J. Infect. Dis. 125*:407–411 (1972).

36. Giannella, R. A., Suckling mouse model for detection of heat-stable *Escherichia coli* enterotoxin: Characteristics of the model. *Infect. Immun. 14*:95–99 (1976).
37. Dallas, W. S., D. M. Gill, and S. Falkow, Cistron encoding *Escherichia coli* heat-labile toxin. *J. Bacteriol. 139*:850–858 (1979).
38. Picken, R. N., A. J. Mazaitis, W. K. Mass, M. Rey, and H. Heyneker, Nucleotide sequence of the gene for heat-stabile enterotoxin II of *Escherichia coli. Infect. Immun. 42*:269–275 (1983).
39. Moseley, S. L., I. Huq, A. R. Alim, M. So, M. Samadpour-Motalebi, and S. Falkow, Detection of enterotoxigenic *Escherichia coli* by DNA colony hybridization. *J. Infect. Dis. 141*:892–898 (1980).
40. Echeverria, P., J. Seriwatana, O. Chityothin, W. Chaicumpa, and C. Tirapat, Detection of enterotoxigenic *Escherichia coli* in water by filter hybridization with three enterotoxin gene probes. *J. Clin. Microbiol. 16*: 1086–1090 (1982).
41. Echeverria, P., J. Seriwatana, D. N. Taylor, C. Tirapat, W. Chaicumpa, and B. Rowe, Identification by DNA hybridization of enterotoxigenic *Escherichia coli* in a longitudinal study of villages in Thailand. *J. Infect. Dis. 151*:124–130 (1985).
42. Georges, M. C., I. K. Wachsmuth, K. A. Birkness, S. L. Moseley, and A. J. Georges, Genetic probes for enterotoxigenic *Escherichia coli* isolated from childhood diarrhea in the Central African Republic. *J. Clin Microbiol. 18*: 199–202 (1983).
43. Echeverria, P., J. Seriwatana, U. Patamaroj, S. L. Moseley, A. McFarland, O. Chityothin, and W. Chaicumpa, Prevalence of heat-stable II enterotoxigenic *Escherichia coli* in pigs, water and people at farms in Thailand as determined by DNA hybridization. *J. Clin. Microbiol. 19*:489–491 (1984).
44. Hill, W. E., J. M. Madden, B. A. McCardell, D. B. Shah, J. A. Jagow, W. L. Payne, and B. K. Boutin, Foodborne enterotoxigenic *Escherichia coli*: Detection and enumeration by DNA colony hybridization. *Appl. Environ. Microbiol. 45*:1324–1330 (1983).
45a. Echeverria, P., D. N. Taylor, J. Seriwatana, A. Chatkaeomorakot, V. Khungvalert, T. Sakuldaipeara, and R. D. Smith, A comparative study of enterotoxin gene probes and tests for toxin production to detect enterotoxigenic *Escherichia coli. J. Infect. Dis. 153*:255–260 (1986).
45b. Murray, B. E., J. J. Mathewson, H. L. DuPont, and W. E. Hill, Utility of oligodeoxyribonucleotide probes for detecting enterotoxigenic *Escherichia coli. J. Infect. Dis. 155*:809–911 (1987).
46. Lanata, C. F., J. B. Kaper, M. M. Baldini, R. E. Black, and M. M. Levine, Sensitivity and specificity of DNA probes with the stool blot technique for detection of *Escherichia coli* enteotoxins. *J. Infect. Dis. 152*:1087–1090 (1985).
47. O'Brien, A. D., G. D. LaVeck, M. R. Thompson, and S. B. Formal, Production of *Shigella dysenteriae* 1-like cytotoxin by *Escherichia coli. J. Infect. Dis. 146*:763–769 (1982).

48. Marques, L. R. M., M. A. Moore, J. G. Wells, I. K. Wachsmuth, and A. D. O'Brien, Production of Shiga-like toxin by *Escherchia coli. J. Infect. Dis. 154*:338–341 (1986).
49. O'Brien, A. D., M. E. Chen, R. K. Holmes, J. B. Kaper, and M. M. Levine, Environmental and human isolates of *Vibrio cholerae* and *Vibrio parahemolyticus* produce a *Shigella dysenteriae* 1 (Shiga)-like cytotoxin. *Lancet 1*:77–78 (1984).
50. O'Brien, A. D., L. R. M. Marques, J. W. Newland, S. F. Miller, N. A. Strockbine, and R. K. Holmes, Shiga and Shiga-like toxins, *Microecol. Ther. 14*:25–30 (1984).
51. Strockbine, N. A., L. R. M. Marques, J. W. Newland, H. W. Smith, R. K. Holmes, and A. D. O'Brien, Two toxin-converting phages from *Escherichia coli* 0157:H7 strain 933 encode antigenically distinct toxins with similar biologic activities. *Infect. Immun. 53*:135–140 (1986).
52. Newland, J. W., N. A. Strockbine, S. F. Miller, A. D. O'Brien, and R. K. Holmes, Cloning of Shiga-like toxin structural genes from a toxin converting phage of *Escherichia coli. Science 230*:179–181 (1985).
53a. Newland, J. W., N. A. Strockbine, and R. N. Neill, Cloning of genes for production of *Escherichia coli* Shiga-like toxin type II. *Infect. Immun. 55*:2675–2680 (1987).
53b. Willshaw, G. A., H. R. Smith, S. M. Scotland, and B. Rowe, Cloning of genes determining the production of Vero cytotoxin by *Escherichia coli. J. Gen. Microbiol. 131*:3047–3053 (1985).
54. Edelman, R., and M. M. Levine, Summary of a workshop on enteropathogenic *Escherichia coli. J. Infect. Dis. 147*:1108–1118 (1983).
55. Rothbaum, R., A. J. McAdams, R. Giannella, and J. C. Partin, A clinicopathologic study of enterocyte-adherent *Escherichia coli*: A cause of protracted diarrhea in infants. *Gastroenterology 83*:441–454 (1982).
56. Moon, H. W., S. C. Whipp, R. A. Argenzio, W. W. Levine, and R. A. Giannella, Attaching and effacing activities of rabbit and human enteropathogenic *Escherichia coli* in pig and rabbit intestines. *Infect. Immun. 41*: 1340–1351 (1983).
57. Takeuchi, A., L. R. Inman, P. D. O'Hanley, J. R. Cantey, and W. B. Lushbaugh, Scanning and transmission electronmicroscopic study of *Escherichia coli* 015 (RDEC-1) enteric infections in rabbits. *Infect. Immun. 19*:686–694 (1978).
58. O'Brien, A. D., and J. P. Nataro, New concepts in the pathogenesis of enteropathogenic *Escherichia coli* diarrhea. In *Microbiology-1985*, L. Leive (ed.). American Society for Microbiology, Washington, pp. 78–82 (1985).
59. Cravioto, A., R. J. Gross, S. M. Scotland, and B. Rowe, An adhesive factor found in strains of *Escherichia coli* belonging to the traditional infantile enteropathogenic serotypes. *Curr. Microbiol. 3*:95–99 (1979).
60. Baldini, M. M., J. B. Kaper, M. M. Levine, D. A. Candy, and H. W. Moon, Plasmid-mediated adhesion in enteropathogenic *Escherichia coli. J. Pediatr. Gastroenterol. Nutr. 2*:534–538 (1983).

61. Levine, M. M., J. P. Nataro, H. Karch, M. M. Baldini, J. B. Kaper, R. E. Black, M. L. Clements, and A. O'Brien, The diarrheal response of humans to some classic serotypes of enteropathogenic *Escherichia coli* is dependent on a plasmid encoding a enteroadhesiveness factor. *J. Infect. Dis. 152*: 550–559 (1985).
62. Nataro, J. P., M. M. Baldini, J. B. Kaper, R. E. Black, N. Bravo, and M. M. Levine, Detection of an adherence factor of enteropathogenic *Escherichia coli* with a DNA probe. *J. Infect. Dis. 152*:560–565 (1985).
63. Evans, D. G., R. P. Silver, D. J. Evans, Jr., D. G. Chase, and S. L. Gorbach, Plasmid-controlled colonization factor associated with virulence in *Escherichia coli* enterotoxigenic for humans. *Infect. Immun. 12*:656–667 (1975).
64. Evans, D. G., and D. J. Evans, New surface associated heat-labile colonization factor antigen (CFA/II) produced by enterotoxigenic *Escherichia coli* of serogroups 06 and 08. *Infect. Immun. 21*:638–647 (1978).
65. Honda, T., M. Arita, and T. Miwatani, Characterization of new hydrophobic pili of human enterotoxigenic *Escherichia coli*: A new possible colonization factor. *Infect. Immun. 43*:959–965 (1984).
66. Thomas, L. V., A. Cravioto, S. M. Scotland, and B. Rowe, New fimbrial antigen type (E 8775) that may represent a new colonization factor in enterotoxigenic *Escherichia coli* in humans. *Infect. Immun. 35*:1119–1124 (1982).
67. Willshaw, G. A., H. R. Smith, M. M. McConnell, and B. Rowe, Expression of cloned plasmid regions encoding colonization factor antigen (CFA/1) in *Escherichia coli. Plasmid 13*:8–16 (1985).
68. Manning, P. A., K. N. Timmis, and G. Stevenson, Colonization factor antigen II (CFA/II) of enterotoxigenic *Escherichia coli*: Molecular cloning of the CS3 determinant. *Mol. Gen. Genet. 200*:322–327 (1985).
69. Mooi, F. R., F. K. DeGraaf, and J. D. A. VanEmbden, Cloning, mapping and expression of the genetic determinant that encodes for the K88 ab antigen. *Nucleic Acids Res. 6*:849–865 (1979).
70. VanEmbden, J. D. A., F. K. DeGraaf, L. M. Schouls, and J. S. Teppema, Cloning and expression of a deoxyribonucleic acid fragment that encodes for the adhesive antigen K99. *Infect. Immun. 29*:1125–1133 (1980).
71. Sereny, B., Experimental keratoconjunctivitis *Shigellosa. Acta Microbiol. Hung. 4*:367–376 (1957).
72. Kopecko, D. J., L. S. Baron, and J. Buysse, Genetic determinants of virulence in *Shigella* and dysenteric strains of *Escherichia coli*: Their involvement in the pathogenesis of dysentery. *Curr. Top. Microbiol. Immunol. 118*:71–95 (1985).
73. Boileau, C. R., H. M. d'Hauteville, and P. J. Sansonetti, DNA hybridization technique to detect *Shigella* species and enteroinvasive *Escherichia coli. J. Clin. Microbiol. 20*:959–961 (1984).
74. Small, P. L. C., and S. Falkow, Development of a DNA probe for the virulence plasmid of *Shigella* spp. and enteroinvasive *Escherichia coli*. In *Microbiology-1986*, L. Leive (ed.). American Society for Microbiology, Washington, pp. 121–124 (1986).

75. Wood, P. K., J. G. Morris, Jr., P. L. C. Small, O. Sethabutr, M. R. F. Toledo, L. Trabulsi, and J. B. Kaper, Comparison of DNA probes and the Sereny test for identification of invasive *Shigella* and *Escherichia coli* strains. *J. Clin. Microbiol. 24*:498–500 (1986).
76. Pal, T., P. Echeverria, D. N. Taylor, O. Sethabutr, and S. Hanchalay, Identification of enteroinvasive *Escherichia coli* by indirect ELISA and DNA hybridization. *Lancet 2*:785 (1985).
77. Maurelli, A. T., H. B. B. d'Hauteville, T. L. Hale, and P. J. Sansonetti, Cloning of plasmid DNA sequences involved in invasion of HeLa cells by *Shigella flexneri. Infect. Immun. 49*:164–171 (1985).
78. Oaks, E. V., T. L. Hale, and S. B. Formal, Serum immune response to *Shigella* protein antigens in rhesus monkeys and humans infected with *Shigella* spp. *Infect. Immun. 53*:57–63 (1986).
79a. Buysse, J. M., C. K. Stover, E. V. Oaks, M. Venkatesan, and D. J. Kopecko, Molecular cloning of invasion plasmid antigens (*ipa*) genes from *Shigella flexneri*: Analysis of *ipa* gene products and genetic mapping. *J. Bacteriol. 169*:2561–2569 (1987).
79b. Taylor, D. N., P. Echeverria, T. Pal, O. Sethabutr, S. Saiborisuth, S. Srichamorn, B. Rowe, and J. Cross, The role of *Shigella* spp., enteroinvasive *Escherichia coli* and other enteropathogens as causes of childhood dysentery in Thailand. *J. Infect. Dis. 153*:1132–1138 (1986).
80. Sethabutr, O., S. Hanchalay, P. Echeverria, D. N. Taylor, and U. Leksomboon, A non-radioactive DNA probe to identify *Shigella* and enteroinvasive *Escherichia coli* in stools of children with diarrhoea. *Lancet 2*: 1095–1097 (1985).
81. Felix, A., S. Bhatnagar, and R. M. Pitt, Observations on the properties of the Vi antigen of *B. typhosus. Br. J. Exp. Pathol. 15*:346–354 (1934).
82. Snellings, N. J., E. M. Johnson, D. J. Kopecko, H. H. Collins, and L. S. Baron, Genetic regulation of variable Vi antigen expression in a strain of *Citrobacter freundii. J. Bacteriol. 145*:1010–1017 (1981).
83a. Rubin, F. A., D. J. Kopecko, R. B. Sack, P. Sudarmono, A. Yi, M. Maurta, R. Meza, M. A. Moechtar, D. C. Edman, and S. L. Hoffman, Evaluation of of a DNA probe for identification of *Salmonella typhi* in Peruvian and Indonesian bacterial isolates. *J. Infect. Dis. 157*:1051–1053 (1988).
83b. Fitts, R., M. Diamond, C. Hamelton, and M. Neri, DNA–DNA hybridization assay for detection of *Salmonella* spp. in foods. *Appl. Environ. Microbiol. 46*:1146–1151 (1983).
84. Horowitz, M. A., R. A. Pollard, M. H. Merson, and S. M. Martin, A large outbreak of foodborne salmonellosis on the Navajo Nation Indian Reservation, epidemiology and secondary transmission. *Am. J. Public Health 67*:1071–1076 (1977).
85. Taylor, D. N., J. M. Bied, J. S. Munro, and R. A. Feldman, *Salmonella dublin* infections in the United States, 1979–1980. *J. Infect. Dis. 146*: 322–327 (1982).
86. Centers for Disease Control. Milk-borne salmonellosis—Illinois. *Morbid. Mortal. Week. Rep. 34*:200 (1985).
87. Spika, J., S. Waterman, M. St. Louis, G. Soo Hoo, R. Pacer, S. James, M.

Potter, K. Greene, L. Mayer, and P. Blake, *Salmonella newport* infections traced through hamburger to dairy farms. *Abstr. 26th Intersci. Conf. Antimicrobiol. Agents Chemother.* p. 153 (1986).
88. Small, R. G., and J. C. M. Sharp, A milk-borne outbreak due to *Salmonella dublin. J. Hyg. 82*:95–100 (1979).
89. Olsvik, O., S. Henning, K. Birkness, K. Wachsmuth, M. Fjolstad, J. Lassen, K. Fossum, and J. C. Feeley, Plasmid characterization of *Salmonella typhimurium* transmitted from animals to humans. *J. Clin. Microbiol. 22*:336–338 (1985).
90. Holmberg, S. D., M. T. Osterholm, K. A. Senger, and M. L. Cohen, Drug resistant *Salmonella* from animals fed antimicrobials. *N. Engl. J. Med. 311*:617–622 (1984).
91. Lysons, R. W., C. L. Samples, H. N. DeSilva, K. A. Ross, E. M. Julian, and P. J. Checko, An epidemic of resistant *Salmonella* in a nursery: Animal-to-person spread. *J. Am. Med. Assoc. 243*:546–547 (1980).
92. Baker, E. F., H. W. Anderson, and J. Allard, Epidemiologic aspects of turtle-associated salmonellosis. *Arch. Environ. Health 24*:1–9 (1972).
93. Tompkins, L. S., N. Troup, A. Labigne-Roussel, and M. L. Cohen, Cloned, chromosomal sequences as probes to identify *Salmonella* species. *J. Infect. Dis. 154*:156–162 (1986).
94. Jones, G. W., D. K. Rabert, D. M. Svinarich, H. J. Whitfield, Association of adhesive, invasive and virulent phenotypes of *Salmonella typhimurium* with autonomous 60-megadalton plasmids. *Infect. Immun. 38*:476–486 (1982).
95. Terakado, N., T. Sekizaki, K. Hashimoto, and S. Naitoh, Correlation between the presence of a fifty-megadalton plasmid in *Salmonella dublin* and virulence in mice. *Infect. Immun. 41*:443–444 (1983).
96. Manning, E. J., G. D. Baird, and P. W. Jones, The role of plasmid genes in the pathogenicity of *Salmonella dublin. J. Med. Microbiol. 21*:239–243 (1986).
97. Helmuth, R., R. Stephan, C. Bunge, B. Hoog, A. Steinbeck, and E. Bulling, Epidemiology of virulence-associated plasmids and outer membrane protein patterns with seven common *Salmonella* serotypes. *Infect. Immun. 48*:175–182 (1985).
98. Smith, H. L., Jr., Serotyping of non-cholerae *Vibrios. J. Clin. Microbiol. 10*:85–90 (1979).
99. Sakazaki, R., and T. Shimada, Serovars of *Vibrio cholerae. Jpn. J. Med. Sci. Biol. 30*:279–282 (1979).
100. Vasil, M. L., R. K. Holmes, and R. A. Finkelstein, Conjugal transfer of a chromosomal gene determining production of enterotoxin *Vibrio cholerae. Science 187*:849–850 (1975).
101. Richards, K. L., and S. D. Douglas, Pathophysiological effects of *Vibrio cholerae* and enterotoxigenic *Escherichia coli* and their exotoxins on eucaryotic cells. *Microbiol. Rev. 42*:592–613 (1978).
102. Moseley, S. L., and S. Falkow, Nucleotide sequence homology between the heat-labile enterotoxin gene of *Escherichia coli* and *Vibrio cholerae* deoxyribonucleic acid. *J. Bacteriol. 144*:444–446 (1980).

103. Kaper, J. B., H. B. Bradford, N. Roberts, and S. Falkow, Molecular epidemiology of *Vibrio cholerae* in the U.S. Gulf Coast. *J. Clin. Microbiol. 16*:129–134 (1982).
104. Echeverria, P., J. Seriwatana, O. Sethabutr, and D. N. Taylor, DNA hybridization in the diagnosis of bacterial diarrhea. *Clin. Lab. Med. 5*:447–462 (1985).
105. Gennaro, M. L., P. J. Greenaway, and D. A. Broadbent, The expression of biologically active cholera toxin in *Escherichia coli. Nucleic Acids Res. 10*:4883–4890 (1982).
106. Kaper, J. B., and M. M. Levine, Cloned cholera enterotoxin genes in study and prevention of cholera. *Lancet 2*:1162–1163 (1981).
107. Pearson, G., and J. J. Mekalanos, Molecular cloning of *Vibrio cholerae* enterotoxin genes in *Escherichia coli* K-12. *Proc. Natl. Acad. Sci. USA 79*:2976–2980 (1982).
108. Morris, J. G., Jr., and R. E. Black, Cholera and other vibrioses in the United States. *N. Engl. J. Med. 312*:343–350 (1985).
109. Kaper, J. B., R. K. Campen, R. J. Seidler, M. M. Baldini, and S. Falkow, Cloning of the thermostable direct or Kanagawa phenomenon-associated hemolysin of *Vibrio parahaemolyticus. Infect. Immun. 45*:290–292 (1984).
110. Nishibuchi, M., M. Ishibashi, Y. Takeda, and J. B. Kaper, Detection of the thermostable direct hemolysin gene and related DNA sequences in *Vibrio parahaemolyticus* and other *Vibrio* species by the DNA colony hybridization test. *Infect. Immun. 49*:481–486 (1985).
111. Nishibuchi, M., W. E. Hill, G. Zon, W. L. Payne, and J. B. Kaper, Synthetic oligodeoxyribonucleotide probes to detect Kanagawa phenomenon-positive *Vibrio parahaemolyticus. J. Clin. Microbiol. 23*:1091–1095 (1986).
112. Wright, A. C., J. G. Morris, Jr., D. R. Maneval, Jr., K. Richardson, and J. B. Kaper, Cloning of the cytotoxin-hemolysin gene of *Vibrio vulnificus. Infect. Immun. 50*:922–924 (1985).
113. Richardson, K., J. Michalski, and J. B. Kaper, Hemolysin production and cloning of two hemolysin determinants from classical *Vibrio cholerae. Infect. Immun. 54*:415–420 (1986).
114. Bottone, E. J., *Yersinia enterocolitica*: A panoramic view of a charismatic microorganism. *Crit. Rev. Microbiol 5*:211–241 (1977).
115. Maki, M., P. Gronroos, and T. Vesikari, In vitro invasiveness of *Yersinia enterocolitica* isolated from children with diarrhea. *J. Infect. Dis. 138*: 677–680 (1978).
116. Agbonlahor, D. E., T. O. Odugbemi, and O. Dosunmu-Ogunbi, Isolation of species of *Yersinia* from patients with gastroenteritis. *J. Med. Microbiol. 16*:93–96 (1983).
117. Portnoy, D. A., and S. Falkow, Virulence-associated plasmids from *Yersinia enterocolitica* and *Yersinia pestis. J. Bacteriol. 148*:877–883 (1981).
118. Gemski, P., J. R. Lazere, and T. Casey, Plasmid associated with pathogenicity and calcium dependency of *Yersinia enterocolitica. Infect. Immun. 27*:682–685 (1980).

119. Gemski, P., J. R. Lazere, T. Casey, and J. A. Wohlhieter, Presence of a virulence-associated plasmid in *Yersinia pseudotuberculosis. Infect. Immun. 28*:1044–1047 (1980).
120. Portnoy, D. A., H. Wolfe-Watz, I. Bolin, A. B. Beeder, and S. Falkow, Characterization of common virulence plasmids in *Yersinia* species and their role in the expression of outer membrane proteins. *Infect. Immun. 43*:108–114 (1984).
121. Isberg, R. R., and S. Falkow, A single genetic locus encoded by *Yersinia pseudotuberculosis* permits invasion of cultured animal cells by *Escherichia coli* K-12. *Nature 317*:262–264 (1985).
122. Hill, W. E., W. L. Payne, and G. C. Aulisio, Detection and enumeration of virulent *Yersinia enterocolitica* in food by DNA colony hybridization. *Appl. Environ. Microbiol. 46*:636–641 (1983).
123. Jagow, J. J., and W. E. Hill, Enumeration by DNA colony hybridization of virulent *Yersinia enterocolitica* colonies in artifically contaminated food. *Appl. Environ. Microbiol. 51*:441–443 (1986).
124. Totten, P. A., K. K. Holmes, H. H. Handsfield, J. S. Knapp, P. L. Perine, and S. Falkow, DNA hybridization technique for the detection of *Neisseria gonorrhoeae* in men with urethritis. *J. Infect. Dis. 148*:462–471 (1983).
125. Perine, P. L., P. A. Totten, K. K. Holmes, E. H. Sng, A. V. Ratnam, R. Widy-Wersky, H. Nsanze, E. Habte-Gabr, and W. G. Westbrook, Evaluation of a DNA hybridization method for detection of African and Asian strains of *Neisseria gonorrhoeae* in men with urethritis. *J. Infect. Dis. 152*:59–63 (1984).
126. Tenover, F. C., Studies of antimicrobial resistance genes using DNA probes. *Antimicrobiol. Agents Chemother. 29*:721–725 (1986).
127. Wallace, R. B., J. Shaffer, R. F. Murphy, J. Bonner, T. Hirose, and K. Itakura, Hybridization of synthetic oligodeoxyribonucleotides ΦX174 DNA: The effect of single base pair mismatch. *Nucleic Acids Res. 6*: 3543–3557 (1979).
128. Datta, A. R., B. A. Wentz, and W. E. Hill, Detection of hemolytic *Listeria monocytogenes* by using DNA colony hybridization. *Appl. Environ. Microbiol. 53*:2256–2259 (1987).
129. Ogle, J. W., J. M. Janda, D. E. Woods, and M. L. Vasil, Characterization and use of a DNA probe as an epidemiological marker for *Pseudomonas aeruginosa. J. Infect. Dis. 155*:119–126 (1987).
130. Honda, T., M. Nishibuchi, T. Miwatani, and J. B. Kaper, Demonstration of a plasmid-borne gene encoding a themostable direct hemolysin in *Vibrio cholerae* non-01 strains. *Appl. Environ. Microbiol. 52*:1218–1220 (1986).
131. Levine, M. M., J. Xu, J. B. Kaper, H. Lior, V. Prado, B. Tall, J. Nataro, H. Karch, and K. Wachsmuth, A DNA probe to identify enterohemorrhagic *Escherichia coli* of 0157:H7 and other serotypes that cause hemorrhagic colitis and hemolytic uremic syndrome. *J. Infect. Dis. 156*:175–182 (1987).

132. Bialkowska-Hobrzanska, H., Detection of enterotoxigenic *Escherichia coli* by dot blot hybridization with biotinylated DNA probes. *J. Clin. Microbiol. 25*:338–343 (1987).
133. Kirii, Y., D. Hirofumi, K. Komase, J. Arita, and J. Yoshikawa, Detection of enterotoxigenic *Escherichia coli* by colony hybridization with biotinylated enterotoxin probes. *J. Clin. Microbiol. 25*:1962–1965 (1987).
134. Seriwatana, J., P. Echeverria, D. N. Taylor, T. Sakuldaipeara, S. Changchawalit, and O. Chivoratanond, Identification of enterotoxigenic *Escherichia coli* with synthetic alkaline phosphatase-conjugated oligonucleotide DNA probes. *J. Clin. Microbiol. 25*:1438–1441 (1987).
135. Oprandy, J. J., S. A. Thornton, C. H. Gardiner, D. Burr, R. Batchelor, and A. L. Bourgeois, Alkaline phosphatse-conjugated oligonucleotide probes for enterotoxigenic *Escherichia coli* in travelers to South America and West Africa. *J. Clin. Microbiol. 26*:92–95 (1988).
136. Chityothin, O., O. Sethabutr, P. Echeverria, D. N. Taylor, U. Vongsthongsri, and S. Tharavanij, Detection of heat-stable enterotoxigenic *Escherichia coli* by hybridization with an RNA transcript probe. *J. Clin. Microbiol. 25*:1572–1573 (1987).

9

DNA Probe Assays for Detection of *Campylobacter* and *Salmonella*

AYOUB RASHTCHIAN* and MICHAEL S. CURIALE
Integrated Genetics, Inc., Framingham, Massachusetts

I. INTRODUCTION

The specificity and quantitative characteristics of nucleic acid hybridizations (DNA-DNA, DNA-RNA, RNA-RNA) have been exploited in several areas of molecular biology and microbial taxonomy for over 2 decades. With the aid of molecular-cloning techniques it is now possible to isolate DNA sequences specific for an organism or organism group (e.g., procaryotic or eukaryotic pathogens, mycoplasmas, viruses), or organism characteristics of clinical relevance (e.g., virulence factors, antimicrobial resistance). The use of DNA probes held promise for aiding in the detection of microorganisms but, until recently, the use of these techniques has been restricted to the research laboratory. A limiting factor in practical application of nucleic acid hybridization has been the development of reliable, rapid formats for sample preparation that will eliminate the time-consuming, labor-intensive, and biochemically sophisticated aspects of the procedure.

A number of characteristics of nucleic acids make DNA hybridization an attractive alternative to traditional growth-based assays. Nucleic acids are found in all microorganisms; unlike antigens or metabolic products, their presence is constant and not a function of variable expression. The physical-chemical relationships of nucleic acid interactions are well characterized, thus, allowing control of hybridization specificity by altering the conditions of the reaction. For specific hybridization to occur, only the target sequence need be present, not necessarily an intact or entire genome. This makes possible sequence detection even in samples in which some degree of nucleic acid degradation has occurred. By use

---

**Present affiliation*: Life Technologies, Inc., Gaithersburg, Maryland.

of molecular-cloning techniques, nucleic acid probes can now be isolated or synthesized inexpensively and with relative ease.

In the following discussion we review the use of nucleic acid hybridization as a diagnostic tool and discuss two specific examples in which this technology has been applied to practical areas.

II. GENERAL CONSIDERATIONS IN DEVELOPMENT OF A DNA PROBE ASSAY

To use a DNA probe-based test as a routine assay with adequate precision, a number of practical issues must be addressed. These include

1. What degree of specificity and sensitivity is required for a given application? Does one wish to detect a genus, a species, or an organism carrying a specific gene? For each specimen type, what number of organisms is significant, i.e., how sensitive must the assay be? In a given sample, will competing organisms be present?
2. Can a practical format be created that allows the processing of samples in a rapid and reproducible manner to expose target nucleic acids for hybridization? To what extent will probe-target interactions be affected by characteristics of each sample type (e.g., foods, feces, sputum)?

As an example, we have addressed such questions for the development of DNA probe-based tests to detect important bacterial pathogens in foods and clinical samples. In the following sections the technical considerations in the development of tests for *Campylobacter* species in fecal samples and *Salmonella* species in food are discussed.

III. HYBRIDIZATION FORMATS

Although nucleic acid hybridizations can be performed in a variety of ways, basically there are three general techniques that are widely used. These include (a) solution hybridization, (b) hybridization on membrane filters, and (c) sandwich hybridization.

A. Solution Hybridization

Historically, solution hybridization was one of the first techniques to be described for hybridization of two nucleic acids. In the solution hybridization format the sample and the probe DNA are mixed in solution and are allowed to hybridize. After hybridization, the single-stranded and double-stranded nucleic acids are separated by a variety of methods. Nygaard et al. (1) described a method for quantitation of DNA-RNA hybrids after hybridization, by filtration through a nitrocellulose membrane. Another method for quantitation of hybridization is hydrolysis of the single-stranded nucleic acids with single-stranded specific nuclease S_1, followed by precipitation of nonhydrolyzed nucleic acids with acid (trichloroacetic acid), and filtration or centrifugation. Double-stranded hybrids formed in solution hybridization can also be detected by hydroxyapatite chromotography (2) or a centrifugation method (3).

The previously described methods for isolation of double-stranded nucleic acids from unhybridized probe are not convenient for routine use in a diagnostic laboratory. These procedures that use pure DNA as samples generally show inefficient separation of hybrids and require specialized instruments. As will be described in the sections that follow, we have used the solution hybridization technique for hybridization of DNA probes complementary to ribosomal RNA (rRNA). Here, however, we have used an antibody specific for RNA-DNA hybrids for separation of double-stranded RNA-DNA hybrids from the single-stranded DNA probe (see the following).

B. Hybridization on Membrane Filters

Membrane filter hybridization [Southern blot (4) and colony blots (5)] is the most widely used hybridization technique. In this method the sample nucleic acids are immobilized on a solid support and incubated with a labeled probe (in solution) for the duration of hybridization. After hybridization the filters are washed several times and are assayed by autoradiography for hybridization of the probe. This method of hybridization has been very successful when pure DNA samples are analyzed. It is also a useful technique with more complex samples, such as those encountered in microbiology laboratories.

The rate of hybridization to filter membranes is generally slower than that for solution hybridization by a factor of 10 (6). In many instances, an improvement of about threefold is possible by addition of dextran sulfate. A limitation of filter membrane hybridization is with its use for complex samples, such as colony blots or other samples encountered in microbiology laboratories. The use of these specimens results in immobilization of cell debris and other sample material on the filter in addition to the test nucleic acids. Although these materials do not affect hybridization per se, they may nonspecifically bind to test reagents and cause high backgrounds.

C. Sandwich Hybridizations

The sandwich hybridization technique is a variation of the traditional membrane filter hybridization. This technique uses two separate DNA reagents. One of the DNA fragments is immobilized onto a solid support, such as nitrocellulose filters, and the other is labeled and functions as probe. When present in the sample, a DNA fragment homologous to both of these DNA fragments, will hybridize both to the solid-phase DNA and to the labeled probe. Therefore, the binding of the labeled probe to the solid phase is determined as the signal (for review see Ref. 7). The advantage of this system is that the sample is kept in solution, and interfering substances, such as cell debris, can be removed by washing the filters after hybridization. The disadvantage of this system is that it relies on the occurrence of two independent hybridization events that might require long hybridization periods or a high concentration of the reagents. The feasibility of the sandwich hybridization technique has been extensively studied by Ranki et al. (7,8). We have recently modified the sandwich hybridization procedure by using plastic or agarose supports in place of membrane filters (9). The DNA fragments of interest were covalently bonded to CnBr-activated sepharose beads or to plastic tubes previously coated with a protein and this served as solid-phase DNA. In a model system the applicability of these solid supports to sandwich hybridizations with radioactive probes was shown. However, the use of nonisotopic probes was not investigated.

IV. DEVELOPMENT OF A DNA HYBRIDIZATION ASSAY FOR *CAMPYLOBACTER*

Campylobacter jejuni is one of the most important newly recognized enteric pathogens from the standpoint of human disease (10,11). This organism is a major cause of bacterial gastroenteritis in humans (11) and is encountered as frequently as *Salmonella* and *Shigella* species (12). Since the implication of *Campylobacter* species as a cause of enteritis in humans, the frequency of isolation of these organisms has increased dramatically. The rather recent appreciation of these organisms as major pathogens has, at least in part, been due to substantial improvements in techniques and the use of selective media. Although culturing followed by microscopy has been satisfactory for the most part, more rapid methods are needed.

For development of a DNA hybridization assay for detection of *Campylobacter*, we have isolated DNA probes that are specific for *Campylobacter*. These probes can be divided into two groups: DNA probes that are specific for *C. jejuni* and *C. coli* and do not recognize other *Campylobacter* species; DNA probes that are genus-specific and recognize all established *Campylobacter* species.

A. Probes for *Campylobacter jejuni* and *Campylobacter coli*

DNA probes specific for *C. jejuni* and *C. coli* were isolated from a clinical isolate of *C. jejuni*. The detailed description of probe preparation has been described elsewhere (13). Briefly, a genomic library of DNA fragments from *C. jejuni* strain N941 was constructed in *Escherichia coli*. The recombinant plasmids from this gene library were screened for homology with *Campylobacter* strains and a panel of non-*Campylobacter* bacteria. For any probe sequence, it is necessary to demonstrate two definite properties: those of exclusivity and inclusivity. Inclusivity of DNA probes was shown by testing 130 strains of *C. jejuni* and *C. coli* and demonstrating that the probes hybridized to a large number of these isolates. The exclusivity of the probes was demonstrated by testing a large panel of non-*Campylobacter* strains which included members of *Enterobacteriacae*, *Pseudomonas*, *Vibrio*, and *Bacteroides*. Extensive testing showed that each of the DNA probes hybridized specifically to *C. jejuni* and *C. coli*, whereas no detectable hybridization was seen to other *Campylobacter* species or to non-*Campylobacter* bacteria. Stool specimens from 25 healthy individuals were examined and only in those inoculated with *C. jejuni* or *C. coli* was hybridization observed. Table 1 shows the characteristics of five of the DNA probes that have been isolated.

B. Genus-Specific Probes

Previous studies have shown that the members of the genus *Campylobacter* are heterogenous with respect to relatedness of DNA among different species (14, 15). Although *C. jejuni* and *C. coli* are closely related (14), other members of the genus have little DNA homology (if any) to each other, to *C. jejuni*, or to *C. coli*. Therefore, we approached isolation of genus-specific probes by examining the DNA sequences coding for ribosomal RNA (rRNA). Because the nucleotide sequence of rRNA are highly conserved (16), they would be expected to be highly similar in different species of *Campylobacter*. In addition, rRNA in each

Table 1 List of DNA Probes Specific for *C. jejuni* and *C. coli*

Probe	Size (bp)	*C. jejuni* and *C. coli* detected[a] (%)	Hybridization to other campylobacters[b]
CJ12A	960	100	–
CJ29B	970	85	–
CJ41A	2150	91	–
CJ44B	1700	99	–
CJ87A	1200	96	–

[a]A total of 130 *C. jejuni* and *C. coli* strains from human and animal sources were tested.
[b]*Campylobacter* species tested were *C. fetus*, *C. fetus* subsp. *venerialis*, *C. laridis*, *C. fecalis*, and *C. hyointestinalis*.
(–) No detectable hybridization

bacterial cell is present in many copies (approximately 10^4) and, therefore, is a good target nucleic acid when small numbers of bacteria are examined.

For isolation of DNA probes complementary to the rRNA of *Campylobacter*, DNA sequences coding for rRNA were cloned from *C. jejuni*. The details of the cloning procedure and characterization of the gene coding for 16S rRNA has been described elsewhere (17). Briefly, DNA fragments coding for 16S rRNA were sequenced and compared with the nucleotide sequences of 16S rRNA from other bacteria, such as *E. coli* (18) and *Proteus vulgaris* (19). Despite the high degree of sequence homology among bacterial 16S rRNA sequences, several regions were identified that appeared to be unique for *C. jejuni*. Comparison of nucleotide sequences of these regions to nucleic acid sequences compiled in the genetic sequence data bank (Genbank, NIH) also revealed no sequence homology to known procaryotic sequences. Given these sequences, synthetic oligonucleotide probes complementary to 16S rRNA of *C. jejuni* were constructed. These probes were shown to be specific for members of the genus *Campylobacter*, and no hybridization was observed to non-*Campylobacter* bacteria or to the fecal flora. Table 2 shows the characteristics of the synthetic probes that have been constructed. It can be seen that whereas some probes hybridize to all *Campylobacter* species, others hybridize to all except *C. fecalis*.

C. Use of DNA Probes for Detection of *Campylobacter*

Although isolation of DNA probes specific for the organism(s) of interest is an integral part of a DNA probe assay, it is not the only requirement for test development. Recent advances in molecular-cloning techniques (20) have made isolation of specific probes straightforward and rather easy (21). A more challenging task is development of rapid and reliable methods in which DNA probes can be used effectively by those untrained in molecular biology. This becomes

Table 2 List and Characteristics of Synthetic Oligonucleotide Probes Complementary to rRNA

Probe	Size	*Campylobacter* spp. detected[a]
AR196	50	All species except *C. fecalis*
AR197	50	All species except *C. fecalis*
AR352	60	All species
AR353	56	All species
AR354	70	All species

[a]Tested species were *C. jejuni, C. coli, C. laridis, C. fecalis, C. fetus, C. fetus* subsp. *venerialis*, and *C. hyointestinalis*.

particularly challenging and difficult if the organism of interest is to be detected in complex biological samples, such as fecal specimens. The difficulty in using DNA probe assays for direct analysis of such specimens spotted on nitrocellulose filters is that variable results are obtained. Material in some specimens binds probe nonspecifically (22); in others the high number of organisms and, therefore, of DNA, overloads the filter resulting in lower sensitivity because of inaccessibility or loss of target DNA. These problems can be overcome by extraction of the samples in various ways to yield relatively pure or clean nucleic acids for hybridization assays. However, most of the current methods for isolation of nucleic acids from complex specimens are cumbersome and require use of noxious chemicals.

The problems associated with detection of microorganisms in complex specimens are solved by including a short enrichment step similar to the microbiology methods currently used in clinical laboratories. We have used this approach for detection of *Campylobacter* in fecal specimens. Current microbiological methods for isolation of *Campylobacter* call for inoculation of agar plates containing a variety of antibiotics for selective enrichment of *Campylobacter* species (23). Although this method is effective, it is slow, and it generally takes 48-72 hr before visible *Campylobacter* colonies appear on the plate for biochemical analysis. We have combined this enrichment method with a DNA hybridization assay that allows us to obtain definitive results within 12-16 hr.

For detection of *Campylobacter* from stool specimens, 0.05-0.1 ml of stool samples is plated on Campy-Bap selective plates (23) and incubated under microaerophilic conditions, at 42°C for 8-12 hr. The growth from these plates is then resuspended in saline solution and the cells are collected on a membrane filter using a GENE-TRAK manifold filtration apparatus (Integrated Genetics, Inc.). The cells are lysed and after fixing of DNA to the membrane filter they are hybridized to ^{32}P-labeled DNA probes. After hybridization the filters are washed and counted in a scintillation counter. Unlike most published DNA probe assays that use autoradiography for analysis of results, we have chosen scintillation counting because of its speed and quantitative nature. We have successfully used this solid-phase DNA hybridization system for detection of *C. jejuni* and *C. coli*. An example in which *Campylobacter*-negative stool specimens were inoculated with *C. jejuni* cells and subjected to hybridization is shown in Table 3. It can be seen that *C. jejuni* can be readily detected using DNA probes.

We have used this DNA hybridization method for rapid detection of *Campylobacter* in specimens from patients with diarrhea. For a period of 3 months all stool samples submitted to clinical laboratory for detection of *Campylobacter* were analyzed. Duplicate plates were inoculated with the stool samples. One plate was incubated for 8-12 hr before hybridization, and the other plate was incubated for 48-72 hr for microbiological analysis (23). The results of this study is shown in Table 4. It appears that the DNA hybridization assay described

Table 3 Hybridization of DNA Probes to Fecal Samples after Enrichment for *Campylobacter*

Specimen[a] no.	No. of *C. jejuni* in inoculum (cfu)	Probe hybridized[b] (cpm)
1	None	210
2	None	345
3	None	156
4	None	195
4	1×10^4	20,412
4	1×10^3	9,100
5	None	267
5	1×10^4	21,652
5	1×10^3	8,206

[a]Culture-negative stool specimens were inoculated with *C. jejuni* strain N933 and plated on CAMPY-BAP medium for 12 hr at 42°C. The entire growth from the plate was resuspended and subjected to hybridization with DNA probes.
[b]Probe consisted of a mixture of several probes, some of which are described in Table 1.

here is more sensitive than the microbiological detection. During the study summarized in Table 4, among the 12 stool specimens that were positive in the DNA hybridization assay, four were initially reported as culture-negative. When these specimens were cultured for a second time *C. jejuni* was indeed recovered from two of the specimens. Given the results of this study, the DNA assay has a sensitivity of 100% and a specificity of 99.5%.

D. Nonradioactive Assays

It is generally thought that if two assays were available that had similar sensitivity, specificity, and ease of use, the assay that did not involve the use of a radioisotopic detection system would be preferred in many laboratories.

Table 4 Comparison of Radioactive DNA Probe Assay with Microbiological Detection of *Campylobacter* in Stool Specimens

	DNA probe	
Culture	Positive	Negative
Positive	10	0
Negative	2	446

Methods for labeling DNA fragment with nonradioactive groups, such as biotin, have been known for some time (24-27). However, the use of these probes has been delayed by the lack of practical formats for these assays. Most of the existing procedures (24) that use biotinylated probes require purification of nucleic acids from the samples because unacceptably high backgrounds are observed when biotin-labeled probes are used for hybridization with bacterial cell lysates on membrane filters. Therefore, the use of biotin-labeled probes has been limited to its use in Southern blots and other procedures generally used in research laboratories.

We have developed a novel diagnostic system based on hybridization of synthetic oligonucleotide probes to highly amplified rRNA target in the test sample. In this system, synthetic oligonucleotide probes are labeled with biotin and hybridized in solution to the rRNA sequences in the cell lysates. In this system large amounts of rRNA present in each bacterial cell (approximately 10^4 copies per cell) makes possible substantially greater sensitivity than is feasible with similarly labeled genomic probes. After the hybridization takes place, the double-stranded probe-rRNA hybrids are captured by using an antibody, specific for RNA-DNA hybrids (28), that is immobilized on a solid support. This step efficiently separates RNA-DNA hybrids from unhybridized probe and other cell material and sample debris. The hybridized biotinylated oligonucleotide probes are then detected with avidin conjugated to an enzyme, such as horseradish peroxidase or alkaline phosphatase, with use of chromogenic substrates.

The suitability and sensitivity of this nonisotopic assay is shown in Table 5. As can be seen, as little as 5×10^4 *C. jejuni* cells can be detected readily, whereas no significant signal is seen from *E. coli* at higher titers. In addition to being very sensitive, this assay is easy to perform and results are obtained within 2.5 hr. Usually, if 10^6 or more organisms are being tested, the results can be interpreted visually. Greater sensitivity is achieved by using a photometer.

Table 5 Detection of *C. jejuni* with Biotin-Labeled Synthetic Oligonucleotides Complementary to rRNA

Organism	Cells (n)	OD 450
C. jejuni	1×10^7	>2
C. jejuni	2.5×10^6	>2
C. jejuni	5×10^5	0.862
C. jejuni	1×10^5	0.194
C. jejuni	5×10^4	0.090
E. coli	2×10^7	0.029

The advantages of the solution format over the solid-phase hybridization for detection of *Campylobacter* are several. Because hybridization is performed in solution, hybridization periods as short as 45 min can be used in place of 2-hr hybridization for the membrane filter system. The membrane filter hybridizations generally require a prehybridization step (30 min), whereas in the solution hybridization format, this step is not needed. The most important advantage of this system is that nonisotopic biotin-avidin method replaces the isotopic detection system used in the filter format. The avidin-biotin system cannot be used in the solid-phase hybridization because, in addition to the test nucleic acids, other cellular material and undesirable components of the sample are immobilized on filters. The use of the antibody capture technique in the format described here solves this problem because only RNA-DNA hybrids are immobilized and other cellular materials are washed away.

As with the radioactive DNA-hybridization format we have extended this nonisotopic format to the testing of stool specimens. In the current format, stool specimens are inoculated on *Campylobacter*-selective plates, and after 8-12

Table 6 Comparison of the Standard Microbiological Assay with Two DNA Hybridization Assays for *Campylobacter* spp. in Clinical Specimens

Microbiological assay[a]	Membrane filter radioactive DNA hybridization	Nonradioactive solution-based DNA hybridization
Incubation on selective medium (47-72 hr)	Incubation on selective medium (8-12 hr)	Incubation on selective medium (8-12 hr)
Microscopic examination of suspect colonies	DNA hybridization[b] (4 hr)	DNA hybridization[c] (2.5 hr)
Biochemical testing (up to 24 hr)		
Total time: 48-96 hr	Total time: 12-16 hr	Total time: 10-14 hr

[a]Standard microbiological procedure described in Ref 23.

[b]In this assay cells from the plates are collected on membrane filters and, after lysis and fixing of DNA to membrane filters, the filters are prehybridized for 30 min. Hybridization for 2 hr follows, after which filters are washed six times (10 min each) and counted. Total time to perform the assay is approximately 4 hr. A similar assay has been developed for salmonellae in foods and is commercially available from Integrated Genetics, Inc. Details follow later in this work.

[c]Cells are lysed in solution and, after addition of the probe, allowed to hybridize for 45 min. RNA-DNA hybrids are then incubated with anti-RNA-DNA antibody-coated tubes for 45 min. After rinsing the tubes, strepavidin-HRP is added and incubated for 20 min. Substrate is then added after rinsing and incubated for 10 min for color development. Optical density is determined after addition of stop solution. Total time for this assay is approximately 2.5 hr.

hr of enrichment, the cell suspensions are subjected to DNA hybridization assay. The total time for definitive detection of *Campylobacter* is 10-14 hr from the time of obtaining the specimen from the patient. Table 6 shows comparison of the standard microbiological assay with the DNA hybridization protocol outlined here. As can be seen, by using DNA hybridization assay, time savings of 1.5-3.5 days can be achieved. Studies of side-by-side comparison of nonradioactive DNA probe assay with the standard microbiological assay using clinical specimens are in progress. Results comparable with the radioactive assay have been observed.

V. DEVELOPMENT OF DNA HYBRIDIZATION ASSAY FOR *SALMONELLA*

Salmonellae are facultatively, anaerobic gram-negative rods and belong to the family *Enterobacteriaceae* which includes *Escherichia, Shigella, Citrobacter, Enterobacter, Klebsiella, Erwinia, Kluyvera, Serratia, Cedecea, Morganella, Hafnia, Edwardsiella, Providencia, Proteus*, and *Yersinia* (29). Members of the genus *Salmonella* are pathogenic for humans and cause enteric fevers, gastroenteritis, and septicemia. There are over 2400 recognized strains based on serological typing of three surface antigens: somatic (O), capsular (Vi), and flagellar (H). Independent of serotypic classification, the salmonellae have been divided into five "subgenera" by a set of nine biochemical tests. Members of subgenera I generally inhabit warm-blooded animals and most of the serovars have individual strain names, whereas others are identified by their antigenic formulae. Salmonellae generally found in cold-blooded animals and in the environment, are classified in subgenera II through V. Some serovars in II, IV, and V have names, but most are identified by their antigenic formulae. Group III strains are all named *S. arizonae* even though they show a high degree of antigenic variation.

DNA hybridization analysis divides the family *Enterobacteriaceae* into genera along the same lines established by microbiological and biochemical classification schemes. Different genera are typically 50% or less related by DNA hybridization. *Salmonella* as a group is about 45-55% related to *E. coli* or *Citrobacter* and are less closely related to the *Klebsiella* at 20-40%, *Erwinia* at 20-25%, and *Edwardsiella* and *Proteus* at less than 20%. The grouping of strains to make up a genus occurs when the DNA is greater than 50% related. Members of *Salmonella* subgenera I-IV are observed to be more than 70% related by DNA hybridization (30,31).

A. *Salmonella* in Foods

Outbreaks of salmonellosis commonly follow the consumption of infected food products that had been contaminated during processing when proper sanitation procedures were not followed. Fortunately, occurrences are rare, because commercially available food products and ingredients are regularly tested for the presence of salmonellae. Food products that are found to be contaminated with salmonella are discarded, as being unfit for human use.

The test for salmonella in foods uses a battery of bacteriological, biochemical, and serological procedures that take 4–7 days to complete. These methods are outlined in the *Official Methods of Analysis*-AOAC (32) and are commonly called the BAM procedures (33). In brief, for each product to be analyzed either a 25-g portion is placed in 225 ml of broth or a composite of 17, 25-g samples is placed in 4 L of broth. The sample is usually made homogeneous by swirling, blending, or kneading, depending on the nature of the sample, and then incubated for 24 ± 2 hr. The enrichment broth is a nonselective medium that allows the growth of most bacteria. Selection for *Salmonella* species is not imposed at this stage, because cells often are heat injured or desiccated and need to be revived; selection would impose stresses that already stressed cells are unable to overcome easily. A 1-ml portion of the enrichment culture is transferred to 9 ml of each of two selective broths (tetrathionite brilliant green and selenite cysteine) and incubated for 24 ± 2 hr. These broths are semiselective and usually increase the salmonella titer relative to many non-salmonella bacteria. A portion of each broth is streaked onto each of three different selective media for colony isolation. The plates are examined after 24 and 48 hr for salmonella-like reactions. Positive appearing colonies are classified as presumptive positives and additional analysis is required because some non-salmonellae, notably *Proteus* and *Citrobacter* species, produce salmonella-like reactions. Another complication is that some strains of salmonella show atypical reactions. Careful attention, colony selection, and analysis greatly increases the workload of the testing laboratory; inattention increases the probability of releasing a contaminated product. A sample of the presumptive salmonella colony is subjected to additional biochemical analysis and serological classification, which may take several days. Only after a suspect colony has passed all tests, is it labeled *Salmonella*. Non-salmonella colonies may drop out of the test procedure at various times. It is not unusual for the food manufacturer to quarantine products during the salmonella test phase; otherwise, there is risk that they will have to be recalled if found to be contaminated. The inconvenience and expense generated by having to wait 4–7 days for a salmonella test result makes it highly desirable to have a quicker assay.

B. DNA Probes Specific for *Salmonella*

The observation that salmonellae are related by DNA hybridization and can likewise be differentiated from other organisms, implies that there may be identifiable sequences of DNA that are unique or at least sufficiently unique to

Salmonella species that an assay for *Salmonella* might be developed based on DNA hybridization. For the DNA probes to be useful, they must identify all target organisms, including atypical strains (inclusive), and not cross-hybridize to non-salmonella organisms likely to be present in the test sample (exclusive). DNA probes for salmonellae that meet these two requirements have been described by Fitts et al. (21) and Fitts (34). The probe set is comprised of ten *Bam*HI restriction fragments of 1400-6000 base pairs. These are random genomic pieces of DNA, characterized solely by their ability to perferentially hybridize to salmonella DNA. Two of the fragments detected all serotypes examined, three fragments detected more than 90% of the serotypes, and five detect fewer than 50%. Although only about 7.5% of all known serotypes were tested, there was no correlation between DNA hybridization and grouping by serotypes. Consequently, only a small number of DNA probes are needed to detect a wide variety of salmonella isolates.

DNA sequence data of particular genes represents another source of salmonella probes. The *trpA* and *oriC* genes of *Salmonella* and *E. coli* are closely related (35-37). The genes have both conserved and nonconserved sequences. Probes based on the nonconserved sequences may be made. It is possible that probes for salmonella rRNA may be developed as has been demonstrated for *Campylobacter* species (13). The rRNA genes of *Salmonella* and *E. coli* are 96% homologous, compared with the total genomic DNA, which exhibits almost 50% relatedness. It may be possible, therefore, to estimate that the rRNAs of *E. coli* and *Salmonella* differ by about 180 bases. Probes that are based upon these sequences or other nonconserved genomic regions need to be tested rigorously for inclusivity and exclusivity as Fitts (34) describes for the random genomic probes for *Salmonella.*

C. DNA Hybridization Assay for the Detection of Salmonellae in Foods

Fitts and co-workers (21,34) demonstrated that a DNA probe assay for the detection of salmonellae in foods was feasible. A prototype version of the GENE-TRAK DNA hybridization assay for *Salmonella* was compared with the standard microbiological method and with a commercially available immunoassay and was found to give superior results (38). The DNA hybridization was more sensitive than either the microbiological assay or the immunoassay for the detection of *Salmonella* species and had a lower false-positive reaction rate than the immunoassay.

DNA probes enable the rapid detection of *Salmonella* species by the elimination of the colony isolation step from enrichment culture and subsequent taxonomic classification and strain identification steps required in the traditional culture method, which accounts for 1-5 days of the 4-7 day assay. The DNA hybridization assay, when applied directly to enrichment culture, determines the presence or absence of *Salmonella* species in 4 hr. When coupled to the 40 hr of enrichment culture, the entire assay is completed within 48 hr.

The GENE-TRAK assay procedure for *Salmonella* combines widely accepted and commonly used culture enrichment methods, which have undergone evolution and development for years by food microbiologists, with filter hybridization methods that date back over 10 years. The procedure initially follows the BAM recommended enrichment step of 24 hr. The salmonella selective enrichment is a two-step procedure. As in BAM, the enrichment culture is diluted into tetrathionite and selenite cysteine broths which are incubated for either 6 or 12 hr depending on food type. These are then diluted into Gram-Negative broth and incubated for an additional 12 or 6 hr for a total time of 18 hr in selective enrichment. One milliliter of each selective broth is collected on a single 25-mm diameter nylon membrane filter by application of vacuum using a GENE-TRAK manifold apparatus. Cells are lysed and the DNA is denatured by application of an alkaline solution. After incubation for 2 min, the solution is removed by filtration. A buffer is added to neutralize the first solution which is removed by filtration. An alcohol solution is then applied to allow rapid drying and fixation of the DNA to the membrane. The filters, including positive and negative control filters, are warmed at 65°C for 30 min in a vessel containing prehybridization solution that blocks nonspecific binding of probe DNA to the filter. Hybridization solution containing salmonella-specific DNA probe that is radiolabeled with ^{32}P is then substituted for the prehybridization solution and allowed to hybridize for 2 hr at 65°C. Excess probe and weakly hybridized probe are then removed by a series of six washes, lasting 5 min each, with low salt buffer. The radioactivity on the filter is determined with a β-counter and samples having a predetermined number of disintegrations per filter over those observed on the average of the negative control filters are considered to be positive for salmonellae.

D. Performance Characteristics of the *Salmonella* Assay

Diagnostic assays typically have limits of detection. The limit is usually governed by the number of target organisms. In the standard salmonella assay a loop of culture is streaked on an agar plate; hence, to detect *Salmonella* species there must be at least one salmonella cell per loop, and when it is streaked it must be sufficiently isolated from other cells to give a distinct colony. The higher the salmonella titer relative to non-salmonella competitors the easier it is to isolate and identify it. The selective enrichment stages used in microbiological assay are designed to optimize the salmonella titers. This level is usually about 10^7 cells/ml or higher (Flowers and Curiale, unpublished data). Consequently, GENE-TRAK salmonella assay was designed with this target in mind. Table 7 shows that the salmonella assay meets this sensitivity requirement.

Table 7 Detection of *S. senftenberg* in Artificially Contaminated Milk Chocolate

Trial	Number of cells added per sample			
	0	4.5×10^6	9.0×10^6	2.3×10^7
1	192	505	929	2166
2	224	794	938	1989
3	205	694	1242	2013
4	185	858	926	1817
5	240	693	990	1935

Cells are concentrated and collected on a filter by vacuum filtration. The sensitivity in the food test is 2–3 $\times$ 10^6 cells/ml when the recommended 2 ml of culture is assayed. Results are expressed as counts per minute (cpm) per filter. Filters with more counts than the control filters (0 cells, i.e., blank filters) indicate the presence of *Salmonella.*

Specificity of the assay begins with the isolation of DNA probes. Probes that are highly specific for salmonella show little background noise resulting from hybridization to non-salmonella organisms. In food cultures, in addition to non-salmonella organisms, food debris is also present. Data in Table 8 show that GENE-TRAK salmonella assay shows little cross-reactivity with food debris or non-salmonella organisms. Extensive testing in our laboratories with a variety of

Table 8 Detection of *Salmonella* in Foods Inoculated with Salmonellae and Non-Salmonella Organisms

Food type	Addition	Total cpm/filter	Final cells/ml culture
Nonfat dry milk	*S. typhimurium*	8,847	5.7×10^8
	S. tennessee	25,676	1.4×10^9
	E. coli	278	1.1×10^7
	K. pneumoniae	192	1.0×10^8
	None	140	3.0×10^3
Wheat flour	*S. eastbourne*	6,306	6.8×10^8
	None	146	1.2×10^6
Macaroni	*S. anatum*	25,974	3.0×10^9
	None	252	Less than 10
Cheese powder	*S. senftenberg*	18,315	1.1×10^9
	P. mirabilis	457	8.9×10^8
	None	526	Less than 10

foods showed that the GENE-TRAK salmonella assay always detected *Salmonella* species when present (Curiale, Neri, Prinz, Bell, Landes, in preparation). In 3.5% of the samples no salmonellae were present, but a positive result was indicated. Upon retesting, those samples that do not contain salmonellae are rarely positive, indicating a random adherence of probe to some filters.

VI. SUMMARY

In this review we have discussed development of practical test formats for DNA probe technology in two major infectious disease settings—clinical laboratories and the food microbiology. The application of this powerful methodology to other practical situations (e.g., other microorganisms, human genetic diseases, diagnosis of malignancy) will involve similar basic steps, as outlined here: the identification of relevant probe sequences, the development of formats appropriate for each contemplated use, and the ability to achieve sensitivity and specificity appropriate for the desired applications.

We have described two strategies for the isolation of probes suitable for hybridization assays: genomic probes and rRNA probes. It is important to point out that each of these probe types have unique features. The rRNA probes have the advantage of a high abundance of target rRNA copies per cell. Chromosomal probes are applicable to many situations in which specific genes, such as antibiotic-resistance markers, virulence-related probe properties, closely related organisms, or viral pathogens, are of interest.

In the food industry, substantial cost reductions will accrue by earlier release of food lots and reduction in labor costs. In diagnostic microbiology, improvements in clinical care should result from successful development of hybridization-based assays for the following reasons: Organism identification will be rapid, and definitive (this especially applies to those agents currently difficult to cultivate, isolate, or identify). Such early identification will facilitate more direct antimicrobial therapy, reducing costly, poorly focused empiric therapy. They will aid in early recognition of nosocomial infections enabling timely intervention and prevention of institution-associated outbreaks. Nucleic acid probes will provide for identification of specific genes of interest (e.g., virulence-related factors, specific antimicrobial-resistance mechanisms, plasmid markers). Finally, hybridization-based assays should readily adapt to automated instrumentation, data-base management, and laboratory/hospital information networking.

REFERENCES

1. Nygaard, A. P., and B. C. Hall, Formation and properties of RNA–DNA complexes. *J. Mol. Biol. 9*:125–142 (1964).
2. Bernardi, G., Chromatography of nucleic acids on hydroxyapatite columns. *Method Enzymol. 21D*:95–139 (1971).

3. Johnson, J. L., Genetic characterization. In *Manual of Methods for General Bacteriology*, P. Gerhart et al. (eds.). American Society for Microbiology, Washington, pp. 450–472 (1981).
4. Southern, E., Detection of specific sequences among DNA fragments separated by gel electrophoresis. *J. Mol. Biol. 98*:503–517 (1975).
5. Grunstein, M., and D. Hogness, Colony hybridization: A method for the isolation of cloned DNAs that contain a specific gene. *Proc. Natl. Acad. Sci. USA 72*:3961 (1975).
6. Flavel, R. A., E. J. Birfelder, J. P. M. Sanders, and P. Borst, DNA–DNA hybridization on nitrocellulose filters, 1. General considerations and non-ideal kinetics. *Eur. J. Biochem. 47*:535–543 (1974).
7. Ranki, M., M. Virtanen, A. Palva, M. Laaksonen, R. Petterson, L. Kaariainen, P. Halonen, and H. Sorderlund, Nucleic acid sandwich hybridization in adenovirus diagnosis. *Curr. Top. Microbiol. Immunol. 104*:307–318 (1983).
8. Ranki, M., M. Virtonen, A. Palva, M. Laaksonen, and H. Soderlund, Sandwich hybridization as a convenient method for the detection of nucleic acids in crude samples. *Genes 21*:77–85 (1983).
9. Polsky-Cynkin, R., G. H. Parsons, L. Allerdt, G. Landes, G. Davis, and A. Rashtchian, Use of DNA immobilized on plastic and agarose supports to detect DNA by sandwich hybridization. *Clin. Chem. 31*:1438–1443 (1985).
10. Skirrow, M. B., Campylobacter enteritis: A "new" disease. *Br. Med. J. 2*: 9–11 (1977).
11. Butzler, J. P., Campylobacter *Infection in Man and Animals*. CRC Press, Boca Raton (1984).
12. Skirrow, M. D., Campylobacter enteritis—the first five years. *J. Hyg. Camb. 89*:175–184 (1982).
13. Rashtchian, A., M. A. Abbott, G. A. Mock, D. Lovern, and M. M. Shaffer, Use of cloned and synthetic DNA probes for detection of *Campylobacter* species. (submitted for publication).
14. Harvey, S. M., and J. R. Greenwood, Relationships among catalase-positive *Campylobacter* determined by deoxyribonucleic acid-deoxyribonucleic acid hybridization. *Int. J. Syst. Bacteriol. 33*:275–284 (1983).
15. Roop, R. M., II, R. M. Smibert, J. L. Johnson, and N. R. Krieg, Differential characteristics of catalase positive *Campylobacter* correlated with DNA homology groups, *Can J. Microbiol. 30*:938–951 (1984).
16. Noller, H. F., Structure of ribosomal RNA. *Ann. Res. Biochem. 53*:119–162 (1984).
17. Rashtchian, A., M. A. Abbott, and M. Shaffer, Cloning and characterization of genes coding for ribosomal RNA in *Campylobacter jejuni. Current Microbiol. 14*:311–317 (1987).
18. Brosius, J., T. J. Dull, D. D. Sleeter, and H. F. Noller, Gene organization and primary structure of a ribosomal RNA operon from *Escherichia coli. J. Mol. Biol. 148*:107–127 (1981).
19. Carbon, P., J. P. Ebel, and C. Ehresmann, The sequence of the ribosomal 16S RNA from *Proteus vulgaris*. Sequence comparison with *E. coli* 16S

RNA and its use in secondary structure model building. *Nucleic Acid Res.* *9*:2325–2333 (1981).
20. Maniatis, T., E. F. Fritsch, and J. Sambrook, *Molecular Cloning: A Laboratory Manual*. Cold Spring Harbor Laboratory, Cold Spring Harbor, N.Y. (1982).
21. Fitts, R., M. Diamond, C. Hamilton, and M. Neri, DNA–DNA hybridization assay for detection of *Salmonella* spp. in foods. *Appl. Environ. Microbiol.* *46*:1146–1151 (1983).
22. Rashtchian, A., Detection of *Campylobacter* species in clinical specimens using nucleic acid probes. In *Campylobacter III*, A. D. Pearson et al. (eds.). PHLS, London, pp. 71–73 (1985).
23. Morris, G. K., and C. M. Patton, *Campylobacter*. In *Manual of Clinical Microbiology*, Lennette, E. H., A. Balows, W. J. Hausler, Jr., H. J. Shadomy (eds.). American Society for Microbiology, Washington (1985).
24. Langer, P. R., A. A. Waldrop, and D. C. Ward, Enzymatic synthesis of biotin-labelled polynucleotides: Novel nucleic acid affinity probes. *Proc. Natl. Acad. Sci. USA* *75*:6633–6637 (1981).
25. Chollet, A., and E. H. Kawashima, Biotin-labeled synthetic oligodeoxyribonucleotides: Chemical synthesis and uses as hybridization probes. *Nucleic Acids Res.* *13*:1529–1541 (1985).
26. Draper, D. E., Attachment of reporter groups to specific, selected cytidine residues in RNA using a bisulfite-catalyzed transamination reactions. *Nucleic Acids Res.* *12*:989–1002 (1984).
27. Draper, D. E., and L. Gold, A method for linking fluorescent labels to polynucleotide: Application to studies of ribosome–ribonucleic acid interactions. *Biochemistry* *19*:1774–1781 (1980).
28. Rudkin, G. T., and B. D. Stollar, High resolution detection of DNA:RNA hybrids in situ by indirect immunofluorescence. *Nature* *265*:472–473 (1977).
29. Brenner, D. J., In *Bergey's Manual of Systematic Bacteriology*, N. R. Krieg, and J. G. Holt (eds.). Williams & Wilkins Co., Baltimore (1984).
30. Crosa, J. H., D. J. Brenner, U. H. Ewing, and S. Falkow, Molecular relationships among the salmonellae. *J. Bacteriol.* *115*:307–315 (1973).
31. Stoleru, L., L. LeMinor, and A. M. Lheritier, Polynucleotide sequence divergence among strains of *Salmonella* subgenus IV and closely related organisms. *Ann. Microbiol.* *127*:477–486 (1976).
32. AOAC. *Official Methods of Analysis of the Association of Official Analytical Chemists*. Association of Official Analytical Chemists, Arlington (1984).
33. FDA, *Bacteriological Analytical Manual*. Food and Drug Administration, Association of Official Analytical Chemists, Arlington (1984).
34. Fitts, R., Development of a DNA–DNA hybridization test for the presence of *Salmonella* in foods. *Food Technol.* *39*:95–102 (1985).
35. Nichols, B. P., and C. Yanofsky, Comparison of the nucleotide sequence of *trp*A and sequences immediately beyond the *trp* operon of *Klebsiella aerogenes, Salmonella typhimurium* and *Escherichia coli. Nucleic Acids Res.* *9*:1743–1755 (1981).

36. Nichols, B. P., and C. Yanofsky, Nucleotide sequences of *trp*A of *Salmonella typhimurium* and *Escherichia coli*: An evolutionary comparison. *Proc. Natl. Acad. Sci. USA 76*:5244–5248 (1979).
37. Cleary, J. M., D. W. Smith, N. E. Harding, and J. W. Zyskind, Primary structure of the chromosomal origins (*ori*C) of *Enterobacter aerogenes* and *Klebsiella pneumoniae*: Comparisons and evolutionary relationships. *J. Bacteriol. 150*:1467–1471 (1982).
38. Flowers, R. S., Comparison of rapid *Salmonella* screening methods and the conventional culture method. *Food Technol. 39*:103–108 (1985).

RECENT REFERENCES ADDED IN PROOF

Rashtchian, A., J. Eldredge, M. Ottaviani, M. Abbott, G. Mock, D. Lovern, J. Klinger, and G. Parsons, Immunological capture of nucleic acid hybrids and application to non-radioactive DNA probe assay. *Clin. Chem. 33*:1526–1530 (1987).

Stollar, B. D., and A. Rashtchian, Immunochemical approaches to gene probe assays. *Anal. Biochem. 161*:387–394 (1987).

10
DNA Probe Diagnosis of Periodontal Disease

CYNTHIA K. FRENCH, SUSANNE L. SIMON, MICHAEL C. CHEN,*
SUZANNE M. EKLUND, LYNN C. KLOTZ,† and KAREN K. VACCARO
BioTechnica Diagnostics, Inc., Cambridge, Massachusetts

EUGENE D. SAVITT‡
Forsyth Dental Center, Boston, Massachusetts

I. PERIODONTAL DISEASE: AN OVERVIEW

Periodontal diseases are among the most common afflictions of mankind and cause loss of more teeth than does any other disease, including dental caries. Evidence of periodontal disease has been observed in the fossil remains of Neanderthal man, and detailed descriptions of the disease have been found in Chinese and Egyptian writings over 5000 years old. Epidemiological surveys conducted during this century suggest that essentially all of the world's adult population have experienced some form of periodontal disease (1). At any one time, 10% of the United States' population shows evidence of destructive periodontal disease, and about 40% of adults over 40 years of age develop the disease (2). Currently, in the United States, periodontal disease causes over 60% of tooth loss in adults over the age of 40 (3).

There are several clinical forms of the disease that can be divided into two broad groups: gingivitis and destructive periodontal disease (periodontitis). *Gingivitis* is generally characterized by inflammation and bleeding of the gums caused by microbial colonization (plaque) of adjacent tooth surfaces. In *periodontitis* the infection involves deeper periodontal tissues, resulting in pocket formation between the teeth and gums, destruction of alveolar bone and periodontal ligament (the supporting structures of the teeth), increased mobility of

Present affiliations:
*American BioTechnologies, Inc., Cambridge, Massachusetts.
†Biotechnology consultant, Cambridge, Massachusetts.
‡BioTechnica Diagnostics, Inc., Cambridge, Massachusetts.

teeth, and finally tooth loss. Gingivitis is usually not a serious condition and can readily be treated by conservative debridement of plaque and improved oral hygiene. Periodontitis, however, requires more extensive professional treatment to arrest the disease and prevent tooth loss. Figure 1a presents a schematic representation of a healthy tooth and surrounding tissue, and in Figure 1b is a representation of moderately advanced periodontitis.

The understanding of the causes of periodontal disease has increased rapidly in the past decade. Before 1970, it was generally thought that supragingival (above the gum line) plaque accumulation led to gingivitis which, in turn, progressed to periodontitis. These conclusions resulted because supra- and subgingival plaque were not differentiated, plaque samples were pooled from different sites, and experimental limitations prevented culturing and classification of many fastidious bacteria. Since about 1970, new findings have changed this view of periodontal disease. Supragingival plaque and gingivitis were found to persist in patients for years without leading to bone erosion and tooth loss (4). The heterogeneity of plaque became apparent in supra- and subgingival plaque, between locations, and over time. In some instances, the host immune response was found to be different with different states of gingival health (5). The current view of periodontal disease is that different states of gingival health are correlated with different subgingival bacterial populations. Some bacteria are benign and are entirely compatible with gingival health, whereas others cause the tissue and bone destruction of periodontitis (6,7).

Included in the general term *periodontitis* are several different disease conditions that can be categorized as either juvenile periodontitis or adult periodontitis:

A. Juvenile Periodontitis

It is generally accepted that juvenile periodontitis consists of two forms: *localized*, which affects the permanent first molars and incisors, and *general*, which may affect most of the dentition. Classically, localized juvenile periodontitis often shows little clinical evidence of gingival inflammation. It spreads rapidly, causing pocket formation, bone destruction, and eventually tooth loss within a few years after onset. Depending on population segment, up to 10% of children aged 12-20 develop localized juvenile periodontitis (8). The National Institute of Dental Research (NIDR) has also reported that, of the general population, 6% of youths 12-17 years of age have some form of destructive periodontal disease (9).

The gram-negative capnophile *Actinobacillus actinomycetemcomitans* (*Aa*) has been shown to be the pathogenic microorganism predominantly associated with localized juvenile periodontitis (LJP), and substantial evidence indicates it to be the primary causative agent (9-12). Furthermore, if *Aa* is eliminated or its

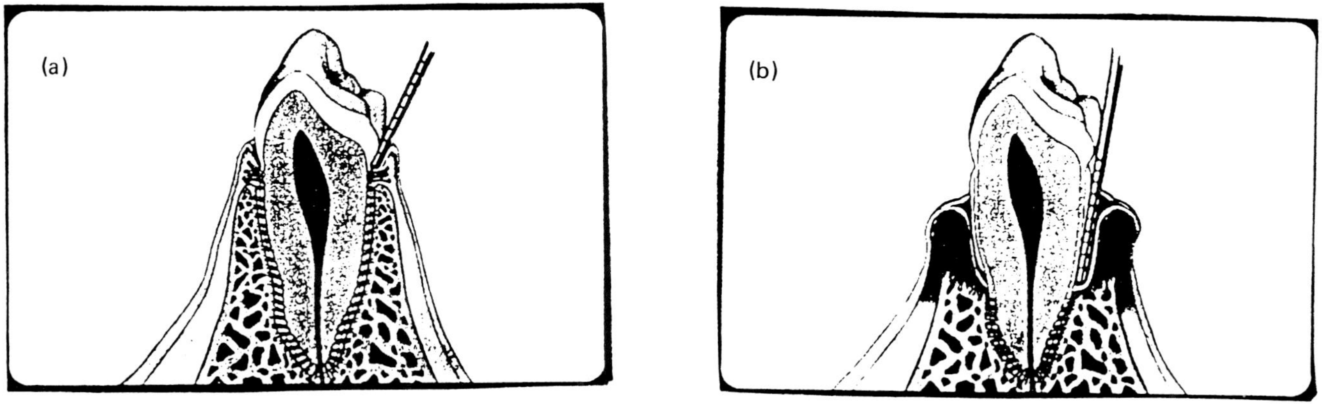

Figure 1 Schematic representation of destructive periodontal disease. (a) A healthy periodontal state. In this state, a metal periodontal probe inserted between the gum and teeth will penetrate only about 2 mm (two marks on the probe). Also, the bone (mottled area) covers the tooth root completely. (b) A tooth with moderate to severe periodontitis. In the tooth pictured, the periodontal probe penetrates about 8 mm, and substantial bone destruction around the tooth root can be observed. In such a state, the tooth may exhibit some mobility. Further bone loss could lead to tooth loss because of loss of anchorage.

numbers greatly reduced in the oral cavity by antibiotics or surgical removal of *Aa*-infected tissue, there is a corresponding remission of symptoms (13).

Optimal treatment for LJP appears to be a combination of surgery and systemically administered tetracycline (13,14). Tetracycline is particularly well suited for this purpose because of its broad effectiveness against gram-negative organisms, such as those involved in periodontitis, and its ability to concentrate in the crevicular fluid between the teeth and gums. The length of time needed and effectiveness of treatment, however, is variable. Therefore, the effectiveness of therapy should be monitored to determine when the infection has been eliminated (12).

B. Adult Periodontitis

Adult periodontitis occurs in patients over 20 years of age and makes up most of the destructive periodontal disease. This form of the disease usually occurs in "bursts" of activity, which may be caused by invasion of pathogenic microorganisms that have developed in the periodontal pockets into the gingival tissue (15). The presence of bacteria often provokes an antibody response, which has been hypothesized to be involved with the pathogenesis. These bursts are separated by periods of stable condition. Both the onset of the disease and the ensuing bursts may be facilitated by host and environmental factors that appear to allow increased levels of pathogenic microorganisms.

Bacteroides gingivalis (*Bg*) is an oral microorganism strongly implicated in the etiology of adult periodontitis (AP). The ability of *Bg* to produce a number of potential virulence factors, such as collagenase and lipopolysaccharides, contributes to its pathogenic potential. *Bacteroides gingivalis* is found in small numbers at a low frequency in the subgingival plaque of periodontally normal patients. However, it is detected at a much higher frequency and in much greater numbers in the subgingival plaque of adult periodontitis patients. It is often a major component of the plaque of these patients. Studies have indicated that the presence or absence of *Bg* correlates well with either the presence or absence of disease (16-19).

Other studies point to the importance of both *Aa* and *Bg*, plus an additional black-pigmented bacteroides, *B. intermedius* (*Bi*), in the etiology of periodontal disease, particularly AP. In one study, at least one of the three bacteria, *Aa, Bi,* and *Bg*, was detected in 99% of 130 oral sites that continued to lose bone and gingival attachment. These three bacteria are sometimes found in nonprogressing sites but in markedly lower proportions (20).

Zambon (21) found that in 92 AP patients, *Bi* and *Bg* were detected in 79% and 67% of patients, respectively, and in high numbers, whereas of 617 clinic patients (representative of the general population) *Bi, Bg*, and *Aa* were found in 8, 4, and 0.4% of patients, respectively, and at low percentages of the total bacteria.

In recent years, researchers using careful anaerobic sampling and culture procedures have identified several other gram-negative anaerobes correlated with destructive adult periodontitis (22-25).

Fusobacterium nucleatum has been commonly isolated in gingivitis (24,26) and in diseased sites in both juvenile periodontitis (27) and adult periodontitis (25,26,28,29). However, despite the consistency of isolation of *F. nucleatum* in periodontal pockets, it is unclear if this species is an opportunistic colonizer of the anaerobic environment of the periodontal pocket (25) or a possible pathogen (30). *Capnocytophaga* species have been associated with both juvenile (27) and adult (24) destructive periodontal disease. However, as with *F. nucleatum, Capnocytophaga* species are ubiquitous in subgingival floras and healthy periodontia (31) and are commonly isolated from chronic gingivitis sites (25). *Capnocytophaga* species have been shown to be present in greater concentrations in samples from healthy sites than from periodontal disease sites (32,33). These studies further suggest that isolation of *C. ochracea* in disease sites (24), is indicative of the stimulatory effect on this species of bacteria and host byproducts located in the pocket.

Selenomonas sputigena represents a species of motile gram-negative rod infrequently associated with destructive periodontal disease. *Selenomonas sputigena* has been isolated from a group of prepubertal periodontitis patients (34). *Selenomonas* have also been found in low percentages from moderately inflamed periodontitis lesions but at levels not significantly elevated above baseline (24). A similar finding of low, but significantly elevated numbers of *S. sputigena*, were noted in overall disease-active sites from 19 subjects (35).

Eikenella corrodens is another species infrequently associated with floras from periodontitis sites. The name is derived from the pitting of agar surfaces by the colonies. *Eikenella corrodens* has been implicated in advancing lesions in adult periodontitis (24) and in the juvenile form of disease (36). This microorganism is of clinical concern, even though the frequency of detection in advancing sites is irregular.

Wolinella recta, a motile vibriolike gram-negative rod, and *Bacteroides forsythus*, a fastidious nonmotile species, have been reported to be present in high proportions in isolated advancing sites in adult patients (29,30) and in slightly increased proportions of juvenile periodontitis sites (37). Because of the difficulty of culturing these microorganisms, coupled with their relatively recent characterization and classification, with *W. recta* registered in 1981 (38) and *B. forsythus* in 1986 (39), data from supporting studies are not yet available.

Spirochetes including *Treponema denticola* are cultivatable only under extremely stringent methods and are not routinely sought even by laboratories oriented toward dealing with the complexities of the subgingival flora. Most data has utilized morphological descriptors and are unable to differentiate among several different species and possibly genera as well. As a result, it is unclear about the importance of spirochetes, as a group or as a particular species, in the disease pockets. Investigators have been unable to establish any decisive role for spirochetes and other motile forms in advancing periodontal lesions compared with controls (37).

II. RATIONALE FOR MICROBIOLOGICAL DIAGNOSIS

In current practice, periodontal disease is usually detected by observing the destruction from the disease. The most common and simplest measurement made by the periodontist or dentist is the loss of attachment between gum and teeth (measured in millimeters of pocket depth by using a metal periodontal probe). Figure 1b illustrates a periodontal probe recording a pocket depth of approximately 8 mm. This clinical measurement has been prone to significant error (40). In controlled clinical trials, probing error is about 1 mm. Because diagnostic determination can differ with even a 1-mm change, relying on probing measurements may lead to an error in diagnosis.

Additionally, destruction, as evidenced by bone loss, can be viewed in dental x-rays. As with nonstandardized probing measurements, radiographic techniques, as they are currently clinically applied, are often misrepresentative of the actual severity of bone loss (41). It has also been shown that radiographically evident bone loss does not appear until 6 months after the burst of active destruction (42).

Other nondestructive clinical signs, easily observed by practitioners and often thought to be associated with active, destructive periodontal disease, are gingival redness, heavy accumulations of supragingival plaque, suppuration (pus), and bleeding of the gums upon probing with the periodontal probe. Unfortunately, these characteristics are of no predictive value in active periodontal disease (43). The lack of correlation between these standard dental observations and active periodontal disease is illustrated in Table 1. For the most part, any one of these measurements has less than a 50% chance of predicting active disease.

Studies have suggested that destructive periodontal disease activity occurs with bursts of acute destruction followed by periods of remission (44). The detection of these active sites is difficult, as outlined previously. Although pocket depths greater than 4 mm may be viewed as evidence of destructive periodontal disease, it is unclear from this measurement alone that the disease is active and, therefore, the patient is at high risk for further near-term destruction, or whether the pockets are evidence of past disease, and causative bacteria are no longer present.

As previously discussed, a number of gram-negative anaerobic bacteria have been correlated with, and are thought responsible for, much of the destruction in periodontal disease. Therefore, identification and enumeration of the bacteria present in subgingival plaque can be a measure of risk of active disease.

Table 1 Clinical Observations As Predictors Of Destructive Periodontal Disease Activity

Clinical measurements (3414 sites)	Proportion of active sites[a] showing condition
Gingival redness	0.32
Plaque	0.42
Suppuration	0.03
Bleeding on probing	0.25
Pocket < 4 mm	0.63
Pocket 4–6 mm	0.21
Pocket > 6 mm	0.10

[a]Determined by tolerance analysis of pairs of probe measurements made bimonthly.
Source: Ref. 43.

Besides prediction of oral sites at risk of active disease, identification of the disease-associated bacteria can provide a basis for monitoring therapy. Such an approach is an alternative to the method of follow-up care dependent on detection of further destruction over time. A semiquantitative enumeration of the disease-associated bacteria before, during, and after therapy can provide a measure for evaluation of treatment success.

For example, in the treatment of juvenile periodontitis, the use of tetracycline has been shown to be a useful adjunct for the elimination of *Aa*, the putative pathogen (45). However, it has been found that the length of time the antibiotic should be administered is variable from patient to patient (12). Thus, a diagnostic test for *Aa* would allow the periodontist to determine the time course of antibiotic treatment, whether to try another antibiotic, or to try some other mode of therapy if the patient is not responding to tetracycline.

III. RATIONALE FOR DNA PROBES

The periodontal disease-associated bacteria are anaerobic and capnophilic. Cells in plaque are typically in a fragile state. Exposure to oxygen (air) for even a short period results in rapid cell death. Historically, these bacteria were first identified and classified by use of strict anaerobic culturing methods. Although most periodontal microbiology is still carried out using methods based on culturing, such methods have limitations for routine patient diagnosis.

For patient diagnosis, the samples must be taken at the periodontist's office and then transported to the laboratory for analysis. The periodontist must ensure that the bacteria are not exposed to oxygen for more than a few seconds. Furthermore, even though media have been developed for transporting these anaerobic bacteria under oxygen-free conditions, the bacteria may die as a result of extremes in temperature during transport or exposure to oxygen because of improperly sealed transport containers, and the like.

Many of these bacteria require a complex growth medium for culturing because they can have fastidious nutritional requirements (24). The selective media and biochemical tests required for speciation of the various bacteria can also be very expensive, and their use time-consuming (25).

For DNA probe diagnosis, the bacteria need not be alive. The DNA should remain intact under all conditions ordinarily encountered in mail delivery. Paper points (a tightly rolled piece of filter paper) commonly used in endodontics, make a useful sampling device. This sampling device would be suitable as a transport surface to the laboratory as well.

Immunoassays with both polyclonal and monoclonal antibodies have been utilized for identification of periodontal disease-associated bacteria (8,46,47). However, DNA probes are potentially superior to antibodies for diagnosis.

Routine detection of as little as 10^3 periodontal disease-associated bacteria has been demonstrated for these microorganisms (48). Typical plaque samples from periodontal pockets taken with paper points contain about 10^6 total bacteria (49). By use of this sampling method, minimum significant quantities of disease-associated bacteria would fall between 10^3 and 10^4 bacteria. Immunoassays, that use polyclonal antibodies against these bacteria, now routinely detect only approximately 10^5 bacteria in the enzyme-linked immunosorbent assay (ELISA) format (3).

In addition, strains of oral species have been shown to change their surface markers (50,51), perhaps as a means of evading the host's immune response. Such surface changes could lead to false-negative diagnostic results, because antibodies are usually prepared against surface antigens.

IV. DNA PROBE DETECTION OF PERIODONTAL PATHOGENS

A. Materials and Methods

1. Preparation of Genomic DNAs

Microorganisms were cultured according to published procedures (52), harvested, and centrifuged to form a cell pellet, and lyophilized. Total cellular DNA was extracted and purified according to the procedure of Marmur (53). The strains used for DNA purification were as follows: Y4 for *A. actinomycetemcomitans*, 581 for *B. intermedius*, and 33277 for *B. gingivalis.*

2. Radioisotope Labeling of DNA Probes

Each genomic DNA or plasmid-insert complex was nick translated with ^{32}P-labeled dCTP (DuPont) to a specific activity of $1\text{-}4 \times 10^8$ cpm/μg of DNA by standard techniques (55).

3. Nonisotopic Labeling of DNA Probes

Actinobacillus actinomycetemcomitans and *Escherichia coli* C600 were labeled by using the Bethesda Research Laboratories (BRL) nick translation kit (Cat. No. 8160SB) with biotin-11-dUTP. Detection was done with the BluGENE nonradioactive nucleic acid detection system (Cat. No. 8279SA).

4. Sample Preparation

Sample strains to be detected were grown as fresh overnight cultures from individually isolated colonies and regrown as log cultures from which optical density measurements were taken to determine cell number. Cultures were appropriately diluted to represent $10^5\text{-}10^2$ cells. Samples for dilutions were also plated for confirmation of the optical density measurements. Samples were prepared for application to nitrocellulose filters with use of a proprietary methodology.

5. Immobilization of Prepared Cell Samples

Aliquots of prepared samples were equilibrated with an equal volume of loading buffer (1:1 3 M NaCl, 0.3 M NaOH–2 M NH_4OAc) and applied to a nitrocellulose filter placed in a Schleicher and Schuell Slot Blot apparatus. Vacuum was applied to bring the samples onto the filters, followed by a 1-ml rinse of loading buffer. After sample application, the nitrocellulose filters were rinsed in 0.5 M NaCl, air dried, and baked at 80°C for 1 hr.

6. Hybridization Conditions

Filters were prehybridized in 5X SSC, 0.5% SDS, 100 μg/ml salmon sperm DNA and 5X Denharts, for 1 hr at 65°C. Hybridization conditions were the same, but hybridization buffer contained 10% dextran sulfate. Nick-translated whole genomic probes (^{32}P) were generated at a specific activity of $1\text{-}4 \times 10^8$ cpm/μg, and 1×10^7 cpms were added per 5-ml hybridization volume. Filters were washed in two steps: 10 min in 1X SSC, 0.5% SDS at room temperature, followed by a 60°C 0.1X SSC, 0.5% SDS wash for 30 min. Filters were then air dried and exposed to Kodak X-OMAT AR film with an intensifying screen at −70°C for 2–18 hr.

7. Generation of Genomic Libraries for Species-Specific (Non-Cross-Reacting) Probes

Bacterial DNA libraries from the previously listed microorganisms can be prepared using (pSP64) as a vector. The following example, using *A. actinomycetemcomitans*, is illustrative of this procedure. The plasmid (pSP64), obtained from Promega Biotec, was digested with *Bam*HI to give a linear double-stranded DNA molecule with GATC cohesive ends. Purified bacterial DNAs, prepared as described earlier, were digested with *Sau*3A, yielding fragments with cohesive ends complementary to the (pSP64) cohesive ends. The restricted bacterial DNA was mixed with the (pSP64) (10:1 M ratio), the mixture was ligated and used to transform competent *E. coli* HB101, and selection for ampicillin-resistant colonies was performed.

To verify that the bacterial DNA had inserted into (pSP64), 20 clones from each bacterial library were randomly selected and analyzed. The plasmids were purified and digested with *Eco*RI and *Pst*I, to detect the presence of inserts by restriction mapping.

8. *Isolation of Actinobacillus actinomycetemcomitans-Specific Probes*

Specific probes were obtained by preabsorbing the ^{32}P-nick-translated *A. actinomycetemcomitans* DNA to the cross-reactive DNAs, which were immobilized onto nitrocellulose. After preabsorption the probe was used to screen the specific library of choice by conventional methods (55).

B. Results

1. *Quantitation and Sensitivity of Genomic DNA Probes*

Factors affecting the rate and efficiency of hybridization with DNA probes were studied with *A. actinomycetemcomitans, B. intermedius*, and *B. gingivalis* isolated pure cell cultures. Pure cultures whose cell numbers were carefully determined by optical density, cell count by Petroff Hausser counter, and culture number confirmation, were diluted, sample processed, and immobilized under our standard conditions. Immobilized samples were hybridized using nick-translated ^{32}P genomic probes for each cell. Surprisingly, detectable levels of 10^4 cells are found after only 2 hr of autoradiographic exposure (data not shown). Figure 2A–C shows routine detection of as few as 10^2–10^3 cells after an 18-hr autoradiographic exposure time for *A. actinomycetemcomitans, B. intermedius*, and *B. gingivalis*. The controls for these experiments included ^{32}P-nick-translated nonhomologous DNA, which showed little or no binding to the immobilized cells but hybridized to the homologous positive control DNAs (data not shown).

2. *Cross-Reactivity Analysis with Whole-Genomic Actimobacillus actinomycetemcomitans, Bacteroides intermedius, and Bacteroides gingivalis DNA Probes*

To determine the specificity and sensitivity of the genomic DNA probes, cross-reactivity studies were performed with immobilized cell panels of other related and predominant oral microbial strains. These experiments consists of sample preparation of isolated laboratory strains of cells, immobilization of each strain

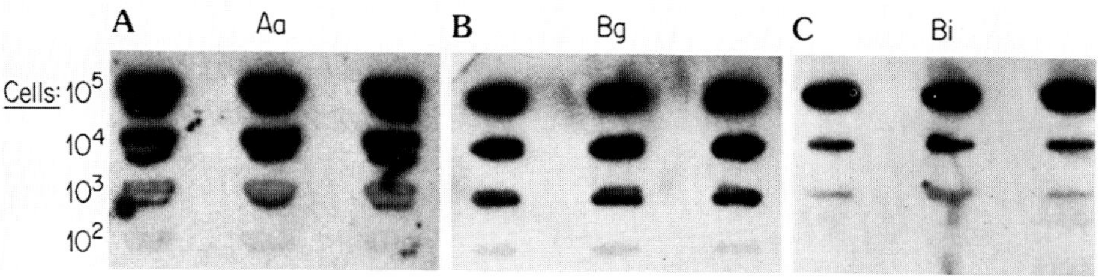

Figure 2 Sensitivity and quantitation of whole genomic DNA probes to immobilized cell samples. Dilutions of cell numbers from 10^5 to 10^2 were probed with ^{32}P-labeled whole-genomic DNA probes and exposed to autoradiographic film. Panels: (A) *A. actinomycetemcomitans*, (B) *B. gingivalis*, (C) *B. intermedius* (exposure time: 18 hr).

in triplicate panels, followed by hybridization with the appropriate genomic DNA probe. These results were correlated with the positive and negative controls and are summarized in Table 2.

The extent of cross-reactivity was assessed on the basis of assuming 100% homology with the positive cell control for each probe and comparing the signal detection level for the cross-reactant candidate strains in which a hybridization signal was found. The *A. actinomycetemcomitans* probe was found to be highly cross-reactive with the *Haemophilus aphrophilus* strains tested but showed no reaction with the subsequent strains (i.e., *Bacteroides*) examined. This result indicated that a refined DNA probe or *A. actinomycetemcomitans*-specific cloned subset would be required for the accurate diagnosis of the presence of *A. actinomycetemcomitans* cells.

The results of the *B. intermedius* panel showed that there is less than 1% cross-reactivity with *B. gingivalis* strains and much less cross-reaction with *B. corporis* and *B. endodontalis* at an 18 hr autoradiographic analysis. Results listed as minus (−) in Table 2 are for strains that showed no visible signal after 4 days of film exposure. The results summarized in Table 2 for both *B. intermedius* and *B. gingivalis* suggest that whole genomic probes are suitable for diagnosis of these two species because of their limited cross-reactivity to other *Bacteroides* strains.

3. *Development of Actinobacillus actinomycetemcomitans-Specific Probes*

After the analysis of the cross-reactivity tests with the whole genomic DNA probes, further development for species-specific *A. actinomycetemcomitans* DNA probe was carried out utilizing recombinant DNA technology. Construction of a genomic library consisting of *A. actinomycetemcomitans* genomic DNA sequences was done in the (pSP64) recombinant DNA vector system. From the results of the cross-reactivity study, it was known that those sequences that con-

Table 2 Summary of Cross-Reactivity Experiments for *A. actinomycetemcomitans*, *B. intermedius*, and *B. gingivalis* Using Whole-Genome Probes

Strain	*A. actinomycetemcomitans*	*B. intermedius*	*B. gingivalis*
A. actinomycetemcomitans			
Y4	++	–	–
29522	++	–	–
29524	++	–	–
2092	++	–	–
511	++	–	–
29523	++	–	–
2112	++	–	–
653	++	–	–
9710	++	–	–
H. aphrophilus			
626	+	–	–
13252	++	–	–
19415	+	–	–
5906	++	–	–

H. paraphrophilus			
H76	+	–	–
29241	++	–	–
B. intermedius I			
581	–	++	1%
25611	–	++	1%
B. intermedius II			
25261	–	++	1%
33563	–	++	1%
B. gingivalis			
381/397	–	1%	++
35277	–	1%	++
B. corporis	–	1%	ND
B. melaninogenicus	–	1%	ND
B. denticola	–	1%	ND
B. loeschii	–	1%	ND
B. asaccharolyticus	ND	–	–
B. endodontalis	–	1%	ND

Scoring is expressed as: –, no hybridization observed; +, light hybridization; 1%, less than 1% hybridization observed; ++, very strong hybridization; ND, not done.

tributed to the *H. aphrophilus* cross-reactivity would need to be removed from the *A. actinomycetemcomitans* DNA sequences that would be used to screen the library.

The following procedure was performed to generate an enriched *A. actinomycetemcomitans*-specific probe. First, *A. actinomycetemcomitans* genomic DNA was nick-translated under our standard conditions. This ^{32}P–nick-translated *A. actinomycetemcomitans* probe was then hybridized to a nitrocellulose filter containing whole genomic *H. aphrophilus* DNA. After the hybridization, the remaining unhybridized probe was removed and used as an enriched non–cross-reactive *A. actinomycetemcomitans* genomic probe.

The probe was then tested against a Southern transfer of *A. actinomycetemcomitans* and *H. aphrophilus* restriction enzyme-digested genomic DNAs. As shown in Figure 3, after an 18-hr exposure almost no detectable *H. aphrophilus*–*A. actinomycetemcomitans* cross-reacting sequences remain, and most of the sequences remaining were *A. actinomycetemcomitans*-specific. The probe was then used to screen the (pSP64) *A. actinomycetemcomitans* library for *A. actinomycetemcomitans*-specific sequences.

Approximately five recombinant DNA clones were obtained in the first round of screening and were subsequently tested as shown in Figure 4 for *A. actinomycetemcomitans* specificity. Clone 5 appeared to show low-level cross-reaction with *H. aphrophilus* when compared with the other selected isolates, which show no cross-reactivity after a 3-4 day autoradiographic exposure. The remaining four clones were pooled and used as specific *A. actinomycetemcomitans* DNA probe material.

4. *Sensitivity of the Actinobacillus actinomycetemcomitans-Specific DNA Probes*

The sensitivity and specificity of the *A. actinomycetemcomitans* DNA clones were analyzed, as previously described, by using the immobilized strain panels. The strain specificity of the probes was verified under standard conditions. Interestingly, the four cloned probes showed only a 10-fold decrease in sensitivity when compared with the whole genomic *A. actinomycetemcomitans* DNA probe (data not shown). The level of detection in 18 hr was between 10^3-10^4 cells compared with the 10^2 level from the whole-genomic probe hybridization. Thus, by obtaining at least 10 more *A. actinomycetemcomitans*-specific DNA clones the sensitivity should be achieved at the optimum 10^2 limit.

5. *Detection and Quantitation of Bacteria in Mixed Cultures*

To study the probe sensitivity in cell culture mixtures we obtained samples of cells that contained a background of non-cross-reacting *Actinomyces viscosus* at 2×10^7 cells/ml, which were then spiked with a mixture of *A. actinomycetemcomitans, H. aphrophilus, B. intermedius*, or *B. gingivalis* cells at 1 or 10% of the total cell number. The cell mixtures were constructed at Forsyth Dental Center

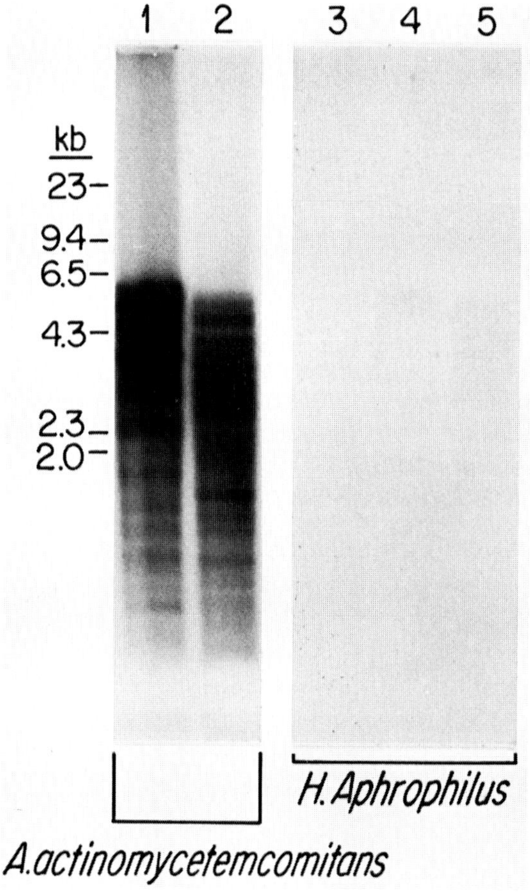

Figure 3 Autoradiogram of a Southern hybridization of *A. actinomycetemcomitans*-enriched specific DNA sequences. Genomic DNAs from *A. actinomycetemcomitans* and the cross-reactive *H. aphrophilus* were digested with restriction enzyme, electrophoresed on gel, and transferred according to Southern (11). The filter was then probed with ^{32}P-labeled *A. actinomycetemcomitans*-specific DNA sequences (prepared as described in materials and methods). Lane 1, *Eco*RI restriction enzyme-digested *A. actinomycetemcomitans*; lane 2, *Hind*III restriction enzyme-digested *A. actinomycetemcomitans*; lane 3, *Eco*RI restriction enzyme-digested *H. aphrophilus*; lane 4, *Hind*III restriction enzyme-digested *H. aphrophilus*; and lane 5, *Eco*RI-*Hind*III digested *H. aphrophilus*.

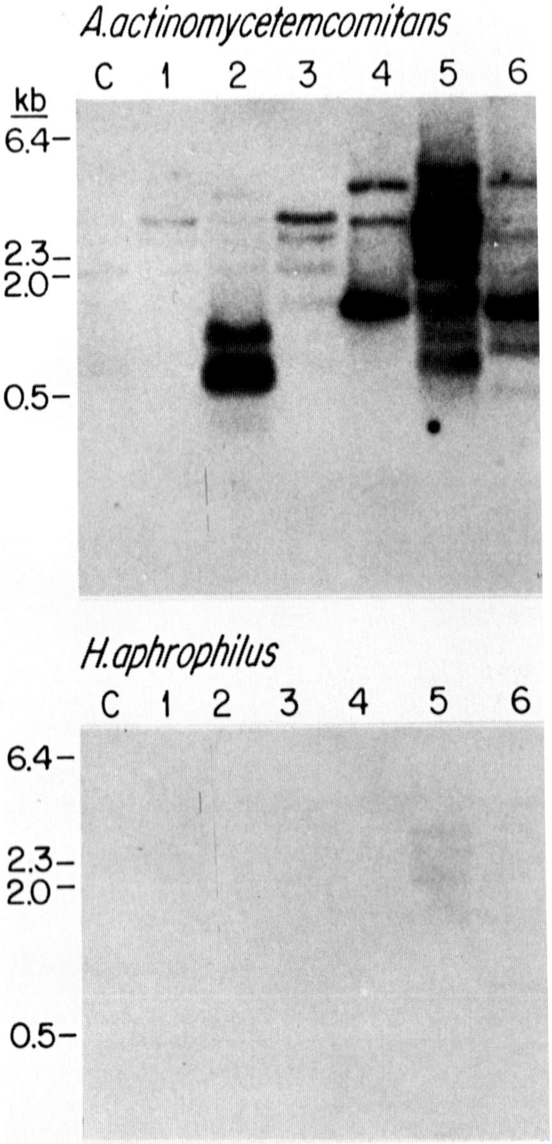

Figure 4 (a) Autoradiogram of Southern hybridization of whole genomic *A. actinomycetemcomitans* DNA to isolated *A. actinomycetemcomitans*-specific clones. (b) Autoradiogram of the Southern hybridization of *H. aphrophilus* DNA to *A.actinomycetemcomitans*-specific clones.

and provided blind to BioTechnica Diagnostics. After sample processing, approximately 5×10^5 total cells were applied to each slot in triplicate (for each unknown sample) to multiple nitrocellulose filters and analyzed as previously described. The results of the sensitivity and accuracy of the probe hybridization are shown in Table 3. Probes used were the *A. actinomycetemcomitans, H. aphrophilus, B. intermedius*, and *B. gingivalis* whole-genomic probes as well as the (pSP64) *A. actinomycetemcomitans*-specific probes.

The presence of a specific microorganism was determined with 100% accuracy at both the 1 and 10% levels in all 13 samples tested within the *A. viscosus* background by hybridization with the specific DNA probes. In addition, the expected false-positive reaction results in samples 2 and 7 when using the whole-genome *A. actinomycetemcomitans* probe, owing to the *H. aphrophilus* spike, were correctly shown as negative with the specific-cloned *A. actinomycetemcomitans* probes.

6. Sensitivity and Reproducibility of Nonradioactive Probes

One of the objectives of this program is to develop a nonradioactive detection system. The feasibility study of the nonisotopic system was determined with control *E. coli* C600 or *A. actinomycetemcomitans* cells. The two bacterial strains were counted by visual cell number assessment, optical density, and culture plating to obtain precise numbers for the amounts of cells immobilized. The panels containing positive and negative controls for either the *E. coli* C600 or *A. actinomycetemcomitans* strains were then hybridized with the nonradioactive probe (BRL) or ^{32}P-nick-translated probe. It was found that, under our standard hybridization conditions, the nonradioisotopic procedure can detect 10^3-10^4 cells after a 60-min development time following processing with the BluGENE detection system (BRL; Fig. 5). The ^{32}P radioactive system requires an 18-hr exposure after washing and processing, to detect 10^3-10^2 cells compared with the BluGENE system, which is resolved within 60 min. The nonradioactive *A. actinomycetemcomitans* probe was also used in the detection of *A. actinomycetemcomitans* cells in mixed cultures, and results were equal to the levels detected by the ^{32}P-labeled probe (data not shown).

C. Discussion

Our approach for the development of a rapid, sensitive, and specific assay for the detection of *A. actinomycetemcomitans, B. intermedius*, and *B. gingivalis* involves the use of whole-genomic DNA probes as well as species-specific DNA probes derived through recombinant DNA technology. By using molecular hybridization techniques and quantitative sample preparation procedures, conditions have been established for the rapid detection of 10^2-10^3 cells with the genomic probes. Species-specific probes for *A. actinomycetemcomitans* cells were also generated and have detectable ranges from 10^3 to 10^4 cells. Further

Table 3 Detection of *A. actinomycetemcomitans*, *B. intermedius*, and *B. gingivalis* in Constructed Mixed Culture Samples by DNA Probes

	Percentage of species in 13 constructed samples (1%) of total cells)[a]												
Species	1	2	3	4	5	6	7	8	9	10	11	12	13
A. actinomycetemcomitans	–	–	–	1	–	1	–	–	–	10	–	–	10
H. aphrophilus	–	10	–	–	–	10	1	–	–	–	1	–	1
B. intermedius	–	–	1	–	–	1	10	1	–	–	–	10	–
B. gingivalis	10	–	–	–	–	10	1	–	10	–	–	10	–

	Results of DNA probe analysis of 13 constructed samples[c]												
DNA probe	1	2	3	4	5	6	7	8	9	10	11	12	13
A. actinomycetemcomitans (wg)[b]	–	+	–	+	–	+	+	–	–	++	–	–	++
H. aphrophilus (wg)	–	++	–	–	–	++	+	–	–	+	+	–	+
A. actinomycetemcomitans (cloned)	–	–	–	+	–	+	–	–	–	+	–	–	+
B. intermedius (wg)	–	–	+	–	–	+	++	+	–	–	–	+	–
B. gingivalis (wg)	++	–	–	–	–	++	+	–	++	–	–	+	–

[a]–, no cells spiked into *A. viscosus* background of 2×10^7 cells/ml.
[b]wg, whole genome
[c]Scoring is expressed as: –, no hybridization; +, light hybridization; ++, strong hybridization.

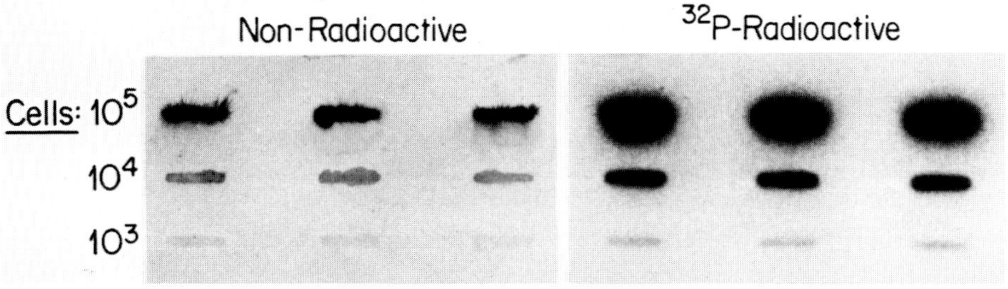

Figure 5 Comparison of BluGENE versus autoradiographic detection. Cell samples were immobilized under standard conditions. Whole genomic DNA was labeled by nick translation with [^{32}P]dCTP, (1.0×10^8 cpm/μg DNA) (right) or biotin-11-dUTP (left). Signal generation time of the nonradioactive detection system was 60 min after processing time. In the radioactive detection at least 18 hr is required to generate sensitivity of 10^3 cells.

isolation of more *A. actinomycetemcomitans*-specific clones is in progress, and the pooling of a larger number of specific clones should provide detection in the range of 10^2-10^3 cells.

The initial experiments with the nonradioactive BluGENE system (BRL) already exhibit detection of 10^3-10^4 cells after only a 60-min signal development time following hybridization and washing. Although the limit of detection of 10^2 cells is achieved with the radioactive probe system, an 18-hr signal generation time is required to generate this sensitivity.

Cloned probes are now used in many clinical research settings for the detection of viral and bacterial pathogens (52). We have found the use of DNA probes rapid, sensitive, and specific for the detection of three primary oral pathogenic microorganisms. The use of this technology should have a wide range of applications for the detection of pathogenic oral microorganisms, and the prospect of being able to quantitative monitor the effectiveness of a course of antibiotic treatment should prove to be an invaluable tool for the future.

REFERENCES

1. NIDR National Institutes of Health, Proceedings from the state of the art workshop on surgical therapy for periodontitis. (May 13-14) 53:475-501 (1981).
2. Institute of Medicine, *Report of a Study: Public Policy Options for Better Dental Health*. Division of Health Care Services, Publication No. 80-06:132, National Academy Press, Washington, (December 1980).
3. Polson, A. M., and J. M. Goodson, Periodontal diagnosis: Current status and future needs. *J. Periodontal 56*:25-34 (1985).

4. Socransky, S. S., A. D. Haffajee, J. M. Goodson, and J. Lindhe, New concepts of destructive periodontal disease. *J. Clin. Periodontol. 11*:21–32 (1984).
5. Ebersole, J. L., M. A. Taubman, and D. J. Smith, Gingival crevicular fluid antibody to oral microorganism. *J. Periodontol. Res. 20*:349–356 (1985).
6. Gibbons, R. J. (personal communication, 1985).
7. Hammond, B. F., The microbiology of periodontal diseases with emphasis on localized juvenile periodontitis. *Alpha Omegan 76*:27–31 (1983).
8. Newman, M. (personal communication, 1985).
9. *NIDR Fact Sheet*, Periodontal (Gum) Disease. NIH Publication #81-1142, reprinted December 1980.
10. Bonta, Y., J. J. Zambon, R. J. Genco, and M. E. Neiders, Rapid identification of periodontal pathogens in subgingival plaque: Comparison of indirect immunofluorescence microscopy with bacterial culture for detection of *Actinobacillus actinomycetemcomitans. J. Dent. Res. 64*:793–798 (1985).
11. Zambon, J. J., L. S. Christersson, and J. Slots, *Actinobacillus actinomycetemcomitans* in human periodontal disease: Prevalence in patient groups and distribution of biotypes and serotypes within families. *J. Periodontol. 54*:707–711 (1983).
12. Slots, J., and B. G. Rosling, Suppression of the periodontopathic microflora in juvenile periodontitis by systemic tetracycline. *J. Clin. Periodontol. 10*:465–468 (1983).
13. Zambon, J. J., *Actinobacillus actinomycetemcomitans* in human periodontal disease. *J. Clin. Periodontol. 12*:1–20 (1985).
14. Socransky, S. S. (personal communication, 1985).
15. Manor, A., M. Lebendigen, A. Shiffer, and H. Tovel, Bacterial invasion of periodontal tissues in advanced periodontitis in humans. *J. Periodontol. 55*:567–573 (1983).
16. National Institute of Dental Research, Periodontal diseases. In *Challanges for the Eighties*. NIH Publication No. 85-860:37–55 (1983).
17. Christersson, L. A., B. G. Rosling, J. J. Zambon, and R. J. Genco, Antimicrobial treatment of periodontal diseases: Predictability of clinical results. *IADR/AADR Abstr.* No. S18 (1985).
18. Christersson, L. A., U. M. E. Wikesjo, R. G. Dunford, and J. J. Zambon, Patterns of periodontal attachment changes and association with *B. gingivalis* in untreated adult periodontitis. *IADR/AADR Abstr.* No. 531 (1985).
19. Rosling, B. G., L. A. Christersson, J. Slots, and R. J. Genco, Predictability of topical and subgingival antimicrobial therapy in the management of advanced adult periodontitis. *IADR/AADR Abstr.* No. 1672 (1985).
20. Slots, J., Bacterial specificity in adult periodontitis—A summary of recent work.
21. Zambon, J. J., Methodology and clinical significance of rapid assays for periodontal pathogens. *IADR/AADR Abstr.* No. S38 (1986).
22. Haffajee, A. D., S. S. Socransky, J. L. Ebersole, and D. J. Smith, Clinical, microbiological and immunological features associated with treatment of active periodontosis lesions. *J. Clin. Periodontol. 11*:600–618 (1984).

23. Socransky, S. S., A. Tanner, A. D. Haffajee, J. D. Hillman, and J. M. Goodson, Present status of studies on the microbial etiology of periodontal diseases. In *Host-Parasite Interactions in Periodontal Diseases*, Genco and Mergenhagen (eds.). American Society for Microbiology, Washington, pp. 1-12 (1982).
24. Tanner, A., C. Haffer, G. T. Bratthall, R. A. Visconti, and S. S. Socransky, A study of the bacteria associated with advancing periodontitis in man. *J. Clin. Periodontol. 6*:278-307 (1979).
25. Savitt, E. D., and S. S. Socransky, Distribution of certain subgingival microbial species in selected periodontal conditions. *J. Periodontol. Res. 19*: 111-123 (1984).
26. Vincent, J. W., W. C. Cornett, W. A. Falkler, and R. G. Montoya, Biologic activity of type I and type II *Fusobacterium nucleatum* isolates from clinically characterized sites. *J. Periodontol. 56*:334-339 (1985).
27. Newman, M. G., S. S. Socransky, E. D. Savitt, D. A. Propas, and A. Crawford, Studies of the microbiology of juvenile periodontis. *J. Periodontol. 47*:373-379 (1976).
28. Slots, J., The predominant cultivable microflora of advanced periodontitis. *Scand. J. Dent. Res. 85*:114-121 (1977).
29. Tanner, A., S. S. Socransky, and J. M. Goodson, Microbiota of periodontal pockets losing crestal alveolar bone. *J. Periodontol. Res. 19*:279-291 (1984).
30. Dzink, J. L., A. Tanner, A. D. Haffajee, and S. S. Socransky, Gram-negative species associated with active destructive periodontal lesions. *J. Clin. Periodontol. 12*:648-659 (1985).
31. Savitt, E. D., K. A. Malament, S. S. Socransky, A. J. Mecer, and K. J. Backman, Effects on colonization of microbiota of cast ceramic crowns compared to natural teeth. *Int. J. Periodont. Restor. Dent. 7*:23-36 (1987).
32. Moore, W. E. C., R. R. Ranney, and L. V. Holdeman, Subgingival microflora in periodontal disease: Cultural studies. In *Host-Parasite Interactions in Periodontal Diseases*, Genco and Mergenhagen (eds.). American Society for Microbiology, Washington, pp. 13-26 (1982).
33. Moore, W. E. C., L. V. Holdeman, E. P. Cato, and R. M. Smibert, Comparative bacteriology of juvenile periodontitis. *Infect. Immun. 48*:507-519 (1985).
34. Sasaki, S., S. S. Socransky, M. Lescerd, and E. Sweeney, Destructive periodontal disease of children II. Microbiological and immunological findings. *AADR Abstr.* No. 422 (1977).
35. Dzink, J. L., A. Tanner, A. D. Haffajee, and S. S. Socransky, Gram-negative species associated with active destructive periodontal lesions. *J. Clin. Periodontol. 12*:648-659 (1985).
36. Mandell, R. L., A longitudinal microbiological investigation of *Actinobacillus acintomycetemcomitans* and *Eikenella corredens* in juvenile periodontitis. *Infect. Immun. 45*:778-780 (1984).
37. Dunham, S. L., J. M. Goodson, P. E. Hogan, and S. S. Socransky, Failure

of darkfield microbiologic parameters to predict periodontal disease activity at periodontal sites. *J. Dent. Res. 64*:IADR Abstr. No. 1657 (1985).
38. Tanner, A., S. Badger, C-H. Lai, M. A. Listgarten, R. A. Visconti, and S. S. Socransky, *Wolinella* gen. nov., *Wolinella succinogenes* (*Vibrio succinogenes* Wolin, et al.) comb. nov., and description of *Bacteroides gracilis* sp. nov., *Wolinella recta* sp. nov., *Campylobacter concisus* sp. nov., and *Eikenella corrodens* from humans with periodontal disease. *Int. J. Syst. Bacteriol. 31*:432–445 (1981).
39. Tanner, A., M. A. Listgarten, J. L. Ebersole, and M. N. Strzempko, *Bacteroides forsythus* sp. nov., a slow-growing fusiform *Bacteroides* species from the human oral cavity. *Int. J. Syst. Bacteriol. 36*:213–221 (1986).
40. Listgarten, M. A., R. Mao, and P. J. Robinson, Periodontal probing and relationship of the probe tip to periodontal tissues. *J. Periodontol. 47*:511–513 (1976).
41. Duckworth, J. E., J. M. Goodson, S. S. Socransky, and P. F. Judy, A method for the geometric and densitometric standardization of intraoral radiographs. *J. Periodontol. 54*:435–440 (1983).
42. Goodson, J. M., A. D. Haffajee, and S. S. Socransky, Relationship between attachment level loss and alveolar bone loss. *J. Clin. Periodontol. 11*:348–359 (1984).
43. Haffajee, A. D., S. S. Socransky, and J. M. Goodson, Clinical parameters as predictors of destructive periodontal disease activity. *J. Clin. Periodontol. 10*:257–265 (1983).
44. Goodson, J. M., A. C. R. Tanner, A. D. Haffajee, G. C. Sornberger, and S. S. Socransky, Patterns of progression and regression of advanced destructive periodontal disease. *J. Clin. Periodontol. 9*:472–481 (1982).
45. Mandell, R. L., L. S. Tripodi, E. D. Savitt, J. M. Goodson, and S. S. Socransky, The effect of treatment on *Actinobacillus actinomycetemcomitans* in localized juvenile periodontitis. *J. Periodontol. 57*:94–99 (1986).
46. Slots, J., C. Hafstrom, B. Rosling, and G. Dahlen, Detection of *Actinobacillus actinomycetemcomitans* and *Bacteroides gingivalis* in subgingival smears by the indirect fluorescent-antibody technique. *J. Periodontol. Res. 20*:613–620 (1985).
47. Zambon, J. J., H. S. Reynolds, P. Chen, and R. J. Genco, Rapid identification of periodontal pathogens in subgingival dental plaque: Comparison of indirect immunofluorescence microscopy with bacterial culture for detection of *Bacteroides gingivalis*. *J. Periodontol. 56* (Suppl.):32–30 (1985).
48. French, C. K., E. D. Savitt, S. L. Simon, S. M. Eklund, M. C. Chen, L. C. Klotz, and K. K. Vaccaro, DNA probe detection of periodontal pathogens. *J. Oral Microbiol. Immunol. 1*:58–62 (1986).
49. Slots, J., and B. G. Rosling, Suppression of the periodontopathic microflora of juvenile diabetes. Culture, immunofluorescence, and serum antibody studies. *J. Periodontol. 54*:420–430 (1983).
50. Beem, J. E., W. B. Clark, and Bleiweis, Antigenic variation of indigenous streptococci. *J. Dent. Res. 64*:1039–1045 (1985).
51. Brathall, D., and R. J. Gibbons, Antigenic variation of *Streptococcus mutans* colonizing grotobiotic rats. *Infect. Immun. 12*:1231–1236 (1975).

52. Dzink, J. L., C. Smith, and S. S. Socransky, Semi-automated technique for identification of subgingival isolates. *J. Clin. Microbiol. 19*:599–605 (1984).
53. Marmur, J., and P. Doty, Determination of the base pair composition of deoxyribonucleic acid from its thermal denaturation. *J. Mol. Biol. 5*:109–122 (1962).
54. Rigby, P. W., M. S. Dieckmann, C. L. Rhodes, and P. Berg, Labeling deoxyribonucleic acid to high specific activity in vitro by nick translation with DNA polymerase I. *J. Mol. Biol. 113*:237–251 (1977).
55. Maniatis, T., E. F. Fritsch, and J. Sambrook, The identification of recombinant clones. In *Molecular Cloning (A Laboratory Manual)*. Cold Spring Harbor Laboratory, Cold Spring Harbor, N.Y., pp. 309–363 (1982).
56. Engleberg, N. C., and B. I. Eisenstein, The impact of new cloning techniques on the diagnosis and treatment of infectious diseases. *N. Engl. J. Med. 311*:892–901 (1985).

11

Detection and Speciation of Mycoplasmas by Use of DNA Probes

STEVEN J. GEARY and MICHAEL G. GABRIDGE
Bionique Laboratories, Inc., Saranac Lake, New York

I. INTRODUCTION

Mycoplasmas are wall-less pleomorphic prokaryotes. They are the smallest (approximately 300 nm) and simplest self-replicating microorganisms, possessing a double-stranded deoxyribonucleic acid (DNA) genome averaging 5×10^8 Da (1-3). This is sufficient to code for approximately 650 proteins (4). Mycoplasmas have a low guanine + cytosine (G+C) ratio, generally ranging between 25 and 40 mol% (5). They typically grow slowly (up to 96 hr) and require a rich medium with sterols, fatty acids, and nucleic acid precursors. The medium must be supplemented with serum, and at least one purine and one pyrimidine base (6). As seen in Table 1, many mycoplasmas are pathogenic for both humans and animals. They also are common contaminants of cell cultures, in which they cause a wide variety of detrimental effects.

Current methods of detecting mycoplasmas include direct culture, DNA fluorochrome staining, and immunofluorescent tests. Direct culture is the most sensitive method. However, some fastidious strains of *Mycoplasma hyorhinis* will not grow in the standard mycoplasma medium, and other mycoplasmas may take as long as 3 weeks to yield visible growth from a direct culture sample. DNA fluorochrome staining is a sensitive method, but this is still less sensitive than direct culture (7). This test is widely used for detecting cell culture contaminants. It involves a nonspecific DNA fluorochrome stain to indicate the presence of nuclear material and is practical because the minute mycoplasma nuclear regions are easily seen between the large eukaryotic nuclei when preparations are viewed in an ultraviolet microscope. Immunofluorescent tests rely on the availability and specificity of antimycoplasmal antibodies. Such tests are

Table 1 Major Pathogenic Mycoplasmas

Mycoplasma species	Host	Disease or effect
M. pneumoniae	Human	Atypical pneumonia
M. hyopneumoniae	Porcine	Pneumonia
M. hyorhinis	Porcine	Pneumonia
M. gallisepticum	Avian	Respiratory disease
M. mycoides subsp. *mycoides*	Bovine	Pneumonia
M. mycoides subsp. *capri*	Ovine	Pneumonia
M. pulmonis	Murine	Respiratory disease
M. bovis	Bovine	Mastitis
M. arthritidis	Murine	Polyarthritis
M. capricolum	Ovine	Polyarthritis
M. hyorhinis	Cell culture	Cytotoxicity[a]
M. orale	Cell culture	Change in rate of growth[a]
M. arginini	Cell culture	Chromosomal aberrations[a]
A. laidlawii	Cell culture	Changes in virus titers[a]
		Interferon induction[a]
		Influence cell fusion[a]
		Competition for nutrients[a]

[a]Varied, but reports include all.

most efficient at speciation, although detection is not practical in clinical samples because of the specificity of the antibodies and the low concentrations of antigen.

Because of the wide array of detrimental effects caused by mycoplasmas, the strict cultural requirements, the extended time required for growth, and the limitations of the current methods of detection, it is appropriate to begin developing rapid, sensitive detection techniques. DNA probes appear to be ideally suited to this task.

DNA probes are labeled sequences of DNA (either a specific gene or the total genome) that recognize and hybridize to specific complementary sequences of DNA. The means of detection of the hybridization reaction depends on the label incorporated during the nick-translation reaction that is used to construct the probe. For example, probes labeled with radioactivity are detected by autoradiography, whereas those labeled with biotinylated nucleotides can be detected by a combination of streptavidin, polyalkaline phosphatase, nitroblue tetrazolium, and 5-bromo-4-chloro-3-indolyl phosphate. Probes provide a high degree of specificity and a sensitivity that is equal to, or greater than, that seen when using the current methods of indirect detection of mycoplasmas. Because of the specificities of the probes, they can be developed into powerful tools for detecting mycoplasmas present in relatively low concentrations in clinical samples and cell cultures without interference by eukaryotic DNA.

II. NUCLEIC ACID PROBES FOR MYCOPLASMAS

For the detection of mycoplasmas, cloned gene probes have been constructed to hybridize with ribosomal ribonucleic acid (rRNA) genes or with genes coding for a specific protein. The mycoplasmal genes coding for rRNAs have been well characterized. Many mycoplasmas have one or two sets of rRNA genes, and these rRNA genes and their products are highly conserved (8). Hence, such a probe would be an efficient indicator of mycoplasmas because the target is present in multiple sets per cell, and any mycoplasma present will trigger a positive response. Amikam et al. (9) cloned the major part of one of the rRNA sets of *M. capricolum* containing the 5S, 23S, and most of the 16S rRNA genes into plasmid (pBR325). This plasmid was designated (pMC5). Southern blot hybridization of this *M. capricolum* probe with *Eco*RI digests of DNAs from various *Mollicutes* (mycoplasmas, plus the closely-related acholeplasmas, ureaplasmas, and spiroplasmas) resulted in patterns specific for the *Mollicutes* tested (9-12). It also hybridized with rRNA genes of *Escherichia coli* but not with eukaryotic DNA. The reaction with *E. coli* is not surprising because there is substantial homology in the rRNA genes of several mycoplasmas and *E. coli* (9). This probe has been used to detect mycoplasmas in cell cultures that were either naturally infected or intentionally inoculated with mycoplasmas (12,13). The DNAs were extracted from the samples and digested with the restriction enzyme *Eco*RI. The samples were electrophoresed in agarose gels, transferred by Southern blot, and then hybridized with (pMC5). The results indicated that this method was capable of detecting 1 ng of mycoplasmal DNA, which is equivalent to about 10^6 colony forming units (cfu) of mycoplasmas.

Taylor et al. (14) used a dot hybridization technique to assess the specificity of (pMC5) hybridization. Five hundred nanograms of DNA from various mycoplasmas was deposited onto nitrocellulose filter paper using a dot-blot apparatus. The (pMC5) probe hybridized to all of the *Mycoplasma* species tested. This demonstrated the utility of this probe as a tool for mycoplasmal detection without the need for restriction endonuclease digestion or electrophoretic separation of the DNA fragments. Göbel and Stanbridge (15) also constructed a probe to detect mycoplasma rRNA genes. This was accomplished by cloning a *Hind*III *M. hyorhinis* DNA digest into bacteriophage M13. This resulted in two clones, M13Mh129 and M13Mh171, which contained 900 and 1200-bp fragments of the 23S rRNA gene, respectively. These fragments were purified and used as probes to detect rRNA genes in mycoplasmas. The M13Mh129 probe hybridized with all *Mycoplasma* species tested but did not react with *E. coli* DNA or with HeLa cell DNA. These investigators were able to detect between 7×10^4 and 1×10^5 cfu in suspensions of mycoplasma-infected cells that had been blotted onto nitrocellulose paper. Researchers at Gen-Probe (San Diego, California) constructed a cDNA probe by using *Acholeplasma laidlawii* and *M. hominis* rRNA and reverse transcriptase. This tritiated probe hybridizes with homologous ribosomal RNA from all *Mycoplasma* and *Acholeplasma* species tested. It does not react with eukaryotic DNA, but it does detect bacterial DNA. However, the reaction with bacterial DNA is much less efficient than that with mycoplasma DNA. This probe is designed specifically to detect mycoplasmal contamination of tissue culture samples. The sensitivity of detection has been reported to range from 9.3×10^3 for *M. arginini* to 2.6×10^6 for *M. orale* (16). Our laboratory has noted similar variations in sensitivity of detection among the different *Mycoplasma* species. We have found the range of detection to vary greatly with the *Mycoplasma* species involved. The commercial probe system was capable of detecting 10^5 *M. hyorhinis*, 10^5 *A. laidlawii*, and 10^6 *M. salivarium* (per ml). Preliminary data indicate that as many as 10^9 *M. gallisepticum* are required to give a positive result. These DNA probes to the highly conserved mycoplasmal rRNA genes or to rRNA have been demonstrated to be useful tools for detection of mycoplasmal contamination of tissue culture samples. However, they lack the specificity necessary to speciate these contaminants.

Another approach to the development of DNA probes is to construct them from DNA fragments that code for specific mycoplasmal proteins. These probes can be used to speciate mycoplasmal contaminants. Taylor et al. (17) constructed two *M. hyorhinis*-specific DNA probes from recombinant clones generated by the incorporation of *Eco*RI digestion fragments of *M. hyorhinis* DNA into bacteriophage λ Charon 4A. The probes were derived from the recombinant clone designated λ Ch4A-MhrGl. The DNA from this clone was extracted and digested with the restriction endonuclease *Eco*RI. This resulted in two *M. hyorhinis* fragments, 3.6 and 9.4 kb, that were subsequently labeled with $[\alpha\text{-}^{32}P]$ATP by nick translation and utilized as specific probes. Southern hybridization analysis revealed that each probe hybridized specifically with only its corresponding-sized fragment in *Eco*RI digests of *M. hyorhinis* strains GDL and BTS-7. Neither probe hybridized with other *Mycoplasma* species or with *E. coli*. These researchers utilized a rapid dot hybridization technique (18) to assess the specificity and sensitivity of the probes. The 9.4-kb probe was capable of detecting approximately 5×10^4 *M. hyorhinis* deposited directly onto nitrocellulose paper. Both probes recognized only *M. hyorhinis* strains GDL and BTS-7 and no other *Mycoplasma* species tested. *Mycoplasma hyorhinis* contamination of BW 5147 lymphoblastoid cells was detected by both probes, even when fewer than 2% of the cells were infected. Neither probe hybridized with the eukaryotic

DNA. This demonstrates the utility of probes such as these for detection of specific mycoplasmas from biological samples.

Chan and Ross (19) constructed a DNA probe to another swine mycoplasma pathogen, *M. hyopneumoniae*. They cloned *M. hyopneumoniae* DNA, which had been cleaved with the restriction endonuclease *Hind*III, into plasmid (pBR322). The DNA from a recombinant clone that contained a 9-kb fragment of *M. hyopneumoniae* DNA was extracted and radioactively labeled by a nick-translation reaction. The probe generated by this procedure was then used in Southern hybridization analysis of broth-grown *M. hyopneumoniae* and *M. flocculare*, as well as clinical samples from pigs. This probe hybridized to both *M. hyopneumoniae* and *M. flocculare* DNAs but to distinctly different DNA fragment sizes. These researchers also were able to detect the presence of mycoplasmas from clinical material. Lung washings were collected from *M. hyopneumoniae*-infected pigs, and total DNA was extracted. Samples of the DNA were cleaved with one of the following restriction endonucleases: *Sau*3A, *Mbo*I, or *Dpn*I, and were analyzed by Southern blot hybridization with the *M. hyopneumoniae* DNA probe. The positive results reported indicate that DNA probes will also be useful to detect mycoplasmal infection in clinical samples from a wide variety of sources. The use of DNA probes in this manner should substantially reduce the time now necessary for the detection of mycoplasmas in clinical samples.

Total genomic DNA has been labeled and used as a probe for the determination of nucleotide sequence homology among *Mollicutes* and as tools for detecting mycoplasmas from tissue culture. Aulakh et al. (20) generated tritiated total-genomic DNA probes to eight species of *Acholeplasma* and utilized these to study the degree of relatedness among the species. This was accomplished by hybridization reactions of each probe with DNA samples from each of the eight *Acholeplasma* species. Each DNA probe hybridized extensively (78–90%) with its homologous DNA, but to a greatly reduced degree ($\leqslant$18%) with DNA from other species. This indicates that the species of acholeplasmas are quite distinct genetically, thereby corroborating the previously noted differences in antigenicity.

Our laboratory has developed total-genome DNA probes to *A. laidlawii, M. hyorhinis*, and *M. gallisepticum*. These were constructed by a standard nick-translation reaction incorporating a biotin-labeled nucleotide, biotin-11-dUTP. These probes were used in a dot-blot hybridization system for which the sample was deposited directly onto nitrocellulose membranes and then hybridized. To assess the sensitivity of the probes, samples of homologous target DNA and broth-grown organisms were immobilized onto nitrocellulose. The hybridization reactions were detected by reaction with streptavidin, biotinylated polyalkaline phosphatase, nitroblue tetrazolium, and 5-bromo-4-chloro-3-indolyl phosphate. As can be seen in Table 2, we were able to detect 15 ng of *A. laidlawii* DNA and 75 ng of *M. gallisepticum* DNA as well as 10^4 broth-grown *M. gallisepticum*. This

Table 2 Limits of Detection Utilizing Biotinylated Total Genome DNA Probes

Probe	Quantity of homologous sample detected: Target DNA	Broth-grown organisms	Organisms from tissue culture samples
A. laidlawii	15 ng	ND	10^4
M. gallisepticum	75 ng	10^4	ND
M. hyorhinis	ND	ND	10^6

ND = not done.

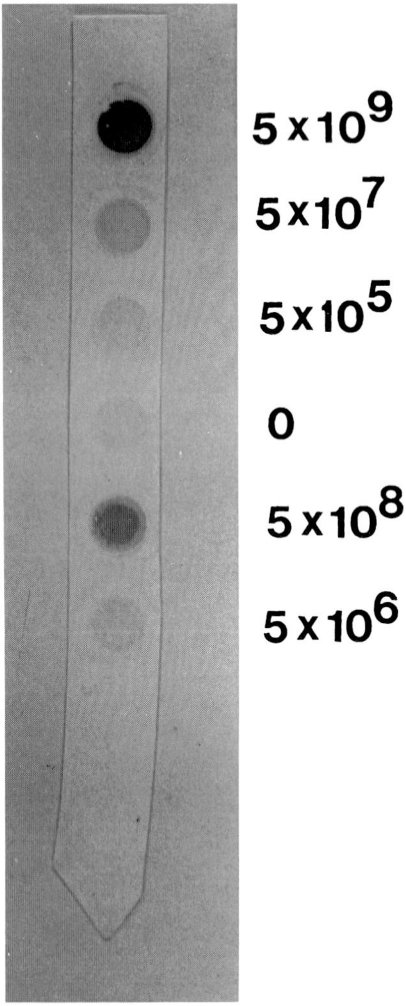

level of sensitivity is equal to that obtained with radiolabeled probes. To determine the applicability of the probes for detecting mycoplasmal contamination in tissue culture, VERO cells were infected with dilutions of either *A. laidlawii* or *M. hyorhinis*. Supernatant fluids were analyzed in the dot hybridization system described earlier. The results (see Table 2) reveal that this system is capable of detecting 10^4 *A. laidlawii* and 10^6 *M. hyorhinis*. No background reaction occurred with supernatant from noninfected VERO cells (Fig. 1). Biotinylated probes, such as these, should become valuable tools for the routine screening of tissue culture for mycoplasmal contamination, especially in laboratories that do not have a radiation license or the necessary equipment and facilities.

III. APPROACHES TO PROBE DETECTION

The work on probes described thus far has included both isotopic and nonisotopic markers or signals to indicate a positive response. Radioisotopes, such as ^{32}P and ^{3}H, are used quite commonly. Although they have a high degree of sensitivity, ^{32}P users must contend with a short half-life for the isotope (14 days) and strict measures on storage, use, and disposal. Tritium (^{3}H) has a fairly low energy level, but users still must deal with special protocols for safe handling. Also, a scintillation counter and potentially hazardous solubilizers and scintillants must be used.

Other alternative means of labeling DNA probes have been developed that, in certain instances, approximate the sensitivity of radiolabeled probes. Both chemiluminescent and fluorescent labels have been developed to label DNA probes (21). Chemiluminescent catalysts, such as peroxidase, luciferases, and microperoxidases, and chemiluminescent reagents, such as luminol, have been used as markers. Catalysts produce light by oxidizing luminol in the presence of hydrogen peroxide. One type of microperoxidase, MP-11, is a particularly sensitive label for DNA probes. In the presence of excess luminol and hydrogen peroxide, as little as 1 fg (10^{-15} g) of MP-11 can be detected in 2 min of counting by a luminometer. When attached to DNA, MP-11 retains considerable chemiluminescence (21). Probes can also be labeled with luminol, which is capable of detection in the tens of femtogram range (22). These means of labeling DNA probes may become very useful for clinical diagnostic purposes.

Another label that has been mentioned previously is a biotinylated derivative, deoxyuridine triphosphate (bio-11-dUTP). Bio-11-deoxyuridine triphosphate is an analogue of thymidine triphosphate that is synthesized (23,24) by the addi-

Figure 1 Detection of *M. hyorhinis* contamination in VERO cell culture by use of a *M. hyorhinis* total-genome biotinylated DNA probe. Samples are in the order in which they were received in a blind study.

tion of a biotin molecule linked to the C-5 position of the pyrimidine ring though a linker-arm 11 atoms long. The biotinylated nucleotide is incorporated into DNA, substituting for thymidine triphosphate, by a standard nick-translation reaction. This biotinylated nucleotide is incorporated at a slightly slower rate but to the same extent as thymidine triphosphate (23,25). The DNA labeled in this manner forms the normal double helix structure with complementary unlabeled DNA. When incubated with avidin or streptavidin, a high-affinity reaction will bind it to the probe. Subsequent addition of the appropriate biotinylated enzyme, substrate, and indicator then will lead to the formation of a colored product.

Figure 2 outlines the basic procedures used in our laboratory to construct an *A. laidlawii* total-genome biotinylated DNA probe. Two hundred and fifty milliliters of broth-grown *A. laidlawii* are harvested by centrifugation and the pellet is resuspended in 10 ml of TE buffer. The DNA is extracted by the procedure of Marmur (26). The cells are lysed in 2.5% sodium dodecyl sulfate (SDS), and then 5 M perchlorate is added to separate proteins from the DNA. The solution is deproteinized with chloroform–isoamyl alcohol (24:1). The DNA is then wound out onto a glass rod. After resuspension in TE, RNA is eliminated by the addition of RNase. The solution is deproteinized and the DNA wound again onto a glass rod. The DNA is then resuspended in a small volume of TE and the purity and concentration determined spectrophotometrically by the 260:280-nm ratio and the 260-nm reading, respectively (27). The probe is constructed by the incorporation of bio-11-dUTP into 1 μg of *A. laidlawii* DNA by a standard nick-translation reaction. The biotin-labeled DNA is separated from unincorporated nucleotides by column chromatography on Sephadex G-50 fine (Pharmacia). Samples are deposited onto 12 cm^2 of nitrocellulose paper by a dot-blot procedure (6 μl of target DNA or 500 μl of cell culture supernatant) using a multiwell manifold. The nitrocellulose paper is removed from the dot-blot apparatus, the organisms are lysed, and the DNA is denatured into single strands. The nitrocellulose is heated at 80°C, placed in 10 ml of prehybridization buffer in a heat-sealable bag and incubated at 42°C for 2.5 hr. The prehybridization buffer is drained and 10 ml of hybridization buffer containing 400 ng of biotinylated probe is added and the bag resealed and incubated at 42°C for 12–16 hr. The nitrocellulose is removed and washed with SSC [0.15 M NaCl, 0.015 M sodium citrate (pH 7.0)]. To detect the sites of probe hybridizations, the multivalent biotin/avidin interaction is used. The nitrocellulose is incubated with 2 μg/ml streptavidin followed by 1 μg/ml biotinylated polyalkaline phosphatase. The reaction is visualized by the addition of the substrate complex, nitroblue tetrazolium, plus 5-bromo-4-chloro-3-indolyl phosphate. This results in an easily visible blue spot at the site of probe hybridization.

Acholeplasma laidlawii

250 ml harvested by centrifugation

↓

DNA extracted (organisms lysed, deproteinized, DNA wound out, RNase treated) and resuspended in TE buffer (0.01 M Tris base, 0.001 M EDTA, pH 8.0)

↓

Biotin-11-dUTP incorporated into 1 μg A. laidlawii DNA in a standard nick translation reaction

↓

Biotin-labeled DNA separated from unincorporated nucleotides by column chromatography on Sephadex G-50 fine

↓

Samples immobilized onto nitrocellulose paper by dot blot procedure

↓

Hybridization reaction using approximatley 400 ng of probe per 10 ml of hybridization buffer to cover 12 sq cm of nitrocellulose

↓

Detection of hybridized probe utilizing:

streptavidin
↓
biotinylated polyalkaline phosphatase
↓
nitroblue tetrazolium and 5-bromo-4-chloro-3-indolyl phosphate
↓
5 mM EDTA to terminate the reaction

↓

Blue spots at the location of hybridized probe

Figure 2 Procedure for the construction and detection of an *A. laidlawii* total-genome biotinylated DNA probe.

IV. OVERVIEW

Mycoplasmas are microbes that cause diseases in wild and domesticated animals, birds, and plants. Hence, they are of economic importance. They also cause human diseases such as pneumonia and nongonococcal urethritis. Their pathogenicity, coupled with the fact that they often contaminate cell cultures and hybridomas used to produce various biologicals in vitro, make detection and speciation critically important.

Methods currently used to detect and identify mycoplasmas are based on direct culture, metabolism, and antigenicity. The latter approach is especially important, and includes growth inhibition or fluorescence with specific antisera, and various enzyme-linked immunosorbent assay (ELISA) systems. The use of DNA probes would be a beneficial supplement to this list of techniques. The DNA probe tests have high sensitivity and the probes can be constructed with various degrees of specificity. A probe to simply detect the presence of any mycoplasma would employ common genes such as those involved in ribosome synthesis. Conversely, probes used to identify a given species would be directed toward unique genes, such as those coding for species-specific proteins.

In the field of mycoplasmology, probes such as these are rapidly becoming useful adjuncts to control contamination and diagnose disease in a reliable, timely, and cost-effective manner. Their importance will likely increase over the next several years.

REFERENCES

1. Ryan, J. H., and H. J. Morowitz, Partial purification of native rRNA and tRNA cistrons from *Mycoplasma* sp (Kid). *Proc. Natl. Acad. Sci. USA 63*: 1282–1289 (1969).
2. Bak, A. L., F. T. Black, C. Christiansen, and E. A. Freundt, Genome size of mycoplasmal DNA. *Nature 224*:1209–1210 (1969).
3. Wallace, D. C., and H. J. Morowitz, Genome size and evolution. *Chromosoma 40*:121–126 (1973).
4. Morowitz, H. J., The genome of mycoplasmas. In *The* Mycoplasmatales *and the L-Phase of Bacteria*, L. Hayflick (ed.). Appleton, New York, pp. 405–412 (1969).
5. Stanbridge, E. J., and M. E. Reff, The molecular biology of mycoplasma. In *The Mycoplasmas*, M. F. Barile and Razin, S. (eds.). Vol. 1, Academic Press, New York, pp. 157–185 (1979).
6. Razin, S., The mycoplasmas. *Microbiol. Rev. 42*:414–470 (1978).
7. Stanbridge, E. J., Mycoplasma detection—an obligation to scientific accuracy. *Isr. J. Med. Sci. 17*:563–568 (1981).
8. Razin, S., Molecular biology and genetics of mycoplasmas (*Mollicutes*). *Microbiol. Rev. 49*:419–455 (1985).
9. Amikam, D., S. Razin, and G. Glaser, Ribosomal RNA genes in mycoplasma. *Nucleic Acids Res. 10*:4215–4222 (1982).

10. Amikam, D., G. Glaser, and S. Razin, Mycoplasmas (*Mollicutes*) have a low number of rRNA genes. *J. Bacteriol. 158*:376–378 (1984).
11. Razin, S., M. F. Barile, R. Harasawa, D. Amikam, and G. Glaser, Characterization of the mycoplasma genome. *Yale J. Biol. Med. 56*:357–366 (1983).
12. Razin, S., G. Glaser, and D. Amikam, Molecular and biological features of *Mollicutes* (mycoplasmas). *Ann. Microbiol. (Paris) 135A*:9–15 (1984).
13. Razin, S., M. Gross, M. Wormser, Y. Pollack, and G. Glaser, Detection of mycoplasmas infecting cell cultures by DNA hybridization. *In vitro 20*: 404–408 (1984).
14. Taylor, M. A., K. S. Wise, and M. A. McIntosh, Selective detection of *Mycoplasma hyorhinis* using cloned genomic DNA fragments. *Infect. Immun. 47*:827–830 (1985).
15. Göbel, U. B., and E. J. Stanbridge, Cloned mycoplasma ribosomal RNA genes for the detection of mycoplasma contamination in tissue cultures. *Science 226*:1211–1213 (1984).
16. McGarrity, G. J., and H. Kotani, Detection of cell culture mycoplasmas by a genetic probe. *Exp. Cell Res. 163*:273–278 (1986).
17. Taylor, M. A., M. A. McIntosh, J. Robbins, and K. S. Wise, Cloned genomic DNA sequences from *Mycoplasma hyorhinis* encoding antigens expressed in *Escherichia coli. Proc. Natl. Acad. Sci. USA 80*:4154–4158 (1983).
18. Brandsma, J., and G. Miller, Nucleic acid spot hybridization: Rapid quantitative screening of lymphoid cell lines for Epstein-Barr viral DNA. *Proc. Natl. Acad. Sci. USA 77*:6851–6855 (1980).
19. Chan, H. W., and R. F. Ross, Restriction endonuclease analysis of two porcine mycoplasma deoxyribonucleic acids: Sequence-specific methylation in the *Mycoplasma hyopneumoniae* genome. *Int. J. Syst. Bacteriol. 34*:16–20 (1984).
20. Aulakh, G. S., E. B. Stephens, D. L. Rose, J. G. Tully, and M. F. Barile, Nucleic acid relationships among *Acholeplasma* species. *J. Bacteriol. 153*: 1338–1341 (1983).
21. Heller, M. J., and L. E. Morrison, Chemiluminescent and fluorescent probes for DNA hybridization systems. In *Rapid Detection and Identification of Infectious Agents*, D. T. Kingsbury and Falkow, S. (eds.). Academic Press, Orlando, FL, pp. 245–256 (1985).
22. Scroeder, H. R., R. C. Boguslaski, R. J. Carrico, and T. T. Buckler, Monitoring specific protein-binding reactions with chemiluminescence. In *Methods in Enzymology—Bioluminescence and Chemiluminescence*, M. A. DeLuca (ed.). Vol. 57, Academic Press, New York, pp. 424–445 (1978).
23. Langer, P. R., A. A. Waldrop, and D. C. Ward, Enzymatic synthesis of biotin-labelled polynucleotides: Novel nucleic acid affinity probes. *Proc. Natl. Acad. Sci. USA 78*:6633–6637 (1981).
24. Brigati, D. J., D. Myerson, J. J. Leary, B. Spalholz, S. Z. Travis, C. K. Y. Fong, G. D. Hsiung, and D. C. Ward, Detection of viral genomes in cultured cells and paraffin-embedded tissue sections using biotin-labelled hybridization probes. *Virology 126*:32–50 (1983).

25. Gardner, L., Non-radioactive DNA labeling of specific DNA and RNA sequences on nitrocellulose in in situ hybridization. *Biotechniques 1*:38 (1983).
26. Marmur, J. A., A procedure for the isolation of deoxyribonucleic acid from microorganisms. *J. Mol. Biol. 3*:208–218 (1961).
27. Maniatis, T., E. F. Fritsch, and J. Sambrook, In *Molecular Cloning, A Laboratory Manual*, Cold Spring Harbor Laboratory, Cold Spring Harbor, N.Y., p. 468 (1982).

12
DNA Probes for Medically Important Yeasts

W. STUART RIGGSBY
University of Tennessee, Knoxville, Tennessee

I. INTRODUCTION

As a number of other chapters in this volume make clear, modern DNA technology has had a strong and increasingly broad impact on solving problems of identification and characterization of bacteria and on assessing interrelationships among various groups of bacteria. Perhaps the most fundamental result has been the now widely accepted proposal for the division of the bacteria into two kingdoms, Eubacteria and Archaebacteria (1). At a less intellectually lofty, but more practical, level, specific probes and characterizations of restriction fragment patterns are being developed for clinical applications in the diagnosis of bacterial and viral infections.

Progress in the applications of these techniques to the fungi—and particularly to the medically important fungi—has been much less rapid. Several recent reviews of the use of diagnostic DNA probes make no mention of fungi (2-4). A recent review volume (5) has adumbrated a science of molecular mycology, but it is significant that the chapter in that volume dealing with the medically important species (6) is entitled "*Toward* Gene Manipulations with Selected Human Fungal Pathogens" (emphasis added), indicating the state of development in this area. There are several factors that have contributed to this relative lag in establishing a discipline of molecular mycology and its biotechnological applications. New technology is typically developed in simple systems, and only subsequently applied to more complex systems. Fungi are more complex than bacteria and viruses in several respects that are important in the application of DNA technology: the compartmentalization of the eukaryotic genome into nuclear and extranuclear components; the larger size of the eukaryotic nuclear genome, even in such "simple" eukaryotes as the yeasts; and the more complex interactions between nucleic acids and proteins in the eukaryotic nucleus. In addition, many fungi appear to be rich in nucleases, compared with bacteria; such nucleases can complicate the purification of DNA with sufficient integrity for restriction analysis, cloning, and probe construction.

Other reasons for the lag are related to the history of research in infectious diseases, particularly in developed countries, and the relative level of interest and funding for mycological research as opposed to research on pathogenic bacteria and viruses. This ratio is clearly revealed by a scanning of the table of contents of any issue of, for example, *The Journal of Clinical Microbiology*. Finally, many of the medically important yeasts are imperfect, so that the rich repertoire of genetic techniques that can be applied to a sexually reproducing organism are not available as adjuncts to the molecular investigation.

There are reasons to believe, however, that mycology is ripe for the application of the new technology. The level of sophistication already developed in bacteria and viruses should, even now, be more than adequate for the purposes of clinical and systematic mycology, so there is no need for extensive new technical development. In fact, the molecular biology of the yeast *Saccharomyces cerevisiae* (7) is probably as sophisticated as that of *Escherichia coli*. As an indication of the success of this technology at an even higher level of complexity, the subject of the 1986 Cold Spring Harbor Symposium was *The Molecular Biology of* Homo sapiens! So the problems of complexity are no longer serious deterents for the application of DNA technology to medical mycology.

At the same time, medical progress has taken a path that is now calling attention to fungal infections. As antibiotics and immunological control of bacterial and viral infections have become more effective and more readily available, fungal pathogens have increased in importance (8). Many fungal pathogens are opportunists, and pose serious problems, often life-threatening, for patients who are undergoing immunosuppressive therapy or who are otherwise immunologically or physiologically compromised. For example, candidal infections are common in patients with acquired immunodeficiency syndrome (AIDS). As a result of this heightened awareness, we may expect to see increased interest and activity in the fundamental molecular biology of these organisms in the near future; what we have now is just a beginning.

II. THE GENUS *CANDIDA*

A. The Human Pathogens

The most common fungal pathogens of humans are yeasts of the artificial genus *Candida*. Authorities differ somewhat in which species of *Candida* ought to be considered as being medical important, but these disagreements are minor

(9-11). The yeast most often responsible for candidosis [or candidiasis; see Odds (10) for a discussion of the disagreement concerning these two terms for infection by *Candida* or *Candida*-like species] is *C. albicans*, followed somewhat distantly by *C. tropicalis*. Although there is no consensus regarding the relative importance of other *Candida* species relative to human disease, those unquestionably involved include *C. stellatoidea, C. pseudotropicalis, C. guilliermondii, C. parapsilosis, C. kruseii*, and *C. glabrata*. One of the difficulties in drawing up a definitive list of medically important candida is the slow but steady appearance of reports in the clinical literature of infections attributable to *Candida* species not generally included in such lists. Be that as it may, this section will be restricted to the preceding, well-established human pathogens. For the most part, the clinical identification of these yeasts is routine, although as will be seen later, present diagnostic techniques may sometimes be misleading. Clinical diagnosis is usually made on the basis of a combination of a number of physiological tests, mostly carbohydrate assimilation tests, accompanied by some morphological observations, and a test for the ability of the yeast to undergo conversion to a mycelial growth pattern under specified growth conditions. Kits that reduce the identification procedure to a rote numerical identification are commercially available.

Limitations on this kind of classic identification procedure are obvious, both from close scrutiny of the details of the identification protocol itself and by abstract consideration of the general principle of identification by means of a few selected phenotypic properties (12-14). Especially in the kinds of biochemical properties used in these tests, a single point mutation in a single gene in a pathway can change the presumptive identification of an isolate from one species to another. To be sure, the design of these diagnostic tests minimizes that possibility; but, in principle, a single base-pair difference can convert a yeast from one species to another by these criteria. A more serious practical problem is the uncertainty of discrimination between certain pairs of *Candida* species. In a significant fraction of cases, the standard identification procedures yield an ambiguous result, that must then be resolved by additional tests not included in the original battery. Additional costs in time and supplies result. The problem of ambiguity in identification is certain to become more important as more and more *Candida* species with overlapping phenotypic properties are implicated as agents of human infection.

There is no a priori assurance that the methods of molecular genetics and DNA technology will supply a less ambiguous identification procedure. Even less certain is the probability that what are now (for the clinical laboratory) rather complicated, unfamiliar, and even arcane procedures will become economically and technically feasible. But there is every reason to hope that such a goal is achievable. What will probably be necessary is not one but a (small) battery of probes, some designed to distinguish between the most distantly related of the

Candida species, and others to distinguish between those that are very closely related or, as discussed later, to distinguish between subgroups within a species.

B. DNA Hybridization Studies

Several laboratories, notably those of Phaff, Meyer, and Kurtzman, have pioneered the use of DNA hybridization measurements in the study of yeast systematics (13,15-18). These studies have included many species of *Candida* (15, 16), including some of the medically important species. The use of DNA hybridization to study the relationships among these medically important *Candida* goes back several years. A number of investigators have reported on the genomic similarity or divergence of different *Candida* species, as judged by DNA-DNA or DNA-RNA hybridization. A compilation of these results is shown in Table 1. A striking feature of this table is its lack of completeness. In addition, the interpretation of these results is complicated by several factors. In most of the studies used in the compilation of this table, the original investigation was not limited to, or even mainly concerned with, the medically important *Candida*; thus, some of these data are simply by-products of studies designed for another purpose. A more important source of difficulty lies in the area of methodology. The results compiled in Table 1 were obtained over a period of several years and resulted from experiments carried out with different techniques. The hybridization determinations of Leth Bak and Stendrup (19) were carried out by using the total DNA of one strain as the driver in each reaction and by using as the testor a complementary RNA that may, or may not, have been an accurate representative of the DNA complement of the species from which it was derived. The other

Table 1 Overall DNA Relatedness of Some Medically Important *Candida* Species (Percentage of Labeled Testor DNA that Hybridizes to an Excess of Unlabeled Driver DNA)

Testor \ Driver	*C. albicans*	*C. stellatoidea*	*C. pseudotropicalis*	*C. tropicalis*
C. albicans	100[b]	81[a]	10[a], 13[a]	9[a]
C. stellatoidea	65[a], 90[b] 100[b], 100[b]			12[a] 6[b], 11[b]
C. pseudotropicalis	5[a], 7[a]			
C. tropicalis	6[a]	6[a]		

[a]Leth Bak and Stendrup, Ref. 19; values are not normalized.
[b]Meyer, Ref. 16; values are normalized to the homologous reaction (same strain used as source for both testor and driver DNA). Multiple entries reflect the use of three *C. albicans* strains and two *C. tropicalis* strains as sources of driver DNA.

measurements were carried out using whole-cell DNA from different species, one (the testor) labeled and in low concentration, the other (the driver) unlabeled and in high concentration. Although this latter approach certainly avoids the problem of representativity, it does not take into account some fundamental problems in using DNA hybridization to assess relatedness in eukaryotes. It does not, for example, take into account the effect of organellar DNA; nor is it sensitive to the effects of repetitive sequences, which can so easily dilute out whatever sensitivity the DNA-DNA hybridization assay might have. It is now becoming clear (20-23) that meaningful systematic inferences based on DNA-DNA hybridization (at least for eukaryotes) should be carried out with single-copy DNA rather than with whole-cell DNA or even purified nuclear DNA, and that the fidelity of the hybrid (thermal stability) is as important a variable as the extent of hybridization for closely related speices. Although these methods have been applied to some fungi (17,23), studies on the relationships among the *Candida*, or of other medically important yeasts, have not yet reached this level of sophistication.

C. Identification Based on Gene-Specific Probes

In attempting to establish a DNA probe for organisms that may be very distantly related, it is necessary to choose a gene, or other DNA element, that is highly conserved. Operationally, this means a DNA element that has a base sequence that has undergone sufficiently few changes among all the species (or other taxa) of interest that it will hybridize with the homologous DNA element of every one. The same principle has been used for many years in comparing protein sequence divergence. In a classic paper, Fitch and Margoliash (24) used cytochrome C as such a "probe", and constructed a phylogeny of the eukaryotes, a part of which is shown in Figure 1. This figure is shown here for two reasons. First, it is a succinct example of the way biomolecular sequence data are used to infer phylogenies. The DNA hybridization methods and protein sequence data methods employ the same strategy; only the tactics are different. More importantly, however, this figure illustrates a key point about the possible relatedness of the organisms under discussion here, namely, that some groups of yeasts diverged from other groups of yeasts a very long time ago. By this criterion, *C. krusei* and *Saccharomyces oviformis* (a close relative of *S. cerevisiae*; 25) are less closely related than are man and moth. Consequently, to have a probe that will be potentially useful for selecting homologous sequences in all these yeasts, one must find a DNA element whose evolution over all this time has been sufficiently slow to allow detection of the hybrid. Our experience, as described later, has shown that the actin gene is satisfactory for this purpose with respect to the medically important *Candida*, and perhaps to some other medically important yeasts as well.

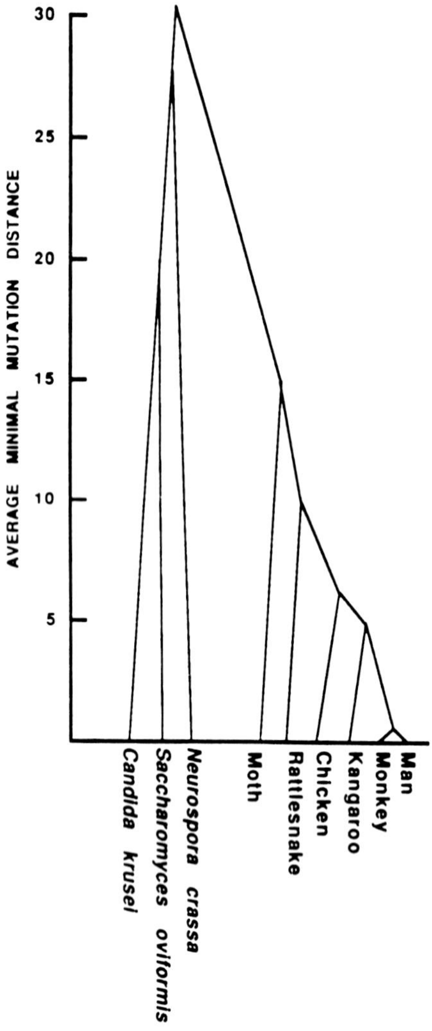

Figure 1 Phylogenetic tree for selected eukaryotes that is based on cytochrome C amino acid sequence comparisons. The ordinate reflects the number of amino acid replacements required to have occurred between taxa and their common postulated ancestors (*Source*: from Ref. 24).

Actin is a protein whose coding sequences are highly conserved over a broad range of eukaryotic taxa. In some higher organisms, there are multiple actin genes, but in *S. cerevisiae* there is a single actin gene (26,27). We have found that *C. albicans* also has a single actin gene, and we have cloned this gene in the plasmid (pBR 322) (B. A. Lasker, J. M. Bagash, and W. S. Riggsby, manuscript in preparation). This cloned gene hybridizes with homologous DNA from *S. cerevisiae* (the *C. albicans* actin gene clone was identified on the basis of this homology), from all of the *Candida* species tested (see following discussion), and from *Drosophila melanogaster* (B. A. Lasker, unpublished observations).

Species identification with specific gene probes depends on a variation from species to species of the positions of restriction endonuclease recognition sites within and near the gene of interest. If a particular restriction endonuclease has no recognition site within a gene, that gene will appear in a single restriction fragment and, therefore, will be found in a single electrophoretic band. The size of the fragment and, thus, the position of the band, will depend solely on the distance between the nearest flanking recognition sites on each side of the gene. If there is a single recognition site within the gene, then the gene will be cleaved internally, and its sequences will be partitioned between two restriction fragments, each bounded by the internal site and one of the flanking sites; consequently, two bands will ordinarily be observed upon electrophoresis. For all of the hexaschizomers (enzymes that cleave at a six-base recognition site) we have used, there is either one internal recognition site or none at all in the actin gene of each of the *Candida* species we have tested. Consequently, each restriction endonuclease digest of DNA from each species, when electrophoresed and probed with the cloned *C. albicans* actin gene, produces either one or two labeled hybrid bands.

Figure 2 is representative of such experiments. For the three species shown, *C. albicans, C. tropicalis*, and *C. guilliermondii*, the actin gene has no internal *Eco*RI recognition site. Both *C. albicans* and *C. tropicalis* have a single internal *Hind*III recognition site, but this site is absent in *C. guilliermondii*. Of the three species, only *C. albicans* has an internal *Cla*I recognition site. (The two fragments generated by digestion of *C. albicans* DNA with *Cla*I are similar in size, 1.3 and 1.5 kb, and are not well resolved in this figure, which was taken from a film that was exposed for a time sufficiently long to reveal the fainter signals produced by the *C. guilliermondii* fragments.) We have analyzed the sizes of actin gene-containing fragments of all of the *Candida* species mentioned in the Introduction of this chapter by using six hexaschizomers (Mason and Riggsby, in preparation). It is usually possible to identify each species on the basis of a single restriction endonuclease fragment pattern, although it would be preferable to use at least two enzymes for this purpose. We have used two or more strains of each species, and more than 20 isolates of *C. albicans*; different isolates of the same species always yield identical patterns. Furthermore, strains of *C. stellatoidea* yield the

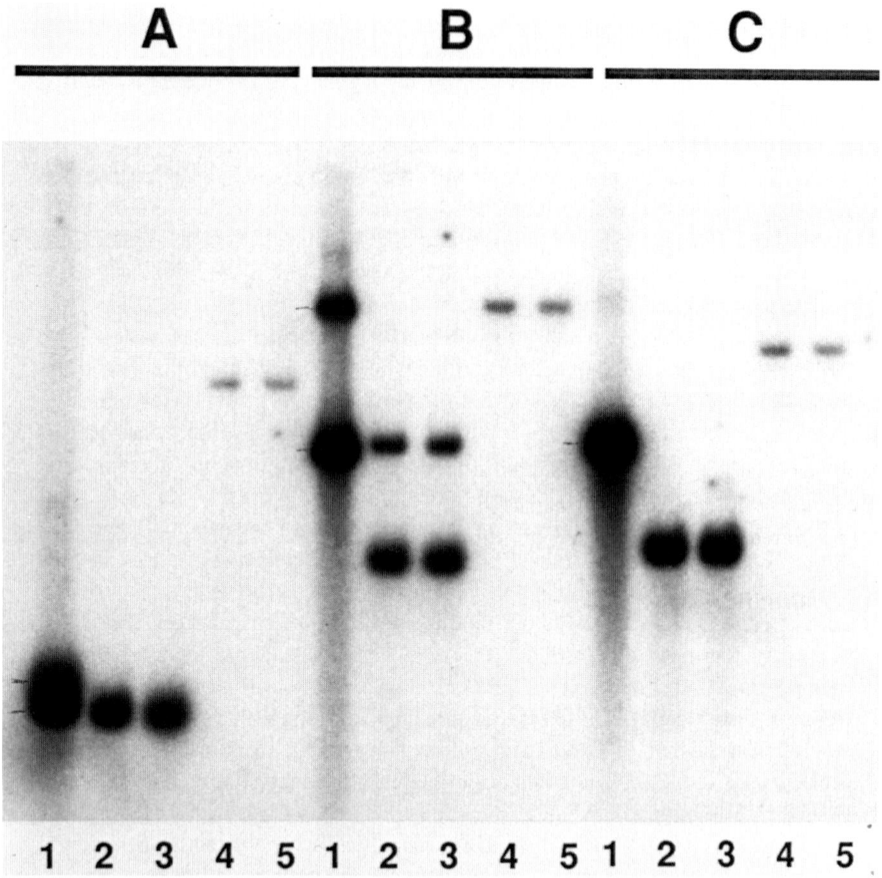

Figure 2 Actin gene-containing fragments from *Candida* species. Whole cell DNA was digested with restriction endonucleases *Cla*I (panel A), *Hind*III (panel B), or *Eco*RI (panel C), electrophoresed on an agarose slab gel, transferred to nitrocellulose, and hybridized to ^{32}P-labeled (by nick translation) *C. albicans* actin gene cloned into plasmid (pBR 322). In each panel the DNAs are from *C. albicans* strain H317 (lane 1), clinical isolates of *C. tropicalis* (lanes 2 and 3), and clinical isolates of *C. guilliermondii* (lanes 4 and 5).

same patterns as obtained for *C. albicans*, in harmony with the view that these two taxa are conspecific (28). The observed constancy within a species, and marked differences between species, suggest that the actin gene restriction fragment patterns might, in principle, provide a one-step differential diagnosis of the medically important *Candida*.

Some other protein-coding genes that have appeared in the literature may prove useful in identifying and characterizing medically important yeasts. Seehaus and co-workers (29) have recently described the use of several probes that code for glycolytic and biosynthetic enzymes in *S. cerevisiae*. They tested these probes against genomic blots of a variety of yeasts, but of those of medical importance, only *C. albicans* was included. However, they did also use *Pichia guilliermondii*, which has been shown to be the perfect form of *C. guilliermondii* (28), and that might, therefore, be expected to produce similar if not identical DNA restriction fragments in a highly conserved gene. Most of these probes give signals of varying degrees of intensity with discrete restriction fragments from both *C. albicans* and *P. guilliermondii* genomic DNA and, therefore, should be considered candidates for specific probes.

D. Identification Based on Ribosomal RNA-Specific Probes

The last decade has seen extensive development of ribosomal RNAs (rRNAs) and their genes as tools in phylogenetic analysis, particularly in the bacteria (30). This technology has been barely touched in work with fungi. The only rRNA that has been studied extensively in any of the organisms of interest here is the 5S rRNA of *C. albicans*. Chen and co-workers (31) compared this RNA with the 5S rRNAs of several other fungi, and showed that *C. albicans* is closely allied with the ascomycetes. This conclusion is consistent with some other biochemical data (28,32), and it is contrary to the view expressed elsewhere (8) that *C. albicans* is allied with the basidiomycetes.

In our analysis of the repetitive DNA of *C. albicans* (33), we identified three strongly fluorescing bands in *Eco*RI digests of whole-cell DNA and in nuclear DNA, which could not be attributed to mitochondrial contamination. These bands are designated nuc-1, nuc-2, and nuc-3 in Figure 3. By analogy with similar results in other fungi, we suggested that these bands represented sequences coding for rRNA. This conjecture has been confirmed (P. T. Magee, personal communication) by experiments that show that these bands hybridize with cloned DNA coding for rRNA. In addition, Magee and his colleagues have analyzed the rRNA-coding sequences from several *Candida* species and conclude that characteristic differences between these species can be detected from the ethidium bromide-stained gels, without the necessity to resort to Southern transfers and probe hybridization. It is of particular interest that these workers report a difference in the rRNA-banding patterns of *C. albicans* and *C. stellatoidea*.

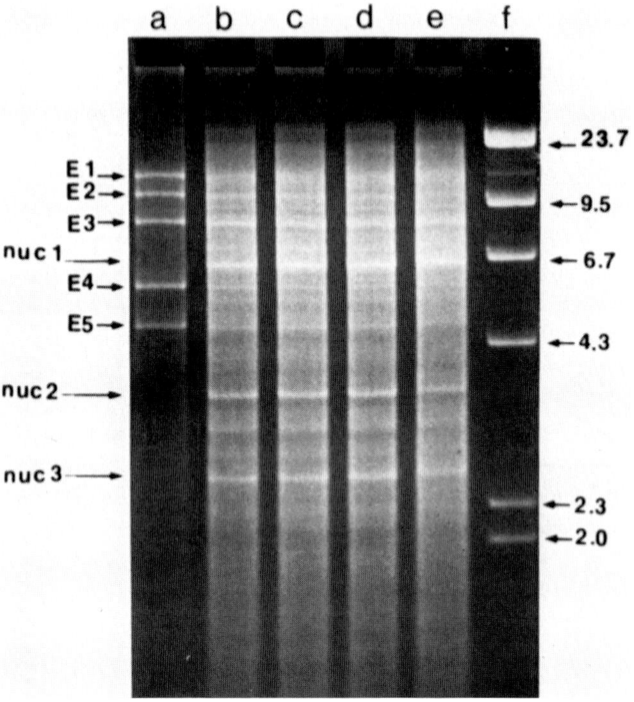

Figure 3 Restriction fragments produced by digestion with restriction endonuclease *Eco*RI of *C. albicans* mitochondrial and partially purified nuclear DNAs. The DNAs were digested with the restriction endonuclease, electrophoresed on an agarose slab gel, and visualized by staining with ethidium bromide and photography under ultraviolet illumination. Lane a, purified mitochondrial DNA; lane b, whole-cell DNA; lanes c–e, DNAs from nuclei partially purified by various methods (Ref. 33); lane f, marker DNA (phage λ DNA digested with restriction endonuclease *Hind*III. The smallest mitochondrial fragment (E6, 350 bp) is not visible on this gel.

III. *CANDIDA ALBICANS*

A. Serotypes and Phenotypes

In Sect. II we dealt with DNA probes designed to differentiate among the medically important *Candida* species and, therefore, sought a probe that gives identical results when tested with different strains or isolates within a species. In this section we deal with the converse problem: finding a test that is sufficiently sensitive to differentiate among different isolates of a single species, *C. albicans*.

Such a test is required for epidemiological studies, as well as for certain systematic and evolutionary investigations.

Traditionally, strain differentiation has relied on morphological, immunological, or physiological techniques, or a combination of these. Cell morphology is not a very sensitive test with yeasts because there is little variation except in the sexual forms, which are nonexistent in many of the medically important species, including *C. albicans*. The only important morphological test in this group is the ability to form germ tubes as a presumptive identification test for *C. albicans* (and for *C. stellatoidea*). However, because *C. albicans* mutants unable to form germ tubes have now been reported from several laboratories (34–36), this test cannot be considered particularly conclusive. Colony morphology variants have been observed in *C. albicans* (37,38). However, these variants appear to be able to undergo spontaneous switching (37), and so again the property is unlikely to be useful in strain differentiation.

Immunologically, *C. albicans* isolates fall into two serogroups, designated A and B. Serotype A is antigenically related to *C. tropicalis*, whereas serotype B is essentially indistinguishable from *C. stellatoidea* (8). At present, immunological techniques offer considerable promise as tools in species identification, but they do not provide a sensitive test for intraspecies variation.

By far the most widely used tests for intraspecies variation are based on physiological properties. A variety of biotyping methods have been introduced to differentiate among *C. albicans* strains. The set of tests devised by Odds and Abbott (39,40) have been used in studies from several laboratories. In a set of cooperative studies (41), Odds and several collaborators have investigated the distribution, relative to geography and associated pathological condition, of over 700 clinical isolates. Although the nine components of the test system are capable, in principle, of resolving 512 phenotypic combinations, these investigators found only 160 of those combinations. These were clustered in a few groups with great internal similarity, and a large fraction of the isolates belonged to relatively few of the groups. Similar clustering was observed in an independent study that used 69 isolates (42). When using the commercial API-20C system, Hellyar (43) found that 89% of 250 isolates fell into three of the 35 assimilation patterns assigned to *C. albicans*. The repeatedly observed clustering of *C. albicans* isolates into relatively few phenotypic classes is the principal motivation for the attempt to develop DNA probes, which are potentially capable of a much finer degree of discrimination than are the phenotypic tests.

B. Mitochondrial DNA Probes

Studies that have used taxa as diverse as yeasts (44) and mammals (45,46) have indicated that the mitochondrial genome evolves more rapidly (on an absolute time scale) than does the nuclear genome. Thus, mitochondrial DNA probes have become popular in studying diversity among closely related groups. For example, mitochondrial DNA probes have been used to elucidate relationships among the higher primates (47), and even among human races (48,49), and to study the divergence of closely related rodents (50). All of these studies (and many more–see Refs 45, 46) confirm the rapid evolution of mitochondrial DNA in higher animals.

The case for such a disparity in rate of evolution of mitochondrial and nuclear genomes in fungi is less well established. The original observation on which this generalization is based (44) involved rRNA-specific sequences, and the fundamental differences between the two types of rRNA (eukaryotelike in the nucleus, but prokaryotelike in the mitochondrion) could conceivably lead to errors in interpretation. Moreover, because different elements in the nuclear genome itself appear to evolve at different rates (genes versus pseudogenes, conserved genes versus plastic genes, introns versus exons, first-position nucleotides versus second- and third-position nucleotides), it is not entirely clear what precise meaning should be attached to statements about the relative rates of evolution of the nuclear and mitochondrial genomes as a whole. Several recent reviews dealing with fungal mitochondrial genomes deal directly or indirectly with the problem of evolutionary rate. Clark-Walker's (51) analysis emphasizes variation in genome size and gene order and the differential presence of genes, and discusses the evolutionary implications of this diversity; but he leaves the resolution of these issues to further study. Wallace (52), whose discussion is in terms of size and gene order in mitochondrial DNAs, documents a number of fundamental differences between the organellar DNAs of unicellular and multicellular eukaryotes, and suggests a model in which the rate of organelle DNA evolution is greater in unicellular organisms. In a different context, Grossman and Hudspeth (53) point out that investigations of mitochondrial DNA rarely use as many as five different strains, thus mitigating against any characterization of whatever diversity may exist in nature. Most of these discussions are speculative, and none explicitly takes into account the very great overall genetic distances separating different fungi (see Fig. 1). A recent study (23) of the nuclear and mitochondrial genomes of two congeneric basidiomycetes revealed much less difference in the evolution of nuclear and mitochondrial DNA. However, because there is no direct evidence to contradict the proposed more rapid evolution of mitochondrial genome, we will adopt it as a working hypothesis. Work in progress in several laboratories should help elucidate this fundamental problem.

Our laboratory is studying mitochondrial DNA polymorphism in strains of *C. albicans* to determine if this particular genetic element lends itself to the problem of distinguishing biotypes. Figure 4 shows a restriction map of the mitochondrial DNA of *C. albicans* strain H317, which we have used as the reference strain in all of our work. This DNA contains a large inverted duplication, indicated by the headed arcs on the right side of the figure. When the DNA is

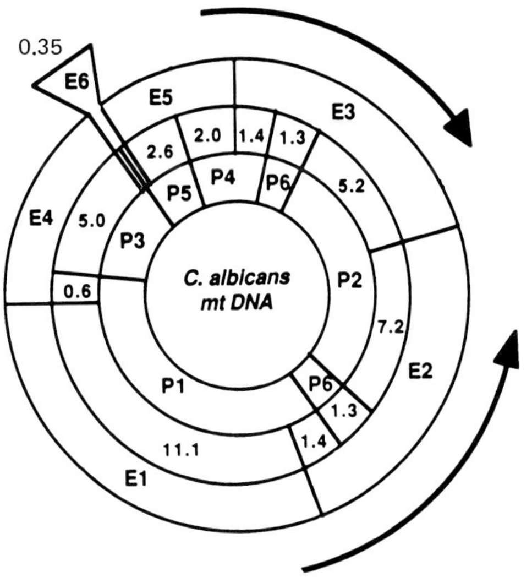

Figure 4 Map of the circular mitochondrial genome of *C. albicans* strain H317. The outer annulus shows the six fragments produced by digestion with restriction endonuclease *Eco*RI, whereas the inner annulus shows the fragments produced by digestion with *Pvu*II. The center annulus indicates the sizes of the fragments produced by simultaneous digestion with both enzymes. The dark arrows indicate the position of a large inverted duplication in this genome.

digested simultaneously with restriction endonucleases *Eco*RI and *Pvu*II, ten fragments with sizes greater than 1 kb are generated. When these fragments are electrophoresed, however, only eight bands are observed, because two fragments in each copy of the duplication are identical, and so comigrate. These eight bands are designated EP1 through EP8, in order of descending size; the lengths of the respective fragments are given in Table 2.

We have examined the restriction fragment sizes produced by simultaneous digestion by these two enzymes for approximately 30 strains of *C. albicans* and *C. stellatoidea*. Representative results from these studies are shown in Figure 5.

Table 2 Lengths of Fragments (in kb Pairs) Produced by Digestion of Mitochondrial DNA of *C. albicans* Strain H317 with Restriction Endonucleases *Eco*RI and *Pvu*II

Fragment	EP1	EP2	EP3	EP4	EP5	EP6	EP7	EP8
Length (kbp)	11.1	7.2	5.2	5.0	2.6	2.0	1.36	1.28

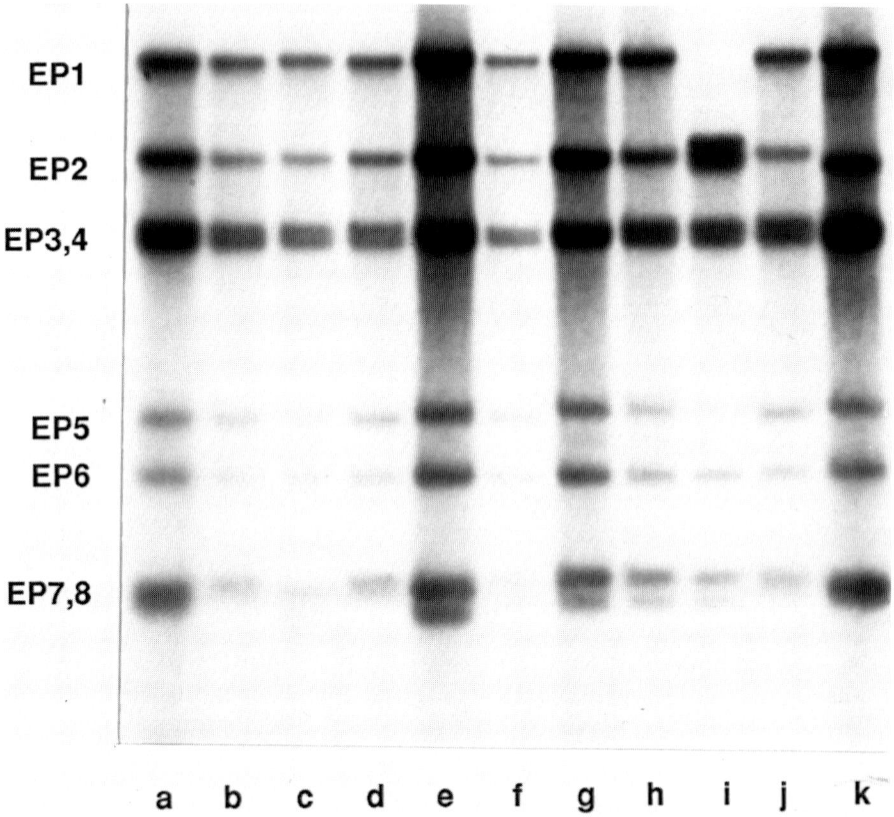

Figure 5 Restriction fragment patterns produced by digestion with restriction endonucleases *Eco*RI and *Pvu*II of DNAs from *Candida* strains. The DNAs were digested simultaneously with both restriction endonucleases, electrophoresed on agarose, and transferred to nitrocellulose. The blot was probed with labeled *C. albicans* mitochondrial DNA and exposed to x-ray film. The DNAs used are: lanes e and g, *C. stellatoidea*; lanes b, d, and j, *C. albicans* reference strain H317; the other lanes are *C. albicans* clinical isolates.

The most obvious fact about this figure is the basic similarity shown by all of the strains. Most of the differences are minor, even though the strains represented here were obtained from geographically widely dispersed locations. In most of the strains we have tested, the order of magnitude of the differences is no more than that observed in lanes j and k of Figure 5; here, in lane k, fragment EP2 migrates somewhat more rapidly, and fragment EP5 migrates somewhat more slowly, than do the corresponding fragments in lane j. The strain used to generate lane i is the most aberrant strain we have found thus far. Here fragments EP1 and EP5 are absent, and two new fragments appear, migrating very close to fragment EP2. (These three fragments are not resolved at the exposure used in this radioautogram.)

Most of the variation we have observed is in the small fragments EP7 and EP8. This variation is real, as can be demonstrated by observing differences between DNAs run in adjacent wells; both the relative separation of the two fragments, and their joint separation from the other fragments, is observed to vary. The fact that there is more variability in these fragments, compared with the other fragments, may result either from our greater ability to detect small differences in small fragments, or from intramolecular recombinational activity in the region of the inverted duplication (54), where these fragments are located.

We have tentatively assigned the strains we have examined to 14 genogroups based on the EP restriction fragment patterns. Most of these groups differ from one or more of the other groups by a single (detected) fragment size difference; for example, genogroup I and genogroup J differ only in the size of fragment EP5, which is 2.7 kb in group I and 2.6 kb in group J. It should also be noted that most of the strains fall into a few genogroups, with genogroup B comprising more than a third of all the strains. This result is reminiscent of the results of phenetic biotyping.

The more rapid evolution of mitochondrial genomes has inspired the concern that "the resolution afforded by mtDNA [restriction fragment] patterns may not be sufficient to recognize the more divergent strains of a species" (18). This possibility may be a concern in some taxa, but it does not seem to be so in *C. albicans*. As shown in Figure 6, virtually all of the *C. albicans* strains we examined differ from at least one other strain by a single fragment size difference. The three *C. stellatoidea* strains differ from one of the *C. albicans* strains in the sizes of only two fragments. The other two *C. albicans* strains examined differ from the strains shown in the Figure 6 by two or more fragment size differences, but they are still clearly identifiable as *C. albicans* simply by visual inspection of the radioautograms.

The basic uniformity of the EP fragment pattern is further emphasized by three cases in which it appeared that *C. albicans* strains had very divergent mitochondrial genomes from those described so far. Three strains sent to our laboratory as *C. albicans* were found to have mitochondrial DNA that reacted only very weakly with the *C. albicans* probe, and that, on extended exposure of the autoradiogram, produced fragment patterns completely different from those shown in Figure 5. When we did genomic blots with DNA from these strains and probed with the actin gene probe, two of these strains produced actin fragment patterns identical with those we had previously found characteristic of *C. tropicalis*. Subsequent analysis of the mitochondrial DNA restriction patterns also showed it to be typical of authentic *C. tropicalis* strains. (The third anomolous

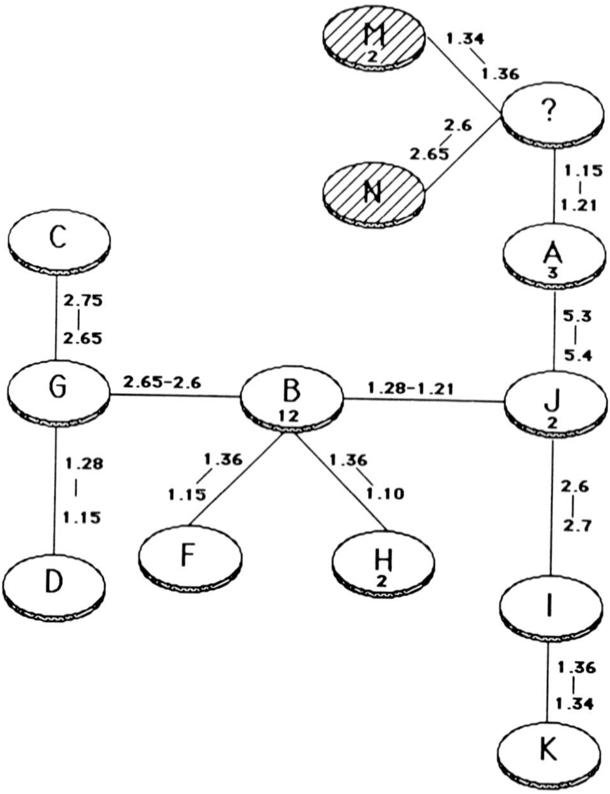

Figure 6 Comparison of various restriction fragment patterns obtained with *C. albicans* and *C. stellatoidea* mitochondrial DNAs digested with *Eco*RI and *Pvu*II. The designations A through N were given arbitrarily to the various EP patterns observed. In this figure, the numeral shown below the group designation indicates the number of strains that yielded that pattern; when no numeral is shown, only one strain yielded that restriction pattern. Groups that differ in the length of only one fragment are joined by lines; the fragment length change between adjacent groups is indicated in kilobase pairs. The shaded groups are *C. stellatoidea*; all of the other groups are *C. albicans*. The "group" designated by a question mark is included to indicate that the *C. stellatoidea* strains all differ from all of the *C. albicans* strains in more than one fragment length.

strain has not yet been identified, but standard identification procedures show that it is not *C. albicans*.) Thus, we conclude that these strains had been misidentified by the routine clinical diagnostic procedures used in the laboratories where they had been isolated. We will return to the misidentification question in Sect. III.C. The significant conclusion here is that *C. albicans* strains isolated from a wide variety of geographical locations (United States, Puerto Rico, United Kingdom, Spain) display very similar overall *Eco*RI/*Pvu*II restriction fragment patterns, with minor variations.

The results obtained with mitochondrial DNA digested with *Eco*RI and *Pvu*II, although they indicate that there was indeed polymorphism among various isolates of *C. albicans*, also indicated that these enzymes alone would not allow us to achieve the level of resolution between strains that we would like for epidemiological and biogeographical purposes. Consequently, we have adopted the strategy of digesting the DNA with enzymes that recognize four-base cleavage sites (tetraschizomers), and probing genomic blots made with these DNAs with cloned mitochondrial DNA fragments, rather than with the whole mitochondrial DNA. The purpose of using the tetraschizomers is to produce a larger number of smaller fragments; small differences between small fragments are much easier to detect on a Southern blot than similar differences between larger fragments, and the tetraschizomers are expected to cleave the DNA much more frequently than do the hexaschizomers *Eco*RI or *Pvu*II. But because so many bands are produced, probing with the whole mitochondrial DNA was expected to produce so many signals that they would be difficult to interpret. We chose, therefore, to probe with cloned fragments that carry the regions that appear to be most variable on the basis of the EP restriction fragment patterns.

Figure 7 shows the results of Southern blot analysis of DNAs from four clinical isolates of *C. albicans*, digested with restriction endonuclease *Msp*I. DNAs from these strains have similar EP fragment patterns, and on ethidium bromide-stained gels, the fluorescent banding patterns produced by the four DNAs after *Msp*I digestion are essentially indistinguishable. However, when the *Msp*I-digested DNA fragments are transferred to nitrocellulose and probed with labeled mitochondrial DNA fragment E2, distinctly different patterns appear. No two of the four patterns are identical. In lane A, fragment 2 is absent; in lane B, fragment B is markedly reduced in intensity; in lane C, both fragments 2 and 3 have slightly higher mobilities compared with lanes B and D. Thus, as expected, the use of a tetraschizomer resolves mitochondrial DNA differences not resolved with enzymes that have more stringent recognition requirements.

An approach similar to that described here has been used in a clinical setting (P. D. Olivo, E. J. McManus, and J. M. Jones, personal communication). They subjected DNA from a number of clinical isolates and repository strains of *C. albicans* to digestion with the restriction nuclease *Hae*III, electrophoresed on polyacrylamide gels, transferred to nitrocellulose, and probed with labeled whole

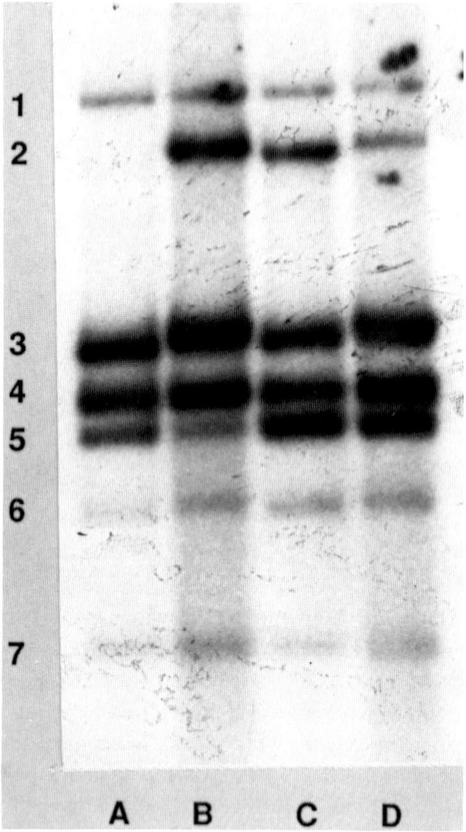

Figure 7 E2-specific restriction fragments produced by digestion of *C. albicans* DNA with restriction endonuclease *Msp*I. The DNAs from four clinical isolates of *C. albicans* were digested with *Msp*I, electrophoresed on an agarose slab gel, and transferred to nitrocellulose. The transferred fragments were hybridized to radiolabeled *C. albicans* mitochondrial DNA fragment E2 cloned into plasmid (pBR 322) (54), and exposed to x-ray film.

mitochondrial DNA (reconstituted from equimolar amounts of the six cloned *Eco*RI fragments). Although the simultaneous probing with the entire mitochondrial genome results in a complex banding pattern (more than 20 bands, some of which probably contain more than one distinct restriction fragment) they were able to distinguish three different patterns among six ATCC strains and eight clinical isolates from four different patients. When *C. albicans* was reisolated from the same patient, an identical restriction fragment pattern was obtained.

We are surveying different tetraschizomers to determine which one or which combination will be most sensitive for determining strain differences. Although this survey is far from complete, our preliminary results indicate that at least *Alu*I, *Msp*I, *Hpa*II, and *Taq*I are candidates for this role.

C. *Candida albicans* and *Candida stellatoidea*

Although *C. albicans* and *C. stellatoidea* have historically been regarded as closely related, but different species, the most recent taxonomic treatises on the yeasts (28,32) regard them as conspecific, largely on the basis of DNA homology data (see Table 1). In addition, protoplast fusions of *C. albicans* and *C. stellatoidea* strains have been accomplished (55). The evidence from molecular probes could be used to argue for either side of this issue, depending on one's allegiance in the splitter/lumper division. We have detected no differences in the actin genes of *C. albicans* and *C. stellatoidea*. In addition, the EP restriction fragment patterns of mitochondrial DNA of *C. stellatoidea* are very close to the consensus pattern of *C. albicans* (see Fig. 5). On the other hand, these *C. stellatoidea* patterns may form a separate subcluster (see Fig. 6) and, as noted, Magee's group reports a difference in the genomic blots of *C. albicans* and *C. stellatoidea* DNAs probed with DNA specific for rRNA.

D. *Candida albicans* and *Candida tropicalis*

From a certain perspective, the question of whether *C. albicans* and *C. stellatoidea* are one species or two may be seen as little more than a whimsical debate over where to draw an arbitrary line. No such triviality can be attached to the question of the relationship between *C. albicans* and *C. tropicalis* because there is an apparent discordance between phenotypic and genotypic assessments of relatedness.

These leading human pathogens share many physiological and immunological properties. With the use of numerical phenetics, Kocková-Kratochvilová and Ondrusová (56) divided the genus into eight clusters, using 70 phenotypic properties; both *C. albicans* and *C. tropicalis* were placed in the same small cluster, designated "albicans/tropicalis." Several studies (57-59) have established that *C. albicans* and *C. tropicalis* form a distinct immunological group among the medically important *Candida*.

The similarity of these organisms extends to the clinical level. *Candida albicans* is regarded as the major causal agent of candidal vulvovaginitis (10) and has been widely associated with the chronic, recurrent form of the infection (60, 61). In one recent report, however, *C. tropicalis* was found to be the most common agent of recurrent vulvovaginitis in the patients studied (62). Some confusion, and possibly some clinical misidentification, may arise from the heavy reliance on germ tube formation in the identification of *C. albicans*. As mentioned earlier, there are mutant strains of *C. albicans* that do not form germ tubes. Conversely, *C. tropicalis* is regarded as not forming germ tubes; but strains that do form germ tubes (63), or structures similar in appearance to germ tubes (36), have been reported.

All of these similarities argue for a close relationship between *C. albicans* and *C. tropicalis*. Yet, as shown in Table 1, these two yeasts appear to have very little genomic similarity. Our investigations of the actin genes and the mitochondrial DNAs of these species are consistent with the DNA hybridization data in arguing against a close genetic relationship. This apparent discordance between genotypic and phenotypic analyses certainly merits more detailed investigation.

IV. OTHER MEDICALLY IMPORTANT YEASTS

A. *Histoplasma capsulatum*

Both mitochondrial DNA and DNA coding for rRNA have been used to study variation in this dimorphic yeast (64). This study revealed the existence of three distinct restriction patterns for both mitochondrial DNA and ribosomal DNA. As in *C. albicans*, the clustering is uneven, with most of the strains (16 out of 22) falling into one of the three classes, designated class 2. The six members of class 3 were isolated from South and Central America, a region geographically distinct from the sites of isolation of the strains of the other two classes. Class 1 comprised a single isolate, which was atypical in that it had reduced virulence and an unusual clinical presentation. The mitochondrial DNA restriction patterns of classes 2 and 3 were recognizably related, sharing several common bands, whereas that of the class 3 isolate showed little similarity to either of the other classes. Within each class, no variation was observed in fragment patterns of either of the DNA elements when cleaved with any of the four enzymes used (all hexaschizomers).

Unlike the *C. albicans* results, these strains exhibit a clear geographical variation in restriction fragment pattern. The results are similar to the *C. albicans* results in that the hexaschizomer-derived patterns are too heavily clustered to be useful for epidemiological purposes. Finer discrimination may be possible with the use of tetraschizomers.

B. *Cryptococcus neoformans*

For molecular genetic studies, *C. neoformans* is attractive because the perfect forms of its various serotypes are known, making it feasible to design genetic experiments to complement molecular studies (65). Although no reports concerning DNA probes for this pathogen have appeared, we have carried out preliminary experiments on the cross-hybridization of *C. neoformans* DNA with the actin gene probe derived from *C. albicans*. As shown in Figure 8, the *C. albicans*

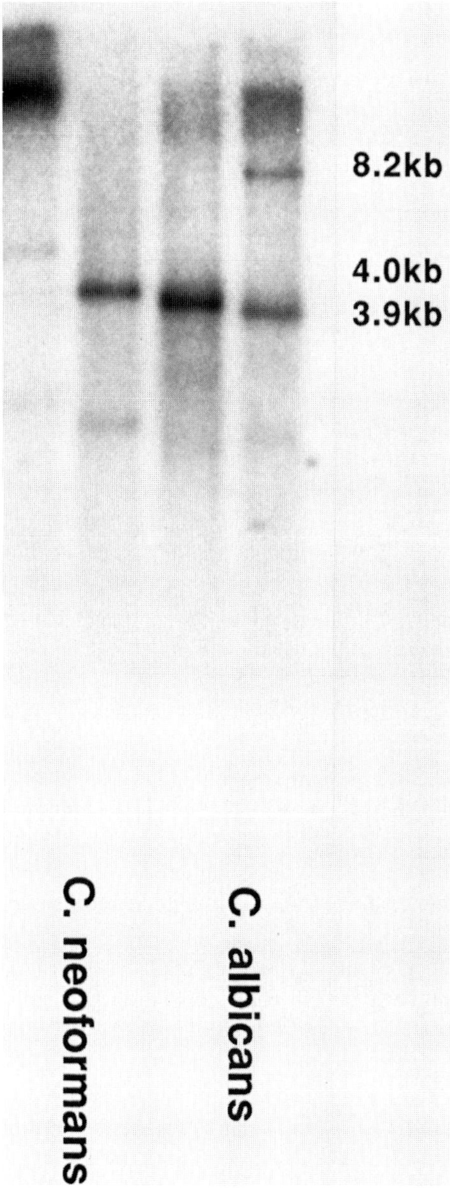

Figure 8 Genomic blots of *C. albicans* and *Cryptococcus neoformans* probed with labeled *C. albicans* actin DNA that was cloned in plasmid (pBR 322). For each species, the DNAs in the left and right lanes were digested, respectively, with *Eco*RI and *Hind*III. The sizes of the *C. albicans* fragments are shown at the right as markers.

probe does have sufficient similarity to a homologous gene in *C. neoformans* to allow the detection of discrete bands in the genomic blot. Detailed analysis of the patterns is not possible because the DNA used in the experiment was incompletely digested by the enzyme *Eco*RI; nevertheless, it is clear that experiments analogous to those described for *Candida* are possible in this pathogen as well.

V. PROSPECTS

As mentioned in the introduction to this chapter, the application of DNA technology to the medically important yeasts has just begun. Nevertheless, there has been a rapid advance in the very recent past, especially for *C. albicans*. Although conventional genetic analysis is precluded by the organism's asexuality, alternative approaches to genetic analysis have proved fruitful; these have been reviewed elsewhere (66,67). More recently, the combination of the development of a cloning vector for *C. albicans* (68) and of methods for isolating *Candida* genes by complementation in *S. cerevisiae* (69) have opened the way for detailed molecular genetic analysis.

The two major research areas considered in this chapter, evolutionary relationships and clinical identification, may be expected to proceed along somewhat different paths. DNA relatedness studies at even finer levels of discrimination have already made a great impact on the study of evolution in general (70). It is now possible to study minor sequence variations in large numbers of strains by using appropriately designed restriction fragment analysis (71). Recently introduced techniques for detecting single base-pair differences (72,73) should provide an even finer level of discrimination. The ultimate measure of genetic relatedness, complete sequence analysis of selected genetic elements, is not an unrealistic prospect, given the current rate of innovation in sequencing technology (74,75).

The future of DNA probes in a clinical setting is less clear. At the very least, it will be interesting to compare DNA probe results with the conventional phenotypic methods and with such newer methods as yeast "killer" profiles (76) and fatty acid analysis (77). Gene-specific probes, such as the actin gene or the DNA coding for rRNA, could, in principle, serve as a single tool for the identification of *Candida* species, and it is not unreasonable to suppose that this analysis could be extended to other yeasts as well. However, clinical mycology is, by its nature, conservative, and the new biotechnology will have to be supported by more extensive testing to justify the investment in equipment and training that would be required. At the appropriate time, the development of commercial diagnostic kits based on DNA probes would accelerate the acceptance of this technology in the clinical mycology laboratory.

ACKNOWLEDGMENTS

Work in our laboratory has been supported by Public Service Grants AI-07123 and GM-07438, and by a Faculty Research Grant from the University of Tennessee. Strains were generously donated by D. G. Ahearn, Jeff Becker, Frank C. Odds, Erroll Reiss, M. Casal Roman, Max Shepherd, and L. J. Torres Bauzá. DNA from *C. neoformans* was a gift of K. J. Kwon-Chung and Y. A. Donkersloot.

Previously unpublished experiments reported here were carried out principally by Mark Mason and Ron Uphoff. The actin gene probe and the mitochondrial DNA probes were developed by Brent Lasker and John Wills, respectively. I am particularly indebted to Luis Torres Bauzá, who convinced me that the molecular biology of *C. albicans* is an area worth studying, and Karl Sirotkin, who has spent many hours teaching me and my students the right way to use recombinant DNA technology.

This chapter is dedicated to Arthur Brown, on the occasion of his retirement as Head of the Department of Microbiology at the University of Tennessee.

REFERENCES

1. Fox, G. E., E. Stackebrandt, R. B. Hespell, J. Gibson, J. Maniloff, T. A. Dyer, R. S. Wolfe, W. E. Balch, R. Tanner, L. Magrum, L. B. Zablen, R. Blakemore, R. Gupta, L. Bonen, B. J. Lewis, D. A. Stahl, K. R. Luehrsen, K. N. Chen, and C. R. Woese, The phylogeny of prokaryotes. *Science 209*: 457–463 (1980).
2. Eisenstein, B. I., and N. C. Engleberg, Applied molecular genetics: New tools for microbiologists and clinicians. *J. Infect. Dis. 153*:416–430 (1986).
3. Sauls, C. D., and C. T. Caskey, Applications of recombinant DNA to pathologic diagnosis. *Clin. Chem. 31*:804–811 (1985).
4. Kingsbury, D. T., and S. Falkow (eds.), *Rapid Detection and Identification of Infectious Agents*. Academic Press, New York (1985).
5. Bennett, J. W., and L. L. Lasure (eds.), *Gene Manipulation in Fungi*. Academic Press, New York (1985).
6. Chaffin, W. L., Toward gene manipulations with selected human fungal pathogens. In *Gene Manipulations in Fungi*, J. W. Bennett and Lasure, L. L. (eds.). Academic Press, New York, pp. 469–490 (1985).
7. Strathern, J. N., E. W. Jones, and J. R. Broach (eds.), *The Molecular Biology of the Yeast Saccharomyces*. Cold Spring Harbor Laboratory, Cold Spring Harbor, N.Y. (1981).
8. Rippon, J. W., *Medical Mycology*, 2nd ed. W. B. Saunders Co., Philadelphia (1982).
9. Ahearn, D. G., Medically important yeasts. *Annu. Rev. Microbiol. 32*:59–68 (1978).

10. Odds, F. C., *Candida and Candidosis*, University Park Press, Baltimore (1979).
11. Hurley, R., The pathogenic *Candida* species and diseases caused by candidas in man. In *Biology and Activities of Yeasts*, F. A. Skinner, Passmore, S. M., and Davenport, R. R. (eds.). Academic Press, New York, pp. 231–248 (1980).
12. Stahl, U., and K. Esser, Inconsistency in the species concept for yeasts due to mutations during vegetative growth. *Eur. J. Appl. Microbiol. 8*:271–278 (1979).
13. Phaff, H. J., and C. W. Price, Strengths and weaknesses of traditional criteria in the systematics of yeasts as revealed by nuclear genome comparison. In *Single Cell Protein*, S. Garattini and Paglialunga, S. (eds.). Pergamon Press, Elmsford, N.Y., pp. 1–12 (1979).
14. Van der Walt, J. P., An evaluation of criteria for yeast speciation. In *Biology and Activities of Yeasts*, F. A. Skinner, Passmore, S. M., and Davenport, R. R. (eds.). Academic Press, New York, pp. 63–78 (1980).
15. Kurtzman, C. P., H. J. Phaff, and S. A. Meyer, Nucleic Acid relatedness among yeasts. In *Yeast Genetics*, J. F. T. Spencer, Spencer, D. M., and Smith, A. R. W. (eds.). Springer-Verlag, New York, pp. 139–166 (1983).
16. Meyer, S. A., DNA relatedness between physiologically similar strains and species of yeasts of medical and industrial importance. In *Single Cell Protein*, S. Garattini and Paglialunga, S. (eds.), Pergamon Press, Elmsford, N.Y., pp. 13–19 (1979).
17. Price, C. W., G. B. Fuson, and H. J. Phaff, Genome comparison in yeast systematics: Delimitation of species within the genera *Schwanniomyces, Saccharomyces, Debaryomyces*, and *Pichia. Microbiol. Revs. 42*:161–193 (1978).
18. Kurtzman, C. P., Molecular taxonomy of the fungi. In *Gene Manipulations in Fungi*, J. W. Bennett and Lasure, L. L. (eds.). Academic Press, New York, pp. 35–63 (1985).
19. Leth Bak, A., and A. Stenderup, Deoxyribonucleic acid homology in yeasts. Genetic relatedness within the genus *Candida. J. Gen. Microbiol. 59*:21–30 (1969).
20. Sibley, C. G., and J. E. Ahlquist, The phylogeny of hominoid primates, as indicated by DNA–DNA hybridization. *J. Mol. Evol. 20*:2–15 (1984).
21. Hall, T. J., J. W. Grula, E. H. Davidson, and R. J. Britten, Evolution of sea urchin non-repetitive DNA. *J. Mol. Evol. 16*:95–110 (1980).
22. Brownell, E., DNA/DNA hybridization studies of muroid rodents: Symmetry and rates of molecular evolution. *Evolution 37*:1034–1051 (1983).
23. Weber, C. A., M. E. S. Hudspeth, G. P. Moore, and L. I. Grossman, Analysis of the mitochondrial and nuclear genomes of two basidiomycetes, *Coprinus cineris* and *Coprinus stercorarius. Curr. Genet. 10*:515–525 (1986).
24. Fitch, W. M., and E. Margoliash, Construction of phylogenetic trees. *Science 155*:279–284 (1967).
25. Bicknell, J. N., and H. C. Douglas, Nucleic acid homologies among species of *Saccharomyces. J. Bacteriol. 101*:505–512 (1970).

26. Gallwitz, D., and I. Sures, Structure of a split yeast gene: Complete nucleotide sequence of the actin gene in *Saccharomyces cerevisiae. Proc. Natl. Acad. Sci. USA 77*:2546–2550 (1980).
27. Ng, R., and J. Abelson, Isolation and sequence of the gene for actin in *Saccharomyces cerevisiae. Proc. Natl. Acad. Sci. USA 77*:3912–3916 (1980).
28. Meyer, S. A., D. G. Ahearn, and D. Yarrow, Genus 4. *Candida* Berkhout. In *The Yeasts*, N. J. W. Kreger-van Rij (ed.). Elsevier, Amsterdam, New York, pp. 585–844 (1984).
29. Seehaus, T., R. Rodicio, J. Heinisch, A. Aguilera, H. D. Schmitt, and F. K. Zimmerman, Specific gene probes as tools in yeast taxonomy. *Curr. Genet. 10*:103–110 (1985).
30. Woese, C. R., E. Stackebrandt, T. J. Macke, and G. E. Fox, Phylogenetic definition of the major eubacterial taxa. *Syst. Appl. Microbiol. 6*:143–151 (1985).
31. Chen, M.-W., J. Anné, G. Volckaert, E. Huysmans, A. Vandenberghe, and R. De Waechter, The nucleotide sequence of the 5S rRNAs of seven molds and a yeast and their use in studying ascomycete phylogeny. *Nucleic Acids Res. 12*:4881–4892 (1984).
32. Barnett, J. A., R. W. Payne, and D. Yarrow, *Yeasts: Characteristics and Classification*. Cambridge University Press, New York (1983).
33. Wills, J. W., B. A. Lasker, K. Sirotkin, and W. S. Riggsby, Repetitive DNA of *Candida albicans*: Nuclear and cytoplasmic components. *J. Bacteriol. 157*:918–924 (1984).
34. Torosantucci, A., and A. Cassone, Induction and morphogenesis of chlamydospores in an agerminative variant of *Candida albicans. Sabouraudia 21*: 49–57 (1983).
35. Buckley, H. R., M. R. Price, and L. Daneo-Moore, Isolation of a variant of *Candida albicans. Infect. Immun. 37*:1209–1217 (1982).
36. Ogletree, F. T., A. T. Abdalal, and D. G. Ahearn, Germ tube formation by atypical strains of *Candida albicans. Antonie Leeuwenhoek J. Microbiol. 44*:15–24 (1978).
37. Slutsky, B., J. Buffo, and D. R. Soll, High-frequency switching of colony morphology in *Candida albicans. Science 230*:666–669 (1985).
38. Pomes, R., C. Gil, and C. Nombela, Genetic analysis of *Candida albicans* morphological mutants. *J. Gen. Microbiol 131*:2107–2113 (1985).
39. Odds, F. C., and A. B. Abbott, A simple system for the presumptive identification of *Candida albicans* and differentiation of strains within the species. *Sabouraudia 18*:301–317 (1980).
40. Odds, F. C., and A. B. Abbott, Modification and extension of tests for differentiation of *Candida* species and strains. *Sabouraudia 21*:79–81 (1983).
41. Odds, F. C., A. B. Abbott, R. L. Stiller, H. J. Scholer, A. Polak, and D. A. Stevens, Analysis of *Candida albicans* phenotypes from different geographical and anatomical sources. *J. Clin. Microbiol. 18*:849–857 (1983).
42. Skorepová, M., and H. Hauck, Differentiation of *Candida albicans* biotypes by the method of Odds and Abbott. *Mykosen 28*:323–332 (1985).

43. Hellyar, A. G., The frequency distribution and consistency of assimilation biotypes of *Candida albicans. J. Hyg. Camb. 96*:89–93 (1986).
44. Groot, G. S. P., R. A. Flavell, and J. P. M. Sanders, Sequence homology of nuclear and mitochondrial DNAs of different yeasts. *Biochim. Biophys. Acta 378*:186–194 (1975).
45. Brown, W. M., Evolution of animal mitochondrial DNA. In *Evolution of Genes and Proteins*, M. Nei and Koehn, R. R. (eds.), Sinauer, Sunderland, MA, pp. 62–88 (1983).
46. Avise, J. C., and R. A. Lansman, Polymorphism of mitochondrial DNA in populations of higher animals. In *Evolution of Genes and Proteins*, M. Nei and Koehn, R. R. (eds.). Sinauer, Sunderland, MA, pp. 147–164 (1983).
47. Ferris, S. D., A. C. Wilson, and W. M. Brown, Evolutionary tree for apes and humans based on cleavage maps of mitochondrial DNA. *Proc. Natl. Acad. Sci. USA 78*:2432–2436 (1981).
48. Brown, W. M., Polymorphism in mitochondrial DNA of humans as revealed by restriction endonuclease polymorphism. *Proc. Natl. Acad. Sci. USA 77*: 3605–3609 (1980).
49. Johnson, M. J., D. C. Wallace, M. C. Rattazzi, and L. L. Cavalli-Sforza, Radiation of human mitochondrial DNA types analysed by restriction endonuclease restriction patterns, *J. Mol. Evol. 19*:255–271 (1983).
50. Brown, G. G., and M. V. Simpson, Intra- and interspecific variation in the mitochondrial genome in *Rattus norvegicus* and *Rattus rattus*: Restriction enzyme analysis of variant mitochondrial DNA molecules and their evolutionary relationships. *Genetics 97*:125–143 (1981).
51. Clark-Walker, G. D., Basis of diversity in mitochondrial DNAs. In *The Evolution of Genome Size*, T. Cavalier-Smith (ed.). John Wiley & Sons, New York, pp. 277–297 (1985).
52. Wallace, D. C., Structure and evolution of organelle genomes. *Microbiol. Revs. 46*:208–240 (1982).
53. Grossman, L. I., and M. E. S. Hudspeth, Fungal mitochondrial genomes. In *Gene Manipulations in Fungi*, J. W. Bennett and Lasure, L. L. (eds.). Academic Press, New York, pp. 65–103 (1985).
54. Wills, J. W., W. B. Troutman, and W. S. Riggsby, Circular mitochondrial genome of *Candida albicans* contains a large inverted duplication. *J. Bacteriol. 164*:7–13 (1985).
55. Poulter, R., and V. Hanrahan, Conservation of genetic linkage in nonisogenic isolates of *Candida albicans. J. Bacteriol. 156*:498–506 (1983).
56. Kocková-Kratochvilová, A., and D. Ondrusová, The grouping of the species within the genus *Candida* Berkhout. In *Yeasts: Models in Science and Technics*, A. Kocková-Kratochvilová and Minárik, E. (eds.). Slovak Academy of Sciences, Bratislava, pp. 313–338 (1972).
57. Bruneau, S. M., R. M. F. Guinet, and I. Sabbagh, Identification and antigenic comparison of enzymes in the genus *Candida* by means of quantitative immunoelectrophoretic methods: Taxonomic significance. *Syst. Appl. Microbiol. 6*:210–220 (1985).

58. Gabriel-Bruneau, S. M., and R. M. F. Guinet, Antigenic relationships among some *Candida* species studied by crossed-line immunoelectrophoresis: Taxonomic significance. *Int. J. Syst. Bacteriol. 34*:227–236 (1984).
59. Brawner, D. L., and J. E. Cutler, Variability in expression of a cell surface determinant on *Candida albicans* as evidenced by an agglutinating monoclonal antibody. *Infect. Immun. 43*:966–972 (1984).
60. Sobel, J. D., Vulvovaginal candidiasis. In *Sexually Related Infectious Diseases: Clinical and Laboratory Aspects*, T. Sun (ed.). Field, Rich and Associates, New York, pp. 143–152 (1986).
61. Sobel, J. D., Vulvovaginal candidiases–What we do and do not know. *Ann. Intern. Med. 101*:390–392 (1984).
62. Horowitz, B. J., S. W. Edelstein, and L. Lippman, *Candida tropicalis* vulvovaginitis. *Obstet. Gynecol. 66*:229–232 (1985).
63. Martin, M. V., and F. H. White, A microbiologic and ultrastructural investigation of germ-tube formation by oral strains of *Candida tropicalis. Am. J. Clin. Pathol. 75*:671–676 (1981).
64. Vincent, R. D., R. Goewert, W. E. Goldman, G. S. Kobayashi, A. M. Lambowitz, and G. Medoff, Classification of *Histoplasma capsulatum* isolates by restriction fragment polymorphism. *J. Bacteriol. 165*:813–818 (1986).
65. Kwon-Chung, K. J., J. E. Bennett, and J. C. Rhodes, Taxonomic studies on *Filibasidiella* species and their anamorphs. *Antonie Leeuwenhoek J. Microbiol. Serol. 48*:25–38 (1982).
66. Riggsby, W. S., Some recent developments in the molecular biology of medically important *Candida. Microbiol. Sci. 2*:257–263 (1985).
67. Shepherd, M. G., R. T. M. Poulter, and P. A. Sullivan, *Candida albicans*: Biology, genetics, and pathogenicity. *Annu. Rev. Microbiol. 39*:579–614 (1985).
68. Kurtz, M. B., M. W. Cortelyou, and D. R. Kirsch, Integrative transformation of *Candida albicans*, using a cloned *Candida* ADE2 gene. *Mol. Cell. Biol. 6*:142–149 (1986).
69. Rosenbluh, A., M. Mevarech, Y. Koltin, and J. A. Gorman, Isolation of genes from *Candida albicans* by complementation in *Saccharomyces cerevisiae. Mol. Gen. Genet. 200*:500–502 (1985).
70. Scott, A. F., and K. D. Smith, Genomic DNA: New approaches to evolutionary problems. In *Macromolecular Sequences in Systematics and Evolutionary Biology*, M. Goodman (ed.). Plenum Press, Elmsford, N.Y., pp. 319–356 (1982).
71. Kreitman, M., and M. Aguadé, Genetic uniformity in two populations of *Drosophila melanogaster* as revealed by filter hybridization of four-nucleotide-recognizing restriction enzyme digests. *Proc. Natl. Acad. Sci. USA 83*: 3562–3566 (1986).
72. Novak, D. F., N. J. Casna, S. G. Fischer, and J. P. Ford, Detection of single base-pair mismatches in DNA by chemical modification followed by electrophoresis in 15% polyacrylamide gel. *Proc. Natl. Acad. Sci. USA 83*: 586–590 (1986).

73. Myers, R. M., Z. Larin, and T. Maniatis, Detection of single base substitutions by ribonuclease cleavage at mismatches in RNA:DNA duplexes. *Science 230*:1242–1246 (1985).
74. Lane, D. J., B. Pace, G. J. Olsen, D. A. Stahl, M. L. Sogin, and N. R. Pace, Rapid determination of 16S ribosomal RNA sequences for phylogenetic analyses. *Proc. Natl. Acad. Sci. USA 82*:6955–6959 (1985).
75. Smith, L. M., J. Z. Sanders, R. J. Kaiser, P. Hughes, C. Dodd, C. R. Connell, C. Heiner, S. B. H. Kent, and L. E. Hood, Fluorescence detection in automated DNA sequence analysis. *Nature 321*:674–677 (1986).
76. Caprilli, F., G. Prignano, C. Latella, and S. Tavarozzi, Amplification of the killer system for differentiation of *Candida albicans* strains. *Mykosen 28*: 569–573 (1985).
77. Kock, J. L. F., P. M. Lategan, P. J. Botes, and B. C. Viljoen, Developing a rapid statistical identification process for different yeast species. *J. Microbiol. Methods 4*:147–154 (1985).

RECENT REFERENCES ADDED IN PROOF

Alsina, A., M. Mason, R. A. Uphoff, W. S. Riggsby, J. M. Becker, and D. Murphy, Catheter-associated *Candida utilis* fungemia in an AIDS patient: Species verification with a molecular probe. *J. Clin. Microbiol. 26*:621–624 (1988).

Magee, B. B., T. M. D'Souza, and P. T. Magee, Strain and species identification by restriction fragment polymorphisms in the ribosomal DNA repeat of *Candida* species. *J. Bacteriol. 169*:1639–1643 (1987).

Mason, M. M., B. A. Lasker, and W. S. Riggsby, Molecular probe for the identification of medically important *Candida* species and *Torulopsis glabrata*. *J. Clin. Microbiol. 25*:563–566 (1987).

Olivo, P. D., E. J. McManus, W. S. Riggsby, and J. M. Jones, Mitochondrial DNA polymorphism in *Candida albicans*. *J. Infect. Dis. 156*:214–215 (1987).

Scherer, S., and D. A. Stevens, A *Candida albicans* dispersed, repeated gene family and its epidemiologic applications. *Proc. Natl. Acad. Sci. USA 85*: 1452–1456 (1988).

13
DNA Probes for Rapid Diagnosis of Malaria

PATRICIA GUERRY
Naval Medical Research Institute, Bethesda, Maryland

MARY A. BUESING
Uniformed Services University of the Health Sciences, Bethesda, Maryland

I. INTRODUCTION TO MALARIA

Malaria is a devastating disease affecting a quarter to a third of the world's population; it annually infects 200 million people and causes more than 1 million deaths in Africa alone (1). Hence, malaria is considered the world's number one infectious disease health problem. The current resurgence of this disease is primarily a result of increased drug resistance of the parasite and a failure of insecticide eradication programs against the mosquito vector. Moreover, it is no longer solely a health threat for Third World countries such as Africa, India, and South America alone. Instead, immigration and tourism have resulted in thousands of cases of "imported" malaria to Europe, North America, and Australia (2).

A. Etiology

Human disease is transmitted by the bite of a female anopheline mosquito which injects one of four species of protozoan belonging to the genus *Plasmodium*: *P. malariae* (Laveran, 1881), *P. vivax* (Grassi and Feletti, 1890), *P. falciparum* (Welch, 1897), and *P. ovale* (Stephens, 1922). There are 120 species of plasmodia causing infection in birds, reptiles, rodents, and primates but, usually, only these four infect humans.

B. Life Cycle

Clinical features of malaria are related to the parasite's life cycle, which is, in general, the same for all four species (Fig. 1). Sporozoites are injected from the

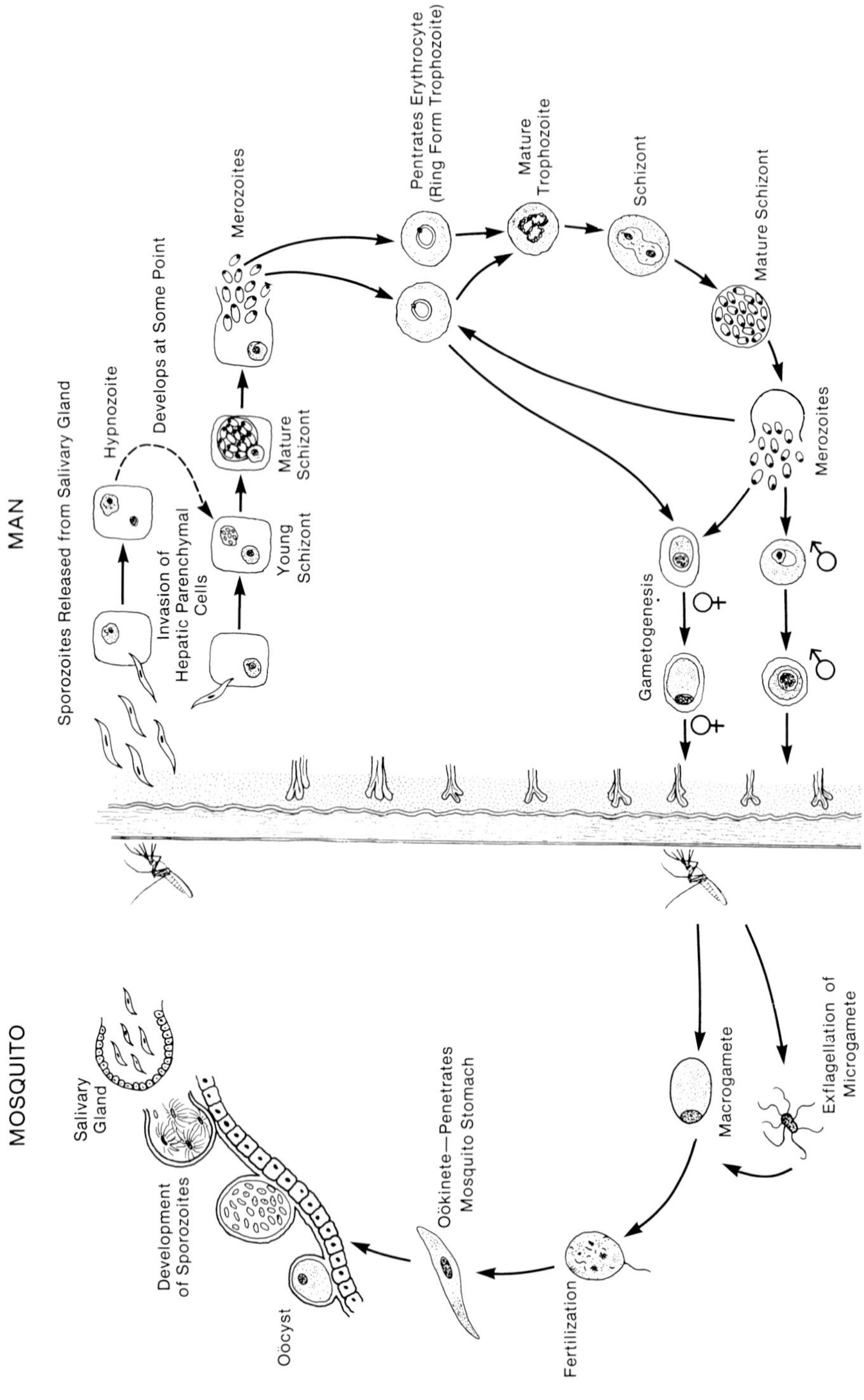

Figure 1 Schematic representation of the *Plasmodium* life cycle.

salivary gland of the mosquito into the blood stream of the host through subcutaneous capillaries. Within 30 min, the sporozoites are either destroyed by the host's immune system or they succeed in invading the liver parenchymal cells. In the liver, the parasites develop into schizonts, which rupture and release merozoites into the bloodstream. In infections with *P. vivax* and *P. ovale*, some liver forms may remain latent, forming hypnozoites that can develop into schizonts months to years later and lead to relapses of malaria. Merozoites invade erythrocytes through parasite receptors. The age of the erythrocyte is an important determinant for the degree of infection: whereas *P. vivax* and *P. ovale* primarily infect young cells and *P. malariae* infects old ones, *P. falciparum* infects all red blood cells, thereby producing extremely high levels of parasitemia. Upon entering the erythrocyte, the merozoites acquire a ringed appearance. These trophozoites then divide to form the schizonts, which rupture to release more merozoites. It is this process that produces the shaking chills, high fevers, and drenching sweats that characterize the malaria paroxysm. The ruptured erythrocytes result in hemolytic anemias and hepatosplenomegaly. Moreover, people with untreated infections with *P.falciparum* have a high mortality, and the infections are associated with serious complications such as acute renal failure, pulmonary edema, and cerebral dysfunction. As the life cycle continues, the merozoites can either reinfect red blood cells or develop into gametocytes, which can be ingested by a female anopheline mosquito during a blood meal. Within the mosquito gut, male gametocytes develop into gametes (through a process called exflagellation) and fertilize female gametes to form zygotes. The latter mature into wormlike oökinetes that migrate through the stomach wall of the mosquito to become oöcysts. Therein, numerous sporozoites develop, rupture, and find their way back to the salivary gland, thus completing the life cycle.

II. CONVENTIONAL METHODS OF DIAGNOSIS

A. Giemsa-Stained Blood Smears

For almost a century, malaria has been diagnosed by observing parasites in stained smears of peripheral blood. To date, Giemsa staining of thin and thick blood smears is the most reliable means of parasite identification. Technical guidance for preparing, staining, and reading blood films for malaria diagnosis is readily available from several laboratory manuals, the classic one being the *Manual for the Microscopic Diagnosis of Malaria in Man*, by Aimee Wilcox (3). Recently, Bruce-Chwatt has outlined the World Health Organization's recommendations for preparation of both smear techniques in his textbook, *Essential Malariology* (2). Although such references provide a framework from which to diagnose the disease, some amount of formal training is required to become accurate in the subtleties of speciation among *P. falciparum, P. vivax, P. malariae,* and *P. ovale*. Some distinguishing features of the species are shown in Table 1.

A skilled technician can detect 100 parasites/μl (2 parasites per 100,000 RBCs, or a 0.002% parasitemia) by examining 200-300 fields of a thin smear under oil immersion for 20-25 min. This sensitivity can be improved to 10-20 parasites/μl (2-4 parasites/1 million RBCs, or a 0.0002% parasitemia), by examining 100 fields of a thick smear for 3-5 min. Practically, however, only 50-60 thick smears can be read by a technician each day (4). Consequently, the ability to identify and speciate malaria parasites is dependent upon the skill and perseverance of the microscopist. This was best illustrated in the Garki project, a major epidemiological study of malaria in the Sudan Savanna of West Africa (5). Each microscopist read an average of 25 slides a day. By examining 400 fields of a thick smear versus the customary 200, the detection of *P. falciparum* increased by 10%. The diagnoses of *P. malariae* and *P. ovale* were improved even more: by 24 and 21%, respectively.

The ability to diagnose malaria by blood smear analysis requires well-stained smears and a microscope, both of which may be temporarily unobtainable in a field situation. It is not surprising then, that most diagnoses of malaria in Third World countries rely exclusively on clinical impressions, which may be correct only one-third to one-half of the time (6).

B. Immunodiagnosis

A wide variety of serological assays have become available for the study of malaria over the past 25 years. Until the 1980s, most of these techniques were directed toward the detection of circulating antibodies and were used primarily for epidemiological surveillance (7-16). However, these assays often missed the early phase of active infection because of the lag period between the appearance of parasitized erythrocytes and the presence of antibodies in the blood. In endemic areas, individuals exposed to recurrent infections often maintain high antibody levels, which further complicates the diagnosis of acute disease. Similarly, microscopic examination is prohibitively labor-intensive in such patients because they often have extremely low parasitemias as a result of their partial immunity. Hence, enzyme-linked immunosorbant assays (ELISA) and radioimmunoassays (RIA) have recently been developed to identify malaria antigens directly (17-21). By using different antibody binding inhibition methodologies, sensitivities ranging from one parasite per 10^4-10^5 erythrocytes (50-500 parasites/μl; 19,20) to eight parasites per 10^6 erythrocytes (40 parasites/μl; 17,18) have been reported for infections with *P. falciparum*. Thus, these studies demonstrated a maximum level of sensitivity intermediate between that of thin and thick smears. However, the serological assays employed to date are not as reliable as microscopic examinations because considerable numbers of false-positive and false-negative reactions occur. By using rats infected with *P. berghei*

Table 1 Speciation of Human Plasmodia in Giemsa-Stained Thin Blood Films

	P. falciparum	*P. vivax*	*P. ovale*	*P. malariae*
Peak parasitemia levels (parasites/μl)	20,000	20,000	9000	6000
Appearance of RBCs				
size	Normal	Enlarged (1.5–2×)	Oval, enlarged (1.25–1.5×)	Normal
Schuffner's stripping (red dots)	–	+	+	–
multiple infections	Common	Occasional	Occasional	Rare
Appearance of parasites				
pigmentation	Brown-black	Yellow-brown	Yellow-brown to dark brown	Brown-black
trophozoites	Small ring forms with double nuclei; also applique forms	Irregular, ameboid	Rounded, occasionally ameboid	Rounded, compact; some band forms
schizonts	8–24 merozoites	12–24 merozoites	4–16 merozoites in rosette pattern	6–12 merozoites in rosette pattern
merozoites	Very small	Small, grapelike clusters	Very large	Large
gametocytes	Banana-shaped	Round	Round	Round
Predominate stage in peripheral blood	Ring form trophozoites	All stages	All stages	All stages

as a model system, Avraham et al. (22) improved parasite detection to five parasites per 10^7 erythrocytes, but similar sensitivities have not yet been achieved in field trials of *P. falciparum* infections (17,18,21).

The main advantages of these immunodiagnostic assays lies in the speed with which many samples can be processed simultaneously. More extensive applications await purification and biochemical characterization of the diverse stage- and species-specific antigens produced during the complex blood stage of the parasite life cycle. Currently, DNA probe technology holds more promise as a rapid and sensitive diagnostic technique because it circumvents the problems of expression of the diverse antigenic repertoire of the parasite.

III. DNA PROBES FOR CLINICAL DIAGNOSIS

The published work on human malaria probes is limited to probes for *P. falciparum* malaria. Because *P. falciparum* is the only species of malaria that can be cultured in vitro, parasite DNA is more readily available than for other species. Thus, work on other species is dependent upon obtaining blood samples from patients or from experimentally infected primates. A summary of diagnostic probes for *P. falciparum* is given in Table 2.

A. Total-Genomic Probes

There has been one study in which total-genomic *P. falciparum* DNA was used as a probe. Pollack and co-workers (23) used total-genomic DNA to probe patient bloods (10 μl samples) that had been lysed directly on nitrocellulose by osmotic shock. Despite the excellent sensitivity of this method (five parasites per microliter or 50 total parasites), total genomic DNA would probably cross-react with other *Plasmodium* species. Thus, although this probe is not suitable for clinical diagnosis, the work does indicate the sensitivity that one can obtain with a very simple, direct lysis technique.

B. Probes Based on the pRepHind Clone

In 1984, Franzen and co-workers (24) published a report on the isolation of a repetitive DNA element from an African strain of *P. falciparum* and demonstrated its potential as a diagnostic probe. The recombinant clone was isolated by screening a λ-phage library of *P. falciparum* with nick-translated total *P. falciparum* DNA. Those plaques that hybridized most strongly with total genomic DNA represented highly repetitive sequences, with approximately 1% of the analyzed clones fitting this criterion. Two of the most strongly hybridizing clones contained overlapping sequences consisting of tandem repeats of 21-base-pair (bp) units. The repeats show slight variations, as demonstrated in the published sequence of two repeat units.

1. 5′ AGG TCT TAA CTT GAC TAA CAT 3
2. 5′ AGG TCT TAA CTT GAC TTA CTT 3′

Table 2 Probes for the Diagnosis of *P. falciparum*

Probe	Total *P. falciparum*	pRepHind clone	pRepHind-oligo	pRepHind-oligo	21-bp genomic clone
Ref.	Pollack et al.	Franzen et al.	McLaughlin et al.	Mucenski et al.	Barker et al.
Sensitivity with DNA (exposure time)	ND	25 pg (10 hr)	100 pg (18 hr) 10 pg (1 wk)	100 pg (18 hr)	10 pg (18 hr)
Blood lysis procedure	Direct osmotic shock lysis on NC	Detergent, phenol; spot lysates to NC	ND	Direct detergent lysis on NC with proteinase K	Detergent, proteinase K, vacuum filter onto NC
Sensitivities (p/μl)					
patients	NA		ND		
lowest detected		5000		70	<40
lowest missed		5000		140	<80
lowest examined		5000		70	<40
chimpanzees	ND	ND	ND		ND
lowest detected				<50	
Cross-reactivities					
P. malariae	ND	ND	ND	–	ND
P. vivax	ND	–	–	–	–
human	–	–	–	–	–

ND, not done; NA, not available; NC, mitrocellulose.

The *P. falciparum* sequence obtained from the recombinant clone (1), called pRepHind, could detect a minimum of 25 pg of purified *P. falciparum* DNA. In addition, parasitemias as low as 0.001% (or approximately 50 parasites per microliter) were detected when 50 μl of cultured *P. falciparum*-infected blood had been lysed with detergent, deproteinized with proteinase K and phenol, precipitated with ethanol, and spotted onto nitrocellulose. In a study of 35 blood samples from patients infected with either *P. falciparum* or *P. vivax*, 16 of 17 *P. falciparum* infections were correctly diagnosed. The patient that was misdiagnosed had a parasitemia of 0.1% or about 500 parasites per microliter of blood, although another patient who had a similar parasitemia level was correctly diagnosed. Blood from one patient diagnosed as having a *P. vivax* infection reacted with the probe, but the authors stated that this patient probably had a mixed infection of *P. vivax* and *P. falciparum*. Overall the strength of the hybridization signal was roughly proportional to the parasitemia level, but there were some deviations. For example, one patient with a 0.5% parasitemia level had a signal equivalent to another with a 1.6% parasitemia. Potential explanations for this discrepancy will be discussed later.

A subsequent study by McLaughlin et al. (25) employed a synthetic oligonucleotide probe (21 bp) based on the malaria sequence of pRepHind. They could detect 0.1 ng of pure *P. falciparum* DNA after overnight autoradiography; this was only fivefold less sensitive than the detection obtained with total *P. falciparum* DNA as the probe. After a week-long exposure the oligonucleotide was as sensitive as total-genomic DNA. These authors also found that the probe failed to cross-react with *P. vivax* DNA.

Our laboratory (26) has evaluated the same synthetic 21-bp oligonucleotide, based on pRepHind, more extensively using a simplified blood lysis procedure. Our goal has been to develop a simple, rapid diagnostic assay that avoids cumbersome techniques such as phenol extraction and vacuum filtration. We developed a technique, adapted from Totten et al. (27), of lysing samples directly on nitrocellulose filters. Blood samples from fingersticks are spotted onto nitrocellulose squares. The nitrocellulose squares are placed sequentially over Whatman papers saturated with saponin (to lyse the erythrocyte membranes), sarkosyl (to lyse the parasites), and proteinase K. Figure 2 shows an autoradiograph of the oligonucleotide probe hybridized to samples of cultured *P. falciparum*. In an initial study, blood from experimentally infected chimpanzees was probed and the results were compared with those obtained with both Giemsa-stained slides and in vitro cultivation of the chimp blood. The probe showed excellent correlation with the Giemsa stain results and was able to detect parasitemias as low as 50 parasites per microliter of blood. After overnight autoradiography, the probe also detected a sample that was read as negative by slide, but that was confirmed to be positive by in vitro cultivation of the sample.

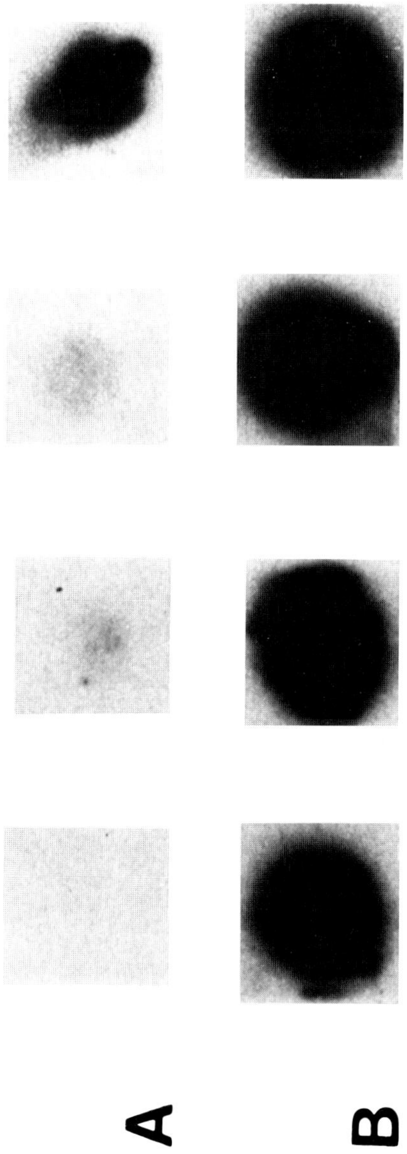

Figure 2 Autoradiograph of pRepHind oligonucleotide probe hybridized to 1 μl spots of blood according to Mucenski et al. (26). Row A, normal blood, cultured blood infected with *P. falciparum* at 0.005%, 0.01%, 0.05% parasitemias; Row B, cultured blood infected with *P. falciparum* at 0.2%, 0.01%, 0.4%, and 0.5% parasitemias. By using 20 μl blood spots lysed under the same conditions, parasitemias of 0.001% are detected.

A clinical study (26) was conducted in which blood samples from 50 febrile patients in the Philippines were spotted directly onto nitrocellulose. Probe analyses carried out in our laboratory were compared with the smear results read in the Philippines. Seventeen of these patients had a diagnosis, by smear, of *P. falciparum* malaria; 7 had a diagnosis of infection with *P. vivax*; 1 had a diagnosis of infection with *P. malariae*. The remaining diagnoses were nonparasitic. In an attempt to shorten the overall length of the procedure, the autoradiographic exposure time was varied. The probe showed the best correlation with the slides after a 4 to 8-hr exposure; after 18-hr film exposure 18 patient bloods that had been read as negative by smear were read as positive by autoradiography. Control blood spots obtained from a United States blood bank remained negative after the same exposure time. Considering the sensitivity observed when using this technique with chimpanzee bloods, it is difficult to ignore the possibility of a low-grade parasitemia in these patients. However, it is virtually impossible to determine if these patients were truly infected. A subsequent study in our laboratory (Buesing et al., manuscript in preparation) examined 175 samples from febrile patients from six countries. When 8-hr autoradiographic exposures were used, the probe again showed excellent sensitivity (90.0%) and specificity (96.9%).

In both the study by Franzen (24) and in our experience, there is an occasional lack of correlation between the strength of the hybridization signal with the probe and the percentage of parasitemia determined by smear. The exact reasons for these occasional discrepancies are not clear. In part, they could be due to the subjective determination of parasite numbers by the smear technique or to difficulties in speciation of mixed infections. Alternatively, they may represent genetic variations in the parasite population. However, our laboratory has screened strains of *P. falciparum* from seven countries and, although the overall pattern of hybridization varies by Southern blot analysis, all strains seem to hybridize with equal overall intensity with the oligonucleotide probe. On the other hand, we have noticed a variation in signal strength dependent on the level of passage of a strain from a human or mosquito host. A strain of *P. falciparum* that was isolated from a human volunteer and passaged a minimal time in vitro showed less overall hybridization than the parental strain that had been maintained in culture for several years. Similar rearrangements have been observed by others (28). The significance of such variation in probe hybridization will undoubtedly require an understanding of the function of this repetitive DNA in the life cycle of *P. falciparum*.

C. Other Probes

Another probe that is based on a repetitive DNA sequence has been reported by Wirth and co-workers (29). The repetitive DNA sequence was isolated from a λ-library of *P. falciparum* using the same method as previously described for pRepHind. The probe consists of 21-bp tandem repeats (30), but its sequence has not been reported and its relationship to the pRepHind clone is not certain. This probe can detect 10 pg of *P. falciparum* DNA after overnight autoradiography. When analyzing patient samples these workers dilute the blood fivefold and then lyse with Triton X-100 and proteinase K. The lysates are then filtered onto nitrocellulose with a vacuum manifold. In clinical studies in Brazil and Thailand, the probe was used to correctly make a diagnosis in 15 of 16 patients who were infected with *P. falciparum*. It did not cross-react with the blood of three patients who had a diagnosis of *P. vivax* or with 19 negative controls. The lowest parasitemia detected was less than 40 parasites per microliter of blood; the only patient confirmed to have malaria by smear whose diagnosis was missed by the probe had less than 80 parasites per microliter of blood. Close examination of the data again shows occasional discrepancies between signal strength with the probe and the percentage of parasitemia, as described earlier for the pRepHind probe.

IV. USE OF PROBES IN STRAIN IDENTIFICATION AND RESEARCH STUDIES

Malaria-specific gene probes are being used in research laboratories to distinguish among strains of *P. falciparum* by Southern blot analysis. The ability to readily identify strains has obvious importance for epidemiological studies. In addition, these probes could be useful for identifying strains with specific markers such as drug resistance.

Goman et al. (31) observed differences in hybridization patterns between two strains of *P. falciparum* isolated in Thailand when using a probe consisting of a moderately repeated DNA sequence. In addition, Coppel et al. (32) have demonstrated that cDNA clones of single-copy genes may be employed as probes to differentiate among isolates from diverse geographical areas. The relatively simple hybridization patterns observed with such probes offer an advantage over the complex patterns observed with repetitive DNAs (26,31). Alternative methods for strain identification include isozyme electrophoresis, two-dimensional gel electrophoresis of total proteins, and reaction with a set of monoclonal antibodies. However, gene probe technology is, in general, simpler than most of the alternatives. In addition, it is probably more specific than even the monoclonal antibody methods because it has been shown to distinguish differences in genes encoding cross-reactive antigens (32).

A recent study (33) reported the use of a repetitive DNA probe specific for the rodent malaria *P. berghei* to study the effect of γ-interferon on the course of sporozoite-induced infections. The probe enabled detection of approximately 1000 parasites just 2 days after infection with sporozoites, when the merozoites had just emerged from the liver. The authors were also able to measure parasite development within the liver by using this probe and to detect inhibition of liver stages in animals treated with γ-interferon. Such studies demonstrate a further usefulness of DNA probes to the study of malaria pathogenesis.

V. FUTURE DIRECTIONS

Work on malaria probes is just beginning to come to fruition. Clearly, the potential usefulness of the technique has been demonstrated, but much work remains to be done. Still, probe diagnosis of malaria offers many promises of improvements over the classic methods of diagnosis. A major advantage is the number of samples that can be processed within a given time. It has been estimated that a trained technician can read 50-60 thick smears per day (4); a single technician could process hundreds of samples by the probe technology in the same time frame. Thus, the probe method is particularly well suited for epidemiological studies and vaccine field trials. Additionally, the ability to accurately read thick and thin blood films and to speciate causative malarial agents, particularly in mixed infections, is a skill requiring considerable training and conscientiousness. Probes offer a less subjective method that can be readily taught to inexperienced personnel.

There are, nonetheless, major problems with malarial probe technology. The first problem is that all work, to date, has been done with radioactive labeling, which restricts the work to sophisticated laboratory settings. The use of non-radioactive-labeling procedures will dramatically increase the usefulness of probe technology to the Third World. Second, of the four species of malaria that infect humans, probes exist only for *P. falciparum*. Although this species is clearly the most virulent form of malaria, probes are needed to more accurately diagnose other species, particularly when mixed infections with *P. falciparum* occur.

ACKNOWLEDGMENTS

This work was supported by the Naval Medical Research and Development Command, Research Work Unit 3M463763A807.AA122.

REFERENCES

1. Strickland, G. T., *Hunter's Tropical Medicine*. W. B. Saunders Co., Philadelphia, pp. 516-552 (1984).
2. Bruce-Chwatt, L. J., *Essential Malariology*. John Wiley & Sons, New York, pp. 103-126 (1985).
3. Wilcox, A., *Manual for the Microscopic Diagnosis of Malaria in Man*. Public Health Service Publication No. 796 (1960).
4. Boyd, M. F., R. Christophers, and L. T. Coggeshal, *Malariology 1*:155 (1949).
5. Molineaux, L., and G. Gramiccia, *The Garki Project*. WHO, Geneva (1980).
6. Fox, J. L., Challenge: Diagnosing malaria in the developing world. *ASM News 52*(3):146-149 (1986).
7. WHO, Serological testing in malaria. *Bull. WHO 50*:527-535 (1974).
8. Label, H. O., and I. G. Kagan, Seroepidemiology of parasitic diseases. *Annu. Rev. Microbiol.* 329-346 (1978).
9. Meuwissen, J. H. E. T., Immunodiagnostic techniques applicable to malaria. Unpublished document WHO/Mal/81.951 (1981).

10. Beaudoin, R. L., J. M. Ramsey, and N. D. Pacheco, Antigens employed in immunodiagnostic tests for the detection of malarial antibodies. Unpublished document, WHO/Mal/81.952 (1981).
11. Siddarth, K. D., I. K. Srivastava, G. P. Dutta, and S. S. Agarwal, Serology and seroepidemiology of malaria. *J. Comm. Dis. 17*(suppl. 1):68–76 (1985).
12. Bayoumi, R. A., A. H. Bashir, and N. H. Abdulhadi, Resistance to falciparum malaria among adults in central Sudan. *Am. J. Trop. Med. Hyg. 35*:45–55 (1986).
13. Edrissan, G. H., M. Ghorbani, and A. Afshar, IFA serological surveys of malaria in north, north-west and south-west of Iran. *Bull. Soc. Pathol. Exot. Filales 78*:349–359 (1985).
14. Cattani, J. A., J. L. Tulloch, H. Vrbova, D. Jolley, F. D. Gibson, J. S. Moir, P. F. Heywood, M. P. Alpers, A. Stevenson, and R. Clancy, The epidemiology of malaria in a population surrounding Madang, Papua, New Guinea. *Am. J. Trop. Med. Hyg. 35*:3–15 (1986).
15. Benzerroug, E. H., P. Demedts, and M. Wery. Detection of antibodies to *Plasmodium vivax* by indirect immunofluoresence: Influence of the geographic origin of antigens and serum samples. *Am. J. Trop. Med. Hyg. 35*: 255–258 (1986).
16. Collins, W. E., W. McWilson, and G. M. Jeffery, Seroepidemiology studies of malaria in Central and South America. Unpublished document, 1–5, WHO/MAL/81.968 (1981).
17. Mackey, L. J., I. A. McGregor, and P. H. Lambert, Diagnosis of *Plasmodium falciparum* infection using a solid-phase radioimmunoassay for the detection of malaria antigens. *Bull. WHO 58*:439–444 (1980).
18. Mackey, L. J., I. A. McGregor, N. Paounova, and P. H. Lambert, Diagnosis of *Plasmodium falciparum* infection in man: Detection of parasite antigens by ELISA. *Bull. WHO 60*:69–75 (1982).
19. Avraham, H., J. Golenser, D. T. Spira, and D. Sulitzeanu, *Plasmodium falciparum*: Assay of antigens and antibodies by means of a solid phase radioimmunoassay with radioiodinated staphylococcal protein A. *Trans. R. Soc. Trop. Med. Hyg. 75*:421–425 (1981).
20. Bidwell, D. E., and A. Voller, Malaria diagnosis by enzyme-linked immunosorbent assay. *Br. Med. J. 282*:1747–1748 (1981).
21. Avraham, H., J. Golenser, D. Bunnag, P. Suntharasamal, S. Tharavanij, K. T. Harinasuta, D. T. Spira, and D. Sulitzeanu, Preliminary field trial of a radioimmunoassay for the diagnosis of malaria. *Am. J. Trop. Med. Hyg. 32*:11–18 (1983).
22. Avraham, H., D. T. Spira, and D. Sulitzeanu, Inhibition of antibody-binding as a radioimmunoassay for *Plasmodium berghei* infection in rats. *J. Parasitol. 68*:177–184 (1982).
23. Pollack, Y., S. Metzger, R. Shemer, D. Landau, D. T. Spira, and J. Golenser, Detection of *Plasmodium falciparum* in blood using DNA hybridization. *Am. J. Trop. Med. Hyg. 34*:663–667 (1985).
24. Franzen, L., R. Shabo, H. Perlman, H. Wigzell, G. Westin, L. Aslund, T. Persson, and U. Pettersson, Analysis of clinical specimens by hybridization

with probe containing repetitive DNA from *Plasmodium falciparum*. A novel approach to malaria diagnosis. *Lancet 1*:525–528 (1984).
25. McLaughlin, T. D., Edlind, G. H. Campbell, R. F. Eller, and G. M. Ihler, Detection of *Plasmodium falciparum* using a synthetic DNA probe. *Am. J. Trop. Med. Hyg. 34*:837–840 (1985).
26. Mucenski, C. M., P. Guerry, M. Buesing, A. Szarfman, R. Sangalong, C. P. Ranoa, M. Tuazon, O. R. Majam, I. Quakyi, L. W. Scheibel, J. H. Cross, J. Olson, and P. L. Perine, Evaluation of a synthetic oligonucleotide probe for diagnosis of *Plasmodium falciparum* malaria. *Am. J. Trop. Med. Hyg.* (in press, 1986).
27. Totten, P. A., K. K. Holmes, H. H. Handsfield, J. S. Knapp, P. L. Perine, and S. Falkow, DNA hybridization technique for the detection of *Neisseria gonorrhoeae* in men with urethritis. *J. Infect. Dis. 148*:462–471 (1983).
28. Bhasin, V. K., C. Clayton, W. Trager, and G. A. M. Cross, Variations in the organization of repetitive DNA sequences in the genomes of *P. falciparum* clones. *Mol. Biochem. Parasitol. 15*:149–158 (1985).
29. Barker, R. H., Jr., L. Suebsaeng, W. Rooney, G. C. Alecrim, H. V. Dourado, and D. F. Wirth, Specific DNA probe for the diagnosis of *Plasmodium falciparum* malaria. *Science 231*:1434–1436 (1986).
30. Barker, R. H., Jr., and D. F. Wirth, Specific probe for the diagnosis of *Plasmodium falciparum* malaria. *J. Cell. Biochem. 10*(suppl. A):124 (1986).
31. Goman, M., G. Langley, J. E. Hyde, N. K. Yankovsky, J. W. Zolg, and J. G. Scaife, The establishment of genomic DNA libraries for the human malaria parasite *Plasmodium falciparum* and identification of individual clones by hybridization. *Mol. Biochem. Parasitol. 5*:391–400 (1982).
32. Coppel, R. L., R. B. Saint, H. D. Stahl, C. J. Langford, G. V. Brown, R. F. Anders, and D. J. Kemp, *Plasmodium falciparum*: Differentiation of isolates with DNA hybridization using antigen gene probes. *Exp. Parasitol. 60*: 82–89 (1985).
33. Ferreira, A., L. Schofield, V. Enea, H. Schellekens, P. van der Meide, W. E. Collins, R. S. Nussenzweig, and V. Nussenzweig, Inhibition of development of exoerythrocytic forms of malaria parasites by gamma interferon. *Science 232*:881–884 (1986).

RECENT REFERENCES ADDED IN PROOF

Aslund, L., L. Franzen, G. Westin, T. Persson, H. Wigzell, and U. Pettersson. Highly reiterated non-coding sequence in the genome of *Plasmodium falciparum* is composed of 21 base-pair tandem repeats. *J. Mol. Biol. 185*:509–516 (1985).

Buesing, M. A., P. Guerry, C. DeiSanti, E. Franke, G. Watt, M. A. Rab, C. N. Oster, J. P. Burans, and P. L. Perine, An oligonucleotide probe for detecting *Plasmodium falciparum*: An analysis of clinical specimens from six countries. *J. Infect. Dis. 155*:1315–1318 (1987).

Wirth, D. F., W. O. Rogers, R. Barker, Jr., Heitor Dourado, L. Suesebang, and B. Albuquerque, Leishmaniasis and malaria: New tools for epidemiologic analysis. *Science 234*:975–979 (1986).

14
DNA Probes in Clinical Microbiology

JOHN A. WASHINGTON and GAIL L. WOODS*
Cleveland Clinic Foundation, Cleveland, Ohio

I. INTRODUCTION

During the past 5 years, a large number of nucleic acid probes have been developed for the detection of many different microorganisms. In this chapter we review the literature and present a perspective on the current state of the use of nucleic acid probes in clinical microbiology and public health laboratories.

II. GASTROENTERITIS

A. Enterotoxigenic *Escherichia coli*

Enterotoxigenic *E. coli* (ETEC) is a major cause of diarrhea in developing countries. Enterotoxin production, which is plasmid mediated, is detected by testing *E. coli* culture supernatants in tissue culture (Chinese hamster ovary or Y1 adrenal cell) assay (1) for heat-labile toxin (LT) and in suckling mouse assay (2) for heat-stable toxin (ST). Both procedures are time-consuming and costly, and neither is suitable for large-scale epidemiological studies.

Genes that code for LT and ST have been characterized, and specific fragments have been cloned and used as probes to identify ETEC. Several investigators have used ^{32}P-labeled DNA probes in a 72-hr dot-blot hybridization assay, comparing results with the standard tissue culture and suckling mouse assays. When two probes, LT and ST, were used to test either colonies or patient specimens directly (stool or rectal swab), the specificity of the hybridization assay was 100%, and the sensitivities of the LT and ST probes were 100% and 71%, respectively (3). The latter observation suggested the existence of two heterologous ST enterotoxins.

Present affiliation:
*University of Nebraska Medical Center, Omaha, Nebraska.

By using three probes, LT, ST (porcine), and ST (human), the overall sensitivity of the hybridization assay was increased to 100% in three separate studies; specificity remained 100%. Mosely et al. (4) examined colonies, canal water, and diarrheal stool or rectal swab specimens from children under 2 years of age. Seriwatana et al. (5) and Georges et al. (6) tested only colonies in cultures.

Other investigators who have tested the three ^{32}P-labeled probes have not observed complete correlation with the standard assays. A study assessing the spotting of specimens "in the field" and performing the hybridization assay in a separate laboratory resulted in major discrepancies. Only 18% of known positive samples hybridized with the probes, indicating the need for an alternative method of fixing the DNA (7). Echeverria et al. (8) tested diarrheal stool from children under 5 years of age, hand imprints from their mothers, rectal swabs from family members, water, and food. Of 221 specimens, 89 were positive by hybridization and 71 by standard assay. The authors suggested possible explanations for this disparity: (a) ETEC toxin plasmids may have been lost during storage, (b) other bacteria may have interfered with isolation of *E. coli*, (c) hybridization is more sensitive than standard assays, or (d) other bacteria may produce enterotoxin.

In a later paper Echeverria et al. (9) tested both colonies and direct stool specimens from children with watery diarrhea of $<$ 48-hr duration in Bangkok, Thailand. In addition to doing spot hybridization with the three ^{32}P-labeled probes described previously, Southern blot analysis, ^{32}P-labeled oligomer probes, and hybridization with the vector plasmid (pBR322) were performed on selected isolates. Of the 5701 *E. coli* tested, 289 produced LT, as determined in the Y1 adrenal cell assay, and 293 hybridized with the LT probe. The four discrepant isolates were shown to contain plasmids that hybridized with the cloned LT probe. These isolates also hybridized with the cloning vector but not with an LT oligonucleotide probe, which suggested nonspecific hybridization. Twenty isolates hybridized with the STp probe and were ST-positive in the suckling mouse assay. Of the 257 *E. coli* that hybridized with the STh probe, however, 184 produced ST in the suckling mouse assay. The discrepant isolates did not hybridize with the ST oligonucleotide probe but did hybridize with the cloning vector (pBR322), again indicating the nonspecificity of the original STh probe.

Recently, the ability of four biotin-labeled DNA probes to detect ETEC in pure *E. coli* cultures has been examined (10). With this method specimens must be treated with phenol-chloroform to extract proteins and RNAse before hybridization. In comparison with the standard assay procedures, the biotin-labeled probes had a specificity of 81.5%. Biotin-labeled probes were able to detect 160 pg of DNA; however, radiolabeled LT and ST probes were eight and five times more sensitive than the respective biotinylated probes.

B. *Shigella* Species and Enteroinvasive *Escherichia coli*

Shigella spp. and enteroinvasive *E. coli* (EIEC) are important causal agents of dysentery throughout the world. Virulence is due to their ability to penetrate the colonic mucosa and, as with toxin production in ETEC, it is plasmid mediated. The responsible plasmids in the four *Shigella* species and EIEC share a high degree of homology; consequently, the plasmid of one species can be cloned and used as a probe to identify enteroinvasive isolates of all *Shigella* species and *E. coli*.

Boileau and co-workers (11) prepared three ^{32}P-labeled probes by using plasmid DNA of *S. flexneri* 5. They first assessed the specificity and sensitivity of each probe by conducting hybridization assays (which took 3 days to complete) with 40 known virulent strains of *Shigella* spp. and EIEC and 40 other gram-negative bacilli. Results were compared with two virulence assays: the Sereny test, which demonstrates the ability of the bacteria to produce keratoconjunctivitis in guinea pigs, and the Hela cell assay, in which virulent strains demonstrate invasion of the cell monolayer. Each probe had a specificity of 100%; the 17-kilobase (kb) probe, however, was the most sensitive. With the 17-kb probe, hybridization assays were performed using colonies from 452 diarrheal stool specimens positive for *Shigella* spp. and 428 other gram-negative bacilli isolated from stool. Only one *S. boydii* serotype 11 failed to hybridize with the probe. It subsequently was shown to lack the plasmid necessary for penetration. There was no hybridization with the other gram-negative bacilli.

A biotinylated probe, also prepared from fragments of *S. flexneri* 5 plasmid DNA, was used by Sethabuter and associates in a 36-hr hybridization assay (12). Specimens were also hybridized with a ^{32}P-labeled probe, and results of both assays were compared with the Sereny virulence assay. Diarrheal stool specimens from children under 10 years of age were directly hybridized and also cultured. From the culture, 10 lactose-positive and all lactose-negative colonies were saved for further testing. Colonies of known EIEC and non-EIEC were also examined. Both biotin- and ^{32}P-labeled probes hybridized with 10 of 12 stool specimens positive for *Shigella* spp. and one positive for EIEC; neither hybridized with specimens negative for these pathogens. Each probe hybridized with all 52 colonies of known EIEC but with none of the 16 non-EIEC colonies. After 4-hr exposure to either dye (biotin-labeled) or x-ray film (^{32}P-labeled), each probe was able to detect 125 pg of DNA. With prolonged exposure, the sensitivity of the ^{32}P-labeled probe improved, detecting 16 pg at 24 hr. In contrast, prolonged exposure of the biotinylated probe resulted in nonspecific staining.

C. *Salmonella typhi*

Typhoid fever continues to be a serious public health problem in developing countries. Rubin and co-workers prepared a ^{32}P-labeled DNA probe that recognized the *viaB* gene locus involved in the synthesis of the Vi antigen and assessed its ability to hybridize with *S. typhi* (13). Colonies of 170 known enteric bacteria were tested. Hybridization assay results were available in 36-72 hr and were in complete agreement with conventional biochemical identification results. Strains of *S. typhi*, Vi-positive *S. paratyphi* C, and Vi-positive *Citrobacter freundii* WR 7004 demonstrated very strong hybridization, whereas there was no hybridization with strains of *E. coli, Salmonella* spp., *S. typhimurium, Shigella* spp., or other *Citrobacter* spp. The sensitivity of the assay was 10^4 Vi-expressing cells.

D. Rotavirus

Rotavirus is a major cause of sporadic and epidemic cases of acute gastroenteritis and diarrhea in infants and young children and, occasionally, in adults. The standard and most sensitive method of diagnosis is direct visualization of the virus by electron microscopy (EM). More commonly, however, diagnosis is based on immunological detection of rotaviral antigens either by enzyme-linked immunosorbent assay (ELISA), by latex agglutination (LA), or by solid-phase radioimmunoassay (RIA). A method used primarily as an epidemiological tool is electrophoretic separation of the segmented viral genomes. In addition, nucleic acid probes that recognize rotaviral RNA have recently been developed and tested with both tissue culture and clinical specimens.

Flores and associates (14) prepared ^{32}P- and ^{125}I-labeled mRNA probes from human rotavirus strains Wa and DS-1 for use in a dot hybridization assay, which required 48 to over 72 hr for completion. They tested stool suspensions directly from symptomatic and asymptomatic children, suspended material from rectal swabs, and material extracted from infected tissue culture. Although the sensitivity of the hybridization assay, as compared with ELISA, was 95%, the probe was able to detect much smaller quantities of virus than was ELISA. No hybridization occurred with samples negative for rotavirus. Weaker signals, however, were observed with heterologous strains when compared with strains homologous with the probe.

A ^{32}P-labeled complementary DNA (cDNA) probe was prepared from simian rotavirus SA11 by Dimitrov et al. (15) and used in a 36-hr dot-blot hybridization assay. Specimens tested were stools, shown to be positive for rotavirus by EM, that had been collected over a 3-year period and frozen. Analyses with ELISA and electrophoresis had also been performed on many of the samples. Compared with EM, the sensitivity of the hybridization assay was 92%; occasionally, however, a positive result was obtained only after the RNA was reextracted and the procedure repeated. The authors postulated that long-term storage and freeze-thawing may have resulted in RNA degradation. No hybridization occurred with specimens negative for rotavirus by EM; in contrast, one EM-negative specimen was positive by ELISA. As noted by Flores and associates, these investigators also found that the amount of RNA detected varied with the strain of rotavirus present in the sample. Furthermore, Dimitrov and co-workers speculated that by using a panel of probes under conditions of high stringency, it may be possible to characterize rotavirus subgroups and serotypes.

E. Enteric Adenoviruses

Included in the enteric adenoviruses are adenovirus group F or type 40 (Ad 40) and adenovirus group G or type 41 (Ad 41). These viruses are associated with gastroenteritis in infants and young children and have been implicated in outbreaks of diarrhea. Diagnosis is difficult because these viruses are fastidious with respect to in vitro growth. They can be variably isolated in Chang conjunctival cells, Graham 293 cells, and tertiary monkey kidney cells; the cytopathic effect, however, may not become apparent for several days. Diagnosis may also be made by direct visualization of adenovirus particles by EM or by adenovirus group-specific ELISA. Typing is done by restriction enzyme analysis. Nucleic acid probes have also been examined as a possible alternative diagnostic method.

Using a ^{32}P-labeled DNA probe prepared from adenovirus type 2 (Ad 2) DNA, Stalhandske and colleagues (16) directly examined 49 blind-coded stool specimens by spot hybridization. Results, obtained within 20–30 hr, were compared with RIA. Three samples positive by RIA demonstrated no hybridization; antigen concentrations in each were low, 0.02–0.2 g/ml. Quantitative differences between RIA and hybridization results were also observed. There was no hybridization with RIA-negative specimens. Discrepancies may have been a result of the limited sequence homology between Ad 2 and Ad 40 and 41.

Takiff et al. (17) prepared a ^{32}P-labeled DNA probe from a fragment of the Ad-41 genome that could detect, but not differentiate, less than 20 pg of Ad 40 or Ad 41 DNA in stool by dot-blot hybridization in 36–48 hr. Stool specimens, previously shown to be positive for adenovirus or enteric adenovirus by ELISA or EM, were tested. The probe sensitivity was 91% but could be increased by culturing the stool sample for 24 hr in Graham 293 cells before performing hybridization. The probe specificity was only 71%, with false-positive reactions occurring in stools with a high titer of adenovirus other than Ad 40 or Ad 41.

With two ^{32}P-labeled probes prepared from fragments of Ad 40 and Ad-41 DNA, Kidd and co-workers (18) were able to detect picogram quantities of enteric adenovirus DNA, distinguishing Ad 40 and Ad 41, by dot-blot hybridization in 48 hr. Stool specimens, positive for adenovirus by EM or ELISA and characterized by restriction enzyme analysis, were tested. Of the 79 specimens, 76 contained either Ad 40 or Ad 41. With no pretreatment of specimens, the sensitivity of the probes was 80%. This, however, increased to 97% when stools were first extracted with phenol–chloroform to remove interfering lipid and protein. One discrepancy in type (Ad 40 versus Ad 41) occurred between probe and enzyme analysis results. There was no hybridization with the three nonenteric adenoviruses.

III. RESPIRATORY TRACT INFECTIONS

A. *Legionella* Species

The most common clinical manifestation of infection with *Legionella* species is pneumonia, which can occur in epidemic form with significant morbidity and mortality if untreated. Increasingly, *Legionella* species are being recognized as an important cause of nosocomial pneumonias and, occasionally, they are responsible for community-acquired pneumonias. Laboratory diagnosis of *Legionalla* infection relies on culture, demonstration of the organism by direct immunofluorescence (DFA), or retrospective detection of an appropriate antibody response. Although DFA is rapid and very specific (99.9%), its sensitivity in relation to culture is low, ranging from 25 to 80%. The serological test is well standardized only for *L. pneumophila* serogroup 1 and has a sensitivity of 60-80% and specificity of 95-99%. Because the prevalence of sporadic Legionnaires' disease is only 5-10%, the positive predictive accuracy of serological testing is low and, thus, more useful in epidemiological studies. The most definitive and sensitive diagnostic method is bacterial isolation, which requires several days, followed by slide agglutination or DFA for species and serogroup identification. Legionellalike organisms, however, may give negative serological test results; therefore, gas-liquid chromatography and DNA hybridization studies may be necessary for final identification.

Nucleic acid probes have been developed for *L. pneumophila* and for the *Legionella* genus in an attempt to simplify detection and identification. Grimont and co-workers (19) prepared a ^{32}P-labeled DNA probe specific for *L. pneumophila* for use in a dot hybridization assay, which required 3-6 days for completion once colonies appeared. Hybridization occurred with colonies of ATCC strains of *L. pneumophila* but not with colonies of other *Legionella* species or with colonies of the many other gram-negative bacilli tested. Colonies from bronchial exudate and from frozen lung tissue, both of which were known to contain *L. pneumophila*, also hybridized with the probe.

An ^{125}I-labeled cDNA probe, which recognizes rRNA of the genus *Legionella*, is presently marketed by Gen-Probe, Inc. (San Diego, California). The hybridization assay is rapid, with results available in 3 hr or less. Evaluations of the product have recently been published. In two studies, colonies were tested. The type strains of all *Legionella* species and serogroups, identified on the basis of growth, biochemical, and serological tests, represented positive specimens, and a variety of nonlegionella gram-negative bacilli were the negative controls. Wilkenson et al. (20) reported a sensitivity of approximately 98% and specificity of 100%. The assay failed to detect four *L. bozemanii*, which demonstrated less than 10% hybridization, the recommended cutoff value for a positive result. Edelstein (21) used a hybridization value of 20%, relative to *L. pneumophila* strain F1772, as the cutoff for a positive value and observed no discrepancies between the legionellae and nonlegionellae tested.

Hogan and associates (22) examined 343 clinical specimens, mainly respiratory, that had been stored at −70°C for 1–8 years. When compared with culture results, the ^{125}I-labeled DNA probe marketed by Gen-Probe had a specificity of 100%, sensitivity of 74%, and an overall agreement of 93%. The authors suggested that the kit be used as a replacement for DFA in conjunction with culture.

B. Adenoviruses

Adenoviruses are a significant cause of febrile upper respiratory tract infections and may be responsible for croup, bronchiolitis, and pneumonia in children under 6 years of age. Serotypes 1, 2, and 5 (group C) and 3 and 7 (group B) are most frequently isolated. Diagnosis is based on recovery of virus in tissue culture, which may take several weeks. Viral isolates can be rapidly identified using antihexon antibody in an immunofluorescent (FA) or ELISA assay. Antihexon antibody is also used in an RIA procedure to detect adenovirus directly in clinical specimens. In addition to RIA, nucleic acid probes have been examined as a possible means of identifying adenovirus directly in clinical specimens.

Two groups of investigators have used sandwich hybridization to detect adenovirus in specimens of nasopharyngeal mucus that were obtained from children with acute respiratory tract infections. In each study ^{125}I-labeled DNA probes were prepared from DNA fragments of Ad 2 and Ad 3. The assay was complete in 20 hr and was able to detect 5×10^6 molecules of adenovirus DNA. Ranki et al. (23) performed both sandwich hybridization and tissue culture on 10 specimens previously tested by RIA. Five of the 10 samples were positive by RIA. Either adenovirus 1 (Ad 1) or 2 was isolated from all five specimens, and each hybridized with the Ad 2 probe. No hybridization occurred with the remaining five specimens; Ad 1, however, was isolated from two of the samples.

Virtanen and co-workers (24) tested 48 specimens that had been previously screened for seven respiratory viruses and blindly coded. Results of sandwich hybridization were compared with RIA detection of hexon antigen, and when discrepancies occurred, tissue culture was done. Of the 25 RIA-positive specimens, 22 were positive by hybridization (19 with Ad 2 and 3 with Ad 3). Adenoviruses (two Ad 1 and one Ad 2) were cultured from each of the three specimens with disparate results. No hybridization occurred with the 23 RIA-negative specimens.

C. Cytomegalovirus

Infection with cytomegalovirus (CMV) is common in the general population. Individuals are usually asymptomatic, but they may present with a mild febrile or mononucleosislike illness. Serious manifestations, however, may develop in two groups of patients: congenitally infected newborn infants and immunocompromised patients. Up to 20% of allogenic bone marrow transplant recipients succumb to CMV pneumonitis. Cytomegaloviral pneumonitis is also a significant cause of morbidity and mortality in renal transplant patients and in individuals with acquired immunodeficiency syndrome (AIDS). Tissue culture of specimens obtained during open-lung biopsy is the most sensitive method to diagnose CMV pneumonitis. Its main disadvantage is that CMV isolation may take as long as 6 weeks. More rapid diagnostic methods are needed. Histological identification of distinctive intranuclear and intracytoplasmic inclusions may provide a rapid diagnosis; the sensitivity, however, is unacceptably low. Other rapid means to identify CMV include EM, which does not distinguish among the herpesviruses, and FA staining of frozen tissue sections. Nucleic acid probes may also prove to be a useful diagnostic tool.

A biotin-labeled CMV DNA probe was used in a 24-hr in situ hybridization procedure to detect CMV in paraffin-embedded sections of 29 consecutive specimens obtained by open-lung biopsy (25). Results of hybridization were compared with both tissue culture and antigen detection in frozen sections by FA monoclonal antibody. Fifteen specimens were negative by all three methods; one of the 14 specimens positive by culture and immunofluorescence, however, was negative by hybridization. In situ hybridization was also used by Myerson and colleagues to study the extent of disseminated CMV infection in two autopsy cases (26).

Procedures less invasive than open-lung biopsy may be useful in the diagnosis of CMV infection. For example, isolation of CMV from buffy coat cultures nearly always correlates with active or impending disease; however, viral isolation is not only time-consuming but also buffy coat specimens are frequently toxic to the cell monolayer used for culture. In an attempt to circumvent these problems, Spector and associates (27) developed a 48-hr hybridization assay for the detection of human CMV DNA in clinical specimens. Two ^{32}P-labeled DNA probes, prepared from cloned fragments of human CMV strain AD 169, were able to detect 10 pg of CMV DNA. Sixty-seven buffy coat specimens from 11 bone marrow transplant recipients were prospectively assessed by DNA–DNA hybridization and tissue culture. Cytomegalovirus was isolated from 14 specimens; hybridization occurred with 13 of these (93%). Of the 53 culture-negative specimens, 32 (60%) were negative by hybridization. Twenty of the 21 culture-negative, hybridization-positive specimens were from patients with documented active CMV infection. The authors, therefore, postulated that culture results were falsely negative and that hybridization was the more sensitive technique.

The complete genome of CMV AD 169 was tritium-labeled and used as a probe by Martin et al. (28) to detect CMV DNA in a buffy coat specimen by reassociation kinetics analysis. The reaction time of the procedure was 4 hr, with results available 36 hr after blood was collected. Cytomegalovirus was also observed in white blood cells by EM and was recovered from the buffy coat culture after 18 days.

Although recovery of CMV from urine is not always indicative of virus-induced disease, a positive result may provide useful information in the appropriate clinical setting. Chou and Merigan (29) prepared a ^{32}P-labeled probe from a fragment of CMV AD 169 DNA for use in a 24-hr DNA spot hybridization assay. Results from 96 urine specimens were compared with standard tissue culture. All 39 culture-positive urines hybridized with the probe, and there appeared to be good correlation between infectivity and intensity of the hybridization signal. The sensitivity limit of the assay was 5 pg, or 10^4 molecules of CMV DNA. No hybridization occurred with the culture-negative specimens.

Spector et al. (27) assessed the ability of their ^{32}P-labeled probes to detect CMV DNA in 50 clinical urine samples, in addition to buffy coat specimens. Results, available in approximately 48 hr, were compared with tissue culture. Of the 24 culture-positive urine specimens, 22 (92%) hybridized with the probes. There were three "false-positive" results with the probes, giving a specificity of 88%.

Sandwich hybridization was used by Virtanen and associates (30) to detect CMV DNA in urine collected from four infants with recent CMV-positive urine cultures. Iodine-125-labeled DNA probes, prepared from the genome of CMV AD 169, were utilized in the 24-hr assay procedure. Hybridization occurred in all four specimens. Although the sensitivity of this assay, 4×10^6 DNA molecules, was lower than that of spot hybridization, an advantage of using sandwich hybridization is that crude samples, containing bacteria, cell lysates, or secretions, can be studied without purification.

Buffone and co-workers (31) developed a biotinylated DNA hybridization assay, which they used for the rapid detection of CMV in urine specimens obtained from hospitalized patients with various medical problems. For the probe the authors used cloned *Bam*HI fragment E of the Towne stain of CMV. The CMV-DNA probe was biotinylated and detected with biotinylated alkaline phosphatase. The procedure, including digestion and extraction of DNA from urine, required approximately 36 hr. The sensitivity of the probe was evaluated by analyzing urine samples to which known quantities of CMV grown in tissue culture had been added. These studies showed an increase in the color intensity with increasing viral titer. The authors also compared the biotinylated probe with the same probe labeled with ^{32}P. The sensitivity of the two probes was similar if autoradiography was carried out for the same time as the enzyme reaction; however, the ^{32}P-labeled probe was approximately 1 log more sensitive with a prolonged exposure time. Eighty urine samples were studied by tissue culture, frozen at −70°C for an unspecified period of time, and later analyzed by DNA hybridization with the biotinylated probe. Cytomegalovirus was isolated from 51 of the specimens, and 47 of these were positive by hybridization. One tissue culture-negative specimen reacted with the probe. When compared with tissue culture, the nonisotopic DNA hybridization assay had a sensitivity of 92.2% and a positive predictive value of 97.9%.

D. *Mycoplasma pneumoniae*

Mycoplasma pneumoniae is a major cause of atypical pneumonia, primarily affecting young adults. Diagnosis is often made either clinically or serologically, because culture can take up to 21 days. In an attempt to rapidly identify *M. pneumoniae*, Shaw et al. (32) (Gen-Probe, San Diego, California) developed an ^{125}I-labeled cDNA probe homologous to *M. pneumoniae* rRNA. The assay was performed in solution with results available in 3 hr or less. Hogan et al. (22) examined 388 throat swab specimens and found the assay specific for *M. pneumoniae* with a sensitivity of 100% compared with culture.

E. *Mycobacterium* Species

Although the incidence of tuberculosis in the United States has declined over the past several years, *M. tuberculosis* remains a significant pulmonary pathogen. Of the mycobacteria other than *M. tuberculosis, M. avium-intracellulare* (MAI) is the most frequent isolate, and in centers where there is a large AIDS population, may equal or exceed those of *M. tuberculosis*. As identification of these organisms is presently both time-consuming and labor-intensive, rapid diagnostic methods would be beneficial. Gen-Probe has developed an ^{125}I-labeled cDNA probe, homologous to rRNA of the genus *Mycobacterium*, which detects all 28 clinically significant mycobacterial species but does not hybridize to other bacterial or mammalian cell rRNA (33). This product has been marketed by Gen-Probe.

Shoemaker and associates (34) prepared a ^{32}P-labeled DNA probe from whole-chromosomal DNA of *M. tuberculosis*, which they used to test known isolates in a 7-day hybridization assay. The probe could detect as little as 10^4 μg of *M. tuberculosis* DNA, or 2×10^4 genomes. No hybridization occurred with a wide variety of nonmycobacterial organisms. Cross-hybridization, however, was observed, although at a lower intensity, with known isolates of MAI, *M. kansasii*, and *M. marinum*. Direct patient specimens were not examined.

Roberts and Coyle (35) prepared ^{32}P-labeled DNA probes from whole-chromosomal preparations of several mycobacteria: *M. bovis* was used for identification of *M. tuberculosis* and three probes were prepared from reference strains of *M. avium, M. intracellulare*, and *M. scrofulaceum* to represent the *M. avium* complex. Stock cultures of 71 clinical isolates were tested by spot hybridization using the four probes, and 67 were correctly identified. Three *M. tuberculosis* and one MAI failed to hybridize. Eight cultures of mycobacteria in BACTEC (Johnston Laboratories, Cockeysville, Maryland) bottles were also subjected to probe analysis; seven reacted correctly. There was no hybridization with one MAI isolate; the authors speculated that this may have been due to an insufficient amount of organism present.

IV. SEXUALLY TRANSMITTED DISEASES

A. *Chlamydia trachomatis*

In the United States, infections with *C. trachomatis* have become the most prevalent of the sexually transmitted diseases. Approximately 3-4 million Americans, including men, women, and infants, are affected each year. Serotypes D through K are associated with sexually transmitted infections (cervicitis, urethritis, neonatal conjunctivitis, and pneumonia); serotypes A, B, and C are the causative agents of hyperendemic trachoma; and lymphogranuloma venereum is a result of infection with serotypes L1, L2, or L3.

The standard method for diagnosis of chlamlydial infections is tissue culture. McCoy cells are inoculated, incubated for 48-72 hr, and then stained either with iodine to detect glycogen-positive inclusions or with fluorescent antibody, which may allow earlier detection of inclusions. Recently developed rapid methods of identification, direct fluorescent monoclonal antibody staining of the specimen, and antigen detection by enzyme immunoassay, provide results in 1-4 hr. The use of nucleic acid probes has also been explored as a means of rapid detection of *Chlamydia* species.

Palva et al. (36) prepared an ^{125}I-labeled DNA probe from DNA fragments of *C. trachomatis* serotype L2 that, in a sandwich hybridization assay, was able to detect 10^6 DNA molecules of serotypes A, C, E, G, H, I, and L2 but did not hybridize with DNA from 41 unrelated organisms commonly encountered in the genitourinary tract. The authors examined 16 endocervical or urethral specimens that had been stored for an unspecified period of time at $-20°C$. Results of the hybridization assay, available in 20 hr, were in complete agreement with culture results, which were finalized at 72 hr. Eight specimens were both culture- and hybridization-positive, and eight were negative by both methods.

Horn and associates (37) prepared an ^{35}S-labeled DNA probe from the 7-kb cryptic plasmid that is harbored by all *C. trachomatis* serovars. With the use of this probe, they performed in situ hybridization (48 hr) on 46 routine Papanicolaou smears obtained from 31 women attending the Johns Hopkins Fertility Control Center. Relative to culture, the sensitivity and specificity of the hybridization assay were 91 and 80%, respectively. The positive predictive value was 79%, and the negative predictive value was 91%. To assess the specificity of the hybridization reaction and decrease the possibility that a positive hybridization reaction was due to the vector (pBR322), duplicate smears were hybridized with a (pBR322) probe. With the (pBR322) probe, only one smear was positive; this specimen was negative by both culture and hybridization with the chlamydial probe.

B. *Neisseria gonorrhoeae*

Neisseria gonorrhoeae remains a significant cause of sexually transmitted disease worldwide. Diagnosis of *N. gonorrhoeae* infection is made by culture isolation of the organism, usually on a selective medium, followed by species identification, which is based on results of carbohydrate utilization reactions. Fluorescent antibody staining or coagglutination methods are alternative means to identify isolated colonies. From genital sites, a presumptive diagnosis of *N. gonorrhoeae* can be made on the basis of colony morphology on a selective medium, Gram stain morphology, and oxidase reaction. Recently, the limulus amebocyte lysate and Gonozyme (Abbott Laboratories, North Chicago, Illinois) assay tests have been developed for rapid direct identification of *N. gonorrhoeae* in patient specimens. Use of a nucleic acid probe, homologous to the 2.6 megadalton (Md) cryptic plasmid present in most *N. gonorrhoeae*, has also been assessed as a possible means of detection of gonococci in clinical specimens.

Totten et al. (38) prepared a ^{32}P-labeled DNA probe from the cryptic plasmid found in *N. gonorrhoeae* strain NRL 8038 which was able to detect 10^2 colony forming units (cfu) of *N. gonorrhoeae* and 0.1 pg of cryptic plasmid DNA in a 5-day spot hybridization assay procedure. The specificity of the probe was determined by testing several gonococcal and nongonococcal strains. Hybridization occurred with all 20 strains of *N. gonorrhoeae* but not with any of the 57 *N. meningitidis*, 20 *Haemophilus* species, or 10 *Branhamella catarrhalis* strains tested. Of the 44 commensal *Neisseria* spp., one *N. mucosa* and one *N. cinerea* hybridized with the probe. Specimens examined were urethral swabs obtained from 130 men attending a sexually transmitted diseases (STD) clinic; a second swab was simultaneously submitted for culture. Species identification of isolated colonies was determined by carbohydrate utilization. When compared with culture, the hybridization assay, using a 3-day x-ray exposure time, had a sensitivity of 89% (63 of 71 culture-positive specimens detected) and specificity of 100% (42 of 42 specimens). Five of the probe-negative, culture-positive *N. gonorrhoeae* lacked the cryptic plasmid; these were frequently of the PCU$^-$ auxotype.

Using a ^{32}P-labeled DNA probe prepared from the cryptic plasmid of *N. gonorrhoeae* strain NRL 8038, Perine et al. (39) studied urethral exudates from 216 males, from four African and two Asian countries, who had a clinical diagnosis of gonococcal urethritis. Urethral exudate was collected on a cotton swab or wire loop, spotted onto a strip of nitrocellulose paper, and air dried. Thayer-Martin medium for *N. gonorrhoeae* isolation was inoculated with separate urethral samples and examined daily for 3 days. Isolates were presumptively identified as *N. gonorrhoeae* by colony morphology and positive oxidase reaction. The *N. gonorrhoeae* isolates from Central African Republic, Singapore, and Japan were lyophilized and mailed to Seattle, together with the nitrocellulose blots. From Zambia, Kenya, and Ethiopia *N. gonorrhoeae* subcultures and nitrocellulose blots were hand-carried to Seattle. As compared with culture results, the sensitivity and specificity of hybridization for detection of *N. gonorrhoeae* were 96% (180 of 187 specimens) and 93% (27 of 29 specimens), respectively. The authors estimated that the cost of testing each specimen by hybridization in their study, "including labor," was less than 2.00 dollars per specimen, compared with the cost of 3.50 dollars each for a *N. gonorrhoeae* culture at the Harborview STD clinic in Seattle. The individual components of each test, however, are not broken down in sufficient detail for objective evaluation. The sensitivity of the probe with specimens from cervix, rectum, or pharynx was questioned by the authors. Their concern was that these sites, as opposed to the male urethra, may harbor sufficient quantities of organisms other than *N. gonorrhoeae* that contain plasmids that would react with the probe. This question has yet to be addressed.

C. Herpes Simplex Virus

During the past several years there has been an increase in the incidence of genital herpes simplex virus (HSV) infection. Of major concern is virus transmission to neonates at the time of delivery. As a means to monitor infection and, consequently, determine the necessity of cesarean section, weekly or biweekly screening of pregnant women with a history of recurrent genital HSV with viral culture or Papanicolaou smear beginning at 32 to 36 weeks of gestation has been suggested. Although the Papanicolaou smear is a rapid means to identify HSV infection, the sensitivity is low, ranging from 26 to 75%. Viral culture is the most sensitive diagnostic technique; however, results may not be available for 2-4 days. Rapid methods that have been examined as possible alternative means of diagnosis include the indirect immunofluorescent assay for detecting HSV antigen and the use of DNA probes to detect HSV DNA.

In 1982, Stalhandske and Pettersson (40) demonstrated that HSV DNA could be detected by membrane filter hybridization. Tissue cultures infected with HSV type 1 (HSV-1) strain C42 and HSV type 2 (HSV-2) C168 were hybridized with a ^{32}P-labeled DNA probe, prepared from a fragment of HSV-1 DNA, in a 20-hr assay procedure. With the probe they were able to detect HSV DNA and, by using more stringent hybridization conditions, distinguish between types 1 and 2.

Redfield and co-workers (41) prepared a ^{32}P-labeled DNA probe from a fragment of HSV-1 DNA which they used to detect HSV DNA in clinical specimens in a 36- to 48-hr hybridization assay. Swab specimens, which collected from patients participating in an evaluation of oral acyclovir for treatment of both primary and recurrent genital HSV, were transported to the laboratory, planted in tissue culture, and then stored at $-70°C$. These frozen samples were later thawed, titrated for infectivity, and hybridized with the DNA probe. Hybridization occurred in 31 of the 40 culture-positive specimens, giving a sensitivity of 78%. Probe signal intensity was positively correlated with the infectivity titer. None of the 30 culture-negative specimens hybridized with the probe.

Fung and associates (42) compared tissue culture, antigen detection (Virgo Antigen Detection System, Electro-Nucleonics, Inc., Columbia, Maryland) by immunofluorescent assay, and DNA detection (Herpes Simplex Virus Patho-Gene Kit, Enzo Biochem, Inc., New York, New York) for identification of HSV in 243 clinical specimens from various sources (206 from genital sites). Specimens were cultured in MRC-5 cells for 2 weeks; isolates were typed on request using MICRO TRAK HSV typing reagents (Syva Co., Palo Alto, California). Cellular pellets were prepared from the specimens and applied to a slide. Only those with two or more intact cells per high-power field (162 specimens) were considered satisfactory for HSV antigen and DNA detection. The Patho-Gene Kit probe is biotin-labeled and consists of a mixture of three sequences, two from the HSV-1 genome and one from the HSV-2 genome. Results of the hybridization procedure were available in 2-3 hr. Of the 162 specimens satisfactory for HSV antigen and DNA detection, 35 were HSV-positive by culture. Twenty-seven of the culture-positive specimens were antigen-positive; all culture-negative specimens were antigen-negative. The sensitivity of the antigen detection method was 77.1%, specificity 100%, positive predictive value 100%, negative predictive value 93.3%. Of the 35 culture-positive specimens, 25 hybridized with the DNA probe, as did 12 of the culture-negative specimens. The sensitivity of this probe in detecting HSV was 71.4%, specificity 90.6%, positive predictive value 67.6%, negative predictive value 92.0%.

The ability of three chemically synthesized, ^{32}P-labeled single-stranded oligonucleotide probes to detect HSV DNA and distinguish types 1 and 2 was assessed by Peterson et al. (43). The probes, each 22 bases in length, were homologous with unique regions of HSV-1 and HSV-2 and a region common to both. Positive HSV cultures from oral or genital clinical specimens were typed with monoclonal antibody (Syva Corporation), frozen, and later processed for hybridization and restriction enzyme digestion. Specific HSV type (1 or 2) determined by probe and monoclonal antibody agreed in 199 of 201 cases. The two isolates that hybridized with the HSV-2 probe but did not react with either monoclonal antibody demonstrated a restriction endonuclease pattern characteristic of HSV-2. Also tested were 45 primary cultures showing typical HSV cytopathic effect. Thirty-two were identified as HSV-2 by monoclonal antibody; 31 of these hybridized with the HSV-2 probe (96.8% sensitivity). Eleven of the 13 identified as HSV-1 by monoclonal antibody hybridized with the HSV-1 probe (84.6% sensitivity). Both negative specimens hybridized with the HSV-1 probe when passed and retested. Hybridization results were available in approximately 48 hr.

D. Human Immunodeficiency Virus Type I

The recently designated (44a) human immunodeficiency virus type I (HIV-I), which was formerly known as human T-cell lymphotropic (leukemic) virus type III (HTLV-III/LAV), has been shown by both serological and viral isolation studies, to be the causative agent of AIDS. The virus is usually transmitted by sexual contact, especially in homosexual males. Other groups at increased risk of acquiring the virus are intravenous drug abusers, Haitains, hemophiliacs receiving factor VIII concentrate, recipients of blood products contaminated with the virus, and children whose mothers belong to one of the above risk groups. In addition to the hallmarks of profound immunodeficiency and accompanying opportunistic infections or malignancies, AIDS is frequently complicated by a progressive encephalopathy. Most commonly, the diagnosis of AIDS is made serologically, the presence of antibody to HIV-I with confirmation by Western blot analysis, because viral culture is an extremely time-consuming and labor-intensive process. Demonstration of virus in tissue by nucleic acid hybridization is, therefore, a valuable diagnostic tool.

By using a ^{32}P-labeled DNA probe representing 2.4 kb of the HIV-I genome, Gelmann et al. (44b) analyzed the peripheral blood lymphocytes of 33 AIDS patients by Southern blot hybridization, a procedure requiring approximately 2 days. Integrated HIV-I proviral sequences were demonstrated in the lymphocytes of two of the 33 patients, both of whom had antibody to HIV-I. In contrast, none of 25 healthy homosexual males had detectable virus. Shaw et al. (45) analyzed lymph node and splenic tissue from 39 patients with AIDS and 26 with AIDS-related complex (ARC) by Southern blot hybridization using a ^{32}P-labeled cDNA probe prepared from HIV-I virions. The HIV-I sequences were detected after 7 days in seven lymph nodes and two spleens from three ARC and six AIDS patients. In most cases, the signal intensity was weak, indicating the presence of a low level of virus. By using in situ hybridization, Harper et al. (46) examined lymph node tissue and peripheral blood from patients with AIDS or ARC for the presence of HIV-I. A ^{35}S-labeled HIV-I-specific RNA probe was hybridized with mononuclear cell preparations, which were then exposed to autoradiographic emulsion for 2 days. Viral RNA was detected at a low frequency ($<$ 0.01% of the total mononuclear cells) in 6 of 7 lymph nodes (86%) and 7 of 14 peripheral blood specimens (50%).

Human immunodeficiency virus has also been demonstrated in brain tissue from AIDS patients with generalized encephalopathy. Shaw et al. (47) examined brains from 15 patients with AIDS encephalopathy by both Southern analysis and in situ hybridization for the presence of HIV-I. A ^{32}P-labeled DNA probe representing an 8.9-kb-long Sst I–Sst I insert from BH-10 was used for Southern blot hybridization, a 4–8 day procedure. For in situ hybridization, a ^{35}S-labeled RNA probe, transcribed from the pBH-10-R3 18.9-kb-long Sst I–Sst I viral insert from HIV-I clone BH-10, which recognizes viral RNA, was used. Five of 15 specimens were positive by Southern analysis; four of the five were positive by in situ hybridization. By both methods, an estimated 1–10% of cells were infected. Epstein et al. (46) performed electron microscopy on brain tissue from three AIDS patients in which HIV-I had been demonstrated by hybridization techniques and observed particles consistent with HIV-I virions within macrophages, both mononuclear and multinucleate, and in the cytoplasm of astrocytes.

E. *Treponema pallidum*

Treponema pallidum subsp. *pallidum* is the causal agent of veneral and endemic syphilis, whereas *T. pallidum* subsp. *pertenue* is the causative agent of yaws. Yaws affects children living in rural, warm, humid areas between the Tropics of Cancer and Capricorn; endemic nonvenereal syphilis occurs in older children and adults in the dry, desert areas of North Africa and Asia Minor; and venereal syphilis is seen worldwide among sexually active individuals. Late, destructive lesions of skin, bone, and cartilage occur in all three diseases; however, visceral, cardiovascular, and central nervous system lesions and congenital transmission are seen only in venereal syphilis.

To date, cultivation of the treponemes has not been accomplished. Visualization by dark-field microscopy of motile, corkscrew-shaped treponemes in exudate of fresh lesions is the most specific diagnostic test for yaws and syphilis. Correct identification of these organisms requires considerable expertise, which is a problem in many clinical microbiology laboratories. Therefore, most frequently these diseases are diagnosed serologically. Because a more specific diagnostic method is needed, both monoclonal antibodies and nucleic acid probes have been developed.

Perine et al. (49) prepared a ^{32}P-labeled DNA probe from a cloned segment (pAW305) of *T. pallidum* (Nichols strain) DNA, which could detect between 1×10^4 and 1×10^5 *T. pallidum* (Nicholos strain) and from 5×10^3 to 1×10^4 *T. pertenue* (CDC1 and CDC2, respectively) extracted from infected rabbit testes. With the probe they tested exudate from dark-field-positive lesions (16 children with yaws from Ghana and 6 adults with primary syphilis from Seattle, Washington) in a 3-day dot hybridization assay. In the adults the probe had a sensitivity of 66% (4 of 16) and in the children, 81% (13 of 16). Specificity was not determined. Because the sensitivity of the probe is dependent upon the numbers of organisms present and because the numbers of treponemes in lesions decreases with time, the authors question whether a treponemal DNA probe could detect organisms during latency or in tertiary lesions.

F. *Mobiluncus* Species

Mobiluncus spp. are motile, curved anaerobic bacteria that have been isolated only from women with bacterial vaginosis. These organisms are exceedingly difficult to culture; an average of 30 days is required for isolation, plus an additional 7 days for speciation. Roberts et al. prepared ^{32}P-labeled DNA probes from whole-chromosomal DNA of *M. curtisii* ATCC 35241 and *M. muleris* ATCC 35243 for use in a spot hybridization procedure (50). Results of the assay were available in an average of 5 days (range of 2-14 days), which is considerably shorter than the time required for isolation and speciation. They then compared Gram stain, culture, and hybridization methods for identification of *Mobiluncus* spp. in female genital specimens. Endocervical, vaginal, urethral, endometrial, urine, and amniotic fluid specimens from 1500 hospitalized women were screened by Gram stain. Curved rods were present in 60 specimens (4%); 56 were further studied by culture and hybridization. *Mobiluncus* spp. were isolated from 77%; 52% hybridized with the DNA probes. There was 100% correlation between species identification obtained with the probes and that from conventional biochemical tests. Vaginal specimens were also randomly collected from 20 college students attending the Student Health Facility and from 92 women attending the STD clinic. From the college students, one specimen was positive by both culture and hybridization; the remaining 19 were negative by all three methods. Thirty specimens from women attending the STD clinic were positive for *Mobiluncus* by any method: 27 by Gram strain, 25 by culture, and 25 by hybridization. Overall, Gram stain detected 90% of the specimens positive by any method, culture detected 77-83%, and hybridization detected 52-83%.

G. Human Papillomavirus

Human papillomaviruses (HPV) are the causative agents of various benign cutaneous and mucosal epithelial proliferations (warts) and have been implicated in the development of cervical carcinoma. Infection of the genital and anal areas with HPV is commonly seen in sexually active populations, and its prevalence appears to be increasing. To date 26 HPV genotypes have been identified, each having $<50\%$ DNA-DNA homology with previously described types. Cell culture systems for either growth of, or transformation by, HPV have not yet been developed. In some lesions, HPV can be visualized electron microscopically.

In other specimens, however, the presence of HPV must be demonstrated by HPV antigen detection or DNA hybridization.

Several investigators have used DNA hybridization in an attempt to determine if specific serotypes are associated with specific lesions. Gissman et al. prepared ^{32}P-labeled DNA probes from cloned fragments of HPV-6 and HPV-11 DNA (51). The DNA was extracted from 14 laryngeal papillomas, 63 condyloma accuminata (exophytic genital warts), and 24 in situ or invasive carcinomas, and assayed by blot hybridization. The HPV-11 DNA was detected in 7 laryngeal papillomas, 13 condyloma accuminata, and 4 in situ or invasive carcinomas, and HPV-6 was found in 41 condyloma accuminata. Durst and co-workers (52) performed spot hybridization on DNA extracted from a variety of genital malignancies by using a ^{32}P-labeled DNA probe prepared from HPV-16 DNA. HPV-16 was detected in 11 of 18 (61.1%) cervical carcinomas from Germany, 8 of 23 (34.8%) cervical carcinomas from Kenya and Brazil, 2 of 7 (28.6%) vulvar carcinomas, 1 of 4 (25%) penile carcinomas, and 2 of 33 condyloma accuminata, both of which were also positive for HPV-6 or HPV-11.

DNA, extracted from colposcopy-directed cervical biopsies from 22 patients referred because of abnormal Papanicolaou smears, was hybridized by Lancaster et al. (53) with a ^{32}P-labeled DNA probe prepared from bovine papillomavirus type 1 (BPV-1) in a 48-hr assay. Specimens were also analyzed for the presence of structural antigens with antiserum against the papillomavirus genus-specific antigens (Dako Corp., Santa Barbara, California) by using a peroxidase-antiperoxidase technique. Selected specimens were also hybridized with ^{32}P-labeled DNA prepared from HPV-11. Of the 11 mild dysplasias, 5 were both BPV-1- and antigen-positive, 3 were only antigen-positive, 1 was only BPV-1-positive, and 2 were negative. There were three cases of moderate dysplasia, all antigen- and BPV-1-positive. The one case of severe dysplasia was BPV-1-positive, as were two of seven specimens demonstrating squamous metaplasia. Tissue from three placentas and six carcinomas were negative. Hybridization with HPV-11 occurred with 1 of 10 dysplasias, 2 of 4 squamous metaplasias, and one of 6 carcinomas.

Beckmann and associates (54) used biotin-labeled DNA probes prepared from HPV-1, -6, -11, and -16 in an 18-hr in situ hybridization procedure to examine tissue from eight condyloma accuminata, one plantar wart, and two laryngeal papillomas. Specimens were also examined for HPV capsid antigen with the avidin-biotin peroxidase complex method. HPV-6 was detected in four vulvar and two anal condylomas, HPV-16 in one perianal condyloma, HPV-1 in the plantar wart, and HPV-11 in one laryngeal papilloma. Capsid antigen was present in all but two of these same specimens.

Cervical scrapings from 83 patients were analyzed by Wagner and co-workers (55) in a 1- to five-day spot hybridization assay using ^{32}P-labeled DNA probes prepared from cloned DNA fragments of HPV-6, -11, and a mixture of HPV-16 and -18 (HPV-16/18). Scrapings were also examined cytologically. Severe dysplasia to carcinoma in situ (CIN III) was diagnosed in 22 cases; 15 of these hybridized with HPV-16/18, 4 with HPV-11, and 2 showed no hybridization. Of the 13 cases of mild to moderate dysplasia, 4 were HPV-16/18 positive, 4 HPV-11 positive, and 3 were negative with all probes. Twelve specimens showed only minor changes; 4 hybridized with HPV-16/18, 4 with HPV-11, and 2 with all three probes, and 2 did not hybridize with the three probes used. Thirty-six normal controls were also analyzed; four hybridized with HPV-11, the remainder were negative.

Wickenden et al. (56) analyzed cervical scrapings with a ^{32}P-labeled DNA probe prepared from HPV-6 in a spot hybridization assay that required 72-168 hr for completion. Two specimens were obtained from each of 78 women; one was examined cytologically, and the other was stored at $-170°C$. Hybridization occurred in 7% of the specimens from patients with normal colposcopic and cytological examination results and in 12.5% of specimens from women with abnormal cervical examination findings, cytologically or colposcopically.

V. MENINGITIS AND ENCEPHALITIS

A. Enteroviruses

Enteroviruses are the most common causative agents of aseptic meningitis and are a significant cause of encephalitis. Clinical illness, which occurs most frequently in infants and young children and in temperate climates, is seen mainly in the summer and fall. Outbreaks occur yearly and are associated with appreciable morbidity. Diagnosis of enterovirus infections is dependent upon tissue culture isolation. Although nearly 70% of positive cultures can be detected within 4 days of specimen inoculation, the cytopathic effect produced by some serotypes may not become apparent for several weeks. For primary isolation of group A coxsackieviruses the method of choice is inoculation into suckling mice, a procedure not available in most laboratories. In addition, serological diagnosis is both time-consuming and expensive because of the multitude of existing serotypes. As a rapid and accurate diagnosis is often critical for appropriate patient care in this clinical setting, an alternative diagnostic method is needed.

Hyypia et al. (57) prepared a ^{32}P-labeled cDNA probe from the genome of coxsackievirus B3 (strain Nancy), which they used in a 2-day spot hybridization assay. To evaluate the probe specificity, viruses grown in tissue cultures were tested initially. Coxsackieviruses A9, B2, B3, B4, and echovirus 17 hybridized strongly with the probe, a weaker signal was seen with poliovirus type 3, and no hybridization occurred with HSV-1, Ad 2, measles virus, or Nebraska calf diarrhea virus. Enteroviruses isolated from 10 stool specimens were examined, and of the 8 culture-positive specimens, 7 hybridized with the probe. The two specimens negative by culture were also negative in the hybridization assay. Finally,

16 direct stool specimens were tested. Enterovirus was isolated from eight specimens, one of which also yielded a positive hybridization signal. No hybridization occurred with the eight specimens from which no virus was isolated.

Three different ^{32}P-labeled cDNA probes, prepared from clones PDS 111, PDS 14, and BAM of poliovirus type 1 (Mahoney strain), were assessed by Rotbart et al. in a 2-day hybridization assay (58). American Type Culture Collection (ATCC) strains of poliovirus types 1, 2, and 3, coxsackieviruses A9 and B1, and echovirus 11 and a clinical isolate of respiratory syncytial virus were grown in tissue culture, frozen at -70°C, and later tested. All enterovirus strains hybridized with probes PDS 14 and PDS 111, whereas only the polioviruses hybridized with the BAM probe. No hybridization occurred with respiratory synctial virus or uninfected controls. The authors also examined the effect of the antiviral drugs arildone and WIN 51711 on the assay. When tissue culture cells were treated with arildone before virus infection and then analyzed by hybridization, no signal occurred with the polioviruses, but there was no effect on remaining enteroviruses tested. Treatment with WIN 51711, on the other hand, resulted in a negative hybridization assay with all enteroviruses assessed.

In a later study, Rotbart and associates (59) conducted reconstruction experiments by adding enteroviruses to cerebrospinal fluid (CSF), which was then analyzed in a 2-day hybridization assay with six different ^{32}P-labeled cDNA probes: pDS 111 and pDS 14 prepared from poliovirus type 1 and pCB 111 51, 29, 33, and 35 prepared from coxsackievirus B3. The goals of the study were to determine the optimal handling and processing conditions for CSF specimens and to evaluate how the presence of an inflammatory response would affect viral RNA detection. The experimental variables tested were temperature (room temperature, 4°C, -20°C, -70°C), length of storage (up to 96 hr), and the addition of white blood cells, red blood cells, and protein (albumin and immunoglobulin) to normal CSF obtained from neurosurgical patients and other patients with noninflammatory illnesses. Of these, only high levels of immunoglobulin had an adverse effect on virus RNA detection; this, however, was corrected with proteinase K treatment of the specimen. The ability of the separate probes to detect the different enteroviruses tested varied. Enteroviruses included in the experiment were ATCC strains of poliovirus types 1, 2, and 3; coxsackieviruses A9, A16, B1, and B6; echoviruses 2, 4, 6, 11, and 22; and clinical isolates of HSV and respiratory syncytial virus, which had all been grown in tissue culture. With the two poliovirus-derived probes, the strongest signals were seen with poliovirus types 1 and 3 (the preparation of poliovirus type 2 was too dilute for detection), echoviruses 2 and 4, and coxsackievirus A9 and A16; echovirus 11 and coxsackievirus B1 were detected only after prolonged autoradiograph exposure. A spectrum of hybridization patterns resulted from the coxsackievirus B3-derived probes. In general, stronger signals were seen when these probes, rather than the poliovirus-derived probes, were tested against the echoviruses, coxsackievirus B, and coxsackievirus A9. By using a combination of probes, such as pDS 14 and pCB111 35 or pCB111 51 and pCB111 35, all enteroviruses tested, with the exception of echovirus 22, could be detected.

B. Herpes Simplex Virus

Herpes simplex virus is the most common cause of sporadic encephalitis in the United States. The age distribution is biphasic, with peaks occurring in the 5-30 year and > 50 year ranges. Most cases are due to HSV-1. An unequivocal diagnosis can be made only by brain biopsy with demonstration of the virus. This is extremely important because antiviral chemotherapy does reduce mortality and may decrease morbidity. Herpes simplex virus may also produce aseptic meningitis. This is most commonly due to HSV-2 and is seen in association with primary genital HSV infection. In contrast to HSV encephalitis, HSV meningitis is an acute, self-limited disease that generally resolves in 2-7 days. The most sensitive means to diagnose HSV is tissue culture isolation, which may not become positive for 2-4 days after inoculation of the specimen. Because a rapid diagnosis is imperative in cases of HSV encephalitis, alternative means of viral detection have been explored. The histological demonstration of intranuclear inclusions, typical of HSV, in tissue sections has an unacceptably low sensitivity. Although the sensitivity of direct IF is higher than that of routine histology, a significant number of cases would go undetected if this method were used alone. The use of nucleic acid probes, therefore, has been examined as a method for diagnosis of HSV encephalitis.

Forghani et al. (60) assessed the ability of the HSV DNA Hybridization Kit (Enzo Biochem, Inc., New York) to detect HSV DNA in 30 autopsy or brain biopsy specimens (17 positive and 13 negative) that had been stored at $-70°C$ for 10-20 years. A biotin-labeled DNA probe consisting of a mixture of three fragments, two from HSV-1 and one from HSV-2, was used in a 2-hour in situ hybridization procedure. Direct FA was repeated at the time of hybridization. Of the 17 originally positive specimens, 16 were positive by hybridization and 11 by FA. All 13 culture-negative specimens were negative in both the hybridization and immunofluorescent assays.

C. Subacute Sclerosing Panencephalitis

Subacute sclerosing panencephalitis (SSPE) is a slowly progressive disease of the central nervous system that occurs in children and young adults and is caused by a persistent infection with measles virus. It is diagnosed by demonstrating high levels of antibodies to measles virus in the serum and cerebrospinal fluid of affected patients. Although measles virus has been isolated from brain tissue, the procedure is very time-consuming and requires cocultivation of brain cells with other cultured cell lines. Measles virus may also be detected in brain tissue by in situ hybridization. By using a virus-specific, tritium-labeled cDNA probe prepared by reverse transcription of full-length measles RNA (Edmonston strain), Haase et al. (61) were able to detect small amounts of measles virus nucleotide sequences in individual cells and in neurophils of four cases of SSPE in which viral antigens were not detectable by FA. This procedure, however, is also time-consuming because in some cases hybridized sections required 2 weeks of autoradiographic exposure.

VI. CUTANEOUS DISEASE

A. Varicella-Zoster Virus

Infections with varicella-zoster virus (VZV) are manifest as varicella (chickenpox) in children and as zoster (shingles) in adults. Because the clinical presentation is classic in most cases, diagnosis is frequently made without virus isolation. In certain situations, however, a specific laboratory diagnosis of VZV is critical. A diagnosis of VZV infection in immunocompromised patients or their contacts is the basis for administration of antiviral agents or immune globulin. Visceral infections, such as varicella pneumonia or encephalitis, are diagnosed only by virus isolation. In addition, it may be necessary to differentiate the vesicular eruptions of VZV from those caused by certain enteroviruses, HSV, bacteria, or hypersensitivity reactions.

Diagnosis of VZV infection is usually made by virus isolation in tissue culture. The typical cytopathic effect generally becomes apparent 4-7 days after specimen inoculation, with a range of 3-14 days. In addition to the problem of time required for isolation of VZV, infected cultures are not always positive because of the labile nature of infectious virus. A more rapid and more sensitive method of diagnosis is immunofluorescence microscopy, which can be performed directly on scrapings from the base of fresh vesicular lesions or from frozen sections of punch biopsies from prevesicular skin lesions. Dot-blot hybridization has also been explored as a means to rapidly detect VZV in clinical specimens.

Seidlin et al. (62) were able to detect 10 pg-10 ng of VZV DNA from clinical specimens in 36-48 hr by using a ^{32}P-labeled DNA probe prepared from fragments of VZV DNA in a spot hybridization assay. Two separate experiments were performed: 49 specimens that had been stored at −70°C for 2-5 years were tested retrospectively by hybridization, and 32 specimens were analyzed prospectively by both hybridization and culture. In the retrospective study, 14 specimens had been culture-positive; 13 of these were positive by hybridization. Hybridization was also observed in two specimens from patients who were thought to have VZV infection clinically but from whom VZV was not recovered. Two false-positive hybridization signals occurred with the remaining culture-negative samples; both were from patients with HSV infection. Of the specimens analyzed prospectively, eight were both culture- and hybridization-positive. Hybridization occurred in specimens from five patients with clinical VZV infection from whom virus was not isolated. A positive signal was also observed in two of four specimens from one patient with an undiagnosed vesicular exanthem. The sensitivity of tissue culture for detection of VZV was 58%, compared with 76% for the hybridization assay ($p = .14$). The specificity of cell culture was 100%, whereas that of hybridization was 94% ($p = .49$).

B. *Leishmania* Species

Infection with *L. tropica, L. mexicana*, or *L. braziliensis* produces a cutaneous lesion at the site of inoculation, which first appears as a macule but becomes ulcerative with time. In addition to the cutaneous lesion, *L. braziliensis* inevitably involves the mucous membranes of the nose and pharynx. These species are endemic in South America; *L. tropica* and *L. mexicana* are also seen in Asia, the Middle East, India, China, Central Africa, the Mediterranean basin, and Mexico. Recently, members of the *L. braziliensis* complex have been identified in Texas. Because facial lesions may become extensive, resulting in serious disfigurement unless appropriately treated, accurate diagnosis is important. Current techniques for isolation and cultivation of the leishmania organism are technically difficult, time-consuming, and often unsuccessful. Serological tests are available, but interpretation of results may be difficult because of cross-reaction in various diseases.

Wirth and Pratt examined the use of nucleic acid hybridization as a means to rapidly and accurately diagnose human leishmaniasis directly from infected tissue (63). DNA probes labeled with ^{32}P were prepared from isolated kinetoplast DNA of *L. braziliensis, L. mexicana*, and *L. tropica*. These probes were tested against stock isolates of *L. tropica* (three strains), *L. mexicana* (three strains), and *L. braziliensis* (five strains) in a spot hybridization assay that required a minimum of 2-3 days for completion. The *L. braziliensis* probe hybridized only with strains of *L. braziliensis*. Cross-hybridization of approximately 1% occurred between strains of *L. tropica* and *L. mexicana*. Either mice or hamsters were infected by subcutaneous injections of promastigotes of *Leishmania* spp.; lesions were excised, and a touch blot was analyzed by hybridization. The *L. mexicana* probe hybridized with blots prepared from lesions caused by *L. mexicana* but not *L. braziliensis*. A field study of seven patients was also conducted. Four specimens hybridized with the probe specific for *L. braziliensis* and three with the probe specific for *L. mexicana*.

VII. HEPATITIS

A. Hepatitis A Virus

Hepatitis A virus (HAV) grows poorly in cell culture; consequently, primary isolation of the virus is slow and unpredictable. Although HAV particles can be detected in fecal specimens by solid-phase RIA, a clinical diagnosis of acute HAV is most frequently made by demonstrating virus-specific immunoglobulin M antibodies in serum.

Jansen et al. (64) recently combined solid-phase immunoabsorption, that utilized a monoclonal anti-HAV capture antibody, with spot hybridization as a means to detect HAV in fecal specimens. By performing the immunoabsorption before hybridization, they eliminated the need to extract or digest interfering nucleic acids and proteins found in crude fecal suspensions that can result in false-positive hybridization signals. By use of a ^{32}P-labeled cDNA probe, prepared from RNA of HAV strain HM-175, Jansen and co-workers were able to detect as little as $1\text{-}2 \times 10^5$ HAV genome copies in a minimum of 3 days. All fecal samples from New World owl monkeys and human volunteers, experimentally infected with HAV, were positive in the assay. Fecal specimens collected during the first week of illness from 23 men with acute hepatitis A were tested by immunoaffinity hybridization and RIA. Thirteen specimens were reproducibly positive for HAV RNA by hybridization; two of these were negative by RIA. The remaining specimens were nonreactive by both procedures.

B. Hepatitis B Virus

Hepatitis B virus (HBV), the prototype agent of the family *Hepadnaviridae*, is a complex double-shelled particle with a diameter of 42 nm consisting of a 27-nm core surrounded by a 7- to 8-nm viral protein coat. Hepatitis B surface antigen (HBsAg) is the complex antigen found on the surface of HBV; it may also be detected on smaller particles and filamentous forms unassociated with the hepatitis B virion. Within the 27-nm core are the hepatitis B core antigen (HBcAg), a small, circular partially double-stranded DNA molecule, and a specific DNA polymerase activity, which has been used as a marker of infectivity. The hepatitis B e antigen (HBeAg) is postulated to be an integral component of the HBV core, and it too has been associated with infectivity. At present, HBV cannot be grown in cell culture; diagnosis is, therefore, made serologically.

Several groups have investigated the use of DNA hybridization for detection of HBV DNA in serum and other tissues. Berninger and co-workers (65) prepared a ^{32}P-labeled DNA probe from recombinant plasmids containing the complete HBV genome. With this probe they tested plasma obtained from known chronic HBsAg carriers and plasma previously titered for infectivity in susceptible chimpanzees in a 2-day spot hybridization assay. Tests for HBsAg, HBcAg,

HBeAg, and DNA polymerase activity were also performed. The HBV infectious dose 50% per milliliter (ID_{50}) in chimpanzees for the five sera studied corresponded to 4×10^7 to 4×10^8 HBV DNA molecules per milliliter. With the probe, however, 1×10^6 DNA-containing Dane particles per milliliter of serum could reliably be detected. In addition, the concentration of HBV DNA molecules correlated well with the titer of infectious HBV and could, therefore, be used as a quantitative marker of infectivity of the serum. Results of the hybridization assay did show some quantitative correlation to DNA polymerase activity; it was, however, more sensitive. There was no quantitative relation between either HBsAg or HBeAg and HBV DNA detected in the hybridization assay.

Scotto et al. (66) assayed HBV DNA in serum by using a ^{32}P-labeled DNA probe by either spot hybridization (181 samples) or Southern blot hybridization. Although the spot hybridization procedure was technically easier, both were potentially time-consuming because two sets of autoradiographs were exposed for 24 hr and 10 days, respectively, before development. Sera were also tested for HBsAg, HBeAg, anti-HBs, anti-HBe, and in select specimens, DNA polymerase activity was also measured. In 96 samples HBV, DNA was detected. Seventy-two specimens were positive for HBV DNA, HABsAg, and HBeAg. Fourteen sera, from eight patients, were HBsAg- and HBeAg-positive, but gave negative or variable HBV DNA results. Serological follow-up was available for seven of these eight patients: five seroconverted to anti-HBe, and four of these became HBeAg negative. Eight sera were positive for HBV DNA and HBsAg but lacked HBeAg; seven contained anti-HBe. The DNA of HBV was demonstrated in six HBsAg-negative sera, suggesting that these sera may have been infective; five of these specimens contained HBV antibodies. Of the 26 sera tested for DNA polymerase activity, three were negative for HBsAg, HBeAg, HBV DNA, and DNA polymerase, and six were positive for all four markers. All remaining 17 specimens were HBsAg positive. Five of the 17 were negative for HBeAg, HBV DNA, and DNA polymerase; 10 were HBeAg and HBV DNA-positive but DNA polymerase-negative; 1 was HBeAg-positive but HBV DNA- and DNA polymerase-negative; and 1 was DNA polymerase-positive but HBeAg- and HBV DNA-negative. The authors concluded that the hybridization assay, used in conjunction with serological HBV markers, provided new information about the serological evolution in various patients and identified potentially infective sera that, by RIA results, would have been considered safe.

Lieberman et al. (67) tested sera from 61 HBsAg carriers and 22 patients with hepatocellular carcinoma for HBV DNA in a 3-7 day spot hybridization assay with a ^{32}P-labeled DNA probe. The assays for HBsAg, anti-HBs, anti-HBc, HBeAg, and anti-HBe were performed by RIA. Of the HBsAg carriers, HBV DNA was detected in 28 of 28 who were HBeAg-positive and 16 of 32 who were anti-HBe-positive. Seven of the 32 patients with hepatocellular carcinoma had detectable HBV DNA in their serum, but in lower levels than were present in

HBsAg carriers without hepatocellular carcinoma. These authors also concluded that serum HBV DNA hybridization analysis may identify potentially infective sera that would not be determined as such by RIA.

By use of a ^{3}H-labeled DNA probe, prepared from a recombinant plasmid (pBR322-HBV) called AM6, in a 24-hr spot hybridization assay, Feinman et al. demonstrated HBV DNA in a dilution of serum shown to be infectious in chimpanzees but negative for HBeAg by RIA (68). Serum from a patient with acute hepatitis B, induced hepatitis B in chimpanzees up to a dilution of 10^{-8}. Hepatitis B viral DNA was also detected up to a 10^{-8} dilution, whereas HBsAg and HBeAg were detected up to dilutions of 10^{-5} and 10^{-3}, respectively. The authors also tested sera from seven normal individuals negative for HBV markers, two patients with HBsAg-negative cryptogenic cirrhosis, and three additional patients with acute hepatitis B. Hybridization occurred in all patients with acute hepatitis B but in none of the normal controls or patients with HBsAg-negative cirrhosis.

In addition to its use for the detection of HBV DNA in serum, hybridization analysis has been used to demonstrate HBV DNA in liver tissue and white blood cells. Lie-Injo et al. (69) analyzed the DNA extracted from primary hepatocellular carcinomas of 23 patients and from white blood cells of 11 of these patients in a 10- to 14-day Southern blot hybridization assay using a ^{32}P-labeled HBV DNA probe cloned in plasmid (pBR325). Sixteen of the hepatocellular carcinomas were positive for HBV DNA and, in 15 of these, HBV DNA was integrated into host liver DNA. In five specimens free HBV DNA in addition to integrated HBV DNA was present, and only free HBV DNA was found in one specimen. In two patients HBV DNA was found in tumor and in adjacent uninvolved hepatic tissue. Hepatitis B viral DNA that was not integrated into host DNA was detected in 2 of the 11 white blood cell DNA samples.

Scotto et al. studied liver tissue of 46 children by spot hybridization using a ^{32}P-labeled cloned HBV DNA probe (70). Fifteen children had chronic hepatitis and were HBsAg and anti-HBc-positive; 24 children had liver disease, which in 11 may have been related to HBV, but were HBsAg-negative; and 7 control children had negative HBV serological markers and liver tissue that was histologically normal or demonstrated dysplasia. Thirteen of the 15 patients with chronic hepatitis were HBeAg- and serum HBV DNA-positive, free HBV DNA was detected in liver tissue of all but one, and in four cases HBV DNA was integrated into host DNA as well. None of the 24 children with HBsAg-negative liver disease had detectable serum HBV DNA; six, however, had integrated HBV DNA sequences in their hepatocyte DNA, and hepatocytes of one patient also contained free HBV DNA. In these six patients hybridization unequivocally established the cause of their disease. No hybridization occurred in specimens from the remaining patients studied.

To define which particular subset of peripheral blood mononuclear cells (PBMC) contain HBV DNA and in what form, Yoffe et al. analyzed PBMC DNA from 14 HBsAg-positive patients by Southern blot hybridization (71). A ^{32}P-labeled probe, prepared from a recombinant plasmid containing a head-to-tail dimer of the complete 3.2-kb pair HBV genome, was used in the assay, which could detect 0.05 HBV genome equivalents per cell after 9–12 days. Hepatitis B viral DNA was detected in five HBsAg-positive carriers, who were also positive for HBeAg, and in three HBeAg-negative carriers, one of whom became HBeAg-positive in follow-up. The most intense hybridization signal occurred in the monocyte population, with a lower level of hybridization detected in the B-cell fraction. The autoradiographic patterns of HBV DNA in PBMC of HBeAg-positive patients (including the patient who seroconverted) and HBeAg-negative patients differed. Fast-migrating, low-molecular-weight species of HBV DNA, consistent with replicative intermediates, were detected in the HBeAg positive group, whereas only high-molecular-weight forms, consistent with nonreplicating extrachromosomal oligomers, were detected in the HBeAg-negative carriers.

Recently, a potential problem with the use of the dot hybridization assay for detection of HBV DNA in serum was reported by Diegutis et al. (72). They encountered false-positive results using the ^{32}P-labeled pHBV CB DNA probe, which consisted of a genome-length copy of HBV subtype *adyw* still inserted in the bacterial plasmid vector (pBR322). The specimens were obtained from three adult Caucasian males who had no previous exposure to hepatitis and who were undergoing preoperative cardiac surgery assessment as part of a posttransfusion hepatitis study. In all three patients HBsAg, anti-HBs, and anti-HBe were negative, bilirubin and aminotransferase values were unremarkable, and none subsequently developed hepatitis after prolonged follow-up. The dot hybridization assay was repeated with the three index specimens using as the probe the plasmid vector (pBR322) lacking any HBV sequences, and all three produced a positive result. It was the opinion of these authors that because the same problem with false-positive results could occur with other diagnostic probes, "the possibility that some positive results might be due to trace contamination of the probe with residual vector sequences should be considered, and appropriate controls should be devised."

VIII. MISCELLANEOUS ORGANISMS OR DISEASES

A. *Bacteroides* Species

Bacteroides species are the most frequently isolated anaerobic bacteria from human clinical specimens. Of these, the *B. fragilis* group are the most commonly recovered, with the species *B. fragilis* accounting for over half of the isolates.

Identification of this group of organisms is important clinically because they are more resistant to antimicrobial agents than are other anaerobic bacteria. The *B. fragilis* group are isolated from most intra-abdominal and pelvic infections and may be recovered from other sites, including blood and abscesses of the lung and brain.

Kuritza et al. (73) prepared several DNA probes by cloning into the *E. coli* plasmid (pBR322) random fragments of chromosomal DNA from various *Bacteroides* species. These probes were labeled with ^{32}P and then tested for specificity with several species of *Bacteroides, Fusobacterium*, and aerobic gram-negative bacilli (obtained from ATCC or Virginia Polytechnic Institute Anaerobe Laboratory) in a 36- to 48-hr hybridization assay. Three probes (pBF11-4, pBF11-5, and pBF11-6) hybridized specifically to *B. fragilis* DNA, one (pBF-3) hybridized to DNA of all members of the *B. fragilis* group, and one (pBO-21) hybridized to DNA from all *Bacteroides* spp. and to DNA from *F. necrophorum* and *F. nucleatum*. The necessary concentration of *Bacteroides* species for detection was 10^6 cfu/ml. Because infections containing *Bacteroides* species are most often polymicrobial, pure cultures of *B. fragilis* and *E. coli* were mixed and then hybridized with pBF11-4. *Bacteroides fragilis* was detected when it accounted for at least 10% of the mixture. To determine if the probes could detect *Bacteroides* spp. in blood cultures, whole rabbit blood was seeded with 10^2 *B. fragilis* per milliliter and inoculated into blood culture medium. *Bacteroides fragilis* DNA was detected with ^{32}P-labeled pBF11-5 in samples removed after 24–42 hr of incubation, by which time there may have been 10^8 cfu/ml. A potential problem with the probes is cross-reactivity with (pBR322) sequences carried in some other bacterium. Although the authors believe this is unlikely because *Bacteroides* are so distant genetically from other gram-negative bacteria, the problem could be avoided by using the cloned insert alone as the probe or by including a labeled (pBR322) (without insert) probe in any screening procedure.

B. Polyomaviruses

The human polyomaviruses, JC, BK, and PML-SV40, are ubiquitous; 40–70% of healthy adults have antibody against these viruses. In general, these viruses cause disease only in immunocompromised patients. Both PML-SV40 and JC viruses have been isolated from brains of patients with progressive multifocal leukoencephalopathy (PML). The JC and BK viruses have been isolated from urine of renal transplant recipients and, in one study, patients with urinary polyomavirus excretion was associated more frequently with the development of drug-requiring diabetes mellitus, arterial occlusive disease, or uretheral stricture (74).

The polyomaviruses can be grown in tissue culture; JC virus, however, has only been grown in fetal glial cell cultures. These viruses can be detected by cytological examination of urine sediment, and they can ve visualized electron microscopically. The BK and JC viral antigens can be identified in brain and renal tissue by immunofluorescent and immunoenzymatic methods; the reagents, however, are not commercially available.

Gibson et al. (75) tested 81 urine samples from 61 patients in a 3- to 4-day spot hybridization assay using a ^{32}P-labeled DNA probe prepared from the complete genome of BK virus. All but 10 of the 61 patients had a known immunological abnormality. Urine samples were initially submitted for virus isolation and electron microscopic examination (20 from 1978–1982, 42 in 1983, and 19 in 1984), after which time they were stored at 4°C until analyzed by hybridization. Hybridization, culture, and EM were performed on 73 specimens. All three tests were positive in 18 specimens and negative in 40 specimens. In 6 of 73 cases, hybridization was the only positive test: three patients had rising serum titers to BK virus, one patient had Hodgkin's disease and neurological symptoms suggestive of PML, and one was a patient with AIDS who had high serum antibody titers to JC virus. Seven specimens were EM-positive and culture-negative. Of these seven, three were negative by hybridization: two had been stored for 4 years and, in one case, only one virus particle had been seen by EM. In two cases only culture was positive. Hybridization plus either EM or culture were performed on the eight remaining specimens. All eight were hybridization-negative, six of which were also EM- or culture-negative. Two of the eight samples were initially positive by EM; these had been stored for 6 years before hybridization.

C. Parvoviruses

Two genera known to infect humans are included in the family *Parvoviridae*. Serotypes 2 and 3 of the adenosatelloviruses naturally infect humans but have not yet been associated with disease. Cossart et al. in 1975 described a serum parvoviruslike virus, SPLV or B19, which has subsequently been associated with aplastic crisis in patients with chronic hemolytic anemias such as sickle cell disease, pyruvate kinase deficiency, and hereditary spherocytosis (76). The SPLV has been seen by EM in serum of patients with aplastic anemia and, upon recovery, the virus particles disappear. Affected patients also develop specific immunoglobulin M (IgM) antibody. Other diseases with which this virus is associated include acute febrile illnesses and erythema infectiosum.

No isolation system is presently available for SPLV. The SPLV particles can be demonstrated in serum by EM or immune EM (IEM), and antigen can be detected by counterimmunoelectrophoresis (CIE). Diagnosis can also by made serologically be demonstrating IgM-specific anti-SPLV with capture RIA.

Anderson et al. (77) used a ^{32}P-labeled DNA probe, prepared from DNA of a parvovirus obtained from plasma of a blood donor, in a spot hybridization assay to test 68 serum specimens from 37 patients with chronic hemolytic anemia (36 sickle cell disease and 1 thalassemia intermedia), each of whom had suffered an aplastic crisis. The exact time required for completion of the assay is not known, as the authors did not specify exposure time. Serum specimens were collected in London (1979-1981), Jamaica (1980-1981), and Chicago (1980-1982), but the means of specimen storage was not indicated. In 36 of the patients the aplastic crisis was associated with primary parvovirus infection, as diagnosed by detection of either parvovirus specific IgM or virus by CIE and IEM. In conjunction with hybridization, solid-phase RIA that used captured IgM from convalescent serum was also performed. Parvovirus DNA was detected in 23 specimens, of which 13 had previously been shown to contain parvovirus by CIE and IEM, 6 by CIE alone, and 3 by IgM antibody. Of these 23 samples, quantities sufficient for further evaluation by RIA were available in 12. Eight of the 12 were positive by RIA; the 4 RIA-negative specimens had previously been found to be positive by CIE and IgM, in addition to being positive in the hybridization assay.

Dot-blot hybridization with a cloned portion of parvovirus genome labeled with ^{32}P was utilized by Plummer et al. in the investigation of an epidemic of an illness resembling erythema infectiosum that occurred in Manitoba, Canada, in 1980 (78). Because the authors investigating the epidemic did not perform the assay, detailed specifics of the procedure were not provided. Serum specimens examined were obtained from 12 patients with unidentified rash, 6 patients with illness but without rash, 19 well family contacts of patients with a rash, and 28 children with measles or rubella. Seventy-one throat, stool, buffy coat, and urine samples were also analyzed. All specimens were stored at -70°C for an unspecified period. Stored serum specimens from 12 patients with rash were also evaluated for the presence of parvovirus-specific IgM; 11 of the 12 were positive. Hybridization was positive in one patient with a rash and one well family contact, and both were confirmed by IEM.

D. *Plasmodium* Species

Malaria is of major significance in the developing world, where 150-200 million cases, resulting in over 1 million deaths, occur yearly. Approximately 300-400 cases of malaria are diagnosed in the United States and reported to the Centers for Disease Control each year. Of the four plasmodia that infect humans, infections due to *P. falciparum* are the most severe and may progress to coma and renal failure within 2-3 days in nonimmune individuals. Rapid diagnosis of *P. falciparum* is also important because this species is often resistant to chloroquine. Diagnosis is generally made by microscopic examination of blood smears from patients with suspected malaria.

Franzen et al. (79) utilized a ^{32}P-labeled DNA probe, prepared from the F-32 strain of *P. falciparum*, in a 24-hr spot hybridization assay. With the probe they could detect parasitemia on the level of 0.001% or a minimum of 25 pg of parasite DNA. Thirty-five blood samples, the pellets of which had been stored at -20°C for an unspecified period, were tested. Boood smears for microscopic examination were prepared in parallel. A positive hybridization signal was obtained in specimens from 16 of the 17 patients infected with *P. falciparum*. The signal intensity gave a general indication of degree of infection. A sample from one patient infected with *P. vivax* also hybridized; the authors postulated that this patient had a dual infection with *P. falciparum*. There was no cross-hybridization with the mouse malaria parasites *P. chabaudi* and *P. yoelii* or with human DNA.

E. Epstein-Barr Virus

Epstein-Barr virus (EBV) is the causal agent of infectious mononucleosis and appears to be an important, if not essential, factor in the development of nasopharyngeal carcinoma and African Burkitt's lymphoma. The diagnosis of infectious mononucleosis is made easily with serological tests. DNA hybridization, however, has been used to detect viral nucleic acid sequences in human tissues, which has provided support for the role of EBV in various neoplastic processes, including nasopharyngeal carcinoma, Burkitt's lymphoma, and certain non-Burkitt's lymphomas.

With a ^{32}P-labeled DNA probe, prepared from a fragment of the genome of EBV strain FF41, Hochberg and co-workers (80) were able to demonstrate, by Southern blot hybridization, EBV DNA in the cerebellar lymphoma of a 48-year-old man but not in the patient's normal brain tissue. Andiman et al. (81) prepared ^{32}P-labeled DNA probes from four different DNA fragments of EBV strain FF41. With these probes they tested a variety of tissues, by either spot or Southern blot hybridization, each requiring 2-4 days for completion. Approximately 10^4 EBV genomes could be detected by spot hybridization, whereas the limit of sensitivity by Southern blot was 10^5 EBV genomes. Clinical specimens analyzed by spot hybridization included suspensions prepared from lymph nodes, tumor tissue, bone marrow, peripheral blood mononuclear cells, and spleens of 54 patients with a variety of diseases. Hybridization occurred in two samples: T lymphocytes of a patient with infectious mononucleosis and a lymph node involved with metastatic nasopharyngeal carcinoma. Of the 92 clinical specimens studied by Southern blot hybridization, eight were positive: pleural fluid cells and spleen of an AIDS patient with Burkitt's lymphoma; metastatic nasopharyngeal carcinoma tissue; immunoblastic sarcoma tissue; histiocytic lymphoma tissue; lymph node tissue from a child with fatal infectious mononucleosis and polyclonal lymphoma; central nervous system lymphoma tissue; lung tissue from a patient with failure to thrive, chronic intersittial pneumonia, and immunological abnormalities; and lymph node of an infant with failure to thrive, organomegaly, and chronic pneumonitis.

Wolf and associates prepared a ^{3}H-labeled DNA probe from EBV DNA that they used to study tonsillar and parotid gland tissue by both in situ hybridization and reassociation kinetics (82). In situ hybridization required 3-4 weeks for completion, whereas reassociation kinetics analysis was complete in less than 1 hr. Frozen sections from five tonsils and three parotid glands, all obtained from adults without clinical or serological evidence of acute EBV infection but who did have antibodies to EBV nuclear antigen, were examined. Parotid gland tissue was positive in both assays, in contrast to sections of tonsillar tissue, all of which were negative. Parotid tissue cell types carrying the EBV genome could not be determined with certainty, for cytoplasmic staining methods did not work following in situ hybridization. The authors, however, believed that the location of grains, representing EBV DNA, was most consistent with ductlike structures and not typical of infiltrating lymphocytes.

The presence of EBV DNA and EBV RNA in oropharyngeal epithelial cells was demonstrated by Sixbey et al. (83), who performed in situ cytohybridization on throat washings using biotin-labeled DNA probes prepared from cloned *Bam*HI V DNA, which contains the large internal repeating sequence within the EBV genome, and *Bam*HI H, which encodes abundant RNA. The hybridization procedure was complete in less than 24 hr. Serial throat washings from 12 patients with clinical and laboratory findings compatible with acute infectious mononucleosis were studied by in situ hybridization and were also assayed for cord-blood lymphocyte-transformation, which if positive indicates the presence of EBV. From 10 of the 12 patients at least one specimen contained epithelial cells positive for EBV DNA; the two negative patients were not excreting infectious EBV. Adequate numbers of epithelial cells for analysis of both infectious virus and cell-associated EBV DNA were present in 28 specimens. In 20 of these, cytohybridization and cord-blood lymphocyte-transformation results agreed: both were positive in 7 specimens and negative in 13. Seven of the remaining eight specimens were positive by hybridization, but the cord-blood lymphocyte transformation assay was negative. One specimen contained infectious virus but lacked EBV-DNA-positive epithelial cells. Epstein-Barr virus RNA was detected in two of the four specimens tested. Because the EBV RNA-negative specimens were obtained from patients who had previously been EBV DNA-positive, the authors postulated that the discrepancy may be due to differences in the preservation of RNA or to sampling variations.

IX. ANTIMICROBIAL SUSCEPTIBILITY TESTING

A. β-Lactamase

Perine et al. (39) prepared a ^{32}P-labeled DNA probe from the 4.4-Md plasmid encoding for β-lactamase found in *N. gonorrhoeae* strain CDC67, which they used to test urethral discharge in a 3- to 5-day spot hybridization assay. Two urethral swab specimens were collected from each of 216 males with a clinical diagnosis of gonococcal urethritis seen in four African and two Asian countries.

One swab was directly applied to nitrocellulose paper, and a Thayer-Martin agar plate was inoculated with the other. Colonies presumptively identified as *N. gonorrhoeae* were tested for β-lactamase production by the chromogenic cephalosporin (Nitrocefin; BBL, Cockeysville, Maryland) method. The nitrocellulose strips, as well as either subcultures or lyophilized preparations of *N. gonorrhoeae* isolates, were transported to Seattle, Washington, for hybridization. For detecting β-lactamase, the sensitivity and specificity of the probe were 91% (59 of 65 samples) and 98% (136 of 139 samples), respectively. The authors, however, question the sensitivity and specificity of the probe in the identification of β-lactamase-positive *N. gonorrhoeae* in specimens other than those from the male urethra. Specimens from cervix, rectum, and pharynx may contain sufficient numbers of organisms other than *N. gonorrhoeae*, such as *Haemophilus influenzae* or *E. coli*, which could harbor β-lactamase-encoding plasmids that would react with the gonococcal β-lactamase probe.

Mendelmen et al. (84) used hybridization analysis as one marker with which to study ampicillin-resistant strains of *H. influenzae*, as determined by disk diffusion, chromogenic cephalosporin, ampicillin hydrolysis with phenol red indicator, and agar dilution. They prepared four ^{35}S-labeled purified plasmids as probes: one isolate each from the northwest, southwest, and south central regions of Alaska, and plasmid (RSF 0885), the TnA sequence that codes for the TEM β-lactamase, for use in a 3-week Southern blot assay. Of the 100 isolates screened, 29 invasive strains and 7 noninvasive strains were ampicillin-resistant. Plasmid analysis of these 36 strains revealed detectable plasmids in 7 (all invasive strains); 4 strains contained one large plasmid of approximately 40-Md, and 3 harbored a small 3-Md plasmid. Both 40-Md and 3-Md plasmids were shown by conjugation or transformation experiments to transfer ampicillin resistance. By DNA-DNA hybridization analysis, each of these probes demonstrated DNA homology with 33 of the 36 resistant strains tested; the three nonhomologous strains harbored the small 3-Md plasmid. Furthermore, restriction endonuclease digestion of the seven isolated plasmids and hybridization of the resulting Southern blots with the ^{35}S-labeled (RSF 0885) probe revealed DNA homology with identical DNA fragments. From these studies the authors concluded that geographically separated strains of ampicillin-resistant *H. influenzae* harbored a common 40-Md conjugative plasmid coding for β-lactamase production. Compared with what has been reported concerning analysis of ampicillin-resistant *H. influenzae* in other areas of the United States, the plasmid isolated from Alaskan strains appeared different from those isolated elsewhere in this country.

B. Aminoglycosides

A radioactive probe coding for gentamicin resistance was constructed by Obbink et al. from a plasmid (IncC) that was common among multiresistance plasmids at the Royal North Shore Hospital, Sydney, Australia (85). The 2-kb probe hybridized only with strains containing the 2"-*O*-adenylyltransferase [AAD(2")] gene. Both colony hybridization and Southern blot assays were performed, with results available in 2-3 days. Eight different bacterial strains were tested. No false-positive signals occurred with colony hybridization, as all were confirmed in Southern hybridizations. Two strains (*Klebsiella pneumoniae* VA273 and *Enterobacter aerogenes* VA274), however, were negative in colony hybridization but positive by Southern hybridization. The authors postulated that this may have been due to the low copy number of large plasmids in some isolates.

Gootz et al. (86) prepared a ^{32}P-labeled DNA probe from the recombinant plasmid (pFCT3103) that was specific for the aminoglycoside 2"-*O*-adenylyltransferase gene [ANT (2")]. With the probe they tested 42 strains of gram-negative bacilli (6 positive controls, 4 negative controls, and 32 clinical isolates) in three different hybridization assays: colony hybridization and spot hybridization were compared with Southern hybridization, which was considered the reference method. The time required for completion of the assays ranged from 3 to 5 days; however, Southern blot was technically more difficult and only one organism could be tested per run, in contrast to 20-25 organisms with the other methods. The phosphocellulose paper binding assay and resistance phenotyping for demonstration of adenylating activity were also performed, and one or the other was negative with three isolates that were positive by Southern blot. These isolates were not further studied. Of the remaining 39 isolates, results of the three hybridization assays agreed for 30 strains. There were six false-positive and three false-negative results with colony hybridization. False-negative results occurred with mucoid strains of *Enterobacteriaceae*, as colony material was lost from nitrocellulose filters during posthybridization washing. All false-positive results were seen with *Pseudomonas aeruginosa* and were thought to be a result of nonspecific binding of the probe to colony material on the filter. Two *P. aeruginosa* strains gave variable results by spot hybridization but were negative by Southern blot. The authors concluded that the spot hybridization procedure was relatively easy, sensitive, and specific, and consequently could be used to screen several clinical isolates for the presence of the ANT(2") gene.

C. Antiviral Agents

Gadler (87) used four ^{32}P-labeled DNA probes, prepared from random fragments of the genome of CMV Ad.169, in a 2- to 3-day spot hybridization assay to determine the effects of three antiviral agents on one laboratory strain and six clinical isolates of CMV. Phosphonoformic acid, acyclovir, and arabinosyladenine were each evaluated at twofold dilutions, with each compound tested being added after the virus had been adsorbed onto human embryonic lung cells. For each drug, virus grown in the absence of that agent and an uninfected cell layer served as controls. The ID_{50} was determined both visually, as the intermediate between two concentrations of three antiviral substances, and by cutting out the spots from the nitrocellulose filter and counting them in a liquid scintillation counter. In the latter method, the ID_{50} was the concentration of antiviral agent at which hybridization signal equaled 50% of the signal obtained with the infected cell control without antiviral substance. The ID_{50} of viral DNA synthesis obtained by hybridization was compared with values published by other authors using different techniques, primarily 50% plaque reduction values. The ID_{50} of CMV Ad-169 determined by hybridization agreed with values published for phosphonoformic acid and acyclovir; however, hybridization values for arabinosyladenine were considerably lower than published values obtained using 50% inhibition of cytopathic effect. The author surmised that the discrepancy was because the detection of inhibition of viral DNA synthesis by hybridization was probably the more sensitive technique.

A mixture of three ^{32}P-labeled DNA probes, prepared from HSV-1 DNA, were used in a 2-day spot hybridization assay by Gadler et al. (88) to test the effects of six antiviral compounds on four laboratory strains of HSV-1. Antiherpes compounds tested were phosphonoformic acid (PFA), (*R*)- and (*S*)-enantiomers of 9-(3,4,-dihydroxybutyl)guanine (DHBG), 9-(4-hydroxybutyl)guanine (HBG), acyclovir (ACV), and (*E*)-5-(2-bromovinyl)-2-deoxyuridine (BVDU). Each agent, in varying concentrations, was added to infected cells after an initial 1-hr adsorption of the virus. Cells were incubated for 16 hr and analyzed immediately or stored at −70°C for an unspecified period. Controls included uninfected cells and HSV-1-infected cells without antiviral substances.

Quantitation was done visually (determining intensity and size of spots) and by counting the spots, which had been cut from the nitrocellulose paper, in a scintillation counter. The mean 50% inhibition value was defined as the concentration of antiviral compound that inhibited viral DNA synthesis by 50%. The mean 50% inhibition value (M) of each of the four strains was compared with reported inhibitory effects on HSV-1 plaque formation (PF) and the effects on HSV-1 (strain C42) DNA synthesis as shown by isopycnic banding of viral and cellular DNA (IB). For BVDU, the mean 50% inhibition value, PF, and IB values are 0.2, 0.1, 0.3; for ADV 0.2, 0.03, 0.5; for HBG 2.3, 0.6, 3.0; for PFA 60, 50, 33. Mean IC_{50} and PF values for (*R*)- and (*S*)-DHBG were 3.6 and 4.0, and 11 and 12, respectively; no IB values were available.

X. CONCLUSIONS

Many nucleic acid probes have been developed for the detection of viruses, bacteria, and parasites. Hybridization assays utilizing these probes have, however, been conducted primarily in research settings, and the techniques employed to date are, for the most part, impractical for use in the clinical microbiology laboratory. With specific technological changes, however, DNA hybridization could play a role in certain areas of clinical microbiology.

To assess the potential of DNA hybridization in clinical microbiology, several issues must be addressed. The first, and perhaps most important, consideration is the potential for utilization of probes as substitutes for culture. Second, the clinical relevance of organism identification and the necessity of antimicrobial susceptibility testing must be considered. Third, any evaluation of DNA hybridization as a replacement for culture should include a comparison between the advantages and disadvantages of the hybridization assay and presently available rapid methods for organism identification. Additional issues are possible uses of hybridization as culture supplements and as a means to detect noncultivable organisms. A final issue is the nature of the probe itself: radioisotopic versus biotinylated.

Can DNA hybridization obviate the need for culture? An objective answer is dependent upon the sensitivity of the assay compared with that of currently available detection methods (i.e., direct examination, culture, serology); however, to date, data in this area are limited. For example, the sensitivity of DNA hybridization for the detection of *M. pneumoniae* in specimens cannot be determined from the available literature. For other organisms, most of the available data are from single publications and, therefore, require confirmation by other investigators. Examples include hybridization assays for detection of the following, with the respective sensitivities: (a) enteroinvasive *E. coli* and *Shigella* spp., 85%; (b) *Mobiluncus* spp., 52-83%; (c) VZV, 76%; and (d) in situ hybridization for detection of HSV, 94%.

Data from two separate studies have been reported for the three organisms most commonly associated with sexually transmitted diseases and for the respiratory adenoviruses. The sensitivities of hybridization assays in each of two published studies for detection of the following organisms in genital specimens were 71.4% and 78% for HSV, 91% and 100% for *C. trachomatis*, and 89% and 96% for *N. gonorrhoeae*. For the identification of respiratory adenoviruses in nasopharyngeal mucus of children, hybridization assay sensitivities were 71.4% and 88%. In two studies, one of *C. trachomatis* and one of adenoviruses, small numbers of specimens were examined. Although the sensitivity of the probe used to detect *N. gonorrhoeae* was high, cross-reactions did occur with two of the saprophytic *Neisseria* spp. in one study. A more important limitation of the *N. gonorrhoeae* hybridization assays that have been studied is that the results from male urethral samples cannot necessarily be extrapolated to specimens from cervical, rectal, or pharyngeal sites.

Probes for detection of ETEC and CMV have been most extensively investigated as potential culture substitutes. The sensitivity of the three probes (LT, STp, and STh) used to identify ETEC in diarrheal stool specimens has approached 100%, and in one study DNA hybridization was more sensitive than the standard tissue culture and suckling mouse assays of isolates of *E. coli* from stool cultures. For CMV hybridization, the assay sensitivities have ranged from 92 to 100%, and with buffy coat specimens, DNA hybridization may be more sensitive than CMV culture.

These data suggest that DNA hybridization could potentially replace culture for ETEC, CMV, and *C. trachomatis* from genital sites, as well as perhaps *N. gonorrhoeae* from male urethral exudates. A test that can be applied to such a limited population as males with gonococcal urethritis, however, is of questionable clinical value, particularly in view of the speed and diagnostic accuracy of the Gram-stained smear in this patient population. Also, because of the low prevalence of disease caused by the ETEC in this country and the limited financial resources in developing countries in which ETEC is endemic, the major use for an ETEC-hybridization assay will probably be as an epidemiological tool for investigation of large outbreaks of diarrhea, rather than as a test routinely performed in a clinical microbiology laboratory.

The clinical utility of hybridization assays for certain other organisms can be problematical. For example, a young, previously healthy, adult who presents with classic atypical pneumonia is frequently treated empirically with erythromycin for *M. pneumoniae* without attempting to isolate the organism or performing specific serological tests. In such cases, a probe would add little except cost to the patient, and use of the probe would be limited to occasional patients with atypical presentations of *M. pneumoniae* infection and in whom a specific diagnosis is necessary. Yet to be examined in this patient population are the sensitivity and specificity of a *M. pneumoniae* probe relative to culture and specific serology. Another example of the potential limited clinical utility of a probe might be bacterial vaginosis which can be easily diagnosed in a physician's office by testing vaginal pH, examining a wet mount preparation for clue cells, and performing the "sniff test" with 10% KOH. *Mobiluncus* spp. need not be identified to make the diagnosis of bacterial vaginosis and, in fact, may not be uniformly present in vaginal secretions of women with otherwise typical presentations of this disease.

If hybridization were to replace culture, the issue of antimicrobial susceptibility testing becomes a significant concern. Although a probe specific for the *N. gonorrhoeae* plasmid encoding β-lactamase production addresses the need to test all isolates of *N. gonorrhoeae* for β-lactamase, such a probe would not detect chromosomally mediated penicillin-resistant strains. Furthermore, the sensitivity of the β-lactamase probe has not been examined with cervical, rectal, or pharyn- specimens. Therefore, hybridization does not now appear to be a feasible alternative to culture of *N. gonorrhoeae*.

Rapid methods of organism detection, such as ELISA and FA using monoclonal antibodies, have recently become commercially available and will undoubtedly have an impact on the utilization of DNA hybridization. With ELISA, results are available in approximately 4 hr. The advantages of ELISA include objectivity in interpretation of end-points and automation, although its cost-effectiveness is very volume sensitive, which in practice virtually prohibits single sample analysis. Fluorescent antibody tests can be used for direct examination of specimens, with results available in less than 1 hr, or for examination of tissue culture cells after 24–48 hr of incubation. It is rapid and applicable to low-volume testing; however, a fluorescence microscope and an experienced microscopist are needed. So to compete favorably with these two methods, DNA hybridization should provide rapid turnaround in a cost-effective manner, objective interpretation, and automation.

In those instances in which probes have comparable sensitivity with existing rapid detection techniques, but that are less sensitive than culture, probes may serve as culture supplements rather than as culture substitutes. An example is the *Legionella* spp.-specific probe. Assuming that the probe is at least as sensitive as DFA for *Legionella*, use of the probe would be desirable if it demanded less technician time, allowed for objective interpretation of results, and were cost effective for laboratories equipped to handle ^{125}I.

Similarly, a probe that generically detects *Mycobacterium* spp. could be used in conjunction with culture and has the potential to replace the acid-fast smear (AFS). For screening specimens, hybridization would have to be at least as sensitive as the AFS, which can detect a minimum of 10^5 acid-fast bacilli per milliliter of sputum. Species-specific mycobacterial probes, such as ones for *M. tuberculosis* and MAI, would be very valuable in rapidly identifying these organisms in screen-positive specimens and in cultures.

A major potential utility of DNA hybridization is to detect noncultivable or difficult-to-culture viruses. For example, a diagnosis of hepatitis B is presently made by immunoassay. It has been demonstrated, however, that hybridization for detection of HBV DNA in serum is more sensitive than immunoassay for detecting either HBsAg or HBeAg. Consequently, an HBV hybridization assay could be a valuable diagnostic tool, if the present technology could be simplified for use in the clinical laboratory setting.

DNA hybridization may also prove to be of value in the diagnosis of viral gastroenteritis. Most microbiology laboratories now detect rotavirus by ELISA or LA, as EM is both labor-intensive and cost-prohibitive. Both LA and ELISA systems, which use polyclonal antibodies, are significantly less sensitive than EM, especially when specimens are obtained later in the illness. A rotavirus-specific probe could, therefore, prove beneficial; however, it would have to compete with ELISA systems that use monoclonal antibodies, because these significantly improve the sensitivity of the ELISA technique. A hybridization assay for enteric adenoviruses would also be useful because restriction enzyme analysis is presently required for definitive identification.

The final issue to be considered is the nature of the probe itself. Most of the probes investigated thus far have been radiolabeled. Problems with radiolabeled probes include the special handling, their short shelf-life, and special counting equipment, all of which limit the application of such probes in clinical microbiology. Although ^{125}I is more manageable than ^{32}P or ^{35}S, a nonisotopic method, such as biotinylation, would be more readily accepted. Unfortunately, biotin-labeled probes, in general, have not been as sensitive as radiolabeled probes.

In summary, DNA hybridization, in its present state of technology, with the possible exception of the ^{125}I-labeled probes soon to be marketed, is not practical for the clinical microbiology laboratory. As methods become more rapid and less cumbersome and as nonradiolabeled probes become available, the role of hybridization will become better defined.

REFERENCES

1. Sack, D. A., and R. B. Sack, Test for enterotoxigenic *Escherichia coli* using Y1 adrenal cells in miniculture. *Infect. Immun. 11*:334–336 (1975).
2. Dean, A. G., Y.-C. Ching, R. G. Williams, and L. B. Harden, Test for *Escherichia coli* exterotoxin using infant mice: Application in a study of diarrhea in children in Honolulu. *J. Infect. Dis. 125*:407–411 (1972).
3. Moseley, S. L., I. Huq, A. R. M. A. Alim, M. So, M. Samadpour-Motalibi, and S. Falkow, Detection of enterotoxigenic *Escherichia coli* by DNA colony hybridization. *J. Infect. Dis. 142*:892–898 (1980).
4. Moseley, S. L., P. Echeverria, J. Seriwatana, C. Tirapat, W. Chaicumpa, T. Sakuldaipeara, and S. Falkow, Identification of enterotoxigenic *Escherichia coli* by colony hybridization using three enterotoxin gene probes. *J. Infect. Dis. 145*:863–869 (1982).
5. Seriwatana, J., P. Echeverria, J. Escamilla, R. Glass, I. Huq, R. Rockhill, and B. J. Stoll, Identification of *Escherichia coli* in patients with diarrhea in Asia with three enterotoxin gene probes. *Infect. Immun. 42*:152–155 (1983).
6. Georges, M. C., J. K. Wachsmuth, K. A. Birkness, S. L. Moseley, and A. J. Georges, Genetic probes for enterotoxigenic *Escherichia coli* isolated from childhood diarrhea in the Central African Republic. *J. Clin. Microbiol. 18*: 199–202 (1983).
7. Echeverria, P., J. Seriwatana, O. Sethabutr, and D. N. Taylor, DNA hybridization in the diagnosis of bacterial diarrhea. *Clin. Lab. Med. 5*:447–462 (1985).
8. Echeverria, P., J. Seriwatana, U. Patamaroj, S. L. Moseley, A. McFarland, O. Chityothin, and W. Chaicumpa, Prevalence of heat-stable II enterotoxigenic *Escherichia coli* in pigs, water, and people at farms in Thailand as determined by DNA hybridization. *J. Clin. Microbiol. 19*:489–491 (1984).

9. Echeverria, P., D. N. Taylor, J. Seriwatana, A. Chat-Kaeomorakot, V. Khungvalert, T. Sakuldaipeara, and R. D. Smith, A comparative study of enterotoxin gene probes and tests for toxin production to detect enterotoxigenic *Escherichia coli. J. Infect. Dis. 153*:255–260 (1986).
10. Bialkowska-Hobrzanska, H., Evaluation of the sensitivity and specificity of biotinylated DNA probe hybridization (BDPH) assay in diagnosis of enterotoxigenic *Escherichia coli* (ETEC). *Abstr. Annu. Meet. Am. Soc. Microbiol.* C-163 (1986).
11. Boileau, C. R., H. M. D'Hauteville, and P. J. Sansonetti, DNA hybridization technique to detect *Shigella* species and enteroinvasive *Escherichia coli. J. Clin. Microbiol. 20*:959–961 (1984).
12. Sethabutr, O., S. Hanghalay, P. Echeverria, D. N. Taylor, and U. Leksomboon, A non-radioactive DNA probe to identify *Shigella* and enteroinvasive *Escherichia coli* in stools of children with diarrhea. *Lancet 2*:1095–1097 (1985).
13. Rubin, F. A., D. J. Kopecko, K. F. Noon, and L. S. Baron, Development of a DNA probe to detect *Salmonella typhi. J. Clin. Microbiol. 22*:600–605 (1985).
14. Flores, J., E. Boggeman, R. H. Purcell, et al., A dot hybridization assay for detection of rotavirus. *Lancet 1*:555–559 (1983).
15. Dimitrov, D. H., D. Y. Graham, and M. K. Estes, Detection of rotavirus by nucleic acid hybridization with cloned DNA of simian rotavirus SA11 genes. *J. Infect. Dis. 152*:293–300 (1985).
16. Stalhandske, P., T. Hyypia, H. Gadler, P. Halonen, and U. Pettersson, The use of molecular hybridization for demonstration of adenoviruses in human stools. *Curr. Top. Microbiol. Immunol. 104*:299–306 (1983).
17. Takiff, H. E., M. Seidlin, and P. Krause et al., Detection of enteric adenoviruses by dot-blot hybridization using a molecularly cloned viral DNA probe. *J. Med. Virol. 16*:107–118 (1985).
18. Kidd, A. H., E. H. Harley, and M. J. Erasmus, Specific detection and typing of adenoviruses types 40 and 41 in stool specimens by dot-blot hybridization. *J. Clin. Microbiol. 22*:934–939 (1985).
19. Grimont, P. A. D., F. Grimont, N. Desplaces, and P. Tchen, DNA probe specific for *Legionella pneumophila. J. Clin. Microbiol. 21*:431–437 (1985).
20. Wilkinson, H. W., J. S. Sampson, and B. B. Plikaytis, Evaluation of a commercial gene probe for identification of *Legionella* cultures. *J. Clin. Microbiol. 23*:217–220 (1986).
21. Edelstein, P. H., Evaluation of the Gen-Probe DNA probe for the detection of *Legionella* in culture. *J. Clin. Microbiol. 23*:481–484 (1986).
22. Hogan, J. J., V. Jonas, C. L. Milliman, et al., Detection of *Legionella* species using a genus specific cDNA probe. *Abstr. Annu. Meet. Am. Soc. Microbiol.* C-166 (1986).
23. Ranki, M., M. Virtanen, A. Palva, et al., Nucleic acid sandwich hybridization in adenovirus diagnosis. *Curr. Top. Microbiol. Immunol. 104*:307–318 (1983).

24. Virtanen, M., A. Palva, M. Laaksonen, P. Halonen, H. Soderlund, and M. Ranki, Novel test for rapid viral diagnosis: Detection of adenovirus in nasopharyngeal mucus aspirates by means of nucleic acid sandwich hybridization. *Lancet 1*:381-383 (1983).
25. Myerson, D., R. C. Hackman, and J. D. Meyers, Diagnosis of cytomegaloviral pneumonia by in situ hybridization. *J. Infect. Dis. 150*:272-277 (1984).
26. Myerson, D., R. C. Hackman, J. A. Nelson, D. C. Ward, and J. C. McDougall, Widespread presence of histologically occult cytomegalovirus. *Hum. Pathol. 15*:430-439 (1984).
27. Spector, S. A., J. A. Rua, D. H. Spector, and R. McMillan, Detection of human cytomegalovirus in clinical specimens by DNA-DNA hybridization. *J. Infect. Dis. 150*:121-126 (1984).
28. Martin, D. C., D. A. Katzenstein, G. S. M. Yu, and M. C. Jordon, Cytomegalovirus viremia detected by molecular hybridization and electron microscopy. *Ann. Intern. Med. 100*:222-225 (1984).
29. Chou, S., and T. C. Merigan, Rapid detection and quantitation of human cytomegalovirus in urine through DNA hybridization. *N. Engl. J. Med. 308*: 921-925 (1983).
30. Virtanen, M., A. Syvanen, J. Oram, H. Soderlund, and M. Ranki, Cytomegalovirus in urine: Detection of viral DNA by sandwich hybridization. *J. Clin. Microbiol. 20*:1083-1088 (1984).
31. Buffone, G. J., C. M. Schimbor, G. J. Demmler, D. R. Wilson, and G. J. Darlinton, Detection of cytomegalovirus in urine by nonisotopic DNA hybridization. *J. Infect. Dis. 154*:163-166 (1986).
32. Shaw, S. B., L. Boas, D. K. Cabanas, et al., A rapid DNA probe assay for the specific detection of *Mycoplasma pneumoniae*. *Abstr. Annu. Meet. Am. Soc. Microbiol.* G-24 (1986).
33. Dean, E. D., W. J. Dalyl, M. S. Hoppe, R. Mikhail, K. A. Murphy, R. D. Smith, and D. L. Kacian, Rapid, specific detection of *Mycobacterium* species by DNA:rRNA hybridization. *Abstr. Annu. Meet. Am. Soc. Microbiol.* C-374 (1986).
34. Shoemaker, S. A., J. H. Fisher, and C. H. Scoggin, Techniques of DNA hybridization detect small numbers of mycobacteria with no cross-hybridization with nonmycobacterial respiratory organisms. *Am. Rev. Respir. Dis. 131*:760-763 (1985).
35. Roberts, M. C., and M. B. Coyle, Whole-chromosomal DNA probes for the rapid identification of *Mycobacterium tuberculosis* and *M. avium* complex. *Abstr. Annu. Meet. Am. Soc. Microbiol.* C-167 (1986).
36. Palva, A., H. Jousimies-Somer, P. Saikku, P. Vaananen, H. Soderlund, and M. Ranki, Detection of *Chlamydia trachomatis* by nucleic acid sandwich hybridization. *FEMS Microbiol. Lett. 23*:83-89 (1984).
37. Horn, J. E., M. L. Hammer, S. Falkow, and T. C. Quinn, Detection of *Chlamydia trachomatis* in tissue culture and cervical scrapings by in situ DNA hybridization. *J. Infect. Dis. 153*:1155-1159 (1986).
38. Totten, P. A., K. K. Holmes, H. H. Handsfield, J. S. Knapp, P. L. Perine,

and S. Falkow, DNA hybridization techniques for the detection of *Neisseria gonorrhoeae* in men with urethritis. *J. Infect. Dis. 148*:4623–471 (1983).
39. Perine, P. L., P. A. Totten, K. K. Holmes, et al., Evaluation of a DNA-hybridization method for detection of African and Asian strains of *Neisseria gonorrhoeae* in men with urethritis. *J. Infect. Dis. 152*:59–63 (1985).
40. Stalhandske, P., and U. Pettersson, Identification of DNA viruses by membrane filter hybridization. *J. Clin. Microbiol. 15*:744–747 (1982).
41. Redfield, D. C., D. D. Richman, S. Albanil, M. O. Oxman, and G. M. Wahl, Detection of herpes simples virus in clinical specimens by DNA hybridization. *Diagn. Microbiol. Infect. Dis. 1*:117–128 (1983).
42. Fung, J. C., J. Shanley, and R. C. Tilton, Comparison of the detection of herpes simplex virus in direct clinical specimens with herpes simplex virus-specific DNA probes and monoclonal antibodies. *J. Clin. Microbiol. 22*:748–753 (1985).
43. Peterson, E. M., S. L. Aarnaes, R. N. Bryan, J. L. Ruth, and L. M. de al Maza, Typing of herpes simplex virus with synthetic DNA probes. *J. Infect. Dis. 153*:757–762 (1986).
44a. Coffin, J., A. Haase, J. Levy et al., Notice: AIDS virus nomenclature. *ASM News 52*:449 (1986).
44b. Gelmann, E. P., M. Popovic, D. Blayney, H. Masur, G. Sidhu, R. E. Stahl, and R. C. Gallo, Proviral DNA of a retrovirus, human T-cell leukemia virus in two patients with AIDS. *Science 220*:862–865 (1983).
45. Shaw, G. M., B. H. Hahn, S. K. Arya, J. E. Groopman, R. C. Gallo, and F. Wong-Staal, Molecular characterization of human T-cell leukemia (lymphotropic) virus type III in the acquired immune deficiency syndrome. *Science 226*:1165–1171 (1984).
46. Harper, M. E., L. M. Marselle, R. C. Gallo, and F. Wong-Staal, Detection of lymphocytes expressing human T-lymphotropic virus type III in lymph nodes and peripheral blood from infected individuals by in situ hybridization. *Proc. Natl. Acad. Sci. USA 83*:772–776 (1986).
47. Shaw, G. M., M. E. Harper, B. H. Hahn, et al., HTLV-III infection in brains of children and adults with AIDS encephalopathy. *Science 227*: 177–182 (1985).
48. Epstein, L. G., L. R. Sharer, E.-S. Cho, M. Myenhofer, B. A. Navia, and R. W. Price, HTLV-III/LAV-like retrovirus particles in the brains of patients with AIDS encephalopathy. *AIDS Res. 1*:447–454 (1984/5).
49. Perine, P. L., J. W. Nelson, J. O. Lewis, et al., New technologies for use in the surveillance and control of yaws. *Rev. Infect. Dis. 7*(Suppl. 2): S295–S299 (1985).
50. Roberts, M. C., S. L. Hillier, F. D. Schoenknecht, and K. K. Holmes, Comparison of Gram stain, DNA probe, and culture for the identification of species of *Mobiluncus* in female genital specimens. *J. Infect. Dis. 152*: 74–77 (1985).

51. Gissman, L., L. Wolnik, H. Ikenberg, U. Koldovsky, H. G. Schnurch, and H. zur Hausen, Human papillomavirus types 6 and 11 DNA sequences in genital and laryngeal papillomas and in some cervical cancers. *Proc. Natl. Acad. Sci. USA 80*:560–563 (1983).
52. Durst, M., L. Gissmann, H. Ikenberg, and H. zur Hausen, A papillomavirus DNA from a cervical carinoma and its prevalence in cancer biopsy samples from different geographic regions. *Proc. Natl. Acad. Sci. 80*:3812–3815 (1983).
53. Lancaster, W. D., R. J. Kurman, L. E. Sanz, S. Perry, and A. B. Jenson, Human papillomaviruses: Detection of viral DNA sequences and evidence for molecular heterogeneity in metaplasias and dysplasias of the uterine cervix. *Intervirology 20*:202–212 (1983).
54. Beckmann, A. M., D. Myerson, J. R. Daling, N. B. Kiviat, C. M. Fenoglio, and J. K. McDougall, Detection and localization of human papillomavirus DNA in human genital condylomas by in situ hybridization with biotinylated probes. *J. Med. Virol. 16*:265–273 (1985).
55. Wagner, D., H. Ikenberg, N. Boehm, and L. Gissman, Identification of human papillomavirus in cervical swabs by deoxyribonucleic acid in situ hybridization. *Obstet. Gynecol. 64*:767–772 (1984).
56. Wickenden, C., A. Steele, A. D. B. Malcolm, and D. V. Coleman, Screening for wart virus infection in normal and abnormal cervices by DNA hybridization of cervical scrapes. *Lancet 1*:65–67 (1985).
57. Hyypia, T., P. Stalhandske, R. Vainionpaa, and U. Pettersson, Detection of enteroviruses by spot hybridization. *J. Clin. Microbiol. 19*:436–438 (1984).
58. Rotbart, H. A., M. J. Levin, and L. P. Villarreal, Use of subgenomic poliovirus DNA hybridization probes to detect the major subgroups of enteroviruses. *J. Clin. Microbiol. 20*:1105–1108 (1984).
59. Rotbart, H. A., M. J. Levin, L. P. Villarreal, S. M. Tracy, B. L. Semler, and E. Wimmer, Factors affecting the detection of enteroviruses in cerebrospinal fluid with coxsackievirus B3 and poliovirus 1 cDNA probes. *J. Clin. Microbiol. 22*:220–224 (1985).
60. Forghani, B., K. W. Dupuis, and N. J. Schmidt, Rapid detection of herpes simplex DNA in human brain tissue by in situ hybridization. *J. Clin. Microbiol. 22*:656–658 (1985).
61. Haase, A. T., P. Ventura, C. J. Gibbs Jr., and W. W. Tourtellotte, Measles virus nucleotide sequences: Detection by hybridization in situ. *Science 212*:672–675 (1981).
62. Seidlin, M., H. E. Takiff, H. A. Smith, J. Hay, and S. E. Straus, Detection of varicella-zoster virus by dot-blot hybridization using a molecularly cloned viral DNA probe. *J. Med. Virol. 13*:53–61 (1984).
63. Wirth, D. F., and D. M. Pratt, Rapid identification of *Leishmania* species by specific hybridization of kinetoplast DNA in cutaneous lesions. *Proc. Natl. Acad. Sci. USA 79*:6999–7003 (1982).
64. Jansen, R. W., J. E. Newbold, and S. M. Lemon, Combined immunoaffinity

cDNA–RNA hybridization assay for detection of hepatitis A virus in clinical specimens. *J. Clin. Microbiol. 22*:984–989 (1985).
65. Beminger, M., M. Hammer, B. Hoyer, and J. L. Gerin, An assay for the detection of the DNA genome of hepatitis B virus in serum. *J. Med. Virol. 9*:57–68 (1982).
66. Scotto, J., M. Hadchouel, C. Hery, J. Yvart, P. Tiollais, and C. Brechot, Detection of hepatitis B virus DNA in serum by a simple spot hybridization technique: Comparison with results for other viral markers. *Hepatology 3*: 279–284 (1983).
67. Lieberman, H. M., D. R. LaBrecque, M. C. Kew, S. J. Hadziyannis, and D. A. Shafritz, Detection of hepatitis B virus directly in human serum by a simplified molecular hybridization test: Comparison to HBeAg/anti-HBe status in HBsAg carriers. *Hepatology 3*:285–291 (1983).
68. Feinman, S. V., B. Berris, A. Guha, R. Sooknanan, D. W. Bradley, W. W. Bond, and J. E. Maynard, DNA:DNA hybridization method for the diagnosis of hepatitis B infection. *J. Virol. Methods 8*:199–206 (1984).
69. Lie-Injo, L. E., M. Balasegaram, C. G. Lopez, and A. R. Herrera, Hepatitis B virus DNA in liver and white blood cells of patients with hepatoma. *DNA 2*:301–308 (1983).
70. Scotto, J., M. Hadchouel, C. Hery, et al., Hepatitis B virus DNA in children's liver diseases: Detection by blot hybridization in liver and serum. *Gut 24*: 618–624 (1983).
71. Yoffe, B., C. A. Noonan, J. L. Melnick, and F. B. Hollinger, Hepatitis B virus DNA in mononuclear cells and analysis of cell subsets for the presence of replicative intermediates of viral DNA. *J. Infect. Dis. 153*:471–477 (1986).
72. Diequtis, P. S., E. Keirnan, L. Burnett, B. N. Nightingale, and Y. E. Cossart, False-positive results with hepatitis B virus DNA dot-hybridization in hepatitis B surface antigen-negative specimens. *J. Clin. Microbiol. 23*:797–799 (1986).
73. Kuritza, A. P., C. E. Getty, P. Shaughnessy, R. Hesse, and A. A. Salyers, DNA probes for identification of clinically important *Bacteroides* species. *J. Clin. Microbiol. 23*:343–349 (1986).
74. Hogan, T. F., E. C. Borden, J. A. McBain, B. L. Padgett, and D. L. Walker, Human polyomavirus infections with JC virus and BK virus in renal transplant patients. *Ann. Intern. Med. 92*:373–378 (1980).
75. Gibson, P. E., S. D. Gardner, and A. A. Porter, Detection of human polyomavirus DNA in urine specimens by hybridot assay. *Arch. Virol. 84*: 233–240 (1985).
76. Cossart, Y. E., A. M. Field, B. Cant, D. Widdows, Parvovirus-like particles in human sera. *Lancet 1*:72–73 (1975).
77. Anderson, M. J., S. E. Jones, and A. C. Minson, Diagnosis of human parvovirus infection by dot-blot hybridization using cloned viral DNA. *J. Med. Virol. 15*:163–172 (1985).
78. Plummer, F. A., G. W. Hammond, K. Forward, et al., An erythema infectiosum-like illness caused by human parvovirus infection. *N. Engl. J. Med. 313*:74–79 (1985).

79. Franzen, L., G. Westin, R. Shabo, et al., Analysis of clinical specimens by hybridization with probe containing repetitive DNA from *Plasmodium falciparum*. *Lancet 1*:525–527 (1984).
80. Hochberg, F. H., G. Miller, R. T. Schooley, M. S. Hirsch, P. Feorino, and W. Henle, Central-nervous system lymphoma related to Epstein-Barr virus. *N. Engl. J. Med. 309*:745–748 (1983).
81. Andiman, W., L. Gradoville, L. Heston, et al., Use of cloned probes to detect Epstein-Barr viral DNA in tissues of patients with neoplastic and lymphoproliferative diseases. *J. Infect. Dis. 148*:967–977 (1983).
82. Wolf, H., M. Haus, and E. Wilmes, Persistence of Epstein-Barr virus in the parotid gland. *J. Virol. 51*:795–798 (1984).
83. Sixbey, J. W., J. G. Nedrud, N. Raab-Traub, R. A. Hanes, and J. S. Pagno, Epstein-Barr virus replication in oropharyngeal epithelial cells. *N. Engl. J. Med. 310*:1225–1230 (1984).
84. Mendelman, P. M., V. P. Syriopoulou, S. L. Gandy, J. I. Ward, and A. L. Smith, Molecular epidemiology of plasmid-mediated ampicillin resistance in *Haemophilus influenzae* type b isolates from Alaska. *J. Infect. Dis. 151*: 1061–1072 (1985).
85. Obbink, D. J. G., L. J. Ritchie, F. H. Cameron, J. S. Mattick, and V. P. Ackerman, Construction of a gentamicin resistance gene probe for epidemiological studies. *Antimicrob. Agents Chemother. 28*:96–102 (1985).
86. Gootz, T. D., F. C. Tenover, S. A. Young, K. P. Gordon, and J. J. Plorde, Comparison of three DNA hybridization methods for detection of the aminoglycoside 2''-*O*-adenylyltransferase gene in clinical bacterial isolates. *Antimicrob. Agents Chemother. 28*:69–73 (1985).
87. Gadler, H., Nucleic acid hybridization for measurement of effects of antiviral compounds on human cytomegalovirus DNA replication. *Antimicrob. Agents Chemother. 24*:370–374 (1983).
88. Gadler, H., A. Larsson, and E. Sølver, Nucleic acid hybridization, a method to determine effects of antiviral compounds on herpes simplex virus type 1 DNA synthesis. *Antiviral Res. 4*:63–70 (1984).

RECENT REFERENCE ADDED IN PROOF

Tenover, F. C., Diagnostic deoxyribonucleic acid probes for infectious diseases. *Clin. Microbiol. Rev. 1*:82–101 (1988).

II
MONOCLONAL ANTIBODY–BASED IMMUNOASSAYS

15
A Practical Guide to Making Hybridomas

RICHARD A. GOLDSBY
Amherst College, and University of Massachusetts at Amherst, Amherst, Massachusetts

I. INTRODUCTION

In the dozen years since Köhler and Milstein published their now classic paper on the derivation of cell lines that produce monoclonal antibodies of predefined specificity, hybridoma technology has made major contributions to many areas of basic and applied biology. In general, where conventional antisera have gone, monoclonal antibodies have followed, and often displaced, their polyclonal cousins. Workers in many different disciplines have found that the making of monoclonal antibodies has become a technology that they must employ from time to time. In this chapter we will provide methods for the derivation, stabilization, and characterization of antibody-producing hybridoma cell lines. It is hoped that this will serve as a basic and practical guide to the production of useful reagents. Further reading may be found in the references at the end of this chapter (1-4).

II. MONOCLONAL VERSUS POLYCLONAL ANTIBODIES: WHY AND WHEN TO MAKE A MONOCLONAL ANTIBODY

Characteristically, antigens bear a variety of antigenic determinants (epitopes). When an animal mounts an immune response to a multideterminant antigen, each immunogenic epitope triggers one or more B-cell clones to divide and differentiate, thereby producing a number of distinct, antibody-secreting, clones of plasma cells. Consequently, the (monoclonal) antibodies made by each activated B-cell clone pool in the circulation and the serum obtained from an animal is a polyclonal mixture of many different monoclonal antibody products from a broad diversity of activated B-cell clones. After a successful immunization, the polyclonal population of serum antibodies will contain a subpopulation (characteristically polyclonal) of antibodies that recognize determinants present on the immunogen. In addition, it should be noted that, even when the same immunization protocol is used, the composition of this polyclonal mixture will change from animal to animal and even from day to day within the same animal. The specificity of polyclonal antiserums can usually be vastly improved by removing unwanted specificities and cross-reactivities by judicious programs of absorption. Also, the antigen-reactive antibodies often can be isolated by affinity purification on immobilized antigen. However, even when this strategy is successful, the affinity-purified polyclonal antibody preparation obtained is a population of immunoglobulin (Ig) molecules whose different subpopulations may bind to different determinants of the antigen, be of different Ig classes or subclasses, and have different variable-region sequences.

This rather chaotic state of affairs can be avoided if one harvests spleen cells from recently immunized donors, constructs hybrids by fusion with a suitable plasmacytoma line, and then identifies those hybridoma clones that secrete antibody to the immunizing antigen. In this approach, each of the hybridomas selected produces a monoclonal antibody that recognizes only a particular epitope of the antigen and offers the following advantages:

1. A stabilized hybridoma cell line, stored under proper cryopreservation, provides a "perpetual" source of a well-defined homogeneous antibody.
2. Large amounts (tens or even hundreds of milligrams) of a particular monoclonal antibody can be obtained with a relatively modest investment of resources and personnel. In fact, the upper limit on the amount of a particular monoclonal antibody that can be obtained is determined only by one's willingness to invest time and resources in its production.
3. Monoclonal antibodies specific for a particular target antigen can be obtained even when the antigen is grossly impure or present in only trace amounts. Only two conditions must be met for one to prepare a monoclonal antibody to a particular antigenic determinant and they are (a) members of at least one B-cell clone reactive with the antigen must be successfully hybridized, and (b) a screening procedure must be devised that is capable of distinguishing monoclonal antibodies that bind the antigen of interest from those that do not.
4. Because monoclonal antibodies react with determinants in an all-or-none fashion, there is no need (in fact, it is impossible) to improve the specificity of a monoclonal antibody by resorting to programs of absorption.

The advantages of monoclonal antibodies are widely appreciated. However, it is extremely useful to be aware of some quirks of monoclonal serology that contrast with the serology of conventional polyclonal preparations. As noted, polyclonal antisera are mixtures of immunoglobulins, and they typically contain antibodies of many different affinities, specificities, antibody classes, and subclasses. On the other hand, a monoclonal antibody is uniform for all of these properties. It is important to be aware of these contrasts because different serological assays depend upon different properties and classes of antibody. Bear in mind that only certain Ig classes (also IgM and certain IgG subclasses in the mouse) activate the complement fixation pathway. Thus, assays that depend upon complement fixation will fail if one tries to use a monoclonal antibody of a non-complement-fixing class or subclass. Precipitation assays, such as Ouchterlony or radial immunodiffusion (RID), require a sufficiently multivalent interaction between antigen and antibody to form a latticework. Radioimmunoassay or enzyme-linked immunosorbent assay (ELISA) of compounds present in trace amounts ($< 10^{-5}$ M) are dependent upon antibodies with high affinities for the antigen of interest. Some screening techniques, such as immunoblotting during Western analysis or searches for the production of a particular polypeptide or peptide fragment by an expression vector-based recombinant DNA library, require that the antibody recognize whatever determinant characteristics of the target antigen happen to be manifested. In contrast to a single monoclonal antibody, a polyclonal serum will often contain a subpopulation of antibody molecules with properties that are appropriate to any one of the assay strategies just described. Certain advantages notwithstanding, a particular monoclonal antibody may lack the properties necessary to function in one or more of these widely used assays. These considerations must be recognized to appropriately design and use monoclonal reagents.

III. MAKING MONOCLONAL ANTIBODIES

The derivation of monoclonal antibodies is an activity that can consume a substantial amount of time and material, and it should not be casually undertaken. Therefore, before starting the production of a collection of antibody-producing hybridomas, one should give careful attention to the following three critical phases that are an essential part of any rational program to produce truly useful monoclonal antibodies:

1. The decision to make monoclonal, instead of polyclonal, reagents
2. The design of an assay that will identify monoclonal antibodies with the desired properties
3. The derivation and stabilization of hybridomas that secrete monoclonal antibodies

These considerations, all of which are important, will be discussed in turn.

A. The Decision to Make Monoclonal, Instead of Polyclonal Reagents

For decades, polyclonal antisera, usually produced in rabbits, sheep, or goats, have made valuable contributions to basic and applied biology. Because of their relative ease of derivation and, in a few instances, their superior suitability, they will continue into the foreseeable future to occupy an important niche in the immunological armentarium. Therefore, before beginning the production of hybridomas, one should determine that monoclonal antibodies are more suitable for the task than are polyclonal reagents. Granting the availability of the antigen in a high state of definition (preferably pure) in amounts that are not limiting for the immunization protocol and a suitable panel of defined absorbents for removing unwanted specificities and cross-reactivities, polyclonal antisera are well suited to the following applications:

1. The detection of any one or several of a number of antigenic determinants, all of which are characteristic of an antigen of interest. Specific examples are provided by techniques such as Western blotting and the screening of expression libraries. In the Western blotting procedure, a particular determinant denatured during the sodium dodecyl sulfate-polyacrylamide electrophoresis (SDS-PAGE) phase may not renature on the blot, or it may be encoded by a polymorphic locus and, hence, not be present on all molecules of the antigen. Although such a determinant may be missed with a monoclonal antibody that is specific for that particular determinant, a polyclonal antibody that incorporates specificities for a number of the antigen's epitopes would be a more reliable probe. Because a particular member of an expression library may express any one, or more, of the many antigenic determinants characteristic of the cloned gene product, a polyclonal antibody probe will be the probe of choice if one wishes to identify the maximum number of clones that express any coding sequence. On the other hand, a selected library of monoclonal antibodies could be used to pick only those colonies producing peptides that display a particular epitope or set of epitopes.
2. Although it is true that each batch must be carefully standardized, appropriately absorbed, and affinity purified, polyclonal antibodies are as suited as monoclonal reagents to most assays of well-defined antigens. The assay of hormones, such as insulin; serum proteins, like immunoglobulins; and drugs, such as digoxin, provide familiar examples of situations in which polyclonal antibody works well.

Nevertheless, as useful as skillfully prepared polyclonal antibodies can be, there are a number of situations in which monoclonal antibodies are clearly the reagents of choice. Consider a situation for which it is a reasonable presumption that either a monoclonal or a polyclonal antibody would behave satisfactorily in an assay or as an immunosorbent for affinity chromatography. If one decides that, for reasons of standardization, it is highly desirable to use an antibody preparation of identical serological and biochemical properties for a number of times and for an indefinite period that extends well into the future, the monoclonal route should be taken. These considerations apply with even greater force when there is a possibility that other laboratories may wish to compare results or standardize protocols for assays or purification procedures.

If pure antibody is desired or required, polyclonal antibody of the required specificity can be obtained only if a suitable immunoabsorbent is available. Regrettably, this too often is not the case. Some antigens are too scarce (consider, for example, a hormone receptor or a minor cell surface antigen). Many times, the target antigen is not available in a purity suitable for use in the affinity purification of antibodies. In contrast with polyclonal serology, hybridoma technology offers alternative pathways to pure antibody. Although monoclonal antibodies can be obtained by affinity purification on the target antigen, they can also be obtained from hybridoma culture fluid without resorting to any program of purification on target antigen, whatsoever.

Finally, we note that there are some problems that can be approached (and solved) only by bringing hybridoma technology to bear. Consider those examples, many of which come from cell biology, in which highly specific and extremely useful monoclonal antibodies have been prepared against antigens whose very existence at the time of immunization was either conjectural or totally unknown. This was the situation for the T (originally OKT) series of antibodies that identify functional human T-cell populations (T4-helper/amplifier subset; T8-cytotoxic subset). Similar successes in identifying and mapping previously unknown cell populations and subpopulations of the nervous system have been achieved. There are also those circumstances for which one wishes to prepare a highly specific antibody against an antigen whose identity is known, but for which suitable absorption programs to render polyclonal antisera specific are impractical or infeasible. Another situation is one in which the antigen of interest is available in only trace amounts, a condition that is often confounded by the presence of substantial amounts of contaminating substances. Techniques of immunization, both in vivo and in vitro, have been devised that require less than a microgram of (not necessarily pure) antigen and enable one to derive antibody-secreting hybridomas. Appropriate members of the library of monoclonal antibodies so derived can be used to isolate useful amounts of the pure antigen.

B. The Design of an Assay That Will Identify Monoclonal Antibodies with the Desired Properties

Assay design is one of the most important determinants of whether or not a monoclonal antibody project will result in the production of reagents that have the desired characteristics of specificity, antibody class, and affinity. *The guiding principle in designing a screen is to make the conditions of screening mirror, as closely as possible, the actual application in which one intends to use the antibody*. Specifically, if one wishes to use the antibodies in cytotoxicity or hemolytic assays, one should design a screen that will select not only for specificity but also for those mouse Igs (IgM, IgG2a and IgG2b) that efficiently fix complement. Monoclonal antibodies that are intended for discrimination among closely related cell types, viruses, or molecules should be subjected to early screening against a number of closely related, as well as irrelevant, antigens. Such a procedure, which should be conducted early in the derivation, not only allows rejection of inappropriate specificities, but can reveal interesting and, sometimes, useful cross-reactivities. High affinity is required when one intends to use the antibodies in radioimmunoassays or ELISA protocols for the measurement of compounds, such as drugs, toxins, or hormones, that are present in only trace amounts. Simple and rapid assay strategies are available for the selection of high-affinity monoclonal antibodies, and two approaches to this problem will be described. Finally, there are some characteristics that, although essential for a particular application, do not readily lend themselves to rapid-screening procedures. Examples include antibodies for use in precipitation assays such as radial immunodiffusion, or antibodies for fluorescence-based analysis of cell populations. In such situations, one constructs a primary screen that allows the rapid and efficient selection of a panel of candidate antibodies, all of which have the desired specificity. The hybridomas secreting these candidate antibodies are stabilized and cryopreserved, and the panel is then screened to determine which members have the required properties.

Some guidelines for obtaining antibodies of a particular class or of high affinity are outlined in the following discussion.

1. Biasing Antibody Class

One may wish to obtain IgMs because they are highly efficient in assays that depend upon complement fixation and may display high avidities. The immunization schedule employed before fusion will be the most important determinant of whether one obtains a high proportion of IgM or of other classes of Ig. Fusion 3 days after the priming of naive animals will yield the highest proportion of IgM antibodies. Alternatively, if, for reasons of tissue distribution, ease of handling, and purification, certain other Ig classes are highly desirable, two steps may be taken. First, the immunization schedule should allow a period of 4 weeks to elapse between priming and boosting. Second, one should take the trouble to screen the library with a class-specific reagent. For example, if mouse IgG2a or IgG2b antibodies are desired, suitably labeled protein A may be used as a screening reagent. It should also be noted that appropriately labeled anti-immunoglobulin class- and subclass-specific reagents can be prepared or purchased.

2. Screening For High Affinity

There are a variety of approaches to identifying monoclonal antibodies of high affinity, and two simple and straightforward methods that are readily conducted in microtiter plates are described here.

Blocking at Low Antigen Concentrations

Step 1:	(a) Incubate 0.1 ml of hybridoma supernatant plus 0.1 ml of antigen (10^{-5}–10^{-7} M) for 2 hr. (b) In parallel, incubate 0.1 ml of hybridoma supernatant plus 0.1 ml of antigen-free solution for 2 hr.
Step 2:	ELISA or solid phase RIA of supernatants.
Step 3:	Comparison of ratio of (+)antigen/(−)antigen.
Judgment criterion:	Highest affinity monoclonal antibodies show smallest ratios.

Antigen Capture at Low Concentrations

Step 1:	16-hr incubation of 0.1 ml of hybridoma supernatant with 0.1 ml of radiolabeled or biotinylated antigen (adjust antigen concentration to between 10^{-7}–10^{-9} M).
Step 2:	Addition of 0.1 ml of Sepharose to which antimouse Ig has been covalently attached.
Step 3:	Agitate for 30 min.
Step 4:	Three sequential washes of the beads.
Step 5:	Determine bead-associated radioactivity or biotin.
Judgment criterion:	Only high-affinity monoclonal antibodies will bind low concentrations of labeled antigen.

C. The Derivation and Stabilization of Hybridomas Secreting Monoclonal Antibodies

A variety of techniques have been applied to the construction of antibody-secreting hybridomas, and different laboratories tend to have a fondness for a particular version of the basic methodology. In the next few sections we present a protocol for hybridoma production and preservation that has served us well. Although we recommend its use as a proven method, the reader should be aware that there are several other protocols that work equally as well. We advise the adoption and consistent use of either this method or (if the reader prefers) another proven method. What will likely result in erratic yields of hybridomas is the frequent switching about from one method to another or the random "hybridization" of one well-worked-out protocol with another equally well-worked-out, but different, protocol. The laboratory that picks one method and sticks with it until true facility is achieved will likely be rewarded by abundant yields of hybrids.

1. Materials and Equipment

1. Water-jacketed CO_2 incubator
2. Bench-top centrifuge
3. Inverted microscope
4. Liquid nitrogen refrigerator
5. -70°C deep-freeze unit
6. Easy access to a gamma-counter or ELISA reader
7. Laminar flow hood
8. A suitable established cell line such as SP 2/0
9. RDG: 50% RPMI/50% Dulbecco's Modified Eagles Medium (DME)
10. RDGS: RDG containing 2 mM L-glutamine, 20% fetal calf serum and 20 μg/ml of gentamicin, if an antibiotic is desired
11. HAT: RDGS containing 1×10^{-4} M hypoxanthine, 4×10^{-7} M aminopterin, and 1.6×10^{-5} M thymidine
12. HT: RDGS containing 1×10^{-4} M hypoxanthine and 1.6×10^{-5} M thymidine
13. Freezing medium: 20% v/v fetal calf serum, 70% v/v phosphate buffered saline (PBS), and 10% v/v dimethyl sulfoxide (DMSO)
14. Freezing ampoules (Nunc or similar design)
15. BALB/c mice
16. Sterile, frosted microscope slides
17. 50- and 15-ml plastic centrifuge tubes
18. 96-well, flat-bottomed, microtiter dishes
19. 40% polyethylene glycol (1/2 ~1500 MW and 1/2 ~4000 MW) in RDG
20. Several gross of sterile cotton-plugged Pasteur pipettes and bulbs
21. 100-mm culture dishes (tissue culture grade or, more economically, bacteriological culture plates)
22. 24-well flat-bottomed cell culture dishes

2. Immunization

Two immunizations (a priming injection and a boost 3-4 weeks later) are usually quite sufficient to ensure the presence of immune spleen cells. Use the following guidelines for immunizations.

1. Prime two or three times more mice than will be required for a hybridization. In the event a mishap or failure to obtain an adequate supply of hybrids makes it necessary to repeat the hybridization, a supply of antigen-primed mice saves time.
2. Immunization with soluble antigens: prime mice by ip injection of 100 μg of the antigen emulsified in complete Freund's adjuvant (CFA). Boost after an interval of 21 days, or more, by iv (the tail vein works well) or ip injection of 0.2 ml of an aqueous solution (i.e., PBS) containing 10-50 μg of the antigen. If the unmodified antigen is unlikely to be antigenic (i.e.,

small peptides, weakly immunogenic proteins, haptens) it is essential to conjugate it to a strongly immunogenic carrier. The mollusk respiratory protein, keyhole limpet hemocyanin (KLH), is a frequently used and highly effective carrier.

3. Note, the recommendation that one prime with 100 μg of antigen should be viewed as a rough guideline. Workers have achieved consistent success with 50 and even 10 μg doses. Successful immunizations have been conducted by immunization with the small amounts of antigen present in a band from an acrylamide gel. When it is necessary to immunize an animal with only a microgram or so of antigen, a route of immunization other than ip should be considered. Injection of the antigen emulsified in CFA into the foot pad, with subsequent harvest and hybridization of lymphocytes from the draining lymph nodes has been successful.
4. Immunization with cells: harvest the cells and wash them twice in PBS. Prime the mice by ip injection of 1-2 $\times 10^7$ eucaryotic cells or 1 $\times 10^8$ bacteria in a volume of 0.2-0.3 ml (the cells may be emulsified in CFA, but this is not necessary for most applications). Boost after 21 days by ip injection of one-tenth the priming dose of cells suspended in PBS.
5. If hyperimmune mice are to be used as spleen cell donors, the animals should be rested 30 days before receiving a second boost.

3. *Spleen Cells and Established Cell Line Fusion Partner*

Spleen Cells. Timing of animal sacrifice is important because it is the actively dividing lymphoblastoid population that gives rise to most of the hybridomas. Three days postimmunization has been found to be an optimal time to harvest spleen cells for a number of antigens. It is useful to exsanguinate the mice by decapitation and to harvest and save the serum to aid in evaluating the success of the immunization protocol and for use as a positive control in assay procedures. Although the appearance of antibody in the serum 3 days after boosting is proof that one has activated lymphocytes, a negative result or a low serum titer does not mean the activation has failed. When there are limitations on the number of immunizations that can be done, it is sometimes advisable to proceed with the fusion of spleen cells from immunized animals that present negative sera. Collect spleen cells as follows:

1. After sacrifice, immerse the mice in 70% ethanol and transfer to a laminar flow hood. Sterilely remove the spleens, place them in 15 ml of RDG and grind them between the faces of frosted microscope slides to release the cells. Transfer the suspension to a 15-ml centrifuge tube, allow the larger tissue fragments to settle (about 1 min), and centrifuge for 10 min at 400x*g*. Resuspend in 25 ml of RDG.
2. Count the cells by diluting 0.5 ml of the suspension with 4.5 ml of 2% acetic acid (to lyse the red blood cells) and immediately count the diluted

suspension with a hemacytometer. A single boosted spleen will yield around 2×10^8 cells.

3. Note that it is usually better to harvest and pool the spleen cells of two mice to increase the potential diversity of the hybridomas obtained.

Established Cell Line Fusion Partner. Over the years, a number of cell lines have been demonstrated to serve as efficient partners for the immortalization of specific antibody production. The lines currently used do not secrete immunoglobulin characteristic of the myeloma parent. Although many of these lines work well, we recommend the use of SP 2/0; actually not a myeloma, but a non-secreting hybridoma variant, selected in the laboratory of George Köhler. This line, like many others, is readily available from the American Type Culture Collection. The following considerations provide guidelines for the growth of SP 2/0.

1. Maintain the cell line in log phase until it is harvested for use in a fusion by seeding each of three or four 100-mm culture dishes with 5×10^5 SP 2/0 cells in 10 ml of RDGS and splitting the cultures each day until use. Do not allow the cultures to overgrow, most workers report a decrease in viability once the culture exceeds a density of $3\text{-}5 \times 10^6$ cells/100-mm culture dish.
2. It is good practice to periodically passage the cells for a week in 2.5 μg/ml 6-thioguanine (6TG) or 25 μg/ml 8-azaguanine to remove any cells that have reverted to $HGPRT^+$. However, it is important to wash cells completely free of the selective agent and grow them in a medium that is free ofselective agent for at least 3 days before their use in a fusion.

4. *Fusion*

1. About 15 min before harvesting the spleen cells, place the following solutions in a 37°C bath to warm up: 40% PEG, RDG, and HAT.
2. Mix 2×10^8 spleen cells with 2×10^8 SP 2/0 cells and copellet the mixture by centrifugation at 400×*g* for 10 min at room temperature.
3. Completely remove all supernatant from the pellet by aspiration with a pipette.
4. Loosen and reslurry the pellet by repeatedly tapping the bottom of the tube on the hard metal surface of the hood. All subsequent steps of the fusion are carried out at 37°C in the confines of the laminar flow hood.
5. Place the tube containing the pellet in a beaker of 37°C water and add 1 ml of 40% PEG, gently agitate the tube and incubate for 1 min and then add 3 ml of 40% PEG. Mix and incubate for an additional 2.5 min with periodic gentle agitation.

6. Add 4 ml of RDG, mix gently and incubate 2 min; then add 8 ml of RDG with gentle mixing and incubate an additional 2 min. Gently mix in 16 ml of RDG, wait 2 min, and then dilute the contents of the tube up to 50 ml.
7. Pellet the cell suspension by centrifugation at 400X*g* for 10 min.
8. Discard the supernatant, and by gentle pipetting, resuspend the cells in 50 ml of HAT.
9. Feeder layers: feeder cells may be mouse or rat thymocytes, mouse macrophages, or irradiated spleen feeder cells. In 96-well, microtiter dishes use 2×10^5 thymocytes per well, 1×10^4 macrophages per well, or 2×10^5 irradiated mouse spleenocytes per well. If spleen cells are used they should be irradiated with 1300-2000 rad or a sufficient dose of UV light to kill 90% of a myeloma cell culture. If unirradiated spleen cells are used, overgrowth of well cultures by fibroblasts is a likely outcome.
10. Use 30 ml of the suspension of fused cells produced in step 8, as is and dilute one 15-ml portion with 45 ml of HAT and dilute the remaining 5 ml with 35 ml of HAT. Add 0.1 ml of the undiluted suspension (4×10^5 spleen cell equivalents) to each of the 96 wells of microtiter dishes that have been previously seeded with 0.1 ml of a feeder cell suspension. The fourfold and eightfold dilutions of the cell suspensions are also seeded in 0.1-ml aliquots into respective sets of feeder layer-containing microtiter dishes.
11. The plates are incubated in a 37°C humidified incubator containing a 7.5% CO_2/92.5% air atmosphere.
12. After 2 days, 1 drop of HT is added to each well. At 4 days postfusion, one-half of the medium is removed and replaced with HT. This process of medium replacement/feeding is repeated every 4 days. When clones have reached a diameter of 1-2 mm (usually 9-14 days) a sample of their culture fluid should be tested to determine if they are secreting antibody of the desired specificity. It should be pointed out that not all hybridomas develop at the same rate. Wells should be maintained for 3 weeks before concluding that they do not contain viable hybridomas. Replace whatever medium is removed for testing with 0.1 ml of HT.
13. Within 24 hr of sampling, replace all of the culture fluid with fresh HT, and transfer the positive clones to the wells of 24-well dishes. Before transfer of hybrid clones, the wells of these dishes should be charged with 1 ml of a feeder cell suspension (1×10^6 thymocytes or irradiated spleen cells) in HT.
14. Two-days posttransfer add 1 ml of HT and test the supernatant for the presence of the desired antibody when the cells are one-half confluent.

15. Immediately clone the positive cultures by limiting dilution onto feeder layers in RDGS in 96-well dishes. Clone at the following cell densities: 24 wells at 1 cell per well; 24 wells at 2 cells per well; 24 wells at 5 cells per well, and 24 wells at 10 cells per well.
16. Expand the balance of the culture by dilution to 5 ml with HT and transfer to a 60-mm tissue culture dish. When these cultures become dense, transfer them to a 100-mm dish. With rapidly growing hybrids it will be necessary to split these cultures within 24-48 hr.
17. After two 100-mm cultures have been obtained, the hybridoma cells should be frozen as a protection against accidental loss. Also, a portion of each hybridoma should be inoculated into a mouse so that substantial quantities of potentially valuable antibodies begin to accrue even before the first cycle of cloning prescribed in the previous step has produced a batch of clones.
18. When positive clones of interest have been identified, two subclones of each of the positive cultures should be expanded, frozen, and then injected into mice as previously indicated.
19. Use the following simple protocol for the freezing of cells: (a) Select healthy logarithmically growing cultures for cryopreservation; (b) centrifuge the cells out of the growth medium (400-600×*g* for 10 min); (c) Resuspend in freezing medium. (d) Put 1 ml of the cell suspension into as many 2-ml Nunc freezing vials as necessary. Immediately place the ampoules on freezing canes and stand these canes in the -70°C deep-freeze unit in such a manner that they do not contact the walls or floor of the unit and leave them overnight. The next day, quickly transfer the ampoules to liquid nitrogen. (e) To return frozen cells to culture, quickly thaw the ampoule in a 37°C water bath; (f) immediately dilute the contents of the ampoules to 10 ml with growth medium, containing 20% serum, and centrifuge at 400×*g* for 10 min. Remove the supernatant and resuspend the cells in 10 ml of RDGS and place in a 100-mm culture dish.

5. *Growth of Hybridomas as Tumors*

For Solid Tumors

1. Harvest cells from growth medium by centrifugation.
2. Resuspend to a cell concentration of 2-5 × 10^7 ml.
3. Inject two or more mice with 0.1-0.2 ml, subcutaneously, just anterior to the right or left rear leg.
4. Tail bleed the mouse when the tumor attains the size of a garden pea (10-20 days) and at 3- to 4-day intervals thereafter until the tumor becomes life-threatening. At this point the mouse should be exsanguinated and the

tumor immediately excised, placed in 10 ml of serum-free growth medium, minced and ground between the faces of frosted glass slides. Aliquots of mince (0.2 ml per mouse) may be reinjected into a number of mice. The balance of the tumor should be harvested by centrifugation, resuspended in freezing medium, and frozen as outlined in the protocol for cryopreservation.

For Ascites

1. Inject the mice intraperitoneally with 0.5 ml of pristane (available from the Aldrich Chemical Co.).
2. Two weeks to 8 weeks after pristane priming, inject (use an 18-gauge needle) $2\text{-}5 \times 10^6$ logarithmically growing hybridoma cells in a volume of 0.5 ml of serum-free medium or PBS.
3. Tap the ascities fluid, when the abdomen becomes swollen, by insertion of a 16 or 18 gauge needle (no syringe) into the lower abdomen. Allow the fluid to drip into a suitable collection vessel (a 15-ml centrifuge tube is convenient). After the initial collection, it is usually possible to tap the animal at 2-day intervals until death.
4. Remove the cells by centrifugation ($400 \times g$ for 10 min). Pipette off the antibody-containing supernatant for assay and freezing. Resuspend the cells to a concentration of 1×10^7/ml in PBS and inoculate as many primed mice as desired with 0.5 ml of the cell suspension. The balance of the cells may be cryopreserved as described earlier.

IV. IN VITRO IMMUNIZATION

Most hybridoma projects will be conducted by harvesting cells from the spleens or lymph nodes of mice that have been immunized with the antigen of interest. However, it is important to realize that in vivo immunization is not the only course available for the production of antigen-activated B-cell blasts for fusion. The technique of in vitro immunization provides a useful alternative to the traditional methods of in vivo immunization. It also offers the following important additional advantages:

Defined levels of antigen can be maintained throughout the immunization.

A useful route to the preparation of human monoclonal antibodies is provided in those many situations when in vivo immunization is infeasible or ill-advised.

Successful immunizations can be accomplished with extremely small antigen concentrations.

Antigens that are toxic at the organismal level may be innocuous in cell culture.

It is possible to produce antibodies to "self" or to highly conserved antigens that are difficult to produce by conventional immunization because of tolerance.

The procedure offers the option of utilizing regulatory lymphokines and monokines to attempt modulation of the response.

Luben and Mohler (5) were the first investigators to use in vitro immunization to obtain antigen-activated B-lineage cells for the hybridization. These investigators wanted to produce monoclonal antibodies to osteoclast activating factor (OAF), a T-cell-derived lymphokine that promotes bone resorption. However, a laborious program of purification was necessary to prepare the antigen in even microgram amounts. Because conventional protocols for the generation of hybridomas routinely involve multiple immunizations of whole animals, they require the expenditure of tens to hundreds of micrograms of antigen. In an experiment that made a crucial advance in the sensitivity and flexibility of the hybridoma technique, Luben and Mohler demonstrated that naive BABL/c spleen cells could be immunized against human OAF in vitro and subsequently fused to a suitable myeloma partner. In fact, these workers showed that significant numbers of hybridomas could be obtained from in vitro immunization with as little as 100 ng of antigen, and some hybrids were obtained when only 10 ng were used. Subsequent to these pioneering studies by Luben and Mohler, Pardue and colleagues (6) demonstrated that the in vitro immunization procedure enabled one to produce monoclonal antibodies against highly conserved, weakly immunogenic antigens. Specifically, these workers used the in vitro approach to prepare monoclonal antibodies against calmodulin, a highly conserved calcium-binding regulatory protein.

Those who want to use the in vitro route to antibody-producing hybridomas may wish to purchase a kit, the IVIS (Dupont, New England Nuclear Division; Boston, Massachusetts). Alternatively, those who prefer to perform do-it-yourself in vitro immunizations will find the following protocol adapted from Reading (7) and McHugh (8) useful as a point of departure:

1. Prepare mixed thymocyte culture-conditioned medium (TCM) by sterilely removing the thymus glands from five young (4-8 weeks old) BALB/c mice and a similar number of C57/B6 mice. The pooled glands are placed in HEPES-buffered (10 mM) Hank's balanced salt solution (HHBSS) and pressed through a stainless steel screen or ground between the frosted surfaces of glass slides. The thymocytes are counted and suspended at a concentration of 5×10^6/ml in DMEM supplemented with 2 mM glutamine and 2% rabbit serum. Fifty-milliliter aliquots of this suspension are cultured in 75-cm^2 flasks at 37°C for 48 hr. The cultures are then harvested by centrifugation and the supernatant, TCM, is filtered (0.2 μm),

divided into 10-ml aliquots and stored at -80°C. Under these conditions TCM preparations retain their activity for at least 6 months and probably indefinitely.

2. Sterilely harvest the spleen from an umprimed BALB/c mouse and prepare a suspension of spleen cells as described in step 1. Allow the clumps of tissue to settle for 30 sec and harvest the spleen cell-containing supernatant.
3. Adjust the volume of the suspension to 20 ml. Add 1-100 μg of soluble antigen to the flask or, if eucaryotic cells are to be used as antigen, 10^7 irradiated (2500 rad) whole cells. A 10-ml vial of TCM is rapidly thawed at 37°C and added to the spleen cell suspension, mixed, and the TCM-supplemented suspension is transferred to a 75-cm^2 flask, flushed with 5% CO_2/95% air, and placed on its culture face in a 37°C incubator. The flask is left undisturbed for 5 days and then harvested and fused to an appropriate established cell line along the outlines of the procedure detailed in Sect. III.C.
4. Note that the amounts of immunogen recommended are merely points of departure. In setting up an in vitro immunization, one should explore a variety of antigen concentrations to obtain optimal responses. One should test antigen concentrations ranging from as little as 5 ng/ml to 5 μg/ml or more. Similarly, if cells are used as antigen, the cell concentration should be varied over 3 or 4 orders of magnitude (i.e., with eucaryotic cells the range of cells added to a 75-cm^2 flask should be varied from 10^5 to 10^8).

V. CONCLUSION

Hybridoma technology is a powerful and constantly evolving approach to the production of highly defined antibody reagents. The procedures detailed in this chapter represent a basic guide to producing useful libraries of monoclonal antibodies. It is hoped that laboratories contemplating or just beginning hybridoma work will find this discussion a helpful introduction and useful point of departure for their work.

REFERENCES

1. Fazekas de St. Groth, S., and D. Scheidegger, Production of monoclonal antibodies: Strategy and tactics. *J. Immunol. Methods 35*:1 (1980).
2. Hammerling, G. J., U. Hammerling, and J. F. Kearney, *Monoclonal Antibodies and T Cell Hybridomas, Perspectives and Technical Advances*. Elsevier, New York (1981).
3. Goding, J. W., *Monoclonal Antibodies: Principles and Practice*. Academic Press, New York (1986).

4. Springer, T. (ed.), *Hybridoma Technology In the Biosciences and Medicine*. Plenum Press, New York (1986).
5. Luben, R. A., and M. A. Mohler, In vitro immunization as an adjunct to the production of hybridomas producing antibodies against the lymphokine osteoclast activating factor. *Mol. Immunol. 17*:635 (1982).
6. Pardue, R. L., R. C. Brady, G. W. Perry, and J. R. Dedman, Production of monoclonal antibodies against calmodulin by in vitro immunization of spleen cells. *J. Cell Biol. 96*:1149 (1983).
7. Reading, C. L., Theory and methods for immunization in culture and monoclonal antibody production. *J. Immunol. Methods 53*:261 (1982).
8. McHugh, Y., In vitro immunization for hybridoma production. In *Hybridoma Technology in Agricultural and Veterinary Research*, N. J. Stern Gamble, H. R. (eds.). Rowman & Allanheld, Totowa, N.J., pp. 216 (1984).

16
Chromatographic Analysis and Purification of Antibodies

DAVID R. NAU
J. T. Baker Inc., Phillipsburg, New Jersey

I. INTRODUCTION

The purification of antibodies is one of the oldest challenges in protein biochemistry. Even today, despite a long history of use of antibodies as tools for analysis, identification, characterization, and purification, the development of rapid and economical purification schemes for antibodies remains a major problem for research and production personnel. Recently, these purification problems have become even more acute. The primary reason for this dilemma is that the development of monoclonal antibody technology (1), by providing an enormous impetus for progress in numerous fields of science and technology (2), has created new and more demanding purification problems. Large quantities of high-purity nonpyrogenic immunoglobulin are now required for in vivo diagnostics (imaging) and immunotherapeutics (3,4), for use in affinity chromatography for the purification of pharmaceuticals (5,6) and, to a lesser extent, for in vitro diagnostic kits (7,8).

Before the advent of monoclonal antibodies, the universal source of immunoglobulins was serum/plasma. Because IgG typically represents a major portion of the total protein complement of serum, rigorous purification was often not essential for the intended use. The major concern was usually the elimination of nonspecific interactions, rather than absolute product purity. Furthermore, the in vivo use of antibody preparations was rarely, if ever, attempted.

However, present trends favor using hybridoma cell cultures to produce large amounts of antibody, despite the fact that the levels of antibody are 2-3 orders of magnitude lower in culture supernatants than in ascites fluid or serum. Furthermore, most cells are cultured in the presence of fetal calf serum or horse serum, which contain high levels of many different types of proteins (9–11). Even in serum-free hybridoma tissue culture medium in which the protein complement is not extremely complex, the purification may be nonetheless difficult because the monoclonal antibody is often present as a minor component. In ascites fluids, contaminating proteins may also consist of "host" polyclonal antibodies, and fetal bovine sera also contain low levels of (40–400 μg/ml) of bovine serum immunoglobulins (9–14).

In addition, antibody purification is no longer limited to IgG, because today, numerous monoclonals of therapeutic interest belong to the IgM and other immunoglobulin classes. Furthermore, as techniques for hybridoma manipulation and immunochemistry become increasingly sophisticated, new purification procedures are needed for cell lines secreting multiple immunoglobulins, bivalent hybrid antibodies, protein-antibody conjugates, antibody fragments, hybrid fragments, and chemically modified monoclonals (15). Therefore, antibody purification is not one problem, but many.

As a result, purification methods are constantly being scrutinized and optimized to achieve more homogeneous antibody preparations. Unfortunately, most of the current approaches to immunoglobulin purification are simply variations of the long-established purification procedures for serum polyclonal antibodies.

II. TRADITIONAL ANTIBODY PURIFICATION

Preparative-scale antibody purification has traditionally been carried out on biopolymer-based DEAE-type soft gels that are based upon polysaccharides and other polymers such as cellulose, dextrans, agarose, or polyacrylamides (16–18). Although anion exchange chromatography on soft gels requires simple equipment, can be scaled-up, to a certain extent, and produces antibodies at a purity that is adequate for a number of end uses, traditional soft gels suffer from a number of serious disadvantages as general preparative-scale purification tools. Most importantly, soft gels provide extremely low resolution between antibodies and contaminating proteins such as albumins and transferrins. Chromatography must typically be combined with an initial ammonium sulfate salt fractionation step (16–18). Soft gels have inherently low capacities for binding proteins. Thus, large quantities of gel and extremely long chromatographic run times are required to achieve adequate separations. In addition, high flow rates, which are desirable in preparative chromatography for high throughput and rapid reequilibration and regeneration, are not possible on soft gels because of high backpressures, poor mechanical strength, and the possibility of column collapse. The life expectancy of soft gels is also rather short. It is often tempting to discard the matrix after each use because of the cumbersome maintenance and regeneration procedures and because polysaccharide-based gels, themselves, support the growth of microorganisms. This biodegradability, as well as chemical instability and difficulty in sterilization, lead to the possibility of endotoxin and degradation product contamination of the final antibody preparation. Finally, soft gels tend to bind immunoglobulin irreversibly, which decreases yield and the overall economics of the purification scheme.

III. MODERN ANTIBODY PURIFICATION

A. High-Performance Anion Exchange Chromatography

Recent work in our laboratories and in those of others, indicates that chromatography on high-performance anion exchange matrices may produce a useful approach to monoclonal antibody purification (9-12,19-23) because these materials overcome many of the problems inherent to soft gels. Most importantly, these high-performance liquid chromatography (HPLC) materials give much higher resolution than traditional matrices. Furthermore, the capacity of these matrices is typically much higher than that of soft gels, binding as much as 150 mg of purified IgG per gram of packing (9-11). These factors, combined with the ability to use higher flow rates, enable substantially higher throughput on much smaller columns. Finally, HPLC supports tend to be much more rugged than soft gels, thus providing a longer column lifetime and an increased chemical stability.

Although substantial progress has recently been made in the area of monoclonal antibody purification, it has become increasingly apparent that the purification of monoclonal antibodies by high-performance anion exchange chromatographic matrices also has serious limitations. The "background" profiles of contaminating nonimmunoglobulin proteins tend to vary considerably depending upon species, strain, and individual hybridoma. More importantly, individual monoclonal antibodies may elute at a wide range of retention times within the gradient (9-11,21,22,24), thereby reducing the possibility of achieving homogeneous antibody and making method development and identification of the antibody peak difficult. The binding of dyes (e.g., phenol red), decreases resolution by broadening the antibody and albumin peaks and changes the elution characteristics of the immunoglobulin peaks, because it changes the mechanism of the separation (9-12,24). These dyes also dramatically reduce the ability of the anion exchange matrices to bind protein to a mere fraction of their original capacity (9-12,24). In fact, anion exchange chromatography is a priori undesirable as an initial step for large-scale murine antibody purification because virtually all of the proteins and dyes are bound, and this drastically reduces the capacity of the matrix that is available to bind immunoglobulin (9-12,24). This reduction in capacity is a particular disadvantage when attempting method development because analytical chromatography on these matrices often does not allow one to correctly predict the chromatographic profile in the presence of the overloaded conditions that are typically used for scale-up to preparative chromatography (9-12,20-22,24). Finally, with very few exceptions (9-11), totally different surface chemistries must be used if preparative chromatography on traditional low pressure, open columns is to be scaled up from methods-development that was conducted on analytical HPLC media.

B. Affinity Chromatography

Another method for antibody purification that has recently gained wide acceptance is the use of protein A, protein G (for IgG), Jacalin (for IgA), concanavalin A (for IgM), anti-immunoglobulin, specific antigen, and other affinity matrices (1,13,25,26). The development of these materials represents a major breakthrough in the field of antibody purification because, in many cases, the monoclonal antibody of interest may be separated in a highly purified state with minimal contamination (9,13,25,26). However, protein A and antigen- or antibody-based affinity matrices usually require harsh elution conditions (25,26) that may denature the monoclonal antibody and reduce immunological activity or may bind immunoglobulins irreversibly. Another drawback of protein A is its inability to bind all antibody classes and subclasses, as well as antibody fragments, antibody conjugates, hybrids, and chemically modified immunoglobulins (9,13,25,26). In addition, protein A or affinity matrices have a low capacity, short column lifetime, and high cost, and as such, are typically uneconomical for preparative or process-scale purifications.

C. Hydroxyapatite Chromatography

Hydroxyapatite has proved to be a useful chromatographic matrix for the purification of monoclonal antibodies (13). The major advantage of hydroxyapatite over conventional nonaffinity methods is that most antibodies appear to be bound more strongly than the majority of the other common protein contaminants. However, hydroxyapatite is fragile, produces high back pressures and low flow rates, requires long chromatographic run times, tends to bind immunoglobulin nonspecifically and irreversibly, and suffers from an extremely short column lifetime (9,13,27). The capacity of hydroxyapatite is up to 15 times lower than that on chromatographic materials that are based on silica or even organic resins (9-11). These problems, along with broad peaks, produce a purified antibody preparation that is substantially dilute.

IV. ABx: ANTIBODY EXCHANGER

As a result of many of the disadvantages that are inherent to many of these separation techniques, protein chemists have been forced to use a series of elaborate purification steps to achieve the desired level of antibody purity. For protein chemists or process engineers faced with the task of purifying numerous antibodies from a number of sources, the most practical approach would involve the use of one or two high-resolving chromatographic matrices that have universal applicability, with minor adjustments in purification protocols to enable the attainment of any desired level of purity at any level, from analytical to process scale.

The requirements for an economical chromatographic matrix that could rapidly purify large quantities of antibodies to homogeneity compelled us to investigate synthetic approaches to construct a chromatographic surface that would bind antibodies more selectively than convention ion exchange matrices. Recent work in this laboratory led to the development of the BAKERBOND ABx (antibody exchanger) matrix (9-11,35); ABx uses mixed mode interactions as the basis (weak cation exchange, "mild" anion exchange, and "mild" hydrophobic interactions), silica gel as the most advantageous support, and a proprietary hydrophilic polymeric coverage to increase stability, eliminate nonspecific protein binding, and maximize recovery (10,11,35). This chromatographic matrix behaves like an ion exchanger, in that immunoglobulins are resolved by the manipulation of buffer species, pH, and ionic strength. However, it also exhibits an affinitylike sensitivity toward all immunoglobulins, indicative of more subtle, more complex interactions.

The major advantage of ABx is that it binds mostly antibodies while exhibiting little or no affinity for albumins, transferrins, proteases, or pH indicator dyes from tissue culture media (Table 1). Therefore, ABx can be used as rapid fractionation media to selectively remove nonpyrogenic immunoglobulins of any class or type from large volumes of ascites fluid, serum-based cell culture media, or serum/plasma from any species. The purity of the immunoglobulin obtained in a single purification step on ABx varies from about 80 to 99+%.

V. GENERAL PROPERTIES OF ABx

The ABx media are made from high-quality, closely sized chromatographic silica which produces high column efficiencies and high resolution (Tables 1 and 2; Fig. 1). This silica base is derivatized with a proprietary hydrophilic polymeric backbone that serves three purposes: (a) It covers up nonspecific interaction sites on the silica surface, leading to quantitative recovery of antibody mass and immunological activity; (b) it protects the entire surface of the silica base, thereby increasing chemical and physical stability and column lifetime; and (c) it increases ligand density and the capacity of the matrix to bind immunoglobulins (see Table 2 and Appendix, Sect. C).

Table 1 Retention Times of Polyclonal and Monoclonal Immunoglobulins on BAKERBOND ABx[a]

Source	Antibody or protein	Antigen if known[b]	Retention time (min)
Mouse ascites	IgG1,k		36
Mouse ascites	IgG2a,k		26
Mouse ascites	IgG2a,k		26
Mouse ascites	IgG2b,k		26
Mouse ascites	IgG2b,k		26
Mouse ascites	IgG3		30
Mouse ascites	IgA		16
Mouse ascites	IgA		25
Mouse ascites	IgM		28
Mouse ascites	IgG	BSA	29
Mouse ascites	IgG	SRBC	40
Mouse ascites	IgG		35
Mouse ascites	IgG		45
Mouse ascites	IgG		22
Mouse ascites	IgD		40
Mouse ascites	IgG		60/67
Mouse ascites	IgG	HCG	25
Mouse ascites	IgG	HCG	27
Mouse ascites	IgG	HCG	27
Mouse ascites	IgM		52/60
Mouse ascites	IgM		60
Mouse ascites	IgM		68
Mouse ascites	IgM		68

Mouse ascites	IgG		30
Mouse ascites	IgG	I1–2	42
Tissue culture	IgG	I1–2	42
Tissue culture	IgG	Int	40
Tissue culture	IgG	Ren	31
Tissue culture	IgG		45
Tissue culture	IgG		50
Tissue culture	IgM		30
Tissue culture	IgM	ODC	27
Tissue culture	IgG		30
Human serum	IgG	Polyclonal	22–35
Human serum	IgG	Polyclonal	22–31
Bovine serum	IgG	Polyclonal	22–31
Guinea pig serum	IgG	Polyclonal	22–31
Mouse serum	IgG	Polyclonal	25–37
Rabbit serum	IgG	Polyclonal	24–32
Bovine serum	Albumin		3
Human serum	Albumin		3
Mouse serum	Albumin		3
Bovine serum	Transferrin		3
Human serum	Transferrin		3
Mouse serum	Transferrin		3
Various	Proteases		3

[a]All retention times were determined on a BAKERBOND ABx analytical column (4.6 × 250 mm). 250 μl of sample was chromatographed over a 60-min linear gradient of 10 mM MES, pH 6.0, to 250 mM KH_2PO_4, pH 6.8, with a flow rate of 1.0 ml/min.

[b]Abbreviations are: BSA, bovine serum albumin; SRBC, sheep red blood cells; HCG, human chorionic gonadotropin; I1–2, interleukin 2; Int, interferon; Ren, renin, ODC, ornithine decarboxylase.

Table 2 BAKERBOND ABx and MAb: General Properties

Matrix	Average silica particle size (μm)	Particle shape	Pore size (A)	Typical backpressures[a] (psi) A	B	Approximate capacity[b]	Stability (pH range)
ABx	5	Spherical	300	200	1,000	0.5 mEq/g[c]	2–10
ABx	15	Spheroidal	300	15	80	0.7 mEq/g	2–10
ABx	40	Irregular	275	3	30	1.0 mEq/g	2–10
MAb*	5	Spherical	300	200	1,000	0.7 mEq/g[c]	2–10
MAb	15	Spheroidal	300	15	80	0.8 mEq/g	2–10
MAb	40	Irregular	275	3	30	2.0 mEq/g	2–10

Mass recovery (%)	Proteins bound	Exchange mechanism	Typical efficiency (plates/m)	Relative resolution	Stability (hr of use)
>97	Immunoglobulins	Mixed mode	50,000	Highest	>1,000
>97	Immunoglobulins	Mixed mode	6,000	High	>1,000
>97	Immunoglobulins	Mixed mode	2,000	Lower[d]	>1,000
>97	Most proteins	Anion	50,000	Highest	>1,000
>97	Most proteins	Anion	6,000	High	>1,000
>97	Most proteins	Anion	2,000	Lower	>1,000

[a] At a flow rate of 1.0 ml/min with (A) 7.75 × 100 mm or (B) 4.6 × 250 mm columns.
[b] Ion exchange groups and milligrams IgG bound per gram of packing.
[c] 150 mg IgG/g ABx.
[d] See Figure 1.

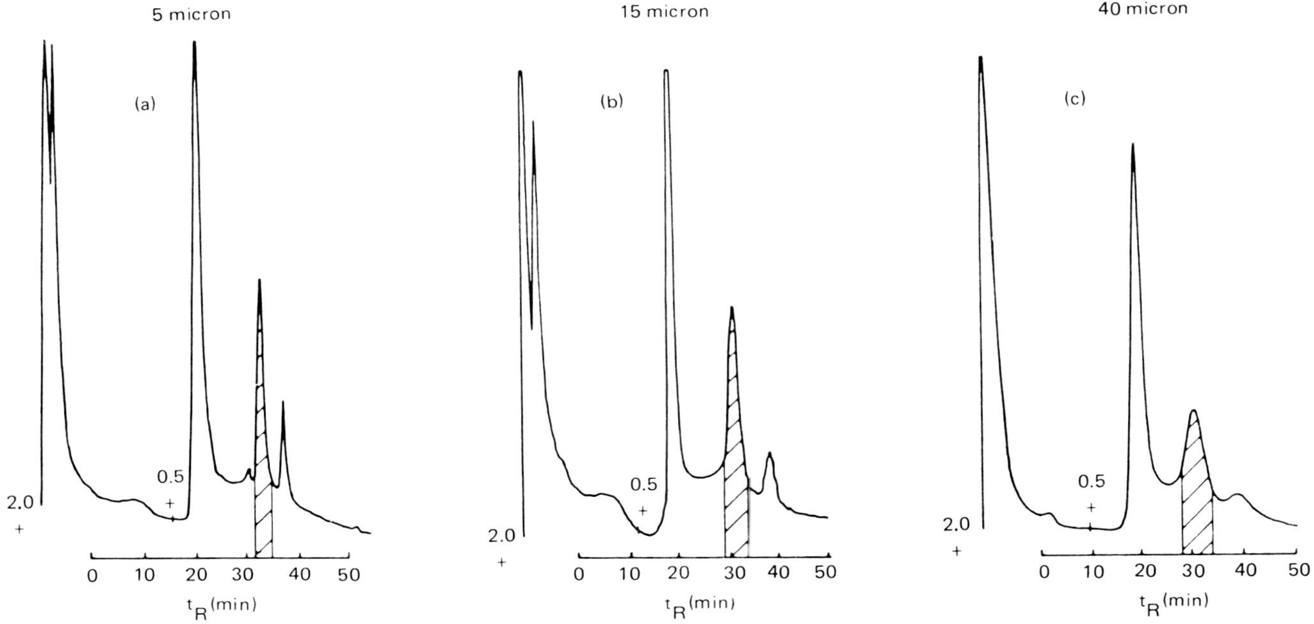

Figure 1 Chromatographic profile of the same cell culture supernatant chromatographed on 5-μm ABx, 15-μm ABx, and 40-μm PREPSCALE ABx. Chromatography was conducted on stainless steel HPLC column (7.75 × 100 mm) containing 5-, 15-, or 40-μm BAKERBOND ABx. The mobile phase consisted of an initial buffer A of 10 mM MES, pH 5.6, and a final buffer B of 1 M NaOAc, pH 7.0, with a linear gradient (100% A to 100% B) over 1 hr. The flow rate was 1.0 ml/min and this produced back-pressures of 200, 15, and 3 psi for the 5, 15, and 40 μm (PREPSCALE) particles, respectively. In each case, 0.3 ml of a 20-fold concentrated tissue culture medium ultrafiltrate (which was diluted to 1.2 ml with buffer A) was loaded onto the column. The void volume peak was composed of albumins, transferrins, proteases, and phenol red, and the cross-hatched peak contained the immunoglobulin. The proteins were detected by UV absorbance at 280 nm and the attenuation (absorbance units full scale; AUFS) was decreased from 2.0 to 0.5 AUFS after the elution of the void volume peak as indicated (see +) in each chromatogram (to better visualize the antibody peak).

VI. USE OF ABx

The selectivity and resolving power of the ABx surface chemistry, plus the rigid, polymer-coated silica support with its high ligand density and high capacity make ABx a versatile tool for analysis and purification.

A. Analysis, Characterization, and Process Monitoring

Traditionally, the analysis of supernatants from large-scale fermentations as well as hybridoma selection and screening processes have been conducted with by sodium dodecyl sulfate-polyacrylamide gel electrophoresis (SDS-PAGE), affinity chromatography, immunological methods, amino acid analysis, gel filtration chromatography, and ion exchange chromatography (9-14,28,29). These methods tend to be rather slow and tedious, give semiquantitative results, or require sophisticated apparatus and techniques. Trace levels of monoclonal antibodies in complex mixtures can be chromatographed on ABx for purpose of hybridoma subcloning, antibody characterization (see Fig. 9), process monitoring, optimization of harvest time, tracking genetic drift, and purification monitoring (11,19,35). Alternatively, if accuracy is less critical, solid-phase extraction of monoclonal antibodies with ABx spe (solid-phase extraction) columns facilitates the rapid spectrophotometric analysis or the simultaneous cleanup of numerous samples (9-11,19,35).

B. Purification

1. Method Development for Scaleup

To facilitate method development for the scale-up to low-pressure preparative antibody purification, the ABx surface chemistry has been bonded to three silica sizes, which produces bonded phases with remarkably similar chromatographic properties (see Fig. 1 and Table 1). These silica sizes include 5-μm spherical silica for methods development, analytical high-performance liquid chromatography (HPLC) or for a high-resolution, high-purity, final-step purification; 15-μm spherical silica for high-resolution, low- to medium-pressure analytical or preparative chromatography; and economical 40 μm (BAKERBOND PREPSCALE) irregularly shaped bulk silica for large-scale batch or solid-phase extraction and traditional low-pressure, open-column chromatography. The differences in resolution on these three particle sizes is minimized because of the similar surface chemistry used in each. Method development is more rapid on 5 or 15 μm ABx because resolution, purity, and specific activities are high, whereas elution volumes and buffer requirements are low; factors that facilitate analysis, assay, and identification (Appendix, Sect. D).

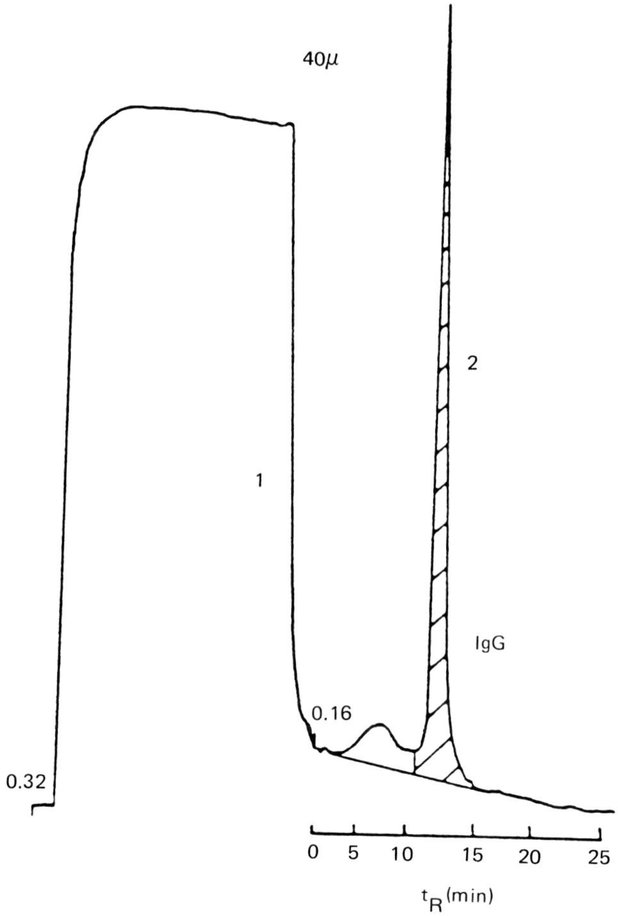

Figure 2 Semipreparative chromatography of serum-based cell culture supernatant on 40-μm BAKERBOND PREPSCALE ABx. 1.5 L of cell culture medium was concentrated 20-fold by ultrafiltration, diluted fivefold with buffer A and chromatographed on a column (21.2 X 150 mm) containing 15 g of 40-μm PREPSCALE ABx. The mobile phase consisted of an initial buffer A of 10 mM MES, pH 5.6, and a final buffer B of 250 mM KH_2PO_4, pH 6.8. The gradient was linear (100% A to 100% B) over 40 min. The flow rate was 20 ml/min and the backpressure was 20 psi. The detector (UV at 280 nm) sensitivity was increased from 0.32 AUFS to 0.16 AUFS after the elution of the void volume, peak 1 eluate, which contained the nonbound proteins (albumins, transferrins, proteases, etc). Peak 2 eluate contained greater than 95% of the original IgG at a purity greater than 90% by SDS–PAGE (cross-hatched area).

2. Preparative Column Chromatography

Because most proteins are not bound by ABx, the relative capacity to bind antibody (150 mg/g ABx) is significantly enhanced. In a typical semipreparative separation (Fig. 2), a small column dry-packed with approximately 15 g of 40 μm BAKERBOND PREPSCALE ABx was used to purify over 100 mg of IgG from 1.5 L of serum-based cell culture medium, with quantitative recovery of highly purified monoclonal antibody. (Further details concerning method development and scale-up are provided in the Appendix, Sect. C and D.)

3. Batch Extraction with ABx

One of the oldest, yet simplest, methods to purify proteins is by batch extraction. The method is quick, easy, and economical and requires no column, pumps, or other chromatographic equipment. The protein of interest is selectively removed from crude biological mixtures by adding a given chromatographic medium or absorbent (Appendix, Sect. F). The ABx matrix operates in many respects as an affinitylike adsorbent toward immunoglobulins by selectively removing antibodies from ascites fluid, cell culture medium, or serum/plasma, making it ideal for use in the batch process (Fig. 3). Whereas soft gels contain fines and require centrifugation or long periods to settle by gravity, 40-μm BAKERBOND PREPSCALE ABx quickly separates from the aqueous phase within minutes. ABx is easily sterilized and regenerated, and it does not support microbial growth (Appendix, Sect. A), making batch extractions with ABx suitable for large-scale applications in which aseptic conditions and extremely high product purity are required. The exact conditions required for aqueous batch extraction of proteins on an ABx medium (Appendix, Sect. F), as well as the kinetics of the process may be conveniently monitored and maximized with ABx analytical HPLC columns (see Fig. 3). Two of the major advantages of batch extraction with ABx are that a given protocol can be scaled up more than 1000-fold, with results that are identical with those achieved on the method-development scale (11,35), and significantly higher capacities can be achieved with batch extractions compared with column chromatography (11, 35). With use of the batch process, up to 250 ml of cell culture fluid can be processed on just 1 g of ABx (35). Recent work suggests that the addition of low levels of ammonium sulfate (approximately 25 mM) during the adsorption step can increase the purity of antibodies extracted from ABx significantly. This data will be presented in detail elsewhere (11,35).

VII. ABx VERSUS ANION EXCHANGE CHROMATOGRAPHY

The high degree of resolution achieved on the ABx is illustrated when a series of collected fractions from the ABx are rechromatographed on a high-performance anion exchanger such as BAKERBOND MAb* (Fig. 4). Although greater than

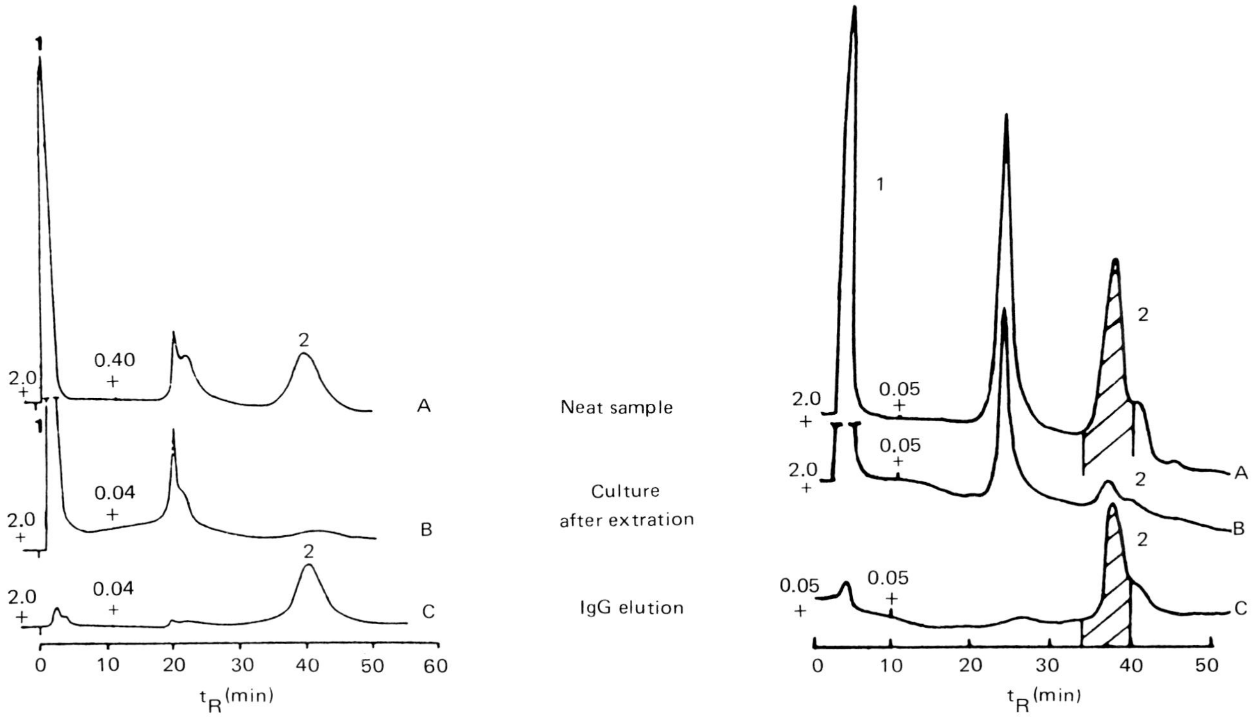

Figure 3 BAKERBOND ABx batch extractions of monoclonal antibodies from two cell culture supernatants as monitored by ABx (5 μm) column chromatography. Analytical conditions for the ABx chromatographic monitoring are the same as in Figure 1 except that buffer B consisted of 125 mM KH_2PO_4, pH 6.8, and the column size was 4.6 × 250 mm. Peak 1 eluate contains the albumins and transferrins, peak 2 eluate is the monoclonal antibody. Note the low levels of antibody being selectively removed by ABx and the increase in recorder sensitivity on the chromatograms after the elution of the unbound proteins (see +). After the extraction procedure, the IgG purity was 85% by SDS–PAGE analysis.

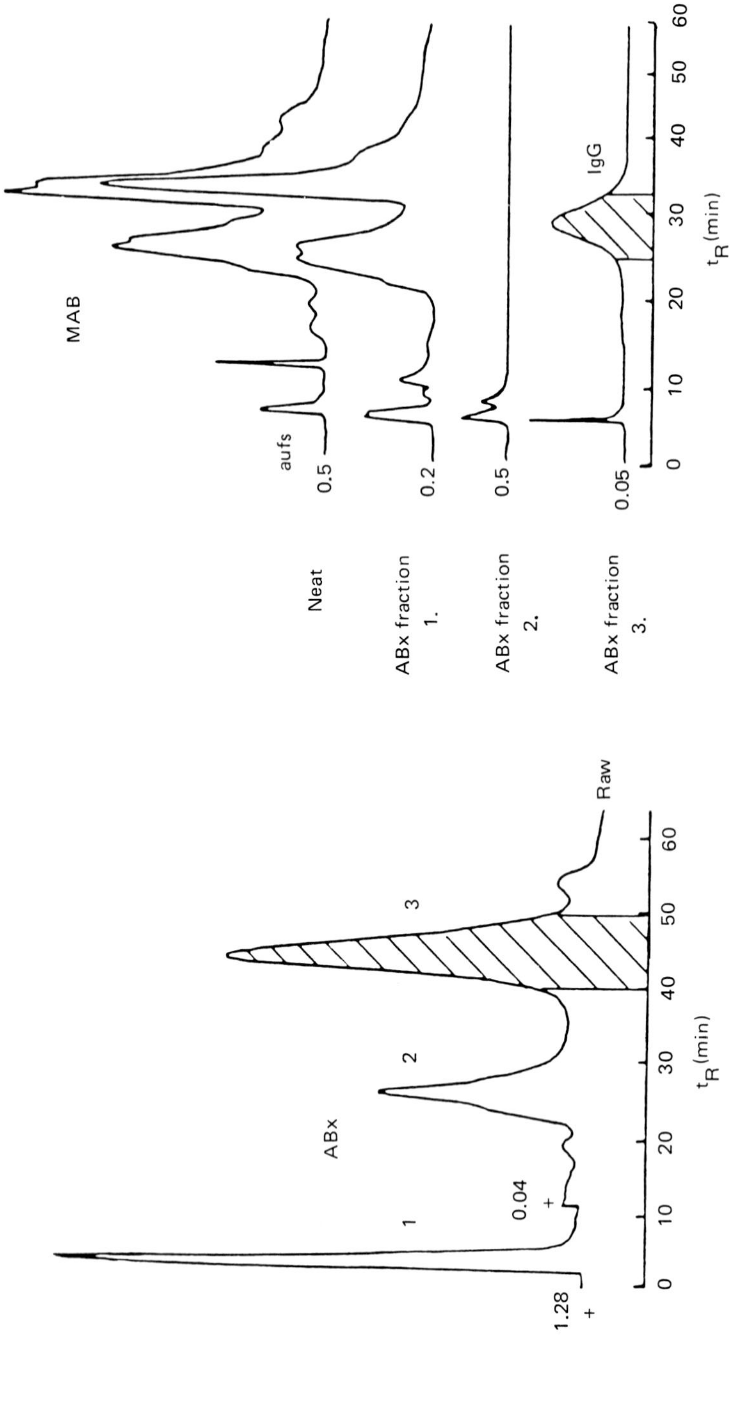
MAB
aufs
0.5
0.2
0.5
0.05
IgG
Neat
ABx fraction 1.
ABx fraction 2.
ABx fraction 3.
0 10 20 30 40 50 60
t_R(min)
ABx
Raw
1.28
+
0.04
+
1
2
3
0 10 20 30 40 50 60
t_R(min)

Figure 4 Rechromatography of ABx fractions of a serum-based cell culture medium on a high-performance anion exchange matrix (BAKERBOND MAb). Neat hybridoma tissue culture medium (0.3 ml) was chromatographed on a 40-μm PREP-SCALE ABx column, and the eluates of three major ABx peaks (1, 2, 3) were rechromatographed on a 15-μm MAb column. More than 95% of the total protein was not bound by the ABx (fraction 1), whereas these nonimmunoglobulin proteins were bound and even coeluted with the monoclonal antibody on the MAb and other anion exchangers. The purity of the monoclonal fraction 3 from the ABx column was greater than 90% IgG by SDS–PAGE. Note the changes in recorder sensitivity (AUFS; see +) and the low levels of monoclonal present. Analytical conditions for the ABx chromatogram were the same as those in Figure 1 except the elution buffer was 125 mM KH_2PO_4, pH 6.0, and the column size was 4.6 × 250 mm. Peak 1 eluate consisted of the albumins, transferrins, proteases, and phenol red, and peak 3 contained the monoclonal antibody (cross-hatched area). Anion exchange chromatography was conducted on an HPLC column (4.6 × 250 mm) packed with 15-μm BAKERBOND MAb matrix. The mobile phase for the MAb consisted of an initial buffer A of 10 mM KH_2PO_4, pH 6.6, and a final buffer B of 500 mM KH_2PO_4, pH 6.6, with a linear gradient (100% A to 50% B) over 1 hr. The flow rate was 1.0 ml/min with a back pressure of 150 psi, with UV (280 nm) detection at various AUFS (see +).

90% of the proteins present within serum-supplemented tissue culture media are not bound by ABx, these proteins are bound by anion exchangers. In fact, rechromatography of the ABx void volume fraction (which contains these nonbound proteins) on an anion exchanger produces a chromatographic profile that is virtually identical with that of the neat sample itself (see Fig. 4). Whereas a substantial portion of contaminating proteins coelute with the monoclonal antibody on high-performance anion exchangers, these proteins are totally resolved from the immunoglobulins on ABx (see Fig. 4).

VIII. EFFECTS OF BUFFERS ON ELUTION PROFILES WITH ABx

Another major advantage of ABx over other chromatographic matrices is the dramatic changes in elution profile and final product purity that occur when various buffer systems are used as the mobile phase. Any given protein has numerous different sites at which physical interactions may occur. Bonded phases such as the ABx which consist of mixed modes of interacting groups (multiple ligand types) offer several types of absorption mechanisms and binding affinities/types. Because of the differences in buffering capacity, pH, ionic strength, and ion composition of various buffers (see Appendix, Sect. D) the physical interactions between any given protein and the ABx chromatographic matrix may differ substantially. As a result, dramatic changes in selectivity and resolution have been achieved by manipulating buffer conditions to optimize resolution in a one-step purification scheme (Fig. 5). Furthermore, rechromatography of the immunoglobulin-containing peak under a second set of buffer conditions typically produces homogeneous antibody (see Fig. 5). Initial results indicate that KH_2PO_4, $(NH_4)_2SO_4$, NaCl, and NaOAc buffer systems each give totally distinct elution profiles (see Appendix, Sect. F, Figs. A1–A4).

In the presence of ammonium sulfate buffer immunoglobulins are bound by ABx more strongly than any other proteins present in mouse ascites fluid, plasma, fetal bovine serum, or serum-based hybridoma tissue culture medium. (see Appendix, Sect. F). This is a particular advantage in the preparative mode. Rather than hindering resolution, column overload actually tends to increase resolution by virtue of displacement chromatography of contaminating proteins that are bound to ABx less tightly than the antibody (compare the analytical with the preparative runs in Figs. 1 and 2). This phenomenon also occurs with other buffer systems. The resolution of IgG was not reduced on a preparative column (see Fig. 2) on which approximately 250 times more tissue culture ultrafiltrate was loaded per gram of resin relative to the comparable analytical column (see Fig. 1).

Antibody Purification and Analysis

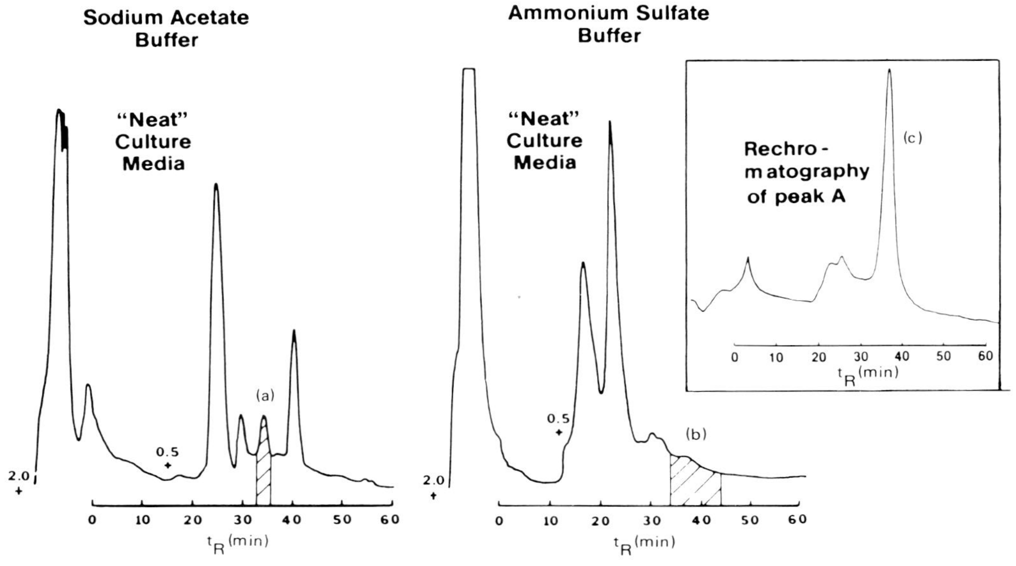

Figure 5 Effects of buffer conditions on the elution profile and the purification of a monoclonal antibody present at low levels in a serum-based cell culture supernatant. A high level of purity (peaks A and B are greater than 85% IgG) was achieved in a single step with either elution buffer system (sodium acetate or ammonium sulfate), despite the low level of monoclonal antibody present (less than 1% of the total protein). Rechromatography of peak A in the presence of ammonium sulfate buffer produced homogeneous antibody (peak C) as determined by SDS–PAGE. Chromatography was conducted on BAKERBOND GOLD HPLC columns (7.75 × 100 mm) packed with 5 μm ABx. The mobile phases consisted of an initial buffer A of 10 mM MES, pH 5.6, and a final buffer B of either 500 mM NaOAc, pH 7.0 (as in chromatogram (*a*), above left), or 250 mM $(NH_4)_2SO_4$ plus 10 mM NaOAc, pH 5.6 (as in chromatograms (*b*), above right, and (*c*) in the insert). The gradient was linear (100% A to 100% B) over 1.5 hr, and the flow rate was 0.7 ml/min with a back pressure of 150 psi. The sample load was 0.4 ml of neat medium (diluted to 1.2 ml with A buffer). The detection was by UV at 280 nm with the sensitivity changed from 2.0 AUFS to 0.5 AUFS after the elution of the void volume peak (see +).

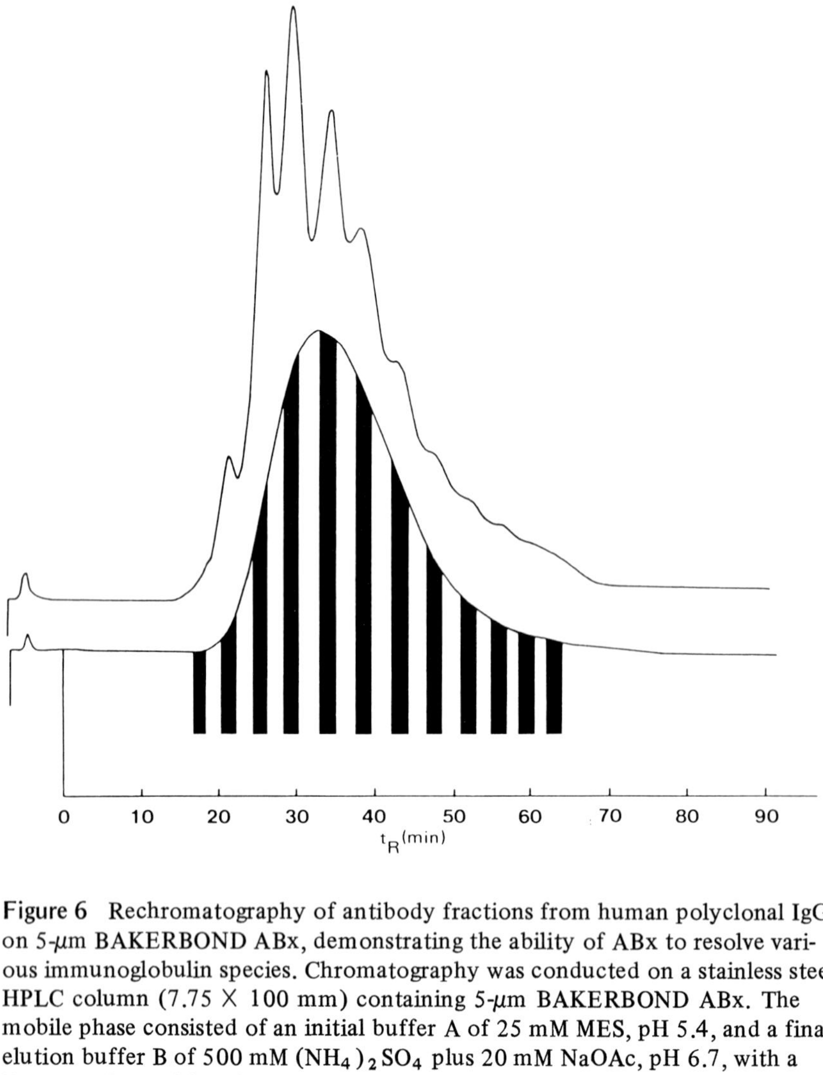

Figure 6 Rechromatography of antibody fractions from human polyclonal IgG on 5-μm BAKERBOND ABx, demonstrating the ability of ABx to resolve various immunoglobulin species. Chromatography was conducted on a stainless steel HPLC column (7.75 × 100 mm) containing 5-μm BAKERBOND ABx. The mobile phase consisted of an initial buffer A of 25 mM MES, pH 5.4, and a final elution buffer B of 500 mM $(NH_4)_2SO_4$ plus 20 mM NaOAc, pH 6.7, with a linear gradient of 100% A to 25% B buffer over 1 hr, at a linear flow rate of 1.0 ml/min, with a back pressure of 200 psi. Proteins were detected by UV absorbance at 280 nm at an attenuation of 2.0 AUFS for the lower chromatogram and 0.02 AUFS for the top chromatogram. The sample injection volume was 4.0 ml consisting of 25 mg of purified human IgG, in the lower chromatogram, or, in the upper chromatogram, 0.04 ml of each of the 12 shaded fractions diluted in 25 mM MES, pH 5.0. Every other fraction was collected from the lower chromatographic run (shaded areas), and was diluted and reinjected under identical conditions to produce the chromatogram on top. These chromatograms illustrate the high degree of resolution that is possible on ABx, and the ability of ABx to separate/fractionate various immunologically active fractions.

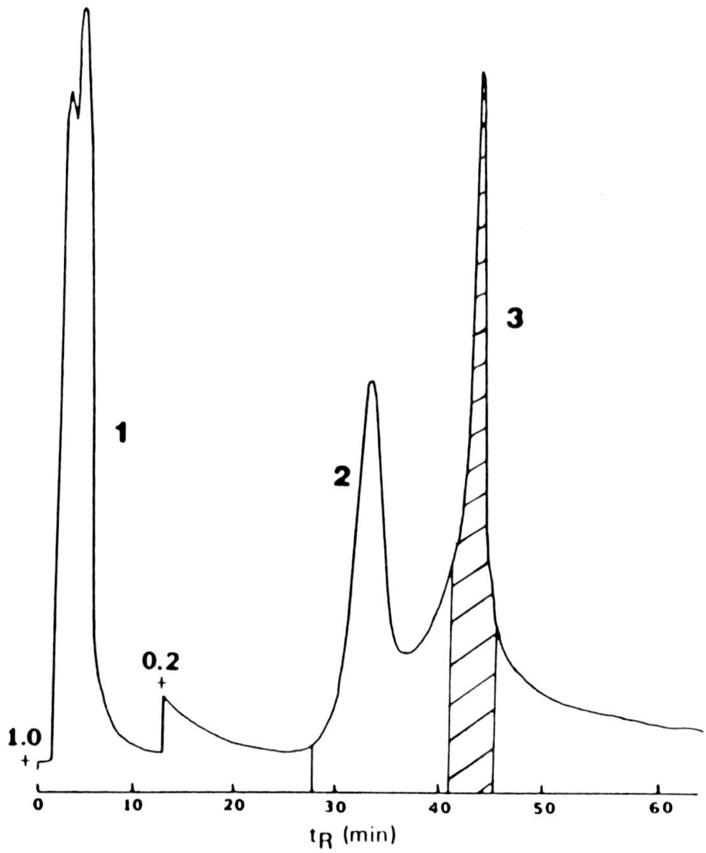

Figure 7 Separation of contaminating, host (mouse serum polyclonal) antibodies from a monoclonal antibody present in mouse ascites fluid on 15-μm BAKERBOND ABx. Analytical conditions were the same as those given in Figure 1, except the column size was 4.6 × 250 mm and buffer A was 10 mM KH_2PO_4, pH 6.0 and buffer B was 125 mM KH_2PO_4, pH 6.8. The sample load was 0.4 ml of neat sample (diluted to 1.2 ml with buffer A). Peak 1 consisted of albumin, transferrin, and proteases, peak 2 was the host IgG, and peak 3 was the monoclonal antibody.

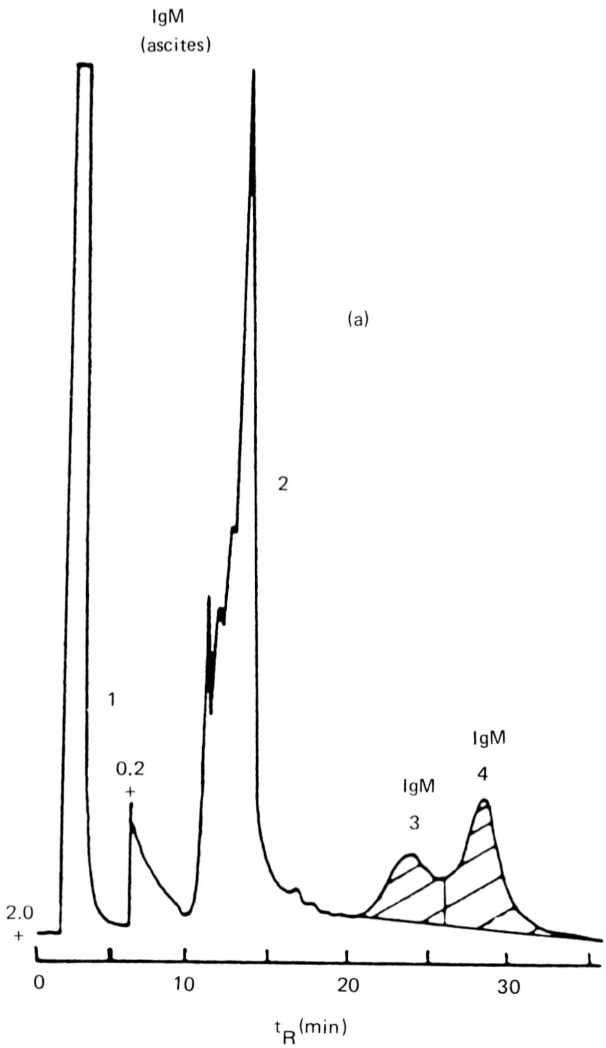

Figure 8 ABx separation of two IgM antibodies produced in the same mouse ascites fluid or two IgG antibodies produced by the same hybridoma grown in culture. Analytical conditions were the same as in Figure 1 except the column size was 4.6 × 250 mm and buffer B was 500 mM KH_2PO_4, pH 6.8. In Figure 8*a*, peak 1 eluate consisted of albumin, transferrin, and proteases, peak 2 eluate contained weakly bound proteins, and peaks 3 and 4 eluates (cross-hatched areas)

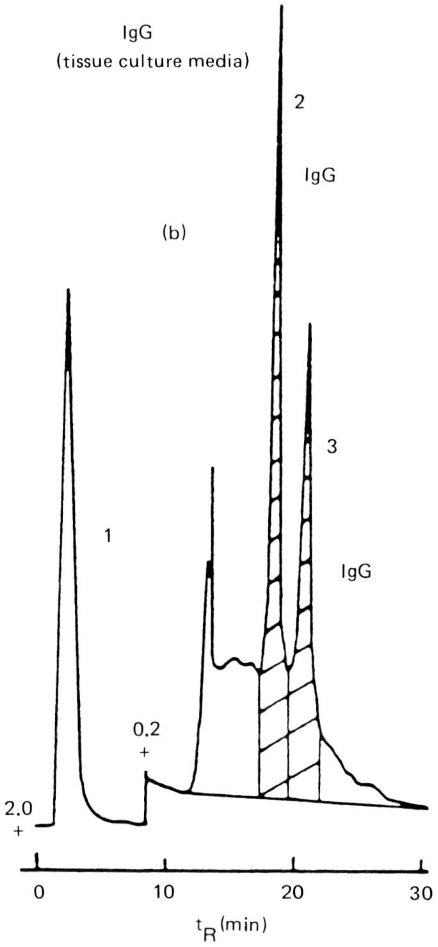

were two distinct IgM species that are both greater than 95% pure by SDS–PAGE. In Figure 8*b*, analytical conditions were the same as in Fgure 1 except the column size was 4.6 X 250 mm and the starting buffer A was 10 mM MES, pH 6.0, and buffer B was 250 mM KH_2PO_4, pH 5.6 and a step-gradient was used (100% A for 10 min, then step to 20% B followed by a linear gradient from 20% B to 50% B over 60 min). In each case, 0.5 ml of neat sample (diluted to 1.5 ml with buffer A) was injected. Peaks 2 and 3 (cross-hatched areas) are IgGs at greater than 95% purity by SDS–PAGE.

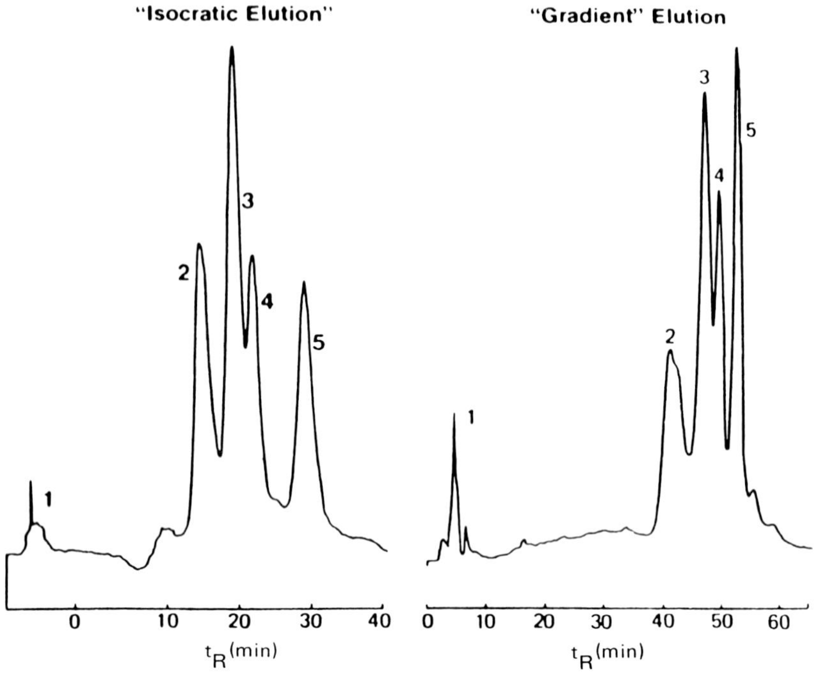

Figure 9 ABx chromatographic profile of a monoclonal antibody that was purified from a serum-free medium by ultrafiltration and two chromatographic steps on a high-performance anion exchange matrix. Although the purified antibody(s) eluted as a single peak on the anion exchanger, ABx was able to resolve small amounts of albumin and transferrin contamination (peak 1) and multiple forms of immunoglobulin (peaks 2, 3, 4, and 5). Immunological methods later revealed the presence of both meyloma IgG1 (65% of the total IgG) and monoclonal IgG2a (35% of the total IgG). The fact that different ABx chromatographic profiles were obtained with different elution buffers (data not shown) further suggests that ABx may be valuable in purifying individual immunoglobulins to total homogeneity. Analytical conditions are the same as in Figure 1 except, in the isocratic run a longer (4.6 × 250 mm) column was used and, after equilibrating the column in the initial buffer (100% A), a step-gradient to 25% final B buffer was employed, and the antibodies were eluted "isocratically."

IX. RESOLUTION OF MULTIPLE ANTIBODY SPECIES ON ABx

Under similar chromatographic conditions antibodies have been found to elute from ABx at a wide range of retention times (see Table 1). Variations in elution times ranging from 16 to 68 min are apparently the result of subtle differences in antibody structure and molecular diversity, which are not, as yet, totally defined.

Whereas the variability in antibody elution times may be a disadvantage on anion exchangers because of possible coelution of the antibody with albumin or transferrin, it represents a distinct advantage of ABx. It facilitates the resolution of different monoclonal antibodies with minimal possibility of contamination by the unbound, nonimmunoglobulin proteins. ABx is capable of fractionating the heterogeneous population of polyclonal antibodies present within serum/plasma (Fig. 6). It is also able to separate monoclonal antibody from host (serum polyclonal) immunoglobulin contaminants in mouse ascites fluid (Fig. 7) and fetal bovine serum (see Appendix, Figs. A3, A5, and A6). Likewise, ABx is capable of high resolution of multiple antibody species secreted from double-producing hybridomas (11,35). Multiple forms of IgG and IgM have been separated in a highly purified state from ascites fluids and cell culture media, respectively, with minimal method development (Fig. 8). Multiple forms of myeloma IgG1 and monoclonal IgG2a from a single hybridoma have also been resolved on ABx with either gradient elution or isocratic elution (Fig. 9). Such a technique will, no doubt, facilitate basic research and development in the field of immunology (15), as well as the purification of truly homogeneous antibody preparations (11,35).

X. TWO-DIMENSIONAL CHROMATOGRAPHY: ABx PLUS MAb CHROMATOGRAPHY OR HYDROPHOBIC INTERACTION CHROMATOGRAPHY

Although many antibodies elute from ABx as electrophoretically pure peaks, those immunoglobulin fractions that contain minor contaminants may yield homogeneous antibody by direct chromatography on an anion exchange matrix, such as the MAb, or on a hydrophobic interaction chromatographic (HIC) matrix, such as BAKERBOND Hi-Propyl* (9–11,19,20). The mechanisms of separation on these complementary-bonded phases are entirely different. Together, ABx plus MAb or ABx plus HIC offer selectivity and resolution that is greatly enhanced relative to any single chromatographic medium alone. Recently, "standardized" procedures have been developed that will enable the purification to near homogeneity of virtually any monoclonal antibody by use of a two-step procedure that comprises ABx and either MAb or HIC, either in the batch mode or in the column chromatographic mode (11,35).

An example of two-dimensional chromatography that uses ABx followed by MAb was carried out with a cell culture supernatant and with human plasma (Figs. 10 and 11). The technique involves collecting the immunoglobulin-containing peak from the ABx column, diluting with two or three parts distilled water or MAb starting buffer, and reinjecting this lower level of protein onto a smaller MAb column. The method can be applied on an analytical or preparative scale to purify virtually any immunoglobulin to homogeneity.

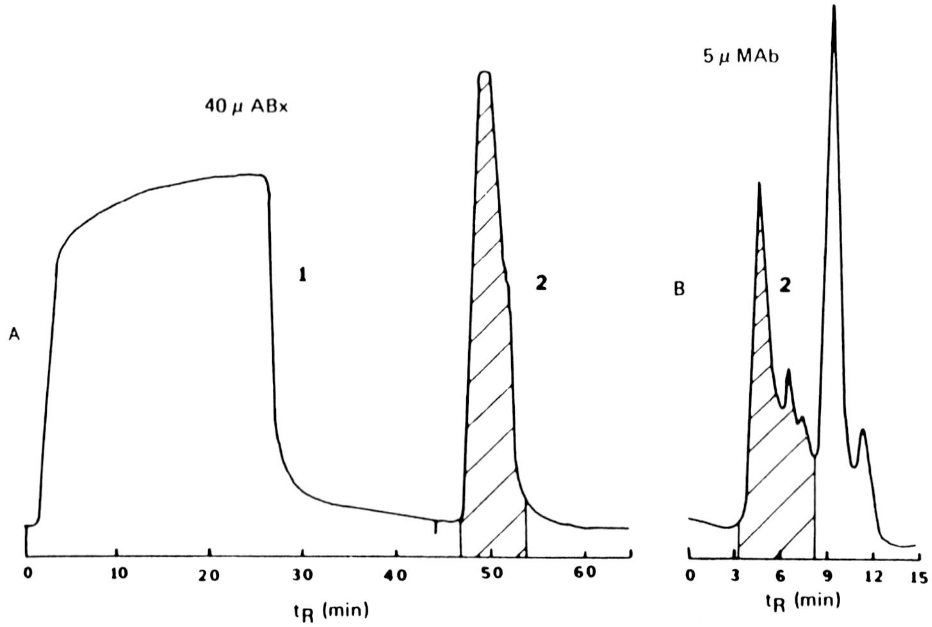

Figure 10 Sequential purification of human polyclonal antibodies from human plasma on ABx and MAb. Neat human plasma was fractionated on 40 μm PREP-SCALE ABx and the peak 2 eluate containing the IgG was diluted twofold and rechromatographed on a 5-μm MAb column to achieve higher immunoglobulin purity. Analytical conditions for ABx were the same as in Figure 1 except that 8 g of protein were loaded onto a 21.1 × 150 mm column of 40-μm PREP-SCALE ABx with a flow rate of 20 ml/min and buffer B was 250 mM KH_2PO_4, pH 6.8. Analytical conditions for anion exchange chromatography on MAb are given in Figure 4, with the peak 2 eluate being greater than 95% pure IgG by SDS–PAGE analysis.

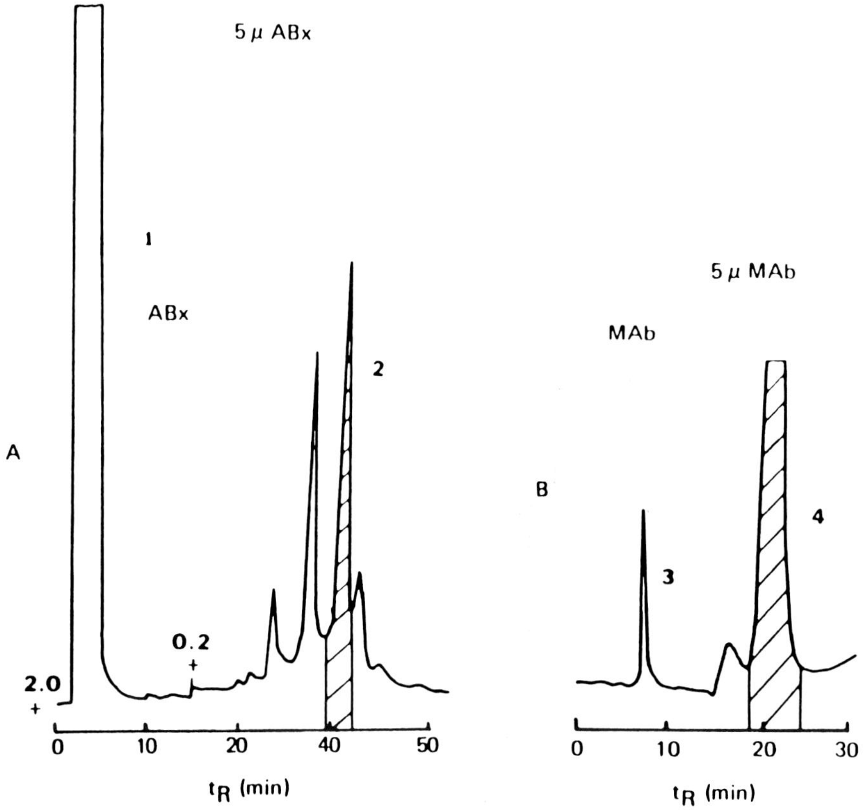

Figure 11 Sequential purification of monoclonal antibody from a cell culture supernatant on ABx and MAb. Neat hybridoma tissue culture medium (0.5 ml of ultrafiltrate diluted to 1.5 ml with buffer A) was chromatographed on ABx and the eluate peak containing the monoclonal antibody was diluted twofold and rechromatographed on an MAb column to give homogeneous IgG. Analytical conditions for the ABx chromatogram are the same as those given in Figure 1, except the initial buffer A was 10 mM MES, pH 6.0, and buffer B was 125 mM KH_2PO_4, pH 6.8. Analytical conditions for the MAb run were the same as in Figure 4, except that the final buffer B was 250 mM KH_2PO_4, pH 6.5. The eluate, peak 1, was composed of albumins and transferrins, peak 2 was IgG (80–90% pure) plus impurities (see peak 3) and peak 4 was homogeneous IgG (by SDS-PAGE analysis).

XI. SUMMARY

By virtue of its ability to bind immunoglobulins rather selectively, ABx is similar to protein A. However, it differs from protein A in a number of important respects. Unlike protein A, ABx is able to bind all antibodies (IgG1, IgG2a, IgG2b, IgG3, IgA, IgD, IgE, and IgM) as well as antibody hybrids, bispecific hybrids, antibody-protein conjugates and chemically modified antibodies (9-11,35). In contrast to affinity matrices, antibodies are eluted from ABx in small volumes of buffer at physiological pH and ionic strength, with quantitative recovery of mass (see Table 2) and immunological activity (30-35). Furthermore, the susceptibility of protein-based chromatographic media to biodegradation, proteolytic digestion, and ligand or endotoxin leakage make chemically defined matrices like ABx an attractive alternative (see Appendix, Sect. A).

ABx has the capacity to bind immunoglobulins (see Table 2) that is approximately 15 times higher than that of protein A or hydroxyapatite (9-11,13). It is also more rugged and durable than hydroxyapatite or affinity matrices, and it has been used for up to 1000 hr, with little loss in resolution or capacity. These advantages, along with the ability to use higher flow rates at lower backpressures (9-11,13), enable substantially higher throughputs for more economical operation.

In any protein purification problem, several factors are of utmost importance. These factors include capacity, resolution, specificity, recovery, hygiene, speed and throughput, durability and longevity, ease and versatility of use, and economy. The ABx bonded phase has been designed with each of these in mind.

As new types of monoclonal antibodies are purified and as purity requirements increase, it becomes increasingly apparent that the molecular diversity inherent within the immunoglobulin family makes a universal single-step purification method less likely (35). However, the development of dedicated chromatographic matrices, such as the ABx, that utilize composite surface interactions to maximize selective antibody binding and resolution, represent a major step in the right direction.

XII. APPENDIX: PROCEDURES FOR USING ABx

This Appendix describes the procedures for using and maintaining the BAKERBOND ABx bulk-bonded phases and HPLC columns. ABx is available from J. T. Baker Inc. (Phillipsburg, New Jersey) in the following configurations: prepacked HPLC columns containing 5-μm ABx, MAb, or Hi-Propyl in the "analytical" (4.6 × 250 mm), "GOLD" (7.75 × 100 mm; FPLC compatible), or custom (any size) column configurations; prepacked HPLC (FPLC compatible) columns containing 15-μm ABx, MAb, or Hi-Propyl in the "semi-prep" (10 × 250 mm) or custom (any size) column configurations; 15 μm bulk matrix for "topping off" old HPLC columns or for slurry (pressure) packing into HPLC columns; 40-μm ABx, MAb, or Hi-Propyl bulk-bonded phases for batch extractions and for traditional open-column (low-pressure) chromatography.

A. Hygiene: Regeneration, Sterilization, and Depyrogenation

If column performance deteriorates unexpectedly, or if column back pressure increases substantially, column clean-up may be required to remove adsorbed proteins or lipids. It should be stressed, however, that normal maintenance procedures (i.e., sample clarification and washing with high salt on a daily basis) should totally eliminate the need for column regeneration for periods of at least 6 months (31-33,35). ABx and MAb may be regenerated with 20 column volumes of high ionic strength salt solutions, such as 2 M sodium acetate or 1 M potassium phosphate buffer, pH 7 or, if necessary, pH 3, and use of high flow rates. If necessary, further washes may be carried out with 10 column volumes of protein-solubilizing agents, such as 6 M urea, 6 M guanidine, or 1 M isothiocyanate; detergents, such as 1% Triton X-100, SDS, or Brij 35; acids, such as 10% acetic acid or 0.2% trifluoracetic acid; or organic solvents, such as methanol, ethanol, or 50:50 DMSO:H_2O (to remove lipids).

At the end of each day it is advisable to store the column in a high salt buffer to clean off contaminants and prevent microbial growth. Unlike most polysaccharide-based soft gels, silica-based bonded phases, themselves, do not support the growth of microorganisms, although microbial growth may occur in low-ionic strength mobile phases. Therefore, if the column is to be stored for long periods (more than a week), the use of pure methanol, ethanol, propanol, or possibly, a preservative such as 0.1% sodium azide or chlorhexidine in high-ionic strength buffer (1 M KH_2PO_4, 2 M $(NH_4)_2SO_4$, or 2 M NaOAc, pH 7.0) is recommended. The column may also be refrigerated. Columns are typically stable for over 1000 hr of use at room temperature. Dry ABx and MAb bulk-bonded phases need not be refrigerated and are extremely stable.

Sterilization of ABx and MAb columns or bulk-bonded phases may be accomplished by washing with any bactericidal organic solvent, such as ethanol, methanol, acetonitrile, a 0.1% aqueous solution of sodium azide or chlorhexidine, or 10% acetic acid. The ABx and MAb matrices are unaffected by, and totally resistant to, these treatments.

Although ABx and MAb media are silica-based, their dense hydrophilic polymeric coverage allows the use of pH extremes (2-10) not permitted with most other silica-based materials. ABx and MAb are totally resistant to most organic solvents, aqueous buffers, chaotropic and protein-solubilizing agents, detergents, antimicrobials, and bactericides.

Endotoxins have been shown not to bind to ABx under normal operational conditions (34,35). "Fouled" ABx can be depyrogenated and cleaned with 10 column volumes of 2 M NaOAc, pH 3.0, or with the same amount of 10% acetic acid (34,35).

B. Sample Preparation and Column Equilibration

Ionic strength must be relatively low to maximize the binding of antibodies to ABx. Any technique may be used for this, including dialysis, membrane ultrafiltration, or dilution. The optimal sample pH may vary with individual samples, but dilution of the sample with, or ultrafiltration or dialysis against, a low ionic strength zwitterionic buffer containing a phosphate or TRIS buffer, at a final pH of 6.0 [e.g., 25 mM 2-(*N*-morpholino)-ethane sulfonic acid (MES) or 3-(*N*-morpholino)-2-hydroxypropane sulfonic acid (MOPSO) plus 15 mM KH_2PO_4 or 20 mM TRIS-OAc, pH 6.0] is generally suitable.

Columns are equilibrated with approximately 20 column volumes of starting buffer (25 mM MES, pH 5.6). This may be accomplished more quickly by beginning the equilibration with 50 mM MES, pH 5.6, and then switching to the lower ionic strength buffer, or simply by increasing the flow rate.

The MES (or MOPSO) buffer is typically used as the starting buffer because zwitterionic buffers facilitate the binding of antibodies to the zwitterionic surface of ABx (11,35). Furthermore, pH equilibration may be conveniently monitored by following absorbance at 220 nm owing to the change in absorptivity (extinction coefficient) of MES that accompanies column pH equilibration. Alternatively, the pH of the column effluent may be monitored to determine when the column is equilibrated.

C. Capacity and Sample Loading

The binding capacity of 5-μm ABx is approximately 150 mg of purified IgG per gram bonded phase (frontal analysis, breakthrough method), corresponding to a titration of about 0.5 mEq/g packing. The ligand titrations of the 15 and 40 μm (PREPSCALE) preparative media are higher (greater than 0.7 mEq/g and 1.0 mEq/g, respectively) because of their higher surface areas and lower densities.

Loading capacities on ABx columns depend upon the nature of the sample. Higher ionic strength, pH, and total protein content, and the concentration of immunoglobulin and other "bound" proteins, all reduce the amount of sample that can be loaded onto the column for quantitative binding of the antibody (11, 35). Therefore, dialysis, dilution or, particularly, ultrafiltration against the ABx starting buffer (containing a low ionic strength of TRIS or phosphate) is recommended to reduce the amount of antibody not bound to the column as a result of displacement chromatography by proteins and salts present within the sample itself (35). Tangential flow ultrafiltration is valuable because of its ability to concentrate large volumes of culture medium into starting buffer (containing a low ionic strength of TRIS or phosphate), and to reduce the concentration of nonimmunoglobulin proteins in the process, decreasing the total amount of protein more than threefold in many cases.

Sample loads as high as 200 ml of serum-based tissue culture medium and up to 10 ml of ascites fluid have been chromatographed on analytical ABx columns (containing less than 5 ml of ABx) with greater than 95% recovery (32,35). An example of preparative sample loading is shown in Figure 2 of the main text, with 1.5 L of culture supernatant being chromatographed on a column containing 15 g of 40 μm ABx, with more than 90% of the total IgG being recovered (11,35). In other cases, 180 mg and 370 mg of a monoclonal antibody in 1.5 L and 3.2 L of serum-supplemented cell culture supernatant (ultrafiltrate) have been purified on columns containing only 10 g and 15 g of ABx, respectively (see Fig. A6). Similar loading capacities have been obtained with rabbit serum, with up to 28 ml of dialyzed serum being proocessed on a column containing less than 10 g of ABx (see Fig. A7). Still higher loading capacities have been obtained in the batch mode, with up to 250 ml of serum-supplemented cell culture supernatant (ultrafiltrate) being processed with only 1 g of ABx (11,35).

D. Optimization of Purification

As indicated previously, each antibody preparation may present different purification problems. Furthermore, not every monoclonal or polyclonal antibody is destined for the same use, and as such, purity requirements differ substantially.

Although many immunoglobulins elute from ABx as electrophoretically pure peaks, the optimization of purification may involve the "fine tuning" of several variables. These include choosing the correct buffer conditions and using step gradients to reduce run time and improve resolution.

As indicated previously, the use of various elution buffers results in different elution profiles from ABx (35). Four basic elution buffers are recommended (Table A1), although a variety of buffer species, concentrations, and pH values may be used successfully (11,35).

Usually, ammonium sulfate elution buffers provide the highest level of antibody purity (Figs. 1-4). This is because sulfate-based mobile phases selectively elute virtually all the bound, contaminating proteins before the immunoglobulins, thereby, enhancing resolution (11,35). As a result, step-gradients with sulfate elution buffers and changes in flow rate can be used to increase purity and peak sharpness, while reducing chromatographic run times and elution volumes (Figs. 5-7). By using these techniques, it is possible to obtain purities greater or equal to those achieved with protein A or affinity chromatography (11,35).

In addition to changing the elution buffer conditions, dramatic changes in elution profile and selectivity can be achieved on ABx by manipulating the ionic strength, the pH, or both, of the MES starting buffer (Fig. A8), or by changing the starting (A) buffer species and pH after sample loading and antibody binding

Table A1 Suggested ABx Gradients[a]

Buffer A	Buffer B
10 mM MES, pH 5.6 (MES acid, pH 5.6 with NaOH)[b]	500 mM $(NH_4)_2SO_4$ plus 10-mM NaOAc, adjusted to pH 5.6 or 7.0
10 mM MOPSO, pH 5.6 (MOPSO acid adjusted to pH 5.6 with NaOH)[b]	500 mM NaOAc, pH 5.6 or 7.0 500 mM KH_2PO_4, pH 5.6 or 7.0 500 mM NaCl plus 10 mM NaOAc, pH 5.6 or 7.0

[a]Other low ionic strength (10–25 mM) MES or MOPSO buffers (pH 5.0–6.0) may also be used.
[b]Gradients may be linear from 1 to 3 hr (with antibodies eluting from 18 to 90 min), or with step gradients to enhance IgG resolution and reduce run times. Alternatively, the ionic strengths of the elution (B) buffer may be reduced.

(35). It should be remembered, however, that increasing the ionic strength or pH of the loading buffer, as well as changing the buffer species itself, may reduce the maximum load tolerated on a given column (11,35).

When extreme product purity is required or large sample volumes are being processed, a purification scheme that comprises several steps on ABx may be required. Initial chromatography may be conducted on a larger column containing 40 μm ABx with use of traditional, open-column chromatographic equipment. Because greater than 95% of the nonimmunoglobulin proteins will be removed in this first step, the second step is typically conducted on a much smaller HPLC column containing high-resolution 5 μm ABx and in the presence of a second buffer system. The active fractions may also be run with a step gradient into 100 mM NaOAc, pH 7.0, to maximize resolution (11,35).

Extremely high antibody purity may also be achieved with a combination of ABx and either a high-performance anion exchanger, such as MAb, or with a high-performance hydrophobic interaction chromatographic (HIC) medium, such as BAKERBOND Hi-Propyl (see Sect. X of the main text). Hydrophobic interaction chromatography is particularly suitable for the chromatographic separation of various immunoglobulin species (including bovine IgG from mouse and human IgG) and as a direct purification step after ABx chromatography or ABx batch extraction, because the antibody is bound in the presence of high salt concentrations and, as such, no sample dialysis/dilution is required. Homogeneous antibody is typically produced by this two-step procedure (11,35).

E. Antibody Fragments, Conjugates, Classes, and Subclasses

It should be pointed out that ABx is capable of separating the major IgG subclasses and the various immunoglobulin classes from numerous species (11,35).

With ABx, it is also possible to resolve both alkaline phosphatase and horseradish peroxidase conjugates of mouse, rabbit, and human polyclonal (and mouse monoclonal) antibodies from both native enzyme and native immunoglobulin (11,35). ABx has also been used to separate native IgG from various antibody fragments, including pepsin and papain digests of mouse, rabbit, and human polyclonal (and mouse monoclonal) IgG (11,35). Details of these investigations are to be presented in detail elsewhere (11,35).

F. ABx Batch Extraction Protocol

The protocol for the batch extraction of cell culture supernatant with 40 μm PREPSCALE ABx (Fig. A9) is as follows: About 40 mg of 40 μm PREPSCALE ABx-bonded phase is stirred in 10 mM MES buffer, pH 5.6. After being stirred for several minutes or sonicated (optional) for 30 sec, the buffer is removed by decanting or filtration. A 5-ml sample of diluted cell culture supernatant [1 ml of supernatant ultrafiltrate or 0.5 ml of crude (undialyzed) supernatant diluted with 4 ml of 10 mM MES, pH 5.60] is added to the bonded phase. After the slurry is gently stirred for 15 min, the supernatant is removed (ABx settles rapidly by gravity). The remaining bonded phase is first washed with 2 ml of 10 mM MES, pH 5.6, and then extracted with 2 ml of 200 mM KH_2PO_4, pH 7.4. At least 70% of the immunoglobulin present in the original sample is now present in this buffer. Supernatants at each step of the extraction procedure may be analyzed for immunoglobulin by using a column packed with 5-μm ABx, or by traditional means of antibody analysis (SDS-PAGE, gel filtration, protein A, ELISA, etc.). The method can be optimized further by using a larger quantity of bonded phase, by using higher molarities to desorb the IgG, by recycling the unadsorbed immunoglobulin from the first extraction in further extraction cycles, or by any combination thereof. Recent evidence suggests that the addition of low concentrations of ammonium sulfate (approximately 20-30 mM) after antibody adsorption can increase product purity substantially by selectively desorbing the contaminating proteins that are bound by ABx (11,35).

(Text continues on p. 429.)

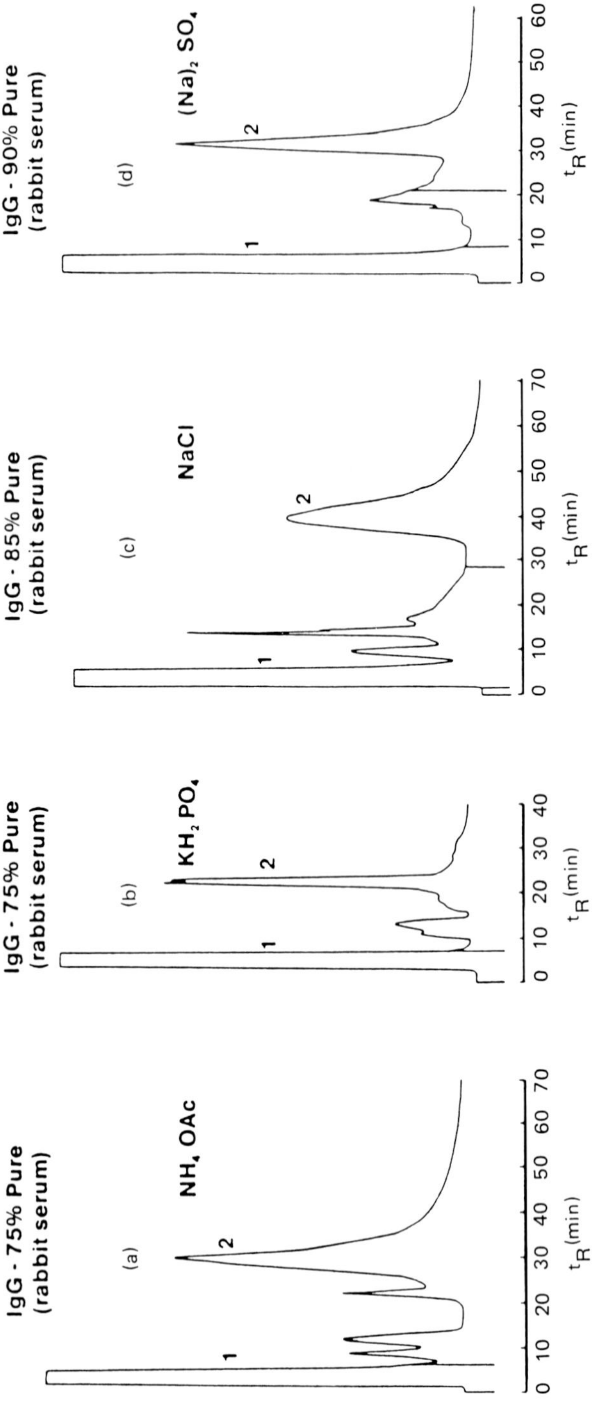

IgG - 75% Pure
(rabbit serum)
(a)
NH$_4$ OAc
1
2
0 10 20 30 40 50 60 70
t_R(min)
IgG - 75% Pure
(rabbit serum)
(b)
KH$_2$ PO$_4$
1
2
0 10 20 30 40
t_R(min)
IgG - 85% Pure
(rabbit serum)
(c)
NaCl
1
2
0 10 20 30 40 50 60 70
t_R(min)
IgG - 90% Pure
(rabbit serum)
(d)
(Na)$_2$ SO$_4$
1
2
0 10 20 30 40 50 60
t_R(min)

Figure A1 Effect of elution buffer on the purification of polyclonal antibodies from rabbit serum on 5-μm BAKERBOND ABx. The chromatographic isolation of the polyclonal IgG from normal rabbit serum was conducted on a BAKERBOND ABx GOLD column (7.75 × 100 mm, containing 5-μm ABx), at a flow rate of 1 ml/min which produced a pressure drop of 180 psi. The starting buffer A was 25 mM MES, pH 5.2, and the elution buffer B was either 500 mM NH_4OAc, pH 7.0 (chromatogram a), 250 mM KH_2PO_4, pH 7.0 (chromatogram b), 500 mM NaCl plus 12 mM KH_2PO_4, pH 7.0 (chromatogram c), or 400 mM $(Na)_2SO_4$ plus 10 mM KH_2PO_4, pH 7.0 (chromatogram d). 1.0 ml of crude serum (which was diluted with 2 ml of buffer A) was loaded onto the column in each run. Proteins were detected by UV absorbance at 280 nm at an attenuation of 2.0 AUFS. The vertical lines in each chromatogram refer to the beginning of the injection, the step-gradient, or the initiation of the linear gradient run. In chromatogram a, a linear gradient from 100% A to 100% B, over 1 hr was begun 7 min after the injection. In chromatogram b, a step gradient from 100% A to 20% B buffer was used 7 min after the injection, followed directly by a linear gradient from 20% B to 100% B over 1 hr. In chromatogram c, a step-gradient from 100% A to 13% B was initiated 2 min after the injection, held at 10% B for 14 min and followed by a linear gradient from 10% B to 100% B over 20 min. SDS–PAGE analysis was conducted on the individual collected fractions (1.5 ml each) that contained immunological activity. In chromatogram a, the purity of the entire IgG peak was approximately 75% (fraction 1 = 50%, 2 = 60%, 3 = 65%, 4 = 65%, 5 = 70%, 6 = 80%, 7 = 75%, 8 = 65%, 9 = 65%, 10 = 70%, 11 = 60% and 12, 13, and 14 were below the levels of accurate quantitation (i.e., the Coomassie Blue-stained bands were too faint to quantitate accurately). In chromatogram b, the IgG peak was approximately 75% pure (fraction 1 = 70%, 2 = 75%, 3 = 80%, 4 = 85%, 5 = 80%, 6 = 70%, 7 = 60%). In chromatogram c, the purity of the entire IgG peak was about 85% (fraction 1 = 70%, 2 = 75%, 3 = 80%, 4 = 90%, 5 = 90%, 6 = 85%, 7 = 80%, 8 = 80%, 9 = 75%, 10 = 70%, and 11 and 12 were below the levels of accurate detection). In chromatogram d, the IgG peak was approximately 90% pure (fraction 1 = 65%, 2 = 80%, 3 = 90%, 4 = 95%, 5 = 90%, 6 = 85%, and 7 was below the detection limits).

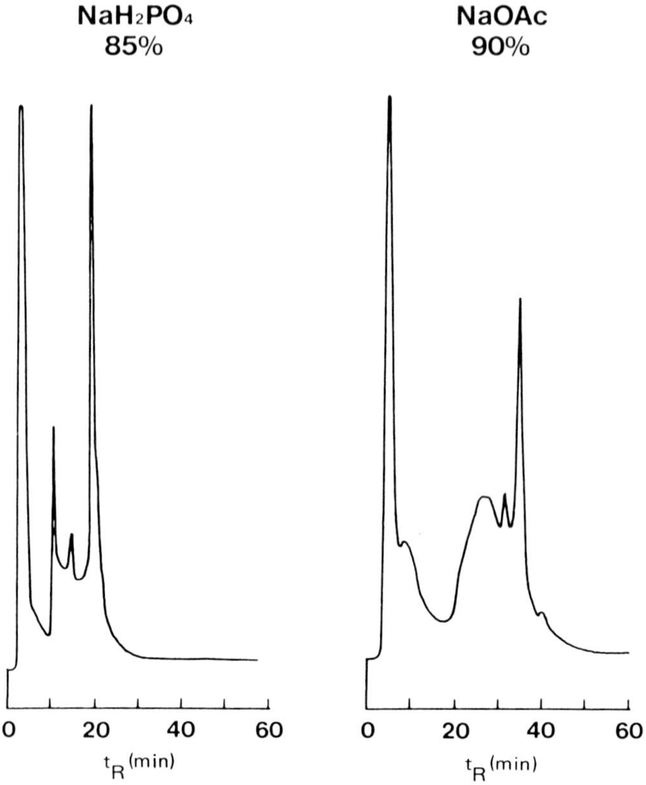

Figure A2 Effects of elution buffer species on the purification of a monoclonal antibody from mouse ascites fluid on a 5-μm BAKERBOND ABx column. Chromatography was conducted on a stainless steel HPLC column (7.75 × 100 mm) containing 5-μm BAKERBOND ABx, which had been destroyed by an improper washing procedure. The mobile phase consisted of an initial buffer A of 25 mM MES, pH 5.4, and a final elution buffer B of either (as indicated above each chromatogram), 500 mM KH_2PO_4, pH 6.8; or 1 M NaOAc, pH 7.0; or 1 M NaCl plus 20 mM NaOAc, pH 5.7; or 500 mM $(NH_4)_2SO_4$ plus 20 mM NaOAc, pH 6.7, with a linear gradient of 100% A to 50% B over 1 hr, at a linear flow rate of 1.0 ml/min with a back pressure of 200 psi. Proteins were detected by UV absorbance at 280 nm at an attenuation of 2.0 AUFS. The sample injection volume was 2.0 ml consisting of 0.4 ml of crude mouse ascites fluid plus 1.6 ml of 25 mM MES, pH 5.4. The monoclonal antibody (MAb) was contained within the last peak in each of the four chromatograms. A substantial proportion of the host (mouse serum polyclonal) IgG was also removed from the monoclonal antibody in the acetate and, particularly, in the chloride and sulfate buffer systems. Higher purities were obtained (data not shown) on columns that passed QC procedures (i.e., greater than 40,000 plates/m); this column failed QC.

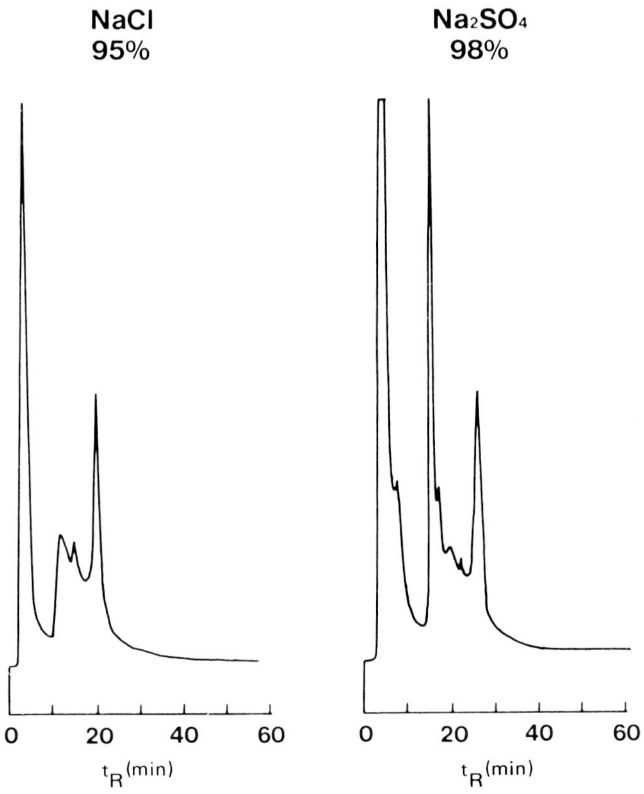
NaCl
95%
Na2SO4
98%
0
20
40
60
tR(min)
0
20
40
60
tR(min)

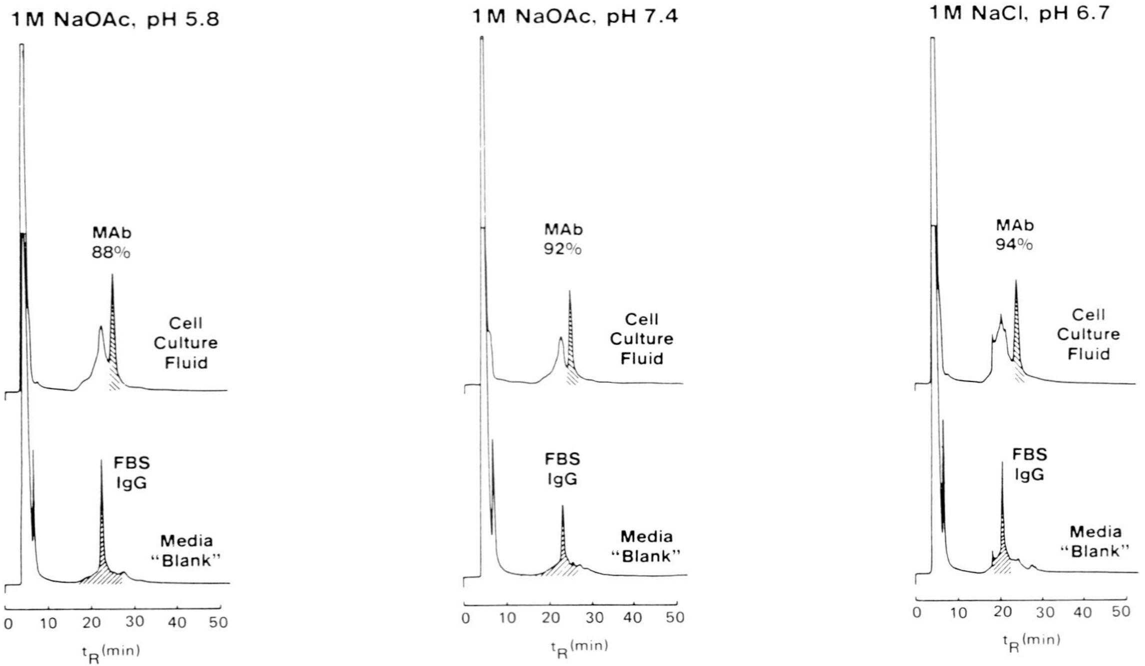

Figure A3 Effect of elution buffer species and pH on the purification of a mouse IgG monoclonal antibody from a serum and hormone/peptide-supplemented cell culture fluid ultrafiltrate on 5-μm BAKERBOND ABx. Chromatography was conducted on a stainless steel HPLC column (7.75 × 100 mm) containing 5-μm BAKERBOND ABx. The mobile phase consisted of an initial starting buffer A of 10 mM MOPSO plus 15 mM MES, pH 5.6, and a final elution buffer B of either (as indicated above each set of the two chromatograms), 1 M NaOAc, pH 5.8; or 1 M NaOAc, pH 7.0; or 1 M NaCl plus 20 mM NaOAc, pH 6.7; or 500 mM KH_2PO_4, pH 5.2; or 500 mM KH_2PO_4, pH 7.4; or 500 mM $(NH_4)_2SO_4$ plus 20 mM NaOAc, pH 6.7. In each chromatogram the gradient consisted of an initial hold at 100% A buffer for 4 min, followed by a rapid linear gradient (begun 4 min after the injection) from 100% A to 100% B buffer over 26 min. The initial flow rate was 0.7 ml/min, and this was increased to 1.0 ml/min at 4 min after the injection; the back pressure was less than 200 psi. Proteins were detected by

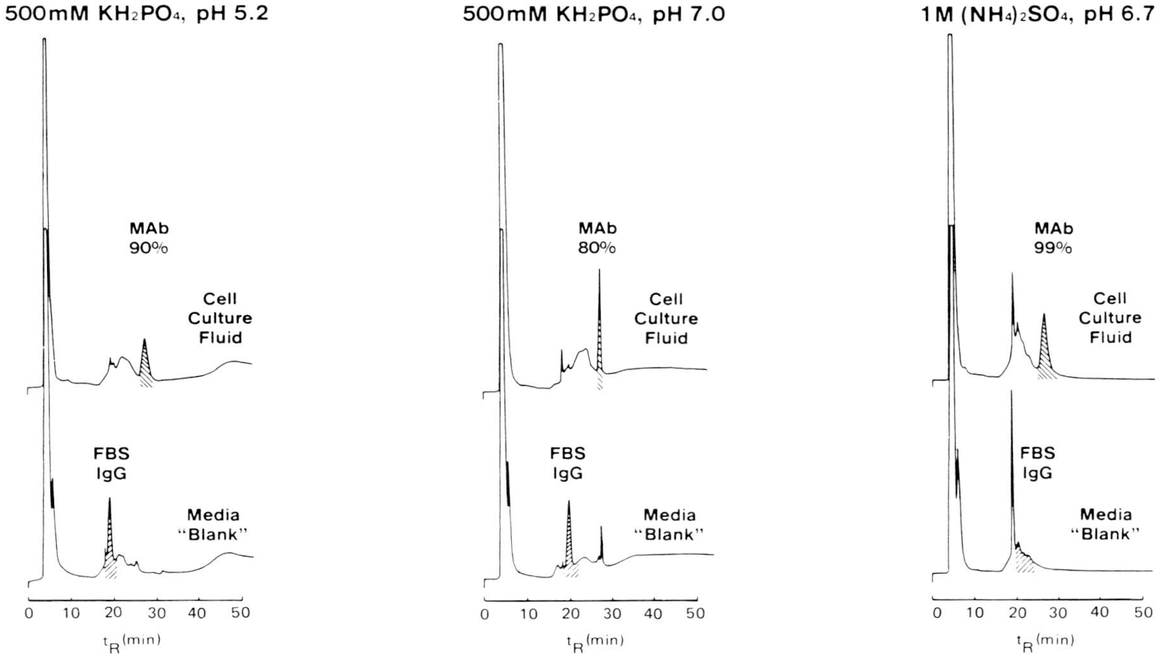

UV absorbance at 280 nm at an attenuation of 2.0 AUFS. The injection volume was 0.8 ml, and consisted of either, 0.4 ml of fetal bovine serum containing the hormone/peptide supplement (cell growth/antibody production stimulators) denoted in each of the lower chromatograms as the "medium blank," or 0.3 ml of the final cell culture ultrafiltrate (20X) denoted in each of the upper chromatograms of each set as "culture fluid." The monoclonal antibody (cross-hatched peaks in the upper chromatograms, labeled as "MAb") eluted as the last major protein peak in the presence of all the elution buffers tested, and was typically well resolved from all protein contaminants, including the host (fetal bovine serum polyclonal) IgG (cross-hatched peaks in the lower set of chromatograms, labeled as "FBS IgG"). The purity of the monoclonal IgG (as determined by SDS–PAGE) and high-performance size-exclusion chromatography is given in each chromatogram, with the highest purity being achieved with the sodium chloride buffer and, in particular, with the ammonium sulfate elution buffer.

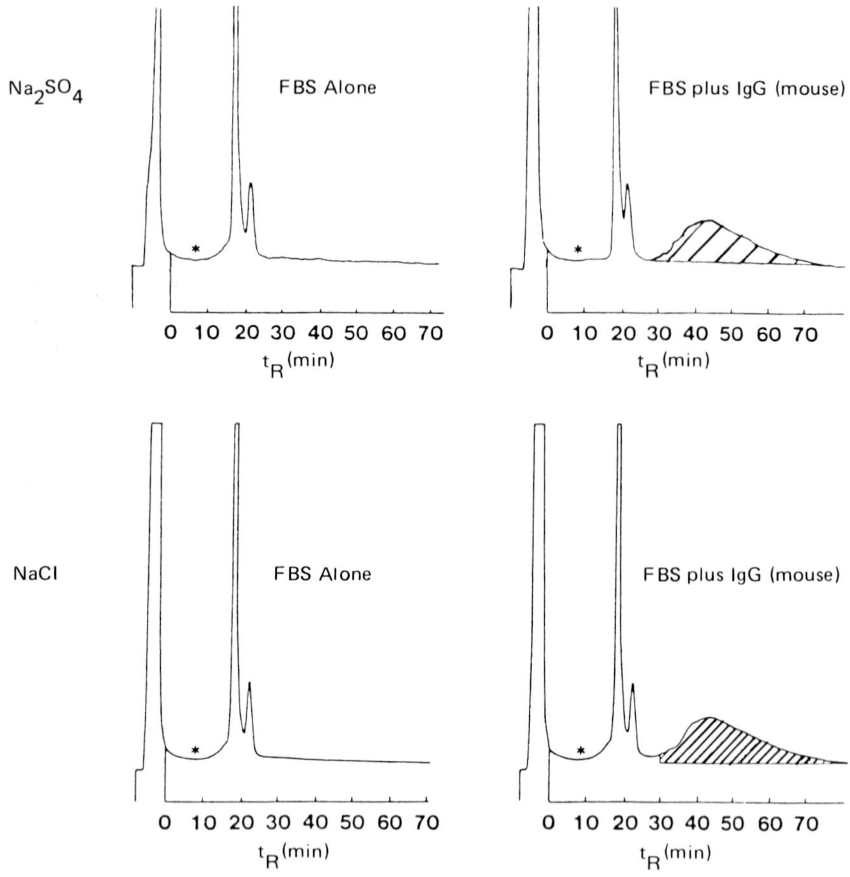

Figure A4 Effect of various elution buffers on the separation of mouse polyclonal IgG from fetal bovine serum proteins on BAKERBOND ABx. Chromatography was conducted on a stainless steel HPLC column (4.6 X 250 mm) dry packed with 5-μm (in the ammonium sulfate chromatogram) or 40-μm BAKERBOND ABx. The mobile phase consisted of an initial buffer A of 25 mM MES, NaOAc, pH 5.8 (or 10 mM MES, pH 5.6, in the $(NH_4)_2SO_4$ run), and a final elution buffer B of either (as indicated in each chromatogram), 250 mM $(NH_4)_2SO_4$ plus 10 mM NaOAc, pH 5.6; or 1 M NaCl plus 20 mM NaOAc, pH 5.7; or 1 M NaOAc, pH 7.0; or 500 mM KH_2PO_4, pH 6.8; at a linear flow rate of 0.5 ml/min, with a back pressure of 20 psi for the 40-μm runs and 500 psi for the 5-μm (top) run, with a linear gradient over 90 min begun at time zero. Proteins were detected by UV absorbance at 280 nm at an attenuation of 2.0 AUFS which was changed (see +) to 1.0 AUFS (for the $(NH_4)_2SO_4$, the NaCl, and the NaOAc elution) or 2.0 AUFS (i.e., no change in sensitivity, for the KH_2PO_4 elution).

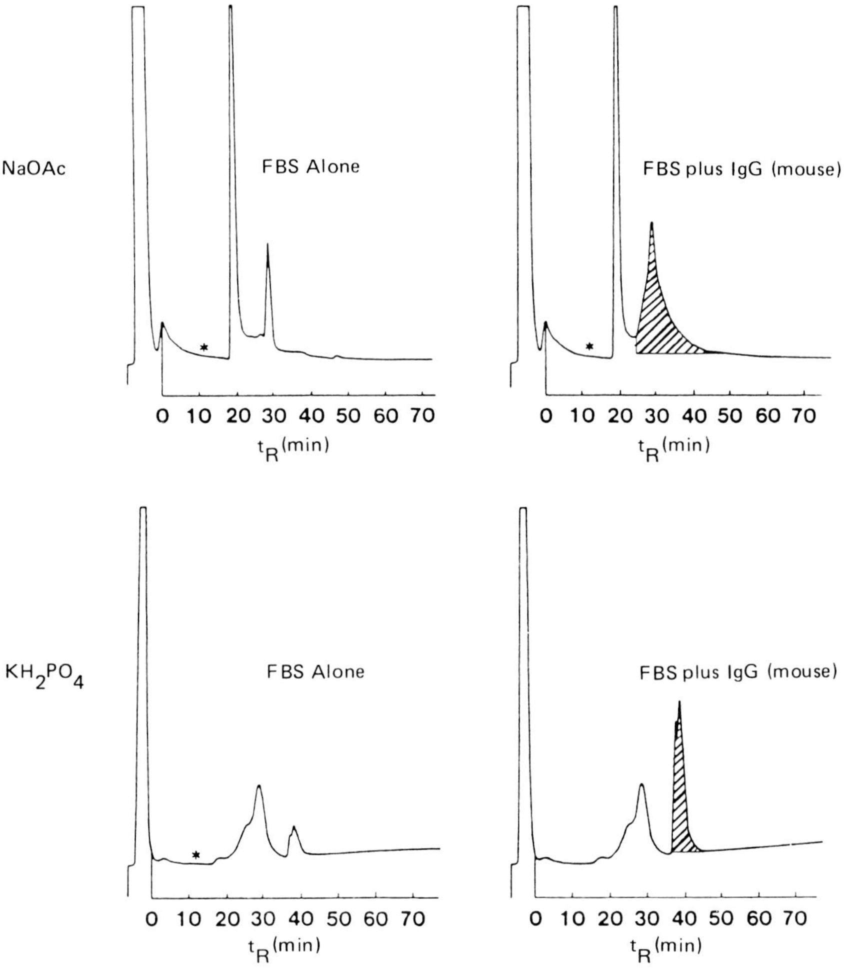

The sample injection volume was 2.0 ml consisting of 0.5 ml of crude fetal bovine serum with or without various amounts of mouse polyclonal IgG, diluted in 25 mM MES, pH 5.4. The cross-hatched peak corresponds to the retention time of the purified mouse polyclonal IgG alone (which is not shown). Similar results were also obtained when the fetal bovine serum was spiked with purified serum polyclonal IgG and IgM from other species (human, rat, rabbit, etc.; data not shown). With ammonium sulfate elution buffer any monoclonal (or polyclonal) antibody(s) can be easily resolved from virtually all the protein contaminants present in fetal bovine serum and tissue culture media containing fetal bovine serum. The antibody peak (cross-hatched area) is broad only because of the polydisperse (polyclonal) nature of the serum antibodies.

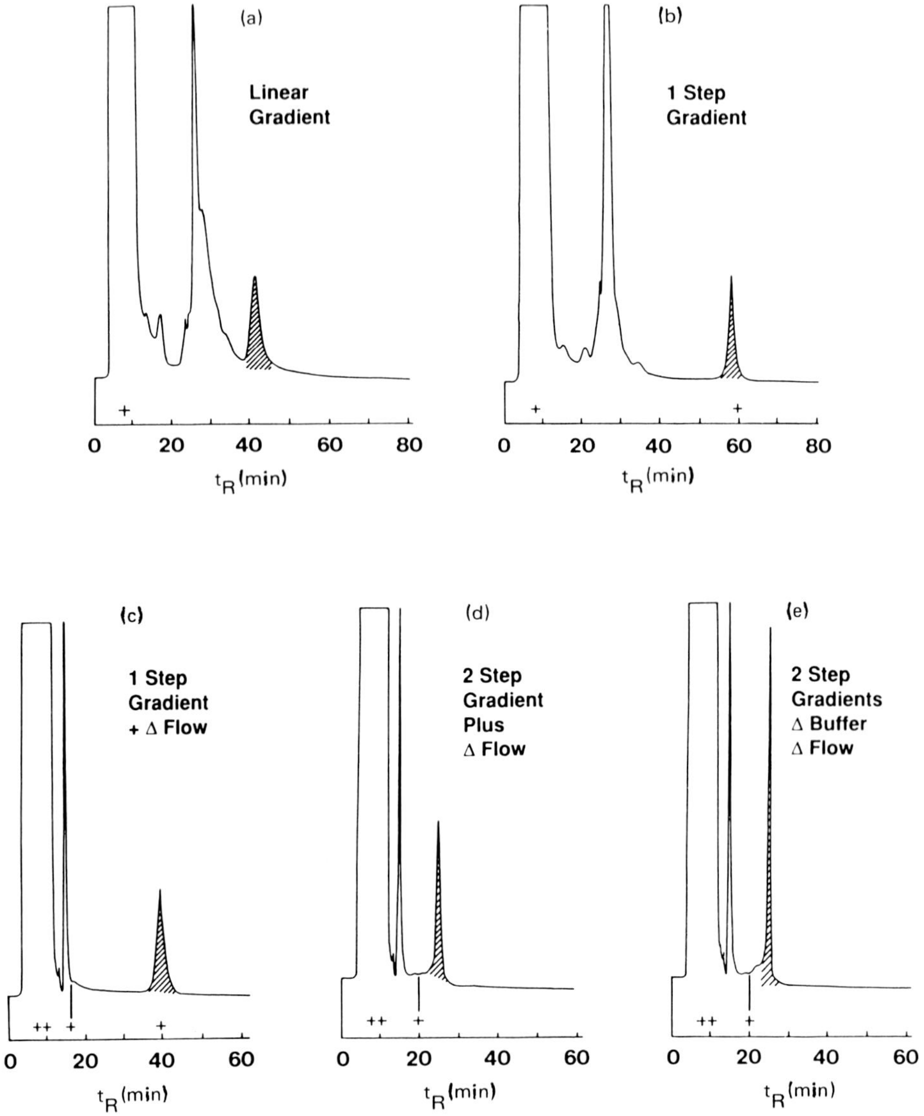
(a)
Linear Gradient
(b)
1 Step Gradient
(c)
1 Step Gradient + Δ Flow
(d)
2 Step Gradient Plus Δ Flow
(e)
2 Step Gradients Δ Buffer Δ Flow
t_R (min)
0
20
40
60
80

Figure A5 Use of step-gradient elution and increased flow rates to improve resolution and purity, and reduce run times and elution volumes on 15-μm BAKERBOND ABx. Chromatography was conducted on a GOLD ABx column (7.75 × 100 mm) packed with 15-μm BAKERBOND ABx. The equilibration buffer A was 25 mM MES, pH 5.5, and the elution buffer was 500 mM $(NH_4)_2SO_4$ plus 20 mM NaOAc, pH 7.0. The flow rate was 1.0 ml/min (except as noted below) at a back pressure of 50 psi, and proteins were detected by UV absorbance at 280 nm, at an attenuation of 1.0 AUFS. The sample consisted of 1.0 ml of a 20-fold concentrated (ultrafiltrate) of a serum-supplemented cell culture supernatant, diluted to 5 ml with buffer A. In each chromatogram, an initial isocratic gradient at 100% A buffer was held for 7 min during sample loading. In chromatogram *a*, a linear gradient over 60 min, from 100% A buffer to 100% B buffer, was begun at 7 min after the injection. In chromatogram *b*, a single step gradient to 17% B buffer was begun at 7 min and held isocratically for 53 min, followed by a linear gradient over 20 min from 60 min to 80 min. In chromatogram *c*, a step gradient to 17% B buffer was begun at 7 min, and held isocratically for 33 min, followed by a 20-min linear gradient from 17% B to 100% B buffer, with the flow rate changed to 4.0 ml/min between 10 min and 16 min (1.0 ml/min from 1 to 10 min, and 16 to 60 min). In chromatogram *d*, a step-gradient to 17% B was begun at 7 min, held isocratically for 13 min and at 20 min a second step-gradient to 100% B buffer was conducted, with the flow rate changed to 4.0 ml/min between 10 and 20 min. In chromatogram *e*, the conditions were the same as those in chromatogram *d*, except the step-gradient at 20 min was into 100% buffer C which was 1 M NaOAc, pH 7.0. The cross-hatched peak in each chromatogram contains the monoclonal antibody at a purity of greater than 99% by SDS–PAGE and high performance size exclusion chromatographic analysis.

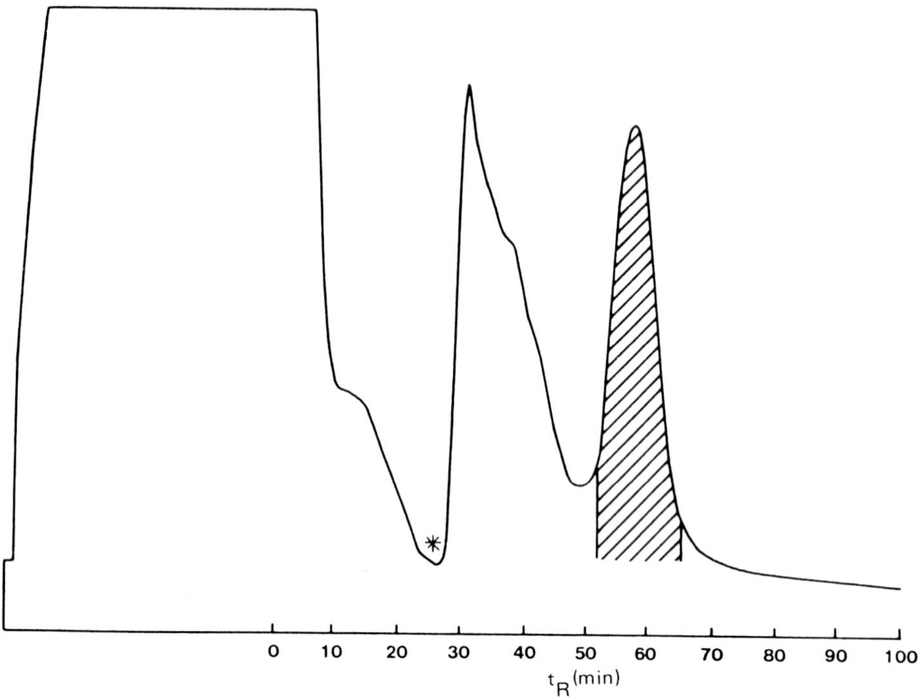

Figure A6 Preparative purification of a monoclonal antibody from a cell culture ultrafiltrate and comparison of linear-gradient elution and step-gradient elution from BAKERBOND ABx. Linear gradient elution was conducted on a semipreparative column (10 × 250 mm) containing approximately 10 g of 15-μm BAKERBOND ABx (chromatogram on the left). The mobile phase consisted of an initial buffer A of 25 mM MES, pH 5.4, and a final elution buffer B of 500 mM $(NH_4)_2SO_4$ plus 20 mM NaOAc, pH 6.7, with an initial isocratic gradient at 100% A buffer for 43 min during the sample loading, followed by a linear gradient from 100% A to 100% B over 120 min at a linear flow rate of 5.0 ml/min, with a backpressure of 200 psi. Proteins were detected by UV absorbance at 280 nm with a preparative flow cell; the attenuation was changed from 2.0 to 1.0 AUFS following the elution of the void volume peak, as indicated (see +), to better visualize the antibody peak. The sample injection volume was 225 ml containing 75 ml of ultrafiltrate (20-fold concentrated cell culture medium ultrafiltrate, or 1500 ml of original cell culture fluid, containing 180 mg of monoclonal antibody) diluted in 25 mM MES, pH 5.4. The cross hatched peak contains the monoclonal antibody at a purity of greater than 90% by SDS-PAGE and high-performance size exclusion chromatographic analysis, with less

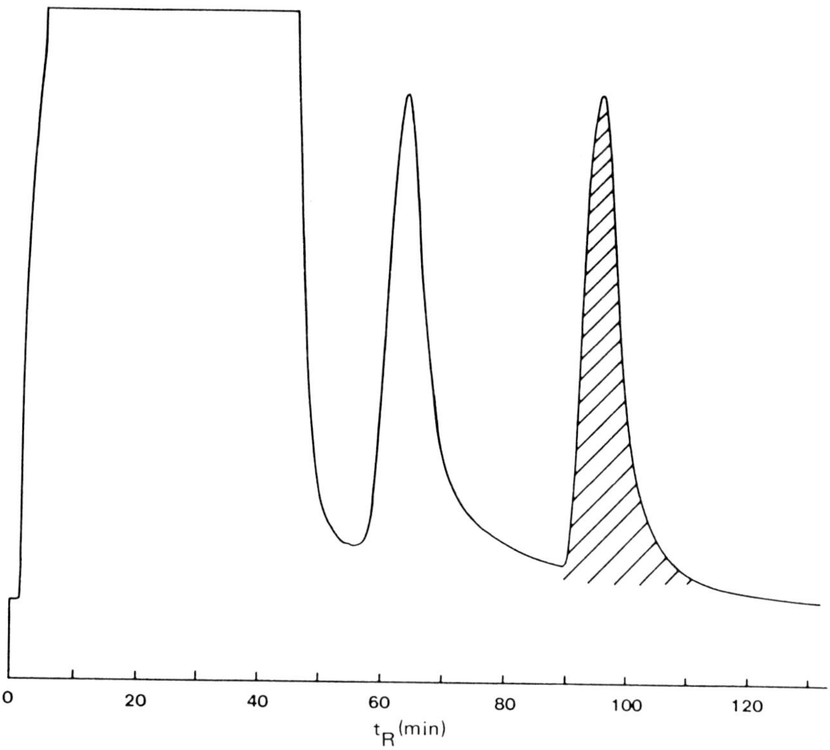

than 15% of the bovine polyclonal immunoglobulins from the fetal bovine serum (chromatogram on the left).

Step-gradient elution was conducted on a "1-in." preparative column (21.2 X 150 mm) which was dry packed with approximately 15 g of 40-μm BAKER-BOND ABx (chromatogram on the right). The chromatographic conditions were the same as detailed above, except, the gradient consisted of a step to 17% buffer (at 42 min after the injection began) which was held isocratically for 36 min; at 78 min, a second step was conducted to 50% B buffer (which was 1 M $(NH_4)_2SO_4$ plus 40 mM NH_4OAc, pH 7.0); the flow rate was 8.0 ml/min and the back pressure was less than 20 psi; detection was at 280 nm at 2.5 AUFS; the injection volume was 320 ml consisting of 160 ml of a 20-fold concentrated serum-supplemented cell culture ultrafiltrate (3.2 L of original supernatant) diluted in buffer A; the cross-hatched peak contains 370 mg of the monoclonal antibody at a purity of greater than 95% by SDS–PAGE and high performance size exclusion chromatographic analysis, with virtually no bovine polyclonal immunoglobulins from the fetal bovine serum (chromatogram on the right).

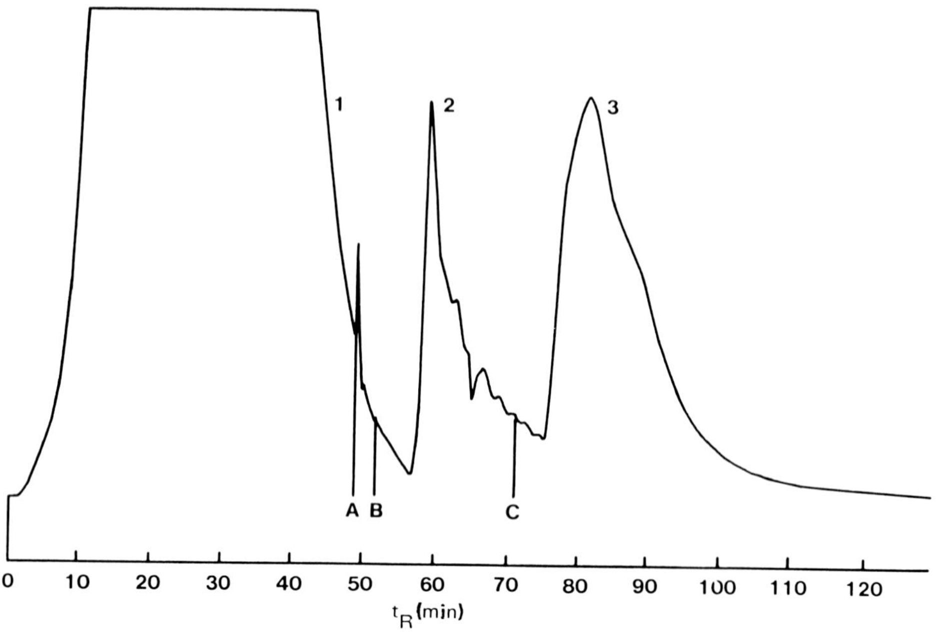

Figure A7 Preparative purification of dialyzed rabbit serum on 40-μm BAKERBOND ABx. Chromatography was conducted on a stainless steel HPLC column (10 × 250 mm) dry packed with less than 10 g of 40 μm BAKERBOND PREPSCALE ABx. The mobile phase consisted of an initial buffer A of 25 mM MES, pH 5.4, and a final elution buffer B of 500 mM $(NH_4)_2SO_4$ plus 20 mM NaOAc, pH 6.7, with an initial isocratic gradient at 100% A buffer for 52 min (during sample loading), followed by a step-gradient (at point B) to 10% B buffer, which was held isocratically for 20 min, followed by a linear gradient (at point C) from 10% B to 100% B over 30 min. The flow rate was 3.0 ml/min during the sample loading and was changed (at point A) to 5.0 ml/min, with a back pressure of 150 psi. Proteins were detected by UV absorbance at 280 nm at an attenuation of 2.0 AUFS. The sample injection volume was 110 ml consisting of 28 ml of crude rabbit serum, which was dialyzed against 20 mM MES plus 15 mM KH_2PO_4, pH 6.0, and then diluted for injection with 25 mM MES, pH 5.4. Peak 1 eluate contained albumins, transferrins, and proteases, peak 2 eluate contained the bound, nonimmunoglobulin proteins, and peak 3 eluate contained the rabbit immunoglobulins (IgG, IgA, and IgM) at a purity of 90% by SDS–PAGE analysis, with a recovery of greater than 90%. Immunological activity from serum samples purified on ABx is typically higher than that obtained by high-performance anion exchange chromatography (31).

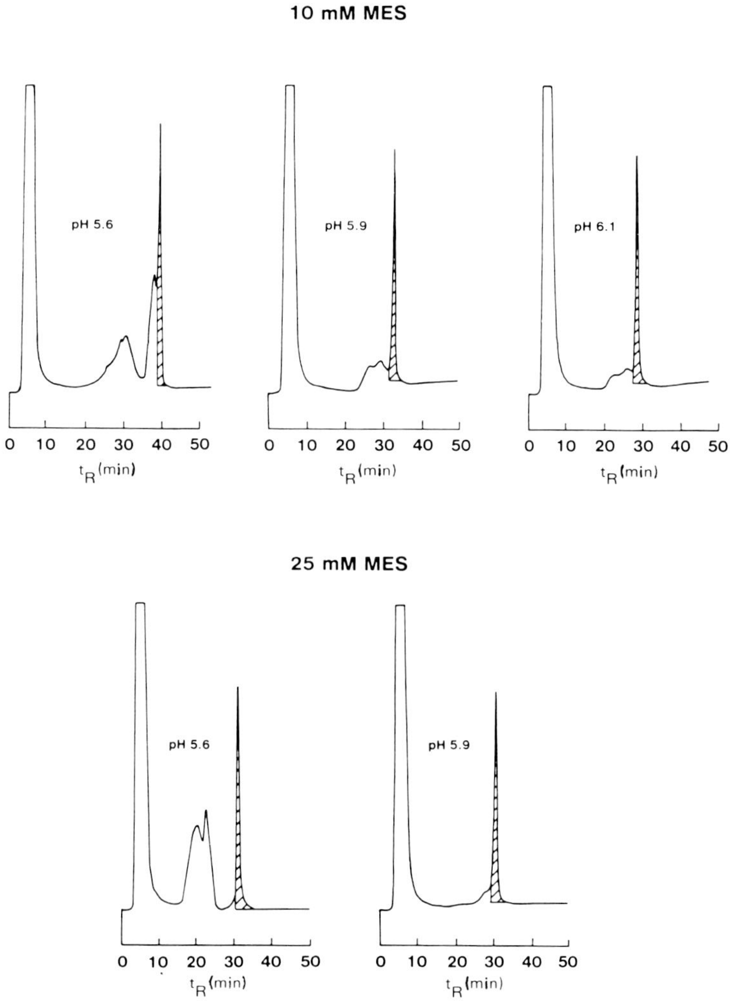

Figure A8 Effect of initial equilibration buffer pH and ionic strength on the purification of a monoclonal antibody from a serum-supplemented cell culture ultrafiltrate, and "rescue" from the poor resolution obtained in the presence of phosphate elution buffer on 15-μm BAKERBOND ABx. Chromatography was conducted on a stainless steel HPLC column (4.6 × 250 mm) containing 15-μm BAKERBOND ABx. The mobile phase consisted of an initial buffer A of 10 or 25 mM MES, at a pH of either 5.6, 5.9, or 6.1 (as indicated in each chromatogram) and a final elution buffer B of 500 mM KH_2PO_4, pH 6.3, with a linear gradient of 100% A to 50% B over 60 min, at a linear flow rate of 1.0 ml/min, with a back pressure of 200 psi. Proteins were detected by UV absorbance at 280 nm at an attenuation of 1.0 AUFS. The sample injection volume was 2.0 ml consisting of 0.5 ml of crude serum-supplemented cell culture ultrafiltrate (10-fold concentrated) plus 1.5 ml of the appropriate A buffer. The cross-hatched peak contains the monoclonal antibody purities of 80, 90, 93, 95, and 97% from top left to bottom right, respectively, as determined by SDS–PAGE.

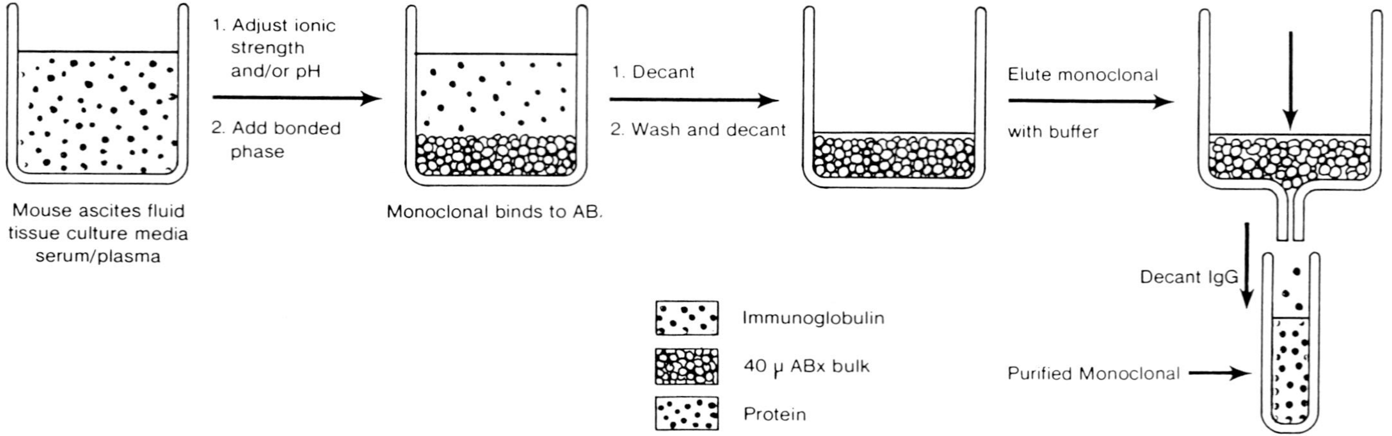

Figure A9 Schematic diagram of the batch or solid-phase extraction of monoclonal antibodies from ascites fluids, cell culture fluids, or from serum/plasma with BAKERBOND ABx.

ACKNOWLEDGMENTS

The author would like to acknowledge the assistance of Dr. Laura Crane and Dr. Sally Seaver for helpful discussions during the preparation of this text, Dr. Michael Henry for his assistance with the batch extractions, Dr. Steve Berkowitz for the two-dimensional chromatography of human plasma, Mrs. Joyce Guenther for her work on the solid phase extraction with ABx, Mrs. Diane Rush for SDS-PAGE analysis and, particularly, Dr. Hugh E. Ramsden and Mr. Joseph Horvath for their eloquent synthesis and help in optimizing the ABx surface chemistry.

BAKERBOND, ABx, PREPSCALE, MAb, HI-Propyl, and spe are all trademarks of J. T. Baker Inc., Phillipsburg, New Jersey. BAKERBOND ABx is registered under U. S. Patent 4,606,825.

REFERENCES

1. Köhler, G., and C. Milstein, *Nature 256*:495–197 (1975).
2. Yelton, D. E., and M. D. Scharff, *Annu. Rev. Biochem. 50*:657–680 (1981).
3. Ritz, J., and S. F. Schlossman, *Blood 59*:1–11 (1982).
4. Larson, S. M., J. P. Brown, P. W. Wright, J. A. Carasquillo, I. Hellstrom, and K. E. Hellstrom, *J. Nucl. Med. 24*:123–129 (1983).
5. Staehelin, T., D. S. Hobbs, H. F. Kung, C. Y. Lai, and S. Pestka, *J. Biol. Chem. 256*:9750–9754 (1981).
6. Calton, G. J., In *Methods in Enzymology*, Vol. 104, W. Jakoby (ed.). Academic Press, New York, pp. 381–387 (1984).
7. Shimizu, F., R. Wang, and M. Varga, *Clin. Chem. 29*:1245–1250 (1983).
8. Wang, R., W. F. Bermudez, R. L. Saunders, W. A. Present, R. M. Bartholomew, and T. Adams, *Clin. Chem. 27*:1063–1069 (1981).
9. Nau, D. R., A unique chromatographic matrix for rapid antibody purification. *BioChromatography 1*:82–94 (1986).
10. Nau, D. R., ABx—A novel chromatographic matrix for the purification of antibodies. In *Commercial Production of Monoclonal Antibodies: A Guide for Scaling-Up Antibody Production*, S. Seaver (ed.). Marcel Dekker, New York (1987).
11. Nau, D. R., Design of chromatographic matrices for antibody purification and analysis. In *HPLC in Biotechnology*, W. Hancock (ed.). Wiley Interscience, New York (submitted, 1988).
12. Gemski, M. J., B. P. Doctor, J. K. Gentry, M. J. Pluskal, and M. P. Strickler, *BioTechniques 3*:378–384 (1985).
13. Brooks, T. L., and A. Stevens, *Am. Lab.* October pp. 54–64 (1985).
14. Bailon, P., N. Drugazima, and A. H. Nichikawa, Paper 210, presented at Biotechnology '86, New Orleans, Louisiana, January 29–30 (1986).
15. Duberstein, R., *Genet. Eng. News* January, pp. 22–25 (1986).
16. Levy, H. B., and H. A. Sober, *Proc. Soc. Exp. Biol. Med. 103*:250 (1960).

17. Garvey, J. S., N. E. Cremer, and D. H. Susodorf, *Methods in Immunology*. Addison Wesley, Reading, Mass., pp. 223–226 (1977).
18. Fahey, J. L., and E. W. Terry, In *Handbook of Experimental Immunology*, Vol. 1, D. M. Weir (ed.). Blackwell, London, pp. 8.1–8.16 (1978).
19. Nau, D. R., ABx—A unique chromatographic matrix for rapid antibody purification. Lecture 3706 presented at HPLC '86—the Tenth International Symposium on Column Liquid Chromatography, May 19, San Francisco (1986).
20. Nau, D. R., Determination of monoclonal antibody concentrations by HPLC. Poster H-15 presented at the Fifth Annual Congress for Hybridoma Research, January 26–29, Baltimore (1986).
21. Deschamps, J. R., J. E. K. Kildreth, D. Derr, and J. T. August, *Anal. Biochem. 147*:451–454 (1985).
22. Clezardin, P., J. L. McGregor, M. Manach, H. Bonkerche, and M. Deschavanne, *J. Chromatogr. 319*:67–77 (1985).
23. Burchiel, S. W., J. R. Billman, and T. R. Alber, *J. Immunol. Methods 69*: 33–44 (1984).
24. Nazareth, A., and S. DiScullio, (Personal communications and personal observation, 1985).
25. Ey, P. L., S. J. Prowse, and C. R. Jenkin, *Immunochemistry 15*:429–436 (1978).
26. Wofsy, L., and B. Burr, *J. Immunol. 103*:380–383 (1969).
27. Hirano, H., T. NIshimura, and T. Iwamura, *Anal. Biochem. 150*:228–234 (1985).
28. Colonick, S. P., N. O. Kaplan, J. J. Langone, and H. Van Vunakis, (eds.), *Methods Enzymol. 73*:21–34 (1981).
29. Frej, A. K., J. G. Gustafsson, and P. Hedman, *Biotechnology* September, pp. 777–781 (1984).
30. Ross, A. H., D. Herlyn, and H. Koprowski, *J. Immunol. Methods 102*: 227–232 (1987).
31. Nazareth, A. (Personal communication, 1987).
32. Seaver, S. S. (Personal communication, 1987).
33. Swaminathan, B., and G. Corleone (Personal communication, 1987).
34. Garg, V. K., Use of Preparative HPLC in Large Scale Purification of Therapeutic Grade Proteins from Mammalian Cell Culture, Poster Number 911, presented at the Seventh International Symposium on HPLC of Proteins, Peptides, and Polynucleotides, November 2–4, Washington, D. C. (1987).
35. Nau, D. R. (Manuscripts in preparation, 1988).

RECENT REFERENCE ADDED IN PROOF

Nau, D. R., Optimization of monoclonal antibody purification. In *Techniques in Protein Chemistry*, papers from 1988 Protein Society (1988, submitted).

17

Monoclonal Antibodies for the Detection of *Chlamydia trachomatis*

RICHARD S. STEPHENS*
University of California–Berkeley, Berkeley, California

I. DISEASES AND PATHOGENESIS

The organisms that the genus *Chlamydia* comprises are some of the most ubiquitous pathogens in the animal kingdom. These organisms were once thought to be viruses because of their small size and obligate intracellular growth. They are now recognized as bacterial parasites and are currently classified into two species, *C. trachomatis* and *C. psittaci* (1). The natural host for *C. trachomatis* is humans, whereas *C. psittaci* is a pathogen of most other mammals and birds.

Chlamydia trachomatis has long been known as the cause of trachoma. This blinding eye disease affects hundreds of millions of people, although it is restricted to endemic areas in arid, developing countries (2). The development of tissue culture systems capable of isolating *C. trachomatis* and the development of seroepidemological assays, have led to an understanding of the wide spectrum of diseases caused by these organisms. Chlamydial infections are now recognized as the most prevalent sexually transmitted infections in the United States, Britain, and other developed countries (3). *Chlamydia trachomatis* colonize the columnar epithelial cells of mucous membranes, such as the conjuctiva, nasopharynx, urethra, cervix, and rectum. Chlamydiae are usually limited to these mucosal surfaces; however, they may spread and are capable of causing systemic diseases.

**Present affiliation*: University of California–San Francisco, San Francisco, California.

Strains of *C. trachomatis* that infect humans consist of two biovars, lymphogranuloma venereum (LGV) and trachoma (1). The LGV strains are invasive, and the site of proliferation is usually the inquinal lymphatic tissues, resulting in lymphogranuloma venereum disease. The incidence of LGV is low in most parts of the world, yet it remains endemic in parts of Africa (4). The trachoma biovar is usually restricted to superficial ocular, respiratory, and urogenital sites and causes trachoma, conjunctivitis, pneumonia in infants, urethritis, and cervicitis. Additionally, trachoma strains often cause epididymitis (5), salpingitis (6), mucopurulent cervicitis (7), and proctitis (8). Other diseases that have been associated with chlamydial infections include endometritis (9), perihepatitis (10), Reiter's syndrome (11), and there may be a role of *C. trachomatis* infections in cervical dysplasia (12). Important complications that arise from chlamydial salpingitis or pelvic inflammatory disease (PID) are tubal infertility (13) and ectopic pregnancy (14).

Trachoma remains the major cause of preventable blindness. However, the severity of disease has been decreasing because of improved sanitation. Endemic areas of blinding trachoma remain in parts of Africa, India, Asia, Australia, South America, and the Pacific Islands (2). Currently, little epidemological data are available on the prevalence and impact of urogenital infection in these areas and on the role of urogenital infection in trachoma. In developed countries, chlamydial infections are the leading sexually transmitted disease, with an estimated incidence in the United States of 3-10 million new cases each year (3). Infants exposed to chlamydia, as they pass through an infected cervix, have a high risk (60-70%) of ocular, nasopharyngeal, vaginal, and rectal colonization (15,16). An estimated 25-50% of these infants will develop conjunctivitis, and 10-20% will develop pneumonia (15,17). If one considers that studies in the United States have found a 8-12% prevalence of chlamydia cervical infection among pregnant women, then chlamydial infections cause a significant infant morbidity (15). Each year in the United States, it is estimated that more than 1 billion dollars are spent in direct and indirect costs, which are primarily attributable to PID and infants with pneumonia (3).

The pathogenesis of chlamydial infection is often marked by chronic and latent infections. The host provides a vigorous immune and inflammatory response, particularly after reactivation of infection or after reinfection. Thus, the progression of disease appears to be a function of the development of immunopathological tissue damage (18). Nevertheless, infections by *C. trachomatis* are treatable, providing the diagnosis can be made. Significantly, symptoms of chlamydial infections are characteristically nonspecific, and asymptomatic infections are common (19). This emphasizes the need for sensitive laboratory diagnostic procedures.

For the past decade, laboratory diagnosis of chlamydial infections has been accomplished by isolation of the organisms in tissue culture systems. A culture for chlamydia takes at least 3 days for growth, followed by a blind passage on negative specimens (20). Consequently, cultures for chlamydia are both time-consuming and expensive. Whereas there has been a history of improvement and standardization of culture techniques, the sensitivity of isolation is estimated to be only 80% (21). Although this is a pressing cause for the development of improved diagnostic techniques, probably more important is that there are a relatively few culture facilities, and these are geographically restricted.

The recent introduction of immunodiagnostic assays that use monoclonal antibodies has offered substantial improvements for culture detection techniques and, more importantly, for culture-independent diagnostic capabilities. The development of suitable monoclonal antibody reagents for *C. trachomatis* exemplifies the challenges involved with accurate detection of this very antigenically diverse group of organisms, as well as providing a model system that challenges the sensitivity capabilities of new immunodiagnostic assay formats.

II. MICROBIOLOGY OF *CHLAMYDIA TRACHOMATIS*

Chlamydiae are obligate intracellular parasites that rely upon host cell ATP because they lack metabolic pathways for their own production of energy-rich compounds (22,23). The two species of *Chlamydia, C. trachomatis* and *C. psittaci*, display very similar biochemical, morphological, and developmental properties (1). It should be emphasized, however, that the genetic relatedness between these two species, as assessed by DNA hybridization studies, shows less than 10% DNA homology (24). Nevertheless, these species do share at least one common surface antigen (25,26), and their protein profiles by polyacrylamide gel electrophoresis (PAGE) are remarkably analagous (27). In this section, I will present a brief review of the important microbiological tissues that relate to the development and characterization of monoclonal antibodies and their use for the detection of *C. trachomatis*. Complete microbiological reviews are available (21, 22,28).

A. Antigenic Relationships

More than 15 *C. trachomatis* serovars have been characterized by the microimmunofluorescent test (MIF) (29,30). In the most utilized form of the micro-IF test, polyvalent antisera produced to a *C. trachomatis* isolate, is tested by immunofluorescence against a panel of prototype reference organisms. At end-point dilutions of antisera, the reactivity patterns among these reference strains define not only serovar specificity but also reveal a complex relationship among serovars (31). Serovars may be classified into either of two groups; B complex or C complex. The serovars that make up the B complex are, as a group, much more closely related than are the serovars that constitute the C complex. In

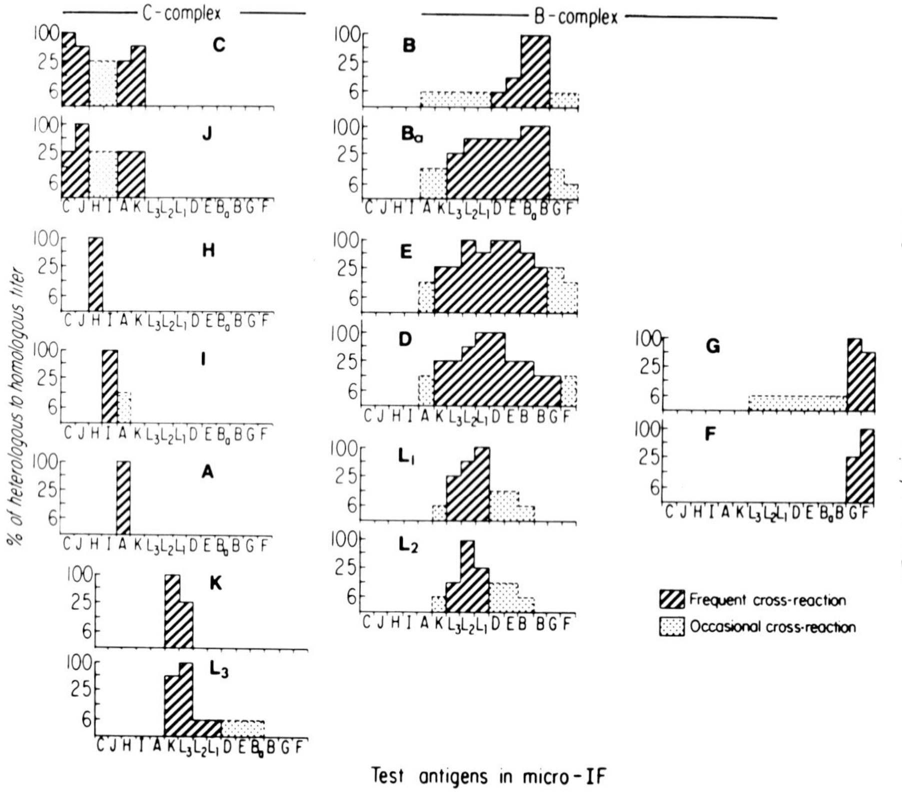

Figure 1 Serological relationships between 15 *C. trachomatis* serovars as shown by polyvalent antisera (*Y* axis). The extent of cross-reactions delineate C-complex and B-complex serogroups and the senior/junior relationships within serogroups (*Source*: reproduced with permission from Ref. 31).

addition, Wang and Grayston (31) have described a junior/senior hierarchy of relationships within each complex (Fig. 1).

Serovars found from eye isolates in trachoma patients are usually A, B, Ba, and C, and serovars found from genital isolates are D through K. However, epidemological studies of the serospecificities of ocular isolates obtained in trachoma endemic areas and isolates obtained from the genital tract are separated in time and geography. Thus, this division may be an artifact because eye isolates and genital isolates have not been evaluated simultaneously in trachoma endemic areas, and virulence determinants that could differentiate ocular from genital strains have not been found. It is likely that any serovar is capable of causing trachoma or genital tract infection (32).

B. Antigenic Structure

Chlamydial antigens that are most useful as targets for immunodiagnostic probes are those antigens that are high in copy number and that display appropriate specificities. Knowledge of antigenic structure is important for consideration of particular assay formats. With some formats, surface exposure may be an essential criterion, whereas with other formats, antigen physicochemical stability after extraction procedures may be necessary. Of the antigenic macromolecular components that have been characterized for chlamydiae, two stand out above others: the chlamydial lipopolysaccharide (LPS) and the major outer membrane protein (MOMP).

1. Lipopolysaccharide

Both *C. trachomatis* and *C. psittaci* share a common LPS that is characterized as a heat-stable, periodate-sensitive, genus-specific antigen. This ethyl ether-soluble antigen was used in the complement fixation test for psittacosis developed by Hillman and Nigg in 1946 (33). Saponification of chlamydial glycolipid extracts enabled Dhir et al. (34) to identify a high-molecular-weight acidic polysaccharide with an antigenic entity similar to the 2-keto-3-deoxyoctonoic acid (KDO) moiety of salmonella LPS.

Genus-specific monoclonal antibodies to chlamydial LPS react to a low-molecular-weight, protease-resistant, and periodate-sensitive antigen (35). Among different genus-specific monoclonal antibodies, we observed differential avidity between some of these antibodies after mild periodate treatment of LPS, which suggested more than one antigenic determinant. Recent studies by Nurminen et al. (36,37), Thornley et al. (38), Brade et al. (39), and Caldwell and Hitchcock (40) have shown that chlamydial LPS is physically similar to entrobacterial LPS, and two antigenic determinants have been described: one that cross-reacts with entrobacterial Re-LPS, and one that is chlamydia-specific.

2. Major Outer Membrane Protein

Although a number of principal outer membrane proteins have been identified for chlamydiae, the MOMP, with a molecular mass of 40,000, is the most quantitatively preponderant protein (27). The MOMP constitutes over 60% of the total protein of the outer membrane, and it is exposed on the surface (41). The chlamydial MOMP displays structural and functional attributes that are uniquely modulated by intra- and intermolecular disulfide bonds. In its oxidized state, it may confer structural integrity to the extracellular form of the organism (42, 43), whereas in the reduced state, found in the intracellular form of the organism (44), it displays porin properties that mediate the transfer of solutes across the chlamydial membrane (45).

In addition to these dual structural and functional roles, MOMP is a strong immunogen and a complex antigen. The antigenic complexity of this protein has been investigated with polyvalent antisera and monoclonal antibodies. Caldwell and Schachter (46) used purified MOMP obtained from different serovars to immunize animals. The elicited antisera showed type specificities at endpoint dilutions, and at lower dilutions recognized other related serovars. The endpoint reactivities for different serovars were the same as those observed in the MIF test. Monoclonal antibodies to MOMP confirmed the presence of these determinants and the relationships observed in the MIF test (35); some antibodies were type-specific, some recognized multiple serovars and, furthermore, some recognized all *C. trachomatis* serovars.

C. Developmental Cycle

An understanding of the developmental cycle of chlamydiae is pertinent for the design, development, and evaluation of diagnostic approaches. Chlamydiae have a complex developmental cycle that is necessary to cope with the harsh realities of extracellular survival and, alternatively, to proliferate in the comfort of the intracellular milieu. The extracellular, infectious form is the elementary body (EB). Elementary bodies are 200-300 nm in diameter, which is very close to the resolving power of light microscopy. This form of the organism is relatively metabolically inert and is resistant to sonic or osmotic lysis. The structural integrity of these dense organisms is maintained by disulfide cross-linked MOMP and other membrane proteins (42,43). This infectious form binds host cell receptors that mediate endocytosis of the EB into a host cell phagosome (47). The entire growth cycle is completed within this membrane-bound vesicle, and chlamydiae are specifically protected because those phagosomes that harbor chlamydia will not fuse with lysosomes (48).

Soon after uptake the EB begins, its transformation into the noninfectious and metabolically active reticulate body (RB) form begins. This transformation results in crucial structural changes of the outer membrane (44). The RB membrane is particularly sensitive to sonic or osmotic lyses and becomes permeable to solutes. It has been proposed that these changes are mediated by the reduction or exchange of disulfide cross-linking. The plastic RBs are 500-1000 nm in diamter, and they divide primarily by binary fission.

Approximately 20 hr after infection, a new transition begins from RB to the infectious EB. Infectivity increases between 20 hr and approximately 48 hr; nevertheless, a mature phagosome also contains substantial numbers of RBs and forms in intermediate stages of transition. Late in the cycle, *C. trachomatis* often produces large quantities of glycogen, which is deposited in the phagosome matrix. The mature phagosome occupies virtually the entire host cell cytoplasmic volume and is called an inclusion. In tissue culture, this intracellular inclusion can be visualized by phase microscopy or by staining with Giemsa or iodine (Fig. 2). By an unknown mechanism, infectious EBs are ultimately released; however, the rate of initiation of infection in vitro is inefficient, thereby prohibiting continous culture of these organisms.

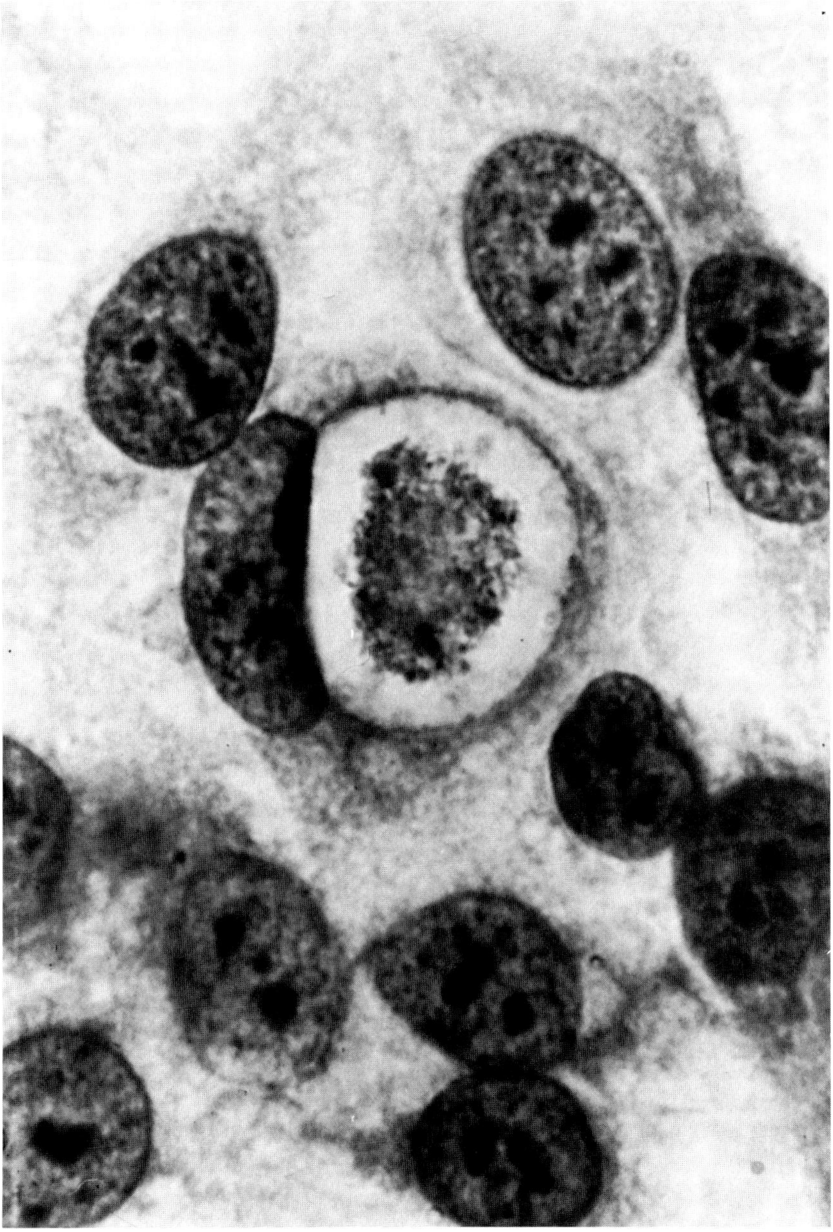

Figure 2 Cell monolayer infected with *C. trachomatis* and stained with Giemsa. The cell in the center of the field shows a large vacular inclusion which has apparently deformed the nucleous (top).

III. MONOCLONAL ANTIBODIES TO *CHLAMYDIA TRACHOMATIS*

Both polyvalent antibodies and monoclonal antibodies to chlamydia have been used for antigenic analysis, immunotyping, and the detection of inclusions. One of the striking features of conventional polyvalent antisera to *C. trachomatis* is its predominant type-specificity (31). Consequently, such antisera have proved very useful for immunotyping chlamydial isolates; however, this attribute has limited the utility of antisera for the detection of inclusions in clinical specimens or in cell culture. Standardized monoclonal antibodies of predefined specificities have applications for probing antigenic composition and structure (35), have greatly simplified immunotyping (49), and have improved inclusion detection systems (50). Most importantly, the high level of specificity of monoclonal antibodies has enabled the direct detection of chlamydia in clinical specimens (51).

The use of monoclonal antibodies as probes of antigenic composition with the immunoblot technique revealed that species-specific, subspecies-specific, and type-specific antibodies are specific to the MOMP, whereas most genus-specific monoclonal antibodies identify LPS (Fig. 3). Appreciation of the complex structure of MOMP was revealed in immunoblots of chlamydial proteins using MOMP-specific monoclonal antibodies (35). In these studies, we identified multiple oligomers of MOMP that were later demonstrated to be mediated by disulfied bonds (42).

A. Specificities

Early antigenic analyses that employed polyvalent antisera in classic precipitation assays (e.g., ouchterlany and immunoelectrophoresis) identified numerous chlamydial antigens (26,52); however, these assays yield little information concerning important macromolecular components without great difficulty. Caldwell and associates undertook this task and selected an interesting antigen identified by two-dimensional (2-D) immunoelectrophoresis (52) that they were able to characterize at the macromolecular level. Nevertheless, this purified component was ultimately a quantitatively minor constituent of unknown function, which frustrated further work (53). The inherent limitations of conventional polyvalent antisera are particularly evident for characterizations of antigentically complex macromolecular components, such as the chlamydial MOMP. Although conventional polyvalent antisera, raised from whole organisms, were of limited value, Caldwell and Schachter (46) purified MOMP from several serovars, and the monospecific antibodies elicited from these preparations showed endpoint reactivities analogous to those observed by the MIF test. They concluded from these studies that MOMP was a complex antigen that displayed multiple and unique specificities. Monoclonal antibodies to MOMP have demonstrated that each

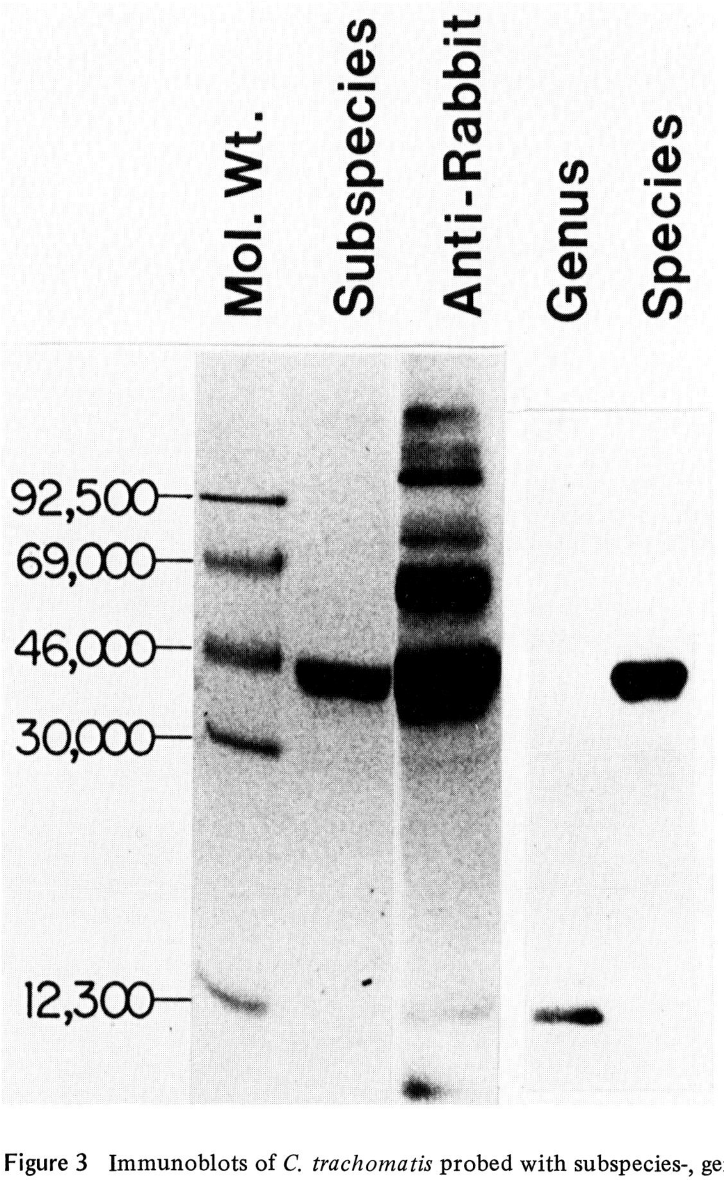

Figure 3 Immunoblots of *C. trachomatis* probed with subspecies-, genus- and species-specific monoclonal antibodies and polyvalent rabbit serum. Subspecies- and species-specific monoclonal antibodies identify the MOMP and the genus-specific antibody identifies LPS (*Source*: reproduced with permission from Ref. 35).

MOMP, in fact, displays multiple and unique epitopes as well as common epitopes (35).

Analysis of the specificities of monoclonal antibodies to *C. trachomatis* revealed four classes of surface reactivity; species-specific, subspecies-specific, type-specific, and genus-specific. These reactivities can be evidenced by using a variety of detection techniques such as enzyme-linked immunosorbent assay (ELISA), immunofluorescence, and immunoblot analyses (35). Examples of the different specificity profiles are presented in Figure 4.

1. Genus-Specific

Genus-specific antibodies are represented by recognition of all strains of *Chlamydia* from both species. The predominant genus-specific antigen is the

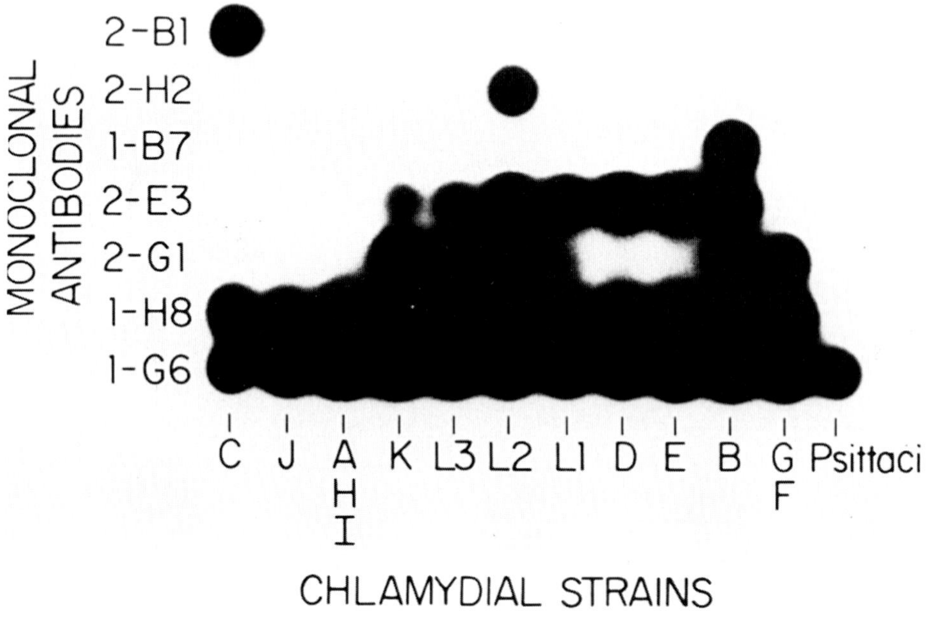

Figure 4 Autoradiograph of antibody-binding assays performed with culture fluids and EB of various chlamydial strains. *C. trachomatis* immunotypes A–K and *C. psittaci* strains Mn were adsorbed to wells of a microtest plate and reacted with monoclonal antibody. Immune reactions were detected with ^{125}I-labeled protein A. Immunotypes A, H, and I, as well as immunotypes G and F, were pooled before adsorption. Representative monoclonal antibody reaction patterns were: type-specific, 2-B1, 2-H2, and 1-B7; subspecies-specific, 2-E3 and 2-G1; species-specific, 1-H8; genus-specific, 1-G6 (*Source*: reproduced with permission from Ref. 35).

LPS. All monoclonal antibodies to this antigen react strongly by immunofluorescence of inclusions, ELISA, and immunoblot techniques (35). By immunofluorescence these antibodies also react strongly to RB developmental forms but equivocally to the condensed EB forms (35). This suggests that these antigens are not readily accessible on the cell surface of the infectious form. This is critical for immunofluorescence assay formats and for the process of monoclonal antibody development and characterization. Recently, Fuentes et al. (54) described some monoclonal antibodies with genus-specific reactivity patterns that reacted to MOMP in immunoblots of *C. trachomatis* proteins; however, it is unclear whether or not these determinants are surface exposed.

2. Species-Specific

Species-specific monoclonal antibodies display essentially equivalent reactivity with all *C. trachomatis* serovars but do not react with *C. psittaci* strains. These monoclonal antibodies recognize a surface-exposed determinant on the chlamydial MOMP; however, considering the number and spectrum of the antibodies produced, they appear to represent a relatively rare event. Although these monoclonal antibodies are the most useful for immunodiagnostic applications, they are also one of the most difficult to produce. Key elements for successful isolation of antibodies with this specificity are careful screening against each serovar and, anecdotally, multiple and long immunization schedules.

3. Subspecies-Specific

Subspecies-specific reactivities are characterized by antibodies that recognize specific groups of two or more serovars but not all seovars. These antibodies are specific to MOMP and demonstrate the complex antigenic relationships among *C. trachomatis* serovars. Immunoblot assays using these monoclonal antibodies may show a broader pattern or reactivity than observed by assays of intact organisms (J. Ma and C. C. Kuo, personal communication). As we have shown, denatured MOMP may expose antigenic determinants that are structurally hidden in the intact organism (55).

4. Type-Specific

Type-specific epitopes are identified by monoclonal antibodies that recognize a single serovar. The exquisite specificity of monoclonal antibodies provides the ability to easily distinguish between closely related serovars. Such an example is represented by monoclonal antibodies that can differentiate between serovars B and Ba (35). The only difference between these serovars when using polyvalent antisera is that the Ba serovar antisera cross-reacts more with other serovars. Type-specific monoclonal antibodies also recognize determinants on the MOMP (35,56). In contrast to species- and subspecies-specific monoclonal antibodies, type-specific antibodies usually react weakly to MOMP by immunoblot (35). These observations may reflect some discontinuous component to that epitope that is disrupted sy sodium doceyl sulfate (SDS) denaturation.

B. Development and Characterization

As in the development of monoclonal antibodies for other systems, a clear understanding of the goals, uses, and assay formats to be utilized streamlines the developmental approach. Monoclonal antibodies may be selected to match these goals early in the process, thereby saving much time and expense. For chlamydia, decisions concerning the desired specificities, morphological location and accessibility, physicochemical stability of determinants, biological activities, and the suitability of the anticipated format will influence the type of immunogen, the type(s) of antigen used for screening, and the screening format. Two particular goals have dominated the direction of monoclonal antibody development for *C. trachomatis*. One is the development of immunodiagnostic reagents, and the second is the development of serovar-specific monoclonal antibodies suitable for rapid immunotyping of chlamydial isolates.

The first critical consideration for the development of monoclonal antibodies is the choice of immunogen and screening antigens. Although there are exhaustive and well-characterized purification procedures for obtaining chlamydial organisms from their host cells, it is clear that considerable host cell immunogenic material is copurified (35). Nevertheless, monoclonal antibodies specific to *C. trachomatis* can readily be selected by using one of two approaches. The first approach consists of testing culture fluids on replicate-plated chlamydial preparations and noninfected host cell material. With this approach, numerous chlamydia-specific antibodies can be selected, but most cell culture fluids react with host cell material (35). The alternative approach employs two different host cell systems. Yolk sac-grown and cell culture-grown organisms are prepared, and one is used for immunization, whereas the other is used as antigen for antibody-binding assays (57). This approach simplifies the screening process, yet, either approach offers the same frequency of specific antibodies.

The next consideration is the material used for immunizations and the immunization schedule. For the development of surface probes suitable for immunodiagnostic uses, chlamydial EBs have been successful. Unfixed EBs are potent immunogens for mice, and adjuvants are unnecessary. The immunization schedule will influence the outcome for obtaining particular specificities to some antigens such as MOMP. For the production of broadly reacting monoclonal antibodies, multiple immunizations with different serovars or mixed serovars is efficacious (35). For the enrichment of serovar-specific monoclonal antibodies, it has been shown that two immunizations within 1-2 weeks increased the frequency of type-specific antibodies by 40-fold (57,58). With either approach, genus-specific monoclonal antibodies are the most common response. The root of immunizations seems to be of little importance, except perhaps for an iv boost just before fusion.

Understanding the complementarity between the immunization material and the material used for screening should be leveraged for maximum effectiveness to speed the development process. The choice of screening antigens can then be selected given the goals of the project; however, one of the earliest evaluations must be the spectrum of serovars identified by each monoclonal antibody because this is such an antigenically diverse group of organisms. Incorporated into this process should be desired attributes, such as physicochemical stability or lability, morphological evaluation, developmental cycle specificity, and identification of the macromolecular component. For example, if the goal is to produce genus-specific antibodies to LPS then the approach might be simply designed as follows:

1. Immunize with yolk sak-grown *C. psittaci* organisms.
2. Replicate screen culture fluids against antigen prepared from tissue-grown organisms, which may include
   a. Purified *C. trachomatis* LPS
   b. Enterobacterial Re-LPS
   c. Periodate-treated chlamydial LPS

The immunization with *C. psittaci* and screening with *C. trachomatis* ensures the selection of antibodies with genus specificity. Evaluation of the reactions to each of the screening preparations will immediately differentiate monoclonal antibodies that are specific for chlamydiae or cross-reactive to enterobacterial LPS and are either specific to the polysaccharide moieties or to lipid-A.

Additional criteria can be built into this primary assay or immediately assessed on only those culture fluids that are of interest from the primary assay. Such considerations as protein A binding or immunoglobulin isotype and other particular format criteria may be addressed. Because much is known about the antigenic, biological, and morphological structure of chlamydiae, monoclonal antibody development can and should be a rapid and efficient endeavor.

IV. STANDARD DIAGNOSTIC METHODS

Clinical laboratory diagnostic methods for *C. trachomatis* have consisted primarily of tissue culture isolation or identification of intracellular inclusions by cytological methods. Unfortunately modern serological tests alone have generally proved to be of low diagnostic value, although in some specific cases they are an important adjunct to diagnosis. Schachter et al. (59) have shown that in cases of infant pneumonia, elevated serum IgM titers are indicative. Furthermore, high titers of IgG are detected in patients presenting with PID, LGV, or perihepatitis (60). The value of the MIF test for seroepidemiological studies cannot be overestimated. These studies have led the way for culture confirmation of the wide spectrum the diseases caused by *C. trachomatis.*

A. Culture

Isolation of *C. trachomatis* is diagnostically definitive. Since the development of tissue culture isolation in 1965 (61), the variables of this method have been thoroughly investigated, and standardized techniques have generally been adopted (15). This method is technically difficult, time-consuming, and expensive; thus, this is not a routine method suitable for implementation in most clinical laboratories. In fact, culture facilities are restricted to relatively few laboratories in large metropolitan areas.

In laboratories with the most experience in isolation techniques, isolation is still believed to be, at best, only 80% sensitive (20). Other factors that may have a crucial impact upon isolation sensitivities are the specimen collection technique and storage and whether or not blind passages are performed (15). Briefly, the culture procedure consists of centrifugation-assisted inoculation of the specimen onto mammalian cell monolayers. Cell monolayers are stained with iodine, Giemsa, or fluorescein-labeled antibody 48-72 hr later. Using Giemsa or iodine staining, usually two-thirds to three-quarters of the positive specimens are identified in the initial assay (15,50), and the remaining positive cultures are identified following a blind passage stained after another 72 hr incubation.

Culture is the "gold standard" for comparison of new diagnostic techniques, thus it is essential to appreciate the effectiveness of a culture system to evaluate comparisons between techniques. For example, inordinately high sensitivities may be deduced for a new assay if the culture procedure uses Giemsa or iodine staining and does not include a blind passage, because comparison is made with culture results that also have low sensitivity. Similarly, a particularly sensitive assay would display unjustifiably low specificity determinations (i.e., "false"-positives) if compared with such a culture evaluation because of many false-negative cultures. A practical approach that may provide internal assessment of the culture system is to employ one or two of the other available culture-independent systems in addition to culture for comparison with a new test format.

B. Inclusion Identification

In the early 1900s Halberstaedter and von Prowazek (62) identified intracellular inclusions in scrapings obtained from the conjunctiva of trachoma patients and infected animals. Although this was the only laboratory diagnostic method available until the introduction of tissue culture, it became clear that the finding of intact inclusions from conjunctival scrapings was only about 20-50% sensitive (63,64). Because this approach requires the visualization of relatively intact inclusions, the low sensitivity may reflect the paucity of such events after the specimen collection procedure. Consequently, inclusion identification from scrapings has generally been superseded by other techniques.

V. IMMUNODIAGNOSTIC MONOCLONAL ANTIBODY PROBES

The exquisite specificities displayed by monoclonal antibodies make them uniquely useful as probes for immunodiagnosis and for immunotyping clinical isolates. The practical use of monoclonal antibody technology for *C. trachomatis* was the initiating example of a successful application of monoclonal antibodies for the immunodiagnosis of a bacterial pathogen. Two assay formats are currently in use. The first evaluation of the use of monoclonal antibodies for improved diagnostic assays was aimed at the cell culture isolation assay. With use of the species-specific monoclonal antibody, the screening of cell cultures for chlamydial inclusions is much more rapid, and cultures can be stained earlier without loss of sensitivity (50). The second format uses monoclonal antibodies to detect organisms directly in clinical specimens (51). This has provided the opportunity for laboratories without tissue culture facilities to engage in diagnostic evaluation for *C. trachomatis* infection.

We also evaluated the use of monoclonal antibodies for the detection of chlamydial antigens using a two-site immunoenzymometric assay (57). This ELISA format antigen-capture method detected 3 ng of protein antigen, which represented approximately 2-4 $\times$ 10^4 chlamydial particles. Assessments of clinical cervical material detected 1 $\times$ 10^2 inclusion-forming units. Although it was clear that substantial improvements would be necessary to develop a useful test, this data supports the conclusion that clinical material contains a large proportion of nonviable chlamydial particles. This provides optimism for the potential development of immunodiagnostic agents with sensitivities greater than culture methods.

A more recent application of monoclonal antibodies for chlamydial infections has been proposed by Thornley et al. (38). They used a genus-specific monoclonal antibody that is specific for chlamydial LPS in a reverse passive hemagglutination assay. Antigens extracted from tissue culture were detected by this assay format, and such an assay may ultimately prove useful for the detection of genus-specific antigen in human and veterinary clinical material. This may be a useful reagent for the detection of *C. psittaci* strains currently under investigation by Grayston and his colleagues as the cause of acute respiratory infections in humans (65). In this section I will provide an overview of the different applications for the detection of *C. trachomatis* by use of monoclonal antibodies.

A. Culture Inclusion Detection

The standard culture methods employ either Giemsa or iodine staining of cell monolayers 72 hr after inoculation. Monolayers are then scanned microscopically for intracellular inclusions. Negative cultures are harvested, passaged, and restained after another 72 hr incubation. By use of a species-specific monoclonal antibody and fluorescent staining, we evaluated the efficacy of this technique compared with Giemsa staining (50). The result showed that immunofluorescent staining, after the second day following inoculation, was superior to Giemsa staining after 3 days following inoculation. The Giemsa technique detected 64% of the positive cultures in the first passage compared with 97% by the immunofluorescent method. The second passage permitted detection of 33% additional Giemsa-positive cultures and one additional culture that was positive by immunofluorescent staining. A third passage was required to detect all the positives with the Giemsa method. Overall, the time required to scan each culture was faster with the monoclonal antibody technique, and in those cultures with fewer inclusions the monoclonal antibody technique averaged three times the inclusion count per specimen. Similar results were obtained with comparisons between the monoclonal antibody technique and the iodine detection system. In this study Stamm et al. (66) also demonstrated higher sensitivity, earlier staining capabilities, and faster evaluations.

These investigations have shown the effective use of monoclonal antibody and fluorescent staining after 48 hr of incubation. Furthermore, if a loss of 2-3% of the positives can be tolerated, this technique represents a single passage test. This offers a savings of 4 days, many fewer cultures, and more rapid microscopic evaluations. These substantial savings in materials and technical time hasten the period required for reporting to 48 hr versus 6 days. For those laboratories equipped for chlamydial culture, monoclonal antibody staining is the method of choice and is gaining wide acceptance (15).

B. Culture-Independent Detection

Before the development of monoclonal antibodies to *C. trachomatis*, it was generally believed that the direct microscopic detection of chlamydial particles in clinical specimens was unlikely. In fact, several studies had employed fluorescein-labeled polyvalent antisera for the detection of chlamydial inclusions in such specimens; however, the nonspecific staining debris precluded the detection of individual organisms. The defined specificity of monoclonal antibodies offers the opportunity to confine staining to a specific antigen, thereby permitting the detection of extracellular organisms in patient specimens. In studies by Tam et al. (51), urethral and cervical smears were stained with a fluorescein-labeled species-specific monoclonal antibody and examined by fluorescent microscopy. We observed extracellular EBs as discrete, evenly staining, smoothed-edged, nearly pinpoint disks approximately 300 nm in diameter. Some RB forms were also seen that were larger and appeared as entire stained organisms. With counterstaining, a contrasting background of mucoid debris and epithelial and inflammatory cells were observed.

In a high-prevalence STD clinic population the direct immunofluorescent test was compared with culture specimens that were stained with either labeled monoclonal antibody or iodine (67). There was 94% agreement among the three evaluations for 926 specimens, and the direct tests was found to be 93% sensitive and 96% specific. In a similar study of a lower-prevalence population, Uyeda et al. (68) reported 6.7% positive results among 401 asymptomatic women attending a family planning clinic in which the direct test resulted in 96% sensitivity and 99% specificity.

The direct test has also been evaluated for specimens from extragenital sites. In a study by Bell et al. (69) of infants presenting with purulent conjunctivities, we found complete correlation of the direct test with monoclonal antibody-stained cultures. Positive nasopharangeal samples were also found. Friis et al. (72) investigated respiratory tract specimens of infants with pneumonia in which 2 of 41 were positive for chlamydia by both the direct test and culture.

Over the last few years there have been numerous studies by diverse groups that have evaluated many different prevalence and geographical populations using the commercially available reagents (Syva Diagnostics, Palo Alto, California). Although there have been a spectrum of sensitivity and specificity determinations, the preponderance of the findings suggests that the direct monoclonal antibody immunofluorescence test can be expected to yield greater than 90% sensitivity and greater than 98% specificity (15). Thus, this test can be utilized for the diagnosis of *C. trachomatis* in genital and in some extragenital infections. The advantages of this test format are (a) ease of specimen collection, preparation and transport; (b) fast processing time per specimen permits rapid diagnosis; (c) evaluation of morphological criteria; and (d) the opportunity to evaluate the specimen adequacy. There are, however, disadvantages to be recognized: (a) an absolute requirement for a high quality fluorescence microscope, (b) an experienced microscopist, and (c) although each specimen takes only a few minutes to evaluate, fluorescent microscopy is a very fatiguing process, thus the attractiveness of this format diminishes with increasing numbers of specimens.

The overall conclusion is the direct test is useful for laboratories without tissue culture facilities that wish to participate in the diagnosis of *C. trachomatis* infections in high-prevalence populations. For laboratories with culture facilities or for examination of low-prevalence groups, the most sensitive diagnostic method remains culture, with inclusion detection using fluorescent monoclonal antibodies. These laboratories may also find that, if the volume is large, it may even be more cost-effective to perform cultures because of the cost of the direct test reagents.

C. Immunotyping Isolates

One of the unique attributes of monoclonal antibodies is their defined and precise specificity. A new application of chlamydial monoclonal antibodies is their use in rapid and direct immunotyping of clinical isolates. Because only the LGV strains demonstrate distinct virulence attributes compared with the trachoma strains, the immunotyping of trachoma strains is, to date, of little clinical importance. Nevertheless, immunotyping is an important epidemiological research tool for the development and evaluation of chlamydial control programs. The standard immunotyping protocol employs the production of antisera to an isolate and antisera to each of the 15 prototype strains. These antisera are evaluated by quantitative MIF assessments of the isolate antisera to each of the prototype antigens and with the prototype antisera to the isolate. This consists of the complete two-way test that is, with considerable technical experience, capable of detecting fine differences between strains. Only a few laboratories type new isolates, and they now generally employ a simplified version of the one-way test that cannot discriminate between closely related serovars (30).

The specificities of type-specific monoclonal antibodies to *C. trachomatis* matched the serotyping data for the MIF test, thus I concluded that monoclonal antibodies might be used for rapid immunotyping without rigorous isolation procedures and immunizations to produce isolate-specific antibody for use in the MIF test (49). This approach was predicated upon the issue of serovar specificity; that is, whether or not a type-specific monoclonal antibody homologous to the prototype strain was a determinant that would be present on all the other strains of the same serovar. Type-specific monoclonal antibodies to serovars F, B, D, and E were used to test over 60 previously typed isolates that had diverse geographical origins. The results demonstrated the requisite breadth of reactivity and specificity of these antibodies for their use in the rapid immunoclassification of chlamydial isolates (49). Wang et al. (58) have recently developed and evaluated a monoclonal antibody approach to immunotyping all serovars. For these evaluations a two-step approach was devised. The first analysis placed isolates into one of three groups, which defined the sets of antibodies used in the second evaluation. This scheme uses six to eight separate monoclonal antibodies, rather than testing 15 different monoclonal antibodies specific for each serovar (Table 1). We evaluated over 300 isolates that had been previously classified by the MIF test. Immunotyping with monoclonal antibodies was rapid and more precise (58). As confidence develops for the usefulness of each serovar-specific monoclonal antibody, antibodies can be pooled to simplify the assay. Combinatorial pooling of appropriate sets of monoclonal antibodies can provide definitive serotyping of chlamydial isolates by the pattern of reactivities observed from as few as four single tests. A schematic example is presented in Table 2.

Once sets of chlamydial serovar-specific monoclonal antibodies have been chosen and made generally available to all investigators, a standardized, reproducible, easy to use, and rapid immunotyping test will provide the opportunity for expanded epidemiological investigations of chlamydial infections. This may provide a more comprehensive evaluation of the distribution and spread of chlamydia that will be valuable for future assessments of basic control programs and, we hope, for vaccine trials.

Table 1 Serological Classification of *C. trachomatis* Isolates with a Two-Step Microimmunofluorescent (MIF) Test with Monoclonal Antibodies

	Serovar or subtype (no. of strains)																	
	C group					Intermediate group					B group							
MA (ascites)	C (15)	J (23)	H (19)	A (12)	I (5)	I′ (17)	G (18)	L_3 (4)	F (29)	K (19)	B (31)	Ba (23)	L_2 (13)	L'_2 (3)	E (55)	L_1 (3)	D (19)	D′ (5)
First step																		
LE-4	+	+	+	+	+	–	–	–	–	–	–	–	–	–	–	–	–	–
G/F/K	–	+	+	–	–	+	+	+	+	+	–	–	–	–	–	–	–	–
BB-11	–	–	–	–	–	–	–	–	–	–	+	+	+	+	+	+	+	+
BB-3	–	–	–	–	–	–	–	–	–	–	+	+	+	+	–	–	–	–
Second step																		
CC-1	+	+	–	–	–													
LA-10	–	–	+	–	–													
AC-11	–	–	–	+	+													
PE-5	–	–	–	–	+													
PE-5						+	–	–	–	–								
GG-11						+	+	+	–	–								
FC-2						–	–	–	+	–								
KK-1						–	–	–	–	+								
DD-1											+	–	–	–				
2F-1											–	–	+	+				
155-35											–	–	+	–				
KB-8															+	+	–	–
JG-9															–	–	+	–

NOTE. + = positive fluorescence; – = negative fluorescence; MA = monoclonal antibodies; G/F/K = pooled GG-11, FC-2, and KK-1 ascites, with equal antibody titer; I′, D′, and L'_2 indicate subtypes of I, D, and L_2, respectively.
Source: Reproduced with permission from Ref. 58.

Table 2 Combinatorial Evaluation of 15 Serovar-Specific Monoclonal Antibodies

Pool	Monoclonal anntibody number														
A	1	2	3	4					9	10				14	15
B		2	3	4	5	6					11		13		15
C			3	4		6	7	8	9		11			14	
D				4		6	7			10		12	13	14	15

Fifteen monoclonal antibodies are distributed into four separate pools (A–D). The reactivity pattern observed, after the testing of an isolate with pools A through D, identifies a reaction by one unique monoclonal antibody. For example, if pools A and B were positive and pools C and D were negative, then only antibody 2 reacted.

VI. CONCLUSIONS

The development of monoclonal antibodies for the detection of *C. trachomatis* has provided an improved culture detection system as well as culture-independent formats. The culture conformation assay using fluorescent monoclonal antibody staining for inclusion detection offers considerable advantages for those laboratories equipped for culture of *C. trachomatis* that wish to increase the sensitivity and speed of culture assessments. The fluorescent monoclonal antibody culture-independent assay provides the ability for those laboratories without culture facilities to participate in chlamydial diagnostic efforts. An ELISA format assay that uses polystyrene beads to nonspecifically capture chlamydial antigens and polyvalent anti-*C. trachomatis* serum for detection has also been developed (70) and has become commercially available (Abbott Diagnostics). Although this test offers objective measurement and is most suitable for testing large numbers of specimens, the sensitivity has varied between 67–90% (15). Nevertheless, this enzyme immunoassay may also be valuable for use with high-prevalence populations.

The first-generation, culture-independent assays have already increased the availability of differential chlamydial diagnostic capabilities, yet the appreciation of the enormous costs and suffering from the damaging sequelae of chronic asymptomatic or latent chlamydial infections heralds the need for more sensitive second-generation assays. Fortunately, there is a substantial market and a growing appreciation of the importance of diagnostic tests that are ideally more sensitive than culture and that are effective for the detection of *C. trachomatis* in low-prevalence and asymptomatic populations (15). Innovative technology should provide assay formats with greater sensitivities than are now available and in formats that meet the demands of speed, cost, and adaptability to evaluations of single tests and large-scale testing.

New horizons for the development of these assays include immunodiagnostic probes that use more-sensitive or amplifying substrates, as well as DNA probe technologies. By using recombinant DNA-cloning techniques, we have expressed portions of the chlamydial MOMP in *E. coli*, and these products are identified by many of the chlamydial monoclonal antibodies (71). Both monoclonal antibodies and antigens of predefined specificity could be used together to develop sensitive homogeneous or competitive assay systems. The challenges of developing laboratory diagnostic tests to this diverse group of organisms that also place stringent demands upon sensitivity as well as having to detect these organisms from complex clinical specimens are clear. Nevertheless, *C. trachomatis* represents an excellent model system for the general development of novel diagnostic systems because of these same attributes, as well as having a clear medical need and a large market for an effective test.

REFERENCES

1. Moulder, J. W., T. P. Hatch, C.-C. Kuo, J. Schachter, and J. Storz, Genus I. *Chlamydia*. In *Bergey's Manual of Systemic Bacteriology*, Vol. 1, N. R. Krieg and Holt, J. G. (eds.). Williams & Williams, Baltimore, pp. 729–739 (1984).
2. World Health Organization Statistics Report, Geneva: WHO *24*:248–329 (1971).
3. Thompson, E. E., and A. E. Washington, Epidemiology of sexually transmitted *Chlamydia trachomatis* infections. *Epidemiol. Rev. 5*:96–123 (1983).
4. Perine, P. L., A. J. Anderson, D. W. Drause, S. Awoke, S.-P. Wang, C.-C. Kuo, and K. K. Holmes, Diagnosis and treatment of lymphogranuloma venereum in Ethiopia. In *Current Chemotherapy and Infectious Disease*, J. D. Nelson and Grassi, C. (eds.). American Society for Microbiology, Washington, pp. 1280–1282 (1980).
5. Berger, R. E., E. R. Alexander, G. O. Monda, J. Ansell, G. McCormick, and K. K. Holmes, *Chlamydia trachomatis* as a cause of acute "idiopathic" epididymitis. *N. Engl. J. Med. 298*:301–303 (1978).
6. Sweet, R. L., J. Schachter, and M. O. Robbie, Failure of β-lactam antibiotics to eradicate *Chlamydia trachomatis* in the endometrium despite apparent clinical cure of acute salpingitis. *J. Am. Med. Assoc. 250*:2641–2645 (1983).
7. Brunham, R. C., J. Paavonen, C. E. Stevens, N. Kiviat, C.-C. Kuo, C. W. Critchlow, and K. K. Holmes, Mucopurulent cervicitis—the ignored counterpart in women of urethritis in men. *N. Engl. J. Med. 311*:1–6 (1984).
8. Quinn, T. C., S. E. Goodell, E. Martichian, M. D. Schuffler, S.-P. Wang, W. E. Stamm, and K. K. Holmes, *Chlamydia trachomatis* proctitis. *N. Engl. Med. 305*:195–200 (1981).
9. Mardh, P. A., B. R. Moller, H. J. Ingerslev, E. Nussler, L. Westrom, and P. Wolner-Hansen, Endometritis caused by *Chlamydia trachomatis*. *Br. J. Vener. Dis. 57*:191–193 (1981).

10. Wolner-Hansen, P., L. Westrom, and P. A. Mardh, Perihepatitis in *Chlamydial salpingitis. Lancet 1*:901–902 (1980).
11. Martin, D. H., S. Pollack, C.-C. Kuo, S.-P. Wang, R. C. Brunham, and K. K. Holmes, Urethral chlamydial infections in men with Reiter's syndrome. In *Chlamydial Infections*, P.-A. Mardh, Holmes, K. K., Oriel, J. D., Piot, P., and Schachter, J. (eds.). Elsevier Biomedical Press, New York, pp. 107–110 (1982).
12. Schachter, J., E. C. Hill, E. B. King, U. R. Coleman, P. Jones, and K. F. Meyer, Chlamydial infection in women with cervical dysplasia. *Am. J. Obstet. Gynecol. 123*:753–755 (1975).
13. Moore, D. E., L. R. Spadoni, H. M. Foy, S.-P. Wang, J. R. Daling, C.-C. Kuo, J. T. Grayston, and D. A. Eschenbach, Increased frequency of serum antibodies to *Chlamydia trachomatis* in infertility due to distal tubal disease. *Lancet 2*:574–577 (1982).
14. Svensson, L., P.-A. Mardh, M. Ahlgren, and F. Nordenskjold, Ectopic pregnancy and antibodies to *Chlamydia trachomatis. Fert. Steril. 44*:313–317 (1985).
15. Centers for Disease Control, *Chlamydia trachomatis* infections. Policy guidelines for prevention and control. *Morbid. Mortal. Week. Rep. Suppl. 34*:53S–74S (1985).
16. Alexander, E. R., and H. R. Harrison, Role of *Chlamydia trachomatis* in perinatal infections. *Rev. Infect. Dis. 5*:713–719 (1983).
17. Beem, M. O., and E. M. Saxon, Respiratory-tract colonization and a distinctive pneumonia syndrome in infants infected with *Chlamydia trachomatis. N. Engl. J. Med. 296*:306–310 (1977).
18. Grayston, J. T., S.-P. Wang, L. J. Yeh, and C.-C. Kuo, Importance of reinfection in the pathogenesis of trachoma. *Rev. Infect. Dis. 7*:717–725 (1985).
19. Stamm, W. E., and K. K. Holmes, *Chlamydia trachomatis* infection of the adult. In *Sexually Transmitted Diseases*, K. K. Holmes, Mardh, P. A., Sparling, P. F., and Wiesner, P. J. (eds.). McGraw-Hill Book Co., New York, pp. 258–270 (1984).
20. Schachter, J., and C. R. Dawson, *Human Chlamydial Infections*, PSG Publishing, Massachusetts (1978).
21. Schachter, J., Biology of *Chlamydia trachomatis.* In *Sexually Transmitted Diseases*, K. K. Holmes, Mardh, P.-A., Sparling, P. F., and Wiesner, P. J. (eds.). McGraw-Hill Book Co., New York, pp. 243–257 (1984).
22. Becker, Y., The chlamydia: Molecular biology of procaryotic obligate parasites of eucaryocytes. *Microbiol. Rev. 42*:274–306 (1978).
23. Hatch, T. P., Utilization of L-cell nucleoside triphosphates by *Chlamydia psittaci* for ribonucleic acid synthesis. *J. Bacteriol. 122*:393 (1975).
24. Kingsbury, D. T., and E. Weiss, Lack of deoxyribonucleic acid homology between species of the genus *Chlamydia. J. Bacteriol. 96*:1421–1423 (1968).
25. Barwell, C. F., Some observations on the antigenic structure of psittacosis and lymphogranuloma venereum viruses. *Br. J. Exp. Pathol. 33*:258–261 (1952).

26. Kuo, C.-C., G. E. Kenny, and S.-P. Wang, Trachoma and psittacosis antigens in agar gel double immunodiffusion. In *Trachoma and Related Disorders Caused by Chlamydial Agents*, R. L. Nichols (ed.). Experpta Medica, Amsterdam, pp. 113–123 (1971).
27. Hatch, T. P., D. W. Vance, and E. Al-Hossainy, Identification of a major envelope protein in *Chlamydia* spp. *J. Bacteriol. 146*:426–429 (1981).
28. Schachter, J., and H. D. Caldwell, Chlamydiae. *Annu. Rev. Microbiol. 34*: 285–309 (1980).
29. Wang, S.-P., and J. T. Grayston, Immunologic relationship between genital TRIC, lymphogranuloma venereum, and related organisms in a new microtiter indirect immunofluorescence test. *Am. J. Ophthalmol. 70*:367–370 (1970).
30. Wang, S.-P., C.-C. Kuo, and J. T. Grayston, A simplified method for immunological typing of trachoma-inclusion conjunctivitis–lymphogranuloma venereum organisms. *Infect. Immun. 7*:356–360 (1973).
31. Wang, S.-P., and J. T. Grayston, Micro immunofluorescence antibody responses in *Chlamydia trachomatis* infection. A review. In *Chlamydial Infections*, P.-A. Mardh, Holmes, K. K., Oriel, J. D., Piot, P., and Schachter, J. (eds.). Elsevier Biomedical Press, New York, pp. 301–316 (1982).
32. Kuo, C.-C., S.-P. Wang, K. K. Holmes, and J. T. Grayston, Immunotypes of *Chlamydia trachomatis* isolates in Seattle, Washington. *Infect. Immun. 41*: 865–868 (1983).
33. Hillman, R. M., and C. Nigg, Studies of lymphogranuloma venereum complement-fixing antigens. *J. Immunol. 59*:349–364 (1946).
34. Dhir, S. P., S. Hakomori, G. E. Kenny, and J. T. Grayston, Immunochemical studies on chlamydial group antigen (presence of a 2-keto-3-deoxycarbohydrate as immunodominant group). *J. Immunol. 109*:116–121 (1972).
35. Stephens, R. S., M. R. Tam, C.-C. Kuo, and R. C. Nowinski, Monoclonal antibodies to *Chlamydia trachomatis*: Antibody specificities and antigen characterization. *J. Immunol. 128*:1083–1089 (1982).
36. Nurminen, M., M. Leinonen, P. Saikku, and P. H. Makela, The genus-specific antigen of *Chlamydia*: Resemblance to the lipopolysaccharide of enteric bacteria. *Science 220*:1279–1281 (1983).
37. Nurminen, M., E. T. Rietschel, and H. Brade, Chemical characterization of *Chlamydia trachomatis* lipopolysaccharide. *Infect. Immun. 48*:573–575 (1984).
38. Thornley, M. J., S. E. Zamze, M. D. Byrne, M. Lusher, and R. T. Evans, Properties of monoclonal antibodies to the genus-specific antigen of *Chlamydia* and their use for antigen detection by reverse passive haemagglutination. *J. Gen. Microbiol. 131*:7–15 (1985).
39. Brade, L., M. Nurminen, P. H. Makela, and H. Brade, Antigenic properties of *Chlamydia trachomatis* lipopolysaccharide. *Infect. Immun. 48*:569–572 (1985).
40. Caldwell, H. D., and P. J. Hitchcock, Monoclonal antibody against a genus-specific antigen of *Chlamydia* species: Location of the epitope on chlamydial lipopolysaccharide. *Infect. Immun. 44*:306–314 (1984).

41. Caldwell, H. D., J. Kromhout, and J. Schachter, Purification and partial characterization of the major outer membrane protein of *Chlamydia trachomatis. Infect. Immun. 31*:1161–1176 (1981).
42. Newhall, W. J., and R. B. Jones, Disulfide-linked oligomers of the major outer membrane protein of chlamydiae. *J. Bacteriol. 154*:998–1001 (1983).
43. Hatch, T. P., M. Miceli, and J. E. Sublett, Synthesis of disulfide-bonded outer membrane proteins during the developmental cycle of *Chlamydia psittaci* and *Chlamydia trachomatis. J. Bacteriol. 165*:379–385 (1986).
44. Hatch, T. P., I. Allan, and J. H. Pearce, Structural and polypeptide differences between envelopes of infective and reproductive life cycle forms of *Chlamydia* spp. *J. Bacteriol. 157*:13–20 (1984).
45. Bavoil, P., A. Ohlin, and J. Schachter, Role of disulfide bonding in outer membrane structure and permeability in *Chlamydia trachomatis. Infect. Immun. 44*:479–485 (1984).
46. Caldwell, H. D., and J. Schachter, Antigenic analysis of the major outer membrane protein of *Chlamydia* spp. *Infect. Immun. 35*:1024–1031 (1982).
47. Byrne, G. I., and J. W. Moulder, Parasite-specified phagocytosis of *Chlamydia psittaci* and *Chlamydia trachomatis* by L and HeLa cells. *Infect. Immun. 19*:598–606 (1978).
48. Eissenberg, L. G., and P. B. Wyrick, Inhibition of phagolysosome fusion is localized to *Chlamydia psittaci*-laden vacuoles. *Infect. Immun. 32*:889–896 (1981).
49. Stephens, R. S., Immunoclassification of *Chlamydia trachomatis* isolates, Doctoral Dissertation, University of Washington, pp. 57–60 (1983).
50. Stephens, R. S., C.-C. Kuo, and M. R. Tam, Sensitivity of immunofluorescence with monoclonal antibodies for detection of *Chlamydia trachomatis* inclusions in cell culture. *J. Clin. Microbiol. 16*:4–7 (1982).
51. Tam, M. R., R. S. Stephens, C.-C. Kuo, K. K. Holmes, W. E. Stamm, and R. C. Nowinski, Use of monoclonal antibodies to *Chlamydia trachomatis* as immunodiagnostic reagents. In *Chlamydial Infections*, P.-A. Mardh, Holmes, K. K., Oriel, J. D., Piot, P., and Schachter, J. (eds.). Elsevier Biomedical Press, New York, pp. 317–320 (1982).
52. Caldwell, H. D., C.-C. Kuo, and G. E. Kenny, Antigenic analysis of chlamydiae by two-dimensional immunoelectrophoresis. I. Antigenic heterogeneity between *C. trachomatis* and *C. psittaci. J. Immunol. 115*:963–975 (1975).
53. Caldwell, H. D., and C.-C. Kuo, Purification of a *Chlamydia trachomatis*-specific antigen by immunoadsorption with monospecific antibody. *J. Immunol. 118*:437–441 (1977).
54. Fuentes, V., J. F. Lefebvre, F. Lema, E. Bissac, and J. Orfila, Establishment of hybrodimas secreting monoclonal antibodies to *Chlamydia psittaci. Immunol. Lett. 10*:325–327 (1985).
55. Stephens, R. S., and C.-C. Kuo, *Chlamydia trachomatis* species-specific epitope detected on mouse biovar outer membrane protein. *Infect. Immun. 45*:790–791 (1984).

56. Terho, P., M. T. Matikainen, P. Arstila, and J. Treharne, In *Chlamydial Infections*, P-A. Mardh, Holmes, K. K., Oriel, J. D., Piot, P., and Schachter, J. (eds.). Elsevier Biomedical Press, New York, pp. 321–324 (1982).
57. Stephens, R. S., C.-C. Kuo, M. R. Tam, and R. C. Nowinski, Monoclonal antibodies to *Chlamydia trachomatis*. In *Chlamydial Infections*, P-A. Mardh, Holmes, K. K., Oriel, J. D., Piot, P., and Schachter, J. (eds.). Elsevier Biomedical Press, New York, pp. 329–332 (1982).
58. Wang, S.-P., C.-C. Kuo, R. C. Barnes, R. S. Stephens, and J. T. Grayston, Immunotyping of *Chlamydia trachomatis* with monoclonal antibodies. *J. Infect. Dis. 152*:791–800 (1985).
59. Schachter, J., M. Grossman, and P. H. Azimi, Serology of *Chlamydia trachomatis* in infants. *J. Infect. Dis. 146*:530–535 (1982).
60. Wang, S.-P., and J. T. Grayston, Micro-immunofluorescence serology of *Chlamydia trachomatis*. In *Medical Virology III*, L. M. de la Mazas and Peterson, E. M. (eds.). Elsevier, New York, pp. 87–118 (1984).
61. Gordon, F. B., and A. L. Quan, Isolation of the trachoma agent in cell culture. *Proc. Soc. Exp. Biol. Med. 118*:354–359 (1965).
62. Halberstadter, L., and S. von Prowazek. Uber Zelleinschlusse parasitarer Natur beim Trachom. *UArb. Kaiserlichen Gesundeitsamte 26*:44–47 (1907).
63. Darougar, S., R. M. Woodland, B. R. Jones, A. Houshmand, and H. A. Farahmandian, Comparative sensitivity of fluorescent antibody staining of conjunctival scrapings and irradiated McCoy cell culture for the diagnosis of hyperendemic trachoma. *Br. J. Ophthalmol. 64*:276–278 (1980).
64. Rowe, D. S., E. Z. Aicardi, C. R. Dawson, and J. Schachter, Purulent ocular discharge in neonates: Significance of *Chlamydia trachomatis*. *Pediatrics 63*:628–632 (1979).
65. Grayston, J. T., C.-C. Kuo, S.-P. Wang, and J. Altman, Isolation of TWAR strains, a human *Chlamydia psittaci*, from acute respiratory disease. *Abst. Intersci. Conf. Antimicrob. Agents Chemother. Am. Soc. Microbiol.* Washington, Abstr. 899 (1985).
66. Stamm, W. E., M. Tam, M. Koester, and L. Cles, Detection of *Chlamydia trachomatis* inclusions in McCoy cell cultures with fluorescein-conjugated monoclonal antibodies. *J. Clin. Microbiol. 17*:666–668 (1983).
67. Tam, M. R., W. E. Stamm, H. H. Hansfield, R. S. Stephens, C.-C. Kuo, K. K. Holmes, K. Ditsenberger, M. Krieger, and R. C. Nowinski, Culture-independent diagnosis of *Chlamydia trachomatis* using monoclonal antibodies. *N. Engl. J. Med. 310*:1146–1150 (1984).
68. Uyeda, C. T., P. Welborn, N. Ellison-Birang, K. Shunk, and B. Tsaouse, Evaluation of the MicroTrak direct·specimen test for identification of *Chlamydia trachomatis* in clinical specimens. *Abstr. Annu. Meet. Am. Soc. Microbiol.* p. 254 (1984).
69. Bell, T. A., C.-C. Kuo, W. E. Stamm, M. R. Tam, R. S. Stephens, K. K. Holmes, and J. T. Grayston, Direct fluorescent monoclonal antibody stain for rapid detection of infant *Chlamydia trachomatis* infections. *Pediatrics 74*:224–228 (1984).

70. Jones, M. F., T. F. Smith, A. J. Houglum, and J. E. Herrmann, Detection of *Chlamydia trachomatis* in genital specimens by the Chlamydiazyme test. *J. Clin. Microbiol. 20*:465–467 (1984).
71. Stephens, R. S., C.-C. Kuo, G. Newport, and N. Agabian, Molecular cloning and expression of *Chlamydia trachomatis* major outer membrane protein antigens in *Escherichia coli. Infect. Immun. 47*:713–718 (1985).
72. Friis, B., C.-C. Kuo, S. P. Wang, C. H. Mordhorst, and J. T. Grayston, Rapid diagnosis of *Chlamydia trachomatis* pneumonia in infants. *Acta Pathol. Microbiol. Immunol. Scand. Sect. B. 92*:139–143 (1984).

RECENT REFERENCE ADDED IN PROOF

Stephens, R. S., E. A. Wagar, and G. K. Schoolnik, High-resolution mapping of serovar-specific and common antigenic determinants of the major outer membrane protein of *Chlamydia trachomatis. J. Exp. Med. 167*:817–831 (1988).

18

Detection and Characterization of *Treponema pallidum* with Monoclonal Antibodies

SHEILA A. LUKEHART
University of Washington School of Medicine, Seattle, Washington

I. BIOLOGY OF *TREPONEMA PALLIDUM*

A. Classification

Four treponemes are recognized as pathogenic for humans: *Treponema pallidum* subsp. *pallidum* (venereal syphilis), *T. pallidum* subsp. *pertenue* (yaws), *T. pallidum* subsp. *endemicum* (endemic syphilis or bejel), and *T. carateum* (pinta). Yaws, bejel, and pinta are nonvenereal treponemal infections found primarily in remote regions of developing countries where the health care systems are inadequate; venereal syphilis, the topic of this chapter, is found throughout the world. A fifth pathogenic treponeme, *T. paraluiscuniculi*, causes venereal spirochetosis in rabbits and is not pathogenic for humans. These five noncultivable organisms are morphologically and, to date, serologically indistinguishable. DNA homology studies have shown virtually 100% homology between *T. pallidum* subsp. *pallidum* and *T. pallidum* subsp. *pertenue* (1). Most investigations of the physiology and antigenic structure of the pathogenic treponemes have been performed with the Nichols strain of *T. pallidum* subsp. *pallidum*, which was isolated from a patient with secondary syphilis in 1912 (2) and has been passaged in rabbits since that time.

Numerous species of commensal treponemes are found as normal flora of the mucosal surfaces of the mouth, genitalia, and gastrointestinal tract; these cultivable anaerobic organisms are antigenically cross-reactive with the pathogenic treponemes, but display no DNA homology with *T. pallidum* subsp. *pallidum* (3).

B. Growth and Morphology

The Nichols strain of *T. pallidum* subsp. *pallidum* (hereafter called *T. pallidum*) is microaerophilic and has been propagated to a limited degree in a tissue culture system (4,5); however, continuous passage has not been achieved. For laboratory examination, *T. pallidum* must be propagated in vivo in rabbit testes and extracted from infected tissue for analysis. The generation time is approximately 33 hr (6).

Treponemes are motile, helically shaped bacteria (Fig. 1), 5-20 μm in length, with multiple endoplasmic flagella (axial filaments) lying within the outer membrane. Beneath the endoflagella are a peptidoglycan layer and a trilaminar cytoplasmic membrane (7). The surface of *T. pallidum* is covered by a layer of loosely associated glycosaminoglycans of unknown origin (8-16), as well as loosely and avidly bound host proteins (17,18).

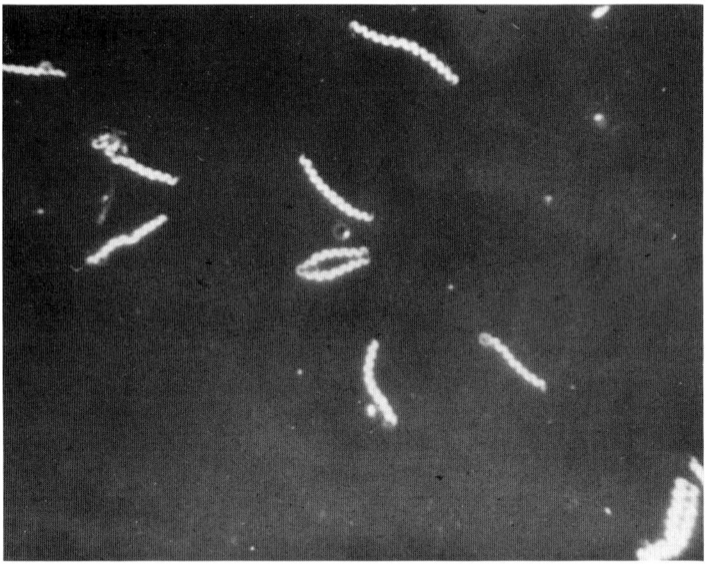

Figure 1 Dark-field microscopic appearance of *T. pallidum* subspecies *pallidum*, Nichols strain [*Source*: Photo courtesy of James N. Miller, Ph.D.; reprinted with permission from Lukehart, S., and S. Baker-Zander, The diagnostic potential of monoclonal antibodies against *Treponema pallidum*. In *Diagnosis of Sexually Transmitted Diseases with Polyclonal and Monoclonal Antibodies*, H. Young and McMillan, A. (eds.). Marcel Dekker, New York (1988)].

II. CLINICAL COURSE AND NATURAL HISTORY OF SYPHILIS

Syphilis is transmitted by direct, usually sexual, contact with the lesion of an infectious person. Because the pathogenic treponemes do not survive for long periods outside of the animal host, fomite transmission is very rare. The organisms penetrate mucosal surfaces or abraded skin and begin to multiply at the site of inoculation. Within a short time, organisms disseminate, through the circulatory and lymphatic systems, to multiple tissue sites. Invasion of the central nervous system, when it occurs, probably takes place in the first days or weeks of infection.

After 10-90 days (average 3-6 weeks), a painless indurated lesion, the chancre, appears at the site of inoculation. Initially a papule, the lesion quickly enlarges and ulcerates. Because syphilis is transmitted sexually, chancres are usually found in the genital, oral, or perianal areas. Histological examination (19) of the primary lesion reveals a marked mononuclear infiltration, with lymphocytes and macrophages predominating. Plasma cells are common, and endothelial proliferation is characteristic. Treponemes are not known to elaborate any toxin or have clear cytopathic activity, and the pathology of treponemal disease is thought to be due to the host's immune response to the organisms. Treponemes, which are abundant in early lesions, are virtually cleared by the immune system, and the lesions resolve spontaneously within 1-2 months.

Following, or concurrent with, the healing of the primary chancre, the disseminated rash of the secondary stage appears as macular, papular, or (occasionally) pustular lesions on the trunk, extremities, scalp, palms, and soles. Other evidence of systemic infection present during the secondary stage includes generalized nontender lymphadenopathy, malaise, fever, and transient liver or kidney dysfunction. Histologically, the lesions resemble the chancre and, like the chancre, the secondary manifestations resolve spontaneously after a period of weeks or months. Recurrent secondary manifestations appear in approximately one-quarter of patients, virtually always within the first year of infection.

After resolution of the secondary manifestations, the patient enters a period of latency, which may last for several years or many decades. Although the patient exhibits no clinical disease during the latent stage, persistent organisms may be identified in various tissues including spleen, lymph nodes, liver, and the central nervous system. Although it is believed that low numbers of organisms occasionally shower from the tissues into the circulatory system during latency, the metabolic status of *T. pallidum* during latency is unknown.

Approximately 30% of untreated patients with syphilis will develop one or more tertiary manifestations which may include gummatous destruction of skin, bone, or internal organs; cardiovascular involvement; or symptomatic or asymptomatic neurosyphilis. The remaining 70% of patients will remain in the latent stage for the remainder of their lives, with no further clinical evidence of syphilis. Although the older literature refers to spontaneous cure in certain individuals, no firm evidence for such exists.

Treponema pallidum subsp. *pallidum* can cross the placenta during any stage of infection in the mother, and fetal infection may result in stillbirth, spontaneous abortion, perinatal death, prematurity, or birth of a live infant with active or latent syphilis.

III. PATHOGENESIS AND THE IMMUNE RESPONSE

The mechanisms of treponemal pathogenesis have not been well defined. One of the early steps in infection is hypothesized to be the attachment of the treponeme to the mammalian cell by specific receptor molecules located on the tip of the bacterium (20-23). Three treponemal molecules with molecular masses of 89,500, 37,000, and 32,000 are thought to mediate cell attachment (24-27). Recent evidence indicates that fibronectin (25-30) may be an important host component in cell-bacteria interaction; laminin and collagen have also been implicated as mediators of attachment (28).

Dissemination of the organisms from the site of inoculation occurs very early through the walls of capillaries and lymphatic vessels, and across the blood-brain and placental barriers. Fitzgerald et al. (23) have suggested that dissemination may be mediated by hyaluronidase (or other mucopolysaccharidase) that destroys the mucopolysaccharide ground substance of vessels walls. This hypothesis is speculative, and there is no firm evidence that *T. pallidum* produces a mucopolysaccharidase or that the enzyme is required for bacterial dissemination.

During active infection, most treponemes are located in extracellular spaces; however, they have been observed by electron microscopy within phagocytic and nonphagocytic cells (31-34). The intracellular location of *T. pallidum* has been hypothesized to contribute to its ability to evade the host's immune response and persist for many years. The passive or active acquisition of host-derived proteins (17,18) or mucopolysaccharides (8-16) may also protect the organisms from antibody, complement, or phagocytic cells, thus contributing to persistence.

The host responds promptly to treponemal infection by producing antibodies, sensitizing T lymphocytes, and activating macrophages. In the experimental rabbit model of syphilis, antitreponemal antibodies can be detected within 1 week of infection (35-38), and neutralizing or immobilizing antibodies are demonstrable within 1 month (39). The contribution of antibody to resistance against treponemal infection has been studied extensively. Antibody, in the presence of complement, can immobilize *T. pallidum* in vitro (40) and render the organism noninfectious (39,41). The passive administration of large amounts of immune serum to naive recipients delays and alters lesion development in challenged rabbits, but does not eliminate the infection (41-47). Recent passive protection studies with *T. pallidum* in the guinea pig model confirm these findings (48). Passively administered immune serum completely protects hamsters from endemic syphilis infection (49), but it does not confer complete protection against yaws (50).

The mechanisms by which immune serum may contribute to protection against treponemal infection are numerous. In addition to complement-dependent immobilization or neutralization, antibody has been shown to significantly enhance the phagocytosis of *T. pallidum* by macrophages (opsonization) in vitro (51). In addition, antibody also has been demonstrated to block the attachment of *T. pallidum* to eukaryotic cells (21,22,52) and extracellular matrix (25). Despite the numerous antitreponemal activities of antibody, the observation that humans develop the rash of secondary syphilis, with a high bacterial load, in the presence of high-titered antitreponemal antibody indicates that antibody is not completely protective.

In the experimental rabbit model of syphilis, specifically sensitized T lymphocytes are demonstrable as early as day 6 after infection (53) and histological studies demonstrate that T lymphocytes are the major infiltrating cell in healing primary lesions (54,55). The sensitization of T lymphocytes results in the production of macrophage-activating factors (56), and the active phagocytosis of *T. pallidum* in healing lesions by infiltrating macrophages is suggested by various histological (55,57) and electron microscopic (58) studies. The phagocytosis of *T. pallidum* by macrophages in vitro has been demonstrated (51).

IV. ANTIGENIC STRUCTURE OF *TREPONEMA PALLIDUM*

Because of the inability to cultivate *T. pallidum* in vitro, the antigenic analysis of the organism has been limited until recent years. Many of the early investigations of treponemal antigens were performed on the cultivable nonpathogenic species such as *T. phagedenis*, biotype Reiter. Antigenic cross-reactivity between the nonpathogens and *T. pallidum* has been demonstrated serologically (59-62), and an extract of the Reiter treponeme has been used for serodiagnosis of syphilis (60-61). Immunization of animals with the nonpathogenic treponemes, however, fails to confer protection against challenge with virulent species, suggesting important antigenic differences between the pathogens and nonpathogens (63-67). Cross-protection studies between the pathogenic treponemes have also revealed antigenic heterogeneity (8,68). Rabbits infected with *T. pallidum* subsp. *pallidum* or immunized with gamma-irradiated *T. pallidum* subsp. *pallidum* are resistant to challenge with the homologous organism but are not protected against infection with *T. pallidum* subsp. *pertenue* (69).

A. Molecular Identification of Treponemal Antigens

Technological developments in the past 6 years have permitted a fairly detailed analysis of the major antigens of *T. pallidum* and the humoral immune response to them. With use of sodium dodecyl sulfate-polyacrylamide gel electrophoresis (SDS-PAGE) and immunoblotting or radioimmunoprecipitation (RIP) techniques with polyvalent antisera, a profile of the major antigens of *T. pallidum* has been defined (24,35-37,70-84), with approximately 22 molecules ranging in molecular mass from 115,000 to 12,000 (Fig. 2). Major bands are seen at approximately 60, 47-48, 37, 35, 33, 30, 14, and 12 kilodaltons (kd). The reported molecular masses of major antigens vary by several thousand in reports from different laboratories, and no standard nomenclature has yet been accepted. This fact makes it difficult to compare the results from different laboratories; however, the use of monoclonal antibodies to define particular molecules may contribute to the emergence of an accepted nomenclature system. Two-dimensional electrophoresis studies have further resolved and characterized many of these antigens (84,86).

Because of the electrophoretic mobility of the identified molecules and their ability to be stained with Coomassie Blue, the antigenic molecules are presumed to be polypeptides; however, the precise chemical composition of the antigenic determinants remains undefined. Using [^{14}C] glucosamine incorporation in vitro, Moskophidis and Muller (82) identified four putative surface glycoproteins, with molecular masses of 59,000, 35,000, 33,000, and 30,500. The molecular masses of the latter three molecules correspond to those hypothesized to comprise the endoflagella.

Labeling of intact *T. pallidum* with ^{125}I has been used to examine surface location of individual molecules (24,37,82-85,87,88). The number of putative surface, or outer membrane, proteins varies from 9 to 20, depending upon the report, but there appears to be consensus that the 47-48-kd molecule and a 37-40-kd molecule are surface-exposed.

As mentioned earlier, Baseman and coworkers have identified three outer membrane molecules that serve as ligands for the attachment of *T. pallidum* to fibronectin and eucaryotic cell surfaces (24-27). Although the reported molecular masses of these molecules has varied in different publications, Thomas et al. (27) recently reported corrected values of 85,500, 37,000, and 32,000. Proteolytic digestion and peptide mapping of these molecules revealed a common 12,000 Da functional domain in these cytadhesins (27).

B. Common and Pathogen-Specific Antigenic Determinants

Although antigenic cross-reactivity between *T. pallidum* and the nonpathogenic treponemes has been recognized for many years, the molecular basis for this cross-reactivity has only recently been examined. Five cross-reactive molecules

Figure 2 Antigenic profile of *Treponema pallidum*, Nichols strain, demonstrated by SDS–PAGE and immunoblotting. Individual molecules were revealed by autoradiography using pooled syphilitic rabbit sera and ^{125}I-labeled protein A. Positions of molecular mass standards are shown. [*Source*: Reprinted with permission from Lukehart, S., and S. Baker-Zander, The diagnostic potential of monoclonal antibodies against *Treponema pallidum*. In *Diagnosis of Sexually Transmitted Diseases with Polyclonal and Monoclonal Antibodies*, H. Young and McMillan, A. (eds.). Marcel Dekker, New York (1988)].

of the Reiter treponeme were identified by Strandberg-Pederson et al. (89,90) by reaction with sera from persons with secondary syphilis; one of these antigens was thought to represent the endoflagella (axial filaments). A number of laboratories have examined *T. pallidum* molecules containing antigens common to the Reiter treponeme: 4 to 15 molecules have been identified using immunoblotting or radioimmunoprecipitation. Molecules of 80, 69, 47-48, 37, 35, 33, and 30 kd were identified by at least three of the laboratories.

Hanff et al. (73) and Lukehart et al. (70) speculated that the 30- and 33-kd or 35- and 33-kd molecules, respectively, represent endoflagellar components. Bharier and Allis (91) reported earlier that the axial filaments of the Reiter treponeme, which contain common antigens, are composed of two to three molecules between 33,000 and 36,500 Da. Limberger and Charon (92) identified two molecules from purified *T. phagedenis* endoflagella with molecular masses of 39,800 and 33,000. Antiserum raised against the 33,000-Da molecule cross-reacted with *T. pallidum*. Recently, Blanco et al. (93) identified three *T. pallidum* endoflagellar molecules (35,000, 33,000, and 30,000) as containing antigenic determinants cross-reactive with Reiter endoflagella. Three molecules with reported molecular masses ranging from 33,500 to 37,000 were also recently identified in purified *T. pallidum* endoflagella by Penn et al. (74).

On the basis of immunoblotting studies using anti-*T. pallidum* antiserum, which had been absorbed extensively with the Reiter treponeme, pathogen-specific determinants were hypothesized by Lukehart et al. (70) to be located on molecules of 14 and 12 kd. Although reactivity of syphilitic rabbit serum to the 47-48- and 37-kd molecules was diminished by absorption with the Reiter treponeme, it was not removed, suggesting that these two molecules contain both common and pathogen-specific determinants. Hanff et al. (75) reported the identification of 14 molecules that contained pathogen-specific components; however, molecular masses were not given. Eight polypeptides containing common antigens were described, including 47-, 45-, 40-, and 35.5-kd molecules. Baughn et al. (78) also showed that cross-reactivity between *T. pallidum* and the Reiter treponeme included a major antigen of approximately 45 kd. Several laboratories (71,73,94) have observed that nonsyphilitic human serum (containing antibodies presumably directed against commensal treponemes) reacts with the major 45-48-kd antigen, supporting the hypothesis of common determinants on that molecule. In contrast, Jones et al. (87) detected only pathogen-specific determinants on the 47-48-kd molecule and refer to that molecule as pathogen-specific. The cumulative evidence, however, suggests that the 47-48-kd antigen probably contains both common and pathogen-specific epitopes. Monoclonal antibody reactivity (discussed in Sect. V) supports this hypothesis.

Cross-reactivity of *T. pallidum* with other related spirochetes, including the rabbit pathogen *T. paraluiscuniculi*, the swine pathogen *T. hyodysenteriae*,

Borrelia hermsii, and *Leptospira interrogans* was examined by immunoblotting *T. pallidum* antigens with antisera raised against the other organisms (79). Although the *T. pallidum* antigens recognized by anti-*T. paraluiscuniculi* were virtually identical with those recognized by human or rabbit syphilitic sera (indicating antigenic near identity), antisera raised against the more distantly related species reacted with fewer *T. pallidum* antigens. The existence of one or more common spirochetal antigens was suggested by this study; the molecules most frequently recognized by heterologous antisera had molecular masses of 80,000, 67,000, 47,000-48,000, 37,000, 35,000, 33,000, and 30,000. These are the same molecules identified as containing antigens in common with the Reiter treponeme.

C. Antigenic Comparisons of Strains and Subspecies of *T. pallidum*

The antigens of *T. pallidum* subsp. *pallidum* and *T. pallidum* subsp. *pertenue* were compared by using one- or two-dimensional gel electrophoresis and immunoblotting or radioimmunoprecipitation (24,37,76,88). Minor molecular mass differences were noted by each laboratory, but consensus was not reached on the identification of individual differences. Stamm and Bassford (37) compared the antigenic profile of *T. pallidum*, Nichols strain with a more recent isolate, Street 14, and again found only minor differences. To date, no molecular or immunological method for distinguishing *T. pallidum* strains or subspecies has been developed.

V. PRODUCTION AND CHARACTERIZATION OF MONOCLONAL ANTIBODIES TO *TREPONEMA PALLIDUM*

A number of laboratories have produced monoclonal antibodies to *T. pallidum*; however, only six have published their results (95-100). This section will include a discussion of the immunization and screening methods for the production of *T. pallidum* monoclonal antibodies, the characterization of those antibodies, and their application to the characterization of *T. pallidum* antigens.

A. Immunization

The optimal method for immunization of mice before hybridoma formation may be suggested by the intended use of the antibodies. If the antibodies will be used to define or characterize treponemal antigens expressed during natural infection, the optimal immunization method may be active infection of the mice. If, on the other hand, the antibodies will be used to detect *T. pallidum* in syphilitic lesions or tissues, any immunization method that presents intact surface antigens to the mouse's immune system would be appropriate. Unfortunately, infection of mice with *T. pallidum* results in only self-limited asymptomatic disease, with low levels of antibody production (102). Hence, most laboratories used a fairly standard immunization protocol, with viable or nonviable organisms emulsified in complete Freund's adjuvant.

The route and duration of immunization, and the total dose of *T. pallidum* varied among the six laboratories (Table 1). Although each laboratory included at least one intraperitoneal injection, no obvious combination of routes appeared to be more successful than others. The duration of immunization averaged 6–8 weeks (range 24 days to 5 months), with no apparent advantage, in terms of numbers or specificities of clones or subclass of immunoglobulins produced, of longer immunization schedules. The total dose of *T. pallidum* used for immunization ranged from 7×10^7 to 1.2×10^9, with no obvious advantage to very large inocula. With the exception of Van Embden et al. (100), numerous clones were produced in the reported fusions. Van Embden et al. (100), who reported a total of only three clones from three fusions, immunized with the lowest number of organisms and did not use complete Freund's adjuvant (CFA). Lukehart et al. (95) immunized with a similar dose of *T. pallidum*, with CFA, and reported a high yield of *T. pallidum* monoclonal antibodies. In another publication (101), Van Embden's laboratory reported the production of more antibodies by using a more intensive immunization scheme, including Freund's adjuvant.

Because of the contamination of *T. pallidum* suspensions with rabbit tissue and the risk of producing antirabbit monoclonal antibodies, the purity of the immunizing preparation must be considered. Immunization has been accomplished using unpurified, freshly extracted *T. pallidum*, as well as density gradient-purified and continuous particle electrophoresis-purified organisms. The degree of purity obtained by different methods and the benefits of higher purity must be weighed against reduced yield and potential loss of labile or loosely associated treponemal surface antigens. In considering the published reports of *T. pallidum* monoclonal antibody production, there appears to be little advantage of one purification method over another. Laboratories using less purified preparations reported a slightly higher number of rabbit-reactive clones, although this did not appear to affect the production of anti-*T. pallidum* clones. Regardless of the purity of the immunizing preparation, all hybridomas must ultimately be tested for reactivity against rabbit tissue, and rabbit reactive clones can be eliminated at that stage.

B. Screening Methods

Most laboratories use an enzyme-linked immunoassay (ELISA) or solid-phase radioimmunoassay (RIA), both with a 96-well configuration, for screening fusion wells and individual clones. Sonicated *T. pallidum* preparations were used by nearly all laboratories in a single-screening assay. The exception was Lukehart

Table 1 Production of Monoclonal Antibodies to *T. pallidum*[a]

Ref.	Preparation	Duration of immunization	Route	Total dose	Adju-vant	Myeloma cell line
97	Freshly isolated *T. pallidum*	87 days	ip, iv	2.7×10^8	CFA	SP2/0
98	Virulent *T. pallidum*	6 weeks	id, ip	Not specified	CFA	P3X63–Ag8.653 or SP2/0
100	Viable, Urografin-purified *T. pallidum*	1, 3, 5 mo.	ip, iv	7×10^7	None	Not specified
88, 99	Freshly extracted, Methocel-Hypaque-purified *T. pallidum*	24 days	im, sc, ip	1.2×10^9	CFA	SP2/0
95	CPE-purified *T. pallidum*	37 days	sc, ip, iv	9×10^7	CFA	NS1/1
96	Freshly extracted, Urografin-purified *T. pallidum*	63 days	ip	2.8×10^8	CFA	SP2/0

[a]CPE, continuous particle electrophoresis; CFA, complete Freund's adjuvant; ip, intraperitoneal; iv, intravenous; id, intradermal; sc, subcutaneous

Source: Reprinted with permission from Lukehart, S., Identification and characterization of *Treponema pallidum* antigens by monoclonal antibodies. In *Monoclonal Antibodies Against Bacteria*, Vol. III, A. J. L. Macario and de Macario, E. C. (eds.). Academic Press, New York (1986).

et al. (95) who used a double-screening method, consisting of initial ELISA testing and confirmatory indirect immunofluorescence (IF) examination, at each screening step. Depending upon the intended use of the antibodies, screening to eliminate antibodies that cross-react with the nonpathogenic treponemes can be performed either as an early screening selection step or as part of the characterization of cloned and established hybridomas.

C. Characterization of *Treponema pallidum* Antigens by Use of Monoclonal Antibodies

Although significant information concerning the antigenic structure of *T. pallidum*, including identification and characterization of cross-reactive and pathogen-specific antigens, has been obtained using polyvalent antisera, the inherent monospecificity of monoclonal antibodies permits the unequivocal definition of particular antigens (Table 2) and the investigation of antigenic cross-reactivity

Table 2 Use of Monoclonal Antibodies for Characterization of *T. pallidum* Antigens

Molecular mass (kd)	Common determinant	Pathogen-specific determinant	Surface location	Immobilization
102	?[a]	+[a]	+[a]	+[a]
84		?[a]	+[a]	
45–48	+[b]	+[b–g]	+[d–g]	+[e,d]
44		+[c,d]	+[d]	+[d]
37		+[b]		
33		+[d]		+[d]
24		+[a]	+[a]	+[a]
15		+[d]		+[d]
12		+[b]		

Sources:
[a]Ref. 104.
[b]Ref. 95.
[c]Ref. 100.
[d]Ref. 96.
[e]Ref. 87.
[f]Ref. 99.
[g]Ref. 103.
Reprinted with permission from Lukehart, S., and S. Baker-Zander, The diagnostic potential of monoclonal antibodies against *Treponema pallidum*. In *Diagnosis of Sexually Transmitted Diseases with Polyclonal and Monoclonal Antibodies*, H. Young and McMillan, A. (eds.). Marcel Dekker, New York, pp. 213–247 (1988).

without the cumbersome, and potentially misleading, absorption steps required for polyvalent antisera.

1. Common Treponemal Antigens

Four laboratories have reported production of monoclonal antibodies that cross-react with the nonpathogenic Reiter treponeme in ELISA, IF, or RIA tests (95–98). Lukehart et al. (95) showed that antibody C2-1, which reacts by immunofluorescence with *T. pallidum*, four nonpathogenic treponemes, as well as species of *Borrelia* and *Leptospira*, binds to the major 47–48-kd molecule of *T. pallidum* in immunoblots. Antibody G2-1, which also has the same broad reactivity pattern as C2-1, does not react in immunoblotting assays, and its molecular specificity is not yet defined. Both of these antibodies, however, fail to react with *Chlamydia trachomatis*-infected HeLa cells and, therefore, appear to recognize a spirochete group determinant, rather than a common bacterial or mammalian cell antigen. Thornburg et al. (99) examined the reactivity of antibody $13F_3$, which identifies an epitope on a 45-kd molecule, against *T. hyodysenteriae* and found no binding.

2. Pathogen-Specific Antigens

Certain pathogen-specific monoclonal antibodies also recognize the 47–48-kd molecule, confirming the presence of both common and pathogen-specific determinants on this molecule. Two-dimensional electrophoresis studies (86) indicate that the 47–48-kd "molecule" can be resolved into two or three discrete molecules with apparent isoelectric points of 6.1, each of which reacts with both common and pathogen-specific monoclonal antibodies. These studies indicate that the broad 47–48-kd band seen by one-dimensional electrophoresis and immunoblotting is actually comprised of several very closely related (precursors? modified?) polypeptides that share significant antigenic identity.

Monoclonal antibodies directed against the 47–48-kd molecule have been reported from every laboratory that determined molecular specificities. With the use of monoclonal antibodies, this molecule has been demonstrated to be located on the surface of the bacterium by immunoelectron microscopy and surface-binding assays (103) and by RIP after surface iodination (87,96). Monoclonal antibodies directed against pathogen-specific determinants of the 47–48-kd antigen can immobilize or neutralize *T. pallidum* in the presence of complement (87,96). There is also preliminary evidence that such monoclonal antibodies can block the attachment of viable *T. pallidum* to eukaryotic cells in tissue culture (reported in Ref. 87) despite the fact that this molecule has not been implicated by others (24–27) in attachment.

Lukehart and co-workers (95) also described monoclonal antibodies with specificities for molecules of 37 and 12 kd. Moskophidis and Muller (96) produced antibodies directed against the 15.5-, 33-, and 44-kd molecules; monoclonal antibodies directed against each of these molecules were reactive in the *T.*

pallidum immobilization (TPI) test. Marchitto et al. (104) recently reported antibodies directed against 84-, 102-, and 24-kd molecules; the latter two are reactive in the TPI test. Although antibodies with specificities for 29-, 32-, and 52–54 kd antigens were also reported, the data were difficult to interpret.

3. *Subspecies Differentiation*

The differentiation of *T. pallidum* subspecies has not yet been accomplished with either polyvalent antisera or monoclonal antibodies. Antigenic differences between subspecies have been demonstrated by cross-protection and cross-immobilization experiments. Although these differences appear to be very important in the production of protective immunity, they are thought to be minor in terms of the total antigenic profile of *T. pallidum*. Monoclonal antibodies, with their exquisite specificity, have the theoretical potential for serological classification of these organisms. To date, three laboratories have examined the reactivity of their monoclonal antibodies with other *T. pallidum* subspecies. Thornburg et al. (88,99) demonstrated that, although *T. pallidum* subsp. *pallidum* and *T. pallidum* subsp. *pertenue* had different surface iodination properties, monoclonal antibody $13F_3$ reacted with both subspecies in immunoblotting. Marchitto et al. (103) compared two anti-47-kd antibodies for reactivity with *T. pallidum* subsp. *pallidum, T. pallidum* subsp. *pertenue*, and *T. pallidum* subsp. *endemicum* by RIA, immunoblotting, and surface-binding assays. Although both antibodies reacted with all subspecies, the degree of binding varied, and the authors concluded that the epitopes were present on all subspecies, but altered in physical orientation or expression, making it more, or less, available for antibody binding.

Preliminary studies by Lukehart and Baker-Zander (unpublished data) also indicate differential reactivity of at least five monoclonal antibodies with *T. pallidum* subsp. *pallidum* and *pertenue* by IF assays. The degree of fluorescence of the antibodies with *T. pallidum* subsp. *pertenue* was significantly lower than with *T. pallidum* subsp. *pallidum*, and titration of the antibodies resulted in preferential loss of reactivity to *T. pallidum* subsp. *pertenue*. These same monoclonal antibodies reacted equally with sonicated preparations of the two subspecies in an ELISA assay. These observations support the hypothesis that certain determinants, although present in both organisms, may be oriented or exposed differently in the two subspecies.

4. *Reactivity with Other Strains of Treponema pallidum subsp. pallidum*

All of the *T. pallidum* monoclonal antibodies reported to date have been produced against the Nichols strain of *T. pallidum*, which has been maintained by rabbit passage since its isolation in 1912. Although the Nichols strain remains virulent for humans, there is the possibility of antigenic shift in the organism during 70 years of animal passage. The practical utility of monoclonal antibodies for diagnostic purposes or identification of protective antigens is dependent upon the relevance of those antibodies to antigenic determinants expressed on modern strains of *T. pallidum*. Lukehart et al. (95) used IF to measure the reactivity of 13 antibodies against four recently isolated (fewer than five passages in rabbits) *T. pallidum*; the degrees of reactivity were comparable with that seen with the Nichols stain. Eight antibodies described by Saunders and Folds (98) reacted by ELISA and RIA equally well with the Nichols and Street 14 strains.

Antibody H9-1, reported by Hook et al. (105), has also been examined for reactivity with *T. pallidum* that have not been passaged in rabbits. The high sensitivity and specificity of this antibody for detection of *T. pallidum* in lesion exudates of syphilis patients from disparate geographical regions (105,106; Sect. VII.A) indicates the conservation of the H9-1-defined antigen to the present. Marchitto et al. (104), with use of monoclonal antibodies directed against the 47- and 84-kd antigens, also recently described the identification of *T. pallidum* in lesion exudate smears from patients with genital ulcers.

VI. LIMITATIONS IN EXISTING METHODS FOR DIAGNOSIS OF SYPHILIS

Syphilis has been called "the great imitator" because of the multiple and varied clinical manifestations that may appear during the various stages of disease. Syphilis is responsible for approximately 20% of genital ulcer disease in the United States and Europe (107), and it must be considered in the differential diagnosis of all genital ulcerations. In addition, syphilis must be considered in the evaluation of many other seemingly disparate clinical presentations including skin rashes, recurrent or persistent headaches, visual or hearing disturbances, neurological abnormalities, aortitis, and liver or kidney dysfunction.

The diagnosis of syphilis is based upon clinical, epidemiological, and laboratory criteria, including history, clinical presentation, serological testing, and microscopic identification of *T. pallidum* in lesion exudate or biopsy tissue. None of these methods alone is sensitive and specific in all stages of syphilis and, frequently, clinicians must synthesize the data from each of these categories into a "best-fit" diagnosis. Culture of the causative agent, which is possible for herpes virus and *Haemophilus ducreyi* (two other common causes of genital ulcer disease), is not available for *T. pallidum*.

A. History

A positive history of exposure to infectious syphilis, or a history of unexplained ulceration or rash, is often helpful in directing diagnostic efforts; however, historical information is frequently incomplete or unreliable. Hence, the diagnosis of syphilis should never be excluded in any patient because of lack of corroborating history.

B. Clinical Presentation

As described previously, the clinical manifestations of syphilis are multiple and varied; virtually any organ system may be involved. Rather than establishing the diagnosis of syphilis, clinical presentations serve largely to heighten the clinician's index of suspicion so that appropriate laboratory tests can be performed.

The most common manifestation of primary syphilis is the chancre, a painless ulcerative lesion, usually in the genital, anal, or oral areas. Although these lesions are characteristically single, well-circumscribed, and indurated, atypical presentations are common; clinical impression may be misleading.

In two evaluations of the reliability of diagnosis of genital ulceration based upon morphological characteristics alone (108,109), the diagnostic accuracy for syphilitic lesions in sexually transmitted diseases clinics was 42-78%; for herpes lesions, 63-75%; and for chancroid lesions, 33-75%. In other settings in which genital ulcer disease is less frequently encountered, the accuracy of morphological diagnosis would be even lower. Rectal chancres may be uncharacteristically painful and superinfected and may cause symptoms of proctitis. Vaginal or cervical chancres frequently go unrecognized by the patient.

The rash of secondary syphilis may be confused with other dermatological conditions including eczema, pityriasis rosea, tinea versicolor, and psoriasis. The rash may also be so subtle as to be unrecognized by the patient or clinician. Condylomata lata may be easily confused with genital warts (condylomata acuminata). The latent stage of syphilis, which may persist for decades, has no clinical manifestations to aid in diagnosis, whereas the tertiary stage may involve one or more organ systems. There are few, if any, tertiary manifestations for which syphilis is the sole diagnostic consideration.

C. Serological Testing

Serological testing for syphilis is based upon a two-test system that includes a nontreponemal screening test [such as the Venereal Disease Research Laboratory (VDRL) or Rapid Plasma Reagin (RPR) test], with confirmation of positive results by a treponemal test [such as the fluorescent treponemal antibody-absorption (FTA-ABS) or a hemagglutination (TPHA, HATTS, MHA-TP) test]. The nontreponemal tests measure antibody directed against a cardiolipin-lecithin-cholesterol antigen and are not specific for treponemal infection; biological false-positive reactions are especially common in persons with autoimmune conditions. These tests are also fairly insensitive; except for the secondary stage, they may be nonreactive in as many as 25% of syphilis patients (Table 3).

The treponemal tests are more sensitive than the nontreponemal tests in latent and tertiary disease, but have variable sensitivities in primary syphilis. Furthermore, the specificities of these tests decrease when they are used for

Table 3 Reactivity of Serological Tests in Untreated Syphilis

	Stage of disease, % positive			
	Primary	Secondary	Latent	Late
VDRL, RPR	59–87	100	73–91	37–94
FTA-ABS	86–100	99–100	96–99	96–100
MHA-TP	64–87	96–100	96–100	94–100

Source: Modified (with permission) from H. Jaffe, Management of the reactive serology. In *Sexually Transmitted Diseases*, K. K. Holmes, Mardh, P. A., Sparling, P. F., and Wiesner, P. J. (eds.). McGraw-Hill Book Co., New York (1984).

routine screening and, therefore, their use is recommended only to confirm a reactive VDRL or RPR result. Another difficulty with the treponemal tests is that they frequently remain reactive after adequate therapy and, therefore, are useless in the evaluation of patients with past syphilis.

D. Microscopic Identification of *Treponema pallidum*

Even when serological testing is useful in establishing the diagnosis of syphilis, the delay in obtaining the result from the laboratory may range from several days to a week; this delay poses a particular problem in patients with potentially infectious primary and secondary lesions. The rapid identification of *T. pallidum* in lesion material is, therefore, useful in establishing the diagnosis early so that appropriate therapy can be initiated. The standard method for detection of *T. pallidum* in primary or secondary lesions is dark-field microscopic examination of lesion exudate. This technique involves the identification of a motile spiral-shaped organism with motility and morphology characteristic of *T. pallidum*. The specificity of the test is dependent upon the skill of the microscopist in differentiating *T. pallidum* from the numerous commensal spirochetes that are found on the genital and rectal mucosal surfaces. Even a skilled microscopist cannot reliably distinguish *T. pallidum* from the very similar-appearing non-pathogenic treponemes found in the mouth.

Although a positive dark-field examination is considered to be sufficient for diagnosis of early syphilis, dark-field-negative lesions may still be treponemal. The sensitivity of the dark-field examination has been shown to be 74% for primary syphilis (110) and can be affected by the common self-administration of topical or systemic antibacterial substances, by natural resolution of the lesion, or by interference with microscopy by refractile tissue debris or erythrocytes. Dark-field microscopes and trained microscopists are rarely found outside of specialty clinics and, for that reason, the practical application of this fast and inexpensive test is limited.

A laboratory alternative to dark-field microscopy involves the identification of *T. pallidum* in lesion exudate by direct immunofluorescent staining (DFA-TP test) using polyvalent anti-*T. pallidum* antiserum (111). In an effort to obtain an antiserum that is specific for the pathogenic treponemes, the antibodies that cross-react with the nonpathogenic commensal treponemes are removed by extensive absorption (112-114). Although these reagents are commercially available and the test has been shown to be very sensitive and specific, compared with dark-field microscopy (110), the DFA-TP is not widely used in the United States.

Both dark-field microscopy and the DFA-TP test can be used for evaluation of primary and secondary skin lesions or for examination of nasal mucus (syphilitic rhinitis) and skin lesions in congenital syphilis. Cerebrospinal fluid, aqueous humor, and endolymph have also been examined for *T. pallidum* by dark-field and immunofluorescence microscopy (115,116), although the sensitivity and specificity of these tests in body fluids have been questioned (117,118).

Occasionally, the diagnosis of syphilis is made by identification of *T. pallidum* in placental, autopsy, or biopsy tissue specimens. Silver staining, the classic method for identification of spirochetes in tissues, is very useful when syphilis is highly suspected, but may be misleading when only few or atypical spirochetes are observed. Normal tissue components, such as collagen, are stained by this procedure and may be confused as spirochetes. Immunohistochemical methods, including immunofluorescence (119,120) and immunoperoxidase (121) staining using polyvalent rabbit anti-*T. pallidum* antiserum have been used for identification of *T. pallidum* in tissue specimens; these methods have the advantage of immunological specificity, but they have been confined primarily to research laboratories (122-124).

VII. USE OF MONOCLONAL ANTIBODIES FOR IDENTIFICATION AND DETECTION OF *TREPONEMA PALLIDUM*

Methods for detection and identification of *T. pallidum* are most applicable in the evaluation of suspicious genital ulcerations or skin lesions. If organisms are identified, a definitive diagnosis of syphilis can be made, even in the absence of reactive serology. The practical and theoretical limitations of existing methods for treponemal detection have been discussed. The recent production of monoclonal antibodies with inherent specificity for *T. pallidum* may permit the development of a new generation of direct diagnostic techniques for early syphilis.

Detection methods that use monoclonal antibodies have several advantages over dark-field microscopy: (a) The specificity of the test is not dependent upon the skill of the microscopist, as with the dark-field examination. Detection methods based upon monoclonal antibody reactions are as specific as the antibody reagent, and results can be accurately interpreted by most laboratory personnel. The specificity of monoclonal antibodies will permit, for the first time, the evaluation of spirochetes in oral lesions. (b) The test is not dependent upon observation of a living treponeme. Dark-field microscopic evaluation of spirochetes is based upon observation of characteristic motility. Because treponemes die rapidly outside the human or animal host, specimens must be examined within 10–30 min to ensure accurate interpretation. Furthermore, organisms that are located on the surface of the lesion (and are accessible for examination) are frequently nonmotile; these organisms are evaluable by immunological methods but not by dark-field microscopy. (c) The test could be made widely available. Dark-field microscopy must be performed "on-site," and frequently physicians must choose to make a diagnosis on the basis of clinical presentation and serological testing or to refer the patient to a specialty clinic for evaluation. As with serological testing, the monoclonal antibody-based tests can be performed on specimens that are submitted to a laboratory. Although monoclonal antibody-based detection methods are unlikely to replace rapid, inexpensive dark-field microscopy in sexually transmitted disease clinics, they would provide a valuable diagnostic tool for most physicians who lack ready access to dark-field microscopy.

A. Identification of *Treponema pallidum* by Direct Immunofluorescence

Lukehart et al. (95) described the identification of *T. pallidum* in exudates derived from syphilitic chancres, by use of a fluorescein-tagged pathogen-specific monoclonal antibody; the sensitivity and specificity of this method have been evaluated in two reported clinical trials. Antibody H9-1, which has been used for these studies, reacts with a pathogen-specific determinant of the 47–48-kd antigen. By ELISA and immunofluorescence assays, H9-1 reacts with *T. pallidum*, but not with five nonpathogenic treponemes or the *Leptospira* and *Borrelia* species tested to date (105). The antibody is partially purified from ascites fluid by ammonium sulfate precipitation and conjugated with fluorescein isothiocyanate. The addition of Evans blue counterstain to the antibody cocktail creates a greater contrast between the cellular debris (red) and the brightly fluorescent treponemes (green). Exudate material is expressed from the lesion, air-dried onto a microscope slide, and fixed with acetone. The slide may be stained immediately, transported to a laboratory for staining, or stored frozen. The entire staining procedure and microscopic examination take approximately 30 min.

Hook et al. (105) reported the examination of specimens from 61 STD clinic patients with genital ulcerations or skin rashes. Using standard methodologies (dark-field microscopy, serological testing) and monoclonal antibody (H9-1) staining *Treponema pallidum* was identified by monoclonal antibody staining in all 30 (100%) patients with syphilis, whereas dark-field microscopy was positive in 29 of 30 (97%). Of 31 patients without syphilis, spirochetes were found by dark-field microscopy in seven (23%); in one case, they were misidentified as *T. pallidum*. Specimens from these patients were negative for *T. pallidum* with the monoclonal antibody test. In this study, then, monoclonal antibody staining was 100% sensitive and specific for *T. pallidum* compared with a diagnosis based upon standard methods.

In a second study, Romanowski et al. (106) examined 128 STD clinic patients using a similar protocol. In this trial, the monoclonal antibody test was 89% sensitive compared with dark-field microscopy and 100% specific for patients with syphilis. Overall, the monoclonal antibody was positive in 73% of syphilis patients, compared with 79% for dark-field microscopy. Both tests were 100% specific in this study.

As with any antigen detection test (including culture), the sensitivity of the test is dependent upon the quality of the specimen collected. Syphilitic lesions contain variable numbers of organisms, depending upon the stage of lesion development or healing and the use of antibacterial substances by the patient. Single specimens from each lesion may not be adequate. Dark-field microscopic examination is not considered to be negative until multiple slides have been examined on several separate occasions. Although the limit of detection of *T. pallidum* by monoclonal antibody staining is one organism on the slide, the examination of multiple slides per lesion ensures maximum sensitivity. In an unpublished study of genital ulcerations in Zambian patients (Hira, Lukehart, Lovett, and Perine, unpublished data), the sensitivity of monoclonal antibody staining of single slides was 76% compared with 61% for dark-field microscopy. The examination of multiple slides by dark-field microscopy increased the sensitivity of that technique to 89%; unfortunately, multiple slides were not available for monoclonal antibody staining.

Direct immunofluorescent staining with monoclonal antibodies can also be used to detect *T. pallidum* in biopsy or autopsy tissue specimens. Although this method has been used in research laboratories, it has not yet been widely applied.

B. Detection of *Treponema pallidum* in a Solid-Phase Radioimmunoassay

Norgard et al. (125) described a solid-phase immunoblot assay for detection of *T. pallidum* using monoclonal antibodies. *Treponema pallidum* is detected, in specimens dried onto nitrocellulose filter paper, using monoclonal antibody, ^{125}I-labeled rabbit antimouse IgG, and autoradiography. The sensitivity of this test for rabbit-derived *T. pallidum* is 500–1000 organisms after a 2- to 4-day autoradiography period. The specificity of this test was determined in the laboratory using *T. phagedenis* biotype Reiter, *Haemophilus ducreyi, Neisseria gonorrhoeae*, herpes simplex virus type 2, and normal rabbit testicular tissue. The sensitivity and specificity of this test with clinical specimens have not yet been examined, and its applicability in a clinical setting has not been determined. With a limit of detection of 500 treponemes, even after prolonged autoradiographic exposure, the practical usefulness of such a test may be limited.

VIII. POTENTIAL VALUE OF MONOCLONAL ANTIBODIES IN SYPHILIS DIAGNOSIS

The most obvious application of monoclonal antibodies in the diagnosis of syphilis is the identification of *T. pallidum* in genital ulcers, skin rashes, and biopsy or autopsy tissues. To date, clinical specimens have been examined using immunofluorescence only. The method is rapid, sensitive, and specific, and it is applicable to many types of specimens. Currently, these reagents are available only in research laboratories. If they should become commercially available, particularly in combination with monoclonal antibodies for detection of other genital ulcer pathogens, the applicability of the test would be wide. The ability of a physician to collect ulcer material on a slide and learn, within a day, whether the patient has syphilis or herpes or both, would be an important advance in the diagnosis of genital ulcer disease.

Monoclonal antibodies may also prove useful in the development of an antigen detection assay for body fluids, particularly cerebrospinal fluid (CSF). The evaluation of syphilitic involvement of the central nervous system presents a major diagnostic problem in syphilis. The solid-phase immunoblot assay described by Norgard et al. (125) might be applicable to body fluids, including CSF, if the sensitivity could be increased 10- to 100-fold, or if the antigen in the sample could be concentrated by filtration onto the nitrocellulose. This assay has the advantage, over immunofluorescence, of not relying upon the intact structure of the treponeme.

IX. CONCLUSIONS

Recent technological advances have permitted the identification and characterization of the major antigens of *T. pallidum*. Monoclonal antibodies to *T. pallidum* have aided in those examinations and have significant potential utility for the diagnosis of syphilis and other treponemal diseases. In addition to *T. pallidum* detection, monoclonal antibodies may serve as useful tools in the isolation and purification of treponemal antigens for the development of new serodiagnostic tests and for vaccine studies.

ACKNOWLEDGMENTS

The author thanks Sally Post and Laurie Johnson for careful manuscript preparation, and Sharon Baker-Zander for helpful discussions. This work was supported by Public Health Service Grants AI-18988 and NS-23677 from the National Institutes of Health.

REFERENCES

1. Miao, R. M., and A. H. Fieldsteel, Genetic relationship between *Treponema pallidum* and *Treponema pertenue*, two noncultivable human pathogens. *J. Bacteriol. 141*:427–429 (1980).
2. Nichols, H. J., and W. H. Hough, Demonstration of *Spirochaeta pallida* in the cerebrospinal fluid. *J. Am. Med. Assoc. 60*:108 (1913).
3. Miao, R., and A. H. Fieldsteel, Genetics of treponema: Relationship between *Treponema pallidum* and five cultivable treponemes. *J. Bacteriol. 133*:101–107 (1978).
4. Fieldsteel, A. H., D. L. Cox, and R. A. Moeckli, Cultivation of virulent *Treponema pallidum* in tissue culture. *Infect. Immun. 32*:908–915 (1981).
5. Norris, S. J., In vitro cultivation of *Treponema pallidum*: Independent confirmation. *Infect. Immun. 36*:437–439 (1982).
6. Cumberland, M. C., and T. B. Turner, The rate of multiplication of *Treponema pallidum* in normal and immune rabbits. *Am. J. Syph. Gon. Vener. Dis. 33*:201–212 (1949).
7. Hovind-Hougen, K., Morphology. In *Pathogenesis and Immunology of Treponemal Infection*, R. F. Schell and Musher, D. M. (eds.). Marcel Dekker, New York, pp. 3–28 (1983).
8. Turner, T. B., and D. H. Hollander, *Biology of the Treponematoses*. World Health Organization, Geneva, pp. 1–278 (1957).
9. Christiansen, S., Protective layer covering pathogenic treponemata. *Lancet 1*:423–425 (1963).
10. Zeigler, J. A., A. M. Jones, R. H. Jones, and K. M. Kubica, Demonstration of extracellular material at the surface of pathogenic *T. pallidum* cells. *Br. J. Vener. Dis. 52*:1–8 (1976).
11. Fitzgerald, T. J., and R. C. Johnson, Surface mucopolysaccharides of *Treponema pallidum*. *Infect. Immun. 24*:244–251 (1979).
12. Strugnell, R. A., C. J. Handley, L. Drummond, S. Faine, D. A. Lowther, and S. R. Graves, Polyanions in syphilis: Evidence that glycoproteins and macromolecules resembling glycosaminoglycans are synthesised by host tissues in response to infection with *Treponema pallidum*. *Br. J. Vener. Dis. 60*:75–82 (1984).
13. Strugnell, R. A., C. J. Handley, D. A. Lowther, S. Faine, and S. R. Graves, *Treponema pallidum* does not synthesize in vitro a capsule containing glycosaminoglycans or proteoglycans. *Br. J. Vener. Dis. 60*:8–13 (1984).
14. Fitzgerald, T. J., J. N. Miller, L. A. Repesh, M. Rise, and A. Urquhart,

Binding of glycosaminoglycans to the surface of *Treponema pallidum* and subsequent effects on complement interactions between antigen and antibody. *Genitourin. Med. 61*:13–20 (1985).
15. Van Der Sluis, J. J., G. Van Dijk, M. Boer, E. Stolz, and T. Van Joost, Mucopolysaccharides in suspensions of *Treponema pallidum* extracted from infected rabbit testes. *Genitourin. Med. 61*:7–12 (1985).
16. Wos, S. M., and K. Wicher, Antigenic evidence for host origin of exudative fluids in lesions of *Treponema pallidum*-infected rabbits. *Infect. Immun. 47*:228–233 (1985).
17. Logan, L. C., Rabbit globulin and antiglobulin factors associated with *Treponema pallidum* grown in rabbits. *Br. J. Vener. Dis. 50*:421–427 (1974).
18. Alderete, J. F., and J. B. Baseman, Surface-associated host proteins on virulent *Treponema pallidum. Infect. Immun. 26*:1048–1056 (1979).
19. Johnson, W. C., Venereal diseases and treponemal infections. In *Dermal Pathology*, J. H. Graham, Johnson, W. C., and Helwig, E. B. (eds.). Harper & Row, Hagerstown, pp. 371–385 (1972).
20. Fitzgerald, T. J., J. N. Miller, and J. A. Sykes, *Treponema pallidum* (Nichols strain) in tissue cultures: Cellular attachment, entry, and survival. *Infect. Immun. 11*:1133–1140 (1975).
21. Hayes, N. S., K. E. Muse, A. M. Collier, and J. B. Baseman, Parasitism by virulent *Treponema pallidum* of host cell surfaces. *Infect. Immun. 17*:174–186 (1977).
22. Fitzgerald, T. J., R. C. Johnson, J. N. Miller, and J. A. Sykes, Characterization of the attachment of *Treponema pallidum* (Nichols strain) to cultured mammalian cells and the potential relationship of attachment to pathogenicity. *Infect. Immun. 18*:467–478 (1977).
23. Fitzgerald, T. J., Attachment of treponemes to cell surfaces. In *Pathogenesis and Immunology of Treponemal Infection*, R. F. Schell and Musher, D. M. (eds.). Marcel Dekker, New York, pp. 195–228 (1983).
24. Baseman, J. B., and E. E. Hayes, Molecular characterization of receptor binding proteins and immunogens of virulent *Treponema pallidum. J. Exp. Med. 151*:573–586 (1980).
25. Thomas, D. D., J. B. Baseman, and J. F. Alderete, Fibronectin mediates *Treponema pallidum* cytadherence through recognition of fibronectin cell-binding domain. *J. Exp. Med. 161*:514–525 (1985).
26. Peterson, K. M., J. B. Baseman, and J. F. Alderete, *Treponema pallidum* receptor binding proteins interact with fibronectin. *J. Exp. Med. 157*:1958–1970 (1983).
27. Thomas, D. D., J. B. Baseman, and J. F. Alderete, Putative *Treponema pallidum* cytadhesins share a common functional domain. *Infect. Immun. 49*:833–835 (1985).
28. Fitzgerald, T. J., L. A. Repesh, D. R. Blanco, and J. N. Miller, Attachment of *Treponema pallidum* to fibronectin, laminin, collagen IV, and collagen I, and blockage of attachment by immune rabbit IgG. *Br. J. Vener. Dis. 60*:357–363 (1984).

29. Fitzgerald, T. J., and L. A. Repesh, Interactions of fibronectin with *Treponema pallidum*. *Genitourin. Med. 61*:147–155 (1985).
30. Steiner, B. M., and S. Sell, Characterization of the interaction between fibronectin and *Treponema pallidum*. *Curr. Microbiol. 12*:157–162 (1985).
31. Azar, H. A., T. D. Pham, and A. K. Kurban, An electron microscopic study of a syphilitic chancre. *Arch. Pathol. 90*:143–150 (1970).
32. Sykes, J. A., and J. N. Miller, Intracellular location of *Treponema pallidum* (Nichols Strain) in the rabbit testis. *Infect. Immun. 4*:307–314 (1971).
33. Lauderdale, V., and J. N. Goldman, Serial ultrathin sectioning demonstrating the intracellularity of *T. pallidum*. *Br. J. Vener. Dis. 48*:87–96 (1972).
34. Sykes, J. A., J. N. Miller, and A. J. Kalan, *Treponema pallidum* within cells of a primary chancre from a human female. *Br. J. Vener. Dis. 50*:40–44 (1974).
35. Hanff, P. A., N. H. Bishop, J. N. Miller, and M. A. Lovett, Humoral immune response in experimental syphilis to polypeptides of *Treponema pallidum*. *J. Immunol. 131*:1973–1977 (1983).
36. Lukehart, S. A., S. A. Baker-Zander, and S. Sell, Characterization of the humoral immune response of the rabbit to antigens of *Treponema pallidum* after experimental infection and therapy. *Sex. Transm. Dis. 13*:9–15 (1986).
37. Stamm, L. V., and P. J. Bassford Jr., Cellular and extracellular protein antigens of *Treponema pallidum* synthesized during in vitro incubation of freshly extracted organisms. *Infect. Immun. 47*:799–807 (1985).
38. Bishop, N. H., and J. N. Miller, Humoral immunity in experimental syphilis. II. The relationship of neutralizing factors in immune serum to acquired resistance. *J. Immunol. 117*:197–207 (1976).
39. Nelson, R. A., Jr., and M. M. Mayer, Immobilization of *Treponema pallidum* in vitro by antibody produced in syphilitic infection. *J. Exp. Med. 89*:369–393 (1949).
40. Blanco, D. R., J. N. Miller, and P. A. Hanff, Humoral immunity in experimental syphilis: The demonstration of IgG as a treponemicidal factor in immune rabbit serum. *J. Immunol. 133*:2693–2697 (1984).
41. Perine, P. L., R. S. Weiser, and S. J. Klebanoff, Immunity to syphilis. I. Passive transfer in rabbits with hyperimmune serum. *Infect. Immun. 8*: 787–790 (1973).
42. Sepetjian, M., D. Salussola, and J. Thivolet, Attempt to protect rabbits against experimental syphilis by passive immunization. *Br. J. Vener. Dis. 49*:335–337 (1973).
43. Turner, T. B., P. H. Hardy Jr., B. Newman, and E. E. Nell, Effects of passive immunization on experimental syphilis in the rabbit. *Johns Hopkins Med. J. 133*:241–251 (1973).
44. Bishop, N. H., and J. N. Miller, Humoral immunity in experimental syphilis. I. The demonstration of resistance conferred by passive immunization. *J. Immunol. 117*:191–196 (1976).
45. Weiser, R. S., D. Erickson, P. L. Perine, and N. N. Pearsall, Immunity to

syphilis: Passive transfer in rabbits using serial doses of immune serum. *Infect. Immun. 131*:1402–1407 (1976).
46. Graves, S., and J. Alden, Limited protection of rabbits against infection with *Treponema pallidum* by immune rabbit sera. *Br. J. Vener. Dis. 55*: 399–403 (1979).
47. Titus, R. G., and R. S. Weiser, Experimental syphilis in the rabbit: Passive transfer of immunity with immunoglobulin G from immune serum. *J. Infect. Dis. 140*:904–913 (1979).
48. Pavia, C. S., C. J. Niederbuhl, and J. Saunders, Antibody-mediated protection of guinea-pigs against infection with *Treponema pallidum. Immunology 56*:195–202 (1985).
49. Azadegan, A. A., R. F. Schell, and J. L. LeFrock, Immune serum confers protection against syphilitic infection on hamsters. *Infect. Immun. 42*: 42–47 (1983).
50. Schell, R. F., J. L. LeFrock, and J. P. Babu, Passive transfer of resistance to frambesial infection in hamsters. *Infect. Immun. 21*:430–435 (1978).
51. Lukehart, S. A., and J. N. Miller, Demonstration of the in vitro phagocytosis of *Treponema pallidum* by rabbit peritoneal macrophages. *J. Immunol. 121*:2014–2024 (1978).
52. Wong, G. H. W., B. Steiner, and S. Graves, Effect of syphilitic rabbit sera taken at different periods after infection on treponemal motility, treponemal attachment to mammalian cells in vitro, and treponemal infection in rabbits. *Br. J. Vener. Dis. 59*:220–224 (1983).
53. Lukehart, S. A., S. A. Baker-Zander, and S. Sell, Characterization of lymphocyte responsiveness in early experimental syphilis. I. In vitro response to mitogens and *Treponema pallidum* antigens. *J. Immunol. 124*: 454–460 (1980).
54. Lukehart, S. A., S. A. Baker-Zander, R. M. C. Lloyd, and S. Sell, Characterization of lymphocyte responsiveness in early experimental syphilis. II. Nature of cellular infiltration and *Treponema pallidum* distribution in testicular lesions. *J. Immunol. 124*:461–467 (1980).
55. Lukehart, S. A., S. A. Baker-Zander, R. M. C. Lloyd, and S. Sell, Effect of cortisone administration on host-parasite relationships in early experimental syphilis. *J. Immunol. 127*:1361–1368 (1981).
56. Lukehart, S. A., Activation of macrophages by products of lymphocytes from normal and syphilitic rabbits. *Infect. Immun. 37*:64–69 (1982).
57. Baker-Zander, S. A., and S. Sell, A histopathologic and immunologic study of the course of syphilis in the experimentally infected rabbit. Demonstration of long-lasting cellular immunity. *Am. J. Pathol. 101*:387–414 (1980).
58. Sell, S., S. A. Baker-Zander, and H. C. Powell, Experimental syphilitic orchitis in rabbits. Ultrastructural appearance of *Treponema pallidum* during phagocytosis and dissolution by macrophages in vivo. *Lab. Invest. 46*:355–364 (1982).
59. Cannefax, G. R., and W. Garson, The demonstration of a common antigen in Reiter's treponeme and virulent *Treponema pallidum. J. Immunol. 82*: 198–200 (1959).

60. D'Allesandro, G., and L. Dardanoni, Isolation and purification of the protein antigen of the Reiter treponeme. A study of its serologic reactions. *Am. J. Syph. 37*:137–150 (1953).
61. Strandberg-Pedersen, N., C. S. Petersen, M. Vejtorp, and N. H. Axelsen, Serodiagnosis of syphilis by an enzyme-linked immunosorbent assay for IgG antibodies against the Reiter Treponeme flagellum. *Scand. J. Immunol. 15*:341–348 (1982).
62. Hunter, E. F., W. E. Deacon, and P. Meyer, An improved FTA test for syphilis, the Absorption Procedure (FTA-ABS). *Public Health Rep. 79*: 410–412 (1964).
63. Gelperin, A., Immunochemical studies of the Reiter spirochete. *Am. J. Syphilol. 35*:1–13 (1951).
64. Miller, J. N., S. I. Whang, and F. P. Fazzan, Studies on immunity in experimental syphilis. I. Immunologic response of rabbits immunized with Reiter protein antigen and challenged with virulent *Treponema pallidum. Br. J. Vener. Dis. 39*:195–198 (1963).
65. Izzat, N. N., W. G. Dacres, J. M. Knox, and R. Wende, Attempts at immunization against syphilis with avirulent *Treponema pallidum. Br. J. Vener. Dis. 46*:451–453 (1970).
66. Al-Samarrai, H. T., and W. G. Henderson, Immunity in syphilis: Studies in active immunity. *Br. J. Vener. Dis. 52*:300–308 (1976).
67. Hindersson, P., C. S. Petersen, and N. H. Axelsen, Purified flagella from *Treponema phagedenis* biotype Reiter does not induce protective immunity against experimental syphilis in rabbits. *Sex. Trans. Dis. 12*:124–127 (1985).
68. Schell, R. F., A. A. Azadegan, S. G. Nitskansky, and J. L. LeFrock, Acquired resistance of hamsters to challenge with homologous and heterologous virulent treponemes. *Infect. Immun. 37*:617–621 (1982).
69. Miller, J. N., Immunity in experimental syphilis. VI. Successful vaccination of rabbits with *Treponema pallidum*, Nichols strain, attenuated by gamma-irradiation. *J. Immunol. 110*:1206–1215 (1973).
70. Lukehart, S. A., S. A. Baker-Zander, and E. R. Gubish Jr., Identification of *Treponema pallidum* antigens: Comparison with a nonpathogenic treponeme. *J. Immunol. 129*:833–838 (1982).
71. Baker-Zander, S. A., E. W. Hook III, P. Bonin, H. H. Handsfield, and S. A. Lukehart, Antigens of *Treponema pallidum* recognized by IgG and IgM antibodies during syphilis in humans. *J. Infect. Dis. 151*:264–272 (1985).
72. Van Eijk, R. V. W., and J. D. A. Van Embden, Molecular characterization of *Treponema pallidum* proteins responsible for the human immune response to syphilis. *Antonie Leeuwenhoek J. Microbiol. Serol. 48*:486–487 (1982).
73. Hanff, P. A., T. E. Fehniger, J. N. Miller, and M. A. Lovett, Humoral immune response in human syphilis to polypeptides of *Treponema pallidum. J. Immunol. 129*:1287–1291 (1982).
74. Penn, C. W., M. J. Bailey, and A. Cockayne, The axial filament antigen of *Treponema pallidum. Immunology 54*:635–641 (1985).

75. Hanff, P. A., J. N. Miller, and M. A. Lovett, Molecular characterization of common treponemal antigens. *Infect. Immun. 40*:825–828 (1983).
76. Baker-Zander, S. A., and S. A. Lukehart, Molecular basis of immunological cross-reactivity between *Treponema pallidum* and *Treponema pertenue. Infect. Immun. 42*:634–638 (1983).
77. Hensel, U., H. J. Wellensiek, and S. Bhakdi, Sodium dodecyl sulfate-polyacrylamide gel electrophoresis immunoblotting as a serological tool in the diagnosis of syphilitic infections. *J. Clin. Microbiol. 21*:82–87 (1985).
78. Baughn, R. E., C. B. Adams, and D. M. Musher, Circulating immune complexes in experimental syphilis: Identification of treponemal antigens in isolated complexes. *Infect. Immun. 42*:585–593 (1983).
79. Baker-Zander, S. A., and S. A. Lukehart, Antigenic cross-reactivity between *Treponema pallidum* and other pathogenic members of the family *Spirochaetaceae. Infect. Immun. 46*:116–121 (1984).
80. Baker-Zander, S. A., R. E. Roddy, H. H. Handsfield, and S. A. Lukehart, IgG and IgM antibody reactivity to antigens of *Treponema pallidum* following treatment of syphilis. *Sex. Trans. Dis. 13*:214–220 (1986).
81. Moskophidis, M., and F. Muller, Molecular analysis of immunoglobulins M and G immune response to protein antigens of *Treponema pallidum* in human syphilis. *Infect. Immun. 43*:127–132 (1984).
82. Moskophidis, M., and F. Muller, Molecular characterization of glycoprotein antigens on surface of *Treponema pallidum*: Comparison with nonpathogenic *Treponema phagedenis* biotype Reiter. *Infect. Immun. 46*:867–869 (1984).
83. Alderete, J. F., and J. B. Baseman, Surface characterization of virulent *Treponema pallidum. Infect. Immun. 30*:814–823 (1980).
84. Alderete, J. F., and J. B. Baseman, Analysis of serum IgG against *Treponema pallidum* protein antigens in experimentally infected rabbits. *Br. J. Vener. Dis. 57*:302–308 (1981).
85. Norris, S. J., and S. Sell, Antigenic complexity of *Treponema pallidum*: Antigenicity and surface localization of major polypeptides. *J. Immunol. 133*:2686–2692 (1984).
86. Fohn, M., S. A. Baker-Zander, and S. A. Lukehart, Resolution of the major common and pathogen-specific antigens of *Treponema pallidum* subspecies *pallidum* by two-dimensional electrophoresis. (submitted).
87. Jones, S. A., K. S. Marchitto, J. N. Miller, and M. V. Norgard, Monoclonal antibody with hemagglutination, immobilization, and neutralization activities defines an immunodominant, 47,000 mol wt, surface-exposed immunogen of *Treponema pallidum* (Nichols). *J. Exp. Med. 160*:1404–1420 (1984).
88. Thornburg, R. W., and J. B. Baseman, Comparison of major protein antigens and protein profiles of *Treponema pallidum* and *Treponema pertenue. Infect. Immun. 42*:623–627 (1983).
89. Strandberg-Pedersen, N., N. H. Axelsen, B. B. Jorgensen, and C. S. Petersen, Antibodies in secondary syphilis against five of forty Reiter treponeme antigens. *Scand. J. Immunol. 11*:629–633 (1980).

90. Strandberg-Pedersen, N., C. S. Petersen, N. H. Axelsen, A. Birch-Andersen, and K. Hovind-Hougen, Isolation of a heat-stable antigen from *Treponema* Reiter, using an immunoadsorbent with antibodies from syphilitic patients. *Scand. J. Immunol. 14*:137–144 (1981).
91. Bharier, M., and D. Allis, Purification and characterization of axial filaments from *Treponema phagedenis* biotype *reiterii* (the Reiter treponeme). *J. Bacteriol. 120*:1434–1442 (1974).
92. Limberger, R. J., and N. W. Charon, Periplasmic flagellar proteins of *Treponema phagedenis. Abstr. Annu. Meet. Am. Soc. Microbiol.* Abstr. D99 (1985).
93. Blanco, D. R., J. D. Radolf, M. A. Lovett, and J. N. Miller, The antigenic interrelationship between the endoflagella of *Treponema phagedenis* biotype Reiter and *Treponema pallidum*, Nichols Strain: I. Treponemicidal activity of cross-reactive endoflagellar antibodies against *T. pallidum. J. Immunol. 137*:2973–2979 (1986).
94. Blanco, D. R., J. D. Radolf, M. A. Lovett, and J. N. Miller, Correlation of treponemicidal activity in normal human serum with the presence of IgG antibody directed against polypeptides of *Treponema phagedenis* biotype Reiter and *Treponema pallidum*, Nichols Strain. *J. Immunol. 137*:2031–2036 (1986).
95. Lukehart, S. A., M. R. Tam, J. Hom, S. A. Baker-Zander, K. K. Holmes, and R. C. Nowinski, Characterization of monoclonal antibodies to *Treponema pallidum. J. Immunol. 134*:585–592 (1985).
96. Moskophidis, M., and F. Muller, Monoclonal antibodies to *Treponema pallidum*: Monoclonal antibodies directed against immunodominant surface-exposed protein antigens of *Treponema pallidum. Eur. J. Clin. Microbiol. 4*:473–477 (1985).
97. Robertson, S. M., J. R. Kettman, J. N. Miller, and M. V. Norgard, Murine monoclonal antibodies specific for virulent *Treponema pallidum* (Nichols). *Infect. Immun. 36*:1076–1085 (1982).
98. Saunders, J. M., and J. D. Folds, Development of monoclonal antibodies that recognize *Treponema pallidum. Infect. Immun. 41*:844–847 (1983).
99. Thornburg, R. W., J. Morrison-Plummer, and J. B. Baseman, Monoclonal antibodies to *Treponema pallidum*: Recognition of a major polypeptide antigen. *Genitourin. Med. 61*:1–6 (1985).
100. Van Embden, J. D., H. J. Van Der Donk, R. V. Van Eijk, H. G. Van Der Heide, J. A. De Jong, M. F. Van Olderen, A. D. Osterhaus, and L. M. Schouls, Molecular cloning and expression of *Treponema pallidum* DNA in *Escherichia coli* K-12. *Infect. Immun. 42*:187–196 (1983).
101. Van De Donk, H. J. M., J. D. A. Van Embden, A. de Jong, M. F. Van Olderen, and A. D. M. E. Osterhaus, Monoclonal antibodies to *Treponema pallidum. Dev. Biol. Stand. 57*:107–111 (1984).
102. Saunders, J. M., and J. D. Folds, Humoral response of the mouse to *Treponema pallidum. Genitourin. Med. 61*:221–229 (1983).
103. Marchitto, K. S., S. A. Jones, R. F. Schell, P. L. Holmans, and M. V. Norgard, Monoclonal antibody analysis of specific antigenic similarities among

pathogenic *Treponema pallidum* subspecies. *Infect. Immun. 45*:660–666 (1984).
104. Marchitto, K. S., C. K. Selland-Grossling, and M. V. Norgard, Molecular specificities of monoclonal antibodies directed against virulent *Treponema pallidum. Infect. Immun. 51*:168–176 (1986).
105. Hook, E. W. III, R. E. Roddy, S. A. Lukehart, J. Hom, K. K. Holmes, and M. R. Tam, Detection of *Treponema pallidum* in lesion exudate with a pathogen-specific monoclonal antibody. *J. Clin. Microbiol. 22*:241–244 (1985).
106. Romanowski, B., E. Forsey, E. Prasad, S. A. Lukehart, M. R. Tam, and E. W. Hook III, Fluorescent monoclonal antibody detection for *Treponema pallidum. Sex. Transm. Dis. 14*:156–159 (1987).
107. Chapel, T. A., Origins of penile ulcerations. *Arch. Androl. 3*:351–357 (1979).
108. Chapel, T. A., W. J. Brown, C. Jeffries, and J. A. Stewart, How reliable is the morphological diagnosis of penile ulcerations? *Sex. Trans. Dis. 4*: 150–152 (1977).
109. Fast, M. V., L. J. D'Costa, H. Nsanze, P. Piot, J. Curran, P. Karasira, N. Mirza, I. W. Maclean, and A. R. Ronald, The clinical diagnosis of genital ulcer disease in men in the tropics. *Sex. Transm. Dis. 11*:72–76 (1984).
110. Daniels, K. C., and H. S. Ferneyhough, Specific direct fluorescent antibody detection of *Treponema pallidum. Health Lab. Sci. 14*:164–171 (1977).
111. Edwards, E. A., Detecting *Treponema pallidum* in primary lesions by the fluorescent antibody technique. *Public Health Rep. 77*:427–430 (1962).
112. Jue, R., J. Puffer, R. M. Wood, G. Schochet, W. H. Smartt, and W. A. Ketterer, Comparison of fluorescent and conventional darkfield methods for the detection of *Treponema pallidum* in syphilitic lesions. *Am. J. Clin. Pathol. 47*:809–811 (1967).
113. Kellogg, D. S., and S. M. Mothershed, Immunofluorescent detection of *Treponema pallidum. J. Am. Med. Assoc. 207*:938–941 (1969).
114. Wilkinson, A. E., and L. P. Cowell, Immunofluorescent staining for the detection of *Treponema pallidum* in early syphilitic lesions. *Br. J. Vener. Dis. 47*:252–254 (1971).
115. Smith, J. L., and C. W. Israel, Spirochetes in the aqueous humor in seronegative ocular syphilis. *Arch. Ophthalmol. 77*:474–477 (1967).
116. Davis, L. E., and S. Sperry, Bell's palsy and secondary syphilis: CSF spirochetes detected by immunofluorescence. *Ann. Neurol. 4*:378–380 (1978).
117. Montenegro, E. N. R., W. G. Nicol, and J. L. Smith, Treponemalike forms and artifacts. *Am. J. Ophthalmol. 68*:197–205 (1969).
118. Brown, B. C., Spiral organisms in body fluids. *Am. J. Ophthalmol. 68*: 945–949 (1969).
119. Yobs, A. R., L. Brown, and E. F. Hunter, Fluorescent antibody technique in early syphilis. *Arch. Pathol. 77*:220–225 (1964).
120. Al-Samarrai, H. T., and W. G. Henderson, Immunofluorescent staining of

Treponema pallidum and *Treponema pertenue* in tissues fixed by formalin and embedded in paraffin wax. *Br. J. Vener. Dis. 53*:1–11 (1977).
121. Beckett, J. H., and J. W. Bigbee, Immunoperoxidase localization of *Treponema pallidum*. *Arch. Pathol. Lab. Med. 103*:135–138 (1979).
122. Yobs, A. R., S. Olansky, D. H. Rockwell, and J. W. Clarke Jr., Do treponemes survive adequate treatment of late syphilis? *Arch. Dermatol. 91*: 379–389 (1965).
123. Handsfield, H. H., S. A. Lukehart, S. Sell, S. J. Norris, and K. K. Holmes, Demonstration of *Treponema pallidum* in a cutaneous gumma by indirect immunofluorescence. *Arch. Dermatol. 119*:677–680 (1983).
124. Quinn, T. C., S. A. Lukehart, S. Goodell, E. Mkrtichian, M. D. Schuffler, and K. K. Holmes, Rectal mass caused by *Treponema pallidum*: Confirmation by immunofluorescent staining. *Gastroenterology 82*:135–139 (1982).
125. Norgard, M. V., C. K. Selland, J. R. Kettman, and J. N. Miller, Sensitivity and specificity of monoclonal antibodies directed against antigenic determinants of *Treponema pallidum* Nichols in the diagnosis of syphilis. *J. Clin. Microbiol. 20*:711–717 (1984).

19
Monoclonal Antibodies for the Characterization and Laboratory Diagnosis of *Neisseria gonorrhoeae*

JOAN S. KNAPP
Center for Infectious Diseases, Centers for Disease Control, Atlanta, Georgia

I. INTRODUCTION

Gonorrhea is a sexually transmitted disease (STD) caused by *Neisseria gonorrhoeae*. In 1987, more than 750,000 cases of gonorrhea were reported in the United States (1). The most common clinical manifestations of gonorrhea are uncomplicated infections of mucous membranes of the urethra, cervix, rectum, and pharynx (2). Some patients develop complicated infections. Approximately 15% of women with gonorrhea may develop pelvic inflammatory disease (PID). This is the most frequent and costly complication of gonorrhea and may also result in sterility (2). Approximately 1-2% of patients with gonorrhea develop disseminated gonococcal infection (DGI), which is manifested either as tenosynovitis or arthritis (2). Newborns infected during birth may develop conjunctivitis, which is a serious infection that may result in blindness if not treated promptly (2). Although gonococcal conjunctivitis is generally rare in adults in the United States, it occurs frequently in adults in some parts of Africa (3). The control of gonorrhea is affected both by the ability to accurately identify the organism in the laboratory and to identify and control infections caused by gonococci that are resistant to antimicrobial agents or cause complicated infections.

Monoclonal antibody reagents have been developed for the laboratory identification of *N. gonorrhoeae* using antibodies that are directed against epitopes on the gonococcal principal outer membrane protein (POMP), protein I (PI). The serological diagnostic tests for *N. gonorrhoeae* have been developed in conjunction with studies to characterize the gonococcus and to determine the diversity within the species and epidemiological correlates of gonorrhea. Therefore, a discussion of the epidemiological studies performed with polyvalent and monoclonal antibodies is presented before a discussion of the laboratory use of these reagents. A discussion of many observations made when using the polyvalent serological-typing systems has also been included because, owing to the direct correlation between polyvalent serogroups and monoclonal serovars or serotypes, these observations may be generally extrapolated to the corresponding monoclonal serovars in instances in which the corresponding studies have not been made at this time.

II. HISTORICAL PERSPECTIVES ON THE PHENOTYPIC CHARACTERIZATION OF *NEISSERIA GONORRHOEAE* ISOLATES

Isolates of *N. gonorrhoeae* have been characterized to determine whether certain strains caused DGI and PID, and whether antimicrobial resistant strains represented the spread of a single clone or many resistant strains. These studies have included surveys of the geographical distribution of strains.

A. Antimicrobial Susceptibility Patterns

Antimicrobial susceptibility patterns were used to characterize gonococcal isolates before the development of typing systems. The use of antibiograms was limited, however, because it was not known whether resistant strains represented resistant variants of strains that were already present in a community or represented the introduction of new resistant strains into a community. It was apparent to investigators (4) that isolates recovered from patients with DGI in Seattle were highly susceptible to antibiotics. The development of auxotyping permitted further phenotypic characterization of these gonococcal isolates.

B. Auxotyping

Auxotyping permitted the classification of gonococcal strains according to their nutritional requirements, i.e., their requirement, in chemically defined medium, for amino acids, vitamins, and pyrimidines (5). A large number of auxotypes of *N. gonorrhoeae* have been described (5-8). Auxotyping studies provided insight into the geographical distribution of strains of *N. gonorrhoeae* as well the association of some strains with specific disease syndromes and antimicrobial resistance.

Gonococcal isolates that required arginine (A;Arg^-), hypoxanthine (H), and uracil (U; AHU), were serum-resistant (9) and highly susceptible to penicillin (10) and were isolated frequently, but not exclusively, from patients with DGI in many cities in the 1970s (10-14); they were also associated with asymptomatic gonorrhea (15). The AHU isolates presented problems for the laboratory diagnosis of gonorrhea because they grew as atypically small colonies and produced a very weak acid reaction from glucose, with the result that they appeared to be glucose-negative (11). The AHU isolates were also susceptible to vancomycin which resulted in their failure to grow on gonococcal selective medium that contained 4 μg/ml of vancomycin (16). Gonococcal isolates that were nonrequiring (Proto) or proline-requiring (P;Pro⁻) were isolated from patients with DGI more frequently than AHU isolates in some cities (12,13) and, in contrast to most Proto and Pro⁻ isolates (10), were also highly susceptible to penicillin and were serum-resistant (12). The similarities between DGI isolates from different geographical areas relative to serum resistance and antimicrobial susceptibilities suggested that, although they belonged to different auxotypes, they might be related in a way that might be easily detected if a different phenotypic-typing system was used.

The distribution of AHU isolates was limited to relatively few areas of the world including Scandinavia, northern Europe, and certain areas of the United States, where these strains accounted for approximately 50% of cases of gonorrhea (17-21). The AHU isolates were recovered more frequently from white than from black patients (17) and more frequently from heterosexual men and women than from homosexual men (22).

Another unusual auxotype that required proline, arginine (citrulline), and uracil (PAU; 7), also attracted much attention. Isolates belonging to this auxotype accounted for approximately 40% of isolates in Ontario, Canada in 1977-1978 (7). The PAU isolates were unusual because they did not possess the 2.6-megadalton (Md) cryptic plasmid that is possessed by most gonococci (23). The PAU isolates now occur frequently in other cities in Canada and in cities in the United States, Europe, and Japan (17,19,24-26).

C. Plasmid Content

Plasmid-mediated resistance to penicillin was first described in *N. gonorrhoeae* in 1976 when β-lactamase-producing strains (PPNG) were isolated from patients who had contracted gonorrhea in the Far East (27) and West Africa (28,29). The PPNG strains have spread to most countries in the world and present problems for gonorrhea control because penicillins cannot be used to treat PPNG infections. The PPNG isolates have been classified according to the size of their β-lactamase plasmids. Isolates from the Far East possessed a 4.4-Md β-lactamase plasmid, whereas those from West Africa possessed 3.2-Md β-lactamase plasmids (30). Spread of the β-lactamase plasmids between gonococci has been facilitated by a 24.5-Md conjugative plasmid (30). The conjugative plasmid has been found frequently in PPNG strains possessing the 4.4-Md β-lactamase plasmid (30), but has only recently been found in PPNG strains possessing the 3.2-Md β-lactamase plasmid (31). Recently, 2.9-, 3.05-, and 4.0-Md β-lactamase plasmids have also been described (32-34).

Plasmid profiles have been used in conjunction with auxotyping to classify PPNG isolates from different geographical areas and to follow temporal changes in PPNG strain populations (35,36). Most of the PPNG strains from the Far East belong to the Proto and Pro^- auxotypes, whereas those initially isolated from Africa were Arg^- (27). Although auxotype and plasmid profile permitted differentiation between PPNG isolates, there was a need for additional phenotypic characters to differentiate among them.

III. SEROLOGICAL CLASSIFICATION OF *NEISSERIA GONORRHOEAE* USING POLYVALENT ANTIBODIES

Practical serotyping systems for *N. gonorrhoeae* were not developed until recently. Typing systems resulting from studies of the PI molecule have proved most successful for serological classification of the gonococcus for epidemiological and diagnostic purposes.

Wang et al. (37) developed a microimmunofluorescent (MIF)-typing system that uses antisera developed against formalinized whole gonococcal cells, which were absorbed to remove cross-reacting antilipopolysaccharide and antiprotein antibodies. Gonococci were divided into three groups, designated A, B, and C (37). Although the MIF test was a laborious and time-consuming test to perform and was not widely used, this typing system became the basis for the subsequent development of gonococcal serotyping systems.

Sandström and Danielsson (38) studied three classes of gonococcal antigens, W, J, and M. Antibodies directed against the W class antigens were used in coagglutination tests to divide gonococci into three serologically distinct groups designated WI, WII, and WIII that corresponded to Wang's serogroups A, B, and C, respectively (38). In contrast to the MIF method, the coagglutination technique was rapid, economical, and easy to perform and has been widely used for serotyping gonococci in research and clinical laboratories.

Buchanan and Hildebrandt (39) developed a serotyping system for *N. gonorrhoeae* using the enzyme-linked immunosorbent assay (ELISA). Gonococci were divided into nine POMP serotypes (39). Serotypes 1-3 corresponded to serogroup WI, serotypes 4-8 generally corresponded to serogroup WII, and serotype 9 generally corresponded to serogroup WIII (40). The antibodies used in both the coagglutination- and ELISA-typing systems were directed against the gonococcal PI molecule (40).

The polyvalent anti-PI antibodies were generally specific for *N. gonorrhoeae* and provided the basis for not only a serological classification system for epidemiology but also for a diagnostic reagent.

IV. W-SEROGROUPING AND -SEROTYPING OF *NEISSERIA GONORRHOEAE* ISOLATES WITH POLYVALENT ANTIBODIES

The AHU isolates belonged to the corresponding serogroups A or WI or to the serotypes 1-3, whereas Proto and Pro^- isolates were classified into each of the three serogroups or nine serotypes (37,40). Many Proto and Pro^- isolates from patients with DGI in Atlanta belonged to the serogroup WI (41). In the United States, although AHU isolates were rarely seen in many cities, approximately one-third of the non-AHU isolates belonged to the serogroup WI (42). In Scandinavia, the proportion of isolates belonging to the different W-serogroups also varied among communities (43). Bygdeman (44) found that serogroup WII isolates were significantly more frequent in homosexual men than were serogroup WI isolates. Bygdeman (44,45) also found that both non-AHU, non-PPNG, and PPNG isolates that belonged to the serogroup WI were significantly more susceptible to antimicrobial agents than those belonging to the serogroups WII and WIII.

The complexity of PPNG epidemics was also demonstrated by W-serogrouping. Handsfield et al. (46) showed that a PPNG outbreak in Seattle in 1981 was caused by several strains because isolates belonging to the Proto and Pro^- auxotypes belonged to each of the three W-serogroups, whereas an outbreak in Shreveport, Louisiana was caused by only one strain. Similar observations were also made by Bygdeman et al. in Sweden (47).

V. STRUCTURE OF THE GONOCOCCAL PROTEIN I MOLECULE USING PEPTIDE MAPPING

Sandström and Danielsson (48) had noted a correlation between the line-rocket immunoelectrophoresis patterns and coagglutination reactions of gonococcal strains. Although isolates belonging to the WI serogroup appeared to be serologically distinct from those belonging to either the WII or WIII serogroups, the latter isolates appeared to be related (48). This relationship had also been suggested previously by Sandström et al. (40) who demonstrated that the W-serogrouping reagents and the POMP-serotyping reagents reacted with PI epitopes. Isolates that belonged to serotypes 1-3 (serogroup WI) did not react with the WII or WIII serogrouping or with the POMP-serotyping reagents 4-9. All isolates that belonged to serotypes 4-7 were also classified as belonging to serogroup WII. In contrast, however, 28 (57%) of 49 isolates that belonged to serotypes 8 or 9 reacted with only the WII reagents; one isolate reacted with both the WII and WIII reagents, and the remaining isolates reacted with the WIII reagent (40). These observations suggested that isolates belonging to the serogroups WII and WIII were antigenically related.

Sandström et al. (49) used two-dimensional peptide mapping of tryptic digests to characterize the PI molecules of strains belonging to the three W-serogroups. Strains belonging to the serogroups WII and WIII possessed similar PI molecules, designated protein IB, whereas strains belonging to the serogroup WI possessed a distinctly different PI molecule, designated protein IA (49).

VI. MONOCLONAL ANTIBODIES AGAINST GONOCOCCAL PROTEIN I EPITOPES

Monoclonal antibodies have been produced against gonococcal outer membrane proteins (50). Antibodies specific for gonococcal PIA or PIB epitopes were selected by screening against W-serogrouping reference strains in coagglutination tests and confirmed by radioimmune precipitation assays (50). The suitability of PI-specific monoclonal antibodies for diagnostic purposes was also demonstrated (51). A "cocktail" of broadly reactive PIA- and PIB-specific antibodies reacted in coagglutination tests with 99.6% of 719 gonococcal isolates from different geographical areas, worldwide (51).

VII. SEROLOGICAL CLASSIFICATION OF GONOCOCCI BY USE OF MONOCLONAL ANTIBODIES

Several serological classification systems have been developed and are currently used to serotype gonococci. Each system uses a panel of monoclonal antibodies directed against gonococcal PI epitopes in coagglutination tests.

A. Serological Classification of *Neisseria gonorrhoeae* Isolates by Serovars

A serological classification system for *N. gonorrhoeae* was developed that uses a standard panel of monoclonal antibody reagents (50,52). Serovars were defined by the pattern of reactions of isolates with either six PIA or six PIB monoclonal antibody reagents (52). The nomenclature used in this classification system combined the prefix IA or IB to indicate the PI molecule possessed by the isolate, with a number that described the reaction pattern of the strain with the corresponding set of reagents (52). For example, an isolate that reacted with the antibodies 3C8, 1F5, 2D6, and 2H1 belonged to the serovar IB-1. In a study of more than 1400 isolates, worldwide, all isolates reacted with at least one of the reagents (52). At this writing, a total of 24 PIA serovars, IA-1 to IA-24, and 32 PIB serovars, IB-1 to IB-32, have been recognized.

The nomenclature for this serotyping system is inflexible because it requires that all isolates be tested with the same standard panel of reagents. If additional resolution within an individual serovar is required, this may be accomplished by using additional monoclonal antibody reagents to subtype within the serovar defined by the standard panel. If this is done, a nomenclature must be devised specifically to accommodate this typing. It was felt, however, that if all isolates were tested with the same standard panel of reagents and subtyping is performed in a limited geographical area, global geographical and temporal comparisons would be possible based on the results obtained with the standard panel. The ability to compare results from different geographical areas was considered paramount to follow the spread of resistant strains so that control strategies could be modified to control outbreaks of gonorrhea caused by a specific strain. An international network among laboratories is being established to exchange information relevant to the spread of resistant strains of *N. gonorrhoeae*, an effort that can be accomplished successfully only if strain data is comparable.

Swedish investigators have used the previously developed monoclonal antibodies (50,52), designated GS antibodies, in some studies (53,54). The serovar nomenclature devised by these investigators uses a combination of upper- and lower-case letters (55). The PI molecule expressed by an isolate is designated A or B and individual monoclonal antibody reagents in the PIA and PIB reagent panels have been assigned lower-case letters. The serovar of an isolate that reacts with the IA-specific monoclonal antibody reagents e, d, i, and h, is *Aedih* (55). Recently, Swedish investigators have developed a set of PI-specific monoclonal antibodies (56), designated Ph-antibodies, whose reaction patterns generally correspond to those of the GS-antibodies (57). A similar nomenclature has been devised for these reagents. The upper-case letters *A* and *B* designate the PI molecule expressed, and a different set of lower-case letters has been used to designate individual reagents. An isolate belonging to the serovar *Aedih* in the GS system belongs to the serovar *Arst* in the Ph system (57).

This serovar nomenclature is flexible because it does not require modification or creation of a supplemental nomenclature if the panel of monoclonal antibody reagents is changed. Swedish investigators have used different panels of reagent to provide greater resolution within serovars that are predominant in some patient populations (53,54,57). Unfortunately, results from different studies cannot be compared with each other or with those from studies performed by other investigators (24,26). Geographical and temporal changes in gonococcal strain populations cannot be compared unless standard panels of antibody reagents are used.

Although the monoclonal antibodies used in these systems are directed against the gonococcal protein I as demonstrated by radioimmune precipitation studies (50), some cross-reactions have occurred between some monoclonal antibody reagents and strains of *N. meningitidis* and *N. lactamica* (56). It is not apparent whether these cross-reactions are specific reactions between the antibody and antigenically related, or identical, epitopes on strains of these nongonococcal *Neisseria* spp., or whether they result from nonspecific Fc binding. These problems are not important in the use of a serotyping system that is designed to type isolates that have been identified as *N. gonorrhoeae*. However, these observations emphasize the need for careful selection of monoclonal antibodies that are used in commercial reagents for the laboratory identification of *N. gonorrhoeae.*

B. Serological Classification of *Neisseria gonorrhoeae* by Serotypes

The POMP-serotyping system, which uses polyvalent antibodies, divided gonococci into nine serotypes (39). Monoclonal antibody reagents are now used to serotype gonococci, and they recognize epitopes of the serotypes 1, 5, 7, 8, and 9 (58). The serotype 9 has been divided into two subtypes, 9a and 9b (58). The serotype of an isolate is described by the pattern of its reaction with the monoclonal antibody reagents. A strain reacting with the serotype reagents 5, 8, and 9a is designated 5,8,9a (58).

Occasional cross-reactions of isolates with PIA and PIB reagents have been observed (59). Cross-reactions were also noted with some isolates typed previously, but that were lost on subculture of the isolates (52). Recently, however, transformation of PI epitopes between IA and IB strains (60) and between different IB strains (61) have been demonstrated in vitro. It is possible that some naturally occurring serovars of *N. gonorrhoeae*, e.g., IB-5 (61) may have evolved by genetic transformation in vivo.

C. Comparison between Serovars and Serotypes

Results obtained using the different serological classification systems for *N. gonorrhoeae* show a good correlation because they were developed based on the same research. The serovar systems permit a greater resolution among gonococcal strains than does the serotyping system. There are advantages and disadvantages in both systems.

The high degree of resolution of the serovar systems permits detailed studies of gonococcal strain populations. The relationships between different serovars are not clear. Thus, it is not readily apparent how the serovars can be grouped for less detailed analyses. There are several serovars that may be grouped. For example, the serovars IA-1 and IA-2 were grouped because it was shown that the reaction of AHU isolates with the reagent 4A12 was not always reproducible (52). This reagent was retained in the standard panel, however, because it reacted strongly with isolates from West Africa (52). Similarly, some isolates belonging to the serovar IB-3 have shown weak reactions with the reagent 2D6 making them difficult to differentiate from IB-1 isolates (62). Although it might be suggested that the serovars IB-1 and IB-3 be grouped, other IB-3 isolates clearly do not react with reagent 2D6, suggesting that this grouping would not be appropriate. For the purpose of comparing serovars with other characteristics, e.g., antimicrobial susceptibilities, isolates that react in different combinations with the IB-specific monoclonal antibody reagents 3C8, 1F5, 2D6, and 2H1, i.e., including the serovars IB-1, IB-3, and IB-2, might be grouped for less detailed analyses. It is clear that, in instances when less detailed analyses of strains are required, the use of the less discriminatory serotyping system may permit a more practical grouping of antigenically related serovars than can be accomplished with the serovar systems.

VIII. AUXOTYPE/SEROVAR CLASSIFICATION OF *NEISSERIA GONORRHOEAE* ISOLATES

Serological classification has been used as an alternative to auxotyping (53,54) because it is more economical and practical for use in many laboratories. However, used by itself, serological classification lacks the discriminatory power to differentiate between gonococcal isolates that was previously recognized as a limitation of auxotyping.

The use of a dual classification system that measures two independent phenotypic characteristics that are stable in vitro provides greater resolution among gonococcal isolates than a system that measures only one phenotypic characteristic, e.g., nutritional type or serovar. Different approaches have been adopted for the dual classification of gonococcal isolates.

An auxotype/serovar (A/S) classification system has been proposed for characterizing gonococcal isolates (52). Serological classification permits considerable resolution among strains that belong to the Proto and Pro⁻ auxotypes internationally and within individual communities (26,52). In the A/S classification system, an isolate that requires proline and belongs to the serovar, IB-4, belongs to the A/S class, Pro⁻/IB-4. The A/S classification has been used alone (24, 26,59,62,63) and in conjunction with plasmid profiles (63-66) and antimicrobial susceptibilities, to perform detailed analyses of gonococcal strain populations (67).

A dual serological classification system has also been recently proposed (68) in which isolates would be tested with both GS- and Ph-antibodies and described by a dual serovar nomenclature that combines the reaction patterns with both panels of monoclonal antibody reagents. Thus, an isolate reacting with the IB GS reagents *a, c,* and *k* (50) and the Ph reagents *r, o, p, y, s,* and *t* (57) would be named *Back/Bropyst* (68).

IX. STUDIES OF *NEISSERIA GONORRHOEAE* STRAIN POPULATIONS

Detailed characterization of gonococcal isolates is important in following changes in gonococcal strain populations that may affect the laboratory diagnosis of the organism and the control of gonorrhea. Serological typing techniques using monoclonal antibodies in conjunction with auxotyping and plasmid typing have permitted detailed analysis of gonococcal strain populations. The distribution of gonococcal strains in different countries and in individual communities, the association of specific strains with specific syndromes, the association of strains with different patient populations, and the epidemiology of gonorrhea caused by strains of *N. gonorrhoeae* with chromosomal- and plasmid-mediated resistance to penicillin (PPNG) and tetracycline (TRNG) have all been studied by using phenotypic-typing techniques.

A. The Diversity of *Neisseria gonorrhoeae* Strain Populations

A total of 107 different A/S classes were identified among more than 1400 strains from countries in Europe; Australasia; Africa; and North, Central, and South America (52). Detailed studies of strain populations have revealed additional A/S classes that were not found in previous studies (26,59,67). It is estimated that more than 200 different A/S classes have now been recognized.

Isolates requiring arginine, hypoxanthine, and uracil (AHU), which previously had been associated with DGI and asymptomatic gonorrhea in some centers in Scandinavia and the United States (10-13), belonged to the serovar, IA-1, 2 (52). The AHU/IA isolates were not isolated before 1946 in Denmark, where they were common in the mid 1970s (21,64). These studies have supported previous studies (69) that suggested that the AHU isolates in different geographical areas have resulted from the spread of a single clone. The origin of these strains, however, remains a mystery.

Isolates requiring proline, arginine, and uracil (PAU) accounted for as many as 40% of isolates in Ontario in the mid 1970s (7) and have been associated with asymptomatic gonorrhea in Winnipeg (24). Isolates of this auxotype have belonged most frequently to closely related serovars, IB-1, IB-2, IB-10, and IB-16 (24,26,63). Serovars IB-2 and IB-16 are probably variants of the same strain for PAU isolates but not for other auxotypes. Thus, the PAU isolates are more diverse than was previously recognized on the basis of auxotype alone.

Proto and Pro$^-$ isolates were differentiated into many A/S classes (26,52). Thus, serological classification has greatly facilitated subdivision within these auxotypes. Proto and Pro$^-$ isolates belonging to some A/S classes have been widely distributed in different geographical areas, whereas others have been limited geographically (24,26,59,65).

B. Geographical Distribution and Temporal Changes in Gonococcal Auxotype/Serovar Classes

Detailed studies of the geographical distribution and temporal changes in gonococcal A/S classes have been made. Among strains isolated from different countries in 1975-1977 (70,71), AHU/IA strains were isolated frequently only in northern Europe and certain parts of the United States. The AHU/IA isolates have now decreased in frequency in Seattle, DesMoines, and Copenhagen, where they once accounted for more than 50% of all isolates (26,63,64). The frequency of isolation of AHU/IA strains has also decreased in Heidelberg, West Germany between 1981 and 1986 (71).

The PAU strains that were geographically limited in the mid-1970s (7,70) have become widespread during the last 10 years. In Seattle and DesMoines, the PAU/IB strains that were not isolated in the mid-1970s now account for approximately 7-10% of all isolates (26,63,67). Isolates requiring proline, hypoxanthine, and uracil (PHU) that also belonged to IB serovars have also emerged (26) as have PAHU/IB strains (26). The PAHU/IB strains are also less susceptible to penicillin and tetracycline than are the AHU/IA strains (24,67). It has been suggested that these strains are hypoxanthine-requiring variants of PAU/IB strains (24). It is also possible that these strains may be arginine-requiring variants of the PHU/IB strains.

Proto and Pro^- strains belonging to some serovars were widely distributed, whereas others were distributed in more limited geographical areas. Strains belonging to the IA serovars IA-5 and IA-9 were isolated more frequently in African countries than elsewhere (70). Strains belonging to the IB serovars IB-5 and IB-7 were isolated more frequently in the Far East than in other areas (70). Longitudinal studies of gonococcal serovars (53) and A/S classes (26,67) have shown that the gonococcal strain populations in communities may constantly undergo change.

C. Distribution of Gonococcal Strains in a Community

The A/S classification has been used to study gonococcal strain populations in which PPNG strains are rarely isolated (26,63,64,71). The distribution of gonococcal strains may change over short periods (26). Gonococcal strain populations may be composed of many different A/S classes. In Seattle, isolates belonging to 57 A/S classes were isolated in January-March 1985 (26). Isolates belonging to each of six to eight A/S classes persisted in 5% of patients during this 3-month period (26). In contrast, isolates belonging to most A/S classes were "transient," i.e., they were isolated from few patients. Isolates belonging to some A/S classes were isolated only from heterosexual patients, and others were isolated only from homosexual men (26), suggesting that different strains may be spread within subgroups of the total population at risk for contracting gonorrhea. A subsequent study of strains isolated in September 1985-February 1986 (72) showed that some A/S classes that had previously been isolated from only heterosexual patients in early 1985 were isolated from homosexual men in 1986. It is to be expected that strains of various A/S classes may be spread between different patient populations with time. The dynamic nature of gonococcal strain populations appears to be a general phenomenon, rather than limited to some geographical areas (26,67,73) and may have important implications for the control of gonorrhea.

It has been hypothesized that strains may become preponderant in a community by a variety of methods (26). A strain may be spread to many sex partners in a short period, i.e., by "high-frequency transmission," and rapidly become preponderant as the result of a single introduction.

In contrast to high-frequency transmission, a strain may be frequently introduced into a community by many persons residing in, or visiting a nearby community where the strain is prevalent. Even if the strain is transmitted to only a few individuals, i.e., by "low-frequency transmission," it may become preponderant in direct proportion to the frequency of its introduction. Although the strain may become preponderant in terms of the total number of cases of gonorrhea attributable to it, there may be no epidemiological connection between the many separate "introductions." Thus, traditional contact tracing efforts to control such an outbreak of gonorrhea may be ineffective in eliminating the strain from the community. Attributes such as antimicrobial resistance may influence the ability of a strain to become preponderant in a community, even if it is introduced rarely from a distant locale. Epidemics caused by strains with chromosomally mediated resistance or plasmid-mediated resistance have also become preponderant in communities as a result of their resistance to therapeutic agents (74-76).

D. Antimicrobial Resistance in *Neisseria gonorrhoeae*

Antimicrobial resistance in *N. gonorrhoeae* has become an increasing problem for gonorrhea control efforts. Antimicrobial resistance in *N. gonorrhoeae* has been associated with both chromosomally mediated (74) and plasmid-mediated resistance (76).

Most strains of *N. gonorrhoeae* are still susceptible to penicillin and tetracycline. However, since the mid-1970s strains have become increasingly resistant to tetracycline when this antibiotic was used as the first-line therapy for gonorrhea in areas where β-lactamase-producing gonococci were more likely to be introduced (67,77). Strains have become resistant to penicillin and tetracycline, and to the first- and second-generation cephalosporins (67,74). A strain that belonged to the A/S class, Pro/IB-1, with chromosomally mediated resistance to penicillin, tetracycline, erythromycin, and cefoxitin, was associated with an epidemic of gonorrhea in North Carolina (74). Subsequent surveillance for strains with chromosomal resistance to multiple antibiotics revealed that strains belonging to several A/S classes exhibited chromosomal resistance to antimicrobial agents (67,75). Thus, the spread of gonorrhea caused by resistant strains is due, not to the spread of a single strain, but to the spread of several different strains. Chromosomal resistance to spectinomycin has also emerged in *N. gonorrhoeae*. Spectinomycin-resistant strains were first isolated in Korea (78) and have since been isolated in England (79) and the United States where spectinomycin-resistant strains, belonging to the A/S class Pro$^-$/IB-1, were isolated in nine different cities (62). These isolates appeared to belong to a single clone that also had chromosomal resistance to several antibiotics and possessed the 24.5-Md conjugative plasmid (62). In addition, a spectinomycin-resistant strain belonging to the A/S class, Proto/IB-5, which was susceptible to other antibiotics was also isolated (62). Although epidemiological links to Korea were established in some cases, others were not linked to foreign travel (62). The emergence of spectinomycin-resistance among strains isolated in the United States is of concern, because it may limit the effectiveness of this antibiotic as an alternative therapy for the control of gonorrhea caused by strains that are resistant to other therapeutic agents.

Auxotype/serovar classification has been used in conjunction with plasmid profiles to make detailed studies of the PPNG strain populations in Miami, Florida, where PPNG infections account for approximately 25-30% of all cases of gonorrhea. Between 1983 and 1986, there has been a shift in the plasmid profile and the A/S classes to which PPNG strains belong (80). In 1983, approximately 90% of PPNG strains possessed the 3.2-Md β-lactamase plasmid, and few possessed the 24.5-Md plasmid. By late 1984, approximately 90% of PPNG strains possessed the 4.4-Md plasmid, and 36% of these also possessed the 24.5-Md plasmid. In January-April 1986, approximately 90% of the PPNG isolates again possessed the 3.2-Md β-lactamase plasmid, and approximately one-third of these isolates also possessed the 24.5-Md conjugative plasmid (80). The PPNG strain population in Miami in each study period was composed of a few dominant strains and many transient strains (80).

Recently, plasmid-mediated, high-level resistance (MICs $\geqslant$ 16.0 μg/ml) to tetracycline has been found in *N. gonorrhoeae* (TRNG) (81). Tetracycline resistance in TRNG strains appears to have resulted from the acquisition of the TetM determinant which became integrated in the 24.5-Md conjugative plasmid, resulting in a 25.2-Md conjugative plasmid (81). The TRNG isolates have belonged to many A/S classes, although $\geqslant$ 50% of TRNG strains have belonged to the Pro$^-$/IB-1 class (82). Genetic studies have shown that the 25.2-Md plasmid may be transmitted between gonococci and related species either by transformation or conjugation (81,83). The TRNG strains have been reported only in the United States (81,82), Canada (84,85), the Netherlands (86), and England (C. Ison, personal communication), but it must be anticipated that these strains will gradually be spread to other countries.

E. Association of Gonococcal Auxotype/Serovar Classes with Disseminated Gonococcal Infection and Pelvic Inflammatory Disease

Previously, it was shown that AHU and non-AHU isolates that were highly susceptible to penicillin and serum-resistant were associated with DGI (9–14) and that many also belonged to the WI serogroup (41) or IA serovars (87). Recently, however, fewer isolates from patients with DGI have belonged to either the AHU auxotype or the IA serovars (41). The frequency of the AHU/IA isolates has decreased from approximately 50% of all isolates in 1974–1975 to approximately 15% of all isolates in 1985 (26). Thus, although certain strains may have a greater propensity to cause DGI, a clear correlation between a specific A/S class and DGI may not be clearly demonstrable in individual communities unless isolates of that class account for a large proportion of isolates that cause uncomplicated gonorrhea in that community. Isolates from patients with PID have not been extensively characterized. However, no clear associations between individual strains and PID have been found.

The AHU strains were associated with asymptomatic infections in Seattle in the mid-1970s (15). Recently, an association between isolates belonging to the PAU/IB serovars and asymptomatic infections has been observed in Winnipeg, Canada (24). In both of these centers, it should be noted that the isolates associated with asymptomatic gonorrhea each accounted for $\geqslant$ 20% of all isolates in the respective communities at the time the correlation was observed.

F. Forensic Applications

Strains isolated from sexual contacts can be compared by A/S classification. The A/S classification has been used to compare strains from victims and alleged assailants in medicolegal cases (70). The isolates must belong to the same A/S class to be used as circumstantial evidence against the alleged assailant. However, it is important that the epidemiological aspects of the cases are managed carefully. In child abuse, all adults who have had private access to the child must be cultured to verify that only the alleged assailant is infected with a strain identical with that isolated from the victim.

G. Limitations of Auxotype/Serovar Classification

The use of A/S classification alone, or in conjunction with plasmid profiles and antimicrobial susceptibility testing, has permitted more detailed analyses of gonococcal strain populations. However, it must be recognized that this classification has some limitations.

For example, isolates belonging to the Pro$^-$/IB-1 class have been associated with an outbreak of resistant gonorrhea in North Carolina, with spectinomycin resistance, and with plasmid-mediated, high-level tetracycline resistance. These isolates have different antimicrobial susceptibility patterns and plasmid profiles which suggest that, although they belong to the same A/S class, they may not be the same strain.

Studies of the arginine biosynthetic pathway (88) and the genetics of the proline requirement (89) have shown that different mutations may result in the same auxotype. Intermediates in the arginine pathway have been used to indicate whether the arginine requirement of an organism can be satisfied either by ornithine or by citrulline (7). However, the nature of the proline requirement can be elucidated only by use of genetic techniques that are not practical for routine typing of gonococci (89).

Similarly, it cannot always be assumed that strains belonging to the same serovar that are isolated in different parts of the world are the same strain. The monoclonal antibodies react with specific epitopes on the PI molecule. These epitopes may be located in different arrangements on the PI but, phenotypically, the strains belong to the same serovar. This must be remembered when data from different geographical areas are interpreted. When it is critical to know the relationship between isolates beloning to the same A/S class, additional characterization of isolates by sophisticated techniques, such as restriction enzyme analyses, may be necessary.

X. LABORATORY DIAGNOSIS OF GONORRHEA WITH USE OF MONOCLONAL ANTIBODIES

The laboratory identification of *N. gonorrhoeae* is dependent on differentiating between strains of this species and closely related species that resemble the gonococcus in cultural and morphological characteristics.

A. Bacteriology

Neisseria gonorrhoeae belongs to the genus *Neisseria* which contains 11 species and biovars that have been isolated from man (90). *Neisseria* spp. and closely related species belonging to the family *Neisseriaceae, Branhamella catarrhalis* and *Kingella denitrificans*, are normal inhabitants of the oro- and nasopharynx and are rarely isolated from other sites infected by the gonococcus (91,92).

Species of the family *Neisseriaceae* are gram-negative, oxidase-positive organisms (90). Strains of the *Neisseria* spp. and *B. catarrhalis* are diplococci. Strains of *K. denitrificans* are coccobacilli, but often appear to be diplococci in Gram-stained smears. Thus, *K. denitrificans* must be considered when a gram-negative, oxidase-positive diplococcus is isolated because this species will grow on selective medium for the gonococcus and produces acid from glucose (91,90). Strains of many species, *N. subflava* biovars *subflava, flava*, and *perflava*; *N. sicca*; *N.*

mucosa; and *B. catarrhalis* produce pigmented or opaque colonies and are easily distinguished from *N. gonorrhoeae*. In contrast, strains of *N. meningitidis, N. lactamica, N. cinerea* (93), *N. polysaccharea* (94), and *K. denitrificans* (91) produce translucent, nonpigmented colonies that closely resemble *N. gonorrhoeae* on an isolation medium. Strains of most *Neisseria* spp. are also easily distinguished from *N. gonorrhoeae* by their ability to produce acid from maltose and other carbohydrates (90). The species that are most difficult to distinguish from *N. gonorrhoeae* by biochemical characteristics are *N. cinerea, B. catarrhalis*, and *K. denitrificans.*

B. Clinical Diagnosis of Gonorrhea

The laboratory identification of *N. gonorrhoeae* from a specimen confirms a clinical diagnosis of gonorrhea. The difficulties in making the laboratory identification, and the implications of this identification, vary according to the patient from whom the specimen was collected, the type of health care sought, the medium on which the specimen is cultured, and the procedures used to identify the organism as *N. gonorrhoeae*.

Patients at high risk for gonorrhea attend STD clinics specifically for the diagnosis of STDs. The laboratory diagnosis of gonorrhea in sexually active patients may be based on a "presumptive" or a confirmed laboratory diagnosis of *N. gonorrhoeae*. A diagnosis of gonorrhea in symptomatic men with urethritis may be made on a presumptive laboratory diagnosis of *N. gonorrhoeae*. A presumptive laboratory diagnosis is based on the observation of intracellular gram-negative diplococci in polymorphonuclear leukocytes in a Gram-stained smear of urethral exudate, or the growth of a gram-negative, oxidase-positive diplococcus on a selective medium for *N. gonorrhoeae*. The Gram stain has a positive predictive value of >95% in men with urethritis, but it is less sensitive in women with cervicitis and may detect only 40-50% of infections that are culture-positive (95). Consequently, the ideal method for the laboratory diagnosis of gonorrhea in women with cervicitis and from all patients with rectal or pharyngeal infections is the isolation and identification of *N. gonorrhoeae* (95).

Specimens for the confirmed identification of *N. gonorrhoeae* in high-risk patients are plated directly on a selective medium that enhances the isolation of colistin-resistant *Neisseria* spp., including *N. gonorrhoeae, N. meningitidis, N. lactamica, K. denitrificans*, and some strains of the sucrose-positive commensal *Neisseria* spp., *N. subflava* biovar *perflava* (96). Some strains of *B. catarrhalis* are also less susceptible to colistin and may grow on a gonococcal selective medium (97). Although strains of *N. cinerea* are colistin-susceptible and will not normally grow on a selective medium, some strains have been isolated from clinical specimens inoculated on selective medium (93).

Patients at low risk for gonorrhea, e.g., children, may be examined in hospital clinics, private doctors offices, and emergency rooms. Except when sexual abuse is suspected and specimens are cultured on selective medium for the gonococcus, low-risk patients are not being examined specifically for STDs. For example, throat or conjunctival specimens may be inoculated on a nonselective medium, such as blood or chocolate agar, as a routine procedure in the diagnosis of a sore throat or conjunctivitis. All *Neisseria* and related species can grow on these media. The presumptive criterion of growth of a gram-negative, oxidase-positive diplococcus cannot be used to diagnose gonorrhea in low-risk patients. When intracellular, gram-negative diplococci are observed in Gram-stained smears or gram-negative, oxidase-positive diplococci are isolated from a specimen, the organism must be identified using confirmatory tests.

C. Laboratory Identification of *Neisseria gonorrhoeae*

In a reference laboratory, isolates of *N. gonorrhoeae* are identified using the following tests: cultural characteristics; growth on gonococcal selective medium (colistin-resistance); production of acid from glucose, maltose, sucrose, fructose, and lactose; reduction of nitrate and nitrite; production of polysaccharide from sucrose; detection of deoxyribonuclease; growth on nutrient agar at 35°C; growth at 22°C; production of catalase (95). Many of these tests must be incubated for 24-48 hr before results can be reported. Although isolates from urethral, cervical, and pharyngeal specimens may be isolated within 24-48 hr, pure cultures from rectal specimens may not be obtained for several days because of contamination of cultures by *Proteus* spp. Thus, a minimum of 48-72 hr may elapse before the laboratory results may be reported to the clinician.

Rapid methods have been developed for the confirmed identification of *N. gonorrhoeae*. These methods include carbohydrate utilization, chromogenic substrates to detect specific enzymes, and serological tests that are designed to permit rapid identification of an isolate. The rapid carbohydrate and chromogenic substrate tests must be inoculated with a pure culture of the organism, and test results may be interpreted within 30 min or 4 hr depending on the test. Because a pure inoculum is required for these tests, the isolate will not be identified until 24 hr after the strain has been isolated.

D. Serological Tests for the Laboratory Identification of *Neisseria gonorrhoeae*

Serological tests that use monoclonal antibodies in coagglutination or fluorescent antibody (FA) tests to identify *N. gonorrhoeae* in primary cultures are commercially available. Monoclonal antibody reagents in these tests generally contain a cocktail of antibodies directed against gonococcal protein IA and IB epitopes. Because these tests can be performed on colonies on the primary isolation plate and do not require the isolation of a pure culture, an isolate may be identified 24 hr earlier than is possible using the rapid carbohydrate or enzyme-substrate tests.

Coagglutination tests for the identification of *N. gonorrhoeae* are the GonoGen (New Horizons Diagnostics, Cockeysville, Maryland) and the Phadebact Monoclonal GC OMNI reagent (Pharmacia, Rahway, New Jersey). The format and test procedures for the coagglutination tests vary. An FA reagent, the Syva Microtrak *Neisseria gonorrhoeae* Culture Confirmation Test, using monoclonal antibodies (SYVA, Palo Alto, California) is also commercially available.

The GonoGen test is performed on glass slides. A suspension of the isolate is prepared to a density equivalent to a MacFarland 3 (10^8 organisms/ml) standard and is heated in a boiling waterbath for 10 min. A drop of the cooled suspension is spread in marked wells on a glass slide with a negative control and a test reagent, respectively. The slide is rotated manually or on a rotating table for 2 min and the agglutination reaction is read in oblique light against a black background. Positive and negative control suspensions should be run with each set of test suspensions. The reaction mixture in the control well should show no agglutination; the result for the test suspension is invalid if agglutination is observed in the control well. If the control well shows no reaction, a positive reaction in the test well confirms that the isolate is *N. gonorrhoeae*.

The Phadebact Monoclonal OMNI test is performed on coated white cards. The reagents contain methylene blue to give a blue agglutinin that can be read against the white background of the card. A suspension of the test organism is prepared to an optical density equivalent to a 0.5 McFarland standard in 0.7% saline and heated in a boiling waterbath for 5 min. A drop of the suspension (cooled to room temperature) is mixed with a drop of a control or test reagent in marked areas on the white card, respectively. The slide is rotated for 1 min, and the agglutination reaction is recorded.

The FA test is performed as follows: A very thin smear is prepared by mixing cells from five colonies presumptively identified as gonococci with 5 μl of distilled or deionized water. The smear is allowed to air dry and is then heat fixed. The smear is then immediately stained with the reagent by placing 30 μl of the *N. gonorrhoeae* reagent on the entire smear. The slide is then incubated at 37°C for 15 min. Excess reagent is removed from the slide without disturbing the smear, which is then rinsed for 5-10 sec with a gentle stream of distilled or deionized water. The smears are then air dried and sealed under a coverslip with mounting fluid. The smear is examined for fluorescent apple-green cells of kidney-shaped diplococci characteristic of *N. gonorrhoeae*. Cells of nongonococcal *Neisseria* spp. should be visible, but will not stain. Comparisons should be made between positive and negative control slides.

The specificities and sensitivities of monoclonal antibody diagnostic reagents have generally been high when they have been used to identify isolates from high-risk patients, i.e., in primary cultures on gonococcal selective medium (98-109). The specificities and sensitivities of the coagglutination reagents quoted for identifying isolates from high-risk patients should not be extrapolated to low-risk patient populations. Some problems have been documented with all reagents. Cross-reactions of the coagglutination reagents with *N. meningitidis* and *N. lactamica* have been reported for both the GonoGen and Phadebact OMNI reagents (102; unpublished observations). In addition, cross-reactions of these reagents with strains of *N. cinerea* and *K. denitrificans* have also been observed (unpublished data). It has also been observed that some *N. gonorrhoeae* isolates failed to react with the GonoGen reagent (102,109); many of these isolates belonged to the IA-4 serovar. The distribution and frequency of isolates belonging to this serovar may vary geographically and temporally. Thus, it is not possible to predict either the geographical areas that may encounter this problem or the proportion of nonreactive isolates that may be encountered.

Difficulties have been encountered with the interpretation of the Phadebact OMNI reagent test. After cross-reactions with nongonococcal isolates were observed, it was recommended that suspensions be prepared to an optical density equivalent to a 0.5 MacFarland standard and that the saline in which the suspensions were prepared be pH 7.0-7.5. Carlson et al. (106) recommended that the saline at pH 7.4 was optimal for the preparation of suspensions. It was further recommended that reactions of $\leqslant$ + + be interpreted as equivocal and that the identity of organisms giving this reaction be confirmed by other tests. We have noted that some *N. gonorrhoeae* isolates gave equivocal reactions under these test conditions (unpublished observations). At the time of writing, the guidelines for the interpretation of this test have not been clarified.

Nongonococcal *Neisseria* spp. have not been observed to react with the Microtrak FA reagent (105,107,109). Occasionally, however, strains of *N. gonorrhoeae* have not reacted in the FA test (108). Nonspecific Fc binding of the FA reagent to other bacterial species, including some strains of *Staphylococcus aureus*, has been noted and has resulted in a notification that consumers should ensure that only gram-negative, oxidase-positive organisms isolated from gonococcal selective medium be tested with the Microtrak reagent. Similarly, it is possible that cross-reactions between the coagglutination reagents and nongonococcal isolates may also result from Fc binding of the reagent to other than *N. gonorrhoeae.*

Some problems associated with the incorrect identification of nongonococcal isolates as *N. gonorrhoeae* have often resulted from the incorrect use of a product. It is important that consumers read the manufacturer's instructions carefully and observe the limitations recommended for the use of the product. It is important that testing be limited to gram-negative, oxidase-positive diplococci. The use of these products is not limited to testing only isolates that have grown on nonselective medium. Test results must be interpreted with caution when the isolate has been grown from the oropharynx which is the normal habitat for most *Neisseria* spp. (92). Strains of *N. lactamica* are rarely isolated from adults (92), but are frequently isolated from children (92,110). It is estimated that 95% of all children have been colonized with *N. lactamica* by the age of 4 years

(110). Strains of *N. meningitidis* may also be carried as normal flora in the oropharynx of adults and children (110,111). *Neisseria cinerea* has been isolated frequently from the oropharynx of both adults and children (92; unpublished data) and, occasionally, strains have cross-reacted weakly with monoclonal antibody reagents (unpublished data). In instances that may have medicolegal implications, e.g., the isolation of *N. gonorrhoeae* from children, it is important that the isolate be purified and the identification confirmed by using several tests based on different principles before the isolate is confirmed as *N. gonorrhoeae*; the isolate should be preserved for further study (112).

It is important that the growth characteristics of the organisms are considered when the results of serological diagnostic tests are interpreted. When the cultural characteristics and the serological test results are not consistent with an identification of *N. gonorrhoeae*, the isolate should also be tested in procedures that use a different diagnostic principle, e.g., acid production or enzyme production.

It is important that monoclonal antibody reagents be evaluated for local stain populations and that the limitations described in the preceding discussion be considered when choosing a serological reagent for diagnostic purposes or making an identification that may have medicolegal implications.

E. Serotyping/Serogrouping Tests

The serotyping or serogrouping tests are available in the GonoType and Phadebact Monoclonal GC tests. The format and test procedures are similar to those for the diagnostic tests described previously. These tests may also be used for the identification of isolates.

The GonoType test divides gonococci into serotypes using six individual reagents, 1, 5, 7, 8, 9a, and 9b. The serotype of an isolate is the reaction pattern of the isolate with these reagents, i.e., an isolate that reacts with the reagents 5, 7, and 8 belongs to the serotype 5,7,8. Isolates that react with reagent 1 will generally not react with the other reagents and vice versa (113). The Phadebact Monoclonal GC test divides isolates into the serogroups WI and WII/III. This test does not use a negative control reagent, but uses the WI-specific reagent as a negative control for the WII/III-specific reagent and vice versa.

The results obtained using these products must be interpreted with caution. Considering the diversity within the WI and WII/III serogroups, it is difficult to make more than the simplest correlations between serogroup and other characteristics such as β-lactamase production, antimicrobial resistance, or syndrome. The GonoGen typing reagents permit greater resolution among gonogoccal isolates and, thus, more legitimate correlations may be made between serotype and other variables. It must be kept in mind, however, that, just as serotyping permitted resolution among auxotypes, serotypes without additional phenotypic characterization may result in misleading interpretations of epidemiological data.

XI. CONCLUSION

The use of monoclonal antibodies for the laboratory identification of *N. gonorrhoeae* must be limited to those antibodies that are specific, but broadly cross-reactive, and are directed against surface-exposed epitopes on the gonococcus. These epitopes must also be stable in vivo and in vitro. Studies performed at this time have demonstrated that monoclonal antibodies that are directed against epitopes on the gonococcal PI molecules have proved most useful for both the laboratory diagnosis of *N. gonorrhoeae* and a practical characterization of gonococcal strains to elucidate the epidemiology of gonococcal infections.

REFERENCES

1. Centers for Disease Control. Cases of specified notifiable diseases, United States, weeks ending January 2, 1988 and December 26, 1986. *Mortal. Morb. Week. Rep. 36*:841 (1988).
2. Hook, E. W., III, and K. K. Holmes, Gonococcal Infections. *Ann. Intern. Med. 102*:229–243 (1985).
3. Odugbemi, T., An open evaluative study of sulbactam/ampicillin with or without probenecid in gonococcal infections in Lagos. *Curr. Ther. Res. 41*:542–551 (1987).
4. Wiesner, P. J., H. H. Handsfield, and K. K. Holmes, Low antibiotic resistance of gonococci causing disseminated gonococcal infection. *N. Engl. J. Med. 288*:1221–1222 (1973).
5. Catlin, B. W., Nutritional profiles of *Neisseria gonorrhoeae, Neisseria meningitidis*, and *Neisseria lactamica* in chemically defined media and the use of growth requirements for gonococcal typing. *J. Infect. Dis. 128*:178–194 (1975).
6. Short, H. B., V. B. Ploscowe, J. A. Weiss, and F. E. Young, Rapid method for auxotyping multiple strains of *Neisseria gonorrhoeae. J. Clin. Microbial. 6*:244–248 (1977).
7. Hendry, A. T., and I. O. Stewart, Auxanographic grouping and typing of *Neisseria gonorrhoeae. Can. J. Microbiol. 25*:512–521 (1979).
8. Hendry, A. T., and J. R. Dillon, Growth inhibition of *Neisseria gonorrhoeae* isolates by phenylalanine and its analogues in defined media. *Can. J. Microbiol. 30*:1319–1325 (1984).
9. Schoolnik, G. K., T. M. Buchanan, and K. K. Holmes, Gonococci causing disseminated gonococcal infection are resistant to the bactericidal action of normal human sera. *J. Clin. Invest. 58*:1163–1173 (1976).
10. Knapp, J. S., and K. K. Holmes, Disseminated gonococcal infections caused by *Neisseria gonorrhoeae* with unique growth requirements. *J. Infect. Dis. 132*:204–208 (1975).
11. Morello, J. A., S. A. Lerner, and M. Bohnhoff, Characteristics of atypical *Neisseria gonorrhoeae* from disseminated and localized infections. *Infect. Immun. 13*:1510–1516 (1976).

12. Eisenstein, B. I., T. J. Lee, and P. F. Sparling, Penicillin sensitivity and serum resistance are independent attributes of strains of *Neisseria gonorrhoeae* causing disseminated gonococcal infections. *Infect. Immun. 15*: 834–841 (1977).
13. Thompson, S. E., G. Reynolds, H. B. Short, C. Thornsberry, J. W. Biddle, N. F. Jacobs, M. F. Rein, A. A. Zaidi, F. E. Young, and J. A. Shulman, Auxotypes and antibiotic susceptibility patterns of *Neisseria gonorrhoeae* from disseminated and local infections. *Sex. Transm. Dis. 5*:127–131 (1978).
14. Turgeon, P. L., and M. J. Granger, Auxotypes of *Neisseria gonorrhoeae* isolated from localized and disseminated infections in Montreal. *Can. Med. J. 123*:381–384 (1980).
15. Crawford, G., J. S. Knapp, J. Hale, and K. K. Holmes, Asymptomatic gonorrhea in men: Caused by gonococci with unique nutritional requirements. *Science 196*:1352–1353 (1977).
16. Mirrett, S., L. B. Reller, and J. S. Knapp, *Neisseria gonorrhoeae* strains inhibited by vancomycin in selective media and correlation with auxotype. *J. Clin. Microbiol. 14*:94–99 (1981).
17. Knapp, J. S., C. Thornsberry, G. A. Schoolnik, P. J. Wiesner, K. K. Holmes, and the Cooperative Study Group, Phenotypic and epidemiologic correlates of auxotype in *Neisseria gonorrhoeae. J. Infect. Dis. 138*:160–165 (1978).
18. LeFaou, A., I. Guy, and J.-Y. Riou, Auxotypes et sensibilité a 6 antibiotiques des souches de *Neisseria gonorrhoeae* isolées a Strasbourg en 1977–78. *Ann. Dermatol. Venerol. (Paris) 106*:267–277 (1979).
19. Knapp, J. S. Typing of gonococci. In *Gonococcal Infection*, G. F. Brooks and Donegan, E. A. (eds.). Edward Arnold, London, pp. 159–167 (1985).
20. Kohl, P. K., J. S. Knapp, H. Hofmann, K. Gruender, D. Petzoldt, M. R. Tam, and K. K. Holmes, Epidemiological analysis of *Neisseria gonorrhoeae* in the Federal Republic of Germany by auxotyping and serological classification using monoclonal antibodies. *Genitourin. Med. 62*:145–150 (1986).
21. Knapp, J. S., and K. K. Holmes, Unique strains of *Neisseria gonorrhoeae* causing epidemic gonorrhea during the penicillin era; a study of isolates from Denmark. *J. Infect. Dis. 154*:363–366 (1986).
22. Handsfield, H. H., J. S. Knapp, P. K. Diehr, and K. K. Holmes, Correlation of auxotype and penicillin susceptibility of *Neisseria gonorrhoeae* with sexual preference and clinical manifestation of gonorrhea. *Sex. Transm. Dis. 7*:1–5 (1980).
23. Dillon, J. R., and M. Pauzé, Relationship between plasmid content and auxotype in *Neisseria gonorrhoeae* isolates. *Infect. Immun. 33*:625–628 (1981).
24. Brunham, R. C., F. Plummer, L. Slaney, F. Rand, and W. DeWitt, Correlation of auxotype and protein I type with expression of disease due to *Neisseria gonorrhoeae. J. Infect. Dis. 152*:339–343 (1985).
25. Noble, R. C., R. R. Reyes, M. C. Parekh, and J. V. Haley, Incidence of disseminated gonococcal infection correlated with the presence of AHU

auxotype of *Neisseria gonorrhoeae* in a community. *Sex. Transm. Dis. 11*: 68–71 (1984).

26. Knapp, J. S., K. K. Holmes, P. Bonin, and E. W. Hook III, Epidemiology of gonorrhea: Distribution and temporal changes in auxotype/serovar classes of *Neisseria gonorrhoeae. Sex. Transm. Dis. 14*:26–32 (1987).
27. Centers for Disease Control, Penicillinase-producing *Neisseria gonorrhoeae. Mortal. Morbid. Week. Rep. 25*:261 (1976).
28. Phillips, I., β-lactamase-producing, penicillin-resistant gonococcus. *Lancet 2*:656–657 (1976).
29. Ashford, W. A., R. G. Golash, and V. G. Hemming, Penicillinase-producing *Neisseria gonorrhoeae. Lancet 2*:657–658 (1976).
30. Roberts, M., P. Piot, and S. Falkow, The ecology of gonococcal plasmids. *J. Gen. Microbiol. 114*:491–494 (1979).
31. Van Embden, J. D. A., B. van Klingerin, M. Dessens-Kroon, and L. J. Wijngaarden, Emergence in the Netherlands of penicillinase-producing gonococci carrying "Africa" plasmid in combination with transfer plasmid. *Lancet 1*:938 (1981).
32. Van Embden, J. D. A., M. Dessens-Kroon, and B. van Klingerin, A new β-lactamase plasmid in *Neisseria gonorrhoeae. J. Antimicrob. Chemother. 15*: 247–258 (1985).
33. Yeung, K.-H., J. R. Dillon, M. Páuze, and M. Wallace, A novel 4.9-kilobase plasmid associated with an outbreak of penicillinase-producing *Neisseria gonorrhoeae. J. Infect. Dis. 153*:1162–1165 (1986).
34. Gouby, A., G. Bourg, and M. Ramuz, Previously undescribed 6.6-kilobase R plasmid in penicillinase-producing *Neisseria gonorrhoeae. Antimicrob. Agents Chemother. 29*:1095–1097 (1986).
35. Van Klingerin, D., M. C. Ansink-Schipper, M. Dessens-Kroon, M. Verheuvel, H. Huikeshoven, and R. K. Woudstra, Relationship between auxotype, plasmid pattern and susceptibility to antibiotics in penicillinase-producing *Neisseria gonorrhoeae. J. Antimicrob. Chemother. 16*:143–147 (1985).
36. Ison, C. A., J. Gedney, J. R. W. Harris, and C. S. F. Easmon, Penicillinase-producing gonococci: A spent force? *Genitourin. Med. 62*:302–307 (1986).
37. Wang, S.-P., K. K. Holmes, J. S. Knapp, S. Ott, and D. Kyzer, Immunologic classification of *Neisseria gonorrhoeae* with immunofluorescence. *J. Immunol. 119*:794–803 (1977).
38. Sandström, E. G., and D. Danielsson, Serology of *Neisseria gonorrhoeae*. Classification by coagglutination. *Acta Pathol. Microbiol. Scand. Sect. B. 88*:27–38 (1980).
39. Buchanan, T. M., and J. F. Hildebrandt, Antigen-specific serotyping of *Neisseria gonorrhoeae* characterization based upon principal outer membrane protein. *Infect. Immun. 32*:985–994 (1981).
40. Sandström, E. G., J. S. Knapp, and T. M. Buchanan, Serology of *Neisseria gonorrhoeae*. W-antigen serogrouping by coagglutination and protein I serotyping by enzyme-linked immunosorbent assay both detect protein I antigens. *Infect. Immun. 35*:229–239 (1982).

41. Sandström, E. G., J. S. Knapp, L. B. Reller, S. E. Thompson, E. W. Hook III, and K. K. Holmes, Serogrouping of *Neisseria gonorrhoeae*: Correlation of serogroup with disseminated gonococcal infection. *Sex. Transm. Dis. 11*:77–80 (1983).
42. Knapp, J. S., E. G. Sandström, T. M. Buchanan, C. Thornsberry, and K. K. Holmes, Correlation between coagglutination (COA) serogroup and auxotype, penicillin susceptibility, and serum resistance of *Neisseria gonorrhoeae. Abstr. Annu. Meet. Am. Soc. Microbiol.* p. 280, C104 (1981).
43. Danielsson, D., S. M. Bygdeman, I. Kallings, and E. G. Sandström, Epidemiology of gonorrhea: Serogroup, antibiotic susceptibility and auxotype patterns of consecutive gonococcal isolates from ten different areas in Sweden. *Scand. J. Infect. Dis. 15*:33–42 (1983).
44. Bygdeman, S., Gonorrhoea in men with homosexual contacts. Serogroups of isolated gonococcal strains related to antibiotic susceptibility, site of infection, and symptoms. *Br. J. Vener. Dis. 57*:320–324 (1981).
45. Bygdeman, S., Antibiotic susceptibility of *Neisseria gonorrhoeae* in relation to serogroups. *Acta Pathol. Immunol. Scand. Sect. B 89*:227–237 (1981).
46. Handsfield, H. H., E. G. Sandström, J. S. Knapp, D. E. Sayers, W. L. Whittington, and K. K. Holmes, Epidemiology of penicillinase producing *Neisseria gonorrhoeae* infections. *N. Engl. J. Med. 306*:950–954 (1982).
47. Bygdeman, S., I. Kallings, and D. Danielsson, Serogrouping and auxotyping for epidemiological study of β-lactamase producing *Neisseria gonorrhoeae* strains isolated in Sweden. *Acta Dermatol. Venereol. 61*:329–334 (1981).
48. Sandström, E. G., and D. Danielsson, Serology of *Neisseria gonorrhoeae*. Characterisation of hyperimmune rabbit sera by line rocket immunoelectrophoresis for use in co-agglutination. *Acta Pathol. Immunol. Scand. Sect. B 88*:17–26 (1980).
49. Sandström, E. G., K. C. S. Chen, and T. M. Buchanan, Serology of *Neisseria gonorrhoeae*: Coagglutination serogroups WI and WII/III correspond to different outer membrane protein I molecules. *Infect. Immun. 38*:462–470 (1982).
50. Tam, M. R., T. M. Buchanan, E. G. Sandström, K. K. Holmes, J. S. Knapp, A. W. Siadak, and R. C. Nowinski, Serological classification of *Neisseria gonorrhoeae* with monoclonal antibodies. *Infect. Immun. 36*:1042–1053 (1982).
51. Nowinski, R. C., M. R. Tam, L. C. Goldstein, L. Stong, C.-C. Kuo, L. Corey, W. E. Stamm, H. H. Handsfield, J. S. Knapp, and K. K. Holmes, Monoclonal antibodies for diagnosis of infectious diseases in humans. *Science 219*:637–644 (1983).
52. Knapp, J. S., M. R. Tam, R. C. Nowinski, K. K. Holmes, and E. G. Sandström, Serological classification of *Neisseria gonorrhoeae* with use of monoclonal antibodies to gonococcal outer membrane protein I. *J. Infect. Dis. 150*:44–48 (1984).
53. Bygdeman, S., D. Danielsson, and E. G. Sandström, Gonococcal serogroups in Scandinavia. A study with polyclonal and monoclonal antibodies. *Acta Pathol. Immunol. Scand. Sect. B 91*:293–305 (1983).

54. Danielsson, D., E. Sandström, S. Bygdeman, M. Bäckman, and H. Gnarpe, W-serogroup (protein I) and serovar patterns of gonococci isolated during two different periods in urban and rural districts of Sweden. In *The Pathogenic Neisseriae*, G. K. Schoolnik, Brooks, G. F., Falkow, S., Frasch, C. E., Knapp, J. S., McCutchan, J. A., and Morse, S. A. (eds.). American Society for Microbiology. Washington, pp. 71–77 (1985).
55. Knapp, J. S., S. Bygdeman, E. Sandström, and K. K. Holmes, Nomenclature for the serological classification of *Neisseria gonorrhoeae*. In *The Pathogenic Neisseriae*. G. K. Schoolnik, Brooks, G. F., Falkow, S., Frasch, C. E., Knapp, J. S., McCutchan, J. A., and Morse, S. A. (eds.). American Society for Microbiology. Washington, pp. 4–5 (1985).
56. Sandström, E., P. Lindell, B. Härfast, F. Blomberg, A.-C. Ryden, and S. Bygdeman, Evaluation of a new set of *Neisseria gonorrhoeae* serogroup W-specific monoclonal antibodies for serovar determination. In *The Pathogenic Neisseriae*. G. K. Schoolnik, Brooks, G. F., Falkow, S., Frasch, C. E., Knapp, J. S., McCutchan, J. A., and Morse, S. A. (eds.). American Society for Microbiology. Washington, pp. 26–30 (1985).
57. Bygdmen, S. M., E.-C. Gillenius, and E. G. Sandström, Comparison of two different sets of monoclonal antibodies for the serological classification of *Neisseria gonorrhoeae*. In *The Pathogenic Neisseriae*, G. K. Schoolnik, Brooks, G. F., Falkow, S., Frasch, C. E., Knapp, J. S., McCutchan, J. A., and Morse, S. A. (eds.). American Society for Microbiology, Washington, pp. 31–36 (1985).
58. Kohl, P. K., and T. M. Buchanan, Serotype-specific bactericidal activity of monoclonal antibodies to protein I of *Neisseria gonorrhoeae*. In *The Pathogenic Neisseriae*, G. K. Schoolnik, Brooks, G. F., Falkow, S., Frasch, C. E., Knapp, J. S., McCutchan, J. A., and Morse, S. A. (eds.). American Society for Microbiology, Washington, pp. 442–444 (1985).
59. Kohl, P. K., T. M. Buchanan, J. S. Knapp, H. Hofmann, D. Petzoldt, and K. K. Holmes, Serological classification of *N. gonorrhoeae* serotypes and serovars. In *The Pathogenic Neisseriae*, G. K. Schoolnik, Brooks, G. F., Falkow, S., Frasch, C. E., Knapp, J. S., McCutchan, J. A., and Morse, S. A. (eds.). American Society for Microbiology, Washington, pp. 78–81 (1985).
60. Catlin, B. W. and E. N. Shinners, Gonococcal protein I: Genetic dissection and linkage associations of *nmp*-1. In *The Pathogenic Neisseriae*, G. K. Schoolnik, Brooks, G. F., Falkow, S., Frasch, C. E., Knapp, J. S., McCutchan, J. A., and Morse, S. A. (eds.). American Society for Microbiology, Washington, pp. 123–129 (1985).
61. Danielsson, D., H. Faruki, D. Dyer, and P. F. Sparling, Recombination near the antibiotic resistance locus *penB* results in antigenic variation of gonococcal outer membrane protein I. *Infect. Immun. 52*:529–533.
62. Zenilman, J. M., L. J. Nims, M. A. Menegus, F. Nolte, and J. S. Knapp. Spectinomycin-resistant gonococcal infections in the United States; 1985–1986. *J. Infect. Dis. 156*:1002–1004 (1987).
63. Whittington, W. L., A. Vernon, J. W. Biddle, W. DeWitt, A. Zaidi, T. Starcher, G. Perkins, R. Rice, B. Anderson, S. Johnson, and W. Albritton,

Serological classification of *Neisseria gonorrhoeae*: Uses at the community level. In *The Pathogenic Neisseriae*, G. K. Schoolnik, Brooks, G. F., Falkow, S., Frasch, C. E., Knapp, J. S., McCutchan, J. A., and Morse, S. A. (eds.). American Society for Microbiology, Washington, pp. 20–25 (1985).

64. Knapp, J. S., M. H. Mulks, I. Lind, H. B. Short, and V. L. Clark, Evolution of gonococcal populations in Copenhagen, 1928–1979. In *The Pathogenic Neisseria*, G. K. Schoolnik, Brooks, G. F., Falkow, S., Frasch, C. E., Knapp, J. S., McCutchan, J. A., and Morse, S. A. (eds.). American Society for Microbiology, Washington, pp. 82–88 (1985).

65. Plummer, J. A., L. J. D'Costa, H. Nzanze, L. Slaney, W. E. DeWitt, J. S. Knapp, J.-A. Dillon, W. L. Albritton, and A. R. Ronald, Development of endemic penicillinase-producing *Neisseria gonorrhoeae* in Kenya. In *The Pathogenic Neisseriae*, G. K. Schoolnik, Brooks, G. F., Falkow, S., Frasch, C. E., Knapp, J. S., McCutchan, J. A., and Morse, S. A. (eds.). American Society for Microbiology, Washington, pp. 101–106 (1985).

66. Knapp, J. S., J. M. Zenilman, J. W. Biddle, G. H. Perkins, W. E. DeWitt, M. L. Thomas, S. R. Johnson, and S. A. Morse, Frequency and distribution in the United States of strains of *Neisseria gonorrhoeae* with plasmid-mediated, high-level resistance to tetracycline. *J. Infect. Dis. 155*:819–822 (1987).

67. Hook, E. W. III, F. N. Judson, H. H. Handsfield, J. M. Ehret, K. K. Holmes, and J. S. Knapp, Auxotype/serovar diversity and antimicrobial resistance of *Neisseria gonorrhoeae* in two mid-sized American cities. *Sex. Transm. Dis. 14*:141–146 (1987).

68. Sandström, E., and S. M. Bygdeman, Serological classification of *Neisseria gonorrhoeae*. Clinical and epidemiologic applications. In *Gonococci and Meningococci*, J. T. Poolman, Zanen, H. C., Meyer, T. F., Heckels, J. E., Mäkelä, P., Smith, R. H., and Beuvery, E. C. (eds.). Kluwer Academic Publishers, Dordrecht, pp. 45–50 (1988).

69. Catlin, B. W., and A. Reyn, *Neisseria gonorrhoeae* isolated from disseminated and localized infections in pre-penicillin era. Auxotypes and antibacterial drug resistances. *Br. J. Vener. Dis. 58*:158–165 (1982).

70. Knapp, J. S., E. G. Sandström, and K. K. Holmes. Overview of epidemiological and clinical applications of auxotype/serovar classification of *Neisseria gonorrhoeae*. In *The Pathogenic Neisseriae*, G. K. Schoolnik, Brooks, G. F., Falkow, S., Frasch, C. E., Knapp, J. S., McCutchan, J. A., and Morse, S. A. (eds.). American Society for Microbiology, Washington, pp. 1–6 (1985).

71. Kohl, P. K., J. S. Knapp, H. Hofmann, K. Gruender, D. Petzoldt, M. R. Tam, and K. K. Holmes, Epidemiological analysis of *Neisseria gonorrhoeae* in the Federal Republic of Germany by auxotyping and serological classification using monoclonal antibodies. *Genitourin. Med. 62*:145–150 (1986).

72. Rice, R. J., E. W. Hook III, K. K. Holmes, and J. S. Knapp, Evaluation of sampling methods for surveillance of *Neisseria gonorrhoeae* strain populations. In *Gonococci and Meningococci*. J. T. Poolman, Zanen, H. C., Meyer, T. F., Heckels, J. E., Mäkelä, P., Smith, R. H., and Beuvery, E. C. (eds.). Kluwer Academic Publishers, Dordrecht, pp. 167–173 (1988).

73. Plummer, F. A., and R. C. Brunham, Gonococcal recidivism, diversity, and ecology, *Rev. Inf. Dis. 9*:846–850 (1987).
74. Faruki, H., R. N. Kohmescher, W. P. McKinney, and P. F. Sparling, A community-based outbreak of infection with penicillin-resistant *Neisseria gonorrhoeae* not producing penicillinase (chromosomally mediated resistance). *N. Engl. J. Med. 313*:607–611 (1985).
75. Rice, R. J., J. W. Biddle, Y. A. JeanLouis, W. E. DeWitt, J. H. Blount, and S. A. Morse, Chromosomally mediated resistance in *Neisseria gonorrhoeae* in the United States: Results of surveillance and reporting. *J. Infect. Dis. 153*:340–345 (1986).
76. Zenilman, J. M., M. Bonner, K. L. Sharp, J. A. Rabb, E. R. Alexander, Penicillinase-producing *Neisseria gonorrhoeae* in Dade County, Florida: Evidence of core-group transmitters and the impact of illicit antibiotics. *Sex Transm. Dis. 15*:45–50 (1988).
77. Centers for Disease Control, Sentinel surveillance system for antimicrobial resistance in clinical isolates of *Neisseria gonorrhoeae. Mortal. Morbid. Week. Rep. 36*:585–593 (1987).
78. Boslego, J. W., E. C. Tramont, E. Takafuji, B. M. Diniega, B. S. Mitchell, J. W. Small, and W. Kahn, Effect of spectinomycin use on the prevalence of spectinomycin-resistant and of penicillinase-producing *Neisseria gonorrhoeae. N. Engl. J. Med. 317*:272–278 (1987).
79. Ison, C. A., J. Gedney, and C. S. F. Easmon, Antibiotic resistance in clinical isolates of *Neisseria gonorrhoeae*. In *The Pathogenic Neisseriae*, G. K. Schoolnik, Brooks, G. F., Falkow, S., Frasch, C. E., Knapp, J. S., McCutchan, J. A., and Morse, S. A. (eds.). American Society for Microbiology, Washington, pp. 116–119 (1985).
80. Zenilman, J. M., J. S. Knapp, W. L. Whittington, R. J. Rice, J. W. Biddle, W. E. DeWitt, M. L. Thomas, M. E. Shepherd, and D. Frazier, Characterization of penicillinase-producing *Neisseria gonorrhoeae* (PPNG) strains from Miami, Florida. *Abstr. Annu. Meet. Am. Soc. Microbiol.* p. 369 C275 (1987).
81. Morse, S. A., S. R. Johnson, J. W. Biddle, and M. C. Roberts, High-level tetracycline resistance in *Neisseria gonorrhoeae* is due to the acquisition of the streptococcal tetM determinant. *Antimicrobial Agents Chemother. 30*: 664–670 (1986).
82. Knapp, J. S., J. M. Zenilman, J. W. Biddle, G. H. Perkins, W. E. DeWitt, M. L. Thomas, S. R. Johnson, and S. A. Morse, Frequency and distribution in the United States of strains of *Neisseria gonorrhoeae* with plasmid-mediated, high-level resistance to tetracycline. *J. Infect. Dis. 155*:819–822 (1987).
83. Roberts, M. C., and J. S. Knapp, Host range of the conjugative 25.2 Mdal tetracycline resistance plasmid from *Neisseria gonorrhoeae* and related species. *Antimicrobial. Agents Chemother. 32*:488–491 (1988).
84. Shaw, C. E., D. G. W. Chan, S. K. Byrne, W. A. Black, and W. R. Bowie, Tetracycline-resistant *Neisseria gonorrhoeae* (TRNG)—British Columbia. *Can. Dis. Week. Rep. 12*:101 (1987).
85. Toma, S., Tetracycline-resistant *Neisseria gonorrhoeae* in Ontario. *Can. Med. Assoc. J. 137*:1109 (1987).

86. Roberts, M. C., J. H. T. Wagenvoori, B. van Klingerin, and J. S. Knapp, *tetM*- and β-lactamase-containing *Neisseria gonorrhoeae* (tetracycline resistance and pencillinase producing) in the Netherlands. *Antimicrob. Agents Chemother. 32*:158 (1988).
87. Cannon, J. G., T. M. Buchanan, and P. F. Sparling, Confirmation of association of protein I serotype of *Neisseria gonorrhoeae* with ability to cause disseminated infections. *Infect. Immun. 40*:816–819 (1983).
88. Catlin, B. W. and E. N. Shinners, Arginine biosynthesis in gonococci isolated from patients. In *Immunobiology of* Neisseria gonorrhoeae, G. F. Brooks, Gotschlich, E. C., Holmes, K. K., Sawyer, W. D., and Young, F. E. (eds.). American Society for Microbiology, Washington, pp. 1–8 (1978).
89. Copley, C. G., *Neisseria gonorrhoeae* subdivision of auxogroups by genetic transformation. *Genitourin. Med. 63*:153–156 (1987).
90. Vedros, N. A., Genus I *Neisseria* Trevisan 1885. In *Bergey's Manual of Systematic Bacteriology*. Vol. I. N. R. Krieg, and Holt, J. G. (eds.). William & Wilkins, Baltimore, pp. 290–296 (1984).
91. Hollis, D. G., G. L. Wiggins, and R. E. Weaver, An unclassified gram-negative rod isolated from the pharynx on Thayer-Martin medium (selective agar). *Appl. Microbiol. 24*:772–777 (1972).
92. Knapp, J. S., and E. W. Hook III, Prevalence and persistence of *Neisseria cinerea* and other *Neisseria* spp. in adults. *J. Clin. Microbiol. 26*:896–900 (1988).
93. Knapp, J. S., P. A. Totten, M. H. Mulks, and B. H. Minshew, Characterization of *Neisseria cinerea*, a nonpathogenic species isolated on Martin-Lewis medium selective for pathogenic *Neisseria* spp. *J. Clin. Microbiol. 19*: 63–67 (1984).
94. Riou, J.-Y., and M. Guibourdenche, *Neisseria polysaccharea* spp. nov. *Int. J. Syst. Bacteriol. 37*:163–165 (1987).
95. Morse, S. A., and J. S. Knapp, Neisserial infections. In *Diagnostic Procedures for Bacterial Infections*, 7th ed., B. B. Wentworth (ed.). American Public Health Association, Washington, pp. 407–432 (1987).
96. Janda, W. M., K. L. Zigler, and J. J. Bradna, API QuadFERM+ with rapid DNase for identification of *Neisseria* spp. and *Branhamella catarrhalis. J. Clin. Microbiol. 25*:203–206 (1987).
97. Doern, G. V., K. G. Siebers, L. M. Hallick, and S. A. Morse, Antibiotic susceptibility of beta-lactamase-producing strains of *Branhamella (Neisseria) catarrhalis. Antimicrob. Agents Chemother. 17*:24–29 (1980).
98. Lawton, W. D., and G. J. Battaglioli, GonoGen coagglutination test for confirmation of *N. gonorrhoeae. J. Clin. Microbiol. 18*:1264–126 (1983).
99. Libonati, J. P., R. L. Leilich, and L. Loomis, Comparison of GonoGen coagglutination test and fluorescent microscopy for the identification of *Neisseria gonorrhoeae. Abstr. Annu. Meet. Am. Soc. Microbiol.* p. 314, C19 (1983).
100. Kanzanis, D., F. Reeves, and J. Frankel, A monoclonal antibody test for confirmation of *Neisseria gonorrhoeae. Abstr. Annu. Meet. Am. Soc. Microbiol.*, p. 315, C20 (1983).

101. Lewis, J. S., Clinical use of GonoGen-monoclonal antibodies directed against *Neisseria gonorrhoeae. Abstr. Annu. Meet. Am. Soc. Microbiol.* p. 315, C22 (1983).
102. Minshew, B. H., J. L. Beardsley, and J. S. Knapp, Evaluation of GonoGen coagglutination test for serodiagnosis of *Neisseria gonorrhoeae*. Identification of problem isolates by auxotyping, serotyping, and with a fluorescent antibody reagent. *Diagn. Microbiol. Infect. Dis. 3*:41–46 (1985).
103. Barth, S. S., C. Tatsch, and S. J. Gibson, Evaluation of Phadebact OMNI test for identification of *Neisseria gonorrhoeae. Abstr. Annu. Meet. Am. Soc. Microbiol.* p. 368, C270 (1987).
104. Elliman, D. L., W. M. Janda, D. Celig, K. L. Ristow, and R. Shone, Evaluation of a fluorescent antibody reagent for identification of Neisseria gonorrhoeae. *Abstr. Annu. Meet. Am. Soc. Microbiol.* p. 368, C273 (1987).
105. Laughon, B. E., J. M. Ehret, T. T. Tanino, B. van der Pol, H. H. Handsfield, R. B. Jones, F. N. Judson, and E. W. Hook III, Fluorescent antibody for confirmation of *Neisseria gonorrhoeae. J. Clin. Microbiol. 25*:2388–2390 (1987).
106. Carlson, B. L., M. B. Calnan, R. E. Goodman, and H. George, Phadebact Monoclonal GC OMNI test for confirmation of *Neisseria gonorrhoeae. J. Clin. Microbiol. 25*:1982–1984 (1987).
107. Welch, D. W., and G. Cartwright, Fluorescent monoclonal antibody compared with carbohydrate utilization for rapid identification of *Neisseria gonorrhoeae. J. Clin. Microbiol. 26*:293–296 (1988).
108. Lewis, J. S., J. W. Biddle, M. E. Shepherd, and J. S. Knapp, Evauation of a fluorescent antibody test for the identification of *Neisseria gonorrhoeae. Abstr. Seventh Int. Meet. Int. Soc. Sex. Transm. Dis.* p. 90 (1987).
109. Dillon, J. R., M. Carballo, and M. Páuze, Evaluation of eight methods for identification of pathogen *Neisseria* species: Neisseria-Kwik, RIM-N, Gonobio-Test, Minitek, Gonochek II, GonoGen, Phadebact monoclonal GC OMNI test, and Syva Microtrak Test. *J. Clin. Microbiol. 26*:493–497 (1988).
110. Gold, R. I. Goldschneider, M. L. Lepow, T. F. Draper, and M. Randolph, Carriage of *Neisseria meningitidis* and *Neisseria lactamica* in infants and children. *J. Infect. Dis. 137*:112–121 (1978).
111. Noble, R. C., R. M. Cooper, and B. R. Miller, Pharyngeal colonization by *Neisseria gonorrhoeae* and *Neisseria meningitidis* in black and white patients attending a venereal disease clinic. *Br. J. Vener. Dis. 55*:14–19 (1979).
112. Whittington, W. L., R. J. Rice, J. W. Biddle, and J. S. Knapp, Incorrect identification of *Neisseria gonorrhoeae* from infants and children. *Pediatr. Infect. Dis. 7*:3–10 (1988).
113. Bohnhoff, M., J. A. Morello, and W. M. Janda, Serotypes of non-AHU gonococci from disseminated and uncomplicated gonorrhea. *Abstr. Annu. Meet. Am. Soc. Microbiol.* p. 322, C136 (1985).

RECENT REFERENCES ADDED IN PROOF

Ansink-Schipper, M. C., S. M. Bygdeman, B. van Klingerin, and E. G. Sandström, Serovars, auxotypes, and plasmid profiles of PPNG strains with Asian type plasmid isolated in Amsterdam. *Genitourin. Med. 64*:152–155 (1988).

Bygdeman, S. M., Polyclonal and monoclonal antibodies applied to the epidemiology of gonococcal infection. In *Immunological Diagnosis of Sexually Transmitted Diseases: Clinical and Biochemical Analysis*, H. Young and A. McMillan (eds.). Marcel Dekker, New York, pp. 117–165 (1988).

20
Monoclonal Antibodies and Mycobacteria

E. PAMELA WRIGHT*
University of Amsterdam, Amsterdam, The Netherlands

AREND H. J. KOLK
Royal Tropical Institute, Amsterdam, The Netherlands

NALIN RASTOGI
Institut Pasteur, Paris, France

I. INTRODUCTION

The causative agents for leprosy and tuberculosis were placed in the genus *Mycobacterium* by Lehmann and Neumann, in 1896 (1), who defined mycobacteria as aerobic, asporogenous, acid-alcohol-fast rods. However, because of an early classification based on a very limited number of their morphological and staining properties, much confusion followed in the classification of the members of the taxa *Corynebacterium-Mycobacterium-Nocardia* (CMN), grouped together in a single family by Lachner-Sandoval in 1898 (2). Today, mycobacterial systematics are based on data derived from cell wall chemistry, numerical taxonomy, in vitro nucleic acid and serological analyses, and cytochemistry. Given this data, Goodfellow and Minnikin (3), more recently, have defined mycobacteria as aerobic, acid-alcohol-fast, rod-shaped actinomycetes with occasional branching. Aerial hyphae are normally absent, and the bacteria are nonmotile, nonsporulating, unencapsulated organisms that contain arabinose, galactose, and *meso*-diaminopimelic acid (DAP) in their walls; have guanine plus cytosine (G+C) DNA base ratios in the range of 62-70 mol%; and have high-molecular-weight mycolic acids (60-90 carbons) lacking components with more than two points of unsaturation in the molecule (an exception has been observed recently for *M. fallax*, which has up to four points of unsaturation; [4] that release C_{22} to C_{26} straight-chain saturated, long-chain acids on pyrolysis.

Present affiliation: Royal Tropical Institute, Amsterdam, The Netherlands.

A. Mycobacteria as Human Pathogens

The major human diseases caused by mycobacteria, with a high death toll in the Third World countries, remain leprosy (caused by *M. leprae*) and pulmonary tuberculosis (*M. tuberculosis* complex comprising *M. tuberculosis, M. africanum*, and *M. bovis*). Both leprosy and tuberculosis are chronic infections of man. According to the World Health Organization (WHO), about 11 million people in the world have leprosy and about 1.4 billion are exposed to the risk of contracting the disease (5), whereas tuberculosis still affects about 30 million people each year (6). Other pathogenic species of mycobacteria include *M. ulcerans, M. marinum, M. avium intracellulare-scrofulaceum* (MAIS) complex, *M. kansasii, M. xenopi, M. szulgai, M. simiae*, and the rapidly growing potential pathogens of the *M. fortuitum* complex.

Prevention of infection with mycobacteria with BCG vaccination against tuberculosis has recently been a subject of controversy because of the failure of the vaccine to protect a South Indian population (7), but it continues to be the only major vaccination program against mycobacteria. Other approaches include vaccination against leprosy, with heat-killed or gamma-irradiated *M. leprae* alone, or in combination with BCG (8); however, an outline of the detailed results of these studies and the conclusions that can be drawn will take a few more years.

B. Diagnostics

The slow growth in vitro of most of the potentially pathogenic mycobacteria (3-4 weeks), including *M. tuberculosis*, and our inability to grow *M. leprae* on artificial media (9), or in cell cultures (10), remain major obstacles to rapid diagnosis. *Mycobacterium leprae* can now be grown only in vivo, in mouse footpads (11), in nine-banded armadillos (12), and in mangabey monkeys (13). The screening of new drugs against *M. leprae* can be performed in the mouse footpad, but this model is not suitable for systematic screening of new drugs nor for routine diagnostics because of the enormous effort and expense involved. Congenitally athymic nude mice, which are highly susceptible to leprosy, and armadillos are alternative sources for *M. leprae* production (14); however, in contrast to armadillos, in which the lesions are extensive and apparent in the internal organs, with a high morphological index (ratio of solid-staining bacilli/total bacterial count under the light microscope; MI) for the bacteria isolated, the internal organs in nude mice do not always show extensive lesions, and fewer bacilli, with a lower MI, are isolated (15).

Current diagnostic tests for routine mycobacterial identification, remain the culture characteristics on various media and the response to various differential biochemical tests (16); drug-susceptibility testing (17); immunologically based diagnostic tests employing tuberculin and other nontuberculous mycobacterial skin-test antigens (18); and the serodiagnostic tests (19,20).

C. Morphology and Ultrastructure

The property of acid-fastness, when mycobacteria are stained by Ziehl-Neelsen or fluorochrome procedures, is useful in discriminating them from most other bacteria, but little information on the structure-function relationship of the wall envelope is revealed. In this field, new insight has been brought in the last few years through electron microscopy.

All of the mycobacteria studied, thus far, possess a trilaminar cell wall composed of a basal peptidoglycan (PG) layer, an intermediate electron-transparent layer (ETL), and an outer dense layer (ODL), which can be visualized further as a polysaccharide outer layer (POL) by ruthenium red staining (21). The components of this POL should now be characterized in more detail by using the fine specificity of monoclonal antibodies.

When observed in situ in sections of infected tissues (Fig. 1A,B), most bacilli appear to be located inside phagocytic cells and are surrounded by a substance that separates the bacterial wall from the phagosomal membrane. This "capsule substance" (22) is now termed the *electron-transparent zone* (ETZ) because of its ultrastructural appearance in transmission electron microscopy. Biochemical and chemical analysis of this capsule-substance around *M. lepraemurium* showed that it contained mycosides (23). Both *M. leprae* and *M. avium* are surrounded by ETZ in host tissues, but after separation from the tissues, the bacilli no longer contain a visible ETZ (Fig. 1C,D). However, when these bacteria are phagocytized in vitro, they are rapidly surrounded by an ETZ (Fig. 1E), which appears to decrease the diffusion of lysosomal enzymes toward the bacterial surface, despite the formation of phagolysosomes (10,24,25; Fig. 1F).

Recently, in a model proposed for the *M. avium* cell envelope (26; Fig. 2), the mycobacterial POL is seen as the amphiphil-protein monolayer, with the hydrophobic ends oriented to face the hydrophobic ends of wall mycolic acids, thus covering the bacterial cell surface partially or completely. The molecular heterogeneity of the amphiphils (glycolipids, peptidolipids, phospholipids, and possibly other complex lipids) and the proteins, relative to their surface charge, structure, and the architectural exposition of reactive groups, may account for the functional diversity of the envelopes from different species of mycobacteria relative to natural multiple drug resistance (27), to their capacity for intracellular growth, and to their pathogenicity and virulence (26), despite the ultrastructural similarity observed for cell envelopes from the 18 species of mycobacteria so far studied (21). These differences would be interesting subjects for the production of monoclonal antibodies because their variation among the species suggests that

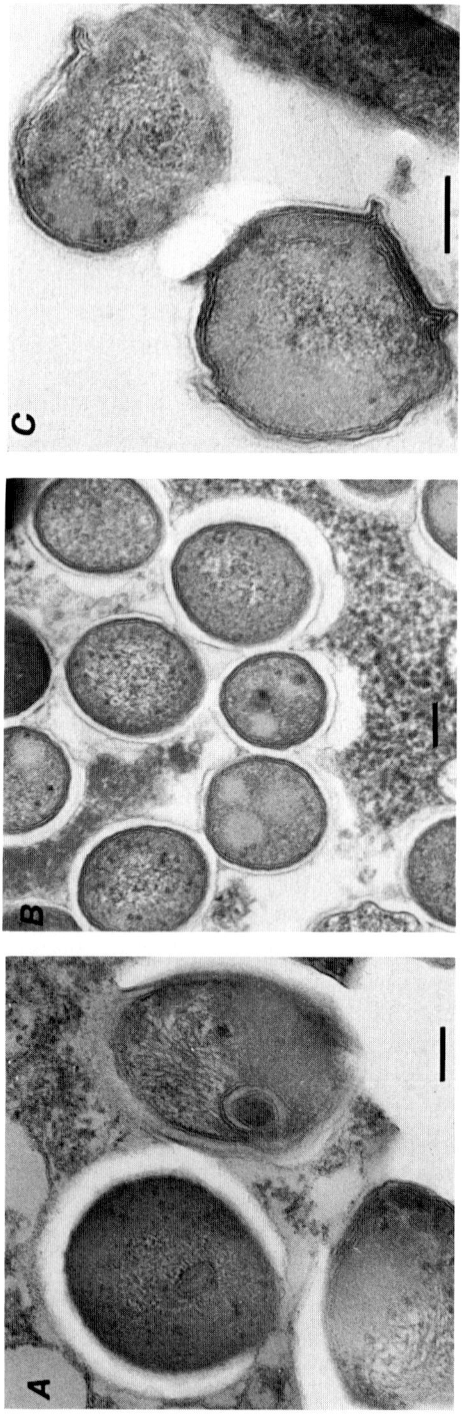
A
B
C

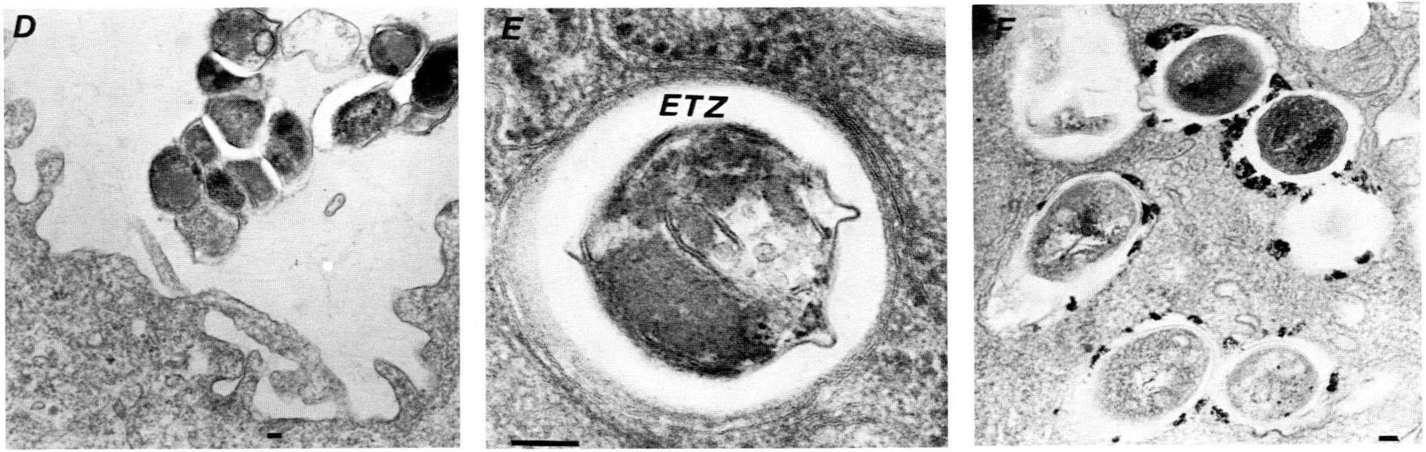

Figure 1 Ulstrastructure of mycobacteria. (A) *M. leprae* observed in situ in infected tissues of armadillo: the bacilli are surrounded by an electron-transparent zone (ETZ). (B) A similar ETZ surrounds *M. leprae* observed in situ in infected tissues of the nude mice. (C) However when isolated and purified from these tissues, the bacteria are no longer surrounded by this ETZ. (D) Purified *M. leprae* bacilli being phagocytized in vitro by a culture of kidney epithelial cells of armadillo. (E) The in vitro phagocytized bacilli are rapidly surrounded by an ETZ which appears to lessen the diffusion of lysosomal enzymes toward the bacterial surface. (F) A similar ETZ is formed around *M. avium* phagocytized by bone marrow-derived macrophages of mouse: the cytochemical demonstration of acid-phosphatase shows the protective role of this ETZ. It does not permit the contact of lysosomal enzymes with the bacterial surface. Bar marker represents 100 nm. (*Sources*: with permission, Figs. A, C, D, and F are from Ref. 10; B is from Ref. 15; and F is from Ref. 24).

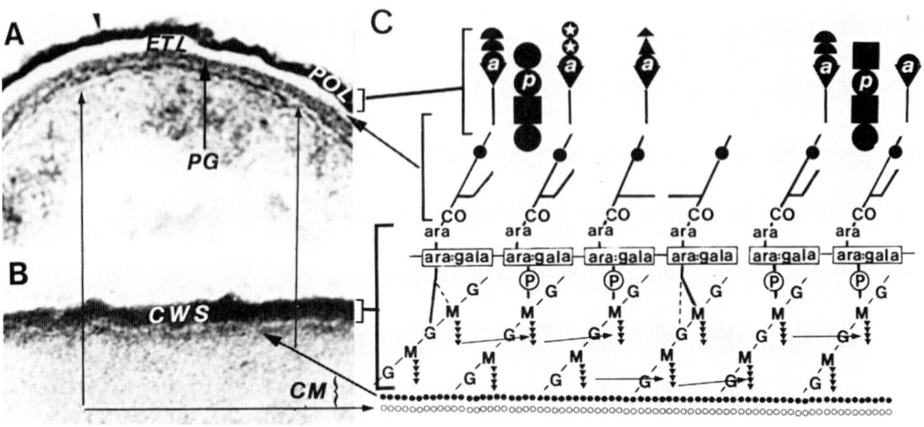

Figure 2 A recent model proposed for the *M. avium* cell envelope. (A) Ruthenium red and lead citrate stained ultrathin section of *M. avium*. (B) Sections stained using the periodic acid–phosphotungstic acid method of Rambourg. (C) Reconstituted model of the *M. avium* cell envelope. In the model, the amphiphils (*a*, extracted using chloroform–methanol) and the proteins (*p*, using Triton X-100) are shown using distinct symbols to illustrate their molecular heterogeneity. The cytoplasmic membrane (CM) is shown as a bilayer with use of closed circles (outer layer) and open circles (inner layer) to illustrate its geometrical and chemical asymmetry. The chemical structure of the cell wall skeleton (CWS) and the amphiphil–protein monolayer are according to the data in the literature. *Abbreviations and symbols*: POL, polysaccharide outer layer; ETL, electron-transparent layer; PG, peptidoglycan; CWS, cell wall structure; CM, cytoplasmic membrane; *a*, amphiphils; *p*, proteins; ⟋●⟍, mycolic acids; ara- [ara-gal] -, arabinogalactan; M, *N*-glycolylmuramic acid; G, *N*-acetylglucosamine; ⇟, tetrapeptide (L-ala, D-glu, *meso*-DAP, D-ala), and intrapeptide linkages (D-ala → *meso*-DAP; *meso*-DAP → *meso*-DAP); *P*, phosphodiester linkages; ⟋⟍, glycosidic linkages. (*Source*: adapted from Ref. 26, with permission.)

they might provide targets for specific reactions related to functions that are clinically important.

D. Immunopathology

Although the host's response to both leprosy and tuberculosis depends upon the cell-mediated immunity (CMI), the pathological evolution of the two diseases is very different. In tuberculosis, a high bacterial load causes rapid death, whereas in leprosy even a bacterial load as high as 10^{10} bacilli per gram of tissue can be maintained in infected subjects. Moreover, tuberculosis is marked by necrotic lesions and necrotic responses to skin-test antigens, whereas in leprosy, necrotic response to soluble leprosy antigens does not occur and necrosis remains rare. Although the toxicity of *M. tuberculosis* may partly be linked to certain cell wall lipids, such as cord-factor and sulfatides (28), the major difference among the immunopathological consequences of the two diseases can be linked to the differential regulatory mechanisms controlling the CMI (29,30).

All pathogenic mycobacteria share the particularity of being intracellular pathogens, with the capacity to grow inside the phagosomes of phagocytic cells. However, the mechanisms involved in the intracellular survival of mycobacteria may differ greatly from one pathogenic species to another (24,25,31).

The Ridley and Jopling classification of leprosy (32), based on the clinical and histopathological spectrum, takes into account two polar developed forms of disease, namely, tuberculoid (TT) and lepromatous (LL) leprosy, and a borderline (BB) or intermediate form of disease. The BB form may later evolve toward borderline tuberculoid (BT) or borderline lepromatous (BL) forms of leprosy, finally developing into the polar TT or LL forms, respectively. In the TT form, the host tissue contains few viable bacilli, but a high number of lymphocytes, compared with the LL form in which the situation is reversed. It is only in the TT form that subjects react positively to a suspension of heat-killed leprosy bacilli (lepromin) or soluble antigens of *M. leprae* in skin-testing (33) and that in vitro response to *M. leprae*, measured by lymphoproliferation and leukocyte migration inhibition, is positive (34).

Two distinct types of acute inflammatory reactions occur in leprosy. In the first type, observed in certain indeterminate forms of the disease, the reaction is marked by an influx of lymphocytes in the lesions (35), whereas in certain LL cases, it is characterized by deposition of circulating immune complexes (CIC), inflammation and activation of the complement cascade, and infiltration of polymorphonuclear cells at the site of macrophage-rich lesions (36,37). The latter type of response, represented by the class II leprosy reaction (erythema nodosum leprosum, ENL) is most probably caused by an Arthus-type hypersensitivity. However, the lack of correlation between circulating and local levels of C_3d (38) and the fact that CIC are found throughout the entire leprosy spectrum in endemic areas, whereas ENL is observed only in LL patients (39), suggest that these observations are not directly related to ENL.

E. Cellular and Humoral Immunity to Mycobacteria

When the mycobacteria, obligate intracellular parasites, attack the host, protection and recovery are mainly dependent on cell-mediated immunity (CMI), in which T-lymphocyte-macrophage interactions predominate. Ingested mycobacteria are carried to regional lymph nodes, spleen, liver, and other organs, where the bacilli are partially digested (processed) inside phagosomes or phagolysosomes, and bacterial antigens are presented on the macrophage membrane in association with the antigens of the major histocompatibility complex (MHC). The human MHC, called the HLA complex, comprises class I (HLA-A, -B, -C) molecules, present on almost all nucleated cells, and class II (HLA-D, -DP, -DQ, -DR) molecules (40), which are present mainly on antigen-presenting cells (APCs), some T and B lymphocytes, and a few other cells. The T lymphocytes are generally capable of recognizing only APC-processed antigen in association with MHC (41).

Functional subsets of T cells include T-helper/inducer (Th/Ti) cells, which are class II, MHC-restricted (41); cytotoxic T cells (Tc), which are usually class I, MHC-restricted (42); and T-suppressor cells (Ts), which may be able to recognize free antigen. Class II MHC restriction of Th/i cells may partly explain the poor response of HLA DR3-negative individuals to *M. leprae* (43).

The binding of the T-cell receptor to foreign antigen plus class II MHC, results in interleukin-1 (IL-1) release by APCs, which causes, in turn, the release of a variety of humoral factors by Th cells. In the T-dependent, B-cell activation responsible for antibody production, the lymphokines implicated are different from those that intervene for macrophage/Tc activation (44). The latter include interleukin-2 (IL-2), macrophage-activating factor (MAF), interferon (IFN), macrophage-migration-inhibiting factor (MIF), and others that certainly play a major role in tuberculosis and leprosy by activating macrophages. The types of lymphokines released depend on the responding Th clone and, consequently, upon the antigenic specificity as read by the repertoire of receptors of the T-cell clones. This T-cell-mediated response is under negative feedback by Ts cells, which may be specific to one antigen only or nonspecific, and which may control either the induction of responses or the effector phase (45,46).

In the past few years the application of monoclonal antibody technology to the identification of these cell types and the factors they produce has permitted a much better analysis of the interactions involved in the generation of immune responses. Some of the information gained with use of these tools in mycobacterial diseases will be described in Sect. II.

As far as humoral immunity to mycobacterial infections is concerned, it appears that antibodies are not protective, and their presence is not always relevant to the stage of infection nor for its diagnosis. Large amounts of antibodies are found in treated as well as in untreated patients, which confirms the persistence of antigenic stimulation despite treatment (29).

II. MONOCLONAL ANTIBODIES

From the foregoing sections, it is clear that many problems concerning mycobacterial taxonomy, immunodiagnosis, and immunopathology of the mycobacterial diseases, and eventual prevention in the form of proper vaccination still remain. In recent years, it has been demonstrated that the judicious application of monoclonal antibodies can contribute to the solution of some of these problems. Although it is accepted that protection and recovery from mycobacterial disease is mainly due to the activities of the cellular branch of the immune system, antibodies can be useful tools, not only to provide evidence of infection, but also to identify molecules that can serve as markers for classifying mycobacteria, or as targets for the preventive or therapeutic manipulation of immunological responses. For mycobacteria, conventional serology has not been able to deliver enough specific information to be particularly useful in any of these areas. Some aspects of this topic have been recently reviewed (46-50), and in this chapter, we will describe our own experience with the production of monoclonal antibodies to mycobacteria and report on the available results of the applications of monoclonal antibody technology to mycobacteriology, as well as discuss the advantages and disadvantages of this approach relative to the requirements of the field, particularly for taxonomy, immunodiagnosis, immunopathology, and future prospects.

A. Making Monoclonal Antibodies to Mycobacteria

Successful production of mouse monoclonal antibodies depends on the immune response of the mouse (about which very little is known for mycobacteria), the immunization scheme, and the assay and selection procedures, once the hybrids have been established in culture.

Most of the mice used for the production of monoclonal antibodies are BALB/c mice. Various immunization schemes have been used to elicit antibody production to mycobacteria in these mice. No systematic studies have been done to compare the schemes, so there is no standard protocol. It is likely that different programs of immunization will be suitable for different antigens and to induce different antibodies.

1. The Choice of Antigen

In general, the antigen injected will also be the desired target of the antibody to be produced. The experience of our group, as well as that of others, indicates that many of the cross-reactive antibodies produced upon immunization with one species react better with another species; this is probably due to quantitative differences in the antigen available for the reaction (51,52). In the several laboratories making antimycobacterial monoclonal antibodies, whole bacteria, culture filtrates, sonicates, and supernatants of sonicated bacteria have been used. Only for the phenolic glycolipid, the PGL of *M. leprae*, has purified antigen been used for immunization (53).

That many antibodies are produced to a few dominant antigens was clearly demonstrated during the WHO workshop on monoclonal antibodies to mycobacteria (54-56). Several laboratories independently produced antibodies against the same antigens, which seem to be dominant. No doubt some preliminary purification of the antigens for immunization would increase the success and variety of the antibody production.

Our own experience with different immunization schemes and the results after fusion, screening, and characterization of the antibodies are shown in Tables 1-4. In all cases, 6- to 8-week-old mice were immunized. It seemed that if the second immunization were given long after the first, the resulting clones grew only slowly and the yield of antibody-producing hybrids was low (fusions 24 and 88).

In contrast to *M. tuberculosis* and *M. bovis* BCG, *M. leprae* was a poor immunogen (compare fusions 23, 29, 67, and 24 with 26, 30, 41). A striking difference was observed between immunization with intact *M. leprae* and with the supernatant of sonicated bacteria. Not only were there many more antibody-producing hybridomas with the supernatant (compare the 10-20% in 26, 30, 31, and 41 with the 20-50% in 47 and 88), but also many more of the clones produced specific antibodies. This, to a lesser extent, was also seen with *M. avium* (compare fusions 75 and 85). Similar findings were reported by Ivanyi et al. (49, 52). It seems that when *M. leprae* cell walls were included in the immunogen, response to species-specific antigens could be suppressed. Immunization with live *M. leprae* (by footpad inoculation) did not improve the results (fusion 31).

With *M. tuberculosis*, more specific clones were detected after immunization with the intact rather than with the sonicated bacteria (compare fusion 29 with 23 and 24).

2. *Fusion*

The spleen cells were removed 3-4 days after the last booster injection and fused with Sp2/0-Ag 14 or NS-1 cells, by standard techniques (51), in a ratio of 4:1. The products of fusion were plated into microtiter plates with a feeder layer of mouse peritoneal macrophages. Generally, hybridomas grew up in 90-100% of the wells. In fusion 41 this was lower (56%) because many young clones were killed by aggressive macrophages, apparently derived from the spleen of the immunized mouse; this problem was greater with mice immunized with Freund's adjuvant. The mouse myeloma line Sp2/0 was less resistant to aggressive macrophages than was NS-1, perhaps because the former grows semiattached and the latter does not. From fusion 75 onward, only NS-1 cells were used for fusion; we found them to give higher fusion frequencies, but lower cloning efficiencies, than Sp2/0.

After fusion, the cells were plated at 2×10^5 spleen cells per well. In some fusions a lower density could be used; this would reduce the chances of two hybrids growing in one well which, if they produce antibodies of different specificities, complicates screening and selection. Hybridomas that gave positive reactions of the desired specificity were cloned by limiting dilution, first at one cell per well then twice at 0.3 cells per well, on macrophage feeder cells. The selected clones were then grown in bulk and injected intraperitoneally into pristane-primed BALB/c mice to obtain ascites rich in the monoclonal antibody.

3. The Assay Technique and Selection Procedures

The first question in this section is always the purpose for which the monoclonal antibody is needed. According to this, the appropriate assay techniques and selection criteria will be chosen. In the beginning, however, it is often necessary to improvise because the sources of variation are not predictable. Usually, we were looking for monoclonal antibodies specific for one species of mycobacteria. In two fusions (67 and 88) we wanted antibodies to a known protein antigen, identified as a band in the polyacryamide gel electrophoresis immunoperoxidase (SGIP) assay. Later, we started looking for antibodies that could be used for identification of mycobacteria in a routine technique, IF (fusions 45, 75, 84, and 85). We also wanted antibodies recognizing (peptido-)glycolipids because it is known, for *M. leprae*, that monoclonal antibodies to the specific phenolic glycolipid-1 could be used for identification in immunofluorescence (IF) (57).

Most of the supernatants were screened in enzyme-linked immunosorbent assay (ELISA) in Terasaki plates, using only 10 μl of supernatant, so that several target antigens could be tested at the same time. Polyvinyl chloride (PVC) plates were used to detect antibodies to glycolipids. When identification was important, the IF test had to be used, so that reagents that would work in that test could be selected.

The screening test contributes to the selection. For example, not all of the *M. tuberculosis* antigens could be detected in ELISA (51). In fusion 29, we tried to select *M. tuberculosis* species-specific antibodies by choosing those positive in ELISA with *M. tuberculosis* H37Rv and negative with *M. bovis* BCG, but this also gave us strain-specific antibodies reacting only with H37Rv (clone F29-3) (51). Similarly, in an *M. avium* fusion, the mice were immunized with seven strains (of serotypes 2, 4, and 8), but because serotype 2 was used for screening, serotype 2-specific antibodies were obtained. In fusion 85, we selected for serotype 4 and *M. intracellulare* and obtained both serotype-specific antibodies and antibodies reacting with *M. avium, M. intracellulare*, and *M. scrofulaceum*.

In some instances, the SGIP assay was used for screening (58). Several clones from fusion 67 were positive in the SGIP, but negative in the ELISA, on Terasaki or polyvinyl chloride plates coated with intact or sonicated bacteria. These were probably antibodies directed against a denatured epitope not present on the plates. Most of the antibodies that were positive in the ELISA on intact mycobacteria were also positive in IF. Because the ELISA is more suitable for screening a large number of samples, it can usually be used to screen for reagents to be used in IF.

Table 1 Immunization Protocol for BALB/C Mice

Fusion number	Mouse immunized with	First immunization day 0		Second and third immunizations			Boosters		
		Amount	Route	Day	Amount	Route	Day	Amount	Route
23	*M. tub.* H37Rv	100 μg sup. (20,000 g)	ip	30 (and 122)	100 μg sup.	ip	140	100 μg sup.	iv
29	*M. tub.* H37Rv	2×10^7 bact.	ip	30	2×10^7 bact.	ip	208	1×10^8 bact.	iv
67	*M. tub.*[d]	100 μg sup. (100,000 g)	ip[c]	—			129	250 μg sup.	iv
24	*M. bovis* BCG	100 μg sup. (20,000 g)	ip	30 (and 341)	100 μg sup.	ip	362	500 μg sup.	iv
86	*M. bovis* BCG	5×10^7 bact.	id/ip	14	5×10^7 bact.	id/ip	256	2.5×10^8	iv[e]
26	*M. leprae*	2×10^8 son.	id/ip[a]	10	2×10^8 son.	id	42	1×10^9 son.	iv
30	*M. leprae*	2.5×10^8 son.	id/ip[a]	17	2.5×10^8 son.	id	47	1×10^8 bact.	iv
31[b]	*M. leprae*	5×10^4 bact.	fp	—			190	1×10^8 bact.	iv
41	*M. leprae*	2×10^7 bact.	id	14	2×10^7 bact.	id	202	1×10^8 bact.	iv
47	*M. leprae*	100 μg sup. (100,000 g)	ip[c]	—			108	250 μg sup.	iv

88	*M. leprae*	100 μg sup. (100,000 g)	ip[c]	304	100 μg sup.	id	347	500 g sup.	iv
75	*M. avium* 7 isolates	5×10^7 bact.	id/ip[a]	14	5×10^7 bact.	ip	63	2.5×10^8 bact.	iv
85	*M. avium* 7 isolates	100 μg sup. (20,000 g)	ip	14	100 μg sup.	ip	249	5×10^8 bact.	iv
84	*M. intra.* *M. scrof.*	5×10^7 bact.	id/ip	14	5×10^7 bact.	ip	213	5×10^8 bact.	ip
45	ADM	5×10^7 bact.	id/ip	–			46	2.5×10^8 bact.	iv

Abbreviations: id, intradermal; ip, intraperitoneal; iv, intravenous; fp, footpad; bact., intact bacilli heated at 80°C for 10 min; μg son., dry weight total sonicated mycobacteria; μg sup., dry weight of the supernatant (20,000 g or 100,000 g) of sonicated mycobacteria; μg pellet, dry weight of the pellet (20,000 g or 100,000 g) of sonicated mycobacteria; *M. tub., M. tuberculosis; M. intra., M. intracellulare; M. scrof., M. scrofulaceum; ADM*, armadillo-derived mycobacteria.

[a]Emulsified in Freund's incomplete adjuvant.

[b]A CBA mouse was inoculated into the footpad with live *M. leprae*.

[c]In aluminum precipitation.

[d]*M. tuberculosis* isolated from a patient with tuberculosis.

[e]Booster was given on each of the 3 days preceeding the fusion.

Table 2 Results of Fusion and First Screening in ELISA

Fusion number	Mouse immunized with	Wells seeded	With hybrids (%)	Day[a]	Positive[b] for antibody (%)	Screened against	Positive[c] (%)	Control antigen	Positive[d] (%)
23	*M. tub.* H37Rv	960	100	10	98	H37Rv sup.	99	None	1
29	*M. tub.* H37Rv	1344	100	10	82	H37Rv sup.	100	ND	
67	*M. tub.*	1008	100	10	34	*M. tub.* sup.	53	ND	
24	*M. bovis* BCG	1404	90	17	8	BCG sup.	100	ND	
86	*M. bovis* BCG	840	100	10	60	BCG int.	37	ND	
26	*M. leprae*	1260	100	10	19	*M. lep.* son.	70	ALH	30
30	*M. leprae*	1500	100	10	10	*M. lep.* son.	100	ND	
31	*M. leprae*	670	13	7	19	*M. lep.* son.	90	ALH	10
41	*M. leprae*	1008	56	12	20	*M. lep.* son.	30	*E. coli*	70
47	*M. leprae*	840	96	17[e]	19	*M. lep.* son.	14	ND	
88	*M. leprae*	1000	100	10	52	*M. lep.* son.	53	ALH	35
75	*M. avium*	1008	100	10	50	*M. avium* int.	71	ND	
85	*M. avium*	840[f]	70[e]	10	78	*M. avium* int. *M. avium* lip.	80	ND	
84	*M. intra.* *M. scrof.*	840	100	10	0.4	*M. scrof.* lip. *M. avium* lip.	0	ND	
45	ADM	1152	58	15	37	ADM int.	78	*E. coli*	57

Abbreviations (see also Table 1): ADM, armadillo-derived mycobacteria; ALH, armadillo liver homogenate; int, intact bacteria; lip, lipid extract of bacteria.

[a]This day is also the day for the first screening of the culture supernatant.

[b]This includes all wells containing hybrids that gave a positive test in one or more of the screening tests.

[c]This is the percentage of the wells positive for antibody that were also positive with the antigen given.

[d]This is the percentage of the wells positive for antibody that were also positive with the control antigen.

[e]The myeloma cells used for the fusion were growing slowly; this was reflected in the retarded growth of the hybridomas.

[f]An infection in the PBS glucose solution used for washing spleen cells resulted in a massive loss of hybridomas. Only 138 wells could be tested.

Table 3 Tests Used in Screening and Selection After Fusion[a]

Fusion number	Mouse immunized with	Desirable hybrids selected by testing in
23	*M. tub.* H37 Rv	ELISA on 20,000 g sup. of *M. tub.*, *M. bovis* BCG, and *M. duv.*
29	*M. tub.* H37 Rv	ELISA on 20,000 g sup. of *M. tub.*, 20,000 g pellet of *M. tub.* and *M. bovis* BCG
67	*M. tub.*	1st: ELISA on 100,000 g sup. of *M. tub.*, *M. bovis* BCG, and *M. kan.* 2nd[b]: SGIP in *M. tub.* with 9% of culture supernatants (77% pos.)
24	*M. bovis* BCG	ELISA on 20,000 g sup. of *M. tub.* and *M. kan.*
86	*M. bovis* BCG	ELISA on intact *M. bovis* BCG and *M. kan.* + lipid extract of *M. bovis* BCG
26	*M. leprae*	ELISA on son. of *M. leprae* + on 20,000 g sup. of *M. tub.* and *M. smeg.* + son. of ALH
30	*M. leprae*	ELISA on son. of *M. leprae* + on 20,000 g sup. of *M. bovis* BCG
31	*M. leprae*	ELISA on son. of *M. leprae* + sup. of *M. bovis* BCG
41	*M. leprae*	1st: ELISA on son. of *M. leprae* + *E. coli* sup. 2nd[b]: IFT on *M. leprae* with 90% of culture supernatants (5% pos.)
47	*M. leprae*	1st: ELISA on *M. leprae* 100,000 sup. + 20,000 sup. of *M. tub.* H37 Rv + son. of ALH 2nd[b]: IFT on *M. leprae* with all supernatants (17% pos.) + SGIP on *M. leprae* with 8% of the supernatants (82% pos.)
88	*M. leprae*	ELISA on son. of *M. leprae* + mixture son. of *M. tub.*, *M. smegm.*, *M. kan.* + son. of ALH
75	*M. avium*	ELISA on intact *M. avium* serotype 2 and *M. tub.* H37Rv + 17% of supernatants tested on lipid extract from *M. avium* and *M. tub.* H37Rv (40% pos.)
85	*M. avium*	ELISA on lipid extracts of intact *M. avium* serotype 4 *M. intra*, and *M. tub.*
84	*M. intra.* *M. scrof.*	ELISA on lipid extracts of *M. scrof.* and *M. avium* serotype 4
45	ADM	IFT on ADM and *E. coli*

[a]See Tables 1 and 2 for abbreviations.
[b]The second screening was done on a part of the total supernatants, according to the results of the first screening.

Table 4 Numbers and Specificities of Clones Finally Derived

Fusion number	Mouse immunized with	Mycobacterial specificity[a]			Nature of antigen		
		Specific	Cross-reactive: Limited	Cross-reactive: Broad	Protein	(glyco)lipid[b]	Unknown
23	*M. tub.* H37Rv	1	–	4	2	–	3
29	*M. tub.* H37Rv	6	5	–	2	–	9
67	*M. tub.*	2	7	4	11	–	2
24	*M. bovis* BCG	1	–	–	1	–	–
86	*M. bovis* BCG	1	–	–	1	–	–
26	*M. leprae*	–	–	4	–	ND	4
30	*M. leprae*	–	–	4	–	ND	4
31	*M. leprae*	–	2	2	–	ND	4
41	*M. leprae*	1	–	1	–	ND	2
47	*M. leprae*	4	–	4[c]	5	1	2
88	*M. leprae*	–	–	6	6	–	–
75	*M. avium*	7	1	3	–	8	3
85	*M. avium*	3	3	–	–	6	–
84	*M. intra.* *M. scrof.*	–	–	–	–	–	–
45	ADM	3	2	–	3	ND	2

[a]Specific, reacted with only one species; limited, reacted with a restricted number of species; broad, reacted with a majority of the species tested. Species tested, *M. tuberculosis* H37Rv, *M. tuberculosis* H37Ra, *M. bovis* BCG, *M. leprae, M. kansasii, M. duvalii, M. vaccae, M. smegmatis, M. avium, M. scrofulaceum, M. fortuitum, M. nonchromogenicum, M. gastrii, M. flavescens, M. terrae*. For fusions 23, 24, 26, 29, 30, 31, 41, 45, 47, also with *M. gadium*. For fusions 45, 47, also with ADM (6 strains). For fusion 47, also with *M. leprae* (5 strains) and *M. lepraemurium*. For fusions 67, 75, 85, 86, 88, also with *M. leprae* (2 strains). For fusions 67, 75, 84, 85, 86, 88 also with ADM, *M. africanum, M. tuberculosis* (patient isolate), *M. intracellulare*, and *M. gordonae*. For fusions 75 and 85, also with *M. avium* (63 strains).
[b]Reaction with lipids was tested on chloroform–methanol extracts of mycobacteria.
[c]Of these, three clones reacted with a 95-kd armadillo protein.

It can be seen that there is no one screening technique that will detect all of the antibodies able to react with antigens present on the mycobacterial target. It would be best to use more than one technique, as far as that is possible. In any event, it is necessary to choose the screening and selection techniques according to the reactions that will be expected from the monoclonal antibody and the use to which it will be put.

As is true for most antigens, there are, as yet, no standard protocols for the production of monoclonal antibodies to mycobacteria. Our experience illustrates the wide range of antibodies that can be obtained, and this is comparable with the experience of others. In the next section we will describe the various monoclonal antibodies to mycobacterial antigens that have been reported and the ways in which they can be used to increase our knowledge of mycobacteria and mycobacterial infections, as well as increase the speed and precision of diagnosis of such infections.

B. Progress in Application of Monoclonal Antibodies to Mycobacteria

1. *Taxonomy*

The problems of mycobacterial taxonomy can be approached from two different directions: that of the specialist who wishes to identify each isolated mycobacterial strain, for example, to define its relationship to all other known species, and that of the clinical mycobacteriologist who needs to identify the isolated strain well enough to be able to predict its importance as a pathogen in a given clinical situation. The former is also necessary for epidemiological studies and to follow transmission, whereas the latter is necessary for rapid decisions in treatment of patients and eventually of contacts (59).

The techniques now used to identify and to classify the mycobacteria have been mentioned previously. The monoclonal antibodies can provide further improvement as labels that may serve as markers for identification and classification, thus, assisting at both levels.

The thrust of monoclonal antibody application to identification has been with an eye to immunodiagnosis, which is discussed more completely in the next section. Many, if not most, of the monoclonal antibodies that arise when mice are immunized with one species or strain of mycobacteria are cross-reactive, either broadly or not. Much of the time, the clones producing cross-reactive antibodies are not selected for further isolation and study because of the interest assigned to specific antibodies. But in addition to the monoclonals with a narrow cross-reactivity, others could be useful to construct panels to establish the relationships among different strains and species of mycobacteria. One area that has not been explored is that in which differences have been identified with other techniques, such as enzymes involved in metabolic pathways. Monoclonal antibodies could be prepared to those components that are known to be different and applied to their detection by using the sensitive immunological tests, thus, exploiting the tools that the taxonomist already has in hand, while saving time and investment.

We have seen that the selection of monoclonal antibodies depends on the antigen and the technique used in screening. Perhaps, for taxonomically useful reagents, whole bacilli must also be used. A limitation of the available data is that the methods used to immunize, screen, and select the antibodies differ greatly among the various reports.

Coates et al. (60) first reported monoclonal antibodies against mycobacteria; they had antibodies that could distinguish among *M. tuberculosis, M. bovis*, and *M. bovis* BCG, as well as among some of the strains of these three species. A year later, Gillis and Buchanan (61,62) reported the production of monoclonals recognizing *M. leprae*; of the 11 antibodies they described, only two were specific for *M. leprae*, whereas others also reacted with *M. flavescens, M. smegmatis, M. gastri, M. tuberculosis*, and *M. gordonae*. One was positive with all of the 19 species tested; it may recognize arabinogalactan. Eight of the cross-reactive antibodies positive with *M. leprae* were positive only with *M. gastri, M. gordonae* and *M. flavescens*, suggesting that these species shared a common antigen. Other relationships would have been missed because the selection of clones for further studies excluded those producing antibodies against *M. smegmatis*. Bach and Hoffenbach (63) produced a broadly cross-reactive antibody by immunizing a mouse with *M. lepraemurium*. This monoclonal antibody, which reacted with 18 species of mycobacteria and two strains of *Nocardia asteroides*, but was negative with three species of corynebacteria, was selected by immunofluorescence and was thought to recognize a cell wall antigen. These authors had intentionally sought antibodies to cross-reactive antigens because of their possible importance in induction of protective responses (64).

Daniel and Olds (65) also reported a broadly cross-reactive monoclonal antibody produced by immunization with *M. tuberculosis*, which recognized a common epitope on two protein antigens and on arabinomannan and arabinogalactan. They emphasized that species-specificity may rest at the level of individual epitopes, rather than at that of intact antigens. This monoclonal antibody, which reacted with the previously reported antigen 5, thought to be limited to *M. tuberculosis* and *M. bovis*, reacted with culture filtrates of many species of mycobacteria, suggesting that the limited distribution of the antigen observed in the earlier studies was a function of the techniques used and did not reflect the true distribution of the antigen among the species studied.

Ivanyi et al. (52) produced antibodies that define both species-specific and cross-reactive determinants on *M. leprae*. The cross-reacting determinants appeared to be on cell wall antigens. Also, Britton et al. (66) identified four antigens with monoclonal antibodies: the two most prominent were common to many species of mycobacteria and were thought to be carbohydrates of the cell wall. The other two were 16-kd and 70-kd protein antigens; the former, found only on *M. leprae*, seemed to be associated with the cell wall, whereas the latter, also detected on *M. tuberculosis* and *M. bovis*, was more likely of cytoplasmic origin.

Our own group (51,57,67,68; see Tables 1-4) has compared the reactions of monoclonal antibodies raised against *M. leprae, M. tuberculosis, M. bovis*, and armadillo-derived mycobacteria (ADM) with a panel of different mycobacterial species using immunofluorescence (IF), ELISA, and sodium dodecyl sulfate (SDS) gels. In the first paper (51), 32 monoclonal antibodies were described according to their reactions with 16 species of mycobacteria. There were various cross-reactions, with a few interesting specific reactions. For example, one antibody F29-29 recognized the *M. tuberculosis* H37Rv strain but not the H37Ra in ELISA and IF; this antibody was positive in SGIP with *M. tuberculosis* complex (i.e., it reacted with a *M. tuberculosis*, virulent or avirulent strains of *M. bovis* BCG, and *M. africanum*). Other *M. tuberculosis* complex-specific monoclonal antibodies were F23-49, F24-2, and several of fusion 67 (68). Antibodies recognizing determinants on *M. tuberculosis, M. kansasii*, and *M. gastri*, but not on other species, were common. From fusions against *M. leprae*, five specific antibodies were produced. The antibodies produced upon immunization with ADM, which had been isolated by culture from armadillo liver that contained *M. leprae* (69) showed that the ADM organisms were little related to *M. leprae* (57); there were some cross-reactions with *M. leprae*, but also with *M. tuberculosis* H37Ra, *M. kansasii, M. duvali, M. gadium*, and *M. nonchromogenicum*. The ADM strains were not equally reactive with all of the anti-ADM antibodies, revealing heterogeneity among these isolates. Antibodies specific for *M. avium* serotypes were also produced (67), some of which reacted strongly with *M. avium, M. intracellulare*, and *M. scrofulaceum* (MAIS complex). These antibodies, however, reacted with only about 80% of the avium strains tested, thus, revealing a heterogeneity among the avium strains. Other monoclonal antibodies were positive in IF with only one or two of the serotypes of *M. avium* tested, again illustrating the variability of antigenic composition within the species.

Because all of these investigations were done to find monoclonal antibodies that could serve as tools for immunodiagnosis, in the first place, and eventually, for identification of antigens that might be useful in vaccines, most of them are not directly useful for taxonomic purposes. Instead of the highly specific, or highly cross-reactive antibodies that were sought, those with limited cross-reactivity would be more interesting for taxonomists. Furthermore, a definite taxonomic assignment would demand consistency of the characters in question, but the expression of antigenic determinants recognized by monoclonal antibodies could be greatly influenced by culture conditions.

Ivanyi et al. (49) observed differences, mainly of a quantitative nature, in the reactions of cultures of *M. bovis* AN5 grown on Sauton's or Reid's medium. Consequently, the methods of immunization and screening are equally important for the interpretation of specificity. Although monoclonal antibodies could be developed to satisfy immediate identification, the techniques of DNA characterization may prove more reliable, consistent and, ultimately, more definitive than serological methods.

2. Immunodiagnosis

Contrary to taxonomy in which all antigens establishing relationships among different species are of interest, in diagnosis, only antigens capable of identifying pathogens specifically have to be considered. This involves making a panel of monoclonal antibodies to interesting species and, then, selecting promising antibodies that can be used to isolate the antigens involved, which can then be characterized and eventually purified or produced on a large scale to develop serological tests. Occasionally, an interesting antigen has been identified with other techniques, the monoclonal antibody that recognizes it is sought, and then is used to develop tests useful for diagnosis, prognosis, or epidemiology. To develop a diagnostic test, it is necessary that the antigens to which monoclonal antibodies are produced appear in the infected host. This will be discussed in more detail along with tests themselves.

Characterization of Antigens. Antigen characterization has resulted in the description of several group-specific or, more or less, species-specific antigens, particularly for *M. leprae* and *M. tuberculosis* in WHO-sponsored exchanges of the available monoclonal reagents to compare results on antigen specificity and on performance in different tests (54–56). The results of these collaborative studies will be summarized after other results that have appeared in the literature.

One of the first antigens to be studied with monoclonal antibodies was the phenolic glycolipid-1 (PGL) of *M. leprae*, which had been isolated earlier and investigated for its specificity and its importance for serodiagnosis (70,71). Monoclonal antibodies were produced to this antigen by Young et al. (53), and later by Kolk and co-workers (51) and Mehra and Bloom (54). Of the nine antibodies to PGL described by Young et al. (53), eight were specific for *M. leprae* and recognized the distinctive trisaccharide, whereas the ninth also reacted with *M. kansasii* glycolipid. In IF, the antibodies brightly labeled whole bacteria, suggesting that the determinants in question were on the bacterial surface and might be promising for the detection of *M. leprae* in tissues.

Another antigen, which had already been recognized using patients' sera and to which monoclonal antibodies are now available, is the 36-kd protein antigen discovered by Klatser et al. (58). This antigen is recognized by sera from both LL and TT patients, but not by that of tuberculosis patients, and it has been shown to bear an *M. leprae*-specific determinant (54). The monoclonal antibody to this antigen has been used to develop a competition ELISA for immunodiagnosis (see later discussion).

Minden et al. (72) used a combination of monoclonal antibodies and affinity chromatography to isolate a 10-kd protein antigen from *M. bovis* BCG (antigen BCG-a) and studied its usefulness for skin testing.

Harboe and associates (73) studied the relationship between the mycobacterial antigens defined by some monoclonal antibodies and the antigens studied earlier in crossed-immunoelectrophoresis (CIE); they were able to identify the antigens assigned to the antibodies of Coates et al. (60) and to relate these to the previously described antigens BCG 56, 78, and 82. Daniel et al. (74) used monoclonal antibodies to demonstrate that antigen 5 of *M. tuberculosis* and its arabinomannan share a single major epitope. When Harboe and co-workers (54) applied CIE to *M. leprae* antigens and antibodies, they reported that the antibodies recognized the previously described *M. leprae* antigens 2, 7, and 11. It is interesting that the antibodies reacting with antigen 7 of *M. leprae* included both specific and cross-reacting antibodies that had been shown to bind to antigens of different composition (protein and carbohydrate/lipid) and different molecular masses (36-, 55-, and 65-kd; 28-50-kd smear). This is an illustration of the ways in which monoclonal antibodies can be used to dissect the antigens known from work with polyclonal antisera. This is further described for *M. leprae* antigen 7 by Reitan et al. (75), including its application to serodiagnosis (discussed later). Papa and associates (76) looked at the interactions between a monoclonal antibody to BCG antigen 60, which also recognized *M. leprae* antigen 7 and the strains of corynebacteria that were isolated from leprosy lesions as well as nonleprosy strains of corynebacteria. In ELISA it was found that the three leprosy-derived strains also bore this determinant common to *M. leprae*, but it could not be concluded that the antigen was related to the origin of the strains in leprosy lesions because two of the seven nonleprosy strains were also found to be positive.

Young et al. (77) produced monoclonal antibodies to *M. leprae* and found that some of them recognized a new 28-kd protein antigen that had cross-reacting as well as specific determinants.

A comparison of the antigenic determinants on cell wall-associated carbohydrates of *M. leprae* and other species was made by Britton et al. (78), using monoclonal antibodies raised against *M. leprae* and *M. bovis* BCG. Evidence from studies from patients' sera had suggested that some of the important antigens in these species were in broad antigen bands on SDS-PAGE and contained carbohydrates and lipids. Their antibodies reacted with antigens of molecular masses 4.5-6 kd, 16-18 kd, 30-40 kd, and 70 kd, present in *M. leprae* and other species. The authors suggested that the target might be the cell wall arabinomannan of *M. leprae*, to which monoclonal antibodies had already been produced by Miller and Buchanan (79), who used mice immunized with arabinomannan purified from *M. smegmatis*. Because there are also quantitative differences in the polysaccharide antigens between the mycobacterial species, even monoclonal antibodies detecting antigens common to one or more species can be functionally specific in diagnostic tests. For example, in a competition assay with the antipolysaccharide monoclonal antibody, the inhibitory titers of sera from tuberculosis patients were not significantly raised above those of controls, whereas sera from leprosy patients were effective inhibitors (66). Carbohydrate antigens of mycobacteria are also of interest in the investigation of cellular immunity, because they have been shown to activate B lymphocytes (80) and may be able to induce suppression of T lymphocytes (81).

Praputpittaya and Ivanyi (82) used a tandem immunoassay with one of their monoclonal antibodies (ML34) to demonstrate that the antigen in question (a repeating epitope on a water-soluble, protease-resistant antigen) was present on several mycobacteria, but because only intact *M. tuberculosis* bacilli bound to the antibody in a capture assay, and not intact *M. leprae*, they suggested that the antigen is expressed on the surface of *M. tuberculosis* and internally in *M. leprae*, which could make a crucial difference for immunodiagnosis.

Morris and co-workers (83) examined the reactions of a panel of monoclonal antibodies with not only the usual laboratory strains but also field isolates. One of the monoclonal antibodies, MB 5, seemed to be specific for *M. bovis*, with the exception of *M. bovis* BCG; it was positive with 44 field isolates and with the pathogenic laboratory strain Valléé. The antibody recognized a 29.8-kd antigen that was not found in extracts of several other species of mycobacteria. The antigens identified by these antibodies were also sought in PPD preparations that were used for immunodiagnosis and epidemiology.

It can be seen that there is a wide variety of monoclonal antibodies that recognize a more restricted variety of antigenic determinants on mycobacteria. The comparative studies organized under the auspices of WHO to standardize the antibodies against *M. leprae* and *M. tuberculosis* have helped to bring order into this array of information. The results of these studies have been published (54–56). In the workshop on *M. leprae*, the monoclonal reagents were shown to react with a range of antigens, including four distinct *M. leprae*-specific determinants on proteins, the *M. leprae*-specific PGL, 55-kd and 65-kd proteins, all of which bore both specific and cross-reactive determinants, and carbohydrate/lipid antigens to which mostly cross-reactive antibodies have been produced. In CIE, at least three different precipitation lines could be detected with this panel of antibodies. When IF was used, the variations in the techniques, as carried out in the different laboratories, may have given rise to some of the variation in the results obtained.

In the workshop of monoclonal antibodies raised against *M. tuberculosis* (56), 31 antibodies were compared. Of these, five were negative in all of the tests used, whereas a further nine were positive in ELISA or dot blot, Western blot, or

SGIP, but negative in IF. It was revealed that most of the antibodies recognized protein antigens with molecular masses of 12, 14, 19, 20-80, 23, 38, 40, 65, and 71 kd, whereas for six antibodies, the nature of the antigen involved remained unknown. None were specific for *M. tuberculosis* alone, although one that reacted well only in dot blot seemed to be specific for an avirulent strain of *M. tuberculosis* IT 30. In the ELISA, seven antibodies reacted with only the complex of *M. tuberculosis, M. bovis* BCG, and *M. africanum*; 10 had limited cross-reactivity within the 23 strains of mycobacteria in the test panel and 6 were broadly cross-reactive. The activities in this workshop also included the screening of DNA libraries of *M. tuberculosis* and the results were in good agreement with the results of the characterization of the antigens by immunochemical methods. This was promising for the efficient identification and isolation of interesting antigens from mycobacteria, for all purposes.

All of the foregoing reports describe results from monoclonal antibody production in mice, the methodology for which is now well developed (although still highly variable). There are several reasons that monoclonal antibodies of human origin would be better tools for many of the studies described. For example, in competitive assays to detect antibodies in patients, antibody probes derived from human cells might be more likely to define their counterparts in human sera. Because the human immune system may perceive the elements of mycobacteria differently than does that of the mouse, the appropriate definition of important antigens is with a human system. In particular, differences in subclinical and clinical infection may be related to differences in antigen recognition. In any event, at this time the technical obstacles with human hybridomas have impeded the progress in this area. There is one report on production of human monoclonal antibodies to *M. leprae* (84), in which a double-mutant human lymphoblastoid cell line (KR4) was used as a fusion partner for Epstein-Barr-transformed B-cell lines from lepromatous leprosy patients to derive human hybridomas that produced antibodies to *M. leprae*. The authors screened 2200 clones and found 359 that were reactive to an extract of *M. leprae* and, furthermore, of 168 screened, there were 21 positive with PG. Nine clones that produce *M. leprae*-specific antibodies were obtained. This report is encouraging by demonstrating that it is possible to obtain human antimycobacterial monoclonal antibodies; it remains to be seen whether their production and specificities will alter any of the concepts about the importance of different antigens derived from the analyses with mouse reagents.

It must be remembered that the cross-reacting antibodies can also contribute to the development of diagnostic tests. First, antibodies to cross-reacting antigens that do not bear specific epitopes can be used to remove such problematic antigens from mixtures to enrich for the antigens of interest. Second, antibodies that in highly sensitive laboratory investigations are cross-reactive, may perform in serological diagnostic tests in a specific manner (e.g., 88). For the development of vaccines, cross-reacting antigens may indeed be the object of the search (64).

Use of Monoclonal Antibodies in Diagnostic Tests. The simplest use of monoclonal antibodies in diagnosis is to use them to detect or to identify the infectious organisms in samples of tissues or fluids from the patients. If conventional staining were to reveal mycobacteria, then monoclonal antibodies in an IF or IP test could be an excellent tool for diagnosis. This kind of exploitation has been worked out for some of the antibodies described earlier (57,67,85). In Figure 3 an example is shown of the identification of *M. tuberculosis* and *M. avium* in patient material using monoclonal antibodies in an immunofluorescent test. When up to three of the *M. tuberculosis* antibodies and four against *M. avium* were used, reaction was seen with 8 of 9 *M. tuberculosis* isolates and 8 of 10 *M. avium* isolates. However, an antibody raised to bacteria obtained from in vitro culture may not react with the antigen in situ or obtained in vivo because of differences in the composition of bacterial components, accessibility of the antigens, or to interference from host components.

Somewhat less directly, the isolated bacteria could be cultured and the cultures then identified with the help of monoclonals. This is less time-saving but smaller amounts of material are required than, for example, for biochemical tests. Morris and Ivanyi (86) used monoclonal antibodies to test field isolates of *M. bovis* and other mycobacteria; the isolates had been cultured for 6 weeks before sonication and testing for binding with radiolabeled monoclonal antibodies. Isolates from 10 badgers and 10 cattle, along with the laboratory strains, were tested. The antibodies were found to be nonspecific for *M. bovis*, but to cross-react with *M. tuberculosis*, necessitating testing with other *M. tuberculosis*-specific antibodies to distinguish between these two groups. The field isolates were all positive and no differences among strains isolated from cattle and badgers were observed.

The next step in the use of monoclonal antibodies for diagnosis would be to use them in techniques, such as ELISA, RIA, agglutination, or hemagglutination, and the like, in which the antibodies could detect small amounts of soluble antigens in blood or fluids obtained from patients. This is routine with polyclonal sera and some antigens such as meningococcus, as well as for detection of hormones with monoclonal antibodies, but it has not been fully investigated in mycobacterial infections. Indeed, it is not clear whether or not such antigens are present in accessible materials in quantities sufficient for detection, even with sensitive methods. This would depend not only upon release of specific antigens from foci of infection but on their persistence, free from antibody, in the circulation. In lepromatous leprosy, in which there is certainly sufficient antigen, it has been difficult to detect antigen in serum, perhaps partly because of high antibody levels and, consequently, high amounts of bacterial antigen bound in immune complexes (39,87). An additional complication to the use of monoclonal antibodies for detection of antigen in vivo was illustrated by the observations of Morris and Thorns (88,89), in the first instance, of the binding of monoclonal antibodies raised against and specific for *M. bovis* to various tissue antigens, such as bovine albumin and collagen and, in the second, of the sharing of epitopes between mycobacteria and many other microorganisms, including *Brucella, Escherichia coli, Salmonella*, and *Trichinella*, among others. Apart from their possible importance for the development of the disease and some of the symptoms, these findings advise great caution in the use of monoclonal antibodies to establish diagnosis by detection of circulating antigen.

A more optimistic note is provided by the possibility of detection, using monoclonal antibodies, of the *M. leprae*-specific glycolopid in tissues (57,61, 62). If this also proves possible for other species, sputum samples could be extracted with chloroform and the glycolipids detected in a capture ELISA with use of plates coated with specific monoclonal antibodies.

When it is difficult to detect either the infections organism itself, or parts thereof, it can be informative to look for antibodies because these do indicate contact with the organism, if they can be identified as specific. The published attempts to develop diagnostic tests for mycobacterial diseases have utilized radio- or enzyme-immunoassays to look for specific antibodies in patients' sera. These do have the advantage that there is no need to purify the antigens from the complex mixtures of bacterial cultures or isolates because the monoclonal antibody provides the specificity. There have been several reports of laboratory investigations of potentially useful immunological tests in which competition between monoclonal antibodies with known specificity and antibodies in patients' sera for binding to crude antigens from mycobacteria has been the measure of exposure to the antigen (bacteria) in question.

Following up on the report on the production of monoclonal antibodies to *M. tuberculosis*, Hewitt et al. (90) tried to exploit these in a radioimmunoassay using *M. tuberculosis* pressate as antigen and different radiolabeled monoclonal antibodies in competition with patients' sera to identify tuberculosis patients. This was not entirely satisfactory because, in the first place, the uninfected controls gave high background levels of inhibition of the monoclonal antibodies' binding, so that there were many false-negative reactions among the patients. In addition, there was the problem of great variability among the patients' sera as to which monoclonal antibody they best inhibited, which implied that none of the antibodies were to "immunodominant" antigens. Even using the combination of the two best monoclonal antibodies, only 71% of the patients gave positive results in the test. This was no improvement on the existing serological tests and did not recommend itself. These antibodies were further tested on a range of tuberculosis patients, treated and untreated, as well as patients with other lung complaints (91). Positive values were found with 74% of all active pulmonary tuberculosis patients and 55% of those untreated. None of the patients with

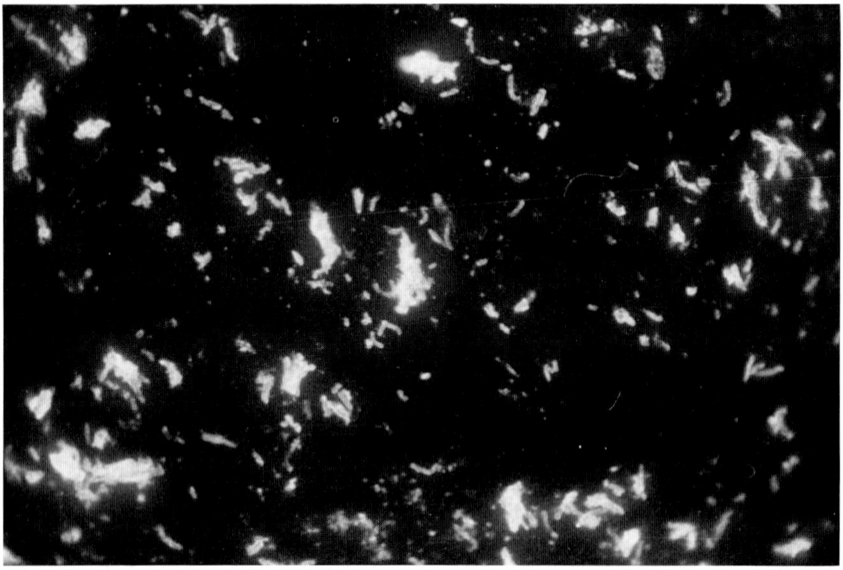

(a)

Figure 3 Indirect immunofluorescence tests showing: (a) Mycobacteria freshly isolated from a patient, stained with monoclonal antibody F29-23, specific for *M. tuberculosis* (magnification 400X). (b) *M. avium* serotype 4, isolated from a patient, stained with monoclonal antibody F85-2, which recognizes *M. avium, M. intracellulare*, and *M. scrofulaceum* complex. (c) Mycobacteria from sputum of a patient stained with monoclonal antibody F29-23, specific for *M. tuberculosis* (the arrows indicate the mycobacteria). In each case the specimens were negative with other monoclonal antibodies, for example, (a) and (c) were negative with antibodies to *M. avium*, and (b) was negative with those against *M. tuberculosis* (data not shown).

suspected, but excluded tuberculosis, were positive. The levels of binding inhibition of the monoclonal antibody were generally correlated with the levels of anti-total-mycobacteria antibodies and it was interesting, but disappointing, that the highest titers were found in patients who had received chemotherapy for 1-8 months before testing. This may be a result of the large quantities of released antigen, to which antibodies may be made; these seem to be the antigens to which the monoclonal antibodies are produced, and other antigens may be important to detect antibody responses in early infection. Because in the test, as it stands, the best results were from treated patients, it is not promising for diagnosis, although it could be useful for follow-up of treatment.

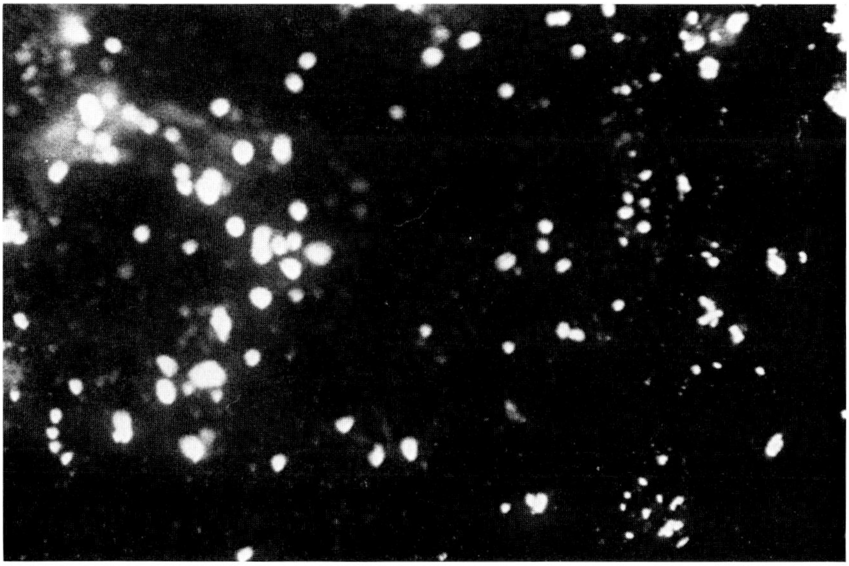

(b)

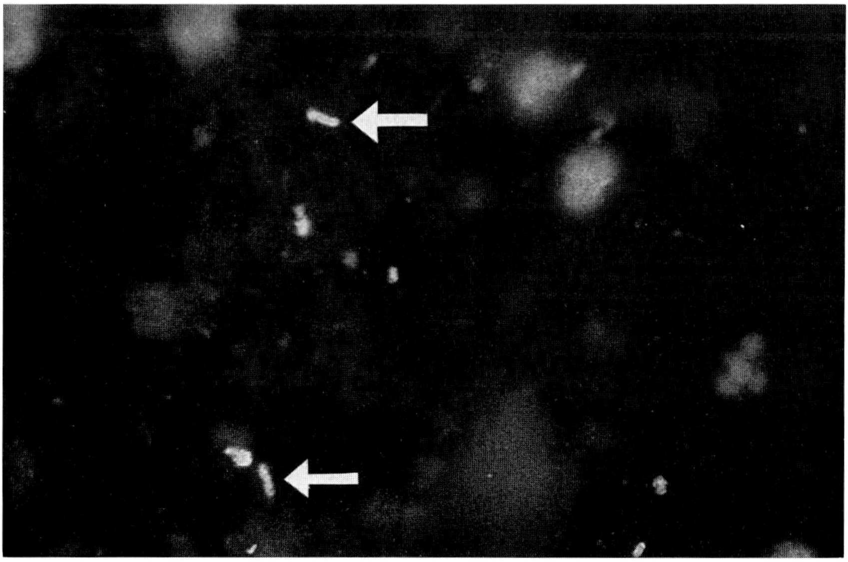

(c)

Several laboratories have been busy with development of tests to detect early infection with *M. leprae*. Sinha et al. (92) used a competitive radioimmunoassay with *M. leprae* sonicate as antigen and had encouraging results with the few patients studied; family contacts as well as tuberculosis patients were negative in this assay. It was interesting that they were able to use one of their specific anti-*M. leprae* antibodies in this test with patients' sera (MY2a, recognizing a 35-kd protein antigen), whereas another monoclonal antibody that was interesting in the laboratory (MY1a) was not inhibited by the patients' sera. This may be a case of the differences in perception of important antigens between mice and men. When this test was used for a larger number of sera in an endemic area (93), the sera from LL and BL patients were all positive, those from BB, 87.5% positive, and from BT-LT, 46.7% positive. Forty-six percent of household contacts of multibacillary patients were also positive, whereas 15 paucibacillary patients and 85 non-leprosy-related subjects were negative (including tuberculosis patients). It was felt that the test was useful for diagnosis, and perhaps for prognosis, of the shift toward LL in borderline patients. To evaluate the meaning of the positive reactions among the contacts, further follow-up would be necessary.

Reitan et al. (75), whose monoclonal antibodies have been described earlier, used these in two kinds of tests to detect antibodies to *M. leprae*: a solid-phase radioimmunoassay and an ELISA on nitrocellulose filters (dot ELISA). Both of these tests gave encouraging results: most of the sera from untreated BT patients and from treated or untreated LL patients gave 25-80% inhibition, whereas sera from treated BT, from treated or untreated TT patients, and from contacts or nonexposed controls gave less than 20% inhibition. The results in the two tests were comparable. Interestingly, there was a correlation between the percentage inhibition and the bacterial index (BI) in the multibacilliferous patients, as well as between the inhibition and the time after the beginning of treatment that the patient had a BI of zero. They concluded that their assay could not discriminate between contacts and unexposed persons, hence, it was not useful for early diagnosis of infection, but that it might be useful for finding the most dangerous highly bacilliferous patients, either early in infection or during relapse. Here, the relative insensitivity could be an advantage, because only the cases more important for control of spread of infection would be detected.

After having identified an *M. leprae* antigen that was promising in its specificity (58), Klatser et al. (94) then produced a monoclonal antibody to it and used that to set up a competitive ELISA to detect antibodies to *M. leprae* in the sera of leprosy patients. The results were very good: seropositivity was found in 100% of the multibacillary patients and in 91% of the paucibacillary patients, whereas only 5% of the control group gave inhibition in the assay. The mean inhibition (%) was also much lower in the controls than in the two patient groups. In contrast to the other serological tests for antibodies in leprosy, this one seems to exploit an antigen that induces response in both multi- and paucibacillary patients; it is, thus, most promising for screening larger groups to detect early infections.

All of the preceding tests have been developed by using sera available in the laboratories and have been tested little, or not at all, on large numbers of sera in endemic areas. This work is now underway and will provide the test of their usefulness as tools for diagnosis and epidemiology.

Although the techniques discussed here are more than acceptable for sensitivity, the specificity is very much dependent on the availability of the appropriate antigens, and it is necessary to define carefully the requirements and to test carefully that the prospective candidates fulfill them under the conditions defined before they can be considered trustworthy aids. However, because in mycobacterial diseases there often are no good alternatives, they will have the chance to prove themselves in the clinical or epidemiological settings.

Theoretically, monoclonal antibodies could also be used to find antigens that could then be prepared in a purified and characterized form for use in skin testing to measure cellular immune responses. However, the antigens picked out with monoclonal antibodies are not necessarily those useful for skin tests. For example, Ivanyi et al. (49) found that only one of seven monoclonal antibodies reacted well with the *M. tuberculosis* or *M. bovis* PPD preparations that elicited good skin test reactions. Thus, it is necessary to combine the use of monoclonal antibodies with studies using cells to investigate the effectiveness of antigens in stimulating cells.

Another way in which monoclonal antibodies could be used to develop a relatively simple diagnostic test, in those cases in whom antibody detection would be useful, would be to produce with this technology an anti-idiotype antibody, which could react with the antibody(ies) produced in response to the infectious agent in the host. Such antibodies could be coated onto plates in the manner of antigens and would, it is hoped, bind only to antibodies that have binding sites specific for the antigen; a positive reaction would then suggest infection. This would have the advantages that monoclonal antibodies can be easier to produce and to standardize than the antigen itself (especially for slow-growing strains or for *M. leprae*), but the specificity of the reaction would have to be well studied, because cross-reacting idiotypes need not be induced by the same antigens.

3. Immunopathology in Mycobacterial Diseases and Monoclonal Antibodies

The applications of monoclonal antibodies, thus far discussed, have been direct, in the sense of exploiting antibodies that react with mycobacterial antigens. There is another way in which monoclonal antibodies are making a contribution, at least to our understanding of the pathogenesis of mycobacterial diseases, if not yet to their treatment and control. This kind of work uses the monoclonal antibodies that identify different types of cells or soluble factors involved in immune responses, to clarify the reactions taking place, especially in the foci of infection. Immunohistochemical techniques that employ monoclonal antibodies can give much more precise information about the nature of the cells in a tissue reaction and about their activities, surface markers, and products, and they permit comparison of studies done in different centers with standard reagents.

Because the responses generated and, to a certain extent, carried out by T lymphocytes are thought to be those most important for resistance to, or recovery from, infection with mycobacteria, among other parasites, investigations into the possible defects in their numbers, types, or functions have been the most common, using material from patients or infected experimental animals. Some of the immunopathology of the mycobacterial diseases is also thought to be due to inappropriate activities of these cells. The functional T-cell subsets in leprosy, as defined by monoclonal antibodies, have been briefly described in an earlier section.

The different subsets of human T cells can be identified by their membrane antigens, as recognized by the OKT-series of monoclonal antibodies. The Th/Ti cells are generally OKT-4$^+$, whereas Ts/Tc cells are usually OKT-8$^+$ (95,96), although exceptions to these observations are not rare (30,97). The study of T-cell subsets using the OKT-series monoclonals has been useful in the histological definition of leprosy granulomas. It has been shown that in TT granulomas the mainly OKT-4$^+$ T-cell population inside the lesion is surrounded by OKT-8$^+$-type cells, whereas in LL granulomas, fewer lymphocytes, with a higher proportion of OKT-8$^+$ cells, are present (30,98). Similar studies in tuberculosis remain difficult to perform because of the paucity of lung biopsies. Although LL is associated with decreased IL-2 production in patients, this defect seems to be secondary, as the quantitative and qualitative deficiencies of T lymphocytes disappear from successfully treated *M. leprae*-free patients (29). Finally, in addition to an impairment of T-cell-mediated immunity specifically against *M. leprae* antigens in leprosy, some degree of nonspecific immunodepression is also observed in the LL form of the disease (29).

In vitro studies of lymphocyte responses to mycobacterial antigens are also assisted by the availability of monoclonal antibodies that define cells and soluble components of the immune response. For example, Kaplan et al. (99) could show that monocytes or their products and not Ts (OKT-8$^+$) lymphocytes were responsible for suppression in leprosy patients' responses to *M. leprae* antigen, whereas in BCG-vaccinated persons, the BCG-induced suppressor cells were T4$^+$, and the effect was, at least, partially histocomptability-restricted (97). Mathew et al. (100) used monoclonal antibodies to identify the types of macrophages in granulomas induced in guinea pigs by *M. bovis* BCG and by *M. leprae*. In the former, there were macrophages bearing the macrophage-specific antigens but not expressing class II histocomptability antigens; they were able to control the microorganisms. In the latter, the macrophages also expressed class II antigens but were not organized and were not successful in controlling the microorganisms. The LL lesions were demonstrated to contain three types of macrophages (101): mainly those bearing typical macrophage antigens, but also 15-30% were identified as interdigitating cells by monoclonal antibodies and a few bore the antigens typical of Langerhans' cells. Alvarenga et al. (102) did not find any differences in the numbers of intraepidermal Langerhans' cells among different types of leprosy patients, despite significant variation among them unrelated to the clinical forms.

In summary, although there are still contradictions in the results reported from investigations, using monoclonal antibody tools, into immune responses, it is obvious that this road will lead to a better understanding of reasons for the lack of effective responses in many persons infected with mycobacteria, once approaches have been chosen to permit comparisons with clear questions to be answered. Nearly all of the investigations thus far have been done with *M. leprae*, perhaps because the biopsy material is more readily accessible than in other infections. In any event, it is clear that the lack of response in lepromatous leprosy may be due to more than one deficiency; those implicated thus far include the recruitment of T lymphocytes, especially of the appropriate type (Th), and the lack, for whatever reason, of production of IL-2 and of response to it, in part because of lack of induction of expression of receptors for that growth factor. The next period will no doubt see the appearance of more information on this subject, which can only help in the improvement of our understanding of the disease and, eventually, to better control its symptoms.

III. PROSPECTS FOR THE FUTURE

Combining the knowledge acquired, in part, through the use of monoclonal antibodies in the fields of antigenic characterization of the composition of the mycobacteria and the functioning of the cellular components of the host immune system, an appreciation of the needs for successful prevention or therapy can evolve and approaches can be considered to realize these.

As we have seen, there are several antigens on all mycobacteria, each with at least several epitopes, which can be identified by the antibody response that they induce. It is known that the generation of an immune response involves a first step in which the macrophage (or a member of the macrophage family) plays the major role; in this step, the mycobacterium is taken up by the phagocytes, killed, and digested. Fragments of the organism later appear on the surface of the macrophage, if this has been successful, and there they are presented to the population of T lymphocytes. Genes controlling these functions of macrophages are known to be important for resistance to mycobacterial infections in mice (reviewed in Ref. 103), but similar genes have not yet been identified in humans. In addition, it is known that the type of leprosy that develops upon successful infection with *M. leprae* in humans is associated with the histocomptability antigens (43). It seems that these two points, when they occur early in the interaction of the infectious organism with the host, are decisive in the development of serious infection or in control and elimination of the mycobacteria.

With the help of T-cell clones and of antigens prepared with monoclonal antibodies, it is now possible to begin to investigate the reactions induced in T cells by mycobacterial antigens and to dissect the immune responses of patients and controls by using known antigens. A review of such experiments is out of place here, but articles such as those by Haanen et al. (104) and Emmrich et al. (105) are examples of possible approaches. We have studied the responses of T-cell clones to purified antigens of *M. tuberculosis* and *M. leprae* (68). We found, for example, a clone from a Mantoux-positive donor that responded to the 64-kd antigens from *M. tuberculosis* but not to the 64-kd antigen from *M. leprae*. On the other hand, the 36-kd antigen, shown to bear *M. leprae*-specific determinants recognized by monoclonal antibodies, was stimulatory for T lymphocytes from both leprosy patients and a Mantoux-positive healthy donor. Thus, the cells recognized some cross-reactive determinants. These observations stress the importance of defining the conditions under which specificity has been assessed and of investigation of T-cell antigen recognition, as well as that by B lymphocytes, if we are to come to understand the inappropriate immune responses that lead to disease.

This kind of approach is promising and may lead to several points of progress. First, it may lead to development of a skin test that could assist in prediction of an infection's outcome, if an antigen were found to which the T-cell response patterns of different categories of patients and contacts were known. Furthermore, an eventual vaccine depends on the identification of an antigen that will induce a protective response in those persons predisposed, either genetically or because of their condition at the moment of infection, to develop disease. This is not necessarily true of any one pure specific or cross-reactive antigen.

In mycobacterial diseases, as in many others, a detailed enough understanding of the events during the time an individual is developing illness could permit manipulation of those events, or at any rate the immunological ones, by interfering in the idiotype network interactions. However, the precision of the knowledge necessary to avoid undesirable and unpredictable side effects puts this possibility well into the future.

If monoclonal antibodies were to be made to other known components of the mycobacteria, for example, enzymes important for drug sensitivity, the antibodies could contribute to their isolation, purification, and study to develop new or to improve existing drugs. By use of monoclonal antibodies directed to membrane and cytoplasmic components, the transport of drugs into the cell could also be studied, as could the roles of receptors for drugs on the surface of the bacterial cell and inside the cell. Because the development of resistance to treatments already long in use is a continuing problem, especially in leprosy, this kind of basic information could accelerate the development of new instruments for control by chemotherapy.

As mentioned in the introduction, the components of the bacterial cell wall have been studied chemically, but many of the fine differences in distribution and function could be elucidated with the help of monoclonal antibodies to the various components. Differences that can be of clinical importance, such as those between "resting" mycobacteria and those active in an infection, could be investigated.

The usefulness of DNA technology in the full exploitation of the new and future immunological knowledge cannot be passed over. It is discussed in Chapter 8 for a related field. It is important here to note the role that monoclonal antibodies can play in the selection of products of cloned genes. They are used, for example, by blotting, to identify the clones producing the interesting antigens so that they can be selected quickly. Once the clones have been selected and grown up, the monoclonal antibody can be used to purify the desired product from the bacterial culture by affinity chromatography. Thus, together, these two areas of strength in biotechnology can be expected to make it possible to deliver the needed materials for control of diseases.

The problem now remains for us to identify what is needed; that is not a simple problem for chronic and complex diseases such as those caused by the mycobacteria. It is necessary to dissect and describe as finely as possible the immune responses and the antigens that induce them, and to relate the information gathered in vitro to the in vivo situation, so that strategies can be worked out and applied.

ACKNOWLEDGMENTS

We are grateful to H.L. David, of the Service de la Tuberculose et des Mycobactéries, Institut Pasteur, for critical reading of the manuscript and to Alice Gilaard for the typing.

REFERENCES

1. Lehmann, K. B., and R. Neumann, *Atlas and Grundes der Bakteriologie und Lehrbuch der speciellen bakteriologischen Diagnostiek*. 1st ed. J. F. Lehmann, Munchen (1896).
2. Lachner-Sandoval, V., Über Strahlenpilze. Inaugural Dissertation Strassburg. Universitäts Buchdrucherei Von Carl Georgi, Bonn (1898).
3. Goodfellow, M., and D. E. Minnikin, Circumscription of the genus. In *The Mycobacteria: A Sourcebook*, Part A, G. P. Kubica and Wayne, L. G. (eds.). Marcel Dekker, New York, pp. 1–24 (1984).
4. Lévy-Frébault, V. E., E. Rafidinarivo, J. C. Promé, J. Grandry, H. Boisvert, and H. L. David, *Mycobacterium fallax* sp. nov. *Int. J. Syst. Bacteriol. 33*: 336–343 (1983).

5. World Health Organization, Report on the Sixth Meeting of the Scientific Working Group on the Immunology of Leprosy. WHO document TDR/IMMLEP-SWG 6 (1982).
6. Styblo, K., Recent advances in epidemiological research in tuberculosis. *Adv. Tuberc. Res. 20*:1-63 (1980).
7. Baily, G. V. J., Trial of BCG vaccines in South India for tuberculosis prevention. *Indian J. Med. Res. 70*:349-363 (1979).
8. Stanford, J. L., Leprosy research, present and future. *Acta Leprol. (Geneve) 2*:421-425 (1984).
9. David, H. L., Classification and identification of *Mycobacterium leprae. Acta Leprol. (Geneve) 2*:137-151 (1984).
10. Rastogi, N., and H. L. David, Phagocytosis of *Mycobacterium leprae* and *M. avium* by armadillo lung fibroblasts and kidney epithelial cells. *Acta Leprol. (Geneve) 2*:267-276 (1984).
11. Shepard, C. C., The experimental disease that follows the injection of human leprosy bacilli into footpads of mice. *Br. J. Exp. Med. 112*:445-454 (1960).
12. Kirchheimer, W. F., and E. E. Storrs, Attempts to establish the armadillo (*Dasypus novemcinctus* Linn.) as a model for the study of leprosy. *Int. J. Lepr. 39*:692-702 (1971).
13. Meyers, W. M., G. P. Walsh, H. L. Brown, Y. Fukunishi, C. H. Binford, P. J. Gerone, and R. H. Wolf, Naturally acquired leprosy in a mangabey monkey (*Cerocebus* sp.). *Int. J. Lepr. 48*:495-496 (1980).
14. Colston, M. J., and K. Kohsaka, The nude mouse in studies of leprosy. In *The Nude Mouse in Experimental and Clinical Research*, Vol. 2, J. Fogh and Giovannella, B. C. (eds.). Academic Press, New York, pp. 247-266 (1982).
15. Ravisse, P., N. Rastogi, H. L. David, and C. C. Guelpa-Lauras, Experimental leprosy in the armadillo and nude mice: Comparative histobacteriology and ultrastructure. *Acta Leprol. (Geneve) 2*:327-339 (1984).
16. Kubica, G. P., Clinical microbiology. In *The Mycobacteria–A Sourcebook*, Part A, G. P. Kubica and Wayne, L. G. (eds.). Marcel Dekker, New York, pp. 133-175 (1984).
17. Hawkins, J. E., Drug-susceptibility testing. In *The Mycobacteria–A Sourcebook*, Part A, G. P. Kubica and Wayne, L. G. (eds.). Marcel Dekker, New York, pp. 177-193 (1984).
18. Chaparas, S. D., Immunologically based diagnostic tests with tuberculin and other mycobacterial antigens. In *The Mycobacteria–A Sourcebook*, Part A, G. P. Kubica and Wayne, L. G. (eds.). Marcel Dekker, New York, pp. 195-220 (1984).
19. Lind, A., and M. Ridell, Immunologically based diagnostic tests–humoral antibody methods. In *The Mycobacteria–A Sourcebook*, Part A, G. P. Kubica and Wayne, L. G. (eds.). Marcel Dekker, New York, pp. 221-248 (1984).
20. Melsom, R., Serodiagnosis of leprosy; the past, the present and some prospects for the future. *Int. J. Lepr. 51*:235-252 (1983).
21. Rastogi, N., C. Frehel, and H. L. David, Triple-layered structure of mycobacterial cell wall: Evidence for the existence of a polysaccharide-rich outer

layer in 18 mycobacterial species. *Curr. Microbiol. 13*:237–242 (1986).
22. Chapman, G. B., J. H. Hanks, and J. H. Wallace, An electron microscope study of the disposition and fine structure of *Mycobacterium lepraemurium* in mouse spleen. *J. Bacteriol. 77*:205–211 (1958).
23. Draper, P., and R. J. W. Rees, Electron-transparent zone of mycobacteria may be a defence mechanism. *Nature 228*:860–861 (1970).
24. Ryter, A., C. Frehel, N. Rastogi, and H. L. David, Macrophage interaction with mycobacteria including *M. leprae. Acta Leprol. (Geneve) 2*:211–226 (1984).
25. Frehel, C., C. de Chastellier, T. Lang, and N. Rastogi, Evidence for inhibition of fusion of lysosomal and prelysosomal compartments with phagosomes in macrophages infected with pathogenic *Mycobacterium avium. Infect. Immun. 52*:252–262 (1986).
26. David, H. L., N. Rastogi, S. Clavel-Sérès, F. Clément, and M. F. Thorel, Structure of the cell envelope of *Mycobacterium avium. Zentralbl. Bacteriol. Hyg. (A) 264*:49–66 (1987).
27. Rastogi, N., C. Frehel, A. Ryter, H. Ohayon, M. Lesourd, and H. L. David, Multiple drug resistance of *Mycobacterium avium*: Is the wall architecture responsible for the exclusions for antimicrobial agents? *Antimicrob. Agents Chemother. 20*:667–677 (1981).
28. Goren, M. B., and P. J. Brennan, Mycobacterial lipids: Chemistry and biological activities. In *Tuberculosis*, G. P. Youmans (ed.). W. B. Saunders Co., Philadelphia, pp. 63–193 (1979).
29. Sansonetti, P., and P. H. Lagrange, The immunology of leprosy: Speculations on the leprosy spectrum. *Rev. Infect. Dis. 3*:422–469 (1983).
30. Rook, G. A. W., The immunology of leprosy. *Tubercle 64*:297–312 (1981).
31. Draper, P., Mycobacterial inhibition of intracellular killing. In *Microbiol Perturbation of Host Defences*: *The Beecham Colloquia*, F. O'Grady and Smith, H. (eds.). Academic Press, London, pp. 143–164 (1981).
32. Ridley, D. S., and W. H. Jopling, Classification of leprosy according to immunity: A five group system. *Int. J. Lepr. 34*:255–273 (1966).
33. Dharmendra, and K. R. Chatterjee, Prognostic value of the lepromin test in contacts of leprosy cases. *Lepr. India 27*:149–157 (1955).
34. Myrvang, B., T. Godal, D. S. Ridley, S. S. Froland, and Y. K. Song, Immune responsiveness to *Mycobacterium leprae* and other mycobacterial antigens throughout the clinical and histopathological spectrum of leprosy. *Clin. Exp. Immunol. 14*:541–553 (1973).
35. Ridley, D. S., Reactions in leprosy. *Lepr. Rev. 40*:77–81 (1969).
36. Wemambu, S. M. G., J. L. Turk, M. F. R. Waters, and R. J. W. Rees, Erythema nodosum leprosum: A clinical manifestation of the Arthus phenomenon. *Lancet 2*:933–935 (1969).
37. Godal, T., Immunological aspects of leprosy—present status. *Prog. Allergy 25*:211–242 (1978).
38. Ridley, M. J., and D. Russel, An immunoperoxidase study of immunological factors in high immune and low resistance granulomas in leprosy. *J. Pathol. 137*:149–157 (1982).
39. Valentijn, R. M., W. R. Faber, R. F. Lai-A-Fat, J. C. Chan Pin Jie, M. R. Daha, and L. A. van Es, Immune complexes in leprosy patients from an

endemic and non-endemic area and a longitudinal study of the relationship between complement breakdown products and the clinical activity of erythema nodosum leprosum. *Clin. Immunol. Immunopathol. 22*:194–202 (1982).
40. Bodmer, J., and W. Bodmer, Histocompatibility 1984. *Immunol. Today 5*: 251–254 (1984).
41. Nixon, D. F., J. P. Ting, and J. A. Frelinger, Ia antigens on nonlymphoid tissues: Their origins and functions. *Immunol. Today 3*:339–342 (1982).
42. Zinkernagel, R. M., and P. C. Doherty, H-2 compatibility requirement for T-mediated lysis of target cells infected with lymphocytic choriomeningitis virus: Different cytotoxic T-cell specificities are associated with structures coded in H2K or H-2D. *J. Exp. Med. 141*:1427–1436 (1975).
43. Van Eden, W., R. R. P. de Vries, J. D'Amaro, G. M. T. Schreuder, D. L. Leiker, and J. L. van Rood, HLA-DR associated genetic control of the type of leprosy in a population from Surinam. *Human Immunol. 4*:343–350 (1982).
44. Howie, S., and W. H. McBride, Cellular interactions in thymus-dependent antibody responses. *Immunol. Today 3*:273–278 (1982).
45. Rook, G. A. W., Suppressor cells of mouse and man: What is the evidence that they contribute to the aetiology of the mycobacterioses? *Lepr. Rev. 53*:306–312 (1982).
46. Closs, O., Antigenic analysis of *Mycobacterium leprae*—A quest for species-specific antigens in hybridoma technique with special reference to parasite diseases. In *Special Programme for Research and Training in Tropical Diseases*, UNDP/World Bank/WHO, Geneva, pp. 151–156 (1980).
47. Seckl, M. J., Monoclonal antibodies and recombinant DNA technology: Present and future uses in leprosy and tuberculosis. *Int. J. Lepr. 53*:618–640 (1985).
48. Ivanyi, J., Application of monoclonal antibodies towards immunlogical studies in leprosy. *Lepr. Rev. 55*:1–9 (1984).
49. Ivanyi, J., J. A. Morris, and M. Keen, Studies with monoclonal antibodies to mycobacteria. In *Monoclonal Antibodies Against Bacteria*, A. J. L. Macario and Macario, E. C. (eds.). Academic Press, New York (1985).
50. Sengupta, U., and S. Sinha, Monoclonal antibodies against *Mycobacterium leprae* and their applications in leprosy research. *Indian J. Lepr. 56*:727–741 (1984).
51. Kolk, A. H. J., M. L. Ho, P. R. Klatser, T. A. Eggelte, S. Kuijper, S. de Jonge, and J. van Leeuwen, Production and characterization of monoclonal antibodies to *Mycobacterium tuberculosis, M. bovis* BCG and *M. leprae*. *Clin. Exp. Immunol. 58*:511–521 (1984).
52. Ivanyi, J., S. Sinha, R. Aston, D. Cussell, M. Keen, and U. Sengupta, Definition of species-specific and cross-reactive antigenic determinants of *Mycobacterium leprae* using monoclonal antibodies. *Clin. Exp. Immunol. 52*: 528–536 (1983).
53. Young, D. B., S. R. Khanolkar, L. L. Barg, and T. M. Buchanan, Generation and characterization of monoclonal antibodies to the phenolic glycolipid of *Mycobacterium leprae*. *Infect. Immun. 43*:183–188 (1984).
54. Engers, H. D., M. Abe, B. R. Bloom, V. Mehra, W. Britton, T. M. Buchanan,

S. K. Khanolkar, D. B. Young, O. Closs, T. Gillis, M. Harboe, J. Ivanyi, A. H. J. Kolk, and C. C. Shepard, Results of a World-Health Organization-sponsored workshop on monoclonal antibodies to *Mycobacterium leprae. Infect. Immun. 48*:603–605 (1985).
55. Engers, H. D., B. B. Bloom, and T. Godal, Monoclonal antibodies against mycobacterial antigens. *Immunol. Today 6*:347–349 (1985).
56. Engers, H. D., V. Houba, J. Bennedsen, T. M. Buchanan, S. D. Chaparas, G. Kadival, O. Closs, J. R. David, J. D. A. van Embden, T. Godall, S. A. Mustafa, J. Ivanyi, D. B. Young, S. H. E. Kaufmann, A. G. Khomenko, A. J. H. Kolk, M. Kubin, J. A. Louis, P. Minden, T. M. Shinnick, L. Trnka, and R. A. Young, Results of a World Health Organization-sponsored workshop to characterize antigens recognized by *Mycobacterium*-specific monoclonal antibodies. *Infect. Immun. 51*:718–720 (1986).
57. Kolk, A. H. J., M. L. Ho, P. R. Klatser, T. A. Eggelte, and F. Portaels, Production of monoclonal antibodies against *Mycobacterium leprae* and armadillo-derived mycobacteria. *Ann. Inst. Pasteur (Paris) 136B*:217–224 (1985).
58. Klatser, P. R., M. M. van Rens, and T. A. Eggelte, Immunochemical characterization of *Mycobacterium leprae* antigens by the SDS-polyacrylamide gel electrophoresis immunoperoxidase technique (SGIP) using patients' sera. *Clin. Exp. Immunol. 56*:537–544 (1984).
59. Collins, C. H., M. D. Yates, and J. M. Grange, Names for mycobacteria. *Br. Med. J. 288*:463–464 (1984).
60. Coates, A. R. M., B. W. Allen, J. Hewitt, J. Ivanyi, and D. A. Mitchinson, Antigenic diversity of *Mycobacterium tuberculosis* and *Mycobacterium bovis* detected by means of monoclonal antibodies. *Lancet 2*:167–169 (1981).
61. Gillis, T. P., and T. M. Buchanan, Production and characterization of monoclonal antibodies to *Mycobacterium leprae*. In *Properties of the Monoclonal Antibodies Produced by Hybridoma Technology and Their Application to the Study of Diseases*. UNDP/World Bank/WHO, Special Programme for Research and Training in Tropical Diseases, Geneva, pp. 125–134 (1982).
62. Gillis, T. P., and T. M. Buchanan, Production and characterization of monoclonal antibodies to *Mycobacterium leprae. Infect. Immun. 37*:172–178 (1982).
63. Bach, M. A., and A. Hoffenbach, A monoclonal antibody against *Mycobacterium lepraemurium* which recognizes a cross-reacting mycobacterial antigen. *Ann. Immunol. 134C*:301–309 (1983).
64. Stanford, J. L., M. J. Shield, and G. A. W. Rook, How environmental mycobacteria may predetermine the protective efficacy of BCG. *Tubercle 62*: 55–62 (1981).
65. Daniel, T. M., and G. R. Olds, Demonstration of a shared epitope among mycobacterial antigens using a monoclonal antibody. *Clin. Exp. Immunol. 60*:249–258 (1985).
66. Britton, W. J., L. Hellquist, A. Basten, and R. L. Raison, *Mycobacterium leprae* antigens involved in human immune responses I. Identification of four antigens by monoclonal antibodies. *J. Immunol. 135*:4171–4177 (1985).

67. Kolk, A. H. J., R. Evers, S. Kuijper, H. Gilis, T. A. Eggelte, D. Groothuis, and P. J. G. Rietra, Identification of *Mycobacterium avium* with monoclonal antibodies. *Antonie Leeuwenhoek J. Microbiol. 52*:460–462 (1986).
68. Kolk, A. H. J., W. van Schooten, R. Evers, J. E. R. Thole, S. Kuijper, M. Y. L. de Wit, T. A. Eggelte, and P. R. Klatser, The use of monoclonal antibodies for the identification of *Mycobacteria* and the diagnosis of mycobacterial diseases; leprosy and tuberculosis. *Proceedings Conference on Mycobacteria of Clinical Interest*, M. Casal (ed.). Excerpta Medica, Amsterdam, pp. 29–33 (1986).
69. Portaels, F., and S. R. Pattyn, Isolation of fastidiously growing mycobacteria from armadillo livers infected with *Mycobacterium leprae. Int. J. Lepr. 50*:370–374 (1982).
70. Cho, Anag-Nae, D. L. Ynagihara, S. W. Hunter, R. H. Belber, and P. J. Brennan, Serological specificity of phenolic glycolipid I from *Mycobacterium leprae* and use in serodiagnosis of leprosy. *Infect. Immun. 41*: 1077–1083 (1983).
71. Brett, S. J., P. Draper, S. N. Payne, and R. J. W. Rees, Serological activity of a characteristic phenolic glycolipid from *Mycobacterium leprae* in sera from patients with leprosy and tuberculosis. *Clin. Exp. Immunol. 52*:271–279 (1983).
72. Minden, P., P. J. Kelleher, J. H. Freed, L. D. Nielsen, P. J. Brennan, L. McPheron, and J. K. McClatchy, Immunological evaluation of a component isolated from *Mycobacterium bovis* BCG with a monoclonal antibody to *M. bovis* BCG. *Infect. Immun. 46*:519–525 (1984).
73. Harboe, M., A. R. M. Coates, and J. Hewitt, Characterization of the specificity of monoclonal antibodies against *Mycobacterium tuberculosis* by crossed immunoelectrophoresis. *Scand. J. Immunol. 22*:93–98 (1985).
74. Daniel, T. M., N. J. Gonchoroff, J. A. Katzmann, and G. R. Olds, Specificity of *Mycobacterium bovis* antigen 5 determined with mouse monoclonal antibodies. *Infect. Immun. 45*:52–55 (1984).
75. Reitan, L. J., O. Closs, and M. Harboe, Characterization of the immune response to an epitope on *Mycobacterium leprae* antigen 7 defined by a monoclonal antibody. *Scand. J. Immunol. 22*:711–720 (1985).
76. Papa, F. P., J. Y. Rauzier, and H. L. David, Occurrence of antigen BCG 60 in leprosy-derived corynebacteria and other coryneforms. *Acta Leprol. (Geneve) 95*:351–358 (1984).
77. Young, D. B., M. J. Fohn, S. R. Khanolkar, and T. M. Buchanan, Monoclonal antibodies to a 28,000 mol. wt. protein antigen of *Mycobacterium leprae. Clin. Exp. Immunol. 60*:546–552 (1985).
78. Britton, W. J., L. Hellqvist, and A. Basten, Separate antigenic determinants on cell wall associated carbohydrated antigens of *Mycobacterium leprae* defined with monoclonal antibodies. *Int. J. Lepr. 54*:545–555 (1986).
79. Miller, R. A., and T. M. Buchanan, Production and characterization of a murine monoclonal antibody recognizing a shared mycobacterial polysaccharide. *Int. J. Lepr. 52*:461–467 (1984).
80. Daniel, T. M., and B. W. Janicki, Mycobacterial antigens: A review of their isolation, chemistry and immunological properties. *Microbiol. Rev. 42*: 84–113 (1978).

81. Ellner, J. J., and T. M. Daniel, Immunosuppression by mycobacterial arabinomannan. *Clin. Exp. Immunol. 35*:250–257 (1979).
82. Praputpittaya, K., and J. Ivanyi, Detection of an antigen (MY4) common to *M. tuberculosis* and *M. leprae* by "tandem" immunoassay. *J. Immunol. Methods 79*:149–157 (1985).
83. Morris, J. A., C. J. Thorns, and J. Woolley, The identification of antigenic determinants on *Mycobacterium bovis* using monoclonal antibodies. *J. Gen. Microbiol. 131*:8125–1831 (1985).
84. Atlaw, T., D. Kozbor, and J. C. Roder, Human monoclonal antibodies against *Mycobacterium leprae. Infect. Immun. 49*:104–110 (1985).
85. Narayanan, R. B., G. Ramu, S. Sinha, U. Sengupta, G. N. Malaviya and K. V. Desikan, Demonstration of *Mycobacterium leprae* specific antigens in leprosy lesions using monoclonal antibodies. *Indian J. Lepr. 57*:258–264 (1985).
86. Morris, J. A., and J. Ivanyi, Immunoassays of field isolates of *Mycobacterium bovis* and other mycobacteria by use of monoclonal antibodies. *J. Med. Microbiol. 19*:367–373 (1985).
87. Chakrabarty, A. K., M. Maire, K. Saha, and P. H. Lambert, Identification of components of immune complexes purified from human sera. II. Demonstration of mycobacterial antigens in immune complexes isolated from sera of lepromatous patients. *Clin. Exp. Immunol. 51*:225–231 (1983).
88. Morris, J. A., and C. J. Thorns, Evidence for the reaction of bovine autoantibodies with *Mycobacterium bovis. Vet. Rec. 117*:169 (1985).
89. Thorns, C. J., and J. A. Morris, Shared epitopes between *Mycobacteria* and other microorganisms. *Res. Vet. Sci. 41*:275–276 (1986).
90. Hewitt, J., A. R. M. Coates, D. A. Mitchinson, and J. Ivany, The use of murine monoclonal antibodies without purification of antigen in the serodiagnosis of tuberculosis. *J. Immunol. Methods 55*:205–211 (1982).
91. Ivanyi, J., E. Krambovits, and M. Keen, Evaluation of monoclonal antibody (TB72) based serological test for tuberculosis. *Clin. Exp. Immunol. 54*: 337–345 (1983).
92. Sinha, S., U. Sengupta, G. Ramu, and J. Ivany, A serological test for leprosy based on competitive inhibition of monoclonal antibody binding to the MY2a determinant of *Mycobacterium leprae. Trans. R. Soc. Trop. Med. Hyg. 77*:869–871 (1983).
93. Sinha, S., U. Sengupta, G. Ramu, and J. Ivanyi, Serological survey of leprosy and control subjects by a monoclonal antibody-based immunoassay. *Int. J. Lepr. 53*:33–38 (1985).
94. Klatser, P. R., M. Y. L. de Wit, and A. H. J. Kolk, An ELISA-inhibition test using monoclonal antibody for the serology of leprosy. *Clin. Exp. Immunol. 62*:468–473 (1985).
95. Reinherz, E. L., P. C. Kung, J. M. Breard, G. Goldstein, and S. F. Schlossman, T-cell requirements for generation of helper factor(s) in man: Analysis of the subsets involved. *J. Immunol. 124*:1883–1887 (1980).
96. Reinherz, E. L., and S. F. Schlossmann, The differentiation and function of human T lymphocytes. *Cell 19*:821–827 (1980).
97. Mustafa, A. S., and T. Godal, In vitro induction of human suppressor T-

cells by mycobacterial antigens. BCG activated OKT-4+ cells mediate suppression of antigen-induced T-cell proliferation. *Clin. Exp. Immunol. 52*: 29–37 (1983).

98. Narayanan, R. B., L. K. Bhutani, A. K. Sharma, and I. Nath, T-cell subsets in leprosy: In situ characterization using monoclonal antibodies. *Clin. Exp. Immunol. 41*:421–429 (1983).
99. Kaplan, G., D. E. Weinstein, R. M. Steinman, W. R. Levis, U. Elvers, M. E. Patarroyo, and Z. A. Kohn, An analysis of in vitro T-cell responsiveness in lepramotous leprosy. *J. Exp. Med. 162*:917–929 (1985).
100. Mathew, R. C., I. Katayama, S. K. Gupta, J. Cutris, and J. L. Turk, Analysis of cells of the mononuclear phagocyte series in experimental mycobacterial granulomas by monoclonal antibodies. *Infect. Immun. 39*: 344–352 (1983).
101. Poulter, L. W., L. A. Collins, K. S. Tung, and M. F. R. Waters, Parasitism of antigen presenting cells in hyperbacillary leprosy. *Clin. Exp. Immunol. 55*:611–617 (1984).
102. Alvarenga, F. de B., E. N. Sarno, A. de A. Figueiredo, C. R. Gattass, and J. A. Porto, Distribuicao de celulas de Langerhans na epiderme de pacientes com hanseniase. *Med. Cutanea Ibero Lat. Am. 13*:187–191 (1985).
103. Fine, P. M., Immunogenetics of susceptibility to leprosy, tuberculosis and leishmaniasis. An epidemiological perspective. *Int. J. Lepr. 49*:437–454 (1981).
104. Haanen, J. B. A. G., T. H. M. Ottenhof, A. Voordouw, B. G. Elferink, P. R. Klatser, H. Spits, and R. R. P. de Vries, HLA class-II-restricted *Mycobacterium leprae* reactive T-cell clones from leprosy patients established with a minimum requirement for autologous mononuclear cells. *Scand. J. Immunol. 23*:101–108 (1986).
105. Emmrich, F., and S. H. E. Kaufmann, Human T-cell clones with reactivity to *Mycobacterium leprae* as tools for the characterization of potential vaccines against leprosy. *Infect. Immun. 51*:879–883 (1986).

RECENT REFERENCES ADDED IN PROOF

De Wit, M. Y. L., and P. R. Klatser, Purification and characterization of a 36kd antigen of *Mycobacterium leprae. J. Gen. Microbiol. 134*:1541–1548 (1988).

Kolk, A. H. J., J. van Leeuwen, H. Gilis, S. Aglibut, S. Sondij, and S. Kuijper, Use of monoclonal antibodies in the identification of mycobacterial antigens. *Quaderni di cooperazione sanitaria/Health Cooperation Papers 7*:101–106 (1988).

Tan Trao, V., P. L. T. Huong, A. T. Thuan, H. T. Long, D. D. Trach, and E. P. Wright, Responses to *Mycobacterium leprae* by lymphocytes from new and old leprosy patients: Role of exogenous lymphokines. *Ann. Inst. Pasteur/ Immunol. 139*:121–133 (1988).

Van Schooten, W. C. A., T. H. M. Ottenhof, P. R. Klatser, J. Thole, R. R. P. de Vries, and A. H. J. Kolk, T Cell epitopes on the 36kd and 65kd *Mycobacterium leprae* antigens defined by human T cell clones. *Eur. J. Immunol. 18*:849–854 (1988).

21

Monoclonal Antibodies Against Group A and Group B Streptococci

Their Use in Immunodiagnosis and Immunoprophylaxis

BHANU P. RAM
Idetek, Inc., San Bruno, California

MARY CATHERINE HARRIS
University of Pennsylvania School of Medicine, and The Children's Hospital of Philadelphia, Philadelphia, Pennsylvania

I. INTRODUCTION

The use of antibodies for detection of extremely low amounts of antigens has been fully documented in the literature (1,2). Some of the problems associated with polyclonal sera, such as antibody heterogeneity, have been circumvented by use of monoclonal antibodies, as first reported by Köhler and Milstein (3).

The use of monoclonal antibody technology (hybridoma technology) for bacteria and their antigens has only begun to accumulate during the last 5 years. Some of the information related to bacterial detection with monoclonal antibodies can be found scattered in various chapters of this book.

For many years, bacteriologists have attempted to understand the mechanisms that are operating in the relationship between a pathogenic bacterial species and the host it has invaded. Some of the mechanisms of the bacterial infection have been unraveled by use of antibodies against the surface antigens or cell wall components.

Monoclonal antibodies against bacteria have enabled workers to gain new insights into the antigenic complexity of several bacterial species. It has permitted the identification and localization of some surface structures that are important for the pathogenicity of some bacteria. The use of monoclonal antibodies in the diagnosis of infectious diseases will probably provide reagents for serological typing and may open new fields for immunoprophylaxis and immunotherapy. This chapter will focus on the production of monoclonal antibodies against group A and group B streptococci and the use of these antibodies in bacterial detection and immunoprophylaxis.

II. BACKGROUND

The streptococci are gram-positive bacteria responsible for a variety of diseases in humans and in certain diseases in lower animals. In addition to the more virulent pathogenic form, relatively harmless parasitic streptococci are, more or less, constantly present in the human throat and in the intestinal tract.

The present classification of streptococci is generally based on their colony morphology and their hemolysis of blood agar; biochemical tests and their resistance to physical and chemical factors; immunological characteristics; and ecological features. Given the preceeding criteria, there are three main categories: (a) β-hemolytic streptococci, (b) non-β-hemolytic streptococci, and (c) peptostreptococci.

The hemolytic streptococci are divided into several serological groups (groups A-U), and some of the groups are subdivided into types that are based on the antigens of the cell wall. These groups are correlated, in a general way, with the epidemiology of the disease concerned.

A. Epidemiology of Group A Streptococci

The most common infection secondary to β-hemolytic streptococci is streptococcal pharyngitis or tonsillitis, which occurs in children between the ages 5 and 15 years (4). The group A streptococci (GAS) are normal inhabitants of the nasopharynx, and the incidence of streptococcal sore throat depends on age, climate, season, and the degree of contact between individuals. The highest incidence of streptococcal pharyngitis is reported between ages 10 and 18 years, whereas skin infections predominate in children under 6 years. Infants do not often acquire GAS disease, and it is felt that they are protected by transplacental antistreptococcal antibodies. Twenty percent of GAS infections are asymptomatic. The transmission of streptococci occurs both by direct contact and by contamination from the environment. Infection may be spread by droplets from the respiratory tract or from the skin and nose of infected individuals. The acquisition of GAS is generally associated with crowding, particularly in the home or institutional setting. Following carriage or infection, type-specific immunity develops to the streptococcal cell substance and to the antigenic soluble products of these organisms. As a consequence, the incidence of streptococcal disease decreases during adult life, once the immunity to the prevalent serotypes has developed.

The most common manifestations of GAS disease include pharyngitis/tonsillitis, and skin, soft-tissue, and blood infections. A unique characteristic of the GAS, however, is their ability to produce nonsuppurative sequelae, such as rheumatic fever, chorea, and acute glomerulonephritis. These complications may follow either overt or subclinical infection.

B. Epidemiology of Group B Streptococci

Streptococcus agalactiae, a group B streptococcus (GBS), was first isolated in 1887 from a case of bovine mastitis (5). It was not until 1938 that GBS was identified as a causal agent for human puerperal sepsis (6). Although there were numerous reports of illnesses caused by GBS during the 1930s and 1940s, most streptococcal isolates belonged to group A. In 1963, Eickoff and co-workers (7) published retrospective data on all GBS isolated during the previous 12 months; 7 infants had GBS sepsis, 4 of whom died, and 10 women had septic abortions, 4 of whom were bacteremic. Since the late 1960s and early 1970s, GBS has emerged as the major pathogen recovered from septicemic infants under 3 months of age (8). Two syndromes of clinical GBS disease, described in 1973, were based upon the age of onset of symptoms (9,10). Early-onset disease typically begins during the first 5 days of life and presents as septicemia with symptoms of apnea, cardiovascular instability, or respiratory distress (11). Exposure of the neonate occurs by the ascending route in utero through ruptured membranes, or by contamination during passage through a colonized birth canal. The infant either inhales or swallows infected amniotic fluid and develops generalized sepsis (12).

Although vertical transmission rates vary from 42 to 72%, several factors increase the neonate's risk of developing invasive disease, including prematurity, premature or prolonged rupture of membranes, heavy maternal GBS inoculum, fever, and lack of protective anti-GBS antibody (13). The attack rate for early-onset infection has varied from 1.3-3:1000 live births, with 100 infants colonized for each one infected. However, despite improvement in neonatal intensive care and use of broad-spectrum antimicrobial agents, the mortality (25-30%) for early-onset GBS disease has not declined during the last decade. Late-onset disease develops, after the first week of life, as meningitis or other focal infection. The mortality from late-onset disease is 20%, although one-third of the survivors may experience hydrocephalus or other neurological sequelae.

The GBS commonly colonize both the gastrointestinal and genitourinary tracts of men and women (14). Reported isolation rates for GBS vary greatly because of the specific methods of culture, the sites chosen for culture (vulva versus cervical, rectal versus vaginal), the number of cultures obtained, and the time interval over which the cultures are collected (15-17). Maternal factors affecting GBS colonization rates include sexual activity, presence of an intrauterine device, parity, age, and ethnic background (18,19). The GBS recovery rates are increased by sampling both the gastrointestinal and genitourinary tracts on multiple occasions. Although the incidence of GBS carriage is relatively stable through pregnancy (20-30%), it is impossible to predict the duration of colonization in any individual. Anthony et al. (20) performed a prospective study of women with sequential vaginal cultures during pregnancy and found that 36%

were chronic carriers, 20% were transient carriers, and 15-20% were intermittent carriers. The remaining 29% had indeterminant GBS carriage. Other studies have confirmed that those women who are culture-negative for GBS during the first and second trimester may acquire the organism around, or shortly before, the time of delivery. Infants born to these women may be at greatest risk to acquire early-onset disease. Therefore, it is most important that GBS cultures be obtained shortly before or at the time of delivery, regardless of previous culture results.

Most infants become colonized with GBS during the first 48 hr of age, and there is evidence for complete concordance of serotypes among isolates from mothers and their neonates. Besides exposure during birth, neonates may also acquire GBS from environmental sources, including other neonates, family members, and nursery personnel (21). However, there is no direct evidence that nosocomially acquired GBS infections are responsible for late-onset disease. Whether acquired by vertical or horizontal transmission, most infants remain asymptomatically colonized with GBS at skin and mucous membrane sites for weeks to months.

III. STREPTOCOCCAL ANTIGENS IMPORTANT IN ANTIBODY PRODUCTION

Antibodies can be produced in animals in response to bacterial antigens. The cell wall antigens of streptococci and many other bacteria have been used to produce polyclonal and monoclonal antibodies. The streptococcal cell wall is mainly composed of peptidoglycan, group-specific carbohydrate, teichoic acid, and proteins such as M, T, and R antigens (22,23). Most group A and group B strains produce capsules composed of hyaluronic acid on their surface (24). We will now describe some of the GAS antigens important in antibody production.

A. Group A Streptococci

1. Carbohydrate Antigens

Group A streptococcal carbohydrate is composed of L-rhamnose and *N*-acetylglucosamine in a ratio of approximately 2:1 (25,26). The group A carbohydrate, which is immunodoominant, is a highly branched structure with β-linked *N*-acetylglucosamine as the terminal unit on the branches or side chains (26,27).

A mutant has been recognized that synthesizes a serologically distinct carbohydrate. Strains of this type of mutant are designated as Av (A variant). Group Av carbohydrate consists of a linear rhamnose polymer with alternating 1-2 and 1-3 linkages (28). Rhamnose is the predominant constituent of Av carbohydrate, and *N*-acetylglucosamine is present in only small amounts, for example, 20 mol of rhamnose per mole of glucosamine (27). There is very little branching of the carbohydrate. The group A and Av carbohydrates are serologically distinct in their reactions with rabbit antisera, although minor cross-reactivity is demonstrable with some specimens of antiserum.

2. Protein Antigens

Other than the carbohydrate moities, protein makes up a substantial portion of the total antigenic substances of the GAS cell wall. The major protein antigens located on the cell wall are termed M, T, and R proteins. Little information is available concerning the actual location of T and R proteins on the cell wall. These antigens are primarily used in epidemiological studies (29) to provide the investigator with additional antigenic markers for following certain streptococcal strains. No biological activity has yet been associated with these two surface antigens.

The M Antigens. The M antigens are located above the peptidoglycan and carbohydrate layers as amorphous irregular projections on the surface of the cell wall (30). The M proteins determine the type (subclass) specificity of GAS as demonstrated by agglutination or precipitation reactions with adsorbed type-specific sera. There are nearly 70 immunologically specific types in GAS.

The M proteins can be removed from the whole streptococci or the cell wall by various methods such as hot acid extraction (31), sonic oscillation (32), alkaline extraction (33), and detergent extraction (34).

The M protein is composed of multiple protein units of which glutamic acid, lysine, alanine, leucine, and aspartic acid are in greatest abundance. The molecule has been found to be low in aromatic amino acids, and cysteine is absent in all types examined (35). The molecule has a frictional ratio of 2:1, indicating that it is asymmetric and not globular (34).

The M proteins are weakly immunogenic, and antisera are not available for all antigens in this series. Therefore, a portion of isolates cannot be typed on this basis. With only rare exceptions, each strain contains only one M antigen; however, some M proteins are similar, and presumably share antigenic determinants, hence, there is cross-reaction between M types (36). The importance of M antigens in the host-parasite relationship is emphasized by the fact that antibodies directed against M antigens are protective in homologous-type infections (35).

Another cell wall antigen, referred to as M-associated protein (MAP) is trypsine-sensitive, and there are at least two antigenic types (type I and II) reported (37). Antibodies to MAP occur widely in human sera, and infection with certain M types elicits a vigorous and frequent anti-MAP response; other types appear to be poorly antigenic (38).

The T Antigens. The T antigens are not so prevalent as the M antigens, and they do not have the same theoretical importance because antibodies against them are not protective. The T antigens are resistant to proteolytic enzymes. They are unstable to heat in dilute acid, but resistant in slightly alkaline solution. The T antigens occur in several types but may be type-specific. For example, types 10 and 12 contain the same M antigens but different T antigens, whereas types 15, 17, 19, 23, and 30 contain similar T antigens but distinct M antigens (39).

The R Antigens. The R antigens occur in GAS types 2, 3, 25, and 48, including some strains of group B, C, and G. They are destroyed by peptic but not by tryptic digestion. They behave similarly to T antigens in their stability in dilute acid and alkali (39).

B. Group B Streptococci

1. Carbohydrate Antigens

The cell wall carbohydrate of GBS is predominantly composed of galactose, hexosamine, and L-rhamnose residues in an α-1,2 linkage as the terminal antigenic moiety (40,41). Typing within the B group is mostly based upon envelope carbohydrate from acid-heat extracts, rather than protein antigens, as for A. There are currently five serotypes of group B, namely Ia, Ib, Ic, II, and III. Further serotypes II/Ic and III/Ic have been recently reported (41).

2. Protein Antigens

Protein antigens are not as important for group B serotyping as they are for group A. Pattison and colleagues (42) utilized Lancefield's classification scheme and introduced two protein moieties, R and X, to reduce the number of strains untypeable with the Lancefield reagent. The R and X protein antigens are unstable and cross-reactive. The R antigen from group B type III is also found in GAS type 28 (43). The Ibc protein composed of two antigenic moieties is stable and present only in group B types Ib, Ic, II, and, rarely, in III (44).

IV. MONOCLONAL ANTIBODIES

A. Anti-Group A Streptococcal Monoclonal Antibodies

Conventional antisera for detection of GAS are heterogeneous mixtures of antibodies of varying specificity and affinity for the antigenic determinants. Furthermore, these assays rely on antisera raised in animals, and they frequently vary in quality between animals and even with each bleed of the same animal. As a consequence, the production and use of antistreptococcal monoclonal antibodies may overcome these difficulties and allow their rapid detection in body fluids.

Hybridoma technology offers the possibility of immunizing the animals with very crude preparations of the antigens against which one wants to raise an antibody, provided there is a rapid and specific screening test available to identify relevant clones. Crude preparation of GAS, referred to as vaccine, to immunize mice for hybridoma production has been described (45,46).

Nahm and associates (47) produced a monoclonal antibody to GAS carbohydrate using mice hyperimmunized with GAS vaccine. By both Lancefield precipitin and fluorescent antibody methods, this antibody proved superior to conventional antisera for the identification of GAS. Although the antibody correctly identified 262 clinical samples, occasional false-positive and false-negative results were observed with conventional antisera.

Nahm et al. (48) characterized the murine anti-GAS carbohydrate monoclonal antibody and found that most of the serum antibody response was composed of IgG3k and IgM κ-classes. When recombinants were examined, it was found that the κ-light-chain variable region did not pair randomly with heavy chains of the IgG subclasses, but associated preferentially with heavy chains of the IgG3 subclasses (49). Other investigators have isolated and characterized the sequence of the variable region of κ-light chains from anti-GAS carbohydrate antibody (50). In this way, monoclonal antibodies have been used as important probes to investigate the diversity of the variable regions of immunoglobulin chains.

By comparing the amino acid sequence variable regions of several heavy and light chains of murine monoclonal antibodies against GAS carbohydrate, Aebershold and colleagues (51) presented evidence suggesting that framework- and specificity-determining regions might be under independent genetic control.

Monoclonal antibodies to GAS carbohydrate have also been reported by Herbst and Braun (52). The basis for similar studies with human monoclonal antibodies has been laid with the establishment of an EBV-transformed human B-cell line secreting antibody to GAS carbohydrate (53).

Hasty and associates (54) developed murine monoclonal antibodies against a purified polypeptide fragment of type 24 streptococcal M protein and compared their opsonic, bactericidal, and protective capacities. None of the antibodies cross-reacted with heart tissue and, with the exception of one antibody directed at a determinant hidden in the interior of the native molecule, all antibodies were protective. These antibodies might provide probes for identifying small peptide fragments still capable of eliciting protective immunity without leading to unwanted cross-reaction with human tissues.

Cunningham and Russell (55) selected murine IgM antibodies that reacted with the M protein of both GAS and human heart tissues, and characterized them as weak or strong reactors. Monoclonal antibodies that were strong reactors with GAS membranes and human heart sarcolemmal antigens were directed against a determinant on a family of proteins. Moreover, when normal and rheumatic heart tissues were compared, the rheumatic sarcolemma reacted more intensely to the monoclonal antibody, but no major immunobanding differences were seen (56). Dale and Beachey (57) demonstrated that antibodies against a single M protein epitope opsonize heterologous serotypes sharing the same epitope, and suggested that monoclonal antibodies could select subpeptides of the M protein that contain protective, rather than cross-reactive, epitopes. Antibodies could be raised, then, to protect the host from many different, possibly rheumatogenic serotypes of GAS.

B. Anti-Group B Streptococcal Monoclonal Antibodies

In 1978, Polin and Kennett (58) prepared the first monoclonal antibodies against bacteria using the technique of somatic cell hybridization (3). These monoclonal antibodies reacted with the polysaccharide determinant unique to each serotype of GBS. Vaccines of formalinized GBS were prepared from type Ia-090/14, type Ib-H36B/60/2, type Ic-1909/14, type II-18RS21/67/2, and type III-D136C reference strains obtained from Dr. Rebecca Lancefield. Female BALB/c mice were given weekly immunizations of formalinized bacteria. The first injection consisted of antigen in incomplete Freund's adjuvant administered intraperitoneally. The second injection, diluted in saline, was given both intraperitoneally and subcutaneously. The final dose was administered into the tail vein.

Cell fusion and cloning were performed according to previously published methods (58). Hybridoma supernatants were screened for binding to GBS types and type 14 pneumococcus in a binding immunoassay in which monoclonal antibody bound to bacteria was detected with a sheep antimouse immunoglobulin conjugated to peroxidase. Following the procedure, four monoclonal antibodies were chosen that exhibited a high affinity in the enzyme immunoassay (EIA) for each of the GBS serotypes. Types Ia and Ic GBS possess a common carbohydrate antigen; therefore, the monoclonal antibody is designated as anti-GBS Ia/Ic. The anti-GBS Ia/Ic, II, and III monoclonal antibodies are $\mu\kappa$, whereas the anti-Ib monoclonal antibody is $\gamma 2_{\alpha}\kappa$. Clones secreting anti-GBS antibody are maintained as cell cultures and routinely tested in enzyme immunoassays. Each clone can also be frozen as a safeguard against loss of cell line. When large quantities of antibodies are desired, the hybridoma clone may be injected into pristane-primed mice, and the ascites harvested after 7–10 days. Alternatively, hybridoma can be mass-produced in a bioreactor.

To detect small quantities of GBS antigens in body fluid specimens (blood, urine, or cerebrospinal fluid), the anti-GBS monoclonal antibodies were incorporated into two enzyme-based assays (59,60).

1. Enzyme-Linked Monoclonal Inhibition Assay

The enzyme-linked monoclonal inhibition assay (ELMIA) is a competitive type enzyme immunoassay for detection of GBS antigens in body fluids (59). The steps involved in performing the assay are depicted in Figure 1.

The ELMIA has been used to identify type III GBS antigen in cerebrospinal fluid specimens from 20 infants with culture-proved GBS meningitis in whom the concentration of antigen ranged from 0.8 to 12.8 μg/ml. The ELMIA also detected GBS antigen in the knee aspirate from an infant with GBS arthritis (61). There were no false-positive reactions noted. In addition, the assay was able to detect native antigen at a concentration of 10 ng/ml, which is 80–100 times lower than that detectable by conventional assays.

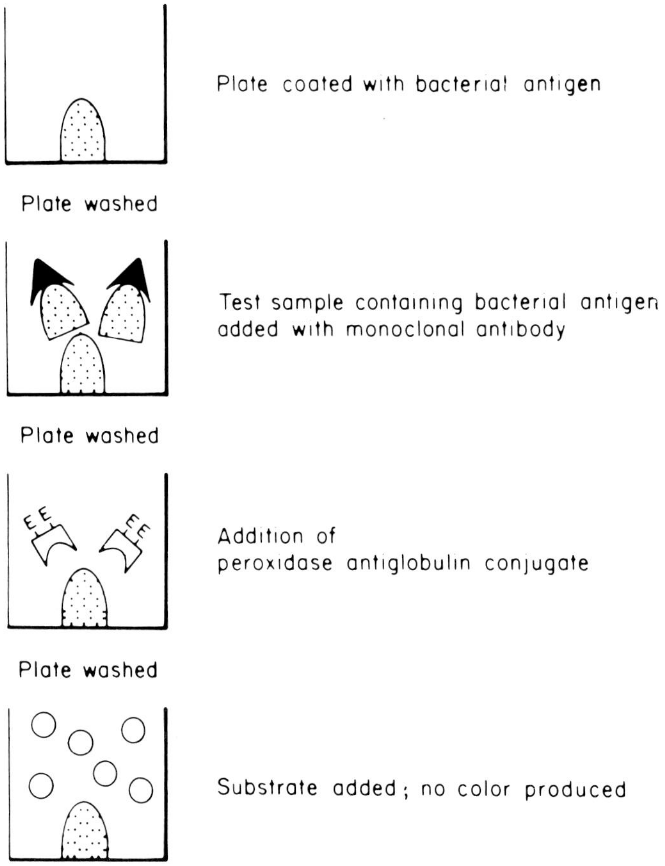

Figure 1 An ELMIA for detection of GBS in body fluids. Microtiter wells were coated with type-specific GBS antigens for 15–20 hr at 4°C. After blocking and washing steps, GBS-suspected body fluid was added to wells followed by type-specific GBS antibody. The plate was incubated for 24 hr at room temperature and then washed. Horseradish peroxidase-labeled antimouse immunoglobulin was added to the wells and the plate incubated for 2 hr at room temperature. After usual washing, peroxidase substrate solution (*o*-phenylenediamine and hydrogen peroxide) was added and the plate incubated at room temperature for color development (20–30 min). The GBS-positive samples did not show color change, contrary to color change in GBS-negative samples. Absorbance of color was monitored in an EIA reader. Proper positive and negative controls were maintained (*Source*: from Ref. 60).

Although the ELMIA was sensitive and specific, there were two drawbacks to this method. As a competitive inhibition assay, the ELMIA was not as easily interpreted visually as assays that have increased color production as their endpoint. In addition, the ELMIA did not prove as reliable for urine as for spinal fluid specimens. Urine from some culture-negative women contained substances that interfered in the assay, producing false-positive reactions.

2. *Sandwich Enzyme Immunoassay*

Polin and co-workers (59) have developed a sandwich EIA in response to some of the criticisms observed with ELMIA. The assay has the advantage of a 10-fold increase in sensitivity over the ELMIA, and a positive endpoint (color production versus color inhibition; Fig. 2).

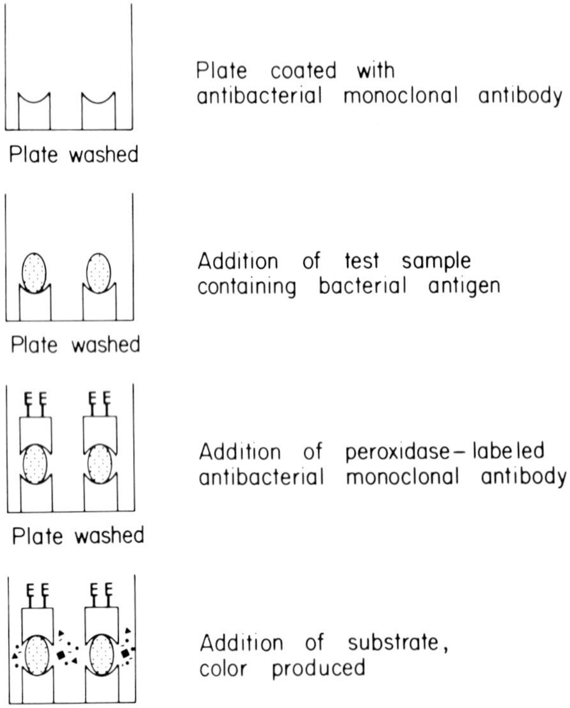

Figure 2 A sandwich EIA for detection of GBS in body fluids or medium from an actively growing culture of GBS. GBS-positive samples caused color production, whereas negative samples did not show color change. Proper positive and negative controls were maintained (*Source*: from Ref. 59).

The sandwich assay proved as reliable as the ELMIA for detection of type III GBS native antigen in cerebrospinal fluid specimens. In contrast to the ELMIA, however, the sandwich assay detected type III GBS purified polysaccharide antigen at a concentration of 1 ng/ml. Results from cerebrospinal fluid specimens from five patients with known GBS type III meningitis are shown in Table 1.

The sandwich assay also proved to be very sensitive and specific for the detection of colony-forming units (CFU/ml) of GBS. No cross-reactions were observed when each of the GBS serotypes was tested with antibodies of different serotype specificities. For types II and III GBS, the sandwich assay detected 5×10^4 CFU/ml; for types Ib and Ic, 10^5 CFU/ml, and for type Ia GBS 5×10^5 CFU/ml.

3. *Characterization of Anti-Group B Streptococcal Monoclonal Antibodies*

More than 2 decades ago, Lancefield and associates demonstrated the role of specific antibody in the protection and treatment of life-threatening infection (62). During the 1970s, Baker and Kasper (63,64) demonstrated a similar protective role for type III GBS antibody in human infections. To demonstrate whether or not the anti-GBS monoclonal antibodies might have therapeutic usefulness, Harris et al. evaluated the functional properties of each monoclonal antibody (agglutination, complement fixation, and opsonization) and tested their ability to protect mice against lethal GBS infection (65). Bacterial agglutination was performed by reacting type-specific GBS with hybridoma supernatant in a microtiter plate. Each of the monoclonal antibodies agglutinated GBS of identical serotype. To evaluate complement fixation, type-specific monoclonal antibodies and serum were incubated in wells of polyvinyl chloride (PVC) plates that were precoated with type-specific GBS of identical specificity to the antibody. The following day, sensitized sheep red blood cells were added to the test

Table 1 Enzyme Immunoassay for Type III GBS Antigen in Patients with Known GBS Meningitis

CSF (sample)	NAC (μg/ml)	Abs.
1	0.8	0.687
2	6.8	0.966
3	1.7	1.004
4	6.8	0.997
5	0.8	0.540

CSF, cerebrospinal fluid; NAC, native antigen concentration; Abs, absorbance.

Table 2 Complement Fixation

Monoclonal antibody	Unreacted sera	Mg-EGTA-treated sera	Heat inactivated sera
Anti-GBS Ia/Ic	+	–	*
Anti-GBS Ib	+	–	*
Anti-GBS II	+	–	*
Anti-GBS III	+	–	*
SP2/0 Ag 14	–	–	*

Symbols: +, antibodies fixed complement, hemolysis absent. –, no complement fixation, hemolysis present. *, both complement pathways inactivated, hemolysis absent.

wells, and hemolysis was assessed macroscopically after 30 min. Each monoclonal antibody fixed complement (Table 2). Pretreatment of the human sera with magnesium-ethylene glycol-*bis*(β-aminoethyl ether) *N,N′*-tetraacetic acid (mg-EGTA) prevented complement fixation and resulted in hemolysis of sensitized sheep red blood cells. These results indicate that the monoclonal antibodies interacted with the classic, but not the alternative complement pathway. Heat inactivation of the serum destroyed complement and prevented hemolysis.

Phagocytosis of GBS was assessed with a radiolabeled bacterial uptake technique (Fig. 3). Anti-GBS Ia/Ic, II, and III monoclonal antibodies were opsonic for GBS of identical serotype specificity. However, anti-GBS Ia/Ic antibody was not opsonic for type Ia GBS, and anti-GBS Ib antibody was not an effective opsonic for that serotype. These results demonstrate that although certain anti-GBS monoclonal antibodies opsonize GBS of identical serotype specificity, others may not possess the functional characteristics to be useful as diagnostic or therapeutic reagents.

Each antibody was also evaluated for its ability to confer protection against fatal GBS infection in mice. BALB/c mice were administered an intrascapular injection of hybridoma cells secreting type-specific monoclonal antibodies. Type III GBS was not studied because it does not easily infect adult BALB/c mice. Tumors were present within 2 weeks. After this period, tumor and control mice were challenged with an intraperitoneal injection of 10^9 live GBS of identical specificity to the tumor. All tumor mice that were challenged with GBS of identical specificity survived and had anti-GBS titers of 1:10,000 (Table 3). All control mice and tumor mice challenged with GBS of a different specificity died.

4. Detection of Group B Streptococci with Monoclonal Antibodies

Because GBS remain the most frequent cause of neonatal sepsis and meningitis, and efforts to reduce neonatal GBS mortality with antimicrobial agents have been largely ineffective, recent work has focused on the use of monoclonal anti-

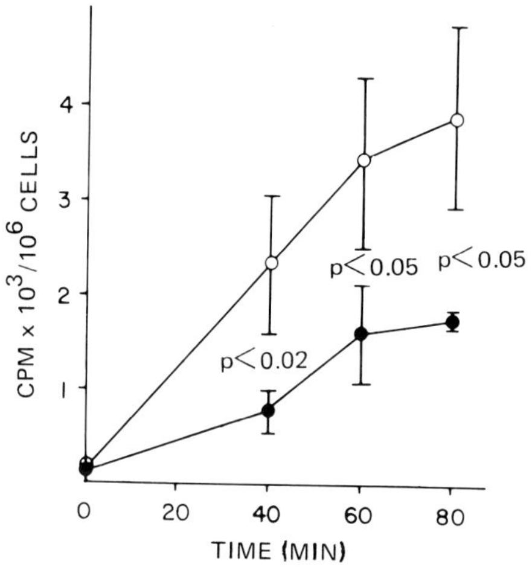

Figure 3 Assessment of phagocytosis of GBS using the radiolabeled bacterial-uptake technique. Preincubation with anti-GBS II monoclonal antibody (0) versus SP2 control (0). [^{3}H] leucine treated type II GBS were opsonized with monoclonal antibody and then incubated with neutrophil monolayers adherent to glass coverslips. At specified time intervals, coverslips were washed and counted. Radiolabeled bacterial uptake was expressed per 10^6 neutrophils and plotted against time. Bars show the standard deviation at each point (*Source*: from Ref. 65).

bodies for the rapid detection of maternal GBS colonization and infant infection. Women at greatest risk to deliver infants with early-onset GBS disease are heavily colonized and lack protective anti-GBS antibody (61-64,66,67). Therefore, assay systems that rapidly detect GBS antigen in mothers or neonates may suggest the high-risk patients who will benefit from antimicrobial therapy.

The identification of GBS on culture swabs or in body fluid specimens requires 24-48 hr for processing. In specific clinical situations, however, a rapid diagnosis of GBS may be preferable. Positive identification of newborn infants with GBS sepsis or meningitis may be lifesaving. Moreover, if rapid detection methods can identify high-risk, culture-positive women during labor, intrapartum administration of ampicillin may prevent vertical transmission of GBS and invasive neonatal disease (68).

Commercial immunodiagnostic assays for the detection of GBS antigens (staphylococcal coagglutination, latex particle agglutination, and counterimmunoelectrophoresis) have enabled the rapid identification of streptococcal antigens in body fluid specimens or culture medium containing mixed flora. The GBS antigen may be identified in 6-18 hr, from specimens containing mixed

Table 3 Mouse Protection Experiment[a]

Antibody[b]	GBS serotypes[c] Ia	Ic	Ib	II
Anti-GBS Ia/Ic	5/5	5/5	–	0/5
Anti-GBS Ib	–	–	5/5	–
Anti-GBS II	0/5	–	–	5/5
Control mice	0/5	0/5	0/5	0/5

[a]Number of survivors/number tested.
[b]Produced by tumor.
[c]Injected intraperitoneally.

flora, by use of these assays. The minimal concentration of GBS detectable by these assays is 800 ng/ml. However, two problems have been associated with the procedures. Variations in the source of antisera used in the assays have decreased the sensitivity and specificity of the tests. In addition, most of the immunoassays detect group-specific determinants; therefore, rapid serotyping of GBS isolates is not possible. Both the ELMIA and sandwich assays offer greatly increased sensitivity and specificity for the detection of GBS antigens in body fluid specimens (59-61).

Ruch and Smith (69) recently developed murine monoclonal antibodies directed against all five GBS serotypes. In this study, monoclonal reagents were equally reliable when compared with conventional diagnostic antisera for the identification of the GBS in urine and cerebrospinal fluid. Rench and co-workers (70) utilized a similar reagent, a murine monoclonal antibody to GBS antigen, and detected GBS antigen in urine and cerebrospinal fluid specimens with 95% sensitivity and specificity. Therefore, monoclonal antibodies directed against both the group and the capsular polysaccharide antigens offer potential as useful reagents for the diagnosis of GBS disease.

V. IMMUNOPROPHYLAXIS AND IMMUNOTHERAPY

Recent attempts to decrease the incidence of streptococcal infection have focused on prophylactic measures. Chemoprophylaxis has been used with fair success by many investigators to treat streptococcal infection (68,71-75). Treatment of a disease by using monoclonal antibodies (immunotherapy/immunoprophylaxis) is being tested in many laboratories and has recently been reviewed with special reference to cancer treatment (76). Antibodies specific for tumor cells are covalently coupled to highly toxic agents. The immunoconjugate is then expected to seek out the malignant cells and destroy them. Be it the treatment of a tumor or of a bacterial infection, this approach will require the greatest care in selecting the appropriate monoclonals. They will have to pass the strictest test for specificity and affinity.

The protective nature of monoclonal antibodies against bacterial infection has been reported in some laboratories (45,77). Because of an antigenic relationship between M protein and tropomyosin, antibodies to the M protein of GAS can protect against infections with bacteria of homologous M serotype (78,79) as well as cross-react with host heart tissues (80,81). In this situation, monoclonal antibodies might be used to define precisely the minimal structure of the M protein molecule required to elicit antibodies of a protective nature.

Immunoprophylaxis may ultimately prove most effective for the prevention or long-term elimination of GBS disease. In a mouse model, Lancefield and co-workers demonstrated that antibody directed against the type-specific surface antigens of GBS protected mice against infection with the homologous serotype (62). Baker and Kasper (64) found that neonatal susceptibility to invasive type III GBS disease resulted from low concentrations of maternal IgG class, type-specific antibody to the capsular polysaccharide. This deficiency was caused by either low levels of type-specific antibody in maternal serum or by failure of sufficient placental transport to confer protection. Baker et al. (82) demonstrated low levels of antibody to the capsular polysaccharide in neonates with early-onset type III GBS and found a significant relationship between maternal and neonatal antistreptococcal antibody titers.

As a preventive measure against streptococcal infection, passive immunization with monoclonal antibodies has been suggested. The monoclonals intended for such purposes must be isologous, e.g., human globulins for clinical use. The improvement of in vitro immunization of human lymphocytes, of finding optimal conditions for constructing human-human hybridomas and, especially, the promise of transfected stable cell lines give hope that monoclonal antibodies can be tested in the near future as prophylactic or therapeutic agents. In one study, Harris and co-workers (65) have demonstrated that the anti-GBS monoclonal antibodies possess the specific functional properties to protect mice against lethal infection. Furthermore, active immunization of either all or selected antibody-deficient women has been suggested as a means to prevent infant disease (62), although safe and effective vaccines have not been developed for all GBS serotypes.

Shigeoka, Hill, and associates (83,84) demonstrated that murine monoclonal IgM antibodies to type III GBS protected neonatal rats from intraperitoneal infection with the homologous group B serotype. The monoclonal antibodies were effective in preventing systemic or respiratory GBS as late as 24 hr after infection. Furthermore, animals that were complement depleted were not protected by the IgM antibody, indicating a critical requirement for complement in the protective and opsonic activity of the monoclonal IgM antibody (85).

Systematically administered monoclonal IgA did not protect against either systemic or respiratory infection, although a tendency to prolong survival was observed (86). Egan et al. (87) also examined the protective properties of mouse hybridoma antibodies directed against the sialated and nonsialated forms of the GBS type III polysaccharide. Mice, receiving monoclonal antibodies specifically directed against the sialated (native) form of the GBS antigen, were protected after a GBS challenge, whereas control mice or those receiving monoclonal antibodies to the nonsialated (core) antigen uniformly died. Antibodies specific for the sialated form of the GBS antigen were highly protective and included IgM, IgG2a, and IgA isotypes.

Christensen and colleagues (88) investigated neutrophil kinetics after hybridoma antibody administration during experimental type III GBS infection in newborn rats. Control rats that received GBS alone developed profound neutropenia, an elevated immature/total neutrophil ratio, and exhaustion of marrow neutrophil reserves. Rats that received both GBS and IgM monoclonal antibodies against type III GBS demonstrated an earlier release of neutrophils from the neutrophil storage pool and maintained an adequate supply of circulating and stored neutrophils. The administration of type-specific antibody prevented the development of neutrophil storage pool exhaustion while preventing mortality. If the administration of monoclonal antibody was delayed 5-7 hr, coincident with neutrophil storage pool depletion, there was no protection against fatal GBS in newborn animals (89).

Fibronectin, an adhesive, high-molecular glycoprotein present on cell surfaces and in plasma, participates in the reticuloendothelial clearance of particles and has been shown to promote the attachment of immunoglobulin-coated erythrocytes to human mononuclear phagocytes (90). Hill and co-workers (83) have demonstrated that fibronectin enhanced the interaction between GBS, monoclonal antibody, and phagocytic cells. Fibronectin increased the opsonic activity of monoclonal antibody directed against type III GBS and resulted in improved survival after bacterial challenge in neonatal rats. Administration of fibronectin with type III GBS monoclonal antibody significantly improved survival. With further refinements in hybridoma technology, GBS monoclonal antibodies (with or without fibronectin) may be useful for the prevention and treatment of invasive GBS disease, particularly as an adjunct to antimicrobial therapy.

VI. PROBLEMS AND FUTURE TRENDS

In recent years, monoclonal antibodies have been used for in vitro clinical laboratory tests. Monoclonal antibody-based clinical tests for in vitro diagnosis are now commercially available. For example, tests for human chorionic gonadotropin (hCG) (Hybritech, San Diego, California), for carcinoembryonic antigen (CEA) (Abbott Laboratories, North Chicago, Illinois), for herpes virus (Genetic

System, Seattle, Washington). Commercial immunoassay tests for bacterial detection are beginning to enter the market. Simple EIA test for GAS antigen is being marketed by Ventrex Laboratories and Abbott Laboratories for detection of GAS. These immunoassay tests take 5 min to complete and have shown specificity and sensitivity of 95% on direct throat swabbing (91).

Currently, there are two major problems in the development of enzyme immunoassay tests for bacteria: (a) the lack of desired specificity and affinity of the antibodies and (b) the desired sensitivity of immunoassay tests. Future research on streptococcal immunodetection needs to focus on these two problems. It has been observed that lack of specificity of an antibody will produce false-positive and false-negative results. Also, such antibody may not be able to screen all serotypes in a particular bacterial group. Some of the foregoing problems may be circumvented by use of newer technology in hybridoma production such as biotin-avidin systems, taggert hybridoma technology (THT), (92) and electrofusion. A cocktail of high-affinity monoclonal antibodies could be used to screen all serotypes belonging to one group, such as GAS or GBS. Furthermore, sensitivity of an enzyme immunoassay may be increased several-fold by using bioluminescent (93), chemiluminescent (94), and fluorogenic (95) substrate systems for the enzyme. Preparation and use of suitable enzyme conjugates is another important factor that affects the sensitivity of enzyme immunoassays.

The use of monoclonal antibodies in vivo as therapeutic agents is still in the offing, although some progress has been made with tumor diagnosis and treatment. Clinical evaluations have been hindered because murine monoclonal antibodies, when used in humans, cause unwanted side effects. They could serve as antigens and stimulate an immune response in the patient; such a response would make the monoclonal antibody therapy impossible. The solution is to use human monoclonal antibodies, which are much less likely to generate an immune response, but there are two major obstacles. First, there is no adequate, genetically marked human myeloma cell line to fuse with immune cells. This makes it very difficult to find a monoclonal antibody-producing cell after the fusion between the immune cells and the parental myeloma cell line has been performed. Second, obtaining immune cells has proved difficult. Obviously, it is not feasible to immunize humans and then remove their spleens as a source of immune cells, as is done with mice.

Another problem with monoclonal antibody in direct human therapy is the potential for generating antigen-antibody complexes. Such complexes have long been known to damage the kidneys, as well as acting as antigens themselves. Although neither problem has yet been demonstrated clinically, experience with monoclonal antibody therapy is still very limited. Until all the preceding questions have been answered, the therapeutic role of monoclonals will stay in the realm of science fiction.

ACKNOWLEDGMENT

The authors are thankful to Dianne M. Ram for proofreading and structuralizing the text.

REFERENCES

1. Ram, B. P., L. P. Hart, A. Suvarnamani, L. Tuck, and J. J. Pestka, Rapid screening for aflatoxin B1 in corn by a direct enzyme-linked immunosorbent assay. 85th Meeting, Am. Soc. Microbiol., Las Vegas, Nevada, March 3–7 (1985).
2. Liu, M.-T., B. P. Ram, L. P. Hart, and J. J. Pestka, Indirect enzyme-linked immunosorbent assay for the mycotoxin zearalenone. *Appl. Environ. Microbiol. 50*:332–336 (1985).
3. Köhler, G., and C. Milstein, Continuous cultures of fused cells secreting antibody of predefined specificity. *Nature 256*:495–497 (1975).
4. Wolin, M. J., The streptococci. In *Textbook of Microbiology*, W. Burrows (ed.). W. B. Saunders Co., Philadelphia, pp. 420–442 (1973).
5. Nocard, E., and A. Mollereau, Sur une mammite contagiense des raches laitieres. *Ann. Inst. Pasteur (Paris) 1*:109–126 (1887).
6. Fry, R. M., Fatal infection by hemolytic streptococcus group B. *Lancet 1*: 199–201 (1938).
7. Eichhoff, T. C., J. O. Klein, A. Daly, D. Ingall, and M. Finland, Neonatal sepsis and other infections due to group B beta-hemolytic streptococci. *N. Engl. J. Med. 271*:1221–1228 (1974).
8. Baker, C. J., Group B streptococcal infections in neonates. *Pediatr. Rev. 1*:5–15 (1979).
9. Franciosi, R. A., J. D. Knostman, and R. A. Zimmerman, Group B streptococcal neonatal and infant infections. *J. Pediatr. 82*:707–718 (1973).
10. Baker, C. J., F. F. Barrett, R. C. Gordon, and M. D. Yow, Suppurative meningitis due to streptococci of Lancefield group B: A study of 33 infants. *J. Pediatr. 82*:724–729 (1973).
11. Harris, M. C., and R. A. Polin, Neonatal septicemia. *Pediatr. Clin. N. Am. 30*:243–258 (1983).
12. Blanc, W. A., Pathways of fetal and early neonatal infection. *J. Pediatr. 59*:473–496 (1961).
13. Baker, C. J., and M. S. Edwards, Group B streptococcal infections. In *Infectious Diseases of the Fetus and Newborn Infant*, J. S. Remington and Klein, J. O. (eds.). W. B. Saunders Co., Philadelphia, pp. 820–881 (1983).
14. Anthony, B. F., R. Eisenstadt, J. Carter et al., Genital and intestinal carriage of group B streptococci during pregnancy. *J. Infect. Dis. 143*:761–766 (1976).
15. McDonald, S. W., F. R. Manuel, and J. A. Embil, Localization of group B beta-hemolytic streptococci in the female urogenital tract. *Am. J. Obstet. Gynecol. 133*:57–59 (1979).

16. Yow, M. D., L. J. Leeds, P. K. Thompson, E. O. Mason Jr., D. J. Clark, and C. W. Beachler, The natural history of group B streptococcal colonization in the pregnant women and her offspring. I. Colonization studies. *Am. J. Obstet. Gynecol. 137*:34–38 (1980).
17. Boyer, K. M., and C. K. Papiernak, Transplacental passage of IgG antibody to group B streptococcus. *J. Pediatr. 104*:618–620 (1984).
18. Baker, C. J., D. K. Goroff, S. Alpert, V. A. Crockett, S. H. Zinner, J. R. Evrard, B. Rosner, and W. M. McCormack, Vaginal colonization with group B streptococcus: A study in college women. *J. Infect. Dis. 135*:392–397 (1977).
19. Embil, J. A., T. R. Martin, N. H. Hansen, S. W. McDonald, and F. R. Manuel, Group B beta-hemolytic streptococci in the female genital tract: A study of four clinic populations. *Br. J. Obstet. Gynaecol. 85*:783–786 (1978).
20. Anthony, B. F., D. M. Okada, and C. J. Hobel, Epidemiology of group B streptococcus: Longitudinal observations during pregnancy. *J. Infect. Dis. 137*:524–530 (1978).
21. Paredes, A., P. Wong, E. O. Mason Jr., L. H. Taber, and F. F. Barrett, Nosocomial transmission of group B streptococci in a newborn nursery. *Pediatrics 59*:679–682 (1976).
22. Krause, R. M., and M. McCarty, Studies on the chemical structure of the streptococcal cell wall I. The identification of a micropeptide in the cell walls of group A and A-variant streptococci. *J. Exp. Med. 114*:127–140 (1961).
23. Krause, R. M., Symposium on relationship of structure of microorganism to their immunological properties. *Bacteriol. Rev. 27*:369–380 (1963).
24. Krause, R. M., A cartographer's survey of streptococcal topography. In *Streptococci and Streptococcal Diseases*, S. E. Read and Zabriskie, J. B. (eds.). Academic Press, New York, pp. 97–109 (1980).
25. McCarty, M., The lysis of group A hemolytic streptococci by extracellular enzymes of *Streptomyces albus* II. Nature of the cellular substrate attacked by the lytic enzymes. *J. Exp. Med. 96*:569–580 (1952).
26. McCarty, M., The streptococcus and its products. In *Streptococcal Diseases and the Immune Response*, S. E. Read and Zabriskie, J. B. (eds.). Academic Press, New York, pp. 97–210 (1980).
27. McCarty, M., and S. I. Morse, Cell wall antigens of gram-positive bacteria. *Adv. Immunol. 4*:249–286 (1964).
28. Coligan, J. E., W. C. Schnute Jr., and T. J. Kindt, Immunochemical and chemical studies on streptococcal group-specific carbohydrates. *J. Immunol. 144*:1654 (1975).
29. Maxted, W. R., and J. P. Widdowson, The protein antigens of group B streptococci. In *Streptococci and Streptococcal Diseases*, L. W. Wannamaker and Matsen, J. M. (eds.). Academic Press, New York, pp. 251–266 (1972).
30. Swanson, J., K. C. Hsu, and E. C. Gotschlich, Electron microscopic studies on streptococci. I. M. antigen. *J. Exp. Med. 130*:1063–1091 (1969).

31. Lancefield, R. C., The antigenic complex of *Streptococcus haemolyticus. J. Exp. Med. 47*:91-103 (1943).
32. Ofek, I., S. Bergner-Rabinowitz, and A. M. Davies, Opsonic capacity of type specific streptococcal antibodies. *Isr. J. Med. Sci. 5*:293-296 (1969).
33. Fox, E. N., and M. K. Wittner, New observation on the structure and antigenicity of the M. proteins of the group A streptococcus. *Immunochemistry 6*:11-24 (1969).
34. Fischetti, V. A., E. C. Gotschlich, G. Siviglia, and J. B. Zabriskie, Streptococcal M protein extracted by nonionic detergent. I. Properties of the antiphagocytic and type-specific molecules. *J. Exp. Med. 144*:32-53 (1976).
35. Fischetti, V. A., Immunochemistry of the proteins of the group A streptococcal cell wall. In *Streptococcal Diseases and the Immune Response*, S. E. Read and Zabriskie, J. B. (eds.). Academic Press, New York, pp. 111-124 (1980).
36. Wittner, M. K., Antigenicity of the M proteins of group A hemolytic streptococci: Further evidence for shared determinants among serotypes. *Infect. Immun. 13*:634-642 (1976).
37. Widdowson, J. P., W. R. Maxted, and A. M. Pinney, An M-associated protein antigen (MAP) of group A streptococci. *J. Hyg. 69*:553-564 (1971).
38. Widdowson, J. P., The M-associated protein antigens of group A streptococci. In *Streptococcal Diseases and the Immune Response*, S. E. Read and Zabriskie, J. B. (eds.). Academic Press, New York, pp. 125-147 (1980).
39. Lancefield, R. C., *Streptococcal Infections*, Columbia University Press, New York, pp. 3-18 (1954).
40. Patterson, M. J., and A. E. B. Hafeez, Group B streptococci in human disease. *Bacteriol. Rev. 40*:774-792 (1976).
41. Wilkinson, H. W., Group B streptococcal infection in humans. *Annu. Rev. Microbiol. 32*:41-57 (1978).
42. Pattison, I. H., P. R. J. Matthews, and D. G. Howell, The type classification of group-B streptococci, with special reference to bovine strains apparently lacking in type polysaccharide. *J. Pathol. Bacteriol. 69*:51-60 (1955).
43. Wilkinson, H. W., Comparison of streptococcal R antigens. *Appl. Microbiol. 24*:669-670 (1972).
44. Wilkinson, H. W., Immunochemistry of polysaccharide type antigens of group B streptococcal types Ia, Ib, and Ic. *Infect. Immun. 11*:845-852 (1975).
45. Briles, D. E., and J. M. Davie, Clonal dominance. I. Restricted nature of the IgM antibody response to group A streptococcal carbohydrate in mice. *J. Exp. Med. 141*:1291-1307 (1975).
46. Cunningham, M., and E. H. Beachey, Immunochemical properties of streptococcal M protein purified by isoelectric focusing. *J. Immunol. 115*:1002-1006 (1975).
47. Nahm, M. H., P. R. Murray, B. L. Clevinger, and J. M. Davie, Improved diagnostic accuracy using monoclonal antibody to group A streptococcal carbohydrate. *J. Clin. Microbiol. 12*:506-508 (1980).

48. Nahm, M. H., B. L. Clevinger, and J. M. Davie, Monoclonal antibodies to streptococcal group A carbohydrate. I. A dominant idotypic determinant is located on V_kI. *J. Immunol. 129*:1513–1518 (1982).
49. Fulton, R. J., M. H. Nahm, and J. M. Davie, Monoclonal antibodies to group A carnohydrate. II. The V_kI GAC light chain is preferentially associated with serum IgG31. *J. Immunol. 131*:1326–1331 (1983).
50. Chang, J., H. Herbst, R. Aebersold, and D. Braun, A new isotype sequence (V_k27) of the variable region of κ-light chains from a mouse hybridoma-derived anti-(streptococcal group A polysaccharide) antibody containing an additional cysteine residue. *Biochem. J. 211*:173–180 (1983).
51. Aebersold, R., J. Y. Chang, R. Knecht, and D. G. Braun, Comparative amino acid sequences of murine monoclonal antibodies against streptococcal group A polysaccharide (A-CHO). *Immunobiology 160*:1 (1981).
52. Herbst, H., and D. G. Braun, Antistreptococcal group A antibodies: Production after in vitro activation and hybridization of mouse spleen cells. *Ann. Immunol. 132C*:87–100 (1981).
53. Steinitz, M., I. Seppala, K. Eichmann, and G. Klein, Establishment of a human lymphoblastoid cell line with specific antibody protection against group A streptococcal carbohydrate. *Immunobiology 156*:41–47 (1979).
54. Hasty, D. L., E. H. Beachy, W. A. Simpson, and J. B. Dale, Hybridoma antibodies against protective and nonprotective antigenic determinants of a structurally defined polypeptide fragment of streptococcal M protein. *J. Exp. Med. 155*:1010–1018 (1982).
55. Cunningham, M. W., and S. M. Russell, Study of heart-reactive antibody on antisera and hybridoma culture fluids against group A streptococci. *Infect. Immun. 42*:531–538 (1983).
56. Cunningham, M. W., K. Krishner, and D. Graves, Murine monoclonal antibodies reactive with human heart and group A streptococcal membrane antigens. *Infect. Immun. 46*:34–41 (1984).
57. Dale, J. B., and E. H. Beachey, Unique and common protective epitopes among different serotypes of group A streptococcal M proteins defined with hybridoma antibodies. *Infect. Immun. 46*:267–269 (1984).
58. Polin, R. A., and R. Kennett, Use of monoclonal antibodies in an enzyme immunoassay for rapid identification of group B streptococcus types II and III. *J. Clin. Microbiol. 11*:332–336 (1980).
59. Morrow, D. L., J. B. Kline, S. D. Douglas, and R. A Polin, Rapid detection of group B streptococcal antigen by monoclonal antibody sandwich assay. *J. Clin. Microbiol. 19*:457–459 (1984).
60. Polin, R. A., and R. Kennett, Use of monoclonal antibodies in an enzyme-linked inhibition assay for rapid detection of streptococcal antigen. *J. Pediatr. 97*:540–544 (1980).
61. Polin, R. A., and M. C. Harris, Monoclonal antibodies for bacterial infection. *Clin. Immunol. Newslett. Am. Soc. Microbiol. 4*:153–159 (1983).
62. Lancefield, R. C., M. McCarty, and W. N. Everly, Multiple mouse protective antibodies directed against group B streptococci. *J. Exp. Med. 142*:165–179 (1975).

63. Baker, C. J., M. S. Edwards, and D. L. Kasper, Immunogenicity of polysaccharides from type III group B streptococcus. *J. Clin. Invest. 61*:1107–1110 (1978).
64. Baker, C. J., and D. L. Kasper, Correlation of maternal antibody deficiency with susceptibility to neonatal group B streptococcal infection. *N. Engl. J. Med. 294*:753–756 (1976).
65. Harris, M. C., S. D. Douglas, G. B. Kolski, and R. A. Polin, Functional properties of anti-group B streptococcal monoclonal antibodies. *Clin. Immunol. Immunopathol. 24*:342–350 (1982).
66. Pass, M. A., B. M. Grey, S. Khare, and H. C. Dillon Jr., Prospective studies of group B streptococcal infection in infants. *J. Pediatr. 95*:437–443 (1979).
67. Ancona, R. J., P. Ferrieri, and P. O. Williams, Maternal factors that enhance the acquisition of group B streptococci by newborn infants. *J. Med. Microbiol. 13*:273–280 (1980).
68. Yow, M. D., E. O. Mason, L. J. Leeds et al., Ampicillin prevents intrapartum transmission of group B streptococcus. *J. Am. Med. Assoc. 241*: 1245–1247 (1979).
69. Ruch, F., and L. Smith, Monoclonal antibody to streptococcal group B carbohydrate: Application in latex agglutination and immunoprecipitin assays. *J. Clin. Microbiol. 16*:145–152 (1982).
70. Rench, M. A., T. A. Metzger, and C. J. Baker, Detection of group B streptococcal antigen in body fluids by a latex-coupled monoclonal antibody assay. *J. Clin. Microbiol. 20*:852–854 (1984).
71. Hall, R. T., W. Barnes, L. Krishnan, D. J. Harris, P. G. Rhodes, J. Fayez, and G. L. Miller, Antibiotic treatment of parturient women colonized with group B streptococci. *Am. J. Obstet. Gynecol. 124*:630–634 (1976).
72. Boyer, K. M., C. A. Gadzala, P. D. Kelly, and S. P. Gotoff, Selective intrapartum chemoprophylaxis of neonatal group B streptococcal early-onset disease. III. Interruption of mother-to-infant transmission. *J. Infect. Dis. 148*:810–816 (1983).
73. Steigman, A. J., E. J. Bottone, and B. A. Hanna, Control of perinatal group B streptococcal sepsis: Efficacy of single injection of penicillin at birth. *Mt. Sinai J. Med. 45*:685–693 (1978).
74. Siegel, J. D., G. H. McCracken, N. Threlkeld, B. Milvernan, and C. R. Rosenfeld, Single-dose penicillin prophylaxis against neonatal group B streptococcal infections. *N. Engl. J. Med. 303*:769–775 (1980).
75. Pyati, S. P., R. S. Pildes, N. M. Jacobs, R. S. Ramamurthy, T. F. Yeh, D. S. Raval, L. D. Lilien, P. Amma, and W. I. Metzger, Penicillin for infants weighing two kilograms or less with early-onset group B streptococcal disease. *N. Engl. J. Med. 308*:1383–1389 (1983).
76. Devita, V. T., Progress in cancer management: Keynote address. *Cancer 51*:2401–2409 (1983).
77. Austrian, R., Pneumococcal vaccine: Development and prospects. *Am. J. Med. 67*:547–549 (1979).

78. Lancefield, R. C., Current knowledge of type-specific M antigens of group A streptococci. *J. Immunol. 89*:307–313 (1962).
79. Stollermann, G. H., *Rheumatic Fever and Streptococcal Infection*. Grune & Stratton, New York (1975).
80. Kaplan, M. H., Immunologic relation of streptococcal and tissue antigens. I. Properties of an antigen in certain strains of streptococcal exhibiting an immunologic cross-reaction with human heart tissue. *J. Immunol. 90*: 595–606 (1963).
81. Van De Rijn, I., J. B. Zabriskie, and M. McCarty, Group A streptococcal antigens cross-reactive with myocardium. Purification of heart-reactive antibody and isolation and characterization of the streptococcal antigen. *J. Exp. Med. 146*:579–599 (1977).
82. Baker, C. J., M. S. Edwards, and D. L. Kasper, Role of antibody to native type III capsular polysaccharide of group B streptococcus in infant infection. *Pediatrics 68*:544–549 (1981).
83. Hill, H. R., A. O. Shigeoka, N. H. Augustine, D. Pritchard, J. L. Lundblad, and R. S. Schwartz, Fibronectin enhances the opsonic and protective activity of monoclonal and polyclonal antibody against group B streptococci. *J. Exp. Med. 159*:1618–1628 (1984).
84. Shigeoka, A. O., S. H. Pincus, N. S. Rote, and H. R. Hill, Protective efficacy of hybridoma type-specific antibody against experimental infection with group B streptococcus. *J. Infect. Dis. 149*:363–372 (1984).
85. Shigeoka, A. O., C. L. Jensen, S. H. Pincus, and H. R. Hill, Absolute requirement for complement in monoclonal IgM antibody-mediated protection against experimental infection with type III group B streptococci. *J. Infect. Dis. 150*:63–70 (1984).
86. Shigeoka, A. O., S. H. Pincus, N. S. Rote, D. G. Pritchard, J. I. Santos, and H. R. Hill, Monoclonal antibody preparations for immunotherapy of experimental GBS infection. *Antibiot. Chemother. 35*:254–266 (1985).
87. Egan, M. L., D. G. Pritchard, H. C. Dillon, and B. M. Gray, Protection of mice from experimental infection with type III group B streptococcus using monoclonal antibodies. *J. Exp. Med. 158*:1006–1011 (1983).
88. Christensen, R. D., G. Rothstein, H. R. Hill, and S. H. Pincus, The effect of hybridoma antibody administration upon neutrophil kinetics during experimental type III group B streptococcal sepsis. *Pediatr. Res. 17*:795–799 (1983).
89. Christensen, R. D., G. Rothstein, H. R. Hill, and S. H. Pincus, Treatment of experimental group B streptococcal infection with hybridoma antibody. *Pediatr. Res. 18*:1093–1096 (1984).
90. Bevilacqua, M. P., D. Amrani, M. W. Mosesson, and C. Bianco, Receptors for cold-insoluble globulin (plasma fibronectin) on human monocytes. *J. Exp. Med. 153*:42–60 (1981).
91. *Med. Lab. Prod. Bull.* May (1986).
92. Taggart, R. T., and I. M. Samloff, Stable antibody-producing murine hybridomas. *Science 219*:1228–1230 (1983).

93. DeLuca, M., and W. McElroy, *Bioluminescence and Chemiluminescence*. Academic Press, New York (1981).
94. Arakawa, H., M. Maeda, and A. Tsuji, Chemiluminescence enzyme immunoassay for thyroxin with use of glucose oxidase and a bis(2,4,6-trichlorophenyl)oxalate fluorescent dye system. *Clin. Chem. 31*:430–434 (1985).
95. Helmmila, I., Fluoroimmunoassays and immunofluorometric assays. *Clin. Chem. 31*:359–370 (1985).

22
Monoclonal Antibodies in the Identification and Characterization of *Brucella* Species

DAVID R. BUNDLE and MALCOLM B. PERRY
National Research Council of Canada, Ottawa, Ontario, Canada

JOHN W. CHERWONOGRODZKY
Agriculture Canada, Animal Diseases Research Institute (Nepean), Nepean, Ontario, Canada

I. INTRODUCTION

Brucellosis is a disease of both humans and animals, the causative agents being members of the bacterial genus *Brucella* (1), which are intracellular parasites. *Brucella* species cause epizootic abortions in a variety of animals and septicemic febril illness or localized infection of bone tissue, or other organ systems, in humans. The species infective for humans are *B. abortus*, which is predominantly a pathogen for cattle; *B. melitensis*, found in cattle, goats, and sheep; *B. suis*, which more usually infects swine; and *B. canis*, a pathogen for dogs. *Brucella neotomae* and *B. ovis*, which are associated with the desert wood rat and sheep, respectively, are not considered human pathogens.

Diagnosis of brucellosis based on recovery and identification of *Brucella* species is usually difficult owing to the length of time required for growth and also because of practical problems involved in the recovery of the infecting bacterium from clinical material. Most diagnoses of brucellosis are made on the basis of serological methods for the detection of antibodies to characteristic cell wall polysaccharide antigens of *Brucella* species (2).

Until very recently, the structural nature of the characteristic specific brucella antigens was unknown so that test systems rested on empirical observations and interpretations of serological findings at the molecular level were limited.

The elucidation of brucella-specific antigenic determinants is essential for the development of sensitive, accurate, and specific serodiagnostic systems. For this reason, a major part of our work on brucellae was directed toward a complete structural analysis of the major polysaccharide antigens of *Brucella* species as well as the antigens of other gram-negative bacteria that had been reported to show varying degrees of cross-serological reactivity with brucella antigens. These studies gave results that provided a sound immunochemical explanation of brucella polyclonal antibody serology and also provided a rationale for our production of monoclonal antibodies together with an understanding of their immunospecificities.

II. BRUCELLAE ANTIGENS

Serological differences between *B. abortus, B. melitensis*, and *B. suis* were found to be attributable to two antigenic determinants, designated A and M (3–5), which were associated with the *O*-polysaccharide chains of their lipopolysaccharide (LPS) cell wall constituents. The A determinant appeared to be predominant, but not exclusive, to *B. abortus*, whereas the M antigen appeared to predominate in *B. melitensis* (6).

Polysaccharide B (poly B), a component of some *B. abortus* and *B. melitensis* strains (7), has been considered a characteristic brucellae antigen; however, our studies indicate that poly B, in purified form, is a nonimmunogenic, nonreducing circular D-glucan composed of 1,2-linked β-D-glucopyranosyl units, the average cyclic structure containing about 20 D-glucose residues (8). It is not clear from the literature whether or not serological activity, such as precipitin reactions, ascribed to poly B, are due to contamination of poly B preparations with material having A or M antigenic activities.

To elucidate the structures of the A and M antigenic determinants present in the LPS O-chains of *Brucella* species, the purified LPS of *B. abortus* 1119-3 (A antigen) and of *B. melitensis* 16M (M antigen) were isolated from the phenol phases of cells extracted with hot aqueous phenol, and the respective *O*-polysaccharides, released from the LPS by mild acid hydrolysis, were obtained by Sephadex gel filtration. The O-chain from *B. abortus* (A antigen) was shown by sodium dodecyl sulfate–polyacrylamide gel electrophoresis (SDS–PAGE) methylation, controlled hydrolysis, periodate oxidation and ^{1}H and ^{13}C nuclear magnetic resonance studies to be an unbranched linear homopolymer of 1,2-linked 4,6-dideoxy-4-formamido-α-D-mannopyranosyl residues (9). A similar analysis of the *O*-chain of *B. melitensis* (M antigen) required more extensive ^{1}H and ^{13}C nuclear magnetic resonance studies involving one- and two-dimensional methods (8,10). This revealed that although the O-chain M antigen, like the O-chain A antigen, was also a linear chain of 4,6-dideoxy-4-formamido-α-D-mannopyranosyl units, it differed from the A antigen in which all the units were 1,2-linked (Fig. 1a) in being composed of a pentasaccharide repeating unit containing one 1,3 linkage and four 1,2 linkages (Fig. 1b).

The elucidation of the A and M antigenic structures makes possible the interpretation of previous serological data obtained using polyclonal antisera. The close structural similarity of the A and M antigens explains the extensive

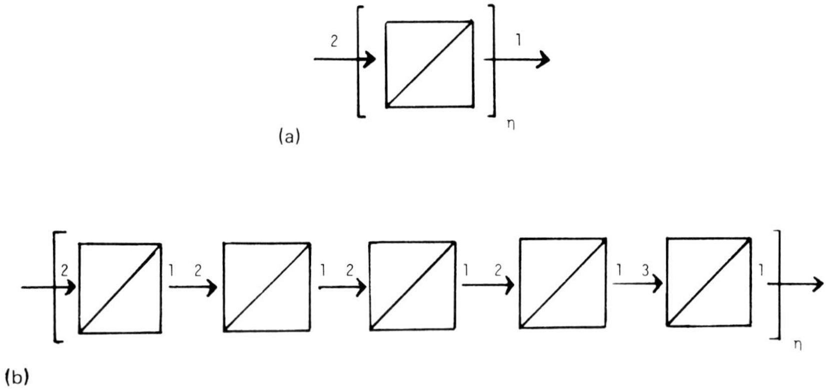

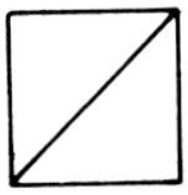

represents 4,6-dideoxy-4-formamido-α-D-mannopyranosyl residues.

Figure 1 Diagrammatic representations of O-polysaccharide chains of the *Brucella* A (*a*) and the M (*b*) antigens.

cross-serological reactions of the two antigens and at the same time explains how, through cross-absorption methods, an antiserum specific for the M antigen could be obtained, which must have specificity for an epitopic feature involving the unique 1,3-linked aminoglycose residue of the M antigen that is absent from the A antigenic structure (Fig. 2).

In the preparation of monoclonal antibodies, it is reasonable now to expect that a specific anti-M hybridoma could be selected in which the antibody would show specificity to an epitope involving a 1,3-linked 4,6-dideoxy-4-formamido-α-D-mannopyranosyl residue, and in fact this goal was achieved.

Because the structures of the brucellae A and M antigens are known, it was also of interest to discover the immunochemical basis for their cross-serological reactivity with *Vibrio cholerae* (11,12), *Yersinia enterocolitica* 0:9 (13–15), *Escherichia coli* 0157 (16,17), *Salmonella landau*, *S. urbana*, and *S. godesberg* (18), and *Pseudomonas maltophilia* (19).

The LPS O-chain of *V. cholerae* (Inaba) has been shown to be a 1,2-linked linear polymer of the same 4-amino-4,6-dideoxy-α-D-mannopyranosyl units, but unlike the A and M brucellae antigen in which the amino group was *N*-formylated, the amino groups of the *V. cholerae* antigen were *N*-acylated with 3-deoxy-L-glycerotetronic acid (20). It appears that bovine brucella A antiserum consists of antibodies that have reactivity to the *V. cholerae* LPS O-chain despite

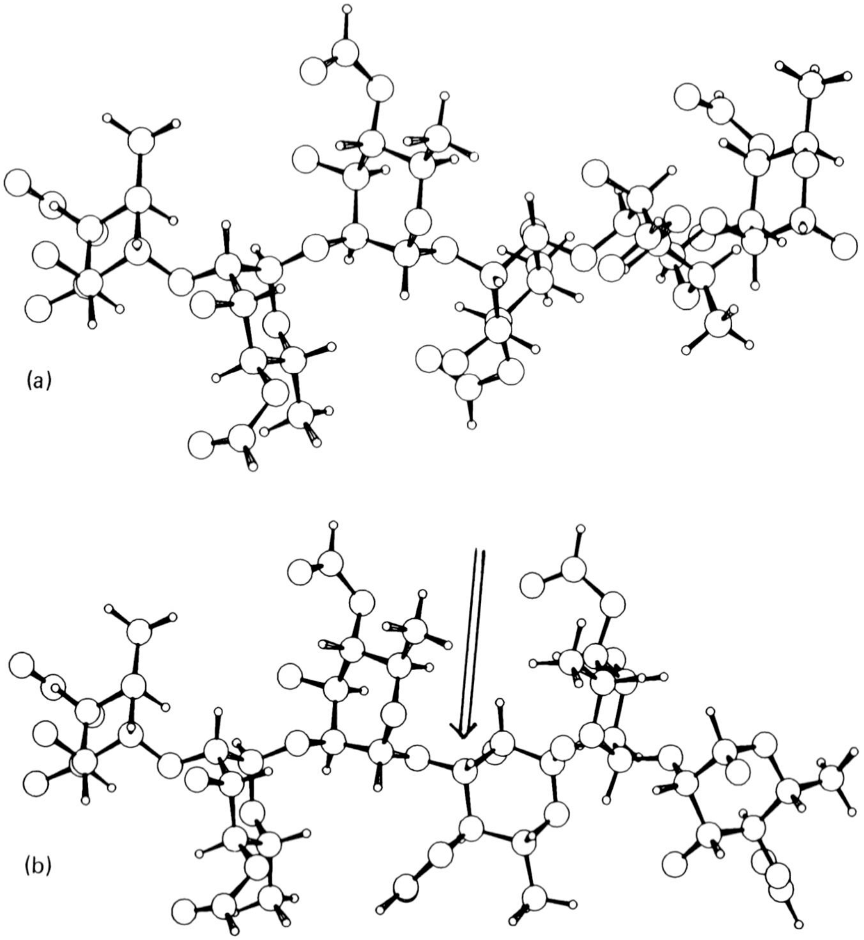

Figure 2 (a) Ball-and-stick molecular model of a hexasaccharide fragment of the brucella A antigen. All glycosidic linkages are α1,2 which leads to a highly regular polymer. (b) Ball-and-stick model of a hexasaccharide fragment of the brucella M antigen. The single α1,3-linkage is highlighted by an arrow, its presence substantially alters the direction of chain propagation, whereas segments containing the α1,2-linkages preserve extensive areas of similarity with the A antigen.

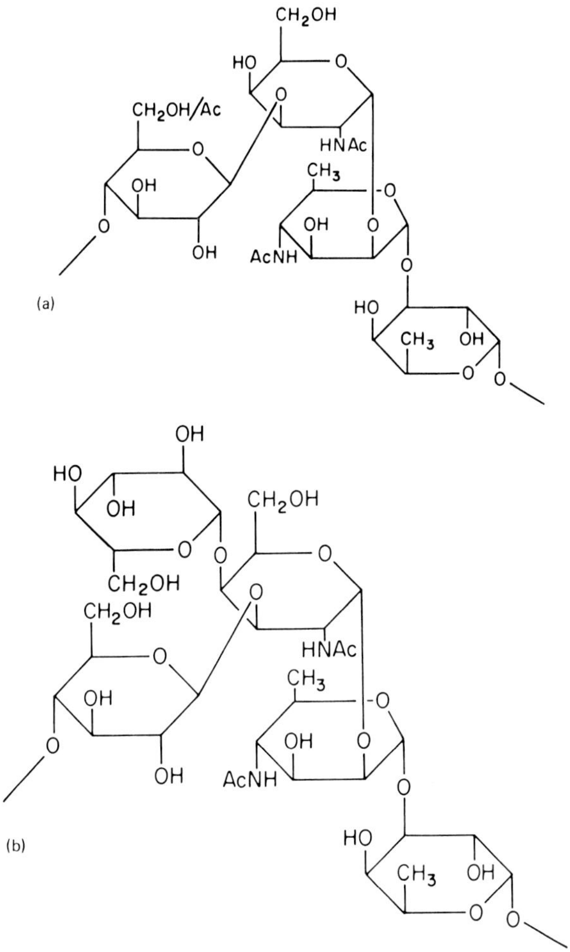

Figure 3 (a) The structure of the chemical repeating unit of the LPS of *E. coli* 0157:H7 and *S. landau*. In *S. landau* one in two glucose residues carry an *O*-6 acetyl substituent. (b) The pentasaccharide repeating unit of *S. godesburg* and *S. urbana* that contains the linear tetrasaccharide of the *E. coli* 0157:H7 O-polysaccharide but, in addition, a branching glucose residue substitutes *O*-4 of the Gal*N*Ac residue.

the different form of *N*-acylation, and we have found that our bovine brucella A antiserum shows only slightly diminished reactivity toward the basic O-chain of *B. abortus* LPS that has been chemically converted to its *N*-acetyl, and *N*-propionyl derivatives (unpublished observations).

Yersinia enterocolitica 0:9 *O*-chain was found to be identical to the A antigen of *Brucella* species except for the minor (2%), residual core constituent at its reducing end (21). The *Y. enterocolitica* 0:9 LPS and O-chain proved to be a valuable alternative source of A antigen, devoid of other possible brucella cross-reacting antigens, for the selection of hybridomas and for defining monoclonal antibody specificities with a greater degree of certainty.

Escherichia coli 0157:H7 LPS O-chain was shown to be a linear polymer of a repeating tetrasaccharide in which one of the four glycose residues is a 1,2 glycosidically substituted 4-acetamido-4,6-dideoxy-α-D-mannopyranosyl unit (22) (Fig. 3a), and it would appear that this glycose must be the residue responsible for cross-reactivity with polyclonal *B. abortus* antiserum. Subsequent studies showed that *S. landau* (23) LPS O-chain had the same basic structure as the *E. coli* 0157 O-chain (see Fig. 3a) and that *S. urbana* and *S. godesberg*, which belong to the same group N (0:30) of the Kauffmann-White classification as *S. landau*, had a related structure. The repeating unit was composed of five sugars (see Fig. 3b) but again glycosidically 1,2-linked 4-acetamido-4,6-dideoxy-α-D-mannopyranosyl units were present in the backbone chain (24).

Preliminary studies on the LPS *O*-chain of *P. maltophilia* 555 indicate that it also contains 4-acetamido-4,6-dideoxy-D-mannopyranosyl residues in linkages unrelated to those described for the *Salmonella* 0:30 species antigens. None of the monoclonal antibodies made against *B. abortus* LPS A antigen or *Y. enterocolitica* 0:9 LPS showed cross-serological reactivity with either the LPS or O-chains from *E. coli* 0157 or *Salmonella* 0:30 species, indicating their greater specificity over polyclonal antisera.

III. GENERATION OF MONOCLONAL ANTIBODIES TO *BRUCELLA*

A previous review (25) covered developments in monoclonal antibodies to brucella antigens prior to the knowledge of the antigen structures. This review is directed toward a molecular interpretation of monoclonal antibodies prepared to defined immunogens and the analysis of their specificities in structural terms.

Once the chemical structure of a particular LPS O-chain of a gram-negative bacterium is known, the task of producing and selecting monoclonal antibodies to this LPS component is considerably simplified. Not only is the purified antigen available for screening assays, in addition, a rational protocol for the rapid screening may be devised to select desirable specificities and eliminate unwanted affinities.

Because chemical and physical analyses indicated that the LPS O-polysaccharides of *B. abortus* 1119-3 and *Y. enterocolitica* 0:9 had identical structures but different core chemotypes (9,21), a dual screening assay employing both of these LPS in enzyme-linked immunosorbent assay (ELISA) with hybridoma culture supernatants was anticipated to yield mainly O-polysaccharide-specific hybrids. This in fact proved to be so, and BALB/C mice immunized with either killed *Y. enterocolitica* 0:9 or *B. abortus* 1119-3 cells subsequently afforded 17 monoclonal antibodies, 15 of which bound the purified A O-polysaccharide antigen (Table 1) (26). Because purified A antigen from LPS was available, it was possible to test the monoclonal antibodies for precipitating activity against this crucial component. Immunodiffusion experiments discriminated antibodies on two levels. Effective precipitation of antigen-antibody complexes is a function of affinity (K_a) and furthermore, requires that the antigen be polyvalent. Thus, in the schematic representation of LPS structure (Fig. 4) the O-chain end determinants (1b) are seen to exist only once per O-polysaccharide chain, whereas internally located determinants (1a) occur in repetitious fashion along the O-chain. Because LPS or alkaline-treated LPS retain elements of their lipid A moiety, aggregation of chains or micelle formation occurs, thereby creating a polyvalent antigen. In contrast, free O-chain does not self-associate and, therefore, terminal determinants would remain univalent. Antibodies precipitated by LPS or alkaline-treated LPS, but not by free O-chains, would be directed toward terminal oligosaccharide determinants, whereas antibodies that precipitated LPS and O-chains would be directed toward internal oligosaccharide determinants. This classification, although undoubtedly an oversimplification, leads to interesting correlations with respect to brucella A and M specificities, especially in light of the structural interrelationship of the two antigens (9,10).

Hybridoma experiments conducted with the spleen cells from two mice immunized with killed cells of *B. melitensis* 16M yielded five stable, recloned cell lines (Table 2). Screening experiments performed with the brucella A and M LPS revealed that only one, Bm-15, exhibited M specificity (Fig. 5a). All the other monoclonals showed similar titers against either A or M antigens (see Fig. 5b,c) (27). Although endpoint titers were similar for antibody Bm-28 (cf. Bm-32 or Bm-10) its titration curve against the A and M antigens showed lower affinity for the A antigen, which is expressed in the failure of this antigen to precipitate the *B. abortus* 1119-3 O-polysaccharide (see Fig. 5c). The low titers of clone Bm-11 reflect low affinity for the A and M antigens, and neither polysaccharide is precipitated by this antibody (27).

Titration curves measured for antibodies Yst9-1, Yst9-2, and Bab-3 and Bab-10 show two types of binding. One exemplified by Yst9-1 and Bab-3 show high titers against the A antigen (*B. abortus* 1119-3 LPS) and negligible or rapidly declining anti-M antigen (*B. melitensis* 16M LPS) activity (Fig. 6a,b). These characteristics mirror the inability of these antibodies to precipitate the

Table 1 Monoclonal Antibodies with *Brucella* A Specificity

		ELISA titer[a]			Immunodiffusion	
Monoclonal antibody	Isotype	*Yersinia* LPS	*B. abortus* LPS	*B. melitensis* LPS	*B. abortus* PS	*B. melitensis* PS
Yst9-1	IgG2b	1×10^5	1×10^5	1×10^4	–	–
Yst9-2	IgG3	5×10^4	1×10^5	1×10^5	+	+
Yst9-3	IgG1	5×10^4	5×10^5	3.16×10^1	–	–
Bab-2	IgG1	1×10^4	1×10^4	0	–	–
Bab-3	IgG3	1×10^3	1×10^3	1×10^1	–	–
Bab-4	IgG3	1×10^5	1×10^5	2×10^4	+	
Bab-6	IgM	5×10^4	5×10^4	ND	+	
Bab-7	IgG3	5×10^4	1×10^4	8×10^3	+	
Bab-8	IgG3	5×10^4	5×10^4	5×10^3	+	
Bab-10	IgG3	5×10^4	5×10^4	3×10^4	+	+

[a]Endpoint titers correspond to the antibody dilution that yields an optical density of 0.1 after 60 min incubation with substrate.

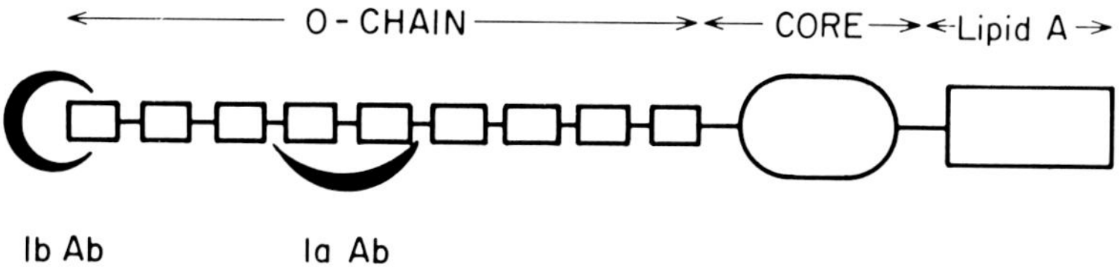

Figure 4 Schematic presentation of a typical smooth lipopolysaccharide. The O-chain repeating units are seen to generate either multiple internal determinants (1a), or univalent chain end epitopes (1b).

homologous O-polysaccharides and their characterization as chain-end specific (26,27). By comparison, antibodies Yst9-2 and Bab-10, thought to bind internal chain oligosaccharide sequences, gave ELISA titration curves indicating that they show no distinction between A and M antigenic structures (see Fig. 6c).

The foregoing data may be interpreted in terms of the recently defined structures of the A and M antigens (9,10). Because the M antigen contains one 4,6-dideoxy-4-formamido-α-D-mannopyranosyl residue that is substituted at *O*-3 in a pentasaccharide repeating unit in which the four remaining 4,6-dideoxy-4-formamido-α-D-mannopyranosyl residues are substituted at *O*-2, it is clear that a substantial common determinant is shared with the A antigen that is composed entirely of 1,2-linked 4,6-dideoxy-4-formamido-α-D-mannopyranosyl units (see Fig. 1). Thus, antibodies Yst9-2, Bab-10, Bm-10, and Bm-32 were probably directed toward this common determinant. Antibody Bm-15 precipitated its homologous O-polysaccharide but not the O-chain derived from the A-LPS. The determinant recognized must be an internal epitope of the M O-polysaccharide involving the unique 1,3-linked aminoglycose moiety because this feature alone distinguishes the M from the A antigen. The O-chain end-binding exhibited by Yst9-1, Yst9-3, and Bab-3 antibodies distinguishes the A from the M antigen because the M antigen possesses a uniquely substituted residue (27). As this latter feature must occur once in the first five residues on the nonreducing end of the M O-chain, its presence will compromise acceptance of any M antigen in the binding site of the anti-A antibodies. The latter effect would be most pronounced for a terminal disaccharide moiety having a 1,3-glycosidic linkage and, at a minimum, if it were the fifth glycosidic linkage from the nonreducing terminal of the O-chain. No clear structural data is available on this point, but the serology would suggest that the 1,3-linkage is located toward the terminal nonreducing end of the biological repeating unit (27).

Table 2 Monoclonal Antibodies with *Brucella* M Specificity

Monoclonal antibody	Isotype	ELISA titer[a]		Immunodiffusion	
		B. melitensis	*B. abortus*	*B. melitensis* PS	*B. abortus* PS
Bm-10	IgG3 κ	1.25×10^4	1.25×10^4	+	+
Bm-11	IgG1 κ	1.58×10^3	2.51×10^2	–	–
Bm-15	IgG3 κ	2.51×10^3	7.9×10^1	(+)	–
Bm-28	IgG3 κ	1.25×10^4	2.51×10^4	+	–
Bm-32	IgG3 κ	1.58×10^4	2.51×10^3	+	+

[a]Endpoint titers correspond to the antibody dilution that yields an optical density of 0.1 after 60 min incubation with substrate.

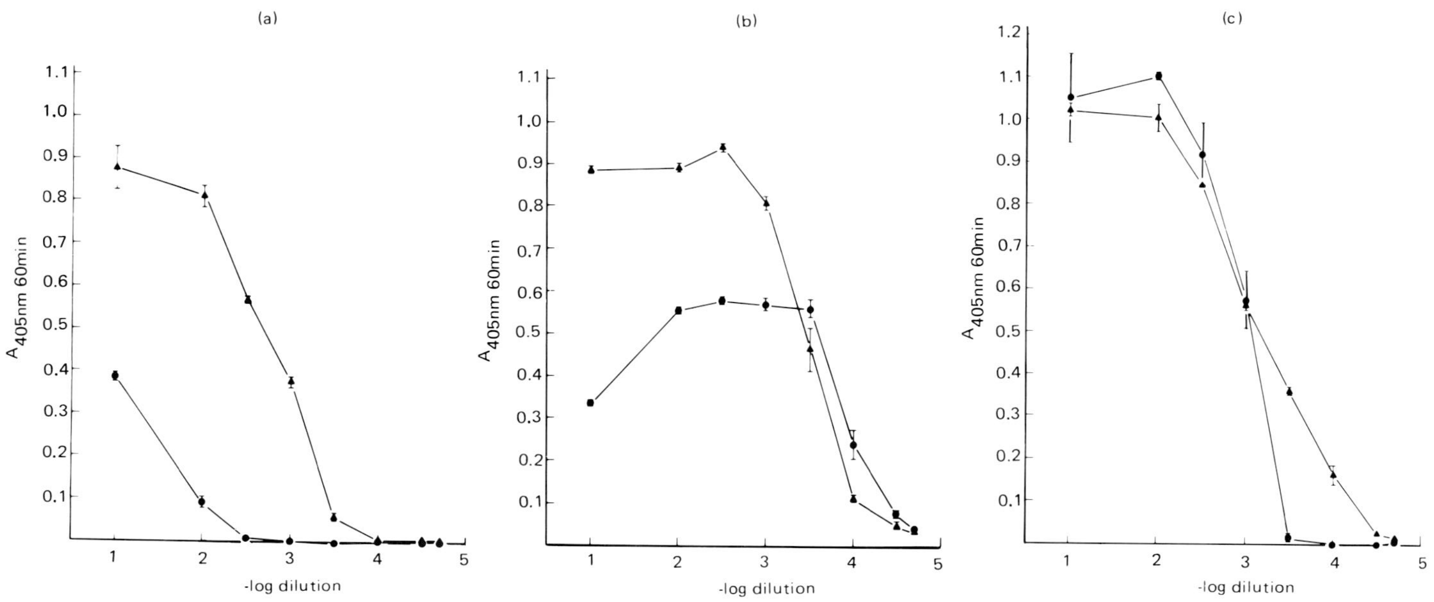

Figure 5 (a) ELISA titration of Bm-15 ascites fluid against the two LPS antigens A (closed circle) and M (closed triangle). (b) Titration curves for the monoclonal antibody Bm-28. Its ability to precipitate the M, but not the A antigen, is mirrored by its ELISA titration against the A-LPS. (c) Monoclonal antibody Bm-32 which gives identical ELISA titration curves with both the homologous M and heterologous A antigens.

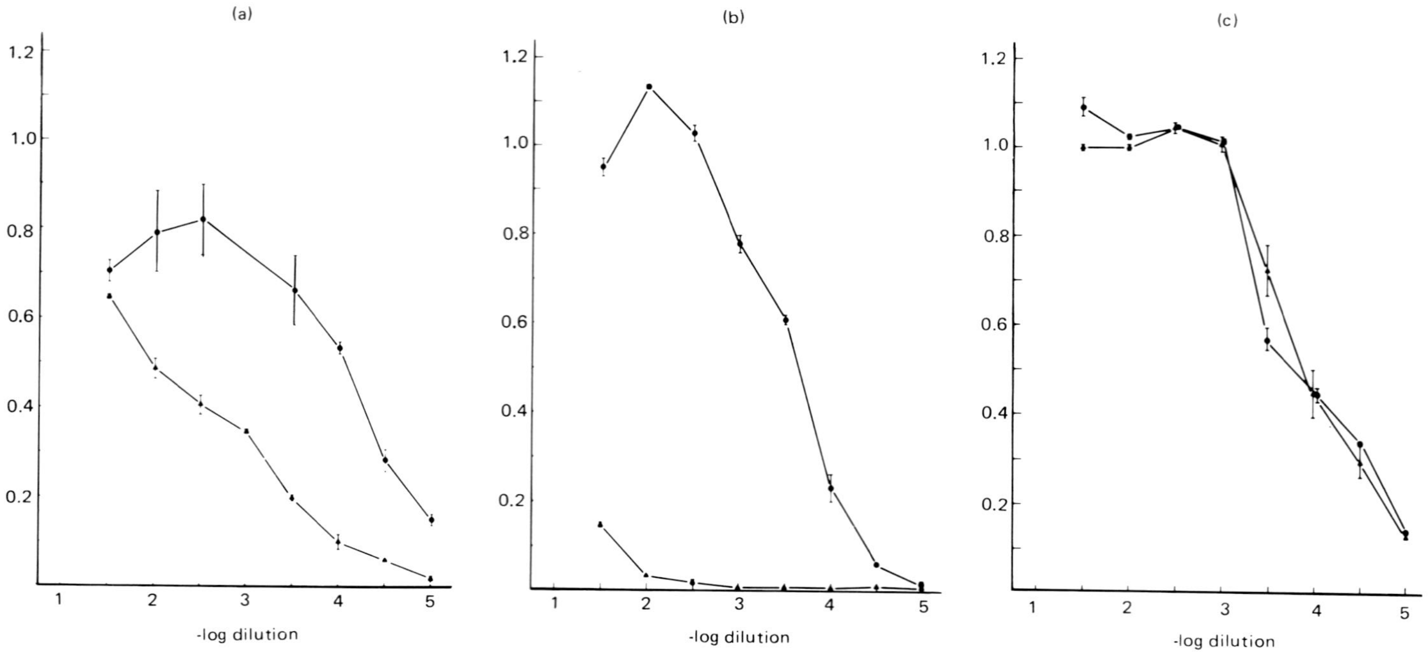

Figure 6 (a) ELISA titration of monoclonal antibody Yst9-1, which exhibits preferential binding toward the A antigen (closed circle). (b) ELISA titration of the A-specific monoclonal antibody Bab-3. Extremely low M activity (closed triangle) was detected compared with good A activity (closed circle). (c) Monoclonal antibody Yst9-2, which binds the A or M antigen with equal affinity. ELISA titration with A (closed circle) and M (closed triangle) LPS.

IV. CONCLUSIONS

The results recorded in the preceding discussion indicate that a detailed understanding of the structures of the brucella A and M antigens (9,10) allowed the construction of three sets of monoclonal antibodies (26,27). Two of the sets correspond to conventional cross-absorbed anti-A and anti-M antisera (28), although monoclonal antibodies possessing a single binding specificity exhibit higher discrimination than polyclonal antibodies, however refined. The third set bind the epitope common to the A and M antigens (9,10). Other antibodies that discriminate between the *Y. enterocolitica* 0:9 LPS and the A antigen of *Brucella* species were also described (26).

Three highly specific brucella reagents are now available to replace conventional typing antisera. The universal anti-A and anti-M set of antibodies bind all smooth brucella LPS, whereas the selective anti-A and anti-M reagents are exclusively specific for their homologous LPS O-antigens (27). The results of conventional tube agglutination experiments (28; Table 3) record endpoint titers for representative antibodies of each class. Clones Yst9-2, Bm-10, Bm-28, and Bm-32 agglutinate all biotypes of *B. abortus* and *B. melitensis*. Monoclonal anti-A antibodies Yst9-1, Yst9-3, and Bab-3 reagents agglutinate only those biotypes known to express the A antigen and not with the *B. abortus* biotype 7 which was previously reported to express both A and M antigens (1). Consistent with these interpretations, the anti-M monoclonal antibody Bm-15 did agglutinate *B. abortus* of biotype 7, indicating the presence of the M antigen and not the A antigen on this bacterium. Although Bm-15 weakly binds the A antigen as indicated by its ELISA titration profile (see Fig. 5a) and has low titers with *Brucella* biotypes of known A serotype, at its operational dilutions of 1:160 or 1:320 it exhibits clear discrimination between A and M *Brucella* serotypes. *Brucella melitensis* biotype 3 is the only bacterium that cannot be clearly assigned to either the A or M serotype by using the monoclonal reagents, a similar result to that observed with the use of conventional polyclonal antisera.

Monoclonal antibodies directed toward the cell wall polysaccharide antigens of *Brucella* species occur with high frequency in hybridoma experiments, and many of the correlations described here may well hold for such antibodies reported by other groups (29–33).

Recent structural and immunochemical data from our laboratory have clarified the structural basis for the serological cross-reactions between the *Brucella* A and M antigen-specific monoclonal antibodies (Tables 2 and 3 and Fig. 5a–c). Fine structural analysis of A antigen preparations by nmr spectroscopy and the use of monoclonal antibodies reveals that they contain one or two 1,3-linked 4,6-dideoxy-4-formamido-α-D-mannopyranosyl residues per O-antigen chain. These minor structural elements can be detected by monoclonal antibodies specific for the M epitope. Thus, the original paradigm of Wilson and Miles in

Table 3 Agglutination Titers[a] of *Brucella* Monoclonal Antibodies with *B. abortus* and *B. melitensis* Biotypes

Brucella strain (biotype)	Classic serological type	Monoclonal antibody							
		Yst9			Bab	Bm			
		−1	−2	−3	−3	−11	−15	−28	−32
B. abortus (1)	A		1280	640	160	160	40	160	160
B. abortus (2)	A		1280	640	160	160	20	160	160
B. abortus (3)	A		1280	320	80	160	40	160	160
B. abortus (4)	M		1280	—	—	160	320	160	160
B. abortus (5)	M		1280	—	—	(160)	320	160	160
B. abortus (6)	A		1280	160	160	160	—	160	80
B. abortus (7)	A&M		1280	—	—	160	320	160	160
B. abortus (9)	M		1280	—	—	160	320	160	160
B. melitensis (1)	M		640	—	—	160	160	160	168
B. melitensis (2)	A		320	640	160	160	10	160	80
B. melitensis (3)	A&M		320	—	80	160	80	160	160

[a]Endpoint titers are expressed as the reciprocal of antibody dilution providing a ++ reaction in conventional tube agglutination tests.

which simultaneous expression of the A and M antigenic determinants is confined to a single antigen is confirmed. The subtle structural differences that differentiate the A and M antigens have been the basis of conflicting interpretations of the occurrence of these antigens on the bacterial cell surface. The use of monoclonal antibodies is able to circumvent these difficulties, but since single O-antigens of *Brucella* biotypes carry both epitopes, the highest affinity monoclonal antibodies are not the best suited to function in bacterial serology of *Brucella*.

REFERENCES

1. Wilson, G., *Brucella*. In *Topley & Wilson's Principles of Bacteriology, Virology and Immunity*, Vol. 2. Edward Arnold, London, pp. 406–421 (1984).
2. Diaz, R., L. M. Jones, D. Leong, and J. B. Wilson, Differences between *Brucella* antigens in indirect haemagglutination tests with normal and tanned red blood cells. *J. Bacteriol. 94*:499–505 (1967).
3. Diaz, R., L. M. Jones, D. Leong, and J. B. Wilson, Surface antigens of smooth *Brucella*. *J. Bacteriol. 96*:893–901 (1968).
4. Wilson, G. S., and A. A. Miles, Serological differential of smooth strains of the *Brucella* group. *Br. J. Exp. Pathol. 13*:1–13 (1932).
5. Corbel, M. J., and B. Morgan, *Int. J. Syst. Bacteriol. 25*:83–89 (1975).
6. Miles, A. A., and N. W. Pirie, The properties of antigenic preparation from *Brucella abortus* and *Brucella melitensis*. I. Chemical and physical properties of bacterial fractions. *Br. J. Exp. Pathol. 20*:83–98 (1939).
7. Diaz, R., P. Garaica, L. M. Jones, and I. Moriyon, Radial immunodiffusion test with a *Brucella* polysaccharide antigen for differentiating infected from vaccinated cattle. *J. Clin. Microbiol. 10*:37–41 (1979).
8. Perry, M. B., D. R. Bundle, and J. W. Cherwonogrodzky, The structures and serology of the A and M antigens of *Brucella abortus* and *Brucella melitensis* and the structure of polysaccharide B of *Brucella* species. (Abstr.) XIV Internat. Congr. Microbiol., Manchester, September (1986).
9. Caroff, M., D. R. Bundle, M. B. Perry, J. W. Cherwonogrodzky, and J. R. Duncan, Antigenic S-Type lipopolysaccharide of *Brucella abortus* 1119-3. *Infect. Immun. 46*:384–388 (1984).
10. Bundle, D. R., M. B. Perry, and J. W. Cherwonogrodzky, The identification and structural elucidation of *Brucella melitensis* M antigen. (Abstr.) XIII International Symp. Carbohydrates. Cornell University, Ithica, New York, August 10–15 (1986).
11. Feeley, J. C., Somatic O-antigen relationships of *Brucella* and *Vibrio cholerae*. *J. Bacteriol. 99*:645–649 (1969).
12. Wong, D. H., and C. H. Chow, Group agglutinations of *Brucella abortus* and *Vibrio cholerae*. *Chin. Med. J. 52*:591–594 (1937).
13. Ahvonen, P., E. Jansson, and K. Ako, Marked cross agglutination between *Brucella* and a subtype of *Yersinia enterocolitica*. *Acta Pathol. Microbiol. Scand. 75*:291–295 (1969).

14. Corbel, M. J., and G. A. Cullen, Differentiation of the serological response to *Yersinia enterocolitica* IX and *Brucella abortus* in cattle. *J. Hyg. 68*: 519–530 (1970).
15. Mittal, K. R., and I. R. Tizzard, Serological cross reactions between *Brucella abortus* and *Yersinia enterocolitica* serotype 9. *Vet. Bull. 51*:501–505 (1981).
16. Stuart, F. A., and M. J. Corbel, Identification of a serological cross reaction between *Brucella abortus* and *Escherichia coli* 0:157. *Vet. Rec. 110*:202–203 (1982).
17. Hurvell, B., Serological cross reactions between *Brucella* species and *Yersinia enterocolitica*. An immunological and immunochemical study. Thesis, Stockholm (1973).
18. Corbel, M. J., The serological relationship between *Brucella* spp., *Yersinia enterocolitica* IX, and *Salmonella* serotype of the Kauffmann White group N. *J. Hyg. Camb. 75*:151–171 (1975).
19. Corbel, M. J., F. A. Stuart, and R. A. Brewer, Observations on serological cross-reactions between smooth *Brucella* species and organisms of other genera. *Dev. Biol. Stand. 56*:341–348 (1984).
20. Kenne, L., B. Lindberg, P. Unger, B. Gustafsson, and T. Holme, Structural studies of the *Vibrio cholerae* antigen. *Carohydr. Res. 100*:341–349 (1982).
21. Caroff, M., D. R. Bundle, and M. B. Perry. Structure of the O-chain of the phenol-phase soluble lipopolysaccharide of *Yersinia enterocolitica* serotype 0:9. *Eur. J. Biochem. 139*:195–200 (1984).
22. Perry, M. B., L. MacLean, and D. W. Griffith, Structure of the O-chain polysaccharide of the phenol-phase soluble lipopolysaccharide of *Escherichia coli* 0:157:H7. *Can. J. Biochem. Cell Biol. 64*:21–28 (1985).
23. Bundle, D. R., M. Gerken, and M. B. Perry, Two-dimensional nuclear magnetic resonance at 500 MHz: The structural elucidation of a *Salmonella* serogroup N polysaccharide antigen. *Can. J. Chem. 64*:255–264 (1986).
24. Perry, M. B., D. R. Bundle, L. MacLean, J. A. Perry, and D. W. Griffith, The structure of the antigenic lipopolysaccharide O-chains produced by *Salmonella urbana* and *Salmonella godesberg*. *Carbohydr. Res. 156*:107–122 (1986).
25. Holman, P. J., G. Schurig, and J. T. Douglas, Development of monoclonal antibodies to *Brucella* cell surface antigens. In *Monoclonal Antibodies Against Bacteria*, Vol. II, A. J. L. Macario and de Macario, E. C. (eds.). Academic Press, New York, pp. 81–110 (1985).
26. Bundle, D. R., M. A. J. Gidney, M. B. Perry, J. R. Duncan, and J. W. Cherwonogrodzky, Serological confirmation of *Brucella abortus* and *Yersinia enterocolitica* 0:9 O-antigens by monoclonal antibodies. *Infect. Immun. 46*:389–393 (1984).
27. Bundle, D. R., J. W. Cherwonogrodzky, M. A. J. Gidney, P. J. Meikle, M. B. Perry, and T. Peters, Definition of brucella A and M epitopes by monoclonal typing reagents and synthetic oligosaccharides. *Infect. Immun.* (submitted 1988).

28. Alton, G. G., L. M. Jones, and D. E. Pietz, Laboratory techniques in brucellosis. *World Health Organization Monograph Series No. 55*, World Health Organization, Geneva (1975).
29. Gorrell, M., G. L. Milliken, B. J. Anderson, and A. Pucci, An enzyme immunoassay for bovine brucellosis using a monoclonal antibody specific for field strains of *Brucella abortus. Dev. Biol. Stand. 56*:491–494 (1984).
30. Quinn, R., A. M. Campbell, and A. P. Phillips, A monoclonal antibody specific for the A antigen of *Brucella*-spp. *J. Gen. Microbiol. 130*:2285–2290 (1984).
31. Cappuccinelli, P., P. L. Fiori, G. Gorgani, and A. M. Pacetti, Antigenic differences between *Brucella abortus* and *Brucella melitensis* recognized by a monoclonal antibody. *Microbiologica 9*:179–188 (1986).
32. Greiser-Wilke, I., V. Moenning, D. Thon, and K. Rauter, Characterization of monoclonal antibodies against *Brucella melitensis. Zentralbl. Veterinaermed. Reihe B 32*:616–627 (1985).
33. Montaraz, J. A., A. J. Winter, D. M. Hunter, B. A. Sowa, A. M. Wu, and L. G. Adams, Protection against *Brucella abortus* in mice with O polysaccharide-specific monoclonal antibodies. *Infect. Immun. 51*:961–963 (1986).

RECENT REFERENCES ADDED IN PROOF

Bundle, D. R., J. W. Cherwonogrodzky, and M. B. Perry, Characterization of *Brucella* polysaccharide B. *Infect. Immun. 56*:1101–1106 (1988).

Bundle, D. R., J. W. Cherwonogrodzky, and M. B. Perry, Structural elucidation of the *Brucella melitensis* M antigen by high-resolution NMR at 500MHz. *Biochemistry 26*:8717–8726 (1987).

Cherwonogrodzky, J. W., M. B. Perry, and D. R. Bundle, Identification of *Brucella* A and M antigens as the O-polysaccharides of smooth lipopolysaccharides. *Can. J. Microbiol. 33*:979–981 (1987).

DiFabio, J. L., M. B. Perry, and D. R. Bundle, Analysis of the lipopolysaccharide of *Pseudomonas maltophilia* 555. *Biochem. Cell Biol. 65*:968–977 (1987).

23
Monoclonal Antibodies to *Salmonella* and *Campylobacter*

BALA SWAMINATHAN*
Purdue University, West Lafayette, Indiana

I. THE GENUS *SALMONELLA*

The genus *Salmonella* is classified into five subgenera on the basis of biochemical characteristics and comprises more than 2000 serovars (1). DNA relatedness tests have shown that all five subgenera of *Salmonella* constitute a single genetic species (2–4). The biochemically typical salmonellae of subgenus I (e.g., *S. typhimurium, S. typhi, S. dublin, S. senftenberg*) are pathogenic to humans and animals. Some serovars are strictly host adapted (e.g., *S. typhi, S. paratyphi* A, *S. sendai*: humans; *S. dublin*: cattle; *S. pullorum* and *S. gallinarum*: fowl; *S. abortus-ovis*: sheep; *S. typhisuis*: swine), although most salmonellae of subgenus I show broad host specificity. Serovars adapted to humans cause typhoid and septicemia, whereas the serovars with broad host specificity cause gastroenteritis in humans. Strains of subgenus II and subgenus III have been isolated mostly from the intestinal contents of cold-blooded animals. Strains of subgenus IV and subgenus V are usually isolated from the environment and are rarely pathogenic for humans (1).

II. SALMONELLOSIS: MAGNITUDE OF THE PROBLEM

Approximately 30,000 cases of human salmonellosis are reported in the United States each year. Surveillance of salmonella infections in the United States is passive, and it is generally agreed that the reported number of salmonella isola-

**Present affiliation*: Center for Infectious Diseases, Centers for Disease Control, Atlanta, Georgia.

tions greatly underestimate the actual incidence (5,6). Hauschild and Bryan estimate that the ratio of actual incidence to number reported (R_j) may range from 29.5:1 to 100:1 in large outbreaks. The reported number of human salmonella isolations in 1983 was 38,881. Applying the factors of Hauschild and Bryan, one may estimate that the actual incidence of human salmonellosis in 1983 was between 1,147,000 cases and 3,888,100 cases. The estimated cost of doctor's visit, hospital care, and productivity loss may range from 1687 to 2025 dollars per person (7). Thus, the annual cost of human salmonellosis to the economy of the United States would be 2 to 8 billion dollars.

III. IMMUNOLOGICAL DETECTION OF *SALMONELLA*

The isolation and identification of salmonella from clinical specimens or foods involve selective enrichment, plating on selective differential agar medium for the isolation of pure cultures, biochemical characterization of suspect colonies, and serological confirmation. Isolation of salmonella from foods involves an additional step of enrichment in a nonselective medium before selective enrichment to resuscitate sublethally damaged cells. The entire procedure takes 4 to 7 days for completion. This labor-intensive and time-consuming cultural procedure impedes prompt commencement of appropriate therapy for patients and prompt sequestering of any contaminated product. Institution of remedial measures in a food processing situation is delayed because of the lengthy detection procedure. Therefore, several attempts have been made to devise rapid detection methods for salmonellae. Immunological tests (immunofluorescence, immunodiffusion, immunoimmobilization, radioimmunoassay, immunoradiometry, and enzyme immunoassay) have been developed for the detection of salmonellae in biological milieu.

Until recently, mixtures of polyclonal antibodies were employed for such assays (8). The Kaufmann-White schema (9) classifies *Salmonella* serovars on the basis of heat-stable somatic "O" antigens and heat-labile flagellar "H" antigens. The O antigens of the genus *Salmonella* are shared among *Salmonella, Citrobacter*, and *Escherichia* species; therefore, cross-reactions are frequently encountered when antibodies directed against O antigens of salmonella are employed in immunological assays. Consequently, many investigators have employed antibodies directed against H antigens in immunological assays for salmonellae because H antigens are not shared outside the genus *Salmonella* (as defined in *Bergey's Manual of Systematic Bacteriology*, which includes *Arizona* as a subspecies of *Salmonella*). Immunological tests employing pooled polyvalent polyclonal flagellar antisera have been found to be more specific than tests employing O antibodies. However, probelms of lack of standardization because of batch-to-batch variation in activity of polyclonal antibodies and heterogeneity of polyclonal antibody preparations have been problematic (8).

IV. MONOCLONAL ANTIBODIES FOR THE DETECTION OF *SALMONELLA*

The advent of monoclonal antibody technology brought new hopes to the realm of immunological detection of *Salmonella*. By employing monoclonal antibodies, one could circumvent problems such as unacceptable batch-to-batch variations in the quality of the antiserum, or cross-reactions with nonsalmonellae owing to the heterogeneity of polyclonal antisera. At the same time, the Kauffmann-White schema indicates the presence of more than 50 distinct flagellar phase 1 and phase 2 antigens in the genus *Salmonella* (10). Of these, 17 different flagellar antigens are represented in the most commonly encountered serovars (11). As monoclonal antibodies are exquisitely specific and are directed against specific defined epitopes, it would be reasonable to conclude that one would have to use a battery of at least 17 monoclonal antibodies to detect the more commonly occurring salmonellae and more than 50 different monoclonal antibodies to detect all but the nonmotile salmonellae. This would be an arduous undertaking, indeed.

Fortunately, the problem has not turned out to be as complicated as originally thought. The Kauffmann-White schema has been an invaluable aid for the serological classification of *Salmonella*. However, excessive reliance on the concept that each serovar of *Salmonella* has a unique flagellar antigenic component has impeded the development of diagnostic immunoassays for *Salmonella*. As early as 1972, Langman (12) reported the presence of shared antigenic determinants in five antigenically distinct serovars of *Salmonella* and suggested that the flagellar antigens contain a variable region (H_v) and a constant region (H_c) and that the Kauffmann-White schema described only the H_v determinants. Langman's suggestion would seem logical given the function of the flagella in the organism. Recent evidence indicates that among serologically unrelated *Salmonella* serovars, there is absolute conservation of amino acid sequences at both ends of the flagellin molecule (locations 1–100 and 421–493 from the NH_2-terminus) with progressively greater variation toward the middle of the molecule (13–16). Wei and Joys (16) suggest that the regions of conservation result from the need for these regions to interact with the various products involved in the regulation of flagellin synthesis.

If there are conserved antigenic determinants, it follows that an antibody directed against such a conserved determinant of a serovar of *Salmonella* would be expected to react with several other *Salmonella* serovars. This is indeed so, as was recently demonstrated with polyclonal antibodies (17) and monoclonal antibodies (discussed in the following).

V. MOPC 467 CELL LINE AND M467 NATURAL MONOCLONAL ANTIBODY

MOPC 467 is a plasmacytoma line isolated from BALB/c mice induced with mineral oil. MOPC 467 produces an IgA (M467) that was found to react with several serovars of *Salmonella, Pasteurella*, and *Herella* (18-20). The reactivity profile of M467 IgA has been determined with more than 600 serovars of *Salmonella* in my laboratory and is shown in Table 1. M467 antibody reacts with

Table 1 Serovars of *Salmonella* that React with M467 Immunoglobulin A

Abadina	Amsterdam
Abaetetuba	Amunigun
Aberdeen	Anatum
Abortusbovis	Anfo
Abortusequi	Apapa
Accra	Aqua
Adabraka	Ardwick
Adamstown	Arechavaleta
Adamstua	Arizonae (19/20)–subgenus III
Adelaide	Artis–subgenus II
Aderike	Atlanta (Mississippi)
Aequatoria	Austin
Agama	Ayinde
Agbeni	Babelsberg
Agona	Bacongo–subgenus II
Ahmadi	Bahati
Ahuza	Bahrenfeld
Akanji	Baildon
Alabama	Ball
Alachua	Baragwanath–subgenus II
Alamo	Bareilly
Albany	Bechuana–subgenus II
Albuquerque	Belem
Allandale	Bellville–subgenus II
Altendorf	Benfica
Amager	Benguella
Amersfoort	Bere
Amherstiana	Bergedorf
Amina	Bergen
Aminatu	Berkeley

Table 1 (Continued)

Bern–subgenus IV	California
Berta	Cambridge
Bietri	Canastel–subgenus II
Bilthoven–subgenus II	Canoga
Binza	Caracas
Birkenhead	Cardiff
Birmingham	Carmel
Bispebjerg	Carno
Blankenese–subgenus II	Carrau
Bleadon–subgenus II	Cerro
Bledgam	Chailey
Blijdorp	Champaign
Blockley	Chandans
Blukwa	Charity
Bodjonegoro	Chester
Boecker	Chicago
Bokanjac	Chingola
Bolton	Choleraesuis
Bonaire–subgenus IV	Christiansborg
Bonariensis	Claibornei
Bonn	Clerkenwell
Bootle	Clifton–subgenus II
Bornum	Coeln
Bovismorbificans	Coleypark
Bracknell	Colindale
Bradford	Colombo
Braenderup	Colorado
Brancaster	Concord
Brandenburg	Coquilhatville
Brazil	Corvallis
Brazzaville	Cubana
Bredeney	Curacao
Brisbane	Dakar
Bristol	Daressalaaem–subgenus II
Bronx	Daytona
Brunei	Decatur (Choleraesuis)
Budapest	Degania–subgenus II
Bukuru	Demerara
Bunnik–subgenus II	Denver
Bury	Derby
Butantan	Dessau
Buzu	Detroit–subgenus II
Cairo (Stanley)	Deversoir
Calabar	Djakarta

Table 1 (Continued)

Djugu	Frintrop
Donna	Fuhlsbuettel–subgenus II
Driffield	Fulica
Drypool	Gamaba
Dublin	Gambaga
Duisburg	Gambia
Durban	Gaminara
Dusseldorf	Garba
Duval	Garoli
Ealing	Gatuni
Eastbourne	Gdansk
Edinburg	Gege
Egusi	Georgia
Egusitoo	Gilbert–subgenus II
Eimsbuettel	Give
Ekotedo	Glasgow
Elisabethville	Glostrup
Elsiesrivier–subgenus II	Gloucester
Emek	Godesberg
Entebbe	Goerlitz
Enteritidis	Goeteborg
Epicrates	Goettingen
Eppendorf	Goldcoast
Escanaba	Gombe
Eschweiler	Greiz
Essen	Grumpensis
Ezra	Guinea
Fandran–subgenus II	Haardt
Fann	Haarlem–subgenus II
Fanti	Hadar
Farcha	Haddon–subgenus II
Fayed	Haifa
Ferlac	Halle
Finchley–subgenus II	Hannover
Finkenwerder	Harburg
Fisherstrasse	Harrisonburg
Flint–subgenus IV	Hartford
Florida	Hato
Flottbek	Havana
Foulpointe–subgenus II	Heidelberg
Frankfurt	Heilbron–subgenus II
Freetown	Hermannswerder
Fresno	Hessarek
Friedenau	Heves
Friedrichsfelde	Hidalgo

Table 1 (Continued)

Hillsborough	Kibusi
Hindmarsh	Kiel
Hisingen	Kilwa–subgenus II
Hofit	Kimuenza
Homosassa	Kingabwa
Horsham	Kingston
Houten–subgenus IV	Kinondoni
Huila–subgenus II	Kinshasa
Humber–subgenus II	Kintambo
Ibadan	Kisangani
Idikan	Kisarawe
Ilala	Kivu
Illinois	Korovi
Indiana	Kottbus
Infantis	Kralendyk–subgenus IV
Inverness	Krefeld
Ipswich	Krugersdorp–subgenus II
Irenea	Kuessel
Irumu	Kumasi
Islington–subgenus II	Kunduchi
Israel	Kuru
Ituri	Labadi
Itutaba	Landala
Jacksonville–subgenus II	Langenhorn
Jaffna	Langford
Jaja (Stanleyville)	Lansing
Jangwani	Leipzig
Javiana	Leith
Jedburgh	Leopoldville
Jericho	Lexington
Jerusalem	Lichtenberg–subgenus II
Jodhpur	Lille
Johannesburg	Limete
Kaapstad	Lindenburg
Kaduna	Lindi
Kalamu	Lisboa
Kaltenhausen–subgenus II	Litchfield
Kamoru	Livingstone
Kandla	Locarno–subgenus II
Kaneshie	Lomalinda
Kapemba	Lomita
Karachi	London
Kasenyi	Luanshya–subgenus II
Kentucky	Luciana
Kiambu	Luckenwalde

Table 1 (Continued)

Luke	Montevideo
Lundby–subgenus II	Morehead
Lyon	Morocco
Macallen	Morotai
Madelia	Moscow
Magumeri	Mountpleasant
Makiso	Mpila–subgenus II
Malstatt	Muenchen
Mampong	Muenster
Manchester	Mundonobo
Manhattan	Nachshonim–subgenus II
Manica–subgenus II	Nagoya
Manila	Nairobi–subgenus II
Mapo	Napoli
Mara	Narashino
Maracaibo	Nashua
Maricopa	Naware
Marseille	Nchanga
Matadi	Ndolo
Mathura	Neasden–subgenus II
Mbandaka	Negev–subgenus II
Mbao	Neukoelln
Meleagridis	Newbrunswick
Memphis	Newington
Mendoza	Newlands
Menston	Newmexico
Mgulani	Newport
Miami	Newrochelle
Michigan	Ngozi–subgenus II
Midhurst–subgenus II	Niamey
Mikawasima	Niarembe
Millesi	Nigeria
Milwaukee	Niloese
Minneapolis	Nima
Minnesota	Nordenham–subgenus II
Mishmarhaemek	Norwich
Mission (Isangi)	Nottingham
Mississippi	Nyanza
Miyazaki	Nyborg
Mjimwema–subgenus II	Ohio
Mobeni–subgenus II	Ohlstedt
Mocamedes	Okerara
Molade	Onderstepoort
Mondeor–subgenus II	Oranienburg
Monschaui	Oregon (Muenchen)

Table 1 (Continued)

Orientalis	Salinatis
Orion	Sandiego
Oritamerin	Sanga
Os	Sanjuan
Oslo	Saphra
Othmarschen	Sarajane
Ouakam	Schleissheim
Overschie	Schoeneberg
Panama	Schwarzengrund
Papuana	Seattle
Patience	Selandia
Penarth	Senegal
Pensacola	Senftenberg
Perth	Seremban
Pharr (weak reaction)	Shamba
Phoenix–subgenus II	Shangani
Pikine (Altona)	Sheffield
Plymouth	Shikmonah
Pomona	Shipley
Poona	Shoreditch
Portland	Shubra
Portsmouth	Simi
Praha	Simsbury
Pretoria	Singapore
Quinhon	Soahanina
Quiniela	Solna
Ramatgan	Solt
Raus	Souza
Reading	Spartel
Redlands	Springs–subgenus II
Richmond	Stanleyville
Ridge	Stellenbosch–subgenus II
Riogrande	Stendal
Rosenthal	Sterrenbos
Rostock	Stockholm
Rotterdam–subgenus II	Stourbridge
Rowbarton–subgenus II	Straengnaes
Rruzizi	Strasbourg
Rubislaw	Suarez–subgenus II
Ruki (Ball)	Suberu
Saarbruecken	Sullivan–subgenus II
Saintpaul	Sundsvall
Saka	Szentes
Sakaraha–subgenus II	Tafo
Salford	Takoradi

Table 1 (Continued)

Taksony	Vancouver
Tallahassee	Vejle
Tananarive	Veneziana
Tanger	Verity–subgenus II
Taunton	Victoriaborg
Tchad	Virchow
Techimani	Virginia
Telaviv	Vleuten
Telelkebir	Wagenia
Telhashomer	Wandsbek–subgenus II
Texas	Wangata
Thomasville	Warragul
Thompson	Wassenaar–subgenus IV
Tilburg	Waycross
Tim (Newington)	Wayne
Tinda	Welikade
Tosamanga–subgenus II	Weltevreden
Tounouma	Wentworth
Tournai	Weslaco
Tranoroa–subgenus II	Westhampton
Treforest	Weston
Tudu	Wichita
Tuebingen	Wien
Typhimurium	Wil
Typhisuis	Wilhelmsburg
Uccle	Winchester–subgenus II
Uganda	Winnipeg
Umbilo	Womba (Altendorf)
Umhlali	Worcester–subgenus II
Umhlatazana	Worthington
Uno	Wuerzburg (Miami)
Uphill–subgenus II	Yalding
Urbana	Zadar
Usumbura	Zanzibar
Utah	Zega
Utrecht	Zongo
Uzaramo	

more than 95% of the *Salmonella* serovars tested. The M467-negative serovars of *Salmonella* identified thus far (21-24) are listed in Table 2. Also, M467 reacts with some strains of *Citrobacter freundii, Proteus mirabilis, Escherichia coli*, and *Yersinia enterocolitica* (24-26). M467 IgA recognizes one or more flagellar antigenic determinants common to many serovars of *Salmonella* and different from the phase 1 and phase 2 antigens of the Kaufmann-White schema (19,27,28). Furthermore, M467 IgA recognizes more than one group of peptide determinants on the flagellin of *Salmonella* species and the number of such M467-reactive determinants in different serovars of *Salmonella* may vary. This leads to a difference in the avidity with wich M467 binds to different serovars of *Salmonella*. For example, M467 binds to *S. milwaukee, S. adelaide*, and *S. mississippi* with greater avidity than to *S. anatum* and *S. typhimurium*. This difference in avidities leads to differences in the sensitivity of enzyme immunoassays for the detection of different serovars of *Salmonella* employing M467 IgA (Table 3).

M467 IgA has been successfully employed for the detection of salmonella by several investigators. Robison et al. (22) used a direct enzyme immunoassay to detect salmonella in 36 hr in artificially inoculated infant formula. It was possible to detect *S. dublin* at a concentration of 10^{-3} cells · ml^{-1} in a competitive immunoassay using polymerized flagellin from *S. milwaukee* as the competing antigen (21). Because of the difference in M467 avidity to salmonella antigens, such sensitivities cannot be achieved with many other serovars (29). A commercially available sandwich enzyme immunoassay that uses M467 IgA for

Table 2 *Salmonella* Serovars that Do Not React with M467 Immunoglobulin A

Abony	Pueris (Newport)
Banalia	Pullorum
Banana	Remo
Emmastad	Rubsilaw
Gallinarum	Sendai
Hvittingfoss	Shangai
Kaposvar (Reading)	Sinstorf
Kirkee	Stanley
Korlebu	Tennessee
Leeuwarden	Tilene
Newington	Tucson
Onarimon	Typhi
Paratyphi A	Willemstad
Paratyphi B	Yaba
Paratyphi C	Zagreb (Saintpaul)
Potsdam	

Table 3 Sensitivity of a Fluorescent Enzyme Immunoassay with M467 Antibody for Five Serovars of *Salmonella*

Serovar	Fluorescence intensity at indicated dilutions				
	10^7 CFU/ml	10^6 CFU/ml	10^5 CFU/ml	10^4 CFU/ml	10^3 CFU/ml
S. milwaukee	1629 ± 93[a]	1041 ± 106	205 ± 7	158 ± 4	120 ± 4
S. montevideo	1735 ± 166	1183 ± 169	209 ± 17	153 ± 8	122 ± 3
S. typhimurium	1299 ± 70	833 ± 65	247 ± 5	149 ± 11	119 ± 4
S. heidelberg	573 ± 26	332 ± 13	133 ± 3	92 ± 7	89 ± 6
S. infantis	539 ± 23	320 ± 15	131 ± 3	88 ± 2	85 ± 4

[a]Values are arbitrary fluorescence units as read on a MicroFluor Reader (Dynatech Laboratories, Chantilly, Va.). The values range from 0 to 4000.

Values are expressed as mean ± standard deviation.

Negative control was a culture of *Escherichia coli* (ca. 10^9 CFU/ml). Mean negative control value was 109 ± 5.

Enzyme immunoassay was performed by direct immobilization of cells in wells of polystyrene microtitration plate (Immulon 2; Dymatech Laboratories) by passive adsorption. M467 antibody labeled with β-galactosidase was used the probe and 4-methylumbelliferyl-β-D-galactopyranoside was the substrate.

capture and detection (30) was evaluated against the conventional cultural procedure for the detection of salmonellae in meat and poultry products. There was 100% agreement between the two methods (31).

VI. MONOCLONAL ANTIBODIES TO COMPLEMENT THE REACTIVITY OF M467

M467 IgA has proved to be a valuable tool in the development of immunoassays for the rapid detection of salmonella. However, several clinically significant serovars of *Salmonella* are M467-negative. For a salmonella immunoassay to be widely accepted in clinical diagnoses and in related health industries (food and dairy processing), it is imperative that the assay detect all clinically significant serovars of *Salmonella*. Therefore, attempts have been made to produce monoclonals that would react with M467-negative *Salmonella* serovars.

One such complementary monoclonal antibody is 6H4 which was produced by a hybridoma derived from the fusion of Sp2/0 Ag-14 mouse myeloma with spleen cells from a BALB/c mouse hyperimmunized with heat extract of *S. typhi*. The monoclonal antibody 6H4 reacts with *S. typhi, S. paratyphi* A, *S. paratyphi* B, *S. newington, S. tennessee*, and *S. potsdam* (29). Thus, it complements the activity of M467 IgA, and the two antibodies, in combination, have been used in sandwich immunoassays for the detection of salmonella in clinical specimens and in foods (29,30).

My laboratory has been engaged in the production and characterization of monoclonal antibodies that would complement the reactivity of M467 IgA. Purified flagellin preparations and heat-extracts of whole cells were used in my laboratory for immunization of BALB/c mice. All fusions were done with Sp2/0 Ag-14 myeloma cells (32). Hybridomas were selected in hypoxanthine-aminopterin-thymidine (HAT) medium and hybridomas of interest were cloned repeatedly by the limiting dilution technique (33). The characteristics and reactivity of selected monoclonal antibodies prepared in my laboratory are shown in Table 4.

A purified flagellin preparation (34a) of *S. typhi* or *S. potsdam* was used for the preparation of monoclonal antibodies JS 133, JS 181, and JS 251. JS 251 exhibits a wide spectrum of reactivity when tested against M467 IgA-negative serovars. It does not react with nonflagellated serovars. JS 251 cross-reacts with *C. freundii* (4/7 strains), *E. coli* (1/3 strains), *Enterobacter aerogenes* (1/1), and *E. cloacae* (1/1). In this respect it is similar to M467 IgA. JS 133 and JS 181 show a much narrower range of activity. However, the reactivity of JS 181 with *S. gallinarum* is interesting and unexpected because a flagellin preparation was used as the immunogen for the preparation of hybridomas from which JS 181 was isolated. It is likely that some outer membrane proteins of salmonella may have been present as contaminants in the flagellar preparation. Also, it is possible

Table 4 Reactivities of Antisalmonella Monoclonal Antibodies with M467 Immunoglobulin A-Negative Serovars

	Monoclonal antibody (isotype)			
Serovar	JS133 (IgG2a)	JS181 (IgG1)	JS251 (IgG1)	4A2D12H9 (IgG2a)
Abony	+	–	+	+
Banalia	–	–	–	+
Banana	–	–	–	+
Emmastad	–	–	+	+
Gallinarum	–	+	–	+
Hvittingfoss	+	–	+	+
Kaposvar (Reading)	–	–	+	ND
Kirkee	+	–	+	+
Korlebu	–	–	+	+
Leeuwarden	–	–	+	+
Newington	+	–	+	+
Onarimon	–	–	–	ND
Paratyphi A	–	+	+	+
Paratyphi B	+	+	+	+
Paratyphi C	–	–	+	+
Potsdam	–	–	+	+
Pueris (Newport)	–	–	+	+
Pullorum	–	–	–	+

ND = Not tested.

that *S. gallinarum* has a processing block in the assembly of flagellin into functional flagellae. Monoclonal antibody 4A2D12H9 has shown the most desirable specific activities with M467 IgA-negative serovars of *Salmonella*. Also, it has not exhibited any cross-reactions with any of the nonsalmonellae tested in our laboratory. Monoclonal antibody 6H4 was also reported to be free from cross-reactions with nonsalmonellae (29).

The protein recognized by 4A2D12H9 monoclonal antibody has been identified by sodium dodecyl sulfate-polyacrylamide gel electrophoresis (SDS-PAGE) of whole-cell extracts of several *Salmonella* serovars, electrophoretic transfer of separated bacterial proteins from gel to nitrocellulose (34b), and probing of the nitrocellulose-bound proteins with 4A2D12H9 antibody. A protein with a relative molecular mass (M_r) of 44–45 kDa (Fig. 1) was recognized by 4A2D12H9 antibody in eight of eight serovars of *Salmonella* (weak reaction in *S. pullorum*).

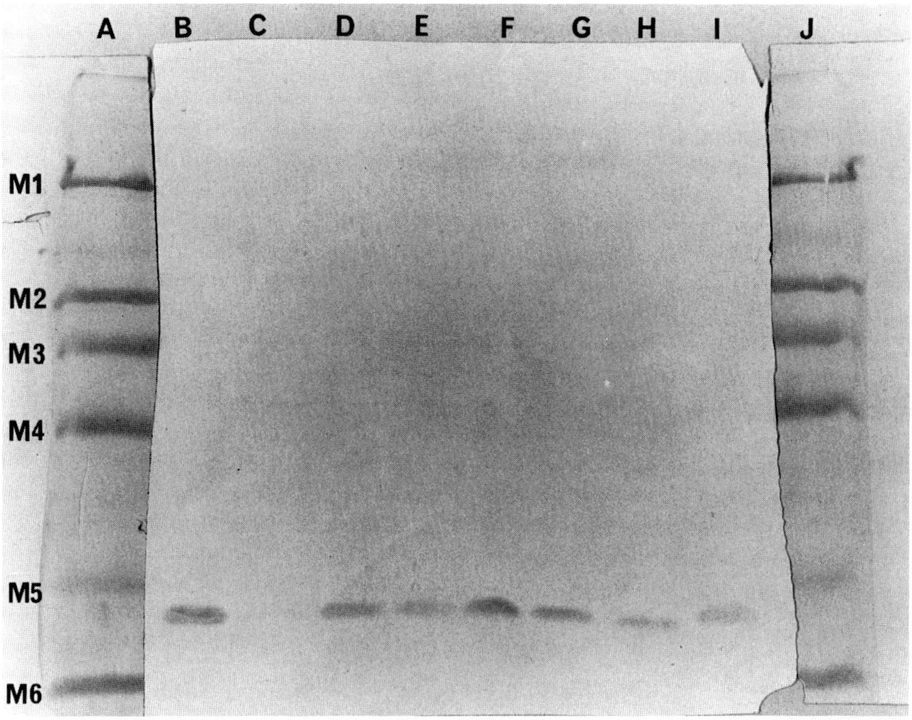

Figure 1 Western blot of outer membrane proteins of eight serovars of *Salmonella* probed with monoclonal antibody 4A2D12H9. M1-M6 are M_r standards (M1, 250 kDa; M2, 116 kDa; M3, 97.4 kDa; M4, 66 kDa; M5, 45 kDa; M6, 29 kDa). Lanes: A, M_r standards; B, *S. typhi*; C, *S. pullorum*; D, *S. gallinarum*; E, *S. typhimurium*; F, *S. paratyphi* C; G, *S. senftenberg*; H, *S. paratyphi* A; I, *S. paratyphi* B; J, M_r standards.

VII. PERSPECTIVES FOR THE FUTURE

Significant progress has been made in the development of antisalmonella monoclonal antibodies for diagnostic purposes since the first publication describing a M467 IgA-based enzyme immunoassay by Robison et al. (22) in 1983. An enzyme immunoassay kit for the rapid detection of salmonellae, based on the work of Dr. Jerome Mattingly, is commercially available (*Salmonella* Bio Enza-Bead, Organon-Teknika). A more interesting aspect of the antisalmonella monoclonal antibody work is the possibility of employing these monoclonal antibodies for the study of common flagellar and outer membrane protein antigens of the genus *Salmonella* and extending our knowledge of the antigens of salmonellae beyond the Kauffman-White schema. Because the 44–45-kDa protein appears to be common across serovars and because of its demonstrated immunogenicity, its vaccinogenic potential may be worth investigating.

VIII. THE GENUS *CAMPYLOBACTER*

The genus *Campylobacter* consists of small, nonsporeforming, oxidase-positive, nitrate-reducing, gram-negative bacteria that have a characteristic curved, S-shaped, or spiral structure. Campylobacters have a single polar flagellum at one or both ends and are highly motile. Virtually all campylobacters are oxygen-sensitive and grow only under reduced oxygen tension (5% oxygen, 10% carbon dioxide, and 85% nitrogen). Campylobacters use the tricarboxylic acid cycle intermediates as their source of energy; they do not utilize carbohydrates. *Campylobacter jejuni, C. coli, C. laridis*, and *C. fetus* subsp. *fetus* have been implicated in human disease. The first three cause acute gastroenteritis in humans, whereas *C. fetus* subsp. *fetus* is an opportunistic pathogen that frequently causes bacteremia in debilitated patients, most of whom have serious underlying problems.

IX. CAMPYLOBACTER ENTERITIS: THE PROBLEM

Since the development of selective isolation procedures for *C. jejuni*, this organism has been recognized as a common cause of diarrhea throughout the world (35,36). A study of 8097 fecal specimens submitted to clinical microbiological laboratories at eight hospitals in the United States showed an isolation rate of 4.6% for *C. jejuni*, which was more than that for *Salmonella* and *Shigella* species combined (2.3% and 1.0%, respectively). During 1983, the first year of national surveillance of campylobacter infections in the United States, *Campylobacter* spp. were isolated from 8593 specimens from humans (37). The R_j for *Campylobacter* is expected to be more than 100:1 because very few laboratories in 1983 had the capability to isolate and identify *Campylobacter* species. Of the 8593 campylobacter-positive clinical specimens, 94% yielded *C. jejuni.*

X. MAJOR ANTIGENS OF *CAMPYLOBACTER*

The major antigens of *Campylobacter* spp. include lipopolysaccharide (LPS), outer membrane (OM) proteins, flagellar protein, and other acid-extractable proteins. Penner and Hennessey (38) have developed a passive hemagglutination-typing scheme based on the soluble, heat-stable antigens of *C. jejuni* and *C. coli.* The Penner-Hennessey serotyping scheme has identified three serotypes that are most common among human isolates. Serological differences observed by the Penner-Hennessey procedure among strains of *C. jejuni* and *C. coli* are due to differences in carbohydrate associated with low-molecular-mass core LPS and are not due to long polysaccharide chain O antigens (39). Lior et al. (40) have developed a serotyping scheme based on the heat-labile antigenic factors of *C. jejuni*. Flagellar protein is a key component of campylobacter heat-labile antigen typing system; also, it is a dominant antigen recognized during infections with *C. jejuni* in humans (36). Mills et al. (41) concluded that campylobacter flagellar protein is a highly conserved protein, from the observation that antibodies prepared against a purified flagellin preparation of *C. jejuni* reacted with 60 strains of *C. jejuni* and *C. coli*. It is interesting that the same antibody did not recognize serotype-specific determinants (Lior scheme) of the organisms.

Outer membrane proteins, other than the flagellar protein, are thought to play major roles in the antigenicity and virulence of campylobacters. The outer membrane proteins purified from a typical human diarrheal isolate of *C. jejuni* have been characterized (42). There are seven major polypeptides in the preparation. The outer membrane protein profiles of *C. jejuni* and *C. coli* are dominated by a 43–46-kDa protein (Omp 1) which is probably the porin, or the matrix protein, responsible for maintaining the hydrophilic size-dependent diffusion channels through the outer membrane (42). The Omp 1 protein and a 92.5-kDa protein reacted strongly in immunoblots performed with antisera against formalinized whole cells of *C. jejuni* (43). A highly immunogenic 27–31-kDa protein is present in glycine-hydrochloride (pH 2.2) extracts of *C. jejuni* and *C. coli* but not in sarcosinate-extracted outer membrane protein preparations of the same strains (43,44). Furthermore, the 27–31-kDa protein is not present in *C. fetus*.

XI. MONOCLONAL ANTIBODIES TO *CAMPYLOBACTER*

Kosunen et al. (45) reported on the development of monoclonal anticampylobacter reagents for antigenic analysis and serotyping purposes. Formalinized cells of a human isolate of *C. jejuni* were used as the immunogen. Twenty-eight hybridoma clones were reported to produce monoclonal antibodies (IgG1-12; IgG2a-5; IgG2b-6; IgG3-3) that reacted with *Campylobacter* spp. None of the monoclonal antibodies reacted with purified lipopolysaccharide antigen of *C. jejuni* but four reacted with a polysaccharide preparation from the immunogen and with autoclaved saline extract of the immunogen. One monoclonal antibody (from clone 6) was reported to react strongly with all of 24 *Campylobacter* strains tested, which included two *C. fetus* subsp. *venerealis* and two *C. fetus* subsp. *fetus*. The same monoclonal antibody was reported to react with an acid glycine (0.2 M glycine–HCl, pH 2.2) extract of *C. jejuni*. A superficial glycoprotein antigen of *C. fetus* subsp. *fetus* was reported to be antiphagocytic and led McCoy et al. to suggest its role as a virulence factor (46). Interestingly, all of 21 rabbits injected with different strains of *Campylobacter* and most humans suffering from campylobacter enteritis had antibodies that reacted with an acid glycine extract of *C. jejuni* (45).

Hart and Nachamkin (47) produced monoclonal antibodies against *C. jejuni* and also characterized them. Three hybridomas produced monoclonal antibodies that reacted with 97% or more of more than 80 clinical isolates of *C. jejuni*. One monoclonal antibody also reacted with *C. coli, C. laridis, C. faecalis, C. fetus* subsp. *venerealis*, and *C. fetus* subsp. *fetus*. One antibody was reported to cross-react with *Staphylococcus aureus* which was attributed by the authors to protein A-IgG interaction. All three antibodies reacted with the flagellar protein of *C. jejuni* and two reacted with a determinant of the outer membrane proteins.

My laboratory has been involved in the development of monoclonal antibodies against *C. jejuni* and *C. coli* for use in immunodiagnostic tests for the rapid identification of these pathogens (48). One such monoclonal antibody is 104-23, produced by the hybridoma line ER 104-23, which was constructed by the fusion (49) of Sp2-0/Ag-14 mouse myeloma cells with spleen cells from BALB/c mice hyperimmunized with formalinized whole cells of *C. coli* (A3315, clinical isolate, courtesy: Dr. D.J. Brenner and Dr. Paul Edmonds, Centers for Disease Control, Atlanta, Georgia). Monoclonal antibody 104-23 reacts strongly with *C. jejuni* (12 of 12 strains representing various Penner serotypes), *C. coli* (5 of 5 strains), and *C. laridis* (2 of 2 strains). Also, 104-23 reacts with *C. fetus* subsp. *fetus, C. fetus* subsp. *venerealis*, and *C. hyointestinalis*. Monoclonal antibody 104-23 belongs to the IgG_{2b} subclass. It appears to recognize a highly conserved antigenic determinant on the flagellar protein of campylobacters. Immunoblots of proteins from whole-cell extracts and outer membrane fractions of *Campylobacter* spp. have shown that 104-23 recognizes a protein of 63 kDa.

Another interesting monoclonal antibody prepared in my laboratory is produced by the clone ER 201. This hybridoma secretes a monoclonal antibody (201) that reacts specifically with *C. coli* in limited tests conducted in my laboratory. If this observation is validated by a substantially larger data base, monoclonal antibody 201 could be used to distinguish between *C. jejuni* and *C. coli*. Discrimination between *C. jejuni* and *C. coli* is presently done by the hippurate hydrolysis test which may not be entirely reliable (49).

XII. PERSPECTIVES

In contrast to the genus *Salmonella*, which has been recognized as a significant human and animal pathogen for several years, *Campylobacter* was recently recognized as a significant cause of human gastroenteritis. Progress in applying monoclonal antibody technology to the detection of *Campylobacter* has been very rapid. At the same time, valuable basic information has been compiled about the antigens of *Campylobacter* and the antigenic relationships among the various species in the genus *Campylobacter*.

REFERENCES

1. Le Minor, L., *Salmonella*. In *The Bergeys Manual of Determinative Bacteriology*, Vol. I, J. G. Holt and Kreig, N. R. (eds.). Williams & Wilkins, Baltimore, pp. 427–458 (1984).
2. Crosa, J. H., D. J. Brenner, W. H. Ewing, and S. Falkow, Molecular relationships among the salmonellae. *J. Bacteriol. 115*:307–315 (1973).
3. Stoleru, G. H., L. Le Minor, and A. M. Lheritier, Polynucleotide sequence divergence among strains of *Salmonella* subgenus IV and closely related organisms. *Ann. Microbiol. (Paris) 127A*:477–486 (1976).
4. Le Minor, L., M. Veron, and M. Popoff, Taxonomie des *Salmonella. Ann. Microbiol. (Paris) 133B*:223–243 (1982).
5. Hauschild, A. H. W., and F. L. Bryan, Estimate of cases of food- and waterborne illness in Canada and the United States. *J. Food Prot. 43*:435–440 (1980).
6. National Research Council, An Evaluation of the *Salmonella* Problem. National Academy of Sciences, Washington, D.C. (1969).
7. Archer, D. L., and J. E. Kvenberg, Incidence and cost of foodborne diarrheal disease in the United States. *J. Food Prot. 48*:887–894 (1985).
8. Swaminathan, B., and S. A. Minnich, Enzyme immunoassay for the detection of *Salmonella*. In *Biotechnology Handbook*, P. N. Chercmisinoff and Ouellette, R. P. (eds.). Technomic Publishing Co., Lancaster, Pa., pp. 526–533 (1985).
9. Kauffmann, F., The Bacteriology of *Enterobacteriaceae*. Williams & Wilkins, Baltimore (1966).
10. Edwards, P. R., and W. H. Ewing, Identification of *Enterobacteriaceae*. 3rd ed. Burgess Publishing Co., Minneapolis (1972).
11. Spicer, C. C., A quick method of identifying *Salmonella* H antigens, *J. Clin. Pathol. 9*:378–379 (1956).
12. Langman, R. E., The occurrence of antigenic determinants common to flagella of different *Salmonella* strains. *Eur. J. Immunol. 2*:582–584 (1972).
13. Iino, T., Genetics of structure and function of bacterial flagella. *Annu. Rev. Genet. 11*:161–182 (1977).
14. Yamaguchi, S., H. Fujita, K. Sugata, T. Taira, and T. Iino, Genetic analysis of H2, the structural gene for phase-2 flagellin in *Salmonella. J. Gen. Microbiol. 130*:255–265 (1984).
15. Joys, T. M., The covalent structure of phase-I flagellar filament protein of *Salmonella typhimurium* and its comparison with other flagellins. *J. Biol. Chem. 260*:15758–15761 (1985).
16. Wei, L. N., and T. M. Joys, The covalent structure of three phase-1 flagellar filament proteins of *Salmonella. J. Mol. Biol. 186*:791–803 (1985).
17. Ibrahim, G. F., G. H. Fleet, M. J. Lyons, and R. A. Walker, Immunological relationships between *Salmonella* flagella and their potential application for salmonellae detection by immunoassay. *Med. Microbiol. Immunol. 174*:87–99 (1985).

18. Potter, M., Antigen-binding myeloma proteins in mice. *Ann. N. Y. Acad. Sci. 190*:306–321 (1971).
19. Potter, M., Mouse IgA myeloma proteins that bind polysaccharide antigens of enterobacterial origin. *Fed. Proc. 29*:85–91 (1970).
20. Smith, A. M., and M. Potter, A BALB/c mouse IgA myeloma protein that binds *Salmonella* flagellar protein. *J. Immunol. 114*:1847–1850 (1975).
21. Smith, A. M., and C. Jones, Use of myeloma protein M467 for detecting *Salmonella* spp. in milk. *Appl. Environ. Microbiol. 46*:826–831 (1983).
22. Robison, B. J., C. I. Pretzman, and J. A. Mattingly, Enzyme immunoassay in which a myeloma protein is used for the detection of salmonellae. *Appl. Environ. Microbiol. 45*:1816–1821 (1983).
23. Mattingly, J. A., and B. J. Robison, Use of monoclonal antibodies to detect *Salmonella*. In *Biotechnology: Applications and Research*, P. N. Cheremisinoff and Oullette, R. P. (eds.), Technomic Publishing, Lancaster, Pa., pp. 519–525 (1985).
24. Swaminathan, B., J. A. G. Aleixo, S. A. Minnich, and V. A. Wallshein, Enzyme immunoassays for *Salmonella*: One-day test is now a reality. *Food Technol. 39*(3):83–89 (1985).
25. Aleixo, J. A. G., and B. Swaminathan, A fluorescent enzyme immunoassay for *Salmonella* detection. *J. Immunoassay 9*:83–95 (1988).
26. Minnich, S. A., B. Swaminathan, and J. A. G. Aleixo, Temperature effects on the processing of *Yersinia enterocolitica* flagellin. *Abstr. Ann. Meet. Am. Soc. Microbiol.* J13, p. 170 (1985).
27. Smith, A. M., J. Slack, and M. Potter, Restriction in the immune response to flagellar proteins in inbred mice. *Eur. J. Immunol. 7*:497–500 (1977).
28. Smith, A. M., J. S. Miller, and D. S. Whitehead, M467: A murine IgA myeloma protein that binds a bacterial protein, I. Recognition of common antigenic determinants on *Salmonella* flagellins. *J. Immunol. 123*:1715–1720 (1979).
29. Mattingly, J. A., An enzyme immunoassay for the detection of all *Salmonella* using a combination of a myeloma protein and a hybridoma antibody. *J. Immunol. Methods 73*:147–156 (1984).
30. Mattingly, J. A., and W. D. Gehle, An improved enzyme immunoassay for the detection of *Salmonella. J. Food Sci. 49*:807–809 (1984).
31. Emswiler-Rose, B., W. D. Gehle, R. W. Johnston, A. Okrend, A. Moran, and B. Bennett, An enzyme immunoassay technique for the detection of salmonellae in meat and poultry products. *J. Food Sci. 49*:1018–1020 (1984).
32. Galfre, G., and C. Milstein, Preparation of monoclonal antibodies: Strtegies and procedures. In *Methods in Enzymology*, J. J. Langone and van Vunakis, H. (eds.), Academic Press, New York, pp. 3–46 (1981).
33. Goding, J. W., Antibody production by hybridomas. *J. Immunol. Methods 39*:285–308 (1980).

34a. Fey, H., and H. P. Wezstein, Production of potent *Salmonella* H antisera by immunization with flagellae, isolated by immunosorption. *Med. Microbiol. Immunol. 161*:73–78 (1973).
34b. Burnette, W. N., "Western blotting": Electrophoretic transfer of proteins from sodium dodecyl sulfate-polyacrylamide gels to unmodified nitrocellulose and radiographic detection with antibody and radioiodinated protein A. *Anal. Biochem. 112*:195–203 (1981).
35. Blaser, M. J., J. G. Wells, R. A. Feldman, R. A. Pollard, J. R. Allen, and The Collaborative Diarrheal Disease Study Group, *Campylobacter* enteritis in the United States. *Ann. Intern. Med. 98*:360–365 (1983).
36. Wenman, W. M., J. Chai, T. J. Louie, C. Goudreau, H. Lior, D. G. Newell, A. D. Pearson, and D. E. Taylor, Antigenic analysis of *Campylobacter* flagellar protein and other proteins. *J. Clin. Microbiol. 21*:108–112 (1985).
37. Riley, L. W., and M. J. Finch, Results of the first year of national surveillance of *Campylobacter* infections in the United States. *J. Infect. Dis. 151*:956–959 (1985).
38. Penner, J. L., and J. N. Hennessy, Passive hemagglutination technique for serotyping. *Campylobacter fetus* subsp. *jejuni* on the basis of soluble heat-stable antigens. *J. Clin. Microbiol. 12*:732–737 (1980).
39. Logan, S. M., and T. J. Trust, Outer membrane characteristics of *Campylobacter jejuni*. *Infect. Immun. 38*:898–906 (1982).
40. Lior, H., D. L. Woodward, J. A. Edgar, L. J. Laroche, and P. Gill, Serotyping of *Campylobacter jejuni* by slide agglutination based on heat-labile antigenic factors. *J. Clin. Microbiol. 15*:761–768 (1982).
41. Mills, S. D., W. C. Bradbury, and J. L. Penner, Isolation and characterization of a common antigen in *Campylobacter jejuni* and *Campylobacter coli*. *J. Clin. Microbiol. 24*:69–75 (1986).
42. Trust, T. J., and S. M. Logan, Outer membrane and surface structure of *Campylobacter jejuni*. In *Campylobacter Infection in Man and Animals*, J. P. Butzler (ed.), CRC Press, Boca Raton, pp. 134–142 (1984).
43. Logan, S. M., and T. J. Trust, Molecular identification of surface protein antigens of *Campylobacter jejuni*. *Infect. Immun. 42*:675–682 (1983).
44. Newell, D. G., H. McBride, and A. D. Pearson, The identification of outer membrane proteins and flagella of *Campylobacter jejuni*. *J. Gen. Microbiol. 130*:1201–1208 (1984).
45. Kosunen, T. U., B. E. Bang, and M. Hurme, Analysis of *Campylobacter jejuni* antigens with monoclonal antibodies. *J. Clin. Microbiol. 19*:129–133 (1984).
46. McCoy, E. C., D. Doyle, K. Burda, L. B. Corbeil, and A. J. Winter, Superficial antigen of *Campylobacter (Vibrio) fetus*: Characterization of an antiphagocytic component. *Infect. Immun. 11*:517–525 (1975).
47. Hart, A. M., and I. Nachamkin, Monoclonal antibodies that define com-

mon antigens of *Campylobacter jejuni. Abstr. Ann. Meet. Amer. Soc. Microbiol.* B233, p. 63 (1986).
48. Riverson, E. A., Monoclonal Antibodies for Detecting *Campylobacter jejuni* and *Campylobacter coli*, M.S. Thesis, Purdue University, West Lafayette, Indiana, p. 93 (1986).
49. Kohler, G., The technique of hybridoma production. In *Immunological Methods*, Vol. II, I. Lefkovits and Pernis, B. (eds.), Academic Press, New York, pp. 285–298 (1981).

24

Monoclonal Antibodies Against *Clostridium difficile* and Its Toxins

Use in Diagnostic Microbiology

DAVID M. LYERLY, CAROL J. PHELPS, and TRACY DALE WILKINS
Virginia Polytechnic Institute and State University, Blacksburg, Virginia

I. INTRODUCTION

Pseudomembranous colitis (PMC) is an antibiotic-associated disease characterized by severe inflammation and necrosis of the colon. Pseudomembranes consisting of fibrin and necrotic material are often observed along the colonic wall. The primary causal agent of the disease is *Clostridium difficile*, a gram-positive anaerobic bacterium. Although the pathogenesis of the disease is not completely understood, it is believed that antibiotic therapy suppresses much of the host's normal intestinal flora. This suppression allows toxigenic strains of *C. difficile* to overgrow in the intestine, resulting in the disease (see Ref. 1-5 for more detailed information on disease caused by *C. difficile*).

The extensive tissue damage that occurs during PMC is thought to result from the action of two toxins produced by *C. difficile* (6-11). These toxins, designated toxin A and toxin B, are large tissue-damaging proteins that are lethal in experimental animals. Toxin A is a potent enterotoxin that causes an intense hemorrhagic fluid response when injected into rabbit ileal loops. The toxin also causes severe diarrhea when given intragastrically to hamsters. In addition to its enterotoxic activity, toxin A is weakly cytotoxic. It is believed that toxin A is responsible for the diarrhea that occurs during PMC. Toxin B is an extremely potent cytotoxin that is active against a wide array of mammalian tissue culture cells. As little as 1 pg, or less, of the toxin is cytotoxic for the cells.

Both of the toxins appear to be involved in the disease. This is based on several observations. First, both toxins are tissue damaging and lethal (6-11). Second, experimental animals must be vaccinated against both of the toxins to protect the animals against *C. difficile* disease; vaccination against just one of the toxins is not protective (12,13). Third, both toxins are consistently detected in fecal specimens from patients with PMC (14,15).

Strains of *C. difficile* produce either both of the toxins or neither toxin (14,15). No strains have been isolated that produce only toxin A or only toxin B. In addition, there appears to be a correlation in the amount of toxins A and B produced; strains that produce high levels of toxin A also produce high levels of toxin B; likewise, strains that produce low levels of toxin A produce low levels of toxin B. Because both toxins are produced by toxigenic strains of *C. difficile*, a test that detects either, or both, of the toxins can be used in the diagnosis of *C. difficile* disease. Currently, a number of reference diagnostic laboratories and hospitals use tissue culture assay for the detection of toxigenic *C. difficile* in patient specimens (16). The assay is based on the detection of cytotoxic activity (toxin B) and the neutralization of the activity by *C. difficile* antiserum. Although this assay is highly sensitive and shows a good correlation with the isolation of toxigenic *C. difficile* from specimens, it has several disadvantages. The assay requires 24-48 hr to complete. In addition, specialized equipment and personnel with experience in tissue culture procedures are needed. As a result, attempts are being made to develop rapid immunoassays [counterimmunoelectrophoresis, enzyme-linked immunosorbent assay (ELISA), latex agglutination] for the detection of *C. difficile* toxin.

A number of reports have been published that describe immunoassays for the detection of *C. difficile* in fecal specimens. In most instances, however, the assays were done with antisera against crude antigen mixtures and were not specific for either of the toxins (17-24). Thus, these assays detected nontoxigenic, as well as toxigenic, strains of *C. difficile*. Our laboratory has been involved in the development of specific antibodies against toxins A and B. We have obtained monospecific antiserum as well as affinity-purified antibodies against each toxin, and these reagents have been useful in the analysis of various types of immunoassays (25). Recently, we have concentrated our efforts on the development of monoclonal antibodies against the toxins, in an effort to obtain large amounts of highly specific antibodies for use in diagnostic tests. Much of our work has centered on obtaining monoclonal antibodies against toxin A. We have concentrated on toxin A for several reasons. We have had more success in preparing large amounts (milligram quantities) of homogeneous toxin A than of toxin B; thus, we have more antigen to work with. Also, toxin A appears to be more stable than toxin B; homogeneous toxin A retains its biological activity and immunoreactivity for months when stored in solution at 4°C. Toxin B, on the other hand, can lose more than 50% of its cytotoxic activity and immunoreactivity when stored under these conditions. Toxin A is also a better antigen than toxin B and elicits higher levels of antibody.

In this chapter, we describe some of the methods that we have used to develop monoclonal antibodies against toxin A. In addition, we list some of the ways in which we characterize the antibodies and utilize them in diagnostic assays.

II. PRODUCTION OF MONOCLONAL ANTIBODIES AGAINST TOXIN A

Most of the methods that we use to produce monoclonal antibodies are based on general procedures developed and described by other investigators, and the reader is referred to those references for additional information (26-31).

A. Preparation of Antigen

We have used two different antigen preparations for the development of toxin A monoclonal antibodies. One of these preparations consists of a crude antigen mixture. The second preparation consists of homogeneous toxin A.

For the crude antigen mixture, a highly toxigenic strain of *C. difficile* is grown in brain-heart infusion dialysis flasks. These flasks consist of a dialysis sac containing saline suspended in brain-heart infusion broth and are similar to those originally described by Sterne and Wentzel (32) for the production of botulinum toxin. The sac is inoculated with the organism and the flasks are incubated at 37°C for 3-4 days. The dialysis sac serves two purposes. First, it minimizes the contamination of the culture with high-molecular-weight components from the brain-heart infusion; only the low-molecular-weight components are able to diffuse into the sac and be utilized as the organism grows. Thus, the large components in the culture supernatant are those produced by the organism. Second, the diffusion of the small components into the sac results in slow growth of the organism. The organism produces much larger amounts of toxin under such a slow-growth condition. After incubation, the cells are removed by centrifugation and the culture supernatant fluid is passed through 0.45-μm membranes and stored at 4°C. We have estimated that toxin A constitutes 5-10% of the total protein in the culture filtrate. Thus,when the culture filtrate is used as the vaccine, enough toxin A (tens of micrograms) is injected to elicit a good antibody response.

Culture filtrate from brain-heart dialysis flasks serves as a good source of starting material for the purification of toxin A because filtrates prepared in this manner contain such high amounts of the toxin. The methods that we have developed for preparing homogeneous toxin A from culture filtrate are listed in Table 1. The homogeneity of the toxin is rigorously tested by using a series of highly resolving techniques such as polyacrylamide gel electrophoresis and crossed-immunoelectrophoresis.

The crude antigen mixture (culture filtrate) and homogeneous toxin A are highly toxic to animals; 50-100 ng of either antigen preparation is lethal when

Table 1 Methods for Purifying Toxin A of *C. difficile*[a]

1. Ultrafiltration, ion-exchange chromatography using a linear NaCl gradient, and precipitation at pH 5.5 (10)
2. Ammonium sulfate precipitation, batch ion-exchange chromatography, and precipitation at pH 5.6 (33)
3. Ultrafiltration or ammonium sulfate precipitation followed by flat-bed electrophoresis
4. Immunoaffinity chromatography using monoclonal antibody-Affi Gel 10 (34)
5. Affinity purification using rabbit erythrocytes (35)

[a]All of these methods yield homogeneous toxin A, as demonstrated by crossed-immunoelectrophoresis and polyacrylamide gel electrophoresis. The starting material in each method is culture filtrate from a highly toxigenic strain of *C. difficile*. References for each procedure are given in parentheses.

injected into mice. Therefore, the antigen preparations must be inactivated before injection. We add formalin to a final concentration of 0.4% (vol/vol) and incubate the mixtures at 37°C for 1-2 hr. This treatment inactivates the toxicity of the preparation; however, much of the antigenicity of the protein is retained, as demonstrated by ELISA.

B. Immunization Protocol

The spleen cells that we use in the fusion protocol are obtained from BALB/c mice vaccinated with the antigen preparations. To boost the antibody response to the antigen, the toxoided preparation is mixed 1:1 with incomplete Freund adjuvant. The mice are injected intraperitoneally or subcutaneously with the toxoid-adjuvant vaccine once every 1-2 weeks over a period of 6-8 weeks. For the crude toxoid vaccine, we administer 0.25-0.5 mg of protein per injection. With the homogeneous toxoid A vaccine, we have found that 0.05-0.1 mg of protein per injection elicits a good antibody titer.

The serum of each mouse used for monoclonal antibody production is screened before and during the immunization by ELISA (Table 2). The amount of toxin A used to coat each well (0.5 μg) gives a high level of sensitivity and low background. We have found that Immulon type 1 microtiter plates from Dynatech Laboratories, Inc., work well in the assay. Serially diluted samples of each serum specimen are screened to determine the highest dilution that gives an absorbance at 405 nm in the range of 0.2-0.4. The reciprocal of this dilution is the ELISA titer. Once the serum of a mouse has reached a titer of 1000 or higher, the animal is injected with crude toxoid or homogeneous toxoid A with-

Table 2 Screening ELISA for the Detection of Hybridomas that Produce Toxin A Antibody

1. Wells of polystyrene microtiter plates are coated with homogeneous toxin A (0.5 μg of toxin per well) overnight at 37°C in carbonate buffer, pH 9.6.
2. Wells are washed with phosphate-buffered saline containing 0.05% Triton X-100 (PBS-T) and samples to be tested for antibody are added.
3. After incubation for 1 hr at 37°C, the wells are washed with PBS-T and rabbit antimouse immunoglobulin–alkaline phosphatase conjugate is added for 1 hr at 37°C.
4. Wells are washed with PBS-T and substrate is added; positive wells are noted after 30 min at room temperature.

out adjuvant. The spleen is then removed from the mouse 3 days later, and the spleen cells are fused to mouse myeloma cells.

C. Cell Fusion

Spleen cells from the vaccinated mouse are collected by teasing apart the spleen and passing the cells through a fine nylon mesh screen. The spleen cells are then combined with mouse myeloma cells that are in exponential growth. Most of our fusions have been done with NS-1 and SP2/0 mouse myeloma cells. Both cell lines have given similar yields of hybridomas.

The fusion is performed in polyethylene glycol. In our experience, Kodak polyethylene glycol 1450 (molecular weight 1300–1600) consistently gives successful fusions. The fusion protocol that we follow is given in Table 3.

Once the fusion is completed, the cells are plated at about 10^6 cells per microtiter well in RPMI 1640 medium (Hazelton Dutchland, Inc.) containing 13% heat-inactivated horse serum (Hyclone Laboratories). The RPMI–horse serum medium is supplemented with thymocytes, which serve as feeder cells. In addition, the medium contains hypoxanthine–aminopterin–thymidine (HAT) to select for hybridoma cells. The cells are maintained for 5–7 days in the selective medium before the medium is changed. This time interval preserves a stable environment for the developing clones and minimizes any possible contamination.

D. Identification and Cloning of Toxin A Antibody-Producing Hybridomas

Once the cells reach 25–50% confluency, the hybridoma supernatant is analyzed for toxin A antibody, using the screening ELISA (see Table 2). If the supernatant tests positive, the hybridomas are cloned using the soft agar method or

Table 3 Protocol for the Fusion of Mouse Spleen Cells to SP2/0 Mouse Myeloma Cells

1. Spleen cells from vaccinated mouse are combined with SP2/0 mouse myeloma cells growing exponentially; the cells are mixed in a 5:1 (spleen cell/myeloma cell) ratio.
2. The cell mixture is centrifuged and the cell pellet is slightly dispersed by gently tapping the bottom of the tube.
3. 1 ml of RPMI medium containing 50% polyethylene glycol (Kodak 1450) is added per 1.6×10^8 spleen cells over 1 min with gentle stirring.
4. 1 ml of RPMI–13% horse serum is added over 1 min with gentle stirring.
5. 8 ml of RPMI-horse serum is added over 3 min with gentle stirring.
6. Cells are collected by centrifugation and suspended in RPMI–horse serum-hypoxanthine-aminopterin-thymidine selective medium containing thymocytes.
7. Cell suspension (5×10^6 spleen cells per ml) is dispensed into microtiter wells.

the limiting dilution method. The cloning is done shortly after the hybridoma has been identified by ELISA so that any overgrowth by unwanted, rapidly developing hybridoma cells is minimized.

For soft agar cloning, 50, 100, and 500 cells of each positive hybridoma are mixed with 0.3% Bacto-Agar (Difco Laboratories) in RPMI-13% horse serum. Each suspension of cells is then poured over a base agar composed of 0.5% Bacto-Agar in RPMI-horse serum. The colonies usually appear 7–10 days after plating. Individual colonies are picked and grown in RPMI-horse serum medium. The supernatant fluid is then screened by ELISA for toxin A antibody to identify positive clones.

In the limiting dilution method, suspensions of 50, 15, and 2.5 hybridoma cells per milliliter of RPMI medium, containing thymocytes, are dispensed into microtiter plates. Usually, about one-third of the wells plated with the 2.5 cell/ml suspension develop hybridomas. The supernatant fluid of these wells is screened by ELISA once the cells reach 25–50% confluency.

Each clone obtained by the soft agar or limiting dilution method is subcloned a second time to ensure monoclonality of the hybridoma.

E. Large-Scale Production of Toxin A Monoclonal Antibodies

In our studies, we use monoclonal antibody from hybridoma supernatant as well as ascites fluid. For the supernatant, we grow the hybridoma cells in 100 ml or 250 ml stir flasks. The supernatant fluid is continuously removed from the flask when the cells are in exponential growth. Ascites fluid is obtained by injecting about 10^6 hybridoma cells into mice treated with pristane. The hybridoma supernatant and ascites are stored as sterile solutions at 4°C or are frozen in aliquots and stored at -20°C until used.

III. CHARACTERIZATION OF MONOCLONAL ANTIBODIES AGAINST TOXIN A

A. Isotypic Characterization

The class, subclass, and light-chain type of our monoclonal antibodies are determined by Ouchterlony double-immunodiffusion against rabbit antimouse immunoglobulin antisera (Litton Bionetics Inc., Kensington, Maryland). For the analysis, we use samples of hybridoma supernatant from wells showing >50% confluency. The toxin A monoclonal antibodies that we have identified thus far include IgG1, IgG2a, and IgM, and contain κ-light chains.

B. Quantitation of Monoclonal Antibodies

The amount of monoclonal antibody in ascites is quantitated by radial immunodiffusion. Serial twofold dilutions of the sample to be quantitated are diffused into agarose that contains rabbit antimouse immunoglobulin (Pel-Freez Biologicals). The concentration of antibody is determined from standard curves obtained with immunoglobulin standards (Sigma Chemical Co.). Ascites obtained from our hybridomas that produce high levels of toxin A antibody contain 1-10 mg/ml of antibody. These ascites have ELISA titers of about 10^6. The supernatant fluids from these hybridomas have ELISA titers of about 10^3. Thus, the supernatant fluids contain approximately 1000-fold less antibody or about 1-10 μg of toxin A antibody per milliliter.

C. Specificity Studies

The monoclonal antibodies that we identify are analyzed using a rigorous series of assays to confirm their specificity. Once a positive hybridoma has been identified by the screening ELISA, the supernatant is tested against toxin B and against culture filtrates from nontoxigenic strains of *C. difficile*. For the assay, wells of microtiter plates are coated with toxin B, or the culture filtrates, instead of toxin A. The nontoxigenic strains used for these assays were selected because they produce all of the same antigens that the toxigenic strains produce, with the exception of toxins A and B. Any monoclonal antibodies that react with toxin B or the culture filtrate are subcloned and reevaluated.

In addition to the ELISA, we analyze our monoclonal antibodies by immunoblotting procedures (34,36-39). A description of these methods is given in

Table 4 Immunoblot Procedure for the Analysis of Toxins A and B

1. Antigen mixture is subjected to polyacrylamide gel electrophoresis or crossed-immunoelectrophoresis. After crossed-immunoelectrophoresis, the immunoprecipitin arcs are solubilized with NaCl-SDS to dissociate antigen–antibody complexes.
2. The proteins are transferred to nitrocellulose membranes in TRIS–glycine–methanol buffer, pH 8.3, at 0.2 A for 18 hr (the long transfer time is necessary because toxin A is a large molecule and does not transfer efficiently).
3. Nitrocellulose membranes are rinsed in TRIS–buffered saline (TBS) and blocked with TBS containing 0.5% casein.
4. Membranes are incubated in monoclonal antibody (diluted in TBS) for 2 hr at room temperature.
5. Membranes are washed (20 min/wash) in three changes of TBS and incubated in goat antimouse immunoglobulin–horseradish peroxidase conjugate (diluted in TBS) for 18 hr at room temperature.
6. Membranes are washed in three changes of TBS and antigens are detected using 4-chloro-1-naphthol.

Table 4. The crude antigen mixture that we use in these tests contains >25 distinct antigens. The specific toxin A monoclonal antibodies that we have characterized bind only to the toxin A band or toxin A immunoprecipitin arc when analyzed by these methods. In addition to evaluating the specificity of the monoclonal antibody, the immunoblot analysis gives some characteristics (e.g., an estimate of size and charge properties) of the antigen being examined.

A third series of tests that we use to analyze our monoclonal antibodies consists of neutralization studies. Serially diluted antibody (either hybridoma supernatant or ascites) is mixed with toxin A. The mixtures are incubated at room temperature for 30 min and then examined for residual cytotoxic and enterotoxic activity. For the cytotoxic assay, the amount of toxin A used is sufficient to cause complete rounding of Chinese hamster ovary K-1 cells (1 TCD_{100}) in the absence of neutralization by the antibody (40). Two different enterotoxin assays are used for the neutralization studies. One is the oral hamster model in which toxin A is administered intragastrically (41). The amount of toxin used (0.16 mg of toxin A per kilogram body weight) causes diarrhea and death of the animal if neutralization does not occur. The second enterotoxin assay consists of injecting toxin A (about 10 μg) into ligated ileal loops in rabbits (11). The toxin causes an intense hemorrhagic fluid response if not neutralized by the antibody. Supernatant fluid and ascites from mouse myeloma SP2/0 cells and from hybridoma cells that produce antibodies against non-*C. difficile* antigens are used as negative controls.

Most of our monoclonal antibodies do not neutralize the biological activity of toxin A. This is not surprising because the toxin is an extremely large molecule and most likely contains a large number of epitopes that are not involved in the active and binding sites. However, we have isolated a group of antibodies that are of different subclass isotypes and that neutralize the enterotoxic but not the cytotoxic activity of toxin A. We have analyzed these antibodies by competitive ELISA to determine if they bind to the same region on the toxin. In the competitive ELISA, toxin A is bound to the wells of microtiter plates as described for the screening ELISA. A saturating amount of unlabeled monoclonal antibody is then added for 1 hr at 37°C to block all of the epitopes to which the antibody binds. For this step, we have found that the sites are effectively blocked with a 10^2 or 10^3 dilution of the ascites. Once the toxin is blocked, enzyme-labeled homologous or heterologous antibody conjugate is added for 1 hr at 37°C. Sigma Chemical Company sells alkaline phosphatase reagents for preparing enzyme–antibody conjugates, and we have found that they are efficiently coupled to antibodies by using the glutaraldehyde method of Voller et al. (42). The dilution of conjugate used in the assay should give an absorbance reading in the upper linear portion of the ELISA curve (usually in the range of 0.4–0.6) in the absence of any blocking by the unlabeled antibody. As a control, the blocking antibody should completely inhibit the binding of the homologous enzyme-labeled antibody. The decrease in binding of the heterologous enzyme-labeled antibody to the blocked toxin is then determined. We have found that the group of antibodies that neutralize the enterotoxic activity of toxin A completely inhibit each other from binding to the toxin. Thus, they bind to the same or similar region of the toxin. Whether this epitope represents the active or the binding site of the toxin remains to be determined. We have also characterized several antibodies that do not inhibit, or only partially inhibit, the binding of this group of antibodies to toxin A, indicating that they bind to different regions on the toxin.

IV. MONOCLONAL ANTIBODY-BASED DIAGNOSTIC TESTS FOR TOXIN A

At the present, the tissue culture assay is the primary assay used for the detection of *C. difficile* toxin in fecal specimens. The assay detects toxin B because this toxin is much more cytotoxic than toxin A and masks the presence of toxin A. In the assay, dilutions of fecal specimens are mixed with buffer or with *C. difficile* antiserum. After an incubation time, the mixtures are assayed for residual cytotoxic activity (noted by rounding of the cells) and specific neutralization of the activity by the *C. difficile* antiserum. Many clinical laboratories assay only low dilutions (a 1:2 or 1:10) of the fecal specimen to test for cytotoxic activity. Feces often contains toxic substances other than *C. difficile* toxin and, at these low dilutions (up to a 1:100 dilution), these substances can cause rounding of the cells. Therefore, we feel that a series of 10-fold dilutions (up to a 10^6 dilution) should be assayed. In our experience, fecal specimens from patients with documented cases of PMC have cytotoxic titers of 10^3 or higher. Given our ELISA results, a cytotoxic titer of 10^3 for toxin B correlates to about 100 ng/ml of toxin A (14). Thus, an immunodiagnostic test for toxin A should have a sensitivity of at least 100 ng/ml.

Monoclonal antibodies against toxins A and B have only recently been developed and, hence, their actual use as diagnostic reagents for the detection of toxin in fecal specimens has not been critically evaluated. However, we have done some preliminary studies to examine their sensitivity in several types of immunodiagnostic assays (25).

For these studies we selected a representative antibody from the group of antibodies that neutralize the enterotoxic activity of toxin A. The antibody that was selected has a very high ELISA titer. Ascites produced with the hybridoma has a titer of $> 10^6$. This antibody precipitates toxin A, indicating that the epitope to which it binds is present in multiple copies on the toxin molecule (25). This multiple binding probably explains the high titer of the antibody.

In the following sections, we describe the use of the antibody in several immunoassays. For each assay, the optimal conditions (i.e., conditions that give the greatest sensitivity and lowest background) were determined to compare the assays.

A. Enzyme-Linked Immunosorbent Assays

The ELISA that we developed using the monoclonal antibody is an indirect double-antibody assay. Toxin A in test samples is trapped by antiserum coating the well. The microtiter plates that we have found to be optimal are Immulon type 2 plates sold by Dynatech Laboratories, Inc. The toxin is then detected with the monoclonal antibody, conjugate, and substrate. A more detailed description of the ELISA is given in Table 5. The sensitivity of the assay (about 20 ng/ml) is sufficient for the detection of toxin A in fecal specimens from patients with PMC. The assay is about fourfold less sensitive than an ELISA that uses polyclonal antibody as the third phase. The decreased sensitivity is to be expected because the monoclonal antibody does not recognize as many toxin epitopes as the polyclonal antibody. In addition to the indirect ELISA, we have developed a direct ELISA, employing just the monoclonal antibody. The wells of microtiter plates (Immulon type 1) are coated with the monoclonal antibody. After coating, the samples to be tested for toxin A are added to the wells for 1 hr at 37°C. The wells are washed and monoclonal antibody–alkaline phosphatase conjugate is added for 1 hr at 37°C. For the preparation of the conjugate, monoclonal antibody is partially purified from ascites by ammonium sulfate precipitation. The antibody is then coupled to alkaline phosphatase with glutaral-

Table 5 Indirect ELISA for the Detection of Toxin A

1. Wells of polystyrene microtiter plate are coated with a 1:1000 dilution (in carbonate buffer, pH 9.6) of rabbit antiserum against culture filtrate of *C. difficile* overnight at 37°C.
2. Wells are washed with phosphate-buffered saline containing 0.05% Triton X-100 (PBS-T) and dilutions of samples to be tested for toxin A are added.
3. After incubation for 1 hr at 37°C, the wells are washed with PBS-T and 1:1000 dilution of toxin A monoclonal antibody (in PBS-T containing 0.1% neutral rabbit serum) is added for 1 hr at 37°C.
4. Wells are washed with PBS-T and 1:1000 dilution of goat antimouse immunoglobulin–alkaline phosphatase conjugate (in PBS containing 0.1% neutral rabbit serum) is added.
5. After incubation for 1 hr at 37°C, the wells are washed with PBS-T and *p*-nitrophenylphosphate substrate is added; the absorbance at 405 nm is measured after 30 min at room temperature.

dehyde (42). After incubation, the wells are washed, substrate is added, and the reaction is measured after 30 min at room temperature. The direct assay requires less time to complete than the indirect assay. However, the sensitivity of the assay (about 50 ng/ml) is lower because there is no amplification step.

B. Countercurrentimmunoelectrophoresis

One of the unusual properties of the toxin A monoclonal antibody used in these studies is that it precipitates the toxin (25). Given this property, we examined the use of the antibody in countercurrentimmunoelectrophoresis. For the analysis, various dilutions of toxin A were run against twofold dilutions of ascites. Our analysis showed that the antibody detected the toxin at a concentration of about 60 μg/ml (about 1 μg of toxin detected). The sensitivity of the assay was increased several-fold if the precipitin bands were stained with Coomassie Blue R-250, although the staining procedure was time-consuming. It appears that countercurrentimmunoelectrophoresis with the monoclonal antibody is not sufficiently sensitive to warrant its use as a diagnostic test. By using polyclonal antibody in place of the monoclonal antibody, we increased the sensitivity about 10-fold; however, even with the polyclonal antibody, the assay still lacks the sensitivity needed for a diagnostic test.

C. Latex Agglutination

Latex agglutination assays are popular as diagnostic tests because the assays are extremely simple and very rapid. In most instances, the results can be read in less than 5 min. Therefore, we examined the use of the monoclonal antibody in latex agglutination tests for toxin A. Monoclonal antibody was partially purified from ascites by ammonium sulfate precipitation. Latex beads were then coated with serial dilutions of the antibody in glycine-NaCl buffer, pH 8.3. The optimal dilution used to coat the beads was a 1:64 to 1:128 (about 0.2 mg/ml of protein). The remaining unbound sites on the beads were blocked with 1% bovine serum albumin.

The sensitivity of the latex agglutination test was about 1 μg/ml of toxin A. At this sensitivity, the test is not adequate for the diagnosis of *C. difficile* disease. We have found that if latex beads are coated with affinity-purified polyclonal antibody instead of the monoclonal antibody, the sensitivity of the assay is increased three- to fourfold. Therefore, we may be able to increase the sensitivity by coating the latex beads with various combinations of the high-titered monoclonal antibody and affinity-purified antibody. In addition, we should be able to increase the sensitivity using a series of monoclonal antibodies that bind to different sites on the toxin.

V. OTHER USES OF MONOCLONAL ANTIBODIES AGAINST TOXINS A AND B

In addition to using monoclonal antibodies for the development of diagnostic assays, we use the antibodies to learn more about the toxins. The monoclonal antibodies have given us important information about the structure of toxin A. Analysis of the toxin with these antibodies indicates that toxin A is an extremely large molecule and that the enterotoxic active site on the toxin is distinct from the cytotoxic site. We have recently found that the monoclonal antibodies that neutralize toxin A do so by binding to the carboxyl-terminus of the molecule. This binding effectively blocks the binding of the toxin to its receptor.

We have been able to purify toxin A from complex antigen mixtures by a single immunoaffinity step by using our monoclonal antibodies (34). We had initially tried this procedure with polyclonal antibody; however, because the polyclonal antibody bound to the toxin at several sites, the toxin was bound too tightly, and harsh denaturing conditions were necessary to elute the toxin.

The monoclonal antibodies that neutralize the enterotoxic activity of toxin A will help us learn more about the mechanism of action of the toxin. Because of their neutralizing activity, these antibodies may be useful as therapeutic agents in patients with PMC.

VI. CONCLUSIONS

The polyclonal antibodies that we have prepared against toxins A and B of *C. difficile* are highly specific and can be used as diagnostic reagents for the detec-

Table 6 Comparison of Monoclonal Antibody-Based Immunoassays for the Detection of *C. difficile* Toxin A

Immunoassay	Sensitivity	Suitability as diagnostic test
Indirect ELISA	20 ng/ml	Sensitivity is sufficient; however, test is time-consuming
Direct ELISA	50 ng/ml	Sensitivity is sufficient; however, test is time-consuming
Countercurrent-immunoelectrophoresis	63 μg/ml	Not suitable
Latex agglutination	1 μg/ml	Test is simple and rapid; however, sensitivity needs to be improved

tion of the toxins in fecal specimens from patients. Monoclonal antibodies against toxin A are less sensitive than polyclonal antibodies in immunoassays; however, they offer several advantages. They can be produced in larger amounts more cheaply than polyclonal antibodies. In addition, they ensure a high degree of specificity with little or no lot-to-lot variation.

We have investigated the potential use of toxin A monoclonal antibodies in various immunoassays, and a comparison of these assays is given in Table 6. The ELISA has the sensitivity needed for the diagnostic assay, but it requires several hours to complete. The latex agglutination assay, on the other hand, is simple and rapid. With additional studies on ways to increase the sensitivity, the latex assay may be useful for the clinical detection of *C. difficile* toxin A.

ACKNOWLEDGMENTS

Much of the work presented in this chapter was supported by Public Health Service grant AI 15749 from the National Institutes of Health and state support grant 1214520 from the Commonwealth of Virginia. We thank Julianna Toth, Roger Van Tassell, Howard Krivan, and David MacDonald for their many contributions to this work.

REFERENCES

1. Bartlett, J. G., Antibiotic-associated pseudomembranous colitis. *Rev. Infect. Dis. 1*:530–539 (1979).
2. Bartlett, J. G., and B. Laughon, *Clostridium difficile* toxins. *Microecol. Ther. 14*:35–42 (1984).
3. George, W. L., Antimicrobial agent-associated colitis and diarrhea: Historical background and clinical aspects. *Rev. Infect. Dis. 6*:S208–S213 (1984).

4. Mulligan, M. E., Epidemiology of *Clostridium difficile*-induced intestinal disease. *Rev. Infect. Dis. 6*:S222–S228 (1984).
5. Borriello, S. P., *Antibiotic Associated Diarrhoea and Colitis*, S. P. Borriello (ed.). Martinus Nijhoff, Boston (1984).
6. Banno, Y., T. Kobayashi, K. Watanabe, K. Ueno, and Y. Nozawa, Two toxins (D-1 and D-2) of *Clostridium difficile* causing antibiotic-associated colitis: Purification and some characterization. *Biochem. Int. 2*:629–635 (1981).
7. Banno, Y., T. Kobayashi, H. Kono, K. Watanabe, K. Ueno, and Y. Nozawa, Biochemical characterization and biologic actions of two toxins (D-1 and D-2) from *Clostridium difficile. Rev. Infect. Dis. 6*:S11–S20 (1984).
8. Taylor, N. S., G. M. Thorne, and J. G. Bartlett, Separation of an enterotoxin from the cytotoxin of *Clostridium difficile. Clin. Res. 28*:285 (1980).
9. Taylor, N. S., G. M. Thorne, and J. G. Bartlett, Comparison of two toxins produced by *Clostridium difficile. Infect. Immun. 34*:1036–1043 (1981).
10. Sullivan, N. M., S. Pellett, and T. D. Wilkins, Purification and characterization of toxins A and B of *Clostridium difficile. Infect. Immun. 35*:1032–1040 (1982).
11. Lyerly, D. M., D. E. Lockwood, S. H. Richardson, and T. D. Wilkins, Biological activities of toxins A and B of *Clostridium difficile. Infect. Immun. 35*:1147–1150 (1982).
12. Libby, J. M., B. S. Jortner, and T. D. Wilkins, Effects of the two toxins of *Clostridium difficile* in antibiotic-associated cecitis in hamsters. *Infect. Immun. 36*:822–829 (1982).
13. Fernie, D. S., R. O. Thomson, I. Batty, and P. D. Walker, Active and passive immunization to protect against antibiotic associated caecitis in hamsters. *Dev. Bio. Stand. 53*:325–332 (1982).
14. Lyerly, D. M., N. M. Sullivan, and T. D. Wilkins, Enzyme-linked immunosorbent assay for *Clostridium difficile* toxin A. *J. Clin. Microbiol. 17*:72–78 (1983).
15. Laughon, B. E., R. P. Viscidi, S. L. Gdovin, R. H. Yolken, and J. G. Bartlett, Enzyme immunoassays for detection of *Clostridium difficile* toxins A and B in fecal specimens. *J. Infect. Dis. 149*:781–788 (1984).
16. Lyerly, D. M., and T. D. Wilkins, Assays for the toxins of *Clostridium difficile*. In *Detection of Bacterial Antigens for the Rapid Diagnosis of Infectious Diseases*, R. B. Kohler (ed.). CRC Press, Boca Raton (1986).
17. Ryan, R. W., I. Kwasnik, and R. C. Tilton, Rapid detection of *Clostridium difficile* toxin in human feces. *J. Clin. Microbiol. 12*:776–779 (1980).
18. Welch, D. F., S. K. Menge, and J. M. Matsen, Identification of toxigenic *Clostridium difficile* by counterimmunoelectrophoresis. *J. Clin. Microbiol. 11*:470–473 (1980).
19. Poxton, I. R., and M. D. Byrne, Detection of *Clostridium difficile* by counterimmunoelectrophoresis: A note of caution. *J. Clin. Microbiol. 14*: 349 (1981).
20. Levine, H. G., M. Kennedy, and J. T. LaMont, Counterimmunoelectrophoresis versus cytotoxicity assay for the detection of *Clostridium difficile* and its toxin. *J. Infect. Dis. 145*:398 (1982).

21. West, S. E. H., and T. D. Wilkins, Problems associated with counterimmunoelectrophoresis assays for detecting *Clostridium difficile* toxin. *J. Clin. Microbiol. 15*:347–349 (1982).
22. Jarvis, W., O. Nunez-Montiel, F. Thompson, V. Dowell, M. Towns, G. Morris, and E. Hill, Comparison of bacterial isolation, cytotoxicity assay, and counterimmunoelectrophoresis for the detection of *Clostridium difficile* and its toxin. *J. Infect. Dis. 147*:778 (1983).
23. Wu, T. C., and J. C. Fung, Evaluation of the usefulness of counterimmunoelectrophoresis for diagnosis of *Clostridium difficile*-associated colitis in clinical specimens. *J. Clin. Microbiol. 17*:610–613 (1983).
24. Yolken, R. H., L. S. Whitcomb, G. Marien, J. D. Bartlett, J. Libby, M. Ehrich, and T. Wilkins, Enzyme immunoassay for the detection of *Clostridium difficile* antigen. *J. Infect. Dis. 144*:378 (1981).
25. Lyerly, D. M., C. J. Phelps, and T. D. Wilkins, Monoclonal and specific polyclonal antibodies for immunoassay of *Clostridium difficile* toxin A. *J. Clin. Microbiol. 21*:12–14 (1985).
26. Coffins, P., and M. D. Scharff, Rate of somatic mutation in immunoglobulin production by mouse myeloma cells. *Proc. Natl. Acad. Sci. USA 68*: 219–223 (1971).
27. Arthur, L. O., and R. J. Massey, *A Workbook on Hybridoma and Monoclonal Antibody Technology*, L. O. Arthur and Massey, R. J. (eds.). (1981).
28. Goding, J. W., *Monoclonal Antibodies: Principles and Practice*. Academic Press, London (1983).
29. Lane, R. D., R. S. Crissman, and M. F. Lachman, Comparison of polyethylene glycols as fusogens for producing lymphocyte-myeloma hybrids. *J. Immunol. Methods 72*:71–76 (1984).
30. Morgan, J., Monoclonal antibody production. In *Modern Methods in Pharmacology*. Alan R. Liss, New York, pp. 29–67 (1984).
31. Westerwoudt, R. J., A. M. Naipal, and C. M. H. Harrisson, Improved fusion technique. II. Stability and purity of hybrid clones. *J. Immunol. Methods 68*:89–101 (1984).
32. Sterne, M., and L. M. Wentzel, A new method for the large scale production of high-titre botulinum Formol-toxoid types C and D. *J. Immunol. 65*: 175–183 (1950).
33. Lyerly, D. M., M. D. Roberts, C. J. Phelps, and T. D. Wilkins, Purification and characterization of toxins A and B of *Clostridium difficile. FEMS Microbiol. Lett. 33*:31–35 (1986).
34. Lyerly, D. M., C. J. Phelps, J. Toth, and T. D. Wilkins, Characterization of toxins A and B of *Clostridium difficile* using monoclonal antibodies. *Infect. Immun. 54*:70–76 (1986).
35. Krivan, H. C., and T. D. Wilkins, Receptor-mediated purification of *Clostridium difficile* enterotoxin (toxin A) in one step. *Abstr. Annu. Meet. Am. Soc. Microbiol.* p. 34 (1986).
36. Towbin, H., T. Stahelin, and J. Gordon, Electrophoretic transfer of proteins from polyacrylamide gels to nitrocellulose sheets. *Proc. Natl. Acad. Sci. USA 76*:4350–4354 (1979).

37. Burnett, W. N., "Western blotting." Electrophoretic transfer of proteins from SDS-polyacrylamide gels to unmodified nitrocellulose and radiographic detection with antibody and radioiodinated protein A. *Anal. Biochem. 112*:195–203 (1981).
38. Hawkes, R., Identification of concanavalin A-binding proteins after sodium dodecyl sulfate-gel electrophoresis and protein blotting. *Anal. Biochem. 123*:143–146 (1982).
39. Bhakdi, S., D. Jenne, and F. Hugo, Electroimmunoassay–immunoblotting (EIA-IB) for the utilization of monoclonal antibodies in quantitative immunoelectrophoresis: The method and its applications. *J. Immunol. Methods 80*:25–32 (1985).
40. Ehrich, M., R. L. Van Tassell, J. M. Libby, and T. D. Wilkins, Production of *Clostridium difficile* antitoxin. *Infect. Immun. 28*:1041–1043 (1980).
41. Lyerly, D. M., K. E. Saum, D. K. MacDonald, and T. D. Wilkins, Effects of *Clostridium difficile* toxins given intragastrically to animals. *Infect. Immun. 47*:349–352 (1985).
42. Voller, A., D. E. Bidwell, and A. Barlett, Enzyme immunoassays in diagnostic medicine. Theory and practice. *Bull. WHO 53*:55–65 (1976).

25
Monoclonal Antibodies to Coliforms and Their Use in Food and Environmental Microbiology

CHARLES KASPAR*
Iowa State University, Ames, Iowa

PETER FENG†
IGEN, Inc., Rockville, Maryland

I. INTRODUCTION

Since the introduction of hybridoma technology by Köhler and Milstein in 1975 (1), the potential of monoclonal antibodies has rapidly become evident in research and in commercial applications. The influence of monoclonal antibodies on clinical microbiology was almost immediate owing to the extensive use of antisera in diagnostics and bacterial serology. In the last several years, however, the use of monoclonal antibodies has expanded into other areas, including food and environmental microbiology. Monoclonal antibodies have been produced and are currently being tested in diagnostic agents for several food- or water-associated bacteria, such as *Salmonella* (2,3); *Shigella* (4); *Vibrio* (5); and *Legionella* (6,7). Diagnostic agents for detection of bacterial toxins with monoclonal antibodies are being developed (8,9). Monoclonal antibodies are also being tested for the detection of rotaviruses in sewage, effluent, and river water samples (10).

One area, yet unexplored, that can benefit from monoclonal antibody technology is in the detection of coliforms in food and water. The coliform group is one of the most commonly used indicators of unsanitary conditions, yet, some current detection methodologies have changed little since their introduction in the early 1900s.

**Present affiliation*: Fisheries Research Branch, U.S. Food and Drug Administration, Dauphin Island, Alabama.
†Present affiliation: U.S. Food and Drug Administration, Washington, District of Columbia.

Antibody-based assays, such as agglutination, radioimmunoassay (RIA), and enzyme immunoassays (EIA), are rapid and highly specific. Antibody-based assays have not been developed for the detection of the coliform group, however, because of the antigenic diversity of the organisms in the group and the problems associated with cross-reactions with noncoliforms with conventional polyclonal antisera. With the advent of monoclonal antibody technology, antibodies to a single antigenic epitope can now be produced, enabling the development of monoclonal antibodies to epitopes common among the coliform group. One, or a pool, of these antibodies may then be applied in an immunoassay for the detection and identification of coliforms.

The aim of this chapter is to briefly summarize current methods used for the detection of coliforms in food and water, and then, discuss several monoclonal antibodies produced in our laboratories that were found to be specific for organisms in the coliform group. The potential application of these monoclonal antibodies to improve current procedures used for the detection or confirmation of coliforms will also be discussed.

II. BACKGROUND

A. Coliform Group as Indicators

Sanitary conditions of food and water have been traditionally monitored by using surrogate systems or bacterial indicators. The use of coliforms as indicators of fecal pollution has been in existence for the past 80 years. Coliforms are defined as aerobic or facultatively anaerobic, asporogenous, gram-negative rods, that ferment lactose to acid and gas in 48 hr at 35°C (11). The four major genera, all from the family *Enterobacteriaceae*, that compose the coliform group *Enterobacter*, *Klebsiella*, *Citrobacter* and, especially, *Escherichia* (12). However, because of the broad definition, a diverse group is encompassed. Several reviews on the use of bacteria as indicators of fecal pollution have been published (13–18).

The standard methods for the enumeration of coliforms in water and wastewater are the multiple-tube fermentation technique and the membrane filtration (MF) test (11). When testing foods, the multiple-tube fermentation method or the direct-plating method in violet red bile (VRB) agar are most commonly used (19). All of these assays are based on the criteria of lactose fermentation by coliforms.

The multiple-tube fermentation test is a statistical assay that estimates the mean density of organisms in a sample. Results are reported as the most probable number (MPN). The assay is a cumbersome procedure that comprises three consecutive parts, the presumptive, the confirmed, and the completed tests. The entire procedure requires 5 to 10 days, and involves extensive labor and material costs. In addition, the MPN test is susceptible to false-negative and false-positive reactions caused by the presence of noncoliforms (20,21), anaerogenic *E. coli* strains (22), and injured or suppressed coliforms (21,23).

Similar drawbacks have been reported for the MF technique and the VRB direct-plating method (24, 25). Noncoliforms have been reported to mimic coliforms in both the M-Endo broth used for the MF test (24), and VRB agar (25). Also, recovery of injured cells in these media have been inadequate, resulting in false-negative reactions (24,25).

Over the years, improvements in media (26-30) and modifications of standard procedures (25,31-34) have improved coliform recovery and accuracy of these tests; however, the assays remain inefficient. A confirmation procedure was recently adopted by the American Public Health Association (11) for the MF technique, whereby individual colonies on the filter are subcultured into nonselective and selective lactose-containing broth media to check for fermentation. Although the numbers of false reactions are reduced, an initially simple, one-step assay is complicated by the requirement of additional tests.

In ideal terms, speciation of bacterial isolates would virtually eliminate most false-negative and false-positive reactions. However, in practical terms, multiple biochemical tests, which are currently used for speciation, are not used routinely because they are time-consuming, labor-intensive, and expensive. The use of monoclonal antibodies may be an attractive alternative to the conventional speciation procedure. A monoclonal antibody-based detection or confirmation assay would not only rapidly confirm the presence of the organisms in question but, by the nature of antibody specificity, could also enable speciation without isolation of organisms.

B. Common Antigens Among Coliforms

The serological classification of the genera within the coliform group is vast and complex. In *E. coli* alone there are 164 somatic (O), 56 flagellar (H), and 103 capsular (K) antigens (35). The task of searching for an epitope common to coliforms among potentially thousands of antigenic determinants is, therefore, both difficult and critical. Fortunately, common antigen epitopes exist and have been reported for genera within a family and even between families (36,37).

The outer surface of gram-negative bacteria contains lipopolysaccharides (LPS), membrane proteins, surface proteins and, frequently, a capsule. These components are often highly immunogenic and potentially useful for producing monoclonal antibodies against coliforms. For example, the enterobacterial common antigen (ECA) component, found on the outer membrane, is well characterized (37,38). A monoclonal antibody produced against ECA reacted with only clinical isolates from the *Enterobacteriaceae*; therefore, the epitope appears to be conserved within the family (37). The lipid A portion of LPS is also conserved structurally (39) and antigenically (36,40). Monoclonal antibodies directed to lipid A from *E. coli* strain J5 reacted with membrane and LPS preparations from numerous gram-negative bacteria from several different families (36,41). In this instance, the epitope appears to be conserved among all gram-negative bacteria and may be too cross-reactive for differential purposes. Interestingly, both anti-ECA and anti-lipid A monoclonal antibodies reacted well with only denatured cell extracts of a wide range of strains but not generally with whole cells. This observation would suggest that these epitopes might not be exposed and readily accessible to antibodies (36,37,41). The requirement for cell extraction or denaturation before antigen detection may be another drawback to using these antigens for producing monoclonal antibodies for the detection of coliforms.

Antibodies to outer membrane proteins are generally strain- or species-specific (36,42), although antigenically related proteins have also been found throughout the *Enterobacteriaceae* (43). Some of the *E. coli* outer membrane proteins to which monoclonal antibodies have been produced include ompA (44), lactose permease (45), phoE (46), and lamB (44,47,48). These antibodies have been used primarily for protein structure analysis; however, some may be potentially useful for coliform diagnostics. As examples, an anti-ompA monoclonal antibody reacted with all *E. coli* and several *Salmonella* species tested (44). Similarly, the cytoplasmic epitopes on the lamB protein appear to be conserved (49), and widely distributed among *E. coli*, and some *Citrobacter*, *Klebsiella*, and *Shigella* species (48,50). Both anti-ompA and anti-lamB monoclonal antibodies may be applicable in coliform detection.

Surface proteins, such as flagella and pili, contain both strain- and species-specific determinants (51-53). A myeloma protein (M467) derived from a mouse plasma cell tumor was found to recognize a common determinant on salmonellae flagellins (51). The antibody reacted with 94% of the salmonellae tested, with no cross-reactivity to enteric organisms (54). Likewise, species- and strain-specific determinants have been identified on *E. coli* pili by using monoclonal antibodies (52). Although pili and flagella have conserved antigenic determinants, some environmental isolates do not have pili or flagellae, nor are these appendages produced under all conditions. For instance, *Klebsiella* species are nonmotile; therefore, an antiflagellar monoclonal antibody alone would not be sufficient to detect all coliforms.

Another potential source of common coliform epitopes may be the cytoplasmic proteins. Biochemical characteristics such as lactose fermentation have been used for many years to define coliforms. Therefore, enzymes, such as β-galactosidase or other cytoplasmic enzymes and proteins, may contain epitopes commonly distributed among the major coliform genera.

From this brief literature survey, it is evident that despite the serological diversity of the organisms in the coliform group, substantial epitope conservation and serological cross-reactions occur among the various antigenic bacterial components. Any of these components should, then, be a suitable immunogen for the production of monoclonal antibodies against coliform organisms.

III. RESULTS AND DISCUSSION

A. Monoclonal Antibody Production

Several approaches to making monoclonal antibodies against coliforms were tested by using different bacterial components as immunogens. Heat-treated cells, cytoplasmic proteins, and surface proteins were all used to induce immune responses. Because the primary purpose of the anticoliform antibodies was to examine their potential application in diagnostics, environmental isolates were almost exclusively used for antibody evaluations.

1. Whole Cells

A pool of five *E. coli* strains was heat-treated for 1 hr at 100°C and used to immunize BALB/c mice. Spleen cells from two immunized mice were then fused with Sp2/0 myeloma cells. To select for broadly reactive antibodies, the resultant hybridomas were screened by enzyme-linked immunosorbent assay (ELISA) by using three heat-treated strains of *E. coli* not used for immunization (55). The reactivity data of the hybridomas are shown in Table 1. Of the 468 antibody-positive hybridomas, 84 reacted with all three test strains and in-

Table 1 ELISA Reactions of Hybridomas to Three Heat-Treated *E. coli* Strains

E. coli strain(s)	No. of positive hybridomas[a]
EC33	75
TM1A	116
EK3C	35
EC33 and TM1A	107
EC33 and EK3C	20
TM1A and EK3C	31
EC33, TM1A and EK3C	84
Total	468

[a]OD measured at 410 nm.

cluded 32 hybridomas which yielded optical densities (OD), at 410 nm, of 1.0 or higher for two or three of the test strains (data not shown). Fifteen of the 32 cell lines were cloned and the cell supernatants tested by ELISA for reactivity against a variety of heat-treated gram-negative bacteria. The specificities of six of the monoclonal antibodies are shown in Table 2. The antibodies are all IgG except for 6H2, which is an IgM. Most of the monoclonal antibodies reacted with a wide spectrum of gram-negative bacteria. This suggested that the antibodies were directed against a determinant conserved among the gram-negative strains tested. Interestingly, these anti-gram-negative antibodies reacted with only heat-treated cells, similar to the observation reported for anti-lipid A monoclonal antibodies (36,41). It is possible that the epitopes of our monoclonal antibodies may also reside in the lipid A portion of the LPS. Because of the strong reactivity of these antibodies to noncoliform strains, such as *Aeromonas* and *Pseudomonas* (see Table 2), they were not suitable for our purpose.

One exception among the group was antibody 6H2, which showed good specificity to *E. coli*, reacting strongly with four of the six *E. coli* strains tested and weakly with *E. agglomerans*. Additional testing with 6H2 showed that 40 of 80 natural isolates of *E. coli* reacted with the antibody preparation. In contrast

Table 2 ELISA Reactivity Pattern of Monoclonal Antibodies to Heat-Treated Gram-Negative Bacteria

Strain	OD at 410 nm with antibody					
	6H9	7A9	4F1	6H2	2H3	9A10
Aeromonas hydrophila	0.75	0.11	0.21	0.00	0.99	0.47
Salmonella arizonae	0.64	0.17	0.13	0.00	1.10	0.45
Enterobacter agglomerans	0.36	0.02	0.04	0.24	0.65	0.21
E. cloacae	0.86	0.06	0.04	0.00	0.51	0.24
Escherichia coli B	0.94	0.20	0.30	1.67	1.40	0.64
E. coli EK3C	0.59	0.06	0.03	0.00	1.36	0.25
E. coli TM1A	0.52	0.22	0.10	0.70	1.67	0.51
E. coli EC31	0.60	0.20	0.36	0.07	1.29	0.60
E. coli EC33	0.66	0.22	0.30	1.05	1.10	0.54
E. coli EC34	0.71	0.18	0.16	1.16	1.66	0.56
Klebsiella pneumoniae	0.51	0.00	0.01	0.00	0.60	0.15
Proteus mirabilis	0.80	0.19	0.03	0.00	1.34	0.49
Pseudomonas aeruginosa	0.97	0.00	0.06	0.00	1.44	0.18
Serratia oderifera	0.78	0.02	0.03	0.00	0.65	0.13
S. liquefaciens	0.75	0.04	0.02	0.00	1.25	0.14
Shigella flexneri	0.90	0.04	0.09	0.00	1.24	0.25
Yersinia enterocolitica	1.18	0.19	0.24	0.00	1.24	0.47

to the other monoclonal antibodies from the same fusion, 6H2 reacted with either heated or unheated cells. This suggests that 6H2 may recognize an epitope on the outer surface of selected *E. coli* strains. The results from this study showed that immunization of mice using heat-treated *E. coli* cells mostly induced antibody production against epitopes that are exposed only upon heat treatment or denaturation. These antibodies usually were broadly cross-reactive. In one instance, however, a monoclonal antibody that apparently recognizes a surface epitope was also isolated. This antibody was specific for selected strains of *E. coli* and may be potentially useful for detection of coliforms.

2. *Cytoplasmic Proteins*

Enzyme activities of β-galactosidase and β-glucuronidase are commonly used biochemical characteristics to define coliforms (29,56,57). These enzymes are, therefore, potentially useful for producing antibodies to coliforms.

For an antibody to effectively detect cytoplasmic proteins, a rapid cell lysis step must be included in the procedure. Such a method was included in a coagglutination assay to detect intracellular enzymes (Kaspar and Hartman, manuscript in preparation). Polyclonal antibodies were prepared against *E. coli* β-galactosidase (GAL), β-glucuronidase (GUD), and glutamate decarboxylase (GAD).

The protocol for the coagglutination assay is as follows: Cultures are grown overnight in lauryl tryptose broth supplemented with 0.05% of L-glutamic acid and 150 μg of 4-methylumbelliferone glucuronide per milliliter to induce enzyme production. The cells are pelleted by centrifugation and resuspended to their original volume in 10 mM sodium phosphate buffer (pH 7.0) containing 25 mM EDTA, 8% sucrose, and 0.05% Triton X-100. Lysozyme is added to a final concentration of 15 μg/ml and the suspension incubated for 30 min at 35°C to lyse the cells. This lysis protocol was found to release enzyme proteins without affecting the antibody reactivity. To prepare the probing antibody reagents, polyclonal antienzyme antibodies are incubated with *Staphylococcus aureus* cells to allow antibody binding to protein A. The antibody–*S. aureus* reagents can then be used to check cell lysates for antigen.

Initial studies that used purified enzyme preparations showed that the antibody reagents reacted specifically with their respective enzymes. Agglutination reactions were rapid, occurring within 30 to 60 sec, with the exception of anti-GUD reagent which required 2 to 3 min. Positive agglutinations appeared as granular clumps that were visible under low magnification. On occasion, the presence of nucleic acids in the lysates obscured the interpretation of the results. However, this problem was remedied by the addition of ribonuclease A to the lysates.

A preliminary study was conducted to examine the reactivity of a number of *E. coli* and non-*E. coli* strains to the anti-GAL, anti-GUD, and anti-GAD reagents.

The results, shown in Table 3, indicate that all *E. coli* strains reacted with either all three (51/55) or two (4/55) of the antienzyme reagents. Of the 42 non-*E. coli* strains tested, only one strain each of *Serratia fonticola* and *Enterobacter cloacae* was able to agglutinate all three reagents. Therefore, by requiring reactions with all three antienzyme reagents, we were able to obtain 93% detection for *E. coli* with only 5% false-positive reaction.

The results from this preliminary study show that it is feasible to develop assays to detect coliforms based on antibodies directed to cytoplasmic proteins. Our agglutination assay, using antienzyme reagents, is simple to perform and rapid.

Although this study was conducted by using polyclonal antienzyme sera, the procedure should be equally applicable to using antienzyme monoclonal antibodies. Purified commercial preparations of GAL and GUD were used to induce an immune response in BALB/c mice. After fusion of spleen cells with Sp2/0 myelomas, monoclonal antibodies to *E. coli* GAL and GUD were isolated. The antibody specificities are currently under evaluation.

3. *Surface Proteins*

Purified flagellin proteins from two *E. coli* strains were used to induce an immune response in BALB/c mice. Spleen cells were then extracted and fused with Sp2/0 myeloma cells. Hybridomas producing the specific monoclonal antibodies were isolated by using the method of Sugasawara et al. (58). The monoclonal antibodies were screened for reactivity by ELISA using either purified flagellin protein or untreated bacteria as antigens. Five monoclonal antibodies were obtained that showed strong reactivity to the parental *E. coli* strains. No reactions were observed to the two *Pseudomonas* strains included in the antibody screenings. These cell lines were cloned and their isotypes are listed in Table 4.

Table 3 Coagglutination of *E. coli* and non-*E. coli* Lysates with Staphylococcal–Antienzyme Reagents

Strain	No. of strain	No. of strains and (%) agglutinated by the following no. of reagents		
		3	2	1
E. coli	55	51 (93)	4 (7)	0 (0)
non-*E. coli*	42	2 (5)[a]	2 (5)[b]	6 (14)[c]

[a]Cultures identified as *Serratia fruticola* and *Enterobacter cloacae*.

[b]*Citrobacter freundii* reacted with anti-GAL and anti-GUD. *Hafnia alvei* agglutinated anti-GAL and anti-GAD reagents.

[c]*Citrobacter freundii, Klebsiella pneumoniae, Enterobacter aerogenes*, and two strains of *Enterobacter cloacae* agglutinated the anti-GAD reagent. One strain of *Acinetobacter calcoaceticus* agglutinated the anti-GUD conjugate.

Table 4 Isotypes of Anti-Surface Protein Monoclonal Antibodies

Monoclonal antibody	Isotype
15D8	IgG1, kappa
32C3	IgG1, kappa
37G12	IgG1, kappa
36F6	IgG2b, kappa
30G4	IgG3, kappa

To determine the specificity of the monoclonal antibodies, 260 gram-negative bacteria, isolated from food, milk, and water, were screened by ELISA. The results from Table 5 show that varying degrees of antibody-specificity were obtained. Antibodies 30G4 and 36F6 were highly specific, reacting with only 5 of the 165 strains of *E. coli* tested. No reactions with any other organisms were observed. Two other antibodies, 37G12 and 32C3, were intermediate in specificity. These two antibodies reacted with a variety of enteric genera; however, fairly low numbers of organisms within each genera were detected by either antibody. The last monoclonal antibody, 15D8 was broadly reactive, detecting a large number of organisms from several genera in the enterobacterial family. Importantly, noncoliforms, such as *Pseudomonas* species which are also found in large numbers in environmental samples, did not react with any of our antibodies.

From the data, it appears that 15D8 antibody may be potentially useful in coliform detection assays. The antibody reacted with *E. coli* and *Enterobacter* species, a variety of other enterobacterial genera (see Table 5), and also *Sal-*

Table 5 Detection of Environmental Isolates by Monoclonal Antibodies

Organism	No. isolated	No. of strains detected by				
		37G12	36F6	32C3	30G4	15D8
Escherichia coli	(165)	59	5	36	5	99
Enterobacter spp.	(31)	7	0	5	0	12
Serratia spp.	(11)	3	0	1	0	8
Proteus spp.	(6)	4	0	2	0	5
Hafnia spp.	(6)	0	0	0	0	1
Klebsiella spp.	(18)	0	0	0	0	0
Shigella spp.	(3)	0	0	0	0	0
Pseudomonas spp.	(20)	0	0	0	0	0

Table 6 Reactivity of 15D8 Monoclonal Antibody to *Salmonella* Species

Bacteria	OD at 414 nm
S. arizonae	2.20
S. potsdam	2.41
S. paratyphi A	1.32
S. paratyphi B	0.62
S. paratyphi C	1.93
S. tennessee	1.75
S. typhi	1.40

monella (Table 6). However, 18 *Klebsiella* and three *Shigella* isolates were not detected by 15D8. Similar results were obtained when laboratory cultures of enterobacterial strains were tested by ELISA with the 15D8 antibody (Table 7). These results suggest that the epitope recognized by the 15D8 antibody is not present on either *Klebsiella* or *Shigella* species. This finding is not surprising, because the antibody was induced using a flagellar antigen and, therefore, non-motile genera such as *Klebsiella* and *Shigella* would not be expected to react with the antibody.

To confirm that the 15D8 antibody is directed to a flagellar epitope, purified flagellin preparations and cell extracts of selected enteric strains were fractionated

Table 7 Reactivity of 15D8 Monoclonal Antibody to American Type Culture Collection (ATCC) Strains Using ELISA

Bacteria	OD at 405 nm
Escherichia coli	2.35
Enterobacter cloacae	2.58
E. aerogenes	0.72
Citrobacter freundii	1.40
Serratia marcescens	1.00
Proteus mirabilis	0.94
P. vulgaris	1.55
Edwardsiella tarda	2.20
Shigella sonnei	0.32
Klebsiella pneumoniae	0.20
Providencia stuartii	0.09
Pseudomonas aeruginosa	0.16
P. fluorescens	0.12

on a SDS-polyacrylamide gel and the proteins transferred onto nitrocellulose paper for immunoblotting using the 15D8 antibody. Figure 1, lanes (*a*) show that the purified flagellin proteins, from each isolate reacted strongly with 15D8. Slight amounts of degradation product were also observed for each flagellin sample. Likewise, blotting of crude cell extracts (lanes *b*) showed that for every isolate tested, the antibody reacted with a single protein band which had an M_r corresponding to that of the purified flagellin (lanes *a*). These results would sug-

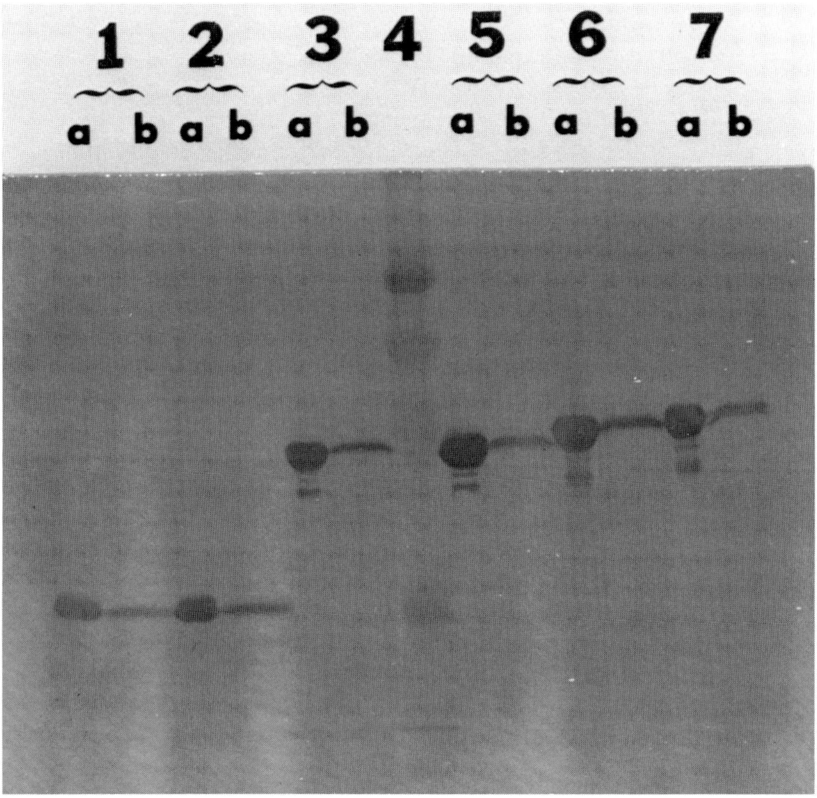

Figure 1 Immuno-blot (Western) of flagellin and cell extracts of isolates using 15D8 monoclonal antibody. The samples are *Enterobacter cloacae* strains MC17 (1) and NJ4 (2); *Serratia rubidaea* strains O10 (3) and O13 (5); *Salmonella arizonae* O8 (6); and *E. coli* EC10 (7). Lanes were loaded with purified flagellin (a) or crude cell extract (b) of each isolate. Sample (4), M_r markers (from top to bottom): phosphorylase B (97,000), bovine serum albumin (65,000), ovalbumin (45,000), carbonic anhydrase (29,000), and trypsin inhibitor (20,000).

gest that 15D8 antibody is recognizing a flagellar epitope. Interestingly, significant variation in flagellar size was noted among the isolates. The M_r variation among bacterial flagellins have previously been reported (59).

From the diagnostic standpoint, the fact that 15D8 antibody was able to detect 60% of the *E. coli* (99/165) and 40% of the *Enterobacter* species (12/31) was encouraging. Both are major genera in the coliform group. The reactivity of 15D8 to other "noncoliform" genera in the family may be undesirable but not critical. Several strains from the genera *Proteus, Serratia*, and *Hafnia* have the ability to ferment lactose (22); hence, by definition can be considered as coliforms. Also, the cross-reactivity of 15D8 to pathogenic species such as *Salmonella* would not detract from the purpose for which these monoclonal antibodies were intended. The absence of *Klebsiella* detection by 15D8, however, is troublesome because *Klebsiella* species are one of the major genera in the coliform group. If 15D8 antibody is to be used for coliform detection, additional anti-*Klebsiella* monoclonal antibody(s) will have to be included to remedy this deficiency. Little is known on the reactivity of 15D8 with *Citrobacter* species, the other major genus of the coliform group, because no *Citrobacter* species were isolated from the environmental samples we collected. However, an ATCC strain of *C. freundii* reacted strongly with 15D8 antibody (see Table 7).

In an effort to determine if we could enhance *E. coli* detection, an antibody cocktail was formulated by using the anti-*E. coli* monoclonal antibody 6H2 and 15D8. When the cocktail was tested by ELISA with environmental isolates, the percentage of *E. coli* detection improved from 60% to 84%.

From these results, it appears that although we have isolated some monoclonal antibodies specific for coliform organisms, the antibodies were deficient in the detection of *Klebsiella* and some *Enterobacter* species. It is also evident that the production of a single monoclonal antibody directed against all coliforms is difficult, if not impossible, because of the antigenic diversity of the genera in the group. However, coliform detection or confirmation with monoclonal antibodies is realistic if panels or cocktails of monoclonal antibodies are used.

B. Potential Applications of Monoclonal Antibodies in Coliform Detection

The use of monoclonal antibodies in coliform assays may be an attractive alternative to conventional methodologies. Antigen-antibody interactions are rapid, antibodies are highly specific, and suitable antibody sensitivities have been demonstrated by both ELISA and coaglutination reactions (60-62). Enzyme immunoassays can efficiently detect 100,000 cells/ml or less (62). Although the application of monoclonal antibodies we presently envision is not in direct detection of coliforms but in the subsequent confirmatory steps, direct detection assays using antibodies are also feasible.

The membrane filtration (MF) test is ideal for incorporating a direct serological testing procedure. The filtration process concentrates organisms from a large sample volume onto a small filter surface area. After 5 to 6 hr of incubation, the microcolonies developing on the filter can be visualized, and individual colonies identified if suitable anticoliform antibodies are available. For instance, antibodies conjugated to an enzyme can be used to probe the microcolonies on the filter. The antibody conjugate would bind specifically to coliform colonies, and the antibody–antigen complex can subsequently be detected by the addition of a fluorogenic enzyme substrate such as a 4-methylumbelliferone derivative. As the substrate is cleaved, the fluorogenic 4-methylumbelliferone radical is released, which can be visualized under UV light. All fluorescent colonies can then be counted as coliforms. Similarly, a fluorescent antibody test that uses an antibody–fluorescein isothiocyanate conjugate to probe for organisms trapped on a black filter membrane (63) could be adapted for coliform detection. Antibody-based assays made directly on developing microcolonies on the filters would provide specificity and sensitivity as well as shorten the required incubation times.

At the present time, the area most suited to antibody-based assay applications is in the confirmatory steps currently required for coliform detection procedures. For instance, the present confirmatory step for the MPN procedure requires subculture of all gas-positive presumptive tubes into BGLB broth and incubation for an additional 24 to 48 hr. With anticoliform monoclonal antibodies, aliquots of bacteria suspensions from presumptive tubes can be removed and tested by either ELISA or coagglutination using specific antibodies to rapidly confirm the presence of coliforms. Moreover, the assays are so simple and inexpensive that even growth-positive, gas-negative tubes might be examined. The presence of anaerogenic coliforms and the suppression of gas formation by coliforms in the presence of other bacteria have been known to yield gas-negative presumptive tests (21–23). Use of a rapid antibody-based test to confirm presumptive MPN tubes would drastically reduce the time and materials required, as well as increase the accuracy of the MPN test.

Monoclonal antibodies could also be used to verify coliform colonies from direct-platings. In direct-plating assays (19), samples are mixed in with molten VRB agar; hence, only subsurface colonies are obtained. Serological confirmation may require subculturing of the colonies in a broth medium for several hours to allow cell density amplification before either ELISA or coagglutination tests can be performed. Still, this could be faster than the conventional confirmation procedure of checking for gas production in BGLB tubes. The problem of subculturing can also be bypassed by using VRB agar spread plates, which would yield surface colonies. These can then be easily picked and confirmed by using ELISA or the cells may be prepared on a glass slide for a coagglutination test. Similarly, in a direct-plating method for *E. coli* developed by Anderson and

Baird-Parker (64), samples are surface-plated on a cellulose membrane placed on top of the agar medium. It should then be possible to remove the membrane without disrupting the colonies and probe the colonies on the membrane surface by using anticoliform monoclonal antibodies.

The present confirmation step for the conventional MF technique requires subculture of at least 10 colonies from the filter into BGLB broth to check for gas evolution. Confirmation is essential because instances of false-negative and false-positive reaction on MF have commonly been reported (24,65,66). The potential application of a monoclonal antibody-based coliform confirmatory test for MF has been examined in our laboratory by use of natural isolates of *E. coli*. Colonies on the filter were picked and either transferred into a microtiter plate for ELISA testing or spotted onto a nitrocellulose membrane for a dot-blot immunoassay using a precipitable substrate. In both assay formats, our monoclonal antibody cocktail was used as the probe. Preliminary results (data not shown) demonstrated that both immunoassays effectively confirmed the colonies as coliforms within a few hours. Residual dyes from the M-Endo medium used in the MF test did not interfere with either the antigen–antibody reaction or the interpretation of results in both assay formats. Comparative studies between the antibody and the conventional BGLB confirmation procedure showed that our tests were equally effective but much simpler and faster in confirming the coliform colonies on the filters.

In this brief discussion, we have pointed out only a few aspects of coliform detection procedures that can benefit from monoclonal antibody technology. The advantages and versatility of these systems, however, are evident. With the production and formulation of a proper combination of anticoliform monoclonal antibodies, it should be possible to incorporate antibody-based confirmation to other tests. With increasing application, new methods of testing for coliforms may be developed that are based on monoclonal antibody technology.

IV. SUMMARY

In this chapter, we have examined several approaches to making monoclonal antibodies against coliforms. From our findings it is apparent that several antibodies will be required to develop a panel of coliform-specific monoclonal antibodies. The potential application and benefits of anticoliform antibodies in diagnostics was illustrated in the examples in which antibody-based confirmation tests could be used to improve and increase the sensitivity of current coliform detection methodologies.

REFERENCES

1. Köhler, G., and C. Milstein, Continuous cultures of fused cells secreting antibody of predefinde specificity. *Nature 256*:495–497 (1975).

2. Mattingly, J. A., An enzyme immunoassay for the detection of all *Salmonella* using a combination of myeloma protein and a hybridoma antibody. *J. Immunol. Methods 73*:147–156 (1984).
3. Mason-Smith, A., and C. Jones, Use of murine myeloma protein M467 for detecting *Salmonella* in milk. *Appl. Environ. Microbiol. 46*:826–831 (1983).
4. Carlin, N. I. A., and A. A. Lindberg, Monoclonal antibodies specific for O-antigen of *Shigella flexneri* and *Shigella sonnei*: Immunological characterization and clinical usefulness. In *Monoclonal Antibodies Against Bacteria*, A. J. L. Macario and de Macario, E. C. (eds.), Academic Press, Orlando, pp. 138–165 (1985).
5. Gustafsson, B., and T. Holme, Monoclonal antibodies against group- and type-specific lipopolysaccharide antigen of *Vibrio cholera* 0:1. *J. Clin. Microbiol. 18*:480–485 (1983).
6. Sethi, K. K., V. Drueke, and H. Brandis, Hybridoma derived monoclonal immunoglobulin M antibodies to *Legionella pneumophila* serogroup 1 with diagnostic potential. *J. Clin. Microbiol. 17*:953–957 (1983).
7. McKinney, R. M., L. Thacker, D. E. Wells, M. C. Wong, M. C. Jones, and W. F. Bibbs, Monoclonal antibodies to *Legionella pneumophila* serogroup 1: Possible applications in diagnostic tests and epidemiologic studies. *Zentralbl. Bakteriol. 1 Abt. Orig. A255*:91–95 (1983).
8. Thompson, N. E., M. J. Ketterhagen, and M. S. Bergdoll, Monoclonal antibodies to staphylococcal enterotoxins B and C: Cross reactivity and localization of epitopes on tryptic fragments. *Infect. Immun. 45*:281–285 (1984).
9. Oguma, K., T. Agui, B. Syuto, K. Kimura, H. Iida, and S. Kubo, Four different monoclonal antibodies against type C1 toxin of *Clostridium botulinum*. *Infect. Immun. 38*:14–20 (1982).
10. Guttman-Bass, N., Problems in detecting water-borne viruses by rapid methods. In *Rapid Methods and Automation in Microbiology and Immunology*, K. O. Habermehl (ed.), Springer-Verlag, Berlin, Heidelberg, New York, Tokyo, pp. 716–726 (1984).
11. American Public Health Association, *Standard Methods for the Examination of Water and Wastewater*. 16th ed. American Public Health Association, Washington (1985).
12. Gavini, F., H. LeClerc, and D. A. A. Mossel, *Enterobacteriaceae* of the "coliform group" in drinking water: Identification and worldwide distribution. *Syst. Appl. Microbiol. 6*:312–318 (1986).
13. Hoadley, A. W., and B. J. Dutka (eds.), *Bacterial Indicators/Health Hazards Associated with Water*. American Society for Testing and Materials, Philadelphia (1977).
14. Berg, G. (ed.), *Indicators of Viruses in Water and Food*. Ann Arbor Science Publishers, Ann Arbor (1978).
15. Mossel, D. A. A., Index and indicator organisms: A current assessment of their usefulness and significance. *Food Technol. Aust. 30*:212–219 (1978).
16. Mossel, D. A. A., Marker (index and indicator) organisms in food and drinking water. Semantics, ecology, taxonomy and enumeration. *Antonie Leeuwenhoek J. Microbiol. 48*:609–611 (1982).

17. Splittstoesser, D. F., R. B. Tompkin, G. W. Reinbold, J. R. Matches, and C. Abeyta, Indicator organisms: A current look at their usefulness. *Food Technol. 37*:105–117 (1983).
18. Hartman, P. A., J. P. Petzel, and C. W. Kaspar, New methods for indicator organisms. In *Foodborne Microorganisms and Their Toxins: Developing Methodology*, M. D. Pierson and Stern, N. J. (eds.), Marcel Dekker, New York, pp. 175–217 (1986).
19. American Public Health Association, *Compendium of Methods for the Microbiological Examination of Food*, 2nd ed. American Public Health Association, Washington (1984).
20. Geldreich, E. E., M. J. Allen, and R. H. Taylor, Interferences in potable water. In *Evaluation of the Microbiology Standards for Drinking Water*, C. W. Hendricks (ed.), Environmental Protection Agency, Washington, pp. 13–20 (1978).
21. Evans, T. M., C. E. Waarvick, R. J. Seidler, and M. W. LeChevallier, Failure of the most probable number techniques to detect coliforms in drinking water and raw water supplies. *Appl. Environ. Microbiol. 41*:130–138 (1981).
22. Edwards, P. R., and W. H. Ewing, *Identification of the* Enterobacteriaceae, 4th ed. Burgess Publishing Co., Minneapolis (1986).
23. Olson, B. H., Enhanced accuracy of coliform testing in seawater by a modification of the most-probable-number method. *Appl. Environ. Microbiol. 36*:438–444 (1978).
24. Evans, T. M., R. J. Seidler, and M. W. LeChevallier, Impact of verification media and resuscitation on accuracy of the membrane filter total coliform enumeration technique. *Appl. Environ. Microbiol. 41*:1144–1155 (1981).
25. Hartman, P. A., P. S. Hartman, and W. W. Lanz, Violet red bile-2 agar for stressed coliforms. *Appl. Microbiol. 29*:537–539 (1975).
26. Andrews, W. H., and M. W. Presnell, Rapid recovery of *Escherichia coli* from estuarine water. *Appl. Microbiol. 23*:521–523 (1972).
27. LeChevallier, M. W., S. C. Cameron, and G. A. McFeters, New medium for improved recovery of coliform bacteria from drinking water. *Appl. Environ. Microbiol. 45*:484–492 (1983).
28. LeChevallier, M. W., P. E. Jakanski, A. K. Camper, and G. A. McFeters, Evaluation of m-T7 as a fecal coliform medium. *Appl. Environ. Microbiol. 48*:371–375 (1984).
29. Wright, R. C., A new selective and differential agar medium for *Escherichia coli* and coliform organisms. *J. Appl. Bacteriol. 56*:381–388 (1984).
30. Damare, J. M., D. F. Campbell, and R. W. Johnston, Simplified direct plating method for enhanced recovery of *Escherichia coli* in food. *J. Food Sci. 50*:1736–1737 (1985).
31. Speck, M. L., B. Ray, and R. B. Read, Jr., Repair and enumeration of injured coliforms by a plating procedure. *Appl. Environ. Microbiol. 29*: 549–550 (1975).
32. Andrews, W. H., A. P. Duran, F. D. McClure, and D. E. Gentile, Use of two

rapid A-1 methods for the recovery of fecal coliforms and *Escherichia coli* from selected food types. *J. Food Sci. 44*:289–293 (1979).
33. Dexter, F., Modification of the standard most-probable-number procedure for fecal coliform bacteria in seawater and shellfish. *Appl. Environ. Microbiol. 42*:184–185 (1981).
34. Hastback, W. G., Short incubation of presumptive media for detection of fecal coliforms in shellfish. *Appl. Environ. Microbiol. 42*:1125–1127 (1981).
35. Orskov, I., F. Orskov, B. Jann, and K. Jann, Serology chemistry, and genetics of O and K antigens of *Escherichia coli. Bacteriol. Rev. 41*:667–710 (1977).
36. Mutharia, L. M., J. S. Lam, and R. E. Hancock, Use of monoclonal antibodies in the study of common antigens of gram-negative bacteria. In *Monoclonal Antibodies Against Bacteria*. A. J. L. Macario and de Macario, E. C. (eds.), Academic Press, Orlando, pp. 131–141 (1985).
37. Peters, H., M. Jurs, B. Jann, K. Jann, K. N. Timmis, and D. Bitter-Suermann, Monoclonal antibodies to enterobacterial common antigen and to *Escherichia coli* lipopolysaccharide outer core: Demonstration of an antigenic determinant shared by enterobacterial common antigen and *E. coli* K5 capsular polysaccharide. *Infect. Immun. 50*:459–466 (1985).
38. Makela, P. H., and H. Mayer, Enterobacterial common antigen. *Bacteriol. Rev. 40*:591–632 (1976).
39. Hase, S., and T. Rietschel, Isolation and analysis of the lipid A backbone. Lipid A structure of lipopolysaccharides from various bacterial groups. *Eur. J. Biochem. 63*:101–107 (1976).
40. Galanos, C., O. Luderitz, and O. Westphal, Preparation and properties of antisera against the lipid-A component of bacterial lipopolysaccharides. *Eur. J. Biochem. 24*:116–122 (1971).
41. Miner, K. M., C. L. Manyak, E. Williams, J. Jackson, M. Jewell, M. T. Gammon, C. Ehrenfreund, E. Hayes, L. T. Callahan III, H. Zweerink, and N. H. Sigal, Characterization of murine monoclonal antibodies to *Escherichia coli* J5. *Infect. Immun. 52*:56–62 (1986).
42. Hancock, R. E. W., A. A. Wieczorek, L. M. Mutharia, and K. Poole, Monoclonal antibodies against *Pseudomonas aeruginosa* outer membrane antigens: Isolation and characterization. *Infect. Immun. 37*:166–171 (1982).
43. Hofstra, H., and J. Dankert, Major outer membrane proteins: Common antigens in *Enterobacteriaceae* species. *J. Gen. Microbiol. 119*:123–131 (1980).
44. Gabay, J., S. Schenkman, C. Desaymard, and M. Schwartz, Monoclonal antibodies and the structure of bacterial membrane proteins. In *Monoclonal Antibodies Against Bacteria*, A. J. L. Macario and de Macario, E. C. (eds.), Academic Press, Orlando, pp. 249–273 (1985).
45. Carrasco, N., S. M. Tahara, L. Patel, T. Goldkorn, and H. R. Kaback, Preparation, characterization, and properties of monoclonal antibodies against the *lac* carrier protein from *Escherichia coli. Proc. Natl. Acad. Sci. USA 79*:6894–6898 (1982).

46. Van der Lay, P., H. Amexz, J. Tommassen, and B. Lugtenberg, Monoclonal antibodies directed against the cell surface exposed part of PhoE pore protein of the *Escherichia coli* K12 outer membrane. *Eur. J. Biochem. 147*: 401-407 (1985).
47. Gabay, J., and M. Schwartz, Monoclonal antibody as a probe for structure and function of an *Escherichia coli* outer membrane protein. *J. Biol. Chem. 257*:6627–6630 (1982).
48. Schenkman, S., E. Couture, and M. Schwartz, Monoclonal antibodies reveal LamB antigenic determinants on both faces of the *Escherichia coli* outer membrane. *J. Bacteriol. 155*:1382–1392 (1983).
49. Block, M. A., and C. Desaymard, Antigenic polymorphism of the LamB protein among members of the family *Enterobacteriaceae. J. Bacteriol. 163*:106–110 (1985).
50. Schwartz, M., and L. LeMinor, Occurrence of the bacteriophage lambda receptor in some *Enterobacteriaceae. J. Virol. 15*:679–685 (1975).
51. Smith, A. M., J. S. Miller, and D. S. Whitehead, M467: A murine IgA myeloma protein that binds a bacterial protein. I. Recognition of common antigenic determinants on *Salmonella* flagellins. *J. Immunol. 123*:1715–1720 (1979).
52. Sonderstrom, T., *Escherichia coli* capsules and pili: Serological, functional, protective, and immunoregulatory studies with monoclonal antibodies. In *Monoclonal Antibodies Against Bacteria*, A. J. L. Macario and de Macario, E. C. (eds.), Academic Press, Orlando, pp. 186–208 (1985).
53. Virji, M., J. E. Heckels, and P. J. Watt, Monoclonal antibodies to gonococcal pili: Studies on antigenic determinants on pili from variant of strain P9. *J. Gen. Microbiol. 129*:1965–1973 (1983).
54. Robison, B. J., C. I. Pretzman, and J. A. Mattingly, Enzyme immunoassay in which a myeloma protein is used for detection of salmonellae. *Appl. Environ. Microbiol. 45*:1816–1821 (1983).
55. Olsen, P. E., and W. A. Rice, Minimal antigenic characterization of eight *Rhizobium meliloti* strains by indirect enzyme-linked immunosorbent assay (ELISA). *Can. J. Microbiol. 30*:1093–1099 (1984).
56. Feng, P. C. S., and P. A. Hartman, Fluorogenic assays for immediate confirmation of *Escherichia coli. Appl. Environ. Microbiol. 43*:1320–1329 (1982).
57. Petzel, J. P., and P. A. Hartman, Monensin-based medium for determination of total gram-negative bacteria and *Escherichia coli. Appl. Environ. Microbiol. 49*:925–933 (1985).
58. Sugasawara, R. J., C. Prato, and J. E. Sippel, Monoclonal antibodies against *Neisseria meningitidis. Infect. Immun. 42*:863–868 (1983).
59. McDonough, M. W., and S. E. Smith, Molecular weight variation among bacterial flagellins. *Microbios 16*:49–53 (1976).
60. Mattingly, J. A., and W. D. Gehle, An improved enzyme immunoassay for the detection of *Salmonella. J. Food Sci. 49*:807–809 (1984).
61. McCarthy, L. R., Latex agglutination tests for the rapid diagnosis of infectious disease. In *Rapid Detection and Identification of Infectious*

Agents, D. T. Kingsbury and Falkow, S. (eds.). Academic Press, Orlando, pp. 165–175 (1985).
62. Minnich, S. A., P. A. Hartman, and R. C. Heimsch, Enzyme immunoassay for detection of salmonellae in foods. *Appl. Environ. Microbiol. 43*:877–883 (1982).
63. Geldreich, E. E., and D. J. Reasoner, Searching for rapid methods in environmental bacteriology. In *Rapid Methods and Automation in Microbiology and Immunology*, K. O. Habermehl (ed.), Springer-Verlag, Berlin, Heidelberg, New York, Tokyo, pp. 696–707 (1984).
64. Anderson, J. M., and A. C. Baird-Parker, A rapid and direct plate method for enumerating *Escherichia coli* biotype 1 in foods. *J. Appl. Bacteriol. 39*:111–117 (1975).
65. Burlingame, G. A., J. McElhaney, M. Bennett, and W. O. Pipes, Bacterial interference with colony sheen production on membrane filters. *Appl. Environ. Microbiol. 47*:56–60 (1984).
66. Schiff, L. J., S. M. Morrison, and J. V. Mayeux, Synergistic false-positive coliform reaction on M-Endo MF medium. *Appl. Environ. Microbiol. 20*:778–781 (1970).

26
Mycotoxin Detection by Immunoassay and Application of Hybridoma Technology

JAMES J. PESTKA and DEBORAH E. DIXON-HOLLAND*
Michigan State University, East Lansing, Michigan

I. INTRODUCTION

Mycotoxins are a chemically diverse group of secondary fungal metabolites that are toxic after ingestion or environmental exposure (1). Mycotoxins have a low molecular weight (300–400): most are nonpolar and have minimal solubility in water. Toxigenic fungi are ubiquitous and can grow and elaborate these toxins in agricultural commodities in the field or after harvest during storage and processing. Examples of some affected agricultural commodities are corn, wheat, peanuts, cotton seed, rice, sorghum, almonds, walnuts, and milk. Factors that affect toxin elaboration include temperature, relative humidity, and moisture content of the substrate. Because of their toxicity, mycotoxins not only cause severe economic losses to farmers and livestock producers but pose a threat to humans consuming contaminated foods.

A number of important questions arise concerning the nature of these mycotoxins: (a) Which ones appear in an agricultural area and to what degree? (b) What specific climatic factors affect this appearance? (c) What is the target tissue of a particular mycotoxin, and what is the biochemical basis for its toxic effect? (d) How much of the toxin finally enters the human food supply? (e) Does the toxin pose a significant threat to humans ingesting it and, if so, how can this threat be minimized? The answers to these questions require a multidisciplinary approach among microbiologists, biochemists, plant pathologists, and toxicologists. Production of specific antibodies to mycotoxins and application of these

**Present affiliation*: Neogen Corporation, Lansing, Michigan.

to immunoassays provide unique tools for this multidisciplinary group of scientists.

The benefits of immunochemical detection of mycotoxins are readily apparent. First, the assays can be employed as rapid and sensitive methods for the toxins in human food and animal feed. Early detection and diversion of contaminated feedstuffs, crops, and foods can prevent contamination of higher-quality materials and, thus, decrease the possibility of human and animal exposure. Second, by using mycotoxin analogues and different conjugation procedures, it is possible to vary the specificity of the resultant antisera. These antisera have potential use for in vitro and in vivo metabolism studies. Third, by using appropriate antibodies, it may be possible to localize bound mycotoxins or mycotoxin metabolites in plant and animal tissue and better elucidate their mode of toxin action. Finally, metabolite assays have general use in epidemiological studies relating mycotoxin occurrence to human and animal disease.

The purpose of this chapter will be to briefly review the biological import of some of the major mycotoxins, to discuss conventional and immunochemical approaches to the detection of mycotoxins, and to review recent advances in the development of antibodies and applications of these to mycotoxin detection.

II. OCCURRENCE AND SIGNIFICANCE OF MYCOTOXINS TO HUMAN AND ANIMAL HEALTH

A. The Problem

Target tissues for mycotoxins include the liver, kidney, spleen, gastrointestinal tract, lymphoid tissue, reproductive organs, skin, and nervous system. Table 1 describes a minimal list of major mycotoxins, the fungi that produce them, and some of their biological effects. The most important of these in the United States are the aflatoxins, zearalenone, the trichothecenes, and ochratoxin.

B. Aflatoxin

The aflatoxins (AF), a family of difuranocoumarins, are of the greatest concern to human health because of their ability to act as hepatotoxins and hepatocarcinogens (2). Of the naturally occurring aflatoxins produced by *Aspergillus flavus* and *A. parasiticus* (Fig. 1), aflatoxin B_1 (AFB_1) is the most prevalent and is also the most toxic. On the basis of the modified salmonella mutagenesis assay, AFB_1 is about 30 times more mutagenic than AFG_1 or AFM_1, which are in turn more mutagenic than AFB_2, AFG_2, or AFM_2 (3). Similar rank order is found for hepatotoxicity and carcinogenicity, indicating the difuran ring with 8:9 vinyl ether is critical for the biological effects of the toxin. The coumarin ring of the aflatoxin molecule absorbs at 360 nm and fluoresces, aiding in its identification and quantification by chromatographic methods.

Table 1 Summary of Major Mycotoxins

Toxin	Fungi	Symptom
Aflatoxins	*Aspergillus flavus* and *A. parasiticus*	Heptotoxic, hepatocarcinogenic, mutagenic, teratogenic
Zearalenones	*Fusarium roseum* and other *Fusarium* spp.	Estrogenic, infertility in cattle, swine, and poultry
Trichothecenes	*Fusarium* spp.	Gastrointestinal tract hemorrhaging, alimentary toxic aleukia, feed rejection, vomiting, and immunotoxicity
Ochratoxins	*Aspergillus* spp. and *Penicillium* spp.	Nephrotoxic, teratogenic, immunotoxic
Sterigmatocystin	*A. versicolor* and other *Aspergillus* spp.	Hepatotoxic, hepatocarcinogenic
Citrinin	*Penicillium* spp. and *Aspergillus* spp.	Nephrotoxic
Rubratoxin	*Penicillium* spp.	Hepatotoxic

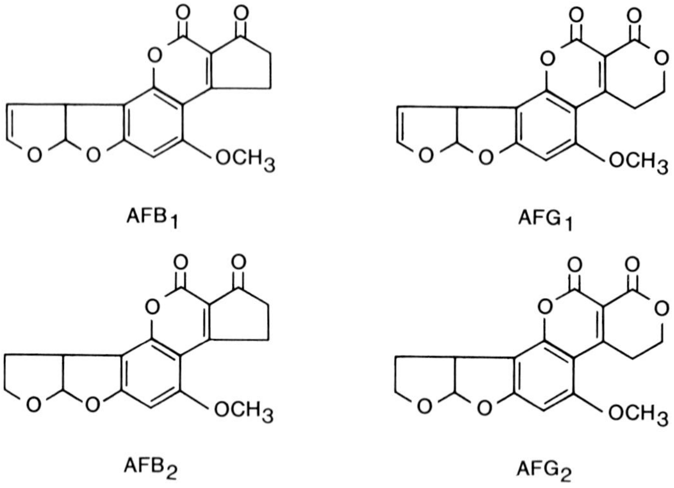

Figure 1 Naturally occurring aflatoxins (*Source*: from Ref. 1 with permission of publishers).

Aflatoxin B_1 is frequently found in corn, peanuts, cottonseed, and various tree nuts (4). Contamination with AFB_1 frequently occurs after drought, after overmaturation of a crop, and after physical or biological damage to a crop. Financial losses resulting after aflatoxin contamination can be devastating to agricultural producers and food processors.

More important are the biological effects of aflatoxin. Aflatoxin B_1 is acutely hepatotoxic with an LD_{50} ranging from 0.3 to 6 mg/kg for several animal species (1,4). At high levels, toxicity can occur in 3–6 hr and is readily evidenced by necrosis of the hepatocytes, followed by widespread hemorrhagic accumulation of blood in the gastrointestinal tract. Chronic levels of aflatoxin can cause marked decreases in growth rate and the immune response of domestic animals. The most notorious effect of aflatoxin is its potency as a liver carcinogen. Rats fed a diet with as little as 1 ppb AFB_1 can develop tumors. Epidemiological studies have demonstrated an association between current exposure to aflatoxins and the incidence in primary liver cancer in humans (2). High aflatoxin risk areas for humans have been designated in Africa, India, Southeast Asia, and China. The U.S. Food and Drug Administration (FDA) currently has a 20-ppb action level for AFB_1 in foods and feeds, and similar regulatory measures exist in most countries.

Aflatoxin B_1 can be metabolized by hepatic enzymes in vitro and in vito to a number of metabolites (1,2). The 8,9-epoxide is believed to be the ultimate carcinogen because it reacts with cellular nucleophiles such as DNA, RNA, and protein. Other pathways involve reduction or hydroxylation steps to more polar compounds and their subsequent removal from the animal. Of the hydroxylated metabolites, AFM_1 is crucial to the food supply because it is also hepatocarcinogenic and can be found in the milk of lactating animals that have ingested it (5). The FDA action level for AFM_1 in dairy products is 0.5 ppb.

C. Zearalenone

The zearalenones are a group of resorcylic lactones, produced by members of the genus *Fusarium*, that have estrogenic effects on a number of animal species (2,4). Zearalenone, the parent compound (Fig. 2), is often found in temperate

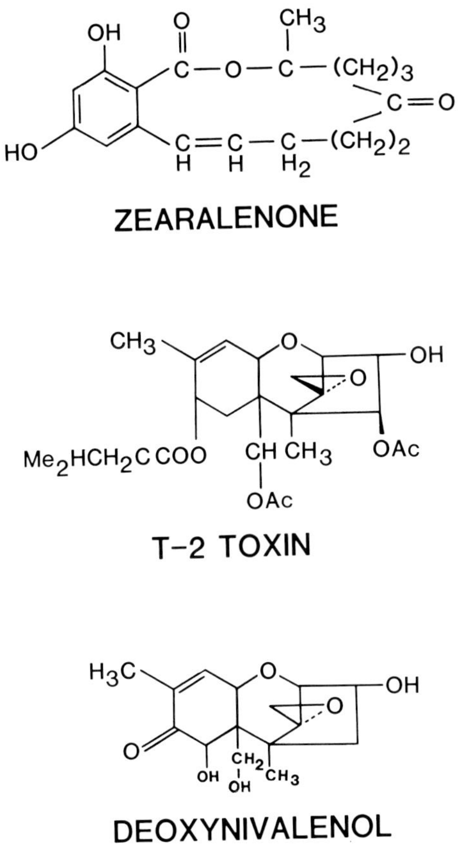

Figure 2 Zearalenone, T-2 toxin, and deoxynivalenol (vomitoxin) (*Source*: from Ref. 1 with permission).

regions in corn, wheat, sorghum, barley, and oats. Although not acutely toxic, high levels of zearalenone interrupt normal estrus in sows and cause vulval swelling and enlargement of the uterus. Testicular atrophy and mammary gland hyperplasia develop in young males exposed to zearalenone. Other species affected by zearalenone include rats, mice, sheep, monkeys, and humans.

D. Trichothecenes

Trichothecene mycotoxins are produced by certain *Fusarium* species and other fungi (6). Among the over 50 known trichothecenes, a group of sesquiterpenoids containing a trichothecane nucleus, T-2 toxin and deoxynivalenol (or vomitoxin) (see Fig. 2) are often identified in fusarium culture filtrates. When applied to the shaved skin of guinea pigs, these compounds produce a strong dermatitic reaction indicated by severe local irritation, inflammation, desquamation, and general necrosis (2). The biological mode of action for trichothecenes is believed to be by the inhibition of initiation or the elongation-termination steps of protein synthesis. Actively dividing tissues, such as bone marrow, lymph nodes, spleen, thymus, testes, ovary, and intestinal mucosa, appear most susceptible to these toxins, with acute LD_{50}s in the milligram per kilogram range. The potential danger of trichothecenes to human health was first realized in the 1940s during several massive outbreaks (1942–1947) of fatal alimentary toxic aleukia in the Soviet Union when overwintered grain containing the toxins was ingested (7).

Members of the genus *Fusarium* occur both as saprophytes in soil and decaying vegetation and as pathogens in cultivated plants. Toxigenic fusaria are often isolated from agricultural commodities such as corn, wheat, barley, and oats (4). Until only recently, actual detection of the trichothecene toxins has been limited by the lack of suitable analytical methods. Vesonder et al. (8) first isolated the trichothecene, deoxynivalenol, as an emetic principle and feed refusal factor in fusarium-infected corn. The toxin has since been found to occur frequently in the parts per million range in grains produced by the United States, Canada, Japan, China, South Africa, and Austria. Many times, zearalenone and deoxynivalenol occur simultaneously.

E. Ochratoxins

The ochratoxins are a group of seven related isocoumarin derivatives linked to phenylalanine that are produced by species of *Aspergillus* and *Penicillium* (2). Ochratoxin A (OA; Fig. 3), the most commonly encountered of the group is highly nephrotoxic to monogastric animals and has also been shown to be immunotoxic and teratogenic. Ochratoxin A is frequently detected in Scandanavian and Balkan countries and, occasionally, in the United States in commodi-

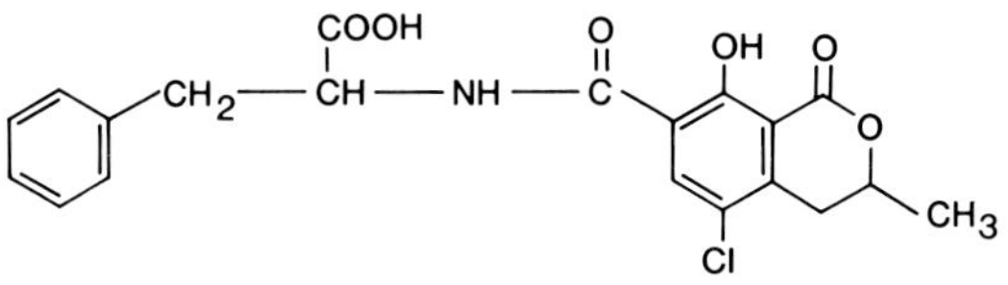

Figure 3 Ochratoxin (*Source*: from Ref. 1 with permission).

ties such as barley, corn, wheat, oats, rye, peanuts, hay, and green coffee beans (4). The toxin has been related to endemic kidney disease in swine and poultry in Denmark and Sweden. During the course of the swine disease, pigs fed diets containing 200 ppb OA develop pale swollen kidneys characterized by atrophy of the proximal tubules and interstitial cortical fibrosis (9). These symptoms are similar to those occurring in humans during the course of endemic Balkan kidney disease of Yugoslavia, Bulgaria, and Romania. The incidence of this human disease can be correlated, by region, to the OA content of foods in the Balkan countries (4).

III. MYCOTOXIN DETECTION

A. Conventional Methods

Numerous analytical methods have been developed, and continuously refined, for the aflatoxins, trichothecenes, zearalenones, and ochratoxins during recent years. The procedures generally utilize thin layer chromatography (TLC), high-performance liquid chromatography (HPLC), gas chromatography (GC), gas chromatography mass spectroscopy (GC-MS) and, in the past, less specific biological assays. Although sensitive methods exist for aflatoxins because of their absorbance and fluorescence characteristics, other mycotoxins do not have these convenient chemical "handles," making their detection at low concentrations difficult. Regardless of the mycotoxin analyzed, problems are always encountered with inherent fluoresence and absorbance by food substrates and biological samples, which substantially decrease attainable sensitivity. Removal of these interfering materials requires lengthy solvent:solvent extractions, column cleanup, and evaporation/concentration steps. These problems have led to the development of immunochemical methods for the routine and rapid screening of mycotoxins.

B. Immunoassay of Mycotoxins

Steps in developing an immunoassay for a mycotoxin include raising of mycotoxin-specific antibody by using mycotoxin carrier protein conjugates, incorporation of antibody into a suitable immunoassay, and demonstration of assay applicability to food and biological sample testing (10).

1. Polyclonal Antibody Production

Because mycotoxins have low molecular weights (300–400) they do not independently induce a hyperimmune response. Usually, they must be derivatized and conjugated to a carrier protein to be rendered immunogenic. Mycotoxin conjugation techniques have been extensively reviewed (10). For example, the approaches for derivatizing aflatoxin, many of which were developed by the laboratory of F. S. Chu at the University of Wisconsin, include conversion of (a) AFB_{2a} or AFB_1 dihydrodiol to its respective dialdehydic phenolate ion form, (b) the putative 8,9 AFB_1 epoxide, (c) carboxymethyloxime, or (d) hemisuccinate or hemiglutarate derivatives. In the first two methods, the molecule reacts directly with carrier protein, whereas in the latter two methods, carboxymethyloxime and hemisuccinate derivatives are conjugated to protein by the carbodiimide, mixed anhydride, or hydroxysuccinimide methods (10,11). The same conjugation techniques can be used to attach the mycotoxin to an enzyme marker.

The most frequently used method for immunizing rabbits against mycotoxin protein conjugates involves multiple-site injection of immunogen (100–1000 μg) in Freund's complete adjuvant, followed by boosting at 5- to 6-week intervals. Usable antisera are usually obtained 12–16 weeks after the initial immunization, but they can take as long as 1 year, as has occurred for T-2 toxin. Antiserum titers can be determined by radioimmunoassay (RIA) or, more conveniently, by enzyme-linked immunosorbent assay (ELISA).

2. Monoclonal Antibody Production

Hybridoma cell lines that secrete monoclonal antibodies against mycotoxins are raised after fusion of hyperimmune spleen cells with myeloma cells (12). BALB/c mice are commonly used as splenocyte donors because most of the myeloma cell lines are of BALB/c origin. Any one of a number of myeloma lines can be chosen (13–17). Myeloma line SP-2 has been used to produce hybridomas that recognize aflatoxins and aflatoxin–DNA adducts (18,19). P3-NS-1-Ag4-1 has been fused for production of hybridomas that secrete antibodies to aflatoxin M_1 (20,21), aflatoxin B_1 (22), T-2 toxin (23), zearalenone (24), and deoxynivalenol (25). Candlish et al. (26) used the cell line PX63Ag8.653 to generate hybridomas that produce monoclonal antibodies to AFB_1.

Immunization protocols for production of monoclonal antibodies to the various mycotoxins vary with respect to route and dosage. In general, repeated subcutaneous, or intraperitoneal injections of 10–500 μg of mycotoxin–carrier protein conjugate have been used to obtain sensitive and specific monoclonal antibodies. Carrier proteins that have been used for immunogen preparation include bovine gamma globulin (BGG; 18,19), keyhole limpet hemocyanin (KIH; 26), goat IgG (GIgG; 27), and bovine serum albumin (BSA; 20-25).

Fusion protocols used to produce mycotoxin monoclonal antibody-secreting hybridomas are similar to well-established methods that employ polyethylene glycol (PEG; 28-30). In general, these PEG-induced fusion methods require addition of warm (37°C) 30-50% PEG to a pellet of mixed spleen cells and myeloma cells, which are usually present in ratios from 10:1 to 1:1.

Enzyme-linked immunosorbent assays are typically used for screening culture supernatants to identify wells that contain cells secreting the desired antibody. A noncompetitive-indirect ELISA has been used to identify cultures containing antibody specific for AFB_1 (19,26) and AFM_1 (20). In our laboratory, a competitive indirect ELISA (CI-ELISA) was developed (21-25) to determine the sensitivity of zearalenone, AFB_1, AFM_1, T-2 toxin, and deoxynivalenol antibodies in mouse sera produced during the course of immunization; to identify culture wells that contain hybridomas secreting desired antibody after fusion and cloning; and to determine the sensitivity and specificity of the monoclonal antibody secreted by the various stabilized cell lines (21-25). In CI-ELISA, free toxin competes with solid-phase-bound toxin for antibody-binding sites; therefore, a decrease in color intensity in assay wells containing free toxin, compared with wells without free toxin identifies antibodies specific for the toxin. This latter method is more useful than the noncompetitive ELISA because the competitive method identifies specific antibody, whereas, in the former, nonspecific cross-reacting antibodies may yield false-positive reactions.

The cloning of cells from wells containing mycotoxin-specific antibody activity is accomplished by limiting dilution (31). Feeder layers of murine splenocytes (1×10^5 cells/ml) have been used (20,26) as well as spent macrophage medium (32) in fresh maintenance medium (21-25). Mycotoxin monoclonal antibodies are typically scaled up in ascites fluid by injecting the appropriate cell line, intraperitoneally (10^6-10^7 cells), into BALB/c mice that have previously been primed with pristane (29).

3. Immunoassay Approaches for Mycotoxin Quantitation

Radioimmunoassay. RIAs are more sensitive then some of the common immunological assays such as precipitation and hemagglutination. In competitive RIAs for mycotoxins, specific antibody is incubated with a constant amount of tritiated toxin in the presence of varying amounts of toxin standard or unknown sample. Ammonium sulfate precipitation or precipitation with a second antibody is used to remove the complex from solution. Toxin content is inversely related to the amount of label present in solution. This method has been applied to aflatoxin (33,34), T-2 toxin (35), zearalenone (36), and ochratoxin (37), but it is subject to several major disadvantages. An RIA requires labeling of the toxins with tritium, which is difficult and expensive. Emission by the isotope will gradually degrade the toxin molecule over time, and the toxin, thus, must be continuously repurified. Also there is the problem of radioactive waste disposal and expense of a scintillation counter. Finally, throughput is low, allowing only a small number of samples to be screened in a short period. Hence, our laboratory has concentrated on ELISA exclusively for mycotoxin quantitation.

Enzyme-Linked Immunosorbent Assay. The ELISA offers perhaps the simplest alternative to the screening of food samples for mycotoxins. The advantages of ELISA over RIA and conventional chemical methods include simplicity, use of stable reagents, high throughput, and absence of radiation hazard. Both competitive indirect and direct ELISAs have been used for mycotoxin detection. In the indirect ELISA (38), mycotoxin-specific antibody competes with free toxin for binding to a solid-phase mycotoxin-protein conjugate. A second antiglobulin enzyme conjugate is then required to determine total bound antibody. In the competitive direct (CD) ELISA (39), mycotoxin-enzyme conjugate (usually horseradish peroxidase) is simultaneously incubated with free toxin over solid-phase-bound mycotoxin antibody. Of the two approaches, we have found the direct assay preferable for analytical purposes because it requires one less incubation step and one less washing step than the indirect ELISA. Generally, polystyrene solid-phases, such as microtiter plates, tubes, and Terasaki plates, have been used for ELISA. Methods for covalent attachment to nylon beads also have been described (40).

C. Application of Immunoassay to the Analysis of Specific Mycotoxins in Foods and Biological Fluids

A large body of literature on the immunochemical analysis of mycotoxins has developed over the past decade. Whereas early studies employ polyclonal antibodies, more recent efforts use monoclonal antisera. The limit of sensitivity for toxin in solution generally ranges between 0.1 and 10 ng/ml. *In interpreting the ultimate sensitivity limits for mycotoxin analysis for a commodity, it is critical to determine whether a published method's sensitivity is reported on an assay volume or on a sample weight basis and whether the sample is diluted or concentrated during preparation for immunoassay*. Although earlier reports employ extraction and cleanup steps before immunoassay, our laboratory has recently demonstrated that simple extraction of a food with 50–70% methanol in water is the only step required before ELISA (24,42–44). Tables 2–6 summarize relevant specificity data on some polyclonal and monoclonal antibodies reported up to this time. Their general properties are described in the following, with specific references to their application to the analysis of aflatoxin, zearalenone, T-2 toxin, and ochratoxin A in food and biological samples.

Table 2 Characteristics of Selected AFB_1 Immunoassays

Type of antibody	Mono[a]	Mono[b]	Poly[b]	Poly[b]	Poly[c]	Poly[d]	Poly[e]	Mono[f]
Type of immuno-assay	CD-ELISA	CD-ELISA	CD-ELISA	RIA	RIA	CD-ELISA	CD-ELISA	RIA
Sensitivity	0.2 ng/ml	0.5 ng/ml	1.0 ng/ml	0.4 ng/ml	10 ng/ml	2 ng/ml	5 ng/ml	3 ng/ml
Crossreactivity with toxin and analogues (%)								
AFB_1	100	100	100	100	100	56	200	100
AFB_2	13	16	100	13	43			100
AFB_{2a}	<1	<2	<1		1	100	100	
AFG_1	14	40	33	10	15	<1	6	5
AFG_2	1	13	1.0	<1	5	<1	4	4
AFM_1	7	16	<1	<1		1	130	100
AFL[g]			80	2.5		20	2	
AFB_1 diol						80	100	
AFB_1 DNA						47	32	3
Ref.	(26)	(21)	(39)	(34)	(33)	(48)	(51)	(19)

[a] AFB_1–oxime–KLH immunogen.
[b] AFB_1–oxime–BSA immunogen.
[c] AFB_1–oxime–PLL immunogen.
[d] AFB_{2a}–BSA immunogen.
[e] AFB_1–diol–ethylenediamine BSA immunogen.
[f] AFB_1–BGG immunogen.
[g] Aflatoxicol.

Table 3 Characteristics of Selected AFM_1 Immunoassays

Type of[a] antibody	Mono	Mono	Poly	Poly
Type of immunoassay	CD-ELISA	CD-ELISA	CD-ELISA	RIA
Sensitivity (ng/ml)	1.0	1.0	0.1	10
Crossreactivity of toxin and analogues (%)				
AFM_1	100	100	100	100
AFB_1	10		1	57
AFB_2	<2.5		<0.1	<1
AFB_{2a}			<0.1	1
AFG_1	10		<0.1	<1
AFG_2	<2.5		<0.1	<1
Ref.	(20)	(21)	(57)	(58)

[a]All were produced against AFM_1 oxime-BSA immunogen.

Table 4 Characteristics of Selected Zearalenone Immunoassays

Type of[a] antibody	Mono	Poly[b]	Poly[c]	Poly[b]	Poly[c]
Type of immunoassay	CD-ELISA	CD-ELISA	CI-ELISA	CI-ELISA	RIA
Sensitivity (ng/ml)	1.0	0.5	10	5.0	5.0
Crossreactivity of toxin and analogues (%)					
Zearalenone	100	100	100	100	100
α-zearalenol	107	280	33	50	100
β-zearalenol	29	35	25	12	44
α-zearalanol	35	22	6	6	53
β-zearalanol	25	10	10	3	44
Ref.	(22)	(42)	(62)	(63)	(64)

[a]All antibodies produced against bovine serum albumin-carboxymethyloxime-zearalenone immunogen (BSA-CMO-ZEA).
[b]Rabbit antiserum.
[c]Pig antiserum.

Table 5 Characteristics of Selected T-2 Toxin Immunoassays

Type of antibody	Mono[a]	Mono[b]	Poly[b]	Poly[b]	Poly[c]
Type of immuno-assay	CI-ELISA	CI-ELISA	CD-ELISA	RIA	RIA
Sensitivity (ng/ml)	100	10	0.5	50	1.0
Crossreactivity of toxin and analogues (%)					
T-2	100	100	100	100	25
T-2HS		32	100		
Acetyl T-2		24.5	100	33	2.2
HT-2	34	2.3	3.4	10.3	2.2
'3OH T-2		128			100
Neosolaniol		1.4	0.1		
T-2 triol		0.1	0.1	0.1	0
T-2 tetraol		<0.1	<0.1	<0.1	0
Deoxynivalenol	0	<0.1	<0.01	0	0
Verrucarol		<0.1	<0.003		
Diacetoxyscirpenol	0		<0.001	0.1	0
Roridin A		<0.1	<0.001	0	
Ref.	(27)	(23)	(69)	(35)	(77)

[a]T-2-HS-GIg immunogen.
[b]T-2-HS-BSA (bovine serum albumin-hemisuccinate-T-2 immunogen).
[c]3'-OH-T-2-HS-BSA immunogen.

Table 6 Characteristics of Selected Ochratoxin A Immunoassays

Type of[a] antibody	Poly	Poly
Type of immunoassay	CD-ELISA	RIA
Crossreactivity of toxin and analogues (%)		
OA	100	100
OB	14	1
OC	44	38
Oα	1	0.01
Ref.	(79)	(37)

[a]Both prepared against OA-BSA immunogen.

1. Aflatoxin B_1

Because of existing regulations in the United States and the rest of the world, major emphasis has been placed on the development of immunoassays for aflatoxins. Varying patterns of cross-reactivity have been observed with antibodies produced against different AFB_1-protein conjugates.

When AFB_1-oxime BSA (22,33,34,39,40) is used as the immunogen, resultant polyclonal and monoclonal antibodies recognize AFB_1 most readily (see Table 2). AFB_2 and AFG_1 are recognized to a smaller degree, whereas, there is negligible cross-reaction observed with the other metabolites. These results suggest that the dihydrofuran moiety of the aflatoxin molecule is most critical in determining the specificity of the antibody raised against the AFB_1-oxime BSA conjugate. Modifications in the dihydrofuran, such as the addition of a hydroxyl in AFM_1, decrease the binding significantly. The lack of a double bond in AFB_2 and AFG_2 also decreases affinity of the antibody for these metabolites. The cyclopentenone ring plays a somewhat minor role in determining specificity because the modification of this structure (e.g., B to G series) does not completelý inhibit binding of the antibody. The overall pattern of cross-reactivity for the RIA developed using antibody obtained from rabbits immunized with AFB_1-oxime-poly-L-lysine (AFB_1-CMO-PLL) (33) is not markedly different, although the antibody appears to cross-react more readily with AFB_2 than the antibodies in the ELISAs. Antibodies prepared against AFB_1-oxime conjugates have been used to detect AFB_1 in corn (38,43,44), wheat (45), cottonseed (43), and peanut butter (38,44,45).

Polyclonal antibodies raised against the AFB_{2a} (46–50) and AFB_1 dihydrodiol (51) immunogens are similar in that the cyclopentenone ring and the methoxy group of the aflatoxin molecule acts as the primary epitope in determining antibody specificity. These antibodies are most useful as diagnostic tools for detection of aflatoxin exposure by ELISA or by immunohistochemical localization in humans and animals (46,49,50,52). The only difference between the two immunogens was the presence of a hydroxyl group at position 9 in AFB_1-dihydrodiol. Although it is likely that this hydroxyl group was located very close to the protein carrier and was not immunodominant, some differences in the relative specificities of the two antibodies are apparent. Table 2 compares the specificity data obtained for AFB_1-diol antibody with those for AFB_{2a} antibody. The AFB_1-diol antiserum has a broader range of specificity for AFB_1 analogues, especially aflatoxicol and AFM_1, than does AFB_{2a} antiserum. Whereas AFB_1 and AFM_1 are more effective than AFB_1-diol and AFB_{2a} in competing in the AFB_1-diol ELISA, lower reactivities for AFB_1 and AFM_1 compared with those for AFB_{2a} and AFB_1-diol are found in the AFB_{2a} ELISA (49,51). In general, the AFB_1-diol antiserum requires higher concentrations of AFB_1-diol, AFB_{2a}, AFB_1-modified DNA, and AFB_1-N^7-guanine than does AFB_{2a} antiserum to inhibit binding of their respective peroxidase conjugates. This suggests that AFB_{2a} antibody has higher affinity for these key metabolites than does AFB_1-diol antibody.

A slightly different pattern of cross-reactivity exists for the monoclonal antibody prepared after immunization of mice with AFB_1 covalently bound (by reaction of its 8,9-epoxide) to bovine gamma globulin (19). Because the toxin was conjugated to the carrier protein on the dihydrofuran side, leaving the coumarin and cyclopentenone rings exposed for antibody recognition, the latter moiety plays the key role in determining the specificity. Similarly, monoclonal antibodies have been made to AFB_1-DNA adducts (18,53). As with the AFB_{2a} and AFB_1-diol antibodies, the monoclonal antibodies to covalently bound AFB_1 are potentially useful for epidemiological investigations (18,19,52-56).

2. Aflatoxin M_1

Monoclonal and polyclonal antibodies have been also developed for specific detection of aflatoxin M_1 (AFM_1). Rabbits and mice were all immunized with the same immunogen (BSA-CMO-AFM_1). Available monoclonal antibodies (20,21) are not as specific or sensitive for AFM_1 as the polyclonal antibody previously employed in the competitive direct ELISA and RIA (57). The AFM_1 antibodies are generally applicable to the analysis of milk directly (21,41,57) or after preliminary extraction and concentration (57-60). Extremely low levels of AFM_1 are detectable in urine after extraction and concentration (60) or after immunoconcentration and separation by HPLC (61).

3. Zearalenone

Both monoclonal and polyclonal antibodies produced against zearalenone oxime-BSA (24,42,62-64) have been successfully used for detection of zearalenone and its analogues in corn (24,42,62), wheat (63), animal feed (3), milk (63), and biological fluids (64). The monoclonal antibody developed in our laboratory (24) reacts as well with α-zearalenol as zearalenone, but it reacts to a lesser extent with β-zearalenol and the zearalanols, suggesting that the C′6 position and double bond at the C′1-C′2 position are dominant in determining the specificity of the monoclonal antibody. Pig antiserum employed in the RIA (64) shows a spectrum of cross-reactivity nearly identical with the monoclonal antibody. In contrast, rabbit and pig antisera in the CI-ELISA show a similar ability to discriminate the double bond at the C′1-C′2 position but lesser cross-reactivity with the α-configuration (62,63). The rabbit antibody developed by Warner et al. (42) has almost three times more cross-reactivity with α-zearalenol than with zearalenone in the direct ELISA. Because α- and β-zearalenol are major animal metabolites of zearalenone and they can occur naturally (65-67), recognition of the α-zearalenol is a useful feature. However, the fact that the polyclonal antibody reacts almost three times as well with the analogue, could bias results for total determination of zearalenone in a sample (42). In contrast, the monoclonal antibody binds both zearalenone and α-zearalenol equally well, so this problem is alleviated (24).

4. T-2 Toxin

Varying degrees of sensitivity have been obtained for T-2 immunoassays by monoclonal and polyclonal antibody systems (see Table 5). In general, polyclonal antibodies (35,68,72) now offer greater sensitivity than monoclonal antibodies developed by our laboratory (23) and in others (27). All antibodies prepared against T-2 hemisuccinate conjugates reveal a fairly similar pattern of cross-reactivity, with one major exception (23). In general, although these bound most readily to T-2 toxin, they also recognize HT-2, a primary metabolite. There was little or no binding to a number of other analogues (neosolaniol, T-2 triol, T-2 tetraol, diacetoxyscirpenol, roridin A, verrucarol, deoxynivalenol). This pattern could be expected because the derivatized toxin was conjugated at the C-3 position, thus making the isovaleryoxy function the immunodominant structure. Antibodies prepared against T-2 hemisuccinate conjugates have been used to detect T-2 toxin in corn (68–70), wheat (68,70), biological fluids (71,72), and tissue (73). In contrast to these reports, Gendloff et al. (23) described a monoclonal antibody that has greater specificity for 3′-OH T-2 than T-2. Cross-reactivity with this and other major metabolites could be especially useful for assaying T-2 in biological systems. These metabolites are considered diagnostic and are present in significant amounts when T-2 toxicosis occurs (74–76). Recently, Wei et al. (77) have reported production of polyclonal antibodies directly to a 3′OH T-2 hemisuccinate BSA conjugate (see Table 5).

5. *Ochratoxin A*

The presence of a carboxylic group on ochratoxin A allows its facile conjugation to carrier proteins for immunogen preparation (37,78–81). Patterns of cross-reactivity and the limits of detection for ochratoxin A immunoassays using these antibodies are similar (Table 6). Levels as low as 1 ng/ml of the nephrotoxin could be detected by either method (RIA and ELISA). The antibody bound most readily to ochratoxin A, followed by ochratoxin C (OC), ochratoxin B (OB), and ochratoxin α. Ochratoxin B differs from OA only in that it lacks a chlorine atom in the dehydroisocoumarin ring. Ochratoxin C most closely resembles OA, whereas the inability of Oα to bind in the assays suggests the amide bond and chlorinated dehydroisocoumarin residues are features that are critical for antibody recognition. Ochratoxin A antibodies have been applied to the analysis of wheat (80) and barley (81).

IV. CONCLUSION

In a recent evaluation of immunoassay procedures, Vogt (82) indicated that ligand binding assays were superior to other analytical procedures for determining the presence of single chemical components, whereas chromatographic methods offered greater advantages in the determination of chemical profiles. The average cost of a mechanized enzyme immunoassay for a single analyte was approximately 5.9% and 2.5% of that required for doing the same analysis by gas chromatography or gas chromatography-mass spectrometry, respectively.

Critical evaluation of the reports discussed herein will undoubtedly lead to the conclusion that the use of ELISA for routine detection of mycotoxins is both desirable and feasible. For smaller research laboratories it is perhaps most convenient to produce polyclonal antisera according to their specific needs. The results presented here demonstrate that it is possible to produce monoclonal antibodies of equivalent quality to polyclonal antibodies after labor-intensive effort. Where production of commercial mycotoxin ELISA kits is the goal, it is most advantageous to invest in hybridoma technology for the production of high-quality, reproducible, reagent antibodies. In 1986, the Neogen company marketed the first ready-to-use monoclonal antibody ELISA kit for AFB_1 (Agri-Screen). Hybridoma-based kits for zearalenone and vomitoxin have followed. Several other companies are currently in the process of developing or marketing similar assay kits based on ELISA. *It is essential that all manufacturers of mycotoxin ELISA kits carefully evaluate the suitability of their methods in terms of extraction efficiency, recovery, commodity to be tested, reproducibility, precision, accuracy, and specificity, and provide these data to the purchaser.* Variability will undoubtedly depend on the type of solid phase, standards, on the shelf-life of the kit, and on whether a monoclonal or polyclonal antibody is used. Ultimately, these methods must be shown to be at least comparable with previously approved Association of Official Analytical Chemists methods to be of value to producers, processors, retailers, and regulators. *As a final note of caution it should be emphasized that the success of any mycotoxin screening procedure will be dependent on obtaining adequate representative samples of the commodity to be tested.*

ACKNOWLEDGMENTS

Michigan State Agricultural Experiment Station No. 000. This work was supported by Neogen Biologics, Lansing, Michigan and USDA Grant No. 83-CRSR-2-2257. We acknowledge the significant contributions of W. Casale, R. L. Warner, L. P. Hart, B. Ram, and E. Gendloff in the work described here and also thank J. Hunt and A. Pearson for assistance with this manuscript.

REFERENCES

1. Pestka, J. J., Fungi and mycotoxins in meats. In *Advances in Meat Research*, Vol 2, A. M. Pearson and Dutson, T. R. (eds.). AVI Publishing Co., Westport, Conn., pp. 277–302 (1986).
2. Busby, W. F., Jr., and G. F. Wogan, Mycotoxins and mycotoxicoses. In *Food-Borne Infections and Intoxications*, 2nd ed, H. Riemann and Bryan, F. L. (eds.). Academic Press, New York, pp. 519–610 (1979).
3. Wong, J. J., and D. D. Hsieh, Mutagenicity of aflatoxins related to their metabolism and carcinogenic potential. *Proc. Natl. Acad. Sci. USA 73*: 2241–2244 (1976).
4. CAST, Aflatoxin and other mycotoxins: An agricultural perspective. Report No. 80 Council for Agricultural Science and Technology. Ames, Iowa (1979).
5. Stoloff, L., Aflatoxin M_1 in perspective. *J. Food Prot. 43*:226–230 (1980).
6. Snyder, A. P., Qualitative, quantitative, and technological aspects of the trichothecene mycotoxins. *J. Food Prot. 49*:544–569 (1986).

7. Joffe, A. Z., *Fusarium poae* and *F. sporotrichoides* as principal causal agents of alimentary toxic aleukia. In *Mycotoxic Fungi, Mycotoxins and Mycotoxicoses, An Encyclopedic Handbook*, Vol. 3, T. D. Wyllie and Morehouse, L. G. (eds.). Marcel Dekker, New York, Chap. 3.10 (1978).
8. Vesonder, R. F., A. Ciegler, and A. Jensen, Isolation of the emetic principle from *Fusarium* infected corn. *Appl. Microbiol. 26*:1008–1010 (1973).
9. Krough, P. Ochratoxins. In *Mycotoxins in Human and Animal Health*, J. V. Rodricks, Hesseltine, C. W., and Mehlman, M. A. (eds.). Pathotox Publishers, Park Forest South, Ill., pp. 489–498 (1977).
10. Chu, F. S., Recent studies on immunochemical analysis of mycotoxins. In *Mycotoxins and Phycotoxins*, P. S. Steyn and Vieggaar, R. (eds.). Elsevier Publishers, Amsterdam, pp. 277–292 (1986).
11. Chu, F. S., Immunoassays for the analysis of mycotoxins, *J. Food Prot. 47*:562–569 (1984).
12. Köhler, G., and C. Milstein, Continuous cultures of fused cells secreting antibody of predefined specificity. *Nature 256*:495–497 (1975).
13. Goding, J. W., Antibody production by hybridomas. *J. Immunol. Methods 39*:285–308 (1980).
14. Fazekas de St. Groth, S., and D. Scheidegger, Production of monoclonal antibodies: Strategy and tactics. *J. Immunol. Methods 35*:1–21 (1982).
15. Kearney, J. F., A. Radbruch, B. Liesegang, and K. Rajewsky, A new mouse myeloma line which has lost immunoglobulin expression but permits the construction of antibody-secreting hybrid cell lines. *J. Immunol. 123*: 1548–1550 (1979).
16. Köhler, G., S. C. Howe, and C. Milstein, Fusion between immunoglobulin secreting and non-secreting myeloma cell lines. *Eur. J. Immunol. 6*:292 (1976).
17. Shulman, M., C. D. Wilde, and G. Köhler, A better cell line for making hybridomas secreting specific antibodies. *Nature 276*:269–170 (1978).
18. Groopman, J. D., A. Haugen, G. R. Goodrich, G. N. Wogan, and C. C. Harris, Quantitation of aflatoxin B_1-modified DNA using monoclonal antibodies. *Cancer Res. 42*:3120–3124 (1982).
19. Groopman, J. D., L. J. Trudel, P. R. Donahue, A. Marshak-Rothstein, and G. N. Wogan, High-affinity monoclonal antibodies for aflatoxins and their application to solid-phase immunoassays. *Proc. Natl. Acad. Sci. USA 81*: 7728–7731 (1984).
20. Woychik, N. A., R. D. Hinsdill, and F. S. Chu, Production and characterization of monoclonal antibodies against aflatoxin M_1. *Appl. Environ. Microbiol. 48*:1096–1099 (1984).
21. Dixon-Holland, D. E., J. J. Pestka, B. A. Bidigare, W. L. Casale, R. L. Warner, B. P. Ram, and L. P. Hart, Production of sensitive monoclonal antibodies to aflatoxin B1 and aflatoxin M1 and their application to ELISA of naturally contaminated foods. *J. Food Prot. 51*:201–204 (1988).
22. Dixon, D. E., L. P. Hart, and J. J. Pestka, (unpublished observations).
23. Gendloff, E. H., J. J. Pestka, D. E. Dixon, and L. P. Hart, Production of a monoclonal antibody to T-2 toxin with strong cross-reactivity to T-2 metabolites. *Phytopathology 77*:57–59 (1987).
24. Dixon, D. E., R. L. Warner, L. P. Hart, and J. J. Pestka, Hybridoma cell line

produces a specific monoclonal antibody to the mycotoxins zearalenone and alpha-zearalenol. *J. Agric. Food Chem. 35*:122–126 (1988).
25. Casale, W. L., J. J. Pestka, and L. P. Hart, Enzyme-linked immunosorbent assay employing monoclonal antibody specific for deoxynivalenol (bomitoxin). *J. Agric. Food Chem. 36*:663–668 (1988).
26. Candlish, A. A. G., W. H. Stimson, and J. E. Smith, A monoclonal antibody to aflatoxin B_1: Detection of the mycotoxin by immunoassay. *Lett. Appl. Microbiol. 1*:57–59 (1985).
27. Hunter, K. W., A. A. Brimfield, M. Miller, F. D. Finkelman, and F. S. Chu, Preparation and characterization of monoclonal antibodies to the trichothecenes mycotoxin T-2. *Appl. Environ. Microbiol. 49*:168–172 (1985).
28. Gefter, M. L., D. H. Margulies, and M. D. Scharff, A simple method for polyethylene glycol-promoted hybridization of mouse myeloma cells. *Somat. Cell Genet. 3*:231–236 (1977).
29. Galfre, G., and C. Milstein, Preparation of monoclonal antibodies: Strategies and procedures. In *Methods in Enzymology*, Vol. 73, J. J. Langone and VanVunakis, H. (eds.). Academic Press, New York, pp. 1–40 (1981).
30. Siraganian, R. P., P. C. Fox, and E. H. Berenstein, Methods of enhancing the frequency of antigen-specific hybridomas. In *Methods in Enzymology*, Vol. 92, Langone, J. J. and VanVunakis, H. (eds.). Academic Press, New York, pp. 17–22 (1983).
31. Oi, V. T., and L. A. Herzenberg, Immunoglobulin-producing hybrid cell lines. In *Selected Methods in Cellular Immunology*, B. B. Mishell and Shugi, S. M. (eds.). Freeman, San Francisco (1980).
32. Sugasawara, R. J., B. E. Cohoon, and A. E. Karu, The influence of murine macrophage-conditioned medium in cloning efficiency, antibody synthesis, and growth rate of hybridomas. *J. Immunol. Methods 79*:263–275 (1985).
33. Langone, J. J., and H. VanVunakis, Aflatoxin B_1-specific antibodies and their use in radioimmunoassay. *J. Natl. Cancer Inst. 56*:591–595 (1976).
34. Chu, F. S., and I. Ueno, Production of antibody against aflatoxin B_1. *Appl. Environ. Microbiol. 33*:1125–1128 (1977).
35. Chu, F. S., S. Grossman, R. D. Wei, and C. J. Mirocha, Production of antibody against T-2 toxin. *Appl. Environ. Microbiol. 37*:104–108 (1979).
36. Thouvenot, D., and R. F. Morfin, A radioimmunoassay for zearalenone and zearalanol in human serum: Production, properties and use of porcine antibodies. *Appl. Environ. Microbiol. 45*:16–23 (1982).
37. Chu, F. S., F. C. C. Chang, and R. D. Hinsdill, Production of antibody against ochratoxin A. *Appl. Environ. Microbiol. 31*:831–835 (1976).
38. Fan, T. S. L., and F. S. Chu, An indirect enzyme-linked immunosorbent assay for the detection of aflatoxin B_1 in corn and peanut butter. *J. Food Prot. 47*:964–968 (1984).
39. Pestka, J. J., P. K. Gaur, and F. S. Chu, Quantitation of aflatoxin B_1 antibody by enzyme-linked immunosorbent microassay. *Appl. Environ. Microbiol. 40*:1027–1031 (1980).
40. Biermann, V. A., and G. Terplan, Nachweis von Aflatoxin B_1 mittels ELISA. *Arch. Lebensmittelhyg. 31*:51–57 (1980).
41. Pestka, J. J., and F. S. Chu, Enzyme-linked immunosorbent assay of myco-

toxins using nylon beads and Terasaki plate solid phases. *J. Food Prot.* *47*:305–308 (1984).

42. Warner, R., B. P. Ram, L. P. Hart, and J. J. Pestka, Screening for zearalenone in corn by competitive direct enzyme-linked immunosorbent assay. *J. Agric. Food Chem.* *34*:714–717 (1986).
43. Ram, B. P., L. P. Hart, O. L. Shotwell, and J. J. Pestka, Analysis of aflatoxin B_1 in naturally contaminated corn and cottonseed by enzyme-linked immunosorbent assay: Comparison with thin layer chromatography and high performance liquid chromatography. *J. Assoc. Off. Anal. Chem.* (in press) *69*:904–907 (1986).
44. Ram, B. P., L. P. Hart, R. J. Cole, and J. J. Pestka, Peanut butter extraction procedure for the enzyme-linked immunosorbent assay of aflatoxin B_1 and its application to a retail survey. *J. Food Prot.* *49*:792–795 (1986).
45. El-Nakib, O., J. J. Pestka, and F. S. Chu, Analysis of aflatoxin B_1 in corn, wheat, and peanut butter by an enzyme-linked immunosorbent microassay and a solid-phase radioimmunoassay. *J. Assoc. Off. Anal. Chem.* *64*: 1077–1082 (1981).
46. Lawellin, D. W., D. W. Grant, and B. K. Joyce, Aflatoxin localization by enzyme-linked immunocytochemical technique. *Appl. Environ. Microbiol.* *34*:88–93 (1977).
47. Lawellin, D. W., D. W. Grant, and B. K. Joyce, Enzyme-linked immunosorbent analysis of aflatoxin B_1. *Appl. Environ. Microbiol.* *34*:94–96 (1977).
48. Gaur, P. K., H. P. Lau, J. J. Pestka, and F. S. Chu, Production and characterization of aflatoxin B_{2a} antibody. *Appl. Environ. Microbiol.* *41*:478–482 (1981).
49. Pestka, J. J., Y. K. Li, and F. S. Chu, Reactivity of aflatoxin B_{2a} antibody with aflatoxin B_1 modified DNA and related metabolites. *Appl. Environ. Microbiol.* *44*:1159–1165 (1982).
50. Pestka, J. J., J. T. Beery, and F. S. Chu, Indirect immunoperoxidase localization of aflatoxin B_1 in rat liver. *Food Chem. Toxicol.* *21*:41–48 (1983).
51. Pestka, J. J., and F. S. Chu, Aflatoxin B_1 dihydrodiol antibody: Production and specificity. *Appl. Environ. Microbiol.* *47*:472–477 (1984).
52. Garner, C., R. Ryder, and R. Montesano, Monitoring of aflatoxin in human body fluids and application to field studies. *Cancer Res.* *45*:922–928 (1985).
53. Haugen, A., J. D. Groopman, I. C. Hsu, G. R. Goodrich, G. N. Wogan, and C. C. Harris, Monoclonal antibody to aflatoxin B_1 modified DNA detected by enzyme immunoassay. *Proc. Natl. Acad. Sci. USA* *78*:4124–4127 (1981).
54. Hertzog, P. J., J. R. Linday Smith, and R. C. Garner, Production of monoclonal antibodies to guanine imidazole-ring opened aflatoxin B_1 DNA, the persistent DNA adduct in vivo. *Carcinogenesis* *3*:825–828 (1982).
55. Martin, C. N., R. C. Garner, P. Tursi, J. V. Garner, H. C. Whittle, R. W. Ryder, P. Sizaret, and R. Montesano, An ELISA procedure for assaying aflatoxin B_1. In *Monitoring Human Exposure to Carcinogenic and Muta-*

56. Tsuboi, S., T. Nakagawa, M. Tomita, T. Seo, H. Ono, K. Kawamura, and N. Iwamura, Detection of aflatoxin B_1 in serum samples of male Japanese subjects by radioimmunoassay and high performance liquid chromatography. *Cancer Res. 44*:1231–1234 (1984).
57. Pestka, J. J., Y. K. Li, W. O. Harder, and F. S. Chu, Comparison of a radioimmunoassay and an enzyme-linked immunosorbent assay for the analysis of aflatoxin M_1 in milk. *J. Assoc. Off. Anal. Chem. 64*:294–301 (1981).
58. Harder, W. O., and F. S. Chu, Production and characterization of antibody against aflatoxin M_1. *Experientia 35*:1104–1105 (1979).
59. Fremy, J. M., and F. S. Chu, A direct ELISA for determining aflatoxin M_1 at ppt levels in various dairy products. *J. Assoc. Off. Anal. Chem. 67*: 1098–1101 (1984).
60. Hu, W. J., N. Woychik, and F. S. Chu, ELISA picogram quantities of aflatoxin M_1 in urine and milk. *J. Food Prot. 47*:126–127 (1984).
61. Wu, S., G. Yang, and T. Sun, Studies on the immunoconcentration and immunoassay of aflatoxins. *Chin. J. Oncol. 5*:81–84 (1983).
62. Pestka, J. J., M.-T. Liu, B. Knudson, and M. Hogberg, Immunization of swine for production of antibody against zearalenone. *J. Food Prot. 48*: 953–957 (1985).
63. Liu, M.-T., B. P. Ram, L. P. Hart, and J. J. Pestka, Indirect enzyme-linked immunosorbent assay for the mycotoxin zearalenone. *Appl. Env. Microbiol. 50*:332–336 (1985).
64. Thouvenot, D., and R. F. Morfin, A radioimmunoassay for zearalenone and zearalanol in human serum: Production, properties and use of porcine antibodies. *Appl. Environ. Microbiol. 45*:16–23 (1983).
65. Mirocha, C. J., S. V. Pathre, and C. M. Christensen, In *Mycotoxins in Human and Animal Health*, J. V. Rodricks, Hesseltine, C. W., and Mehlman, M. A. (eds.). Pathotox Publishers, Park Forest, Ill., pp. 345–364 (1977).
66. Hagler, W. M., C. J. Mirocha, S. V. Pathre, and J. C. Behrens, Identification of the naturally occurring isomer of zearalenol produced by *Fusarium roseum* "Gibbosum" in rice culture. *Appl. Environ. Microbiol. 37*:849–853 (1979).
67. Richardson, K. E., W. M. Hagler, and C. J. Mirocha, Production of zearalenone, alpha and beta-zearalenol and alpha and betz-zearalanol by *Fusarium* sp. in rice culture. *J. Agric. Food Chem. 33*:862–866 (1985).
68. Pestka, J. J., S. S. Lee, H. P. Lau, and F. S. Chu, Enzyme-linked immunosorbent assay for T-2 toxin. *J. Am. Oil Chem. Soc. 58*:940a–944a (1981).
69. Gendloff, E. H., J. J. Pestka, S. P. Swanson, and L. P. Hart, Detection of T-2 toxin in *Fusarium sporotrichioides*-infected corn by enzyme-linked immunosorbent assay. *Appl. Environ. Microbiol. 47*:1161–1163 (1984).
70. Lee, S., and F. S. Chu, Radioimmunoassay of T-2 toxin in corn and wheat. *J. Assoc. Off. Anal. Chem. 64*:156–161 (1981).
71. Lee, S., and F. S. Chu, Radioimmunoassay of T-2 toxin in biological fluids. *J. Assoc. Off. Anal. Chem. 64*:684–688 (1981).
72. Fan, T. S. L., G. S. Zhang, and F. S. Chu, An indirect enzyme-linked immunosorbent assay for T-2 toxin in biological fluids. *J. Food Prot. 47*: 964–967 (1984).

73. Lee, S. C., J. T. Beery, and F. S. Chu, Immunoperoxidase localization of T-2 toxin. *Toxicol. Appl. Pharmacol. 72*:228–235 (1984).
74. Yoshizawa, T., C. J. Mirocha, J. C. Behrens, and S. P. Swanson, Metabolic fate of T-2 toxin in a lactating cow. *Food Cosmet. Toxicol. 19*:31–39 (1981).
75. Yoshizawa, T., T. Sakamoto, Y. Ayano, and C. J. Mirocha, 3′-Hydroxy T-2 and 3′hydroxy HT-2 toxins: New metabolites of T-2 toxin, a trichothecene mycotoxin, in animals. *Agric. Biol. Chem. 46*:2613–2615 (1982).
76. Yoshizawa, T., T. Sakamoto, and K. Okamoto, In vitro formation of 3′-hydroxy T-2 and 3′-hydroxy HT-2 toxins from T-2 toxin by liver homeogenates from mice and monkeys. *Appl. Environ. Microbiol. 47*:130–134 (1984).
77. Wei, R. D., W. Bischoff, and F. S. Chu, Production and characterization of antibody against 3′-OH-T-2 toxin. *J. Food Prot. 49*:267–271 (1986).
78. Aalund, O., K. Brundeldt, B. Hald, P. Krogh, and K. Poulsen, A radioimmunoassay for ochratoxin A: A preliminary investigation. *Acta Pathol. Microbiol. Scand. (Sect. C) 83*:390–392 (1975).
79. Pestka, J. J., B. W. Steinart, and F. S. Chu, Enzyme-linked immunosorbent assay for the detection of ochratoxin A. *Appl. Environ. Microbiol. 41*: 1472–1474 (1981).
80. Lee, S. S., and F. S. Chu, Enzyme-linked immunosorbent assay of ochratoxin A in wheat. *J. Assoc. Off. Anal. Chem. 67*:45–49 (1984).
81. Morgan, M. R. A., R. McNerney, and H. W. S. Chan, Enzyme-linked immunosorbent assay of ochratoxin A in barley. *J. Assoc. Off. Anal. Chem. 66*:1481–1484 (1983).
82. Vogt, W., An evaluation of immunological methods based on the requirements of the clinical chemist. *J. Clin. Chem. Clin. Biochem. 22*:927–934 (1984).

RECENT REFERENCES ADDED IN PROOF

Dragsted, L. O., I. Bull, and H. Autrup, Substances with affinity to a monoclonal aflatoxin B1 antibody in Danish urine samples. *Food Chem. Toxicol. 26*:233–242 (1988).

Mortimer, D. N., M. J. Shepherd, J. Gilbert, and M. R. A. Morgan, A survey of the occurrence of aflatoxin B1 in peanut butters by enzyme-linked immunosorbent assay. *Food Addit. Contam. 5*:127–132 (1987).

Pauly, J. U., D. Bitter-Suermann, and K. Dose, Production and characterization of a monoclonal antibody to the trichothecene mycotoxin diacetoxyscirpenol. *Biol. Chem. Hoppe-Seyler 369*:487–492 (1988).

Pestka, J. J., Enhanced surveillance of foodborne mycotoxins by immunochemical assay. *J. Assoc. Off. Anal. Chem. 71* (Nov/Dec) (1988).

Warner, R. L., and J. J. Pestka, ELISA survey of retail grain-based food products for zearalenone and aflatoxin B_1. *J. Food Prot. 50*:502–503 (1987).

27
Progress in the Use of Monoclonal Antibodies for Diagnostic Microbiology

JOHN E. HERRMANN
University of Massachusetts Medical School, Worcester, Massachusetts

I. INTRODUCTION

Traditional approaches for the investigation of infectious diseases in terms of their etiology, epidemiology, and pathogenesis have relied, for the most part, on methods such as culture techniques, serological tests, and biochemical assays. Since the description of hybridoma technology for production of monoclonal antibodies by Köhler and Milstein in 1975 (1), the use of these antibodies is becoming central for both diagnosis and basic studies of infectious agents.

Polyclonal antibodies have been successfully used as diagnostic reagents in infectious disease for decades, but tests that use these reagents have often had difficulties in standardization and reproducibility. This has especially been a problem for tests run by different laboratories. There is generally less variation in commercially produced test kits that utilize polyclonal antisera, but the inherent difficulties with antisera produced in animals remain. One such difficulty is preexisting antibodies in a given antisera, which may react with antigens unrelated to the immunogen given. If the unrelated antigen is a human pathogen, a false-positive result could be obtained. For example, most animals used for antisera production usually have antibodies to rotaviruses that, because of the group reactive antigen present in all rotaviruses, can react with human strains as well. Thus, if an antisera was prepared against an unrelated viral or bacterial enteric pathogen, for use in an antigen-detecting immunoassay, there can be reactivity with rotaviruses present. This could occur even if two species of animals were used to prepare antisera, such as might be done for a sandwich-type solid-phase enzyme-linked immunosorbent assay (ELISA) or radioimmunoassay (RIA). Another inherent difficulty is the large number of antigenic determinants in most immunogens used in infectious disease research and the resulting broad specificity of antisera produced that can lead to undesired cross-reactions with other antigens. Variation in the ability among animals of the same species to react to a given antigen is also a problem inherent in the production of antisera, and it is sometimes difficult to reproduce an antiserum with the characteristics originally obtained. This often limits availability of a given immunoreagent. Monoclonal antibodies would appear to offer solutions to these problems because of their defined specificity and the potential for greater sensitivity as well, in part, because of higher signal/noise ratios that may be obtained when used in immunoassays.

II. CHARACTERISTICS OF MONOCLONAL ANTIBODIES TO MICROBIAL ANTIGENS

A. Affinity and Sensitivity

In practice, the use of monoclonal antibodies has been highly successful in some applications, but it has been limited in others, because of the nature of the antibodies. Table 1 lists the primary advantages of monoclonal antibodies and also some of the limiting factors to their use. For detection of microbial antigens in clinical specimens either directly or after cultivation, the utility of a monoclonal

Table 1 Advantages and Limitations of Monoclonal Antibodies for Diagnosis of Infectious Diseases

Advantages	Limitations
Permanent supply	Intrinsic cross-reactions
Chemical reproducibility	same determinant on different carriers
Defined specificity	similar determinants: related chemical structures
Problems of limited antigen supply for antisera production are eliminated	unrelated structures: coincidental expression of determinant shape
Potential for greater sensitivity	May be difficult to obtain high affinity antibody
	Cost may be greater than antisera against the same antigen
	Narrow range of reactivity may result in low sensitivity

Source: Table based in part on material presented in Refs. 2 and 3.

antibody prepared against only one epitope may depend on the nature of the microbial species as well as the antibody itself. If a particular species does not have a group antigen, or contains several variants, the monoclonal antibody may not be sufficiently broadly reactive. For direct detection of antigens in clinical specimens, high-affinity antibody is needed to give the sensitivity that is usually needed for a successful assay. Production of monoclonal antibodies with high-affinity constants has been found to be more difficult by most investigators than preparation of high-affinity polyclonal antibodies. Pools of two or more monoclonal antibodies have the potential to circumvent this, in that mixtures of monoclonal antibodies directed against different determinants on a given antigen should give greater sensitivity than one monoclonal antibody when used alone.

The nature of a particular antigen in terms of percentage of the total antigenic mass, the number of repeating epitopes it has, whether or not it is a group antigen, and how accessible the antigen is to antibody is also more critical for sensitivity in detection by monoclonal antibodies. For example, a major antigen of rotaviruses, VP6, is contained in all human and animal strains, and is accessible to antisera for RIA or ELISA. We have found that a single monoclonal antibody to this VP6 group antigen (4) is more sensitive in detecting human rotavirus antigen by ELISA (5) than polyclonal sera, and it is also far more specific. In contrast, there is sufficient variation in the large number of strains (isolates) of *Neisseria gonorrhoeae* that detection, and even identification after cultivation, has been difficult with monoclonal antibodies. In one study utilizing a commercial coagglutination test (GonoGen, Micro-Media Systems, Potomac, Maryland), only 86% of the clinical isolates could be correctly identified by the test (6). Others have found that this test, which uses a pool of monoclonal antibodies directed against gonococcal protein I antigens, to be more effective (7). But the results overall demonstrate that monoclonal antibodies, even when there are pools consisting of several different ones, can be overly specific and, thus, result in lowered sensitivity for identification or detection of some microbial antigens.

B. Specificity

What is considered to be the major characteristic of monoclonal antibodies is their unique specificity. However, as monoclonal antibodies have become more widely used, it has become apparent that there may be intrinsic cross-reactions of the nature listed in Table 1. These types of cross-reactions indicate that multispecificity may be found with monoclonal antibodies as well as with polyclonal ones, although to a far smaller degree. An example of multispecificity found in infectious agents is reported by Saegusa et al. (8) for coxsackievirus B4, in which monoclonal antibody to this virus reacted with heart tissue. Another example is that found by Thornley et al. (9) for monoclonal antibodies to *Chlamydia trachomatis*. Some of the monoclonal antibodies prepared against chlamydial genus-specific lipopolysaccharide (LPS) also reacted with antigens of heat-treated *Acinetobacter calcoaceticus* and with the LPS from *Salmonella minnesota* strain R595.

In both of these examples, however, specific antibodies against either the viral antigen or the chlamydial antigen could be obtained by appropriate screening of the hybridomas. However, this assumes that the potential cross-reactions are known. Because polyclonal sera have been prepared against virtually all of the infectious agents that would be of interest for diagnostic tests, many of the cross-reactions that might be encountered have been described. For example, in the cross-reaction seen with some of the monoclonal antibodies to the chlamydial LPS group antigen just described, antigens from the same bacterial species have been found to cross-react with polyclonal antisera to chlamydial LPS (10–12).

It has long been noted in the preparation of hyperimmune sera that as higher-affinity antibody is produced in a given antiserum, there is lower specificity of that serum because of a polyclonal response. Studies on antibody produced by spleen cells in vitro demonstrated that homogenously reactive antibody from different spleen cell foci can give different affinities (13). This would suggest that monoclonal antibodies with different affinity constants would retain their specificity. However, in one study that examined the relation of avidity to specificity, it was found that there was an inverse relationship between avidity and specificity of monoclonal antichlamydial antibodies (14). The authors suggested that nonspecific monoclonal antibodies become less specific with increasing avidity because avid nonfitting bonds would not be capable of differentiating between different determinants. Whether this will prove to be a universal finding, or will be restricted to certain antigens, remains to be determined.

Another possibility that could result in false-positive results with monoclonal antibodies is binding to Fc receptors present in a given specimen, as has been found with polyclonal antibodies. This is more likely to occur in monoclonal antibodies of murine origin that are not of class IgG1, as this isotype does not bind well to Fc receptors. The problem of this type of nonspecific binding for detection of microbial antigens has not proved to be a major one in polyclonal systems and, thus, is not likely to cause difficulties in monoclonal antibodies used for diagnosis.

When compared with polyclonal sera, the potential for undesired cross-reactivity with monoclonal antibodies is minor and has not yet been shown to be a problem in immunodiagnosis. Rather, the ones that have been found, and are likely to be found, will be useful in studying basic mechanisms and can be used to better define the closeness of related antigens.

III. PRODUCTION OF MONOCLONAL ANTIBODIES

The production of monoclonal antibodies to microbial antigens is, in general, similar to that used for other antigens, and the methodology is available to numerous books and reviews. However, there are some specialized techniques that may be particularly applicable to microbial species in some instances. These include use of in vitro immunization techniques, and use of Epstein-Barr virus (EBV)-transformed B lymphocytes from infected patients for use alone, or for fusion to either murine or human cell lines.

A. Alternative Methods

1. In vitro Immunization

Most monoclonal antibodies to microbial antigens have been produced by standard in vivo techniques, but for situations in which it is difficult to obtain sufficient microbial antigen, the in vitro technique may be advantageous. An example for which this technique has been found useful is in the production of antibodies to Snow Mountain virus. This virus is one of the 27-nm gastroenteritis viruses that cannot be cultivated, and the only source of antigen is that purified from stools of infected patients. By use of in vitro immunization techniques (15) it was possible to produce monoclonal antibodies to this virus with 1-10 ng of antigen.

The method is particularly useful for obtaining hybridomas that produce IgM monoclonal antibodies, but it can also be used to obtain those of class IgG. Immunization of animals in vivo before in vitro immunization increases the likelihood of obtaining IgG-producing hybridomas. The methods for production of monoclonal antibodies by this technique have been given in detail (16,17).

2. Epstein-Barr Virus Hybridoma Techniques

Other techniques that are applicable to situations for which there is limited antigen availability are the EBV transformation technique and the EBV hybridoma technique. The use of these techniques is generally for producing human monoclonal antibodies that could have potential for use in treatment or prophylaxis of disease. They also offer the possibility for production of monoclonal antibodies when antigen supply is limited. There are two general methods of producing human monoclonal antibodies: by transformation of human B cells with EBV and by fusion of these or nontransformed human B cells with either mouse or human myeloma and human lymphoblastoid cell lines. The EBV transformation technique for producing specific antibody, first described by Steinitz et al. (18), has been used to produce monoclonal antibodies to herpes simplex glycoprotein D (19) and to influenza virus nucleoprotein (20). The major disadvantage of using the EBV-transformation technique alone, without subsequent fusion, is that the cell lines may lose their ability to produce specific antibody after long-term culture. Another disadvantage to this technique is that low quantities of antibody (< 1 μg/ml) are usually produced (21). Because the efficiency of direct human B-cell–human or mouse cell line fusion is low, combined EBV transformation–cell fusion techniques were developed. Production of monoclonal antibodies to *Mycobacterium leprae* by fusion to human plasmocytomas (21) or to human × mouse (human B-lymphocytes–mouse myeloma line; 22) has been accomplished. Details of the methods for each of these techniques are available (21,22). Whether or not these techniques will be useful in the production of monoclonal antibodies to human pathogens that cannot be produced by either standard fusion or in vitro techniques remains to be determined.

B. Labeling and Other Factors for Use in Immunoassays

Much of the experience gained with polyclonal antibody immunoassays, in terms of labeling with enzymes or isotopes, is applicable to monoclonal antibody immunoassays as well (3). There are differences among various monoclonal antibodies that need to be evaluated when developing an immunoassay, and some modifications from conditions used for polyclonal antibodies may be necessary. For example, in a study concerned with the binding of monoclonal antibodies to influenza virus proteins on a solid phase, it was found that binding capacity was greatly influenced by the pH of the antigen-coating buffer (23).

As mentioned earlier, it may be difficult to produce a monoclonal antibody to some antigens that has a sufficiently high affinity to be useful in immunoassays requiring high sensitivity. One approach to circumvent this problem is to use mixtures of two or more monoclonal antibodies to increase the sensitivity of the assay (24). The increase in sensitivity is more than additive, which is apparently primarily due to the formation of a circular complex consisting of one of each type of monoclonal antibody and two antigen molecules (24). The specificity may also be affected, so this should be examined carefully when utilizing the mixed monoclonal antibody approach.

IV. SPECIFIC APPLICATIONS IN INFECTIOUS DISEASE DIAGNOSTIC ASSAYS

The application of monoclonal antibodies to diagnosis of infectious diseases is rapidly increasing and includes both direct detection of microbial antigens in clinical specimens and identification of the agent, or a specific component of the agent, after cultivation. There have been monoclonal antibodies produced to many bacterial and viral antigens (25), and the number is continually expanding. The ones that have been used for direct detection in clinical samples will be discussed here.

A. Detection of Bacterial Antigens

The monoclonal antibody tests that have been developed for the direct detection of bacterial antigens in clinical specimens are given in Table 2. Although most of the assays cited here are quite specific, the sensitivity of the tests vary significantly, and some will need to be improved considerably to supplant standard isolation procedures.

One area that has received considerable commercial attention is diagnosis of sexually transmitted disease (STD), especially that caused by *Chlamydia trachomatis*. Tests based on monoclonal antibody to both genus-specific and species-specific *C. trachomatis* antigens are available. Monoclonal antibodies to genus-specific antigens (chlamydial LPS) have been used in both immunofluorescence (IF) and ELISA formats. Tests in the ELISA format (26,27) have not been as extensively evaluated as polyclonal ELISA tests for the same antigen (28), but they appear to give approximately 90% sensitivity in detecting *C. trachomatis* in cervical swabs. The genus-specific monoclonal antibody in the IF format was 90% (37/41) sensitive (29). The first monoclonal antibody-based test that was developed utilized antibody against the chlamydial species-specific membrane protein in an IF format (30). The test was reported to be 93% sensitive and 96% specific, and a number of subsequent evaluations have found the test to give approximately the same results. Some laboratories, however, have reported much lower sensitivities of 49-70% (31,32).

Species-specific monoclonal antibodies have not been used in the ELISA format because of lack of sensitivity. The sensitivity of the genus-specific test is

Table 2 Monoclonal Antibody-Based Immunoassays for the Direct Detection of Bacterial Antigens in Clinical Specimens

Antigen	Type of assay	Ref.
Chlamydia trachomatis		
genus-specific antigen	ELISA	26, 27
	IF	29
species-specific antigen	IF	30-32
Haemophilus influenzae b	ELISA	35, 36
Legionella pneumophila	ELISA	37
	ELISA, IF	38
Neisseria meningitidis group B	ELISA	39
Streptococcus group B	ELISA	40
	Agglutination	41
Treponema pallidum	IF	42
Yersinia pestis	ELISA	43

purportedly enhanced by enzyme amplification, a technique that has been reported to enhance both the speed and the sensitivity of enzyme immunoassays (33).

Tests for other bacterial antigens listed in Table 2 have not been extensively evaluated, but some look promising. There are also ones being developed, such as those for detection of *Escherichia coli* enterotoxin (34), that are able to detect target antigen in fresh clinical isolates but not directly in the clinical sample. As these are developed more fully, more monoclonal antibody-based assays should become available for direct sample testing.

Haemophilus influenzae serotype b was detected by a monoclonal antibody ELISA in all of the cerebrospinal fluid, serum, and urine specimens tested that were positive by other means, with no false-positive results (35). Another study, in which a monoclonal antibody immunoperoxidase stain was used, found a sensitivity of 98% for detecting *H. influenzae* in 169 positive sputum smears (36).

Monoclonal antibody ELISA tests for *Legionella pneumophila* soluble antigen have been developed for detection of antigen in urine and serum (37), although only a few specimens were tested. By IF and ELISA, the same monoclonal antibodies were found useful for identifying legionella antigens in autopsy specimens of lung tissues obtained from the original outbreak of Legionnaires' disease (38).

An ELISA test that used monoclonal antibody specific for *Neisseria meningitidis* pili was able to detect group A meningococcal antigens in 21 of 25 cerebrospinal fluids (CSF) that were positive by a polyclonal ELISA (39). For identification of group B streptococcal antigens in body fluids, an ELISA assay (40) and a commercially available agglutination test (41) have been used. The ELISA evaluation was preliminary, but it was able to detect specific antigen in all of five positive CSF samples. The evaluation of the agglutination test found a sensitivity of 94% (18/19) in one or more admission specimens (urine, CSF, serum) of infants with symptomatic infections.

Another STD infectious agent for which monoclonal antibodies have been developed is *Treponema pallidum*. The use of the antibody in an IF test permitted direct detection of the pathogen in lesion exudates in 100% (30/30) of patients with early syphilis (42). Dark-field microscopy was positive for 29, but this test is not specific for *T. pallidum*, and it is not widely used. The monoclonal IF was negative for 31 patients without syphilis, whereas dark-field microscopy was positive for spiral organisms in 7 of the same 31 patients.

The last study cited in Table 2 concerns an ELISA for *Yersinia pestis*, using a monoclonal antibody to the F1 antigen (43). The sensitivity of the assay for detecting antigen in sera from patients with acute plague, however, was only 20% (2/10).

B. Detection of Viral Antigens

There has been a more rapid development of monoclonal antibody-based assays for viral antigens (Table 3) than for those of bacteria, and commercially produced ELISA or IF tests are available for detection of rotavirus, respiratory syncytial virus, and hepatitis virus antigens.

Monoclonal antibodies to the hexon group antigen of adenoviruses detected virus by ELISA in both stool samples and respiratory tract specimens, but the assay was ineffective when compared with isolation of virus in tissue culture (44). Tests using group-reactive adenovirus antibodies may be useful in screening respiratory specimens, but for diagnosis of adenoviral gastroenteritis, antibodies to the enteric types, 40 and 41, are needed. Monoclonal antibodies to a 17-kd component (apparently polypeptide VII) of enteric adenoviruses reacted in an ELISA with samples containing enteric types, but the efficacy of the test was not established (45). We found that monoclonal antibodies that we developed to

Table 3 Monoclonal Antibody-Based Immunoassays for the Direct Detection of Viral Antigens in Clinical Specimens

Antigen	Type of assay	Ref.
Adenovirus group antigen	ELISA	44
Adenovirus, enteric types	ELISA	45, 46
Cytomegalovirus	IF	47, 48
Hepatitis A virus	ELISA, RIA	49
Hepatitis B virus surface antigen	RIA	50–52
Hepatitis B virus e antigen	RIA	53
Herpes simplex virus	Filtration ELISA	54
	IF	55–58
Influenza A and B viruses	IF	60, 61
	Time-resolved fluoroimmunoassay	62
Parainfluenza 3 virus	IF	63
Respiratory syncytial virus	ELISA	64–66
	IF	67–73
Rotavirus		
group antigen	ELISA	5, 74, 75
	RPHA	76
type-specific antigen	ELISA	77, 78
Varicella-zoster virus	Filtration ELISA	54

the type-specific hexon component of enteric adenoviruses were highly effective when used in an ELISA for detecting the enteric types directly in stools (46). The sensitivity for type 40 was 96% (23/24) and for type 41 was 97% (34/35) when compared with culture results of electron microscopy (EM)-positive stool samples.

Cytomegalovirus (CMV) has been detected with monoclonal antibodies by IF both in lung biopsies (47) and in cells obtained by bronchoalveolar lavage (48). In the latter study, the test permitted rapid diagnosis of CMV infection in 6 of 19 patients by direct IF of cells, and in 18 of 19 after short-term (16-hr) culture isolation.

Hepatitis A virus was detected directly in stool samples by both ELISA and RIA in 9 of 25 patients with acute hepatitis A infection (49). Whether the 16 immunoassay-negative samples contained virus as shown by other means, such as EM or culture, was not determined. Monoclonal antibodies to hepatitis B surface antigen, both IgM (50) and IgG (51,52), may detect surface antigen by RIA in sera, from patients, that are unreactive by polyclonal RIA tests. In some samples, the enhanced sensitivity may be due to monoclonal antibody reactivity with "surface antigenlike" polypeptides (52). Monoclonal antibodies have also been successfully used to monitor hepatitis B virus e antigen in serum by RIA or ELISA (53).

A number of tests, both ELISA and IF, have been developed to detect herpes simplex virus in clinical samples, especially for diagnosis of virus in genital lesions. Monoclonal antibody-based tests have been used mostly in an IF format, but an enzyme-immunofiltration staining assay (54) was found to be 93% (66/71) sensitive in detecting virus obtained from lesions by vigorous swabbing. The format also was effective for detection of varicella-zoster virus in lesions. Among the IF tests for herpes virus antigen (55–58), the sensitivity obtained for direct detection varies. As might be expected, if the virus is cultured there is little difference in the sensitivity of various antibodies from different commercial sources used for typing isolates (59). The most important factor in sensitivity obtained by direct testing may be the status of the clinical lesion. Lafferty et al. (56) were able to detect antigen by IF in 87% of vesicular lesions but in only 67% of pustular lesions, 32% of ulcerative lesions, and 17% of crusted lesions. The isolation rate followed a similar pattern to the IF detection rate.

Rapid tests for diagnosis of respiratory infections have also been a major goal for application of monoclonal antibodies, and immunoassays have been developed and applied for detection of influenza virus (60–62), parainfluenza virus 3 (63), and respiratory syncytial virus (64–73). The tests for respiratory syncytial virus have been the most extensively evaluated and are mostly in the IF format. The sensitivities of the tests have been quite variable, although most have concluded that the tests are comparable with culture. For this virus, no ELISA has yet been reported that uses monoclonal antibodies either solely or as a detector antibody. The commercially available test that uses a monoclonal antibody for coating the solid phase does not appear to be as effective as a polyclonal antibody-based test (66).

Rotavirus infections have been a major target for commercial tests for some time and, at this writing, there were 13 tests available, mostly polyclonal antibody based. In a monoclonal antibody ELISA test we developed (5), the assay was more sensitive than a polyclonal test and detected virus in all of the EM-positive samples from neonates, children, and adults (40 total positives). The specificity was greater than 97%. An evaluation of a commercial assay, using the same monoclonal antibody, reported a sensitivity of 95% compared with 73% for a commercial polyclonal test (74). There is also a one-step ELISA reported (75), and a reverse passive hemagglutination (RPHA) has also effectively utilized monoclonal antibodies (76). Although specific serotyping of rotaviruses is not necessary for routine diagnosis, serotyping is useful in studying the epidemiology of rotaviruses in connection with vaccine development. The development of monoclonal antibodies to four human serotypes and their use for direct, type-specific detection in stools (77,78) will facilitate such studies.

V. CONCLUSIONS

In the past few years the rapid gains in monoclonal antibody technology and the resulting expanded numbers of monoclonal antibodies being developed to microbial antigens are rapidly becoming applied in diagnostic microbiology. The applications discussed here have been primarily those directed at antigen detection. It should be noted that monoclonal antibodies are becoming more frequently used in developing more specific serological assays as well, generally for antigen capture in antibody ELISA tests or for capture of human IgM. These should prove useful for diagnosis, especially when antigen detection is not feasible.

Only a short time ago, almost all techniques for detection of microbial antigens and antibodies were polyclonal antibody based (79). If the current progress continues, it can be expected that most of the assays, both for direct detection of microbial antigens and for their specific identification, will be based on the use of monoclonal antibodies.

REFERENCES

1. Köhler, G., and C. Milstein, Continuous cultures of fused cells secreting antibody of predefined specificity. *Nature 256*:495–497 (1975).
2. Milstein, C., Overview: Monoclonal antibodies. In *Handbook of Experimental Immunology*, 4th ed., Vol. 4, D. M. Weir (ed.). Blackwell Scientific Publications, Oxford, pp. 107.1–107.2 (1986).
3. Yolken, R., Use of monoclonal antibodies for viral diagnosis. In *New De-*

velopments in Diagnostic Virology, P. A. Bachmann (ed.). *Curr. Top. Microbiol. Immunol. 104*:177–196 (1983).
4. Cukor, G., D. M. Perron, R. Hudson, and N. R. Blacklow, Detection of rotavirus in human stools by using monoclonal antibody. *J. Clin. Microbiol. 19*:888–892 (1984).
5. Herrmann, J. E., N. R. Blacklow, D. M. Perron, G. Cukor, P. J. Krause, J. S. Hyams, H. J. Barrett, and P. L. Ogra, Enzyme immunoassay with monoclonal antibodies for the detection of rotavirus in stool specimens. *J. Infect. Dis. 152*:830–832 (1985).
6. Minshew, B. H., J. L. Beardsley, and J. S. Knapp, Evaluation of GonoGen coagglutination test for serodiagnosis of *Neisseria gonorrhoeae*: Identification of problem isolates by auxotyping, serotyping, and with a fluorescent antibody reagent. *Diagn. Microbiol. Infect. Dis. 3*:41–46 (1985).
7. Lawton, W. D., and G. J. Battaglioli, GonoGen coagglutination test for confirmation of *Neisseria gonorrhoeae. J. Clin. Microbiol. 18*:1264–1265 (1983).
8. Saegusa, J., B. S. Prabkakar, K. Essani, P. R. McClintock, Y. Fukuda, V. J. Ferrans, and A. L. Notkins, Monoclonal antibody to coxsackievirus B4 reacts with myocardium. *J. Infect. Dis. 153*:372–373 (1986).
9. Thornley, M. J., S. E. Zamze, M. D. Byrne, M. Lusher, and R. T. Evans, Properties of monoclonal antibodies to the genus-specific antigen of chlamydia and their use for antigen detection by reverse passive haemagglutination. *J. Gen. Microbiol. 131*:7–15 (1985).
10. Brade, H., and H. Brunner, Serological cross-reactions between *Acinetobacter calcoaceticus* and chlamydiae. *J. Clin. Microbiol. 10*:819–822 (1979).
11. Nurminen, M., E. Wahlström, M. Kleemola, M. Leinonen, P. Saikku, and P. Mäkelä, Immunologically related ketodeoxyoctonate-containing structures in *Chlamydia trachomatis, Re* mutants of *Salmonella* species, and *Acinetobacter calcoaceticus* var. *anitratus. Infect. Immun. 44*:609–613 (1984).
12. Caldwell, H. D., and P. J. Hitchcock, Monoclonal antibody against a genus-specific antigen of *Chlamydia* species: Location of the epitope on chlamydial lipopolysaccharide. *Infect. Immun. 44*:306–314 (1984).
13. Klinman, N. R., Antibody with homogeneous antigen binding produced by splenic foci in organ culture. *Immunochemistry 6*:757–759 (1969).
14. Matikainen, M.-T., and O.-P. Lehtonen, Relation between avidity and specificity of monoclonal anti-chlamydial antibodies in culture supernatants and ascitic fluids determined by enzyme immunoassay. *J. Immunol. Methods 72*:341–347 (1984).
15. Treanor, J. J., H. P. Madore, and R. Dolin, Development of a monoclonal antibody to the Snow Mountain agent of gastroenteritis. *Abstr. Intersci. Cong. Antimicro. Agents Chemother.* p. 116 (1986).
16. Reading, C. L., In vitro immunization for the production of antigen-specific lymphocyte hybridomas. In *Methods in Enzymology*, Vol. 121, J. J. Langone and Van Vanakis, H. (eds.). Academic Press, Orlando, pp. 18–26 (1980).

17. Boss, B. D., An improved in vitro immunization procedure for the production of monoclonal antibodies. In *Methods in Enzymology*, Vol. 121, J. J. Langone and Van Vanakis, H. (eds.). Academic Press, Orlando, pp. 27–32 (1986).
18. Steinitz, M., G. Klein, S. Koskimies, and O. Makela, EB virus induced B lymphocyte cell lines producing specific antibody. *Nature 269*:420–422 (1977).
19. Seigneurin, J. M., C. Desgranges, D. Seigneurin, J. Paire, J. C. Renversez, B. Jacquemont, and C. Micouin, Herpes simplex virus glycoprotein D: Human monoclonal antibody produced by bone marrow cell line. *Science 221*:173–175 (1983).
20. Muggeridge, M. I., D. M. Mitchell, E. D. Zanders, and P. C. L. Beverley, Production of human monoclonal antibody to X31 influenza virus nucleoprotein. *J. Gen. Virol. 64*:697–700 (1983).
21. Roder, J. C., S. P. C. Cole, and D. Kozbor, The EBV-hybridoma technique. In *Methods in Enzymology*, Vol. 121, J. J. Langone, and Van Vanakis, H. (eds.). Academic Press, Orlando, pp. 140–167 (1986).
22. Foung, S. K. H., E. G. Engleman, and F. C. Grumet, Generation of human monoclonal antibodies by fusion of EBV-activated B cells to a human-mouse hybridoma. In *Methods in Enzymology*, Vol. 121, J. J. Langone and Van Vanakis, H. (eds.). Academic Press, Orlando, pp. 168–173 (1986).
23. Kammer, K., Monoclonal antibodies to influenza A virus FM1 (H1N1) proteins require individual conditions for optimal reactivity in binding assays. *Immunology 48*:799–808 (1983).
24. Ehrlich, P. H., and W. R. Moyle, Cooperative immunoassays: Ultrasensitive assays with mixed monoclonal antibodies. *Science 221*:279–281 (1983).
25. Porterfield, J. S., and J. O'H. Tobin, Viral and bacterial infectious diseases. *Br. Med. Bull. 40*:283–290 (1984).
26. Caul, E. O., and I. D. Paul, Monoclonal antibody based ELISA for detecting *Chlamydia trachomatis. Lancet 1*:279 (1985).
27. Mohanty, K. C., J. J. O'Neill, and H. H. Hambling, Comparison of enzyme immunoassays and cell culture for detecting *Chlamydia trachomatis. Genitourin. Med. 62*:175–176 (1986).
28. Howard, L. V., P. F. Coleman, B. J. England, and J. E. Herrmann, Evaluation of Chlamydiazyme for the detection of genital infections caused by *Chlamydia trachomatis. J. Clin. Microbiol. 23*:329–332 (1986).
29. Alexander, I., I. D. Paul, and E. O. Caul, Evaluation of a genus reactive monoclonal antibody in rapid identification of *Chlamydia trachomatis* by direct immunofluorescence. *Genitourin. Med. 61*:252–254 (1985).
30. Tam, M. R., W. E. Stamm, H. H. Handsfield, R. Stephens, C.-C. Kuo, K. K. Holmes, K. Ditzenberger, M. Krieger, and R. C. Nowinski, Culture-independent diagnosis of *Chlamydia trachomatis* using monoclonal antibodies. *N. Engl. J. Med. 310*:1146–1150 (1984).
31. Tjiam, K. H., R. V. W. van Eijk, B. Y. M. van Heijst, G. J. Tideman, T. van Joost, E. Stolz, and M. F. Michel, Evaluation of the direct fluorescent antibody test for diagnosis of chlamydial infections. *Eur. J. Clin. Microbiol. 4*: 548–552 (1985).

32. Lipkin, E. S., J. V. Moncada, M.-A. Shafer, T. E. Wilson, and J. Schacter, Comparison of monoclonal antibody staining and culture in diagnosing cervical chlamydial infection. *J. Clin. Microbiol. 23*:114–117 (1986).
33. Stanley, C. J., A. Johannsson, and C. H. Self, Enzyme amplification can enhance both the speed and the sensitivity of immunoassays. *J. Immunol. Methods 83*:89–95 (1985).
34. Svennerholm, A.-M., M. Wikström, M. Lindblad, and J. Holmgren, Monoclonal antibodies against *Escherichia coli* heat-stable toxin (Sta) and their use in a diagnostic ST ganglioside GM-1-enzyme-linked immunosorbent assay. *J. Clin. Microbiol. 24*:585–590 (1986).
35. Belmaaza, A., J. Hamel, S. Mousseau, S. Montplaisir, and B. R. Brodeur, Rapid diagnosis of severe *Haemophilus influenzae* serotype b infections by monoclonal antibody enzyme immunoassay for outer membrane proteins. *J. Clin. Microbiol. 24*:440–443 (1986).
36. Groeneveld, K., L. van Alphen, N. J. Geelen-van den Broeck, P. P. Eijk, H. C. Zanen, and R. J. van Ketel, Detection of *Haemophilus influenzae* with monoclonal antibody. *Lancet 1*:441–442 (1987).
37. Bibb, W. F., P. M. Arnow, L. Thacker, and R. M. McKinney, Detection of soluble *Legionella pneumophila* antigens in serum and urine specimens by enzyme-linked immunosorbent assay with monoclonal and polyclonal antibodies. *J. Clin. Microbiol. 20*:478–482 (1984).
38. Brown, S. L., W. F. Bibb, and R. M. McKinney, Use of monoclonal antibodies in an epidemiological marker system: A retrospective study of lung specimens from the 1976 outbreak of Legionnaires' disease in Philadelphia by indirect fluorescent-antibody and enzyme-linked immunosorbent assay methods. *J. Clin. Microbiol. 21*:15–19 (1985).
39. Sugaswara, R. J., C. M. Prato, and J. E. Sippel, Enzyme-linked immunosorbent assay with a monoclonal antibody for detecting group A meningococcal antigens in cerebrospinal fluid. *J. Clin. Microbiol. 19*:230–234 (1984).
40. Morrow, D. L., J. B. Kline, S. Douglas, and R. A. Polin, Rapid detection of group B streptococcal antigen by monoclonal antibody sandwich enzyme assay. *J. Clin. Microbiol. 19*:457–459 (1984).
41. Rench, M. A., T. G. Metzger, and C. J. Baker, Detection of group B streptococcal antigen in body fluids by a latex-coupled monoclonal antibody assay. *J. Clin. Microbiol. 20*:852–854 (1984).
42. Hook, E. W. III, R. E. Roddy, S. A. Lukehart, J. Hom, K. K. Holmes, and M. R. Tam, Detection of *Treponema pallidum* in lesion exudate with a pathogen-specific monoclonal antibody. *J. Clin. Microbiol. 22*:241–244 (1985).
43. Williams, J. E., M. K. Gentry, C. A. Broden, F. Leister, and R. H. Yolken, Use of an enzyme-linked immunosorbent assay to measure antigenemia during acute plague. *Bull. WHO 62*:463–466 (1984).
44. Anderson, L. J., E. Godfrey, K. McIntosh, and J. C. Hierholzer, Comparison of a monoclonal antibody with a polyclonal serum in an enzyme-linked immunosorbent assay for detecting adenovirus. *J. Clin. Microbiol. 18*:463–468 (1983).

45. Singh-Naz, N., and R. J. Naz, Development and application of monoclonal antibodies for specific detection of human enteric adenoviruses. *J. Clin. Microbiol. 23*:840–842 (1986).
46. Herrmann, J. E., D. M. Perron-Henry, and N. R. Blacklow, Antigen detection with monoclonal antibodies for the diagnosis of adenovirus gastroenteritis. *J. Infect. Dis. 155*:1167–1171 (1987).
47. Hackman, R. C., D. Myerson, J. D. Meyers, H. M. Shulman, G. E. Sale, L. C. Goldstein, M. Rastetter, N. Flournoy, and E. D. Thomas, Rapid diagnosis of cytomegaloviral pneumonia by tissue immunofluorescence with a murine monoclonal antibody. *J. Infect. Dis. 151*:325–329 (1985).
48. Martin, W. J. II, and T. F. Smith, Rapid detection of cytomegalovirus in bronchoalveolar lavage specimens by a monocldnal antibody method. *J. Clin. Microbiol. 23*:1006–1008 (1986).
49. Coulepis, A. G., M. F. Veale, A. MacGregor, M. Kornitschuk, and I. D. Gust, Detection of hepatitis A virus and antibody by solid-phase radioimmunoassay and enzyme-linked immunosorbent assay with monoclonal antibodies. *J. Clin. Microbiol. 22*:119–124 (1985).
50. Wands, J. R., R. I. Carlson, H. Schoemaker, K. J. Isselbacher, and V. R. Jurawski, Immunodiagnosis of hepatitis B with high affinity IgM monoclonal antibodies. *Proc. Natl. Acad. Sci. USA 78*:1214–1218 (1981).
51. Ben-Porath, E., J. Wands, M. Gruia, and K. Isselbacher, Clinical significance of enhanced detection of HbsAg by a monoclonal radioimmunoassay. *Hepatology 4*:803–807 (1984).
52. Karayiannis, P., A. H. Goodall, J. A. Waters, S. Galpin, A. Lok, R. Thorp, and H. C. Thomas, Clinical evaluation of a monoclonal assay for hepatitis B surface antigen: Identification of "HBsAg-like" polypeptides non-reactive in conventional radioimmunoassays. *J. Med. Virol. 15*:291–303 (1985).
53. Ferns, R. B., and R. S. Tedder, Detection of both hepatitis B e antigen and antibody in a single assay using monoclonal reagents. *J. Virol. Methods 11*: 231–239 (1985).
54. Cleveland, P. H., and D. D. Richman, Enzyme immunofiltration staining assay for immediate diagnosis of herpes simplex virus and varicella-zoster virus directly from clinical specimens. *J. Clin. Microbiol. 25*:416–420 (1987).
55. Goldstein, L. C., L. Corey, J. K. McDougall, E. Tolentino, and R. Nowinski, Monoclonal antibodies to herpes simplex viruses: Use in antigenic typing and rapid diagnosis. *J. Infect. Dis. 147*:829–837 (1983).
56. Lafferty, W. E., S. Krofft, R. Remington, R. Giddings, C. Winter, A. Cent, and L. Corey, Diagnosis of herpes simplex virus by direct immunofluorescence and viral isolation from samples of external genital lesions in a high prevalence population. *J. Clin. Microbiol. 25*:323–326 (1987).
57. Pouletty, P., J. J. Chomel, D. Thouvenot, F. Catalan, V. Rabillon, and J. Kadouche, Detection of herpes simplex virus in direct specimens by immunofluorescence assay using a monoclonal antibody. *J. Clin. Microbiol. 25*:958–959 (1987).
58. Fung, J. C., J. Shanley, and R. C. Tilton, Comparison of the detection of herpes simplex virus in direct clinical specimens with herpes simplex virus-

specific DNA probes and monoclonal antibodies. *J. Clin. Microbiol. 22*: 748–753 (1985).

59. Lipson, S. M., T. E. Schutzbank, and K. Szabo, Evaluation of three immunofluorescence assays for culture confirmation and typing of herpes simplex virus. *J. Clin. Microbiol. 25*:391–394 (1987).
60. Shalit, I., P. A. McKee, H. Beauchamp, and J. L. Waner, Comparison of polyclonal antiserum versus monoclonal antibodies for the rapid diagnosis of influenza A virus infections by immunofluorescence in clincial specimens. *J. Clin. Microbiol. 22*:877–879 (1985).
61. Ray, C. G., and L. L. Minnich, Efficiency of immunofluorescence for rapid detection of common respiratory viruses. *J. Clin. Microbiol. 25*:355–357 (1987).
62. Walls, H. H., K. H. Johansson, M. W. Harmon, P. Halonen, and A. Kendal, Time-resolved fluoroimmunoassay with monoclonal antibodies for rapid diagnosis of influenza infections. *J. Clin. Microbiol. 24*:907–912 (1986).
63. Waner, J. L., N. J. Whitehurst, T. Downs, and D. G. Graves, Production of monoclonal antibodies against parainfluenza 3 virus and their use in diagnosis by immunofluorescence. *J. Clin. Microbiol. 22*:535–538 (1985).
64. Lauer, B. A., H. A. Masters, C. G. Wren, and M. J. Levin, Rapid detection of respiratory syncytial virus in nasopharyngeal secretions by enzyme-linked immunosorbent assay. *J. Clin. Microbiol. 22*:782–785 (1985).
65. Ahluwalia, G., J. Embree, P. McNicol, B. Law, and G. W. Hammond. Comparison of nasopharyngeal aspirate and nasopharyngeal swab specimens for respiratory syncytial virus diagnosis by cell culture, indirect immunofluorescence assay, and enzyme-linked immunosorbent assay. *J. Clin. Microbiol. 25*:763–767 (1987).
66. Chonmaitree, T., B. J. Bessette-Henderson, R. E. Hepler, and H. L. Lucia, Comparison of three rapid diagnostic techniques for detection of respiratory syncytial virus from nasal wash specimens. *J. Clin. Microbiol. 25*: 746–747 (1987).
67. Bell, D. M., E. E. Walsh, J. F. Hruska, K. C. Schnabel, and C. B. Hall, Rapid detection of respiratory syncytial virus with a monoclonal antibody. *J. Clin. Microbiol. 17*:1099–1101 (1983).
68. Freke, A., E. J. Stott, A. P. C. H. Roome, and E. O. Caul, The detection of respiratory syncytial virus in nasopharyngeal aspirates: Assessment, formulation, and evaluation of monoclonal antibodies as a diagnostic reagent. *J. Med. Virol. 18*:181–191 (1986).
69. Cheeseman, S. H., L. T. Pierik, D. Leombruno, K. E. Spinos, and K. McIntosh, Evaluation of a commercially available direct immunofluorescent staining reagent for the detection of respiratory syncytial virus in respiratory secretions. *J. Clin. Microbiol. 24*:155–156 (1986).
70. Routledge, E. G., J. McQuillin, A. C. R. Samson, and G. L. Toms, The development of monoclonal antibodies to respiratory syncytial virus and their use in diagnosis by indirect immunofluorescence. *J. Med. Virol. 15*:305–320 (1985).

71. Pothier, P., J. C. Nicolas, G. P. DeSaintMaur, S. Ghim, A. Kazmierczak, and F. Bricout, Monoclonal antibodies against respiratory syncytial virus and their use for rapid detection of virus in nasopharyngeal secretions. *J. Clin. Microbiol. 21*:286–287 (1985).
72. Kadi, Z., S. Dali, S. Bakouri, and A. Boughermouh, Rapid diagnosis of respiratory syncytial virus infection by antigen immunofluorescence detection with monoclonal antibodies and immunoglobulin M immunofluorescence test. *J. Clin. Microbiol. 24*:1038–1040 (1986).
73. Freymuth, F., M. Quibriac, J. Patitjean, M. L. Amiel, P. Pothier, A. Denis, and J. F. Duhamel, Comparison of two new tests for rapid diagnosis of respiratory syncytial virus infections by enzyme-linked immunosorbent assay and immunofluorescence techniques. *J. Clin. Microbiol. 24*:1013–1016 (1986).
74. Knisley, C. V., A. J. Bednarz-Prashad, and L. K. Pickering, Detection of rotavirus in stool specimens with monoclonal and polyclonal antibody-based systems. *J. Clin. Microbiol. 23*:897–900 (1986).
75. Pothier, P., and E. Drouet, Development and evaluation of a rapid one-step ELISA for rotavirus detection in stool specimens using only monoclonal antibodies. *Ann. Inst. Pasteur. (Paris) 137E*:401–410 (1986).
76. Cranage, M. P., A. D. Campbell, J. L. Venters, S. Mawson, R. R. A. Coombs, and T. H. Flewett, Detection and quantitation of rotavirus using monoclonal antibody coupled red blood cells: Comparison with ELISA. *J. Virol. Methods 11*:273–287 (1985).
77. Coulson, B. S., L. E. Unicomb, G. Pitson, and R. Bishop, Simple and specific enzyme immunoassay using monoclonal antibodies for serotyping human rotaviruses. *J. Clin. Microbiol. 25*:509–515 (1987).
78. Taniguchi, K., T. Urasawa, Y. Morita, H. B. Greenberg, and S. Urasawa, Direct serotyping of human rotavirus in stools by an enzyme-linked immunosorbent assay using serotype 1-,2-,3-, and 4-specific monoclonal antibodies to VP7. *J. Infect. Dis. 155*:1159–1165 (1987).
79. Herrmann, J. E., Enzyme-linked immunoassays for the detection of microbial antigens and their antibodies. In *Advances in Applied Microbiology*, Vol. 31, A. I. Laskin (ed.). Academic Press, Orlando, pp. 271–292 (1986).

III
UPDATE

28

Recent Developments in Nucleic Acid and Monoclonal Antibody Probe Technologies

BALA SWAMINATHAN* and CHANGMIN KIM
Purdue University, West Lafayette, Indiana

GYAN PRAKASH
MESA Diagnostics, Inc., Albuquerque, New Mexico

I. INTRODUCTION

In compiling a multiauthored text of this size on two rapidly evolving subjects, the editors and the publisher ran the risk of the book becoming obsolete by the time it was published. Fortunately, topics covered in this volume are still current in September 1988, when we wrote this final chapter. Many more tests utilizing nucleic acid or monoclonal antibody probes have been introduced for a wide spectrum of microorganisms since this book was initiated and some of the available assays have been improved. However, it was not the intent of this book to describe every available nucleic acid or monoclonal antibody probe assay. There have, however, been some significant developments which are expected to significantly impact the nucleic acid and monoclonal antibody technologies. These developments are briefly discussed in this concluding chapter.

II. THE POLYMERASE CHAIN REACTION AND QBETA REPLICASE TECHNOLOGY

The most significant development in nucleic acid probe technology which has the potential to impact diagnostic microbiology in a very significant way is the Polymerase Chain Reaction (PCR). PCR, first described in 1985 by Saiki et al. (1) is an elegant, simple method for amplifying a target DNA sequence in vitro by a factor of 100,000 to 1,000,000 within a few hours. The method consists of

**Present affiliation*: Meningitis and Special Pathogens Branch, Division of Bacterial Diseases, Center for Infectious Diseases, Centers for Disease Control, Atlanta, Georgia.

repetition of three steps which constitute one cycle: denaturation of target DNA, annealing of specific primers to target DNA, and extension of the desired DNA fragment in the range of 1000 bp by DNA polymerase (2-4). PCR amplification requires two synthetic oligonucleotide primers, target DNA with the desired sequence, four deoxyribonucleotides, and DNA polymerase. In the first step, heat denaturation of the target DNA provides two single strands of DNA to act as templates. In the second step, the primers, which have sequences complementary to one or the other strand of the template, anneal to sites flanking the region to be amplified. Extension of the two annealed primers in opposite directions by DNA polymerase in the third step produces complementary strands of a segment of desired sequence flanked by the two primers on the target DNA at the end of one cycle. The number of amplified target DNA sequences produced by PCR doubles after each cycle. Use of a thermostable DNA polymerase (*Taq* polymerase, from *Thermus aquaticus*) obviates the need to add enzyme after each cycle thus simplifying the procedure and rendering it amenable to automation (3,4).

PCR has been applied to rapidly detect proviral sequences of the human immunodeficiency virus (HIV)-1 in blood samples from patients (5,6), in infected T cells (7,8), and even in formalin-fixed, paraffin-embedded tissues (9a).

The Qbeta replicase technology employs a synthetic recombinant RNA molecule which serves the dual function of specific probe and amplifiable reporter. The recombinant RNA contains RNA sequences specific for target (e.g., DNA from *Plasmodium falciparum*) and sequences specific for MDV-1 RNA which is a natural substrate for the RNA-directed DNA polymerase, Qbeta replicase (9b). The target-specific RNA hybridizes to the target DNA in a conventional hybridization reaction. The hybridized recombinant RNA may be amplified up to a billion-fold in about 30 min by Qbeta replicase, thus allowing detection of extremely low levels of target DNA. The Qbeta replicase technology has not yet been applied directly in diagnostic microbiology; however, the technique has the potential to greatly enhance the sensitivity of nucleic acid hybridization reactions.

III. FREE-SOLUTION HYBRIDIZATION

Progress continues to be made in making nucleic acid probe technology simpler, safer, and more efficient. Instead of carrying out hybridization reactions with the target DNA immobilized on a solid matrix (e.g., nitrocellulose membrane) where the kinetics of hybridization are not optimal, free-solution hybridization procedures have been devised to accelerate hybridization reactions (J. Hogan, this volume; 10). Recently developed techniques allow capture of target DNA from a sample by a synthetic oligonucleotide (which has sequences complementary to target DNA) immobilized on Sepharose (11), latex (10,12,13), or other inert particle. In one format, the capture oligonucleotide has a poly dA tail which allows its capture by a poly dT coating on a polystyrene dipstick. Subsequent reactions are carried out by removing the dipstick and dipping it sequentially in a series of solutions. Gene-Trak systems (Framingham, MA) has introduced a second-generation nonisotopic nucleic acid hybridization assay for *Listeria* in foods using the poly dA–poly dT technology (J. D. Klinger, 10th International Symposium on Problems of Listeriosis, Pecs, Hungary, August 1988).

IV. FLUORESCENT DNA PROBES

One exciting development in non-isotopically labeled probes is in the area of fluorescent probes. A family of fluorescent tags (succinyl fluoresceins) have been synthesized and these can be coupled to dideoxynucleoside triphosphates by standard chemical techniques (14). These dyes have absorption maxima close to 488 nm, a wavelength that is suitable for excitation by an argon laser. The dyes can be distinguished by small differences in their absorption and emission spectra. This technology has thus far been applied to automated sequencing of DNA (14). There are problems in direct primary labeling of oligodeoxynucleotides with fluorophores (see Leary and Ruth, this volume). However, the potential of this technology for diagnostic microbiology is evident and will soon be exploited. Using oligonucleotide probes specific for different infectious disease agents and labeling each oligonucleotide with a different fluorescent tag (with a common excitation wavelength), it should be possible to examine a specimen for several infectious agents at the same time in one test.

V. HUMAN IMMUNODEFICIENCY VIRUS 1-RED CELL AGGLUTINATION ASSAY

HIV continues to be the primary research focus in infectious disease research and several immunological methods have been developed for the detection of HIV-1 antibodies in infected patients (15–18). In a competitive enzyme immunoassay format, use of a monoclonal antibody instead of a human anti-HIV-1 polyclonal antibody to capture HIV-1 *gag* antigen and immobilize it on a solid matrix resulted in several improvements: need for less viral antigen in the test to achieve the same level of final response in the enzyme immunoassay, slight increase in sensitivity with no loss of specificity, decrease in false-negative reactions because of the absence of cross-linking of capture human antibody with antihuman antibody probe through mediation of rheumatoid-type factors (19). A simple, ingeniously devised test for the detection of anti-HIV antibodies in whole blood was reported by Kemp et al. (20). The test employs a non-agglutinating antibody to human red blood cells which is conjugated to a synthetic peptide antigen (residues 579–601 of HIV-1 envelope precursor). When this conjugate (30 μl) is added to 10 μl of patient blood, specific agglutination of patient blood cells occurs only if antibodies to the HIV-1 synthetic peptide antigen are present. False-positive rates of 0.1% ($n = 874$) and false negative rates of 1% ($n = 81$) were reported for the red cell agglutination test. Such a test has excellent features for use in the field as a first-line screening test, particularly in developing countries.

VI. HYBRIDOMA TECHNOLOGY—NEW DEVELOPMENTS

Hybridoma technology has also undergone further refinements to make the antibodies more useful for diagnostic immunoassay purposes. For example, a monoclonal antibody that has the desirable specificity and affinity for antigen of interest may not be the right isotype required for a configuration of enzyme immunoassay. In such instances, class and subclass switching (e.g., IgM to IgG or IgG1 to IgG2a) may be accomplished either by the use of fluorescent activated cell sorter (21) or by the Sib selection technique (22). Some antiidiotypes (antibodies directed against the variable region of an antibody) may mimic the spatial arrangement of the antigenic determinant of the antigen and have been used in competitive immunoassays as surrogate antigen. This is particularly useful when the supply of pure antigen is very limited or if the antigen has a limited number of epitopes, making it unsuitable for use in a competitive assay format (23). In vitro mutagenesis and DNA transfection procedures have been used to produce recombinant antibodies called chimeric antibodies in which the antigen-binding portion of an immunoglobulin is fused with a protein having enzymatic activity (24). Such a chimeric molecule will perform the role of antigen binding and substrate modification to reveal the endpoint of an immunoassay. Bifunctional antibodies (a single antibody molecule having two distinct variable regions) have also been constructed in vivo by fusion of two separate hybridomas (quadroma) or in vitro by chemical dissociation of two antibodies, mixing them, allowing the formation of hybrids, and purifying the products by chromatographic techniques (25,26). Researchers have just started exploring the catalytic activities inherent in antibodies (27-29). Progress in this area may significantly impact the application of monoclonal antibodies in immunodiagnostic assays.

VII. CONCLUSIONS

Diagnostic microbiology is successfully being transformed from a collection of labor-intensive, subjective tests to test formats with a high degree of objectivity and great potential for automation. The transition has just begun. Nucleic acid and monoclonal antibody probe technologies, following a phase of confinement to the research laboratory, are on the verge of explosive commercial development. A large number of diagnostic kits have been approved by the U.S. Food and Drug Administration, which is currently approving new kits at the rate of one per month. This trend is expected to accelerate as new product applications are identified.

Further developments in and refinements to nucleic acid and monoclonal probe technologies are expected to result in routine microbiological tests which are simple, have a high degree of precision and reproducibility, and in which most repetitive steps will be automated. We are not predicting the demise of cultural techniques. They will continue to be used whenever additional studies on isolates are warranted. Also, conventional cultural procedures are and will be used as the "gold standard" for evaluation of the new tests. However, most routine microbiological tests will be significantly transformed by the new technologies in coming years.

REFERENCES

1. Saiki, R. K., S. Scharf, F. Faloona, K. B. Mullis, G. T. Horn, H. A. Erlich, and N. Arnheim. Enzymatic amplification of beta-globin genomic sequences and restriction site analysis for diagnosis of sickle cell anemia. *Science 230*:1350–1354 (1985).
2. Mullis, K. B., and F. A. Faloona. Specific synthesis of DNA in vitro via a polymerase-catalyzed chain reaction. In *Methods in Enzymology*, Vol. 155, R. Wu (Ed.), Academic Press, New York, pp. 335–350 (1987).
3. Oste, C. Polymerase chain reaction. *Biotechniques 6*:162–167 (1988).
4. Saiki, R. K., D. H. Gelealand, S. Stoffel, S. J. Scharf, R. Higuchi, G. T. Horn, K. B. Mullis, and H. A. Erlich. Primer-directed enzymatic amplification of DNA with a thermostable DNA polymerase. *Science 239*:487–491 (1988).
5. Ou, C-Y., S. Kwok, S. W. Mitchell, D. H. Mack, J. N. Sninsky, J. W. Krebs, P. Feorino, D. Warfield, and G. Schochetman. DNA amplification for direct detection of HIV-1 in DNA of peripheral blood mononuclear cells. *Science 239*:295–297 (1988).
6. Murakawa, G. J., J. A. Zaia, P. A. Spallone, D. A. Stephens, B. E. Kaplan, R. B. Wallace, and J. J. Rossi. Direct detection of HIV-1 RNA from AIDS and ARC patient samples. *DNA 7*:287–295 (1988).
7. Kwok, S., D. H. Mack, K. B. Mullis, B. Poiesz, G. Ehrlich, D. Blair, A. Friedman-Kien, and J. J. Sninsky. Identification of human immunodeficiency virus sequences by in vitro enzymatic amplification and oligomer cleavage detection. *J. Virol. 61*:1690–1694 (1987).
8. Byrne, B. C., J. J. Li, J. Sninsky, and B. J. Poiesz. Detection of HIV-1 RNA sequences by in vitro DNA amplification. *Nucl. Acids Res. 16*:4165 (1988).

9a. Lai-Goldman, M., E. Lai, and W. W. Grody. Detection of human immunodeficiency virus (HIV) infection in formalin-fixed, paraffin-embedded tissues by DNA amplification. *Nucl. Acids Res. 16*:8191 (1988).

9b. Lizardi, P. M., C. E. Guerra, H. Lomeli, I. Tussie-Luna, and F. R. Kramer. Exponential amplification of recombinant-RNA hybridization probes. *Biotechnology 6*:1197–1202 (1988).
10. Wolf, S. F., L. Haines, J. Fisch, J. N. Kremsky, J. P. Dougherty, and K. Jacobs. Rapid hybridization kinetics of DNA attached to submicron latex particles. *Nucl. Acids Res. 15*:2911–2926 (1987).
11. Polsky-Cynkin, R., G. H. Parsons, L. Allerdt, G. Landers, G. Davis, and A. Rashtchian. Use of DNA immobilized on plastic and agarose supports to detect DNA by sandwich hybridization. *Clin. Chem. 31*:1438–1443 (1985).
12. Kremsky, J. N., J. L. Wooters, J. P. Dougherty, R. E. Meyers, M. Collins, and E. L. Brown. Immobilization of DNA via oligonucleotides containing an aldehyde or carboxylic acid group at the 5′ terminus. *Nucl. Acids Res. 15*:2891–2909 (1987).
13. Urdea, M., B. D. Warner, J. A. Running, M. Stempien, J. Clyne, and T. Horn. A comparison of non-radioisotopic hybridization assay methods using fluorescent, chemiluminescent and enzyme labeled synthetic oligodeoxynucleotide probes. *Nucl. Acids. Res. 11*:4937–4956 (1988).
14. Prober, J. M., G. L. Trainor, R. J. Dam, F. W. Hobbs, C. W. Robertson, R. J. Zagursky, A. J. Cocuzza, M. A. Jensen, and K. Baumeister. A system for rapid DNA sequencing with fluorescent chain-terminating dideoxynucleotides. *Science 238*:336–341 (1987).
15. Gurtler, L. G., J. Eberle, B. Lorbeer, and F. Deinhardt. Sensitivity and specificity of commercial ELISA kits for screening anti-LAV/HTLV III. *J. Virol. Methods 15*:11–23 (1987).
16. Reesink, H. W., J. G. Huisman, M. Gonsalves, I. N. Winkel, A. C. Hekker, P. N. Lelie, W. Schaasberg, C. Aaij, J. J. van der Does, J. Desmyter, and J. Goudsmit. Evaluation of six enzyme immunoassays for antibody against human immunodeficiency virus. *Lancet 2*:483–486 (1986).
17. Sarangadharn, M. G., M. Popovic, L. Bruch, J. Schupbach, and R. C. Gallo. Antibodies reactive with human T-lymphotrophic retroviruses (HTLV-III) in the serum of patients with AIDS. *Science 224*:504–506 (1984).
18. Stute, R. Comparison in sensitivity of 10 HIV antibody detection tests by serial dilutions of Western blot-confirmed samples. *J. Virol. Meth. 20*: 269–273 (1988).
19. Ferns, R. B., R. S. Tedder, and J. L. Donoghue. Comparison of a monoclonal anti-HIV 1 *gag* solid phase with a polyclonal anti-HIV solid phase for detecting anti-HIV 1 in a competition ELISA. *J. Virol. Meth. 20*:143–153 (1988).
20. Kemp, B. E., D. B. Rylatt, P. G. Bundesen, R. R. Doherty, D. A. McPhee, D. Stapleton, L. E. Cottis, K. Wilson, M. A. John, J. M. Khan, D. P. Dinh, S. Miles, and C. J. Hillyard. Autologous red cell agglutination assay for HIV-1 antibodies: Simplified test with whole blood. *Science 241*:1352–1354 (1988).
21. Kipps, T. J. Switching the isotype of monoclonal antibodies. In *Hybridoma Technology in the Biosciences and Medicine*, T. A. Springer (Ed.), Plenum Press, New York, pp. 89–101 (1985).

22. Spira, G., A. Bargellesi, J. L. Teillaud, and M. D. Scharff. The identification of monoclonal class switch variants by Sib selection and an ELISA assay. *J. Immunol. Meth. 74*:307–315 (1984).
23. Sakaguchi, K., R. Ono, M. Tsujisaki, P. Richiardi, A. Carbonara, M. S. Park, R. Tonai, P. I. Terasaki, and S. Ferrone. Anti-HLA B7, B27, BW42, BW55, BW56, BW67, BW73 monoclonal antibodies: Specificity, idiotypes, and application for a double determinant immunoassay. *Hum. Immunol. 21*: 193–208 (1988).
24. Neuberger, M. S., G. T. Williams, and R. O. Fox. Recombinant antibodies possessing novel effector functions. *Nature 312*:604–608 (1984).
25. Milstein, C., and A. C. Cuello. Hybrid hybridoma and their use in immunochemistry. *Nature 305*:537–540 (1983).
26. Reading, C. L. Quadroma cells and trioma cells and methods for the production of the same. U.S. Patent No. 4714681—December 22, 1987. *Off. Gaz. U.S. Pat. Trademark Off. Pat. 1085*:1870–1871 (1987).
27. Napper, A. D., S. J. Benkovic, A. Tramontano, and R. A. Lerner. A stereospecific cyclization catalyzed by an antibody. *Science 237*:1041–1043 (1987).
28. Pollack, S. J., J. W. Jacobs, and P. G. Schultz. Selective chemical catalysis by an antibody. *Science 234*:1570–1573 (1986).
29. Tramontano, A., K. D. Janda, and R. A. Lerner. Catalytic antibodies. *Science 234*:1566–1569 (1986).

Index

D

P

Q

R

S

V

Y

Z

About the Editors

BALA SWAMINATHAN was on the faculty of Purdue University (Food Sciences Institute from 1977 to 1986. In addition to other courses in Food Microbiology and Toxicology, he developed and taught a graduate course on Rapid Methods and Automation in Applied Microbiology at Purdue University. He has been actively engaged in the development of rapid methods for the detection of food-borne pathogenic bacteria since 1974. Dr. Swaminathan's contributions to the development of rapid detection methods for *Salmonella* are nationally and internationally recognized. The author or coauthor of more than 40 research papers, reviews, and book chapters, he has made numerous presentations at national and international conferences and has organized workshops on diagnostic applications of nucleic acid probes and monoclonal antibodies. Dr. Swaminathan's professional affiliations include memberships in the American Society for Microbiology, Institute for Food Technologists, and the American Association for the Advancement of Science. Dr. Swaminathan received his undergraduate degree in chemistry (1966) from Delhi University in India, and his M.S. and Ph.D. degrees in food microbiology from the University of Georgia. Dr. Swaminathan is currently Chief, Epidemic Investigations Laboratory, Meningitis and Special Pathogens Branch, Division of Bacterial Diseases, Center for Infectious Diseases, Centers for Disease Control, Atlanta, Georgia.

GYAN PRAKASH is Director of Microbiology at Mesa Diagnostics, Inc., in Albuquerque, New Mexico. The author or coauthor of some 23 articles, abstracts, and proceedings papers, he has conducted several workshops on the clinical applications of nucleic acid probes and monoclonal antibodies. In 1984 he was a UNESCO Biotechnology Fellow at Rutgers University's Waksman Institute of

Microbiology in Piscataway, New Jersey, and he served as a Guest Lecturer at the University of Bonn and University of Mainz in the Federal Republic of Germany. He is a member of the American Society for Microbiology, Society for Industrial Microbiology, and American Association for the Advencement of Science. Dr. Prakash received the B.S. (1975) degree in biology and M.S. (1977) degree in biochemistry from the University of Allahabad in India, and M.S. (1981) and Ph.D. (1984) degrees in microbiology from the University of Illinois in Urbana-Champaign.